Mastery Learning. Proven Results.

ALEKS® 360
With *eBook* Integration

ALEKS is a unique, online program that uses artificial intelligence and adaptive questioning proven to raise student proficiency and success rates in math.

ALEKS Delivers a Unique Math Experience:

- **Research-Based, Artificial Intelligence** precisely measures each student's knowledge
- **Individualized Learning** presents the exact topics each student is most **ready to learn**
- **Adaptive, Open-Response Environment** includes comprehensive tutorials and resources
- **Detailed, Automated Reports** track student and class progress toward course mastery
- **Course Management Tools** include textbook integration, custom features, and more

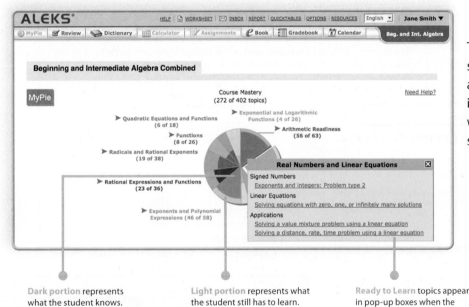

The ALEKS Pie summarizes a student's current knowledge and then delivers an individualized learning path with the exact topics the student is most ready to learn.

Dark portion represents what the student knows.

Light portion represents what the student still has to learn.

Ready to Learn topics appear in pop-up boxes when the student scrolls over a pie slice.

With ALEKS 360, the multimedia eBook is connected to every problem so students can quickly review the exact section they are working on.

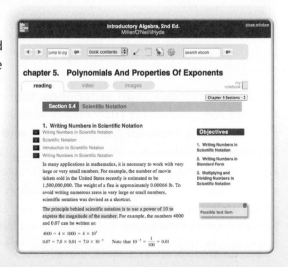

ALEKS is a registered trademark of ALEKS Corporation.

To see ALEKS in action, please visit: **www.SuccessInMath.com**

W9-CSY-643

Visualizing Math Concepts

Dynamic Math Animations

The Miller/O'Neill/Hyde author team has developed a series of Flash animations to illustrate difficult concepts where static images and text fall short. The animations leverage the use of on-screen movement and morphing shapes to enhance conceptual learning. For example, one animation "cuts" a triangle into three pieces and rotates the pieces to show that the sum of the angular measures equals 180° (below).

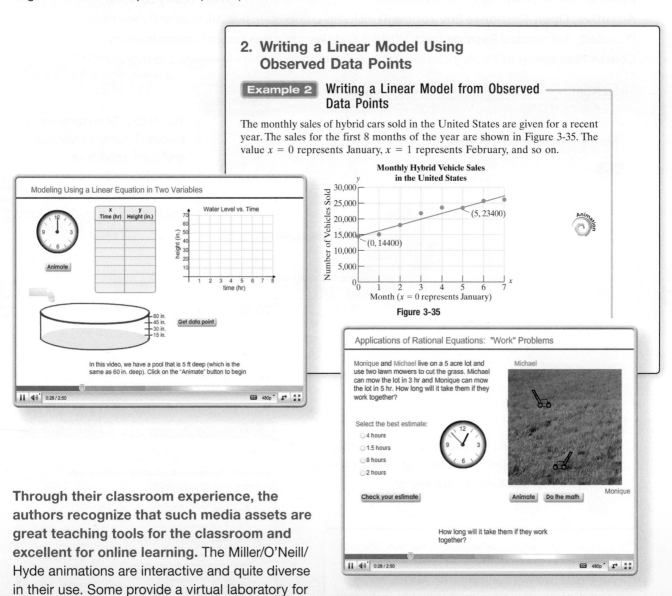

Through their classroom experience, the authors recognize that such media assets are great teaching tools for the classroom and excellent for online learning. The Miller/O'Neill/Hyde animations are interactive and quite diverse in their use. Some provide a virtual laboratory for which an application is simulated and where students can collect data points for analysis and modeling. Others provide interactive question-and-answer sessions to test conceptual learning. For word problem applications, the animations ask students to estimate answers and practice "number sense."

Basic College Mathematics

THIRD EDITION

Julie Miller
Daytona State College

Molly O'Neill
Daytona State College

Nancy Hyde
Broward College—
Professor Emeritus

McGraw Hill Education

BASIC COLLEGE MATHEMATICS, THIRD EDITION

Published by McGraw-Hill Education, 2 Penn Plaza, New York, NY 10121. Copyright © 2015 by McGraw-Hill Education. All rights reserved. Printed in the United States of America. Previous editions © 2013, 2009, and 2007. No part of this publication may be reproduced or distributed in any form or by any means, or stored in a database or retrieval system, without the prior written consent of McGraw-Hill Education, including, but not limited to, in any network or other electronic storage or transmission, or broadcast for distance learning.

Some ancillaries, including electronic and print components, may not be available to customers outside the United States.

This book is printed on acid-free paper.

2 3 4 5 6 7 8 9 0 DOW/DOW 1 0 9 8 7 6 5 4

ISBN 978–0–07–338441–2
MHID 0–07–338441–0

ISBN 978–0–07–734284–5 (Annotated Instructor's Edition)
MHID 0–07–734284–4

Senior Vice President, Products & Markets: *Kurt L. Strand*
Vice President, General Manager, Products & Markets: *Marty Lange*
Vice President, Content Production & Technology Services: *Kimberly Meriwether David*
Managing Director: *Ryan Blankenship*
Brand Manager: *Mary Ellen Rahn*
Director of Marketing: *Alex Gay*
Director of Development: *Rose Koos*
Development Editor: *Emily Williams*
Director of Digital Content: *Nicole Lloyd*
Director, Content Production: *Terri Schiesl*
Content Project Manager: *Peggy Selle*
Buyer: *Nicole Baumgartner*
Senior Designer: *Laurie B. Janssen*
Cover Illustration: *Imagineering Media Services, Inc.*
Lead Content Licensing Specialist: *Carrie K. Burger*
Compositor: *Aptara®, Inc.*
Typeface: *10/12 Times Ten Roman*
Printer: *R. R. Donnelley*

Library of Congress Cataloging-in-Publication Data

Miller, Julie, 1962-
   Basic college mathematics / Julie Miller, Daytona State College, Daytona Beach, Molly O'Neill, Daytona State College, Daytona Beach, Nancy Hyde. – Third edition.
      pages cm
   Includes index.
   ISBN 978–0–07–338441–2 — ISBN 0–07–338441–0 (hard copy : alk. paper) — ISBN 978–0–07–734284–5 — ISBN 0–07–734284–4 (hard copy : alk. paper) 1. Mathematics–Textbooks. I. O'Neill, Molly, 1953-
II. Hyde, Nancy. III. Title.

QA37.3.M55 2015
510 — dc23
                        2013020014

www.mhhe.com

About the Authors

Julie Miller is from Daytona State College, where she has taught developmental and upper-level mathematics courses for 20 years. Prior to her work at Daytona State College, she worked as a software engineer for General Electric in the area of flight and radar simulation. Julie earned a bachelor of science in applied mathematics from Union College in Schenectady, New York, and a master of science in mathematics from the University of Florida. In addition to her textbooks in developmental mathematics, Julie has authored a college algebra textbook and several course supplements for college algebra, trigonometry, and precalculus.

"My father is a medical researcher, and I got hooked on math and science when I was young and would visit his laboratory. I can remember using graph paper to plot data points for his experiments and doing simple calculations. He would then tell me what the peaks and features in the graph meant in the context of his experiment. I think that applications and hands-on experience made math come alive for me and I'd like to see math come alive for my students."

—*Julie Miller*

Molly O'Neill is also from Daytona State College, where she has taught for 22 years in the School of Mathematics. She has taught a variety of courses from developmental mathematics to calculus. Before she came to Florida, Molly taught as an adjunct instructor at the University of Michigan–Dearborn, Eastern Michigan University, Wayne State University, and Oakland Community College. Molly earned a bachelor of science in mathematics and a master of arts and teaching from Western Michigan University in Kalamazoo, Michigan. Besides this textbook, she has authored several course supplements for college algebra, trigonometry, and precalculus and has reviewed texts for developmental mathematics.

"I differ from many of my colleagues in that math was not always easy for me. But in seventh grade I had a teacher who taught me that if I follow the rules of mathematics, even I could solve math problems. Once I understood this, I enjoyed math to the point of choosing it for my career. I now have the greatest job because I get to do math every day and I have the opportunity to influence my students just as I was influenced. Authoring these texts has given me another avenue to reach even more students."

—*Molly O'Neill*

Nancy Hyde served as a full-time faculty member of the Mathematics Department at Broward College for 24 years. During this time she taught the full spectrum of courses from developmental math through differential equations. She received a bachelor of science degree in math education from Florida State University and a master's degree in math education from Florida Atlantic University. She has conducted workshops and seminars for both students and teachers on the use of technology in the classroom. In addition to this textbook, she has authored a graphing calculator supplement for *College Algebra*.

"I grew up in Brevard County, Florida, where my father worked at Cape Canaveral. I was always excited by mathematics and physics in relation to the space program. As I studied higher levels of mathematics I became more intrigued by its abstract nature and infinite possibilities. It is enjoyable and rewarding to convey this perspective to students while helping them to understand mathematics."

—*Nancy Hyde*

Dedications

To my great-niece, Aly	To my grandson, Nick	To Sandy and Jim
—Nancy Hyde	—Molly O'Neill	—Julie Miller

Contents

Chapter 4 **Decimals 219**

Chapter 5 **Ratio and Proportion 293**

Chapter 6 **Percents 337**

Chapter 7   Measurement   413

Chapter 8   Geometry   467

Additional Topics Appendix   A-1

How Will Miller/O'Neill/Hyde Help Your Students *Get Better Results?*

Better Clarity, Quality, and Accuracy!

Julie Miller, Molly O'Neill, and Nancy Hyde know what students need to be successful in mathematics. Better results come from clarity in their exposition, quality of step-by-step worked examples, and accuracy of their exercises sets; but it takes more than just great authors to build a textbook series to help students achieve success in mathematics. Our authors worked with a strong mathematics team of instructors from around the country to ensure that the clarity, quality, and accuracy you expect from the Miller/O'Neill/Hyde series was included in this edition.

> "The most complete text at this level in its thoroughness, accuracy, and pedagogical soundness. The best developmental mathematics text I have seen."
>
> —Frederick Bakenhus, *Saint Phillips College*

Better Exercise Sets!

Comprehensive sets of exercises are available for every student level. Julie Miller, Molly O'Neill, and Nancy Hyde worked with a board of advisors from across the country to offer the appropriate depth and breadth of exercises for your students. **Problem Recognition Exercises** were created to improve student performance while testing.

Practice exercise sets help students progress from skill development to conceptual understanding. Student tested and instructor approved, the Miller/O'Neill/Hyde exercise sets will help your students *get better results*.

▶ **Problem Recognition Exercises**

▶ **Skill Practice Exercises**

▶ **Study Skills Exercises**

▶ **Concept Connections**

▶ **Mixed Exercises**

▶ **Expanding Your Skills Exercises**

▶ **Vocabulary and Key Concepts Exercises**

> "This series was thoughtfully constructed with students' needs in mind. The Problem Recognition section was extremely well designed to focus on concepts that students often misinterpret."
>
> —Christine V. Wetzel-Ulrich, *Northampton Community College*

Better Step-By-Step Pedagogy!

Basic College Mathematics provides enhanced step-by-step learning tools to help students *get better results*.

▶ **Worked Examples** provide an "easy-to-understand" approach, clearly guiding each student through a step-by-step approach to master each practice exercise for better comprehension.

▶ **TIPs** offer students extra cautious direction to help improve understanding through hints and further insight.

▶ **Avoiding Mistakes** boxes alert students to common errors and provide practical ways to avoid them. Both of these learning aids will help students get better results by showing how to work through a problem using a clearly defined step-by-step methodology that has been class tested and student approved.

> "The book is designed with both instructors and students in mind. I appreciate that great care was used in the placement of 'Tips' and 'Avoiding Mistakes' as it creates a lot of teachable moments in the classroom."
>
> —Shannon Vinson, *Wake Tech Community College*

Get Better Results

Formula for Student Success

Step-by-Step Worked Examples

► Do you get the feeling that there is a disconnection between your students' class work and homework?

► Do your students have trouble finding worked examples that match the practice exercises?

► Do you prefer that your students see examples in the textbook that match the ones you use in class?

Miller/O'Neill/Hyde's *Worked Examples* offer a clear, concise methodology that replicates the mathematical processes used in the authors' classroom lectures!

Skill Practice

4. To estimate the number of fish in a lake, the park service catches 50 fish and tags them. After several months the park service catches a sample of 100 fish and finds that 6 are tagged. Approximately how many fish are in the lake?

Classroom Example: p. 323, Exercise 40

Example 4 Applying a Proportion to Environmental Science

A biologist wants to estimate the number of elk in a wildlife preserve. She sedates 25 elk and clips a small radio transmitter onto the ear of each animal. The elk return to the wild, and after 6 months, the biologist studies a sample of 120 elk in the preserve. Of the 120 elk sampled, 4 have radio transmitters. Approximately how many elk are in the whole preserve?

Solution:

Let n represent the number of elk in the preserve.

$$\begin{array}{c}\text{number of elk in the sample} \\ \text{radio transmitters} \longrightarrow \dfrac{4}{120} \\ \text{total elk in the sample} \longrightarrow \end{array} = \dfrac{25}{n} \begin{array}{c}\leftarrow \text{Sample} \\ \leftarrow \text{Population}\end{array}$$

$4 \cdot n = (120)(25)$ Equate the cross products.

$4n = 3000$

$\dfrac{\overset{1}{\cancel{4}}n}{\underset{1}{\cancel{4}}} = \dfrac{3000}{4}$ Divide both sides by 4.

$n = 750$ Divide $3000 \div 4 = 750$.

There are approximately 750 elk in the wildlife preserve.

"As always, MOH's Worked Examples are so clear and useful for the students. All steps have wonderfully detailed explanations written with wording that the students can understand. MOH is also excellent with arrows and labels making the Worked Examples extremely clear and understandable."

—Kelli Hammer, *Broward College–South*

"Easy to read step-by-step solutions to sample textbook problems. The 'why' is provided for students, which is invaluable when working exercises without available teacher/tutor assistance."

—Arcola Sullivan, *Copiah-Lincoln Community College*

Classroom Examples

To ensure that the classroom experience also matches the examples in the text and the practice exercises, we have included references to even-numbered exercises to be used as Classroom Examples. These exercises are highlighted in the Practice Exercises at the end of each section.

Classroom Example: p. 323, Exercise 40

40. Laws have been instituted in Florida to help save the manatee. To establish the number in Florida, a sample of 150 manatees was marked and let free. A new sample was taken and found that there were 3 marked manatees out of 40 captured. What is the approximate population of manatees in Florida?
There are approximately 2000 manatees in Florida.

Better Learning Tools

Concept Connection Boxes

Concept Connections help students understand the conceptual meaning of the problems they are solving—a vital skill in mathematics.

> "This feature is one of my favorite parts in the textbook. It is useful when trying to get students to think critically about types of problems."
> —Sue Duff, *Guilford Technical Community College*

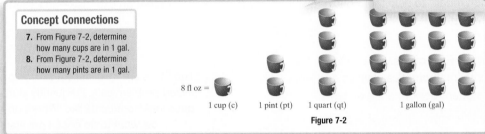

Concept Connections

7. From Figure 7-2, determine how many cups are in 1 gal.
8. From Figure 7-2, determine how many pints are in 1 gal.

8 fl oz =   1 cup (c)   1 pint (pt)   1 quart (qt)   1 gallon (gal)

Figure 7-2

TIP and Avoiding Mistakes Boxes

TIP and **Avoiding Mistakes** boxes have been created based on the authors' classroom experiences—they have also been integrated into the **Worked Examples.** These pedagogical tools will help students get better results by learning how to work through a problem using a clearly defined step-by-step methodology.

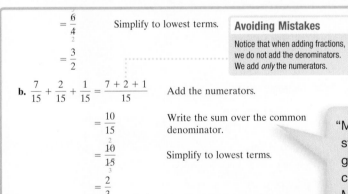

$$= \frac{\overset{2}{6}}{\underset{2}{4}}$$ Simplify to lowest terms.

Avoiding Mistakes

Notice that when adding fractions, we do not add the denominators. We add *only* the numerators.

$$= \frac{3}{2}$$

b. $\dfrac{7}{15} + \dfrac{2}{15} + \dfrac{1}{15} = \dfrac{7+2+1}{15}$ Add the numerators.

$$= \frac{10}{15}$$ Write the sum over the common denominator.

$$= \frac{\overset{2}{10}}{\underset{3}{15}}$$ Simplify to lowest terms.

$$= \frac{2}{3}$$

Avoiding Mistakes Boxes:

Avoiding Mistakes boxes are integrated throughout the textbook to alert students to common errors and how to avoid them.

> "MOH presentation of reinforcement concepts builds students' confidence and provides easy to read guidance in developing basic skills and understanding concepts. I love the visual clue boxes 'Avoiding Mistakes.' Visual clue boxes provide tips and advice to assist students in avoiding common mistakes."
> —Arcola Sullivan, *Copiah-Lincoln Community College*

TIP Boxes

Teaching tips are usually revealed only in the classroom. Not anymore! TIP boxes offer students helpful hints and extra direction to help improve understanding and provide further insight.

TIP: To use the prefix line effectively, you must know the order of the metric prefixes. Sometimes a mnemonic (memory device) can help. Consider the following sentence. The first letter of each word represents one of the metric prefixes.

kids	have	doughnuts	until	dad	calls	mom.
kilo-	hecto-	deka-	unit	deci-	centi-	milli-

↑
represents the main
unit of measurement
(meter, liter, or gram)

Get Better Results

Better Exercise Sets! Better Practice! Better Results!

▶ Do your students have trouble with problem solving?

▶ Do you want to help students overcome math anxiety?

▶ Do you want to help your students improve performance on math assessments?

Problem Recognition Exercises

Problem Recognition Exercises present a collection of problems that look similar to a student upon first glance, but are actually quite different in the manner of their individual solutions. Students sharpen critical thinking skills and better develop their "solution recall" to help them distinguish the method needed to solve an exercise—an essential skill in developmental mathematics.

Problem Recognition Exercises were tested in the authors' developmental mathematics classes and were created to improve student performance on tests.

"The PREs are an excellent source of additional mixed problem sets. Frequently students have questions/comments like 'Where do I start?' or 'I know what to do once I get started, but I have trouble getting started.' Perhaps with these PREs, students will be able to overcome this obstacle."

—Erika Blanken, *Daytona State College*

Problem Recognition Exercises

Operations on Decimals

For Exercises 1–20, perform the indicated operations.

1. a. $123.04 + 100$
    b. 123.04×100
    c. $123.04 - 100$
    d. $123.04 \div 100$
    e. $123.04 + 0.01$
    f. 123.04×0.01
    g. $123.04 \div 0.01$
    h. $123.04 - 0.01$

2. a. $5078.3 + 1000$
    b. 5078.3×1000
    c. $5078.3 - 1000$
    d. $5078.3 \div 1000$
    e. $5078.3 + 0.001$
    f. 5078.3×0.001
    g. $5078.3 \div 0.001$
    h. $5078.3 - 0.001$

3. a. $4.8 + 2.391$
    b. $2.391 + 4.8$

4. a. $632.46 + 98.0034$
    b. $98.0034 + 632.46$

5. a. 32.9×1.6
    b. 1.6×32.9

6. a. 74.23×0.8
    b. 0.8×74.23

7. a. $4(21.6)$
    b. $0.25(21.6)$

8. a. $2(92.5)$
    b. $0.5(92.5)$

10. a. $496.8 \div 9.2$
    b. 54×9.2

12. 20×0.05

14. $6400 \div 0.001$

16. $2000\overline{)5,400,000}$

18. $(340)(5000)$

20. $\begin{array}{r} 124.7 \\ -47.9999 \end{array}$

"These are so important to test whether a student can recognize different types of problems and the method of solving each. They seem very unique—I have not noticed this feature in many other texts or at least your presentation of the problems is very organized and unique."

—Linda Kuroski, *Erie Community College*

Student Centered Applications!

The Miller/O'Neill/Hyde Board of Advisors partnered with our authors to bring the *best applications* from every region in the country! These applications include real data and topics that are more relevant and interesting to today's student.

87. The drug cyanocobalamin is prescribed by one doctor in the amount of 1000 μg. How many milligrams is this? 1 mg

88. An injection of naloxone is given in the amount of 800 μg. How many milligrams is this? 0.8 mg

89. A nurse must administer 45 mg of a drug. The drug is available in a liquid form with a concentration of 15 mg per milliliter of the solution. How many milliliters of the solution should the nurse give? 3 mL

90. A patient must receive 500 mg of medication in a solution that has a strength of 250 mg per 5 milliliter of solution. How many milliliters of solution should be given? 10 mL

Expanding Your Skills

91. A normal value of hemoglobin in the blood for an adult male is 18 g/dL (that is, 18 grams per deciliter). How much hemoglobin would be expected in 20 mL of a males's blood? 3.6 g

92. A normal value of hemoglobin in the blood for an adult female is 15 g/dL (that is, 15 grams per deciliter). How much hemoglobin would be expected in 40 mL of a female's blood? 6 g

Group Activities!

Each chapter concludes with a Group Activity to promote classroom discussion and collaboration—helping students not only to solve problems but to explain their solutions for better mathematical mastery. Group Activities are great for both full-time and adjunct instructors—bringing a more interactive approach to teaching mathematics! All required materials, activity time, and suggested group sizes are provided in the end-of-chapter material.

Group Activity

Constructing a Golden Rectangle

Materials: Ruler, calculator, and a compass

Estimated time: 20 minutes

Group Size: 2

Consider a line segment divided into two pieces and labeled as follows.

The Golden Ratio is given by: $\dfrac{A}{B} = \dfrac{A+B}{A}$

Although it is beyond the scope of this text, we can solve for A to get, $A = \left(\dfrac{\sqrt{5}+1}{2}\right) \cdot B$ or $A \approx 1.618 \cdot B$.

1. This ratio occurs in nature and is used in art and architecture. Measure the length of the hand. Then multiply by 1.618 and compare to the length of the arm.

n Paris, France, is constructed with many golden ratios. Check by blue lines and multiplying each length by 1.618 to get the measure of ding white lines.

gle is one in which the ratio of length to width is approximately 1.618. hought to be most pleasing to the eye.

> "MOH's group activity involves true participation and interaction; fun with fractions!"
>
> —Monika Bender, *Central Texas College*

> "This is one part of the book that would have me adopt the MOH book. I am very big on group work for *Basic College Mathematics* and many times it is difficult to think of an activity. I would conclude the chapter doing the group activity in the class. Many books just have problems for this, but the MOH book provides an actual activity."
>
> —Sharon Giles, *Grossmont College*

Get Better Results

Additional Supplements

Media Suite

NEW Lecture Videos Created by the Authors

Julie Miller began creating these lecture videos for her own students to use when they were absent from class. The student response was overwhelmingly positive, prompting the author team to create the lecture videos for their entire developmental math book series. In these new videos, the authors walk students through the learning objectives using the same language and procedures outlined in the book. Students learn and review right alongside the author! Students can also access the written notes that accompany the videos.

Dynamic Math Animations

The authors have constructed a series of Flash animations to illustrate difficult concepts where static images and text fall short. The animations leverage the use of on-screen movement and morphing shapes to give students an interactive approach to conceptual learning. Some provide a virtual laboratory for which an application is simulated and where students can collect data points for analysis and modeling. Others provide interactive question-and-answer sessions to test conceptual learning.

NEW Exercise Videos

The authors, along with a team of faculty who have used the Miller/O'Neill/Hyde textbooks for many years, have created new exercise videos for designated exercises in the textbook. These videos cover a representative sample of the main objectives in each section of the text. Each presenter works through selected problems, following the solution methodology employed in the text.

The video series is available online as part of Connect Math hosted by ALEKS as well as in ALEKS 360. The videos are closed-captioned for the hearing impaired, and meet the Americans with Disabilities Act Standards for Accessible Design.

Student Resource Manual

The *Student Resource Manual (SRM),* created by the authors, is a printable, electronic supplement available to students through Connect Math hosted by ALEKS. Instructors can also choose to customize this manual and package with their course materials. With increasing demands on faculty schedules, this resource offers a convenient means for both full-time and adjunct faculty to promote active learning and success strategies in the classroom.

This manual supports the series in a variety of different ways:

- NEW Additional Group Activities developed by the authors to supplement what is already available in the text.
- Discovery-based classroom activities written by the authors for each section
- Worksheets for extra practice written by the authors including Problem Recognition Exercise Worksheets
- NEW Lecture Notes designed to help students organize and take notes on key concepts
- Materials for a student portfolio

Annotated Instructor's Edition

In the *Annotated Instructor's Edition* (*AIE*), answers to all exercises appear adjacent to each exercise in a color used *only* for annotations. The *AIE* also contains Instructor Notes that appear in the margin. These notes offer instructors assistance with lecture preparation. In addition, there are Classroom Examples referenced in the text that are highlighted in the Practice Exercises. Also found in the *AIE* are icons within the Practice Exercises that serve to guide instructors in their preparation of homework assignments and lessons.

Powerpoints

The Powerpoints present key concepts and definitions with fully editable slides that follow the textbook. An instructor may project the slides in class or post to a website in an online course.

Instructor's Solutions Manual

The *Instructor's Solutions Manual* provides comprehensive, worked-out solutions to all exercises in the Chapter Openers, the Practice Exercises, the Problem Recognition Exercises, the end-of-chapter Review Exercises, the Chapter Tests, and the Cumulative Review Exercises.

Student's Solutions Manual

The *Student's Solutions Manual* provides comprehensive, worked-out solutions to the odd-numbered exercises in the Practice Exercise sets, the Problem Recognition Exercises, the end-of-chapter Review Exercises, the Chapter Tests, and the Cumulative Review Exercises. Answers to the Chapter Opener Puzzles are also provided.

Instructor's Test Bank

Among the supplements is a computerized test bank using the algorithm-based testing software TestGen® to create customized exams quickly. Hundreds of text-specific, open-ended, and multiple-choice questions are included in the question bank.

With **McGraw-Hill Create™**, you can easily rearrange chapters, combine material from other content sources, and quickly upload content you have written like your course syllabus or teaching notes. Find the content you need in Create by searching through thousands of leading McGraw-Hill textbooks. Arrange your book to fit your teaching style. Create even allows you to personalize your book's appearance by selecting the cover and adding your name, school, and course information. Assemble a Create book and you'll receive a complimentary print review copy in 3—5 business days or a complimentary electronic review copy (eComp) via email in minutes.

To learn more contact your sales rep or visit www.successinmath.com.

Get Better Results

|MATH

Hosted by ALEKS Corp.

- McGraw-Hill's Connect Math hosted by ALEKS® is a web-based assignment and assessment platform that helps students connect to their coursework and prepares them to succeed in and beyond the course.
- Connect Math hosted by ALEKS enables math and statistics instructors to create and share courses and assignments with colleagues and adjuncts with only a few clicks of the mouse. All exercises, learning objectives, and activities are developed by mathematics instructors to ensure consistency between the textbook and the online resources.
- Connect Math hosted by ALEKS also links students to an interactive eBook with a variety of media assets and a place to study, highlight, and keep track of class notes.

To learn more contact your sales rep or visit www.successinmath.com.

ALEKS®360 ALEKS® ALEKS® Prep Products

With *eBook* Integration

- **ALEKS** is a Web-based program that uses artificial intelligence to assess a student's knowledge and provide personalized instruction on the exact topics the student is most ready to learn. By providing individualized assessment and learning, ALEKS helps students to master course content quickly and easily. ALEKS allows students to easily move between explanations and practice, and provides intuitive feedback to help students correct and analyze errors. ALEKS also includes a powerful instructor module that simplifies course management so instructors spend less time with administrative tasks and more time directing student learning.
- **ALEKS 360** includes a fully integrated, interactive eBook and textbook-specific lecture and exercise videos combined with ALEKS personalized assessment and learning.
- **ALEKS Prep** focuses on prerequisite and introductory material, and can be used during the first six weeks of the term to ensure student success in math courses ranging from Beginning Algebra through Calculus. ALEKS Prep quickly fill gaps in prerequisite knowledge by assessing precisely each student's preparedness and delivering individualized instruction on the exact topics students are most ready to learn.

To learn more contact your sales rep or visit www.successinmath.com.

LEARNSMART® SMARTBOOK™

- **LearnSmart®** is an adaptive study tool that strengthens student understanding and retention of your course's fundamental concepts. LearnSmart efficiently prepares your students for class so you can review *less* and teach *more*. The LearnSmart methodology is simple: determine the concepts students don't know or understand, and then teach those concepts using personalized plans designed for each student's success.
- **SmartBook®** is the first and only adaptive reading experience available for the higher education market. Powered by the intelligent and adaptive LearnSmart engine, SmartBook facilitates the reading process by identifying what content a student knows and doesn't know. As a student reads, the material continuously adapts to ensure the student is focused on the content he or she needs the most to close specific knowledge gaps.

To learn more contact your sales rep or visit www.successinmath.com.

Faculty Development and Digital Training

McGraw-Hill is excited to partner with our customers to ensure success in the classroom with our course solutions.

Looking for ways to be more effective in the classroom? Interested in the best practices for using available instructor resources?

Workshops are available on these topics for anyone using or considering the Miller, O'Neill, and Hyde Math Series. Led by the authors, contributors, and McGraw-Hill Learning consultants, each workshop is tailored to the needs of individual campuses or programs.

New to McGraw-Hill Digital Solutions? Need help setting up your course, using reports, and building assignments?

No need to wait for that big group training session during faculty development week. The McGraw-Hill Digital Implementation Team is a select group of advisors and experts in Connect Math. The Digital Implementation Team will work one-on-one with each instructor to make sure you are trained on the program and have everything you need to ensure a good experience for you and your students.

Are you redesigning a course or expanding your use of technology? Are you interested in getting ideas from other instructors who have used ALEKS™ or Connect Math in their courses?

Digital Faculty Consultants (DFCs) are instructors who have effectively incorporated technology such as ALEKS and Connect Math hosted by ALEKS in their courses. Discuss goals and best practices and improve outcomes in your course through peer-to-peer interaction and idea sharing.

Contact your local representative for more information about any of the faculty development, training, and support opportunities through McGraw-Hill. http://catalogs.mhhe.com/mhhe/findRep.do

Our Commitment to Market Development and Accuracy

McGraw-Hill's Development Process is an ongoing, never-ending, market-oriented approach to building accurate and innovative print and digital products. We begin developing a series by partnering with authors that desire to make an impact within their discipline to help students succeed. Next, we share these ideas and manuscript with instructors for review for feedback and to ensure that the authors' ideas represent the needs within that discipline. Throughout multiple drafts, we help our authors adapt to incorporate ideas and suggestions from reviewers to ensure that the series carries the same pulse as today's classrooms. With any new edition, we commit to accuracy across the series and its supplements. In addition to involving instructors as we develop our content, we also utilize accuracy checks through our various stages of development and production. The following is a summary of our commitment to market development and accuracy:

1. 2 drafts of author manuscript
2. 2 rounds of manuscript review
3. Multiple focus groups
4. 2 accuracy checks
5. 2 rounds of proofreading and copyediting
6. Toward the final stages of production, we are able to incorporate additional rounds of quality assurance from instructors as they help contribute toward our digital content and print supplements

This process then will start again immediately upon publication in anticipation of the next edition. With our commitment to this process, we are confident that our series has the most developed content the industry has to offer, thus pushing our desire for quality and accurate content that meets the needs of today's students and instructors.

Acknowledgments and Reviewers

Paramount to the development of this series was the invaluable feedback provided by the instructors from around the country who reviewed the manuscript or attended a market development event over the course of the several years the text was in development.

Albert Groccia, *Valencia College–Osceola*

Albert Guerra, *Saint Philips College*

Alexander Kasiukov, *Suffolk County Community College–Brentwood*

Alice Pollock-Cangemi, *Lone Star College*

Amber Smith, *Johnson County Community College*

Amtul Mujeeb Chaudry, *Rio Hondo College*

Anabel Darini, *Suffolk County Community College–Brentwood*

Andrea Blum, *Suffolk County Community College–Brentwood*

Angela Mccombs, *Illinois State University*

Ann Mccormick, *Lone Star College Kingwood*

Anne Prial, *Orange County Community College*

Antonnette Gibbs, *Broward College–North*

Anuradha Vadrevu, *Prince George's Community College*

Arlene Atchison, *South Seattle Community College*

Ashley Fuller, *John Tyler Community College–Chester*

Azzam Shihabi, *Long Beach City College*

Barbara Lott, *Seminole State College*

Barbara Purvis, *Centura College*

Barry Gibson, *Daytona State College–Daytona Beach*

Bashar Zogheib, *Nova Southeastern University*

Becky Schuering, *Blue River Community College–Independence*

Bernadette Turner, *Lincoln University*

Beverly Pepe, *Community College of Rhode Island–Warwick*

Bill Morrow, *Delaware Technical Community College*

Billie Shannon, *Southwestern Oregon Community College*

Brannen Smith, *Central Georgia Technical College*

Brenda Brown, *University of the District of Columbia*

Brent Pohlmann, *California Maritime Academy*

Bruce Legan, *Century Community & Technical College*

Carl Moxey, *Anna Maria College–Paxton*

Carol Curtis, *Fresno City College*

Carol Elias, *John Tyler Community College–Chester*

Carol Marinas, *Barry University*

Carol Mckillip, *Southwestern Oregon Community College*

Carol Rich, *Wallace Community College*

Carol Weideman, *Saint Petersburg College–Gibbs*

Carolyn Chapel, *Western Technical College*

Cassandra Johnson, *Robeson Community College*

Cassie Firth, *Northern Oklahoma College*

Cassandra Thompson, *York Technical College*
Chad Lower, *Pennsylvania College of Technology*
Christina Morian, *Lincoln University*
Cristi Whitfield, *Wallace Community College*
Cylinda Bray, *Yavapai College–Prescott*
Darla Aguilar, *Pima Community College*
Darlene Hatcher, *Metro Community College–South Campus–Omaha*
David Nusbaum, *Cypress College*
Dawn Chapman, *Columbus Technical College*
Deanna Hardy, *Bossier Parish Community College*
Deborah Logan, *Florida State College–South Campus*
Deborah Wolfson, *Suffolk County Community College–Brentwood*
Denise Nunley, *Glendale Community College*
Diana Dwan, *Yavapai College–Prescott*
D'marie Carver, *Portland Community College*
Don Anderson, *Northwest College*
Don Groninger, *Middlesex County College*
Don Solomon, *University of Wisconsin–Milwaukee*
Donald Robertson, *Olympic College*
Dot French, *Community College of Philadelphia*
Ed Thompson, *River Parishes Community College*
Edelma Simes, *Phillips Community College–Helena*
Eden Donahou, *Seminole State College*
Edith Lester, *Volunteer State Community College*
Edward Migliore, *University of California–Santa Cruz*
Eileen Diggle, *Bristol Community College*
Elaine Fitt, *Bucks Community College*
Eldon Baldwin, *Prince Georges Community College*
Elecia Ridley, *Durham Technical Community College*
Elena Litvinova, *Bloomsburg University of Pennsylvania*
Eleni Palmisano, *Centralia College*
Elisha Van Meenan, *Illinois State University*
Emily Simmons, *Centura College*
Eric Bennett, *Michigan State University–East Lansing*
Eric Kaljumagi, *Mt. San Antonio College*
Evon Lisle, *Seminole State College*
Gary Kersting, *North Central Michigan College*
Gene Ponthieux, *River Parishes Community College*
Gerald J. Lepage, *Bristol Community College*
Geri Philley, *Monterey Peninsula College*
Ghytana Goings, *Wallace Community College*
Ginger Eaves, *Bossier Parish Community College*
Gladys Bennett, *Centura College*
Gloria Hernandez, *Louisiana State University*
Greg Longanecker, *Leeward Community College*

Greg Rosik, *Century Community & Technical College*
Hadley Pridgen, *Gulf Coast Community College*
Heather Gallacher, *Cleveland State University*
Heidi Howard, *Florida State College–South Campus*
Heidi Kiley, *Suffolk County Community College–Brentwood*
Heidi Lyman, *South Seattle Community College*
Ignacio Alarcon, *Santa Barbara City College*
Irma Bakenhus, *San Antonio College*
J. Patrick Malone, *Victor Valley Community College*
James Dorn, *Barstow College*
James Miller, *Hillsborough Community College–Dale Mabry*
James Weeks, *Durham Technical Community College*
Jessica Lopez, *Saint Philips College*
Jean-Marie Magnier, *Springfield Technical Community College*
Jennifer Crowley, *Wallace Community College*
Jennifer Lempke, *North Central Michigan College*
Jian Zou, *South Seattle Community College*
Jill Wilsey, *Genesee Community College*
Joanne Strickland, *California Maritime Academy*
Jody Balzer, *Milwaukee Area Technical College*
Joe Jordan, *John Tyler Community College–Chester*
Joe Joyner, *Tidewater Community College–Norfolk*
Jonathan Cornick, *Queensborough Community College*
Joni Dugan, *Johnson County Community College*
Jordan Neus, *Suffolk County Community College–Brentwood*
Joshua Fontenot, *Louisiana State University*
Joyce Davis, *Heart of Georgia Technical College*
Judith Falk, *North Central Michigan College*
Judith Holbrook, *Yavapai College–Prescott*
Justin Dunham, *Johnson County Community College*
Karan Puri, *Queensborough Community College*
Karen Brown, *Wallace Community College*
Khaled Al-Agha, *Wiley College*
Karen Donnelly, *Saint Joseph's College*
Karen Estes, *Saint Petersburg College–Gibbs*
Karl Viehe, *University of the District of Columbia*
Kathleen Kane, *Community College Allegheny County–Pittsburgh*
Ketsia Chapman, *Centura College Online*
Ken Anderson, *Chemeketa Community College*
Kenneth Mead, *Genesee Community College*
Kenneth Williams, *Albany Technical College*

Kim Johnson, *Mesa Community College*
Kristin Good, *Washtenaw Community College*
Lakisha Holmes, *Daytona State College*
Laura Carroll, *Santa Rosa Junior College*
Laura Perez, *Washtenaw Community College*
Laura Stapleton, *Marshall University*
Lee Raubolt, *Yavapai College–Prescott*
Linda Schott, *Ozarks Technical Community College*
Linda Shackelford, *Tidewater Community College–Portsmouth*
Liz Delaney, *Grand Rapids Community College*
Lorena Goebel, *University of Arkansas–Fort Smith*
Loris Zucca, *Lone Star College Kingwood*
Lynette King, *Gadsden State Community College*
Lynn Irons, *College of Southern Idaho*
Mahshid Hassani, *Hillsborough Community College–Brandon*
Marc Campbell, *Daytona State College*
Marcial Echenique, *Broward College–North*
Maria Rodriguez, *Suffolk County Community College–Brentwood*
Marianna Mcclymonds, *Phoenix College*
Marilyn Peacock, *Tidewater Community College–Norfolk*
Marilyn S. Jacobi, *Gateway Community–Technical College*
Mark Anderson, *Durham Technical Community College*
Mark Batell, *Washtenaw Community College*
Mark Billiris, *St. Petersburg College*
Mark Littrell, *Rio Hondo College*
Marwan Abusawwa, *Florida State College–South Campus*
Mary Deas, *Johnson County Community College*
Mary Hito, *Los Angeles Valley College*
Mary Legner, *Riverside Community College*
Mary Wolyniak, *Broome Community College*
Matthew Pitassi, *Rio Hondo College*
Matthew Utz, *University of Arkansas–Fort Smith*
Maureen Loiacano, *Lone Star College*
Mauricio Marroquin, *Los Angeles Valley College*
Michael Cance, *Southeastern Community College*
Michelle Garey, *Delaware Technical & Community College–Dover*
Myrta Groeneveld, *Manchester Community College*
Nancy Eschen, *Florida State College–South Campus*
Natalie Weaver, *Daytona State College–Daytona Beach*
Nataliya Gavryshova, *College of San Mateo*
Nekeith Brown, *Richland College*
Nicole Francis, *Linn-Benton Community College*
Pam Ogaard, *Bismarck State College*
Pat Jones, *Methodist University*
Patricia Arteaga, *Bloomfield College*
Patricia Jones, *Methodist University*

Paula Looney, *Saint Philips College*
Paula Potter, *Yavapai College–Prescott*
Pavel Sikorskii, *Michigan State University–East Lansing*
Penny Marsh, *Johnson County Community College*
Philip Nelson, *Barstow College*
Phillip Taylor, *North Florida Community College*
Ramona Harris, *Gadsden State Community College*
Randey Burnette, *Tallahassee Community College*
Rhoda Oden, *Gadsden State Community College*
Richard Baum, *Santa Barbara City College*
Richard Hobbs, *Mission College*
Richard Pellerin, *Northern Virginia Community College*
Rick Downs, *South Seattle Community College*
Robbert Mckelvy, *Cossatot Community College*
Robert Cohen, *University of District of Columbia*
Robert Evans, *Monterey Peninsula College*
Robert Fusco, *Bergen Community College*
Rodney Oberdick, *Delaware Technical & Community College–Dover*
Roger Mccoach, *County College of Morris*
Roland Trevino, *San Antonio College*

Ron Powers, *Michigan State University–East Lansing*
Rosa Kontos, *Bergen Community College*
Rose Toering, *Kilian Community College*
Ruby Martinez, *San Antonio College*
Ruth Dellinger, *Florida State College Kent Campus*
Ryan Baxter, *Illinois State University*
Sally Jackman, *Richland College*
Sandra Cox, *Kaskaskia College*
Sandra Jovicic, *University of Akron*
Sandra Leabough, *Centura College*
Shanna Goff, *Grand Rapids Community College*
Shannon Miller-Mace, *Marshall University*
Sharon Hudson, *Gulf Coast Community College*
Shawn Krest, *Genesee Community College*
Sherri Kobis, *Erie Community College Northcamp–Williamsville*
Sima Dabir, *Western Iowa Technical Community College*
Spiros Karimbakas, *Fresno City College*
Stanley Hecht, *Santa Monica College*
Stephen Toner, *Victor Valley Community College*
Susan Metzger, *North Central Michigan College*
Susanna Gunther, *Solano Community College*
Suzette Goss, *Lone Star College Kingwood*

Sylvia Brown, *Mountain Empire Community College*
Tammy Potter, *Gadsden State Community College*
Tian Ren, *Queensborough Community College*
Timothy L. Warkentin, *Cloud County Community College*
Toni Houtteman, *Baker College of Clinton Township*
Tonya Michelle Davenport, *Rowan University*
Vernon Bridges, *Durham Technical Community College*
Wayne Barber, *Chemeketa Community College*
William Kirby, *Gadsden State Community College*
Yon Kim, *Passaic County Community College*

The author team most humbly would like to thank all the people who have contributed to this project.

Special thanks to our team of digital authors for their thousands of hours of work: Jody Harris, Linda Schott, Lizette Hernandez Foley, Michael Larkin, Alina Coronel, and to the masters of ceremonies in the digital world: Donna Gerken, Nicole Lloyd, and Steve Toner. We also offer our sincerest appreciation to the video talent: Jody Harris, Alina Coronel, Didi Quesada, Tony Alfonso, and Brianna Kurtz. You folks are the best! To Mitchel Levy our exercise consultant, thank you for your watchful eye fine-tuning the exercise sets and for your ongoing valuable feedback. To Gene Rumann, thank you so much for ensuring accuracy in our manuscripts.

Finally, we greatly appreciate the many people behind the scenes at McGraw-Hill without whom we would still be on page 1. To Emily Williams, the best developmental editor in the business and steady rock that has kept the train on the track. To Mary Ellen Rahn, our executive editor and overall team leader. We're forever grateful for your support and innovative ideas. To our copy editor Pat Steele who has watched over our manuscript for many years and has been a long-time mentor for our writing. You're amazing. To the marketing team Alex Gay, Peter Vanaria, Tim Cote, John Osgood, and Cherie Harshmann, thank you for your creative ideas in making our books come to life in the market. To the director of digital content, Nicole Lloyd, we are most grateful for your long hours of work and innovation in a world that changes day-to-day. And many thanks to the team at ALEKS that oversees quality control in the digital content. To the production team Peggy Selle, Laurie Janssen, and Carrie Burger for making the manuscript beautiful and bringing it all together. And finally to Ryan Blankenship, Marty Lange, and Kurt Strand, thank you for supporting our projects over the years and for the confidence you've always shown in us.

Most importantly, we give special thanks to the students and instructors who use our series in their classes.

Julie Miller   *Molly O'Neill*   *Nancy Hyde*

New and Updated Content for Miller, O'Neill, and Hyde's *Basic College Mathematics*, Third Edition:

Updates throughout the Text:

- New Vocabulary and Key Concept Exercises
- New Tips and Avoiding Mistakes boxes
- Updated Applications and Data in all instances where appropriate
- Writing has been reworked throughout the text in order to improve clarity and understanding

Chapter by Chapter Changes:

Chapter 1
- 31 New Exercises
- 18 New PRE Exercises
- 2 New Examples

Chapter 2
- New Chapter Opener
- 19 New Exercises
- 1 New Example
- New narrative in Section 2.3 on the Greatest Common Factor
- New narrative in Section 2.5 on the Explanation of Dividing by a Fraction

Chapter 3
- New Chapter Opener
- New Procedure Box in Section 3.4 on Rounding a Fraction
- 12 New Exercises
- 2 New Examples

Chapter 4
- New Chapter Opener
- Updated the Procedure Box in Section 4.1 on Rounding Decimals
- New Procedure Box in Section 4.5 on Writing Fractions as Decimals
- 23 New Exercises
- 1 New Example

Chapter 5
- New Chapter Opener
- New Problem Recognition Exercise Section on Operations on Fractions versus Solving Proportions
- 33 New Exercises
- 2 New Examples

Chapter 6
- New Chapter Opener
- New Group Activity on Credit Card Interest
- 28 New Exercises
- 6 New PRE Exercises
- 3 New Examples

Chapter 7
- 17 New Exercises

Chapter 8
- New Chapter Opener
- New Group Activity on Golden Rectangles
- New Tip Box in Section 8.3 on Perfect Squares
- Reorganized the Examples in Section 8.3
- 20 New Exercises

Chapter 9
- New Chapter Opener
- New narrative in Section 9.2 on Frequency Distribution
- 32 New Exercises
- 6 New Examples

Chapter 10
- New Chapter Opener
- New narrative in Section 10.1 on The Set of Real Numbers
- 1 New Concept Connection
- 22 New Exercises
- 3 New Examples

Chapter 11
- New Chapter Opener
- New Definition of a Linear Equation in One Variable
- New Group Activity on Deciphering Coded Messages
- 18 New Exercises
- 8 new PRE exercises

Additional Topics Appendix
- 3 New Exercises

Whole Numbers

Chapter 1

Chapter 1 begins with adding, subtracting, multiplying, and dividing whole numbers. We also include rounding, estimating, and applying the order of operations.

Review Your Skills

Before you begin this chapter, check your skill level by trying the exercises in the puzzle. If you have trouble with any of these problems, come back to the puzzle as you work through the chapter, and fill in the answers that gave you trouble.

Across

1. $123 + 38$

3. $3866 - 2345$

5. $4 + 7 \times 8$

6. $51 \div 17$

7. $6^2 + 3^3$

Down

2. 125×50

3. 4^2

4. $575 + 89 + 722$

8. $372 \div 12$

Concepts

1. Place Value
2. Standard Notation and Expanded Notation
3. Writing Numbers in Words
4. The Number Line and Order

1. Place Value

Numbers provide the foundation that is used in mathematics. We begin this chapter by discussing how numbers are represented and named. All numbers in our numbering system are composed from the **digits** 0, 1, 2, 3, 4, 5, 6, 7, 8, and 9. In mathematics, the numbers 0, 1, 2, 3, 4, 5, 6, 7, 8, 9, 10, 11, 12, . . . are called the *whole numbers*. (The three dots are called *ellipses* and indicate that the list goes on indefinitely.)

For large numbers, commas are used to separate digits into groups of three called **periods**. For example, the number of live births in the United States in a recent year was 4,058,614. (*Source: The World Almanac*) Numbers written in this way are said to be in **standard form**. The position of each digit determines the place value of the digit. To interpret the number of births in the United States, refer to the place value chart (Figure 1-1).

Concept Connections

1. Explain the difference between the two 3's in the number 303.

Instructor Note: At this time you may want to introduce the definition of a numeral and explain the difference between a number and a numeral.

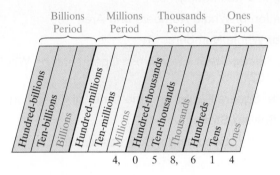

Figure 1-1

The digit 5 in 4,058,614 represents 5 ten-thousands because it is in the ten-thousands place. The digit 4 on the left represents 4 millions, whereas the digit 4 on the right represents 4 ones.

Skill Practice

Determine the place value of the digit 4.

2. 547,098,632
3. 1,659,984,036
4. 6,420

Classroom Example: p. 6, Exercise 10

Example 1   Determining Place Value

Determine the place value of the digit 2.

   a. 417,216,900   **b.** 724   **c.** 502,000,700

Solution:

   a. 417,216,900   hundred-thousands

   b. 724   tens

   c. 502,000,700   millions

Answers

1. First 3 (on the left) represents 3 hundreds, while the second 3 (on the right) represents 3 ones.
2. Ten-millions
3. Thousands
4. Hundreds

Example 2 **Determining Place Value**

Mount Everest, the highest mountain on Earth, is 29,035 feet (ft) tall. Give the place value for each digit.

Solution:

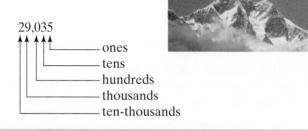

29,035
ones
tens
hundreds
thousands
ten-thousands

2. Standard Notation and Expanded Notation

A number can also be written in an expanded form by writing each digit with its place value unit. For example, 287 can be written as

$$287 = 2 \text{ hundreds} + 8 \text{ tens} + 7 \text{ ones}$$
$$= 2 \times 100 + 8 \times 10 + 7 \times 1$$
$$= 200 + 80 + 7$$

This is called **expanded form**.

Example 3 **Converting Standard Form to Expanded Form**

Convert to expanded form.

  a. 4,672   **b.** 257,016

Solution:

a. 4,672   4 thousands + 6 hundreds + 7 tens + 2 ones

                $= 4 \times 1,000 + 6 \times 100 + 7 \times 10 + 2 \times 1$

                $= 4,000 + 600 + 70 + 2$

b. 257,016   2 hundred-thousands + 5 ten-thousands +

                7 thousands + 1 ten + 6 ones

                $= 2 \times 100,000 + 5 \times 10,000 + 7 \times 1,000 + 1 \times 10 + 6 \times 1$

                $= 200,000 + 50,000 + 7,000 + 10 + 6$

Example 4 Converting Expanded Form to Standard Form

Convert to standard form.

a. 2 hundreds + 5 tens + 9 ones

b. 1 thousand + 2 tens + 5 ones

Solution:

a. 2 hundreds + 5 tens + 9 ones = 259

b. Each place position from the thousands place to the ones place must contain a digit. In this problem, there is no reference to the hundreds place digit. Therefore, we assume 0 hundreds. Thus,

$$1 \text{ thousand} + 0 \text{ hundreds} + 2 \text{ tens} + 5 \text{ ones} = 1{,}025$$

3. Writing Numbers in Words

The word names of some two-digit numbers appear with a hyphen, while others do not. For example:

Number	Number Name
12	twelve
68	sixty-eight
40	forty
42	forty-two

To write a three-digit or larger number, begin at the leftmost group of digits. The number named in that group is followed by the period name, followed by a comma. Then the next period is named, and so on.

Example 5 Writing a Number in Words

Write 621,417,325 in words.

Solution:

621,417,325

six hundred twenty-one million,

four hundred seventeen thousand,

three hundred twenty-five

Notice from Example 5 that when naming numbers, the name of the ones period is not attached to the last group of digits. Also note that for whole numbers, the word *and* should not appear in word names. For example, 405 should be written as four hundred five.

Example 6   **Writing a Number in Standard Form**

Write the number in standard form.

      Six million, forty-six thousand, nine hundred three

Solution:

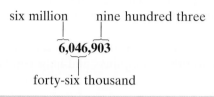

Classroom Example: p. 8, Exercise 58

We have seen several examples of writing a number in standard form, in expanded form, and in words. Standard form is the most concise representation. Also note that when we write a four-digit number in standard form, the comma is often omitted. For example, 4,389 is often written as 4389.

4. The Number Line and Order

Whole numbers can be visualized as equally spaced points on a line called a *number line* (Figure 1-2).

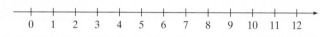

Figure 1-2

The whole numbers begin at 0 and are ordered from left to right by increasing value.

    A number is graphed on a number line by placing a dot at the corresponding point. For any two numbers graphed on a number line, the number to the left is less than the number to the right. Similarly, a number to the right is greater than the number to the left. In mathematics, the symbol $<$ is used to denote "is less than," and the symbol $>$ means "is greater than." Therefore,

$3 < 5$	means	3 is less than 5
$5 > 3$	means	5 is greater than 3

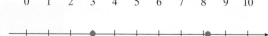

Instructor Note: Point out to students that $3 < 5$ and $5 > 3$ are equivalent expressions.

Example 7   **Determining Order Between Two Numbers**

Fill in the blank with the symbol $<$ or $>$.

a. 7 ☐ 0   **b.** 30 ☐ 82

Solution:

a. 7 $\boxed{>}$ 0

b. 30 $\boxed{<}$ 82

To visualize the numbers 82 and 30 on the number line, it may be necessary to use a different scale. Rather than setting equally spaced marks in units of 1, we can use units of 10. The number 82 must be somewhere between 80 and 90 on the number line.

Classroom Example: p. 8, Exercise 80

Answers

11. 14,609   **12.** $>$ **13.** $<$

Section 1.1   Practice Exercises

Study Skills Exercise

For additional exercises, see Classroom Activities 1.1A–1.1C in the *Student's Resource Manual* at www.mhhe.com/moh.

In this text, we provide skills for you to enhance your learning experience. Each set of practice exercises begins with an activity that focuses on one of eight areas: learning about your course, using your text, taking notes, doing homework, taking an exam (test and math anxiety), managing your time, recognizing your learning style, and studying for the final exam.

Each activity requires only a few minutes and will help you to pass this class and become a better math student. Many of these skills can be carried over to other disciplines and help you become a model college student.

To begin, write down the following information.

a. Instructor's name

b. Instructor's office number

c. Instructor's telephone number

d. Instructor's email address

e. Instructor's office hours

f. Days of the week that the class meets

g. The room number in which the class meets

h. Is there a lab requirement for this course? If so, where is the lab located and how often must you go?

Vocabulary and Key Concepts

1. a. For large numbers, commas are used to separate digits into groups called _____periods_____.

b. The place values of the digits in the ones period are the ones, tens, and _____hundreds_____ places.

c. The place values of the digits in the _____thousands_____ period are the thousands, ten-thousands, and hundred-thousands places.

Concept 1: Place Value

2. Name the place value for each digit in 36,791.
1: ones; 9: tens; 7: hundreds; 6: thousands; 3: ten-thousands

3. Name the place values for each of the digits in 8,213,457.
7: ones; 5: tens; 4: hundreds; 3: thousands; 1: ten-thousands; 2: hundred-thousands; 8: millions

4. Name the place values for each of the digits in 103,596.
6: ones; 9: tens; 5: hundreds; 3: thousands; 0: ten-thousands; 1: hundred-thousands

For Exercises 5–24, determine the place value for each underlined digit. **(See Example 1.)**

5. 3<u>2</u>1   Tens

6. 6<u>8</u>9   Tens

7. 21<u>4</u>   Ones

8. 73<u>8</u>   Ones

9. 8,<u>7</u>10   Hundreds

10. 2,<u>2</u>93   Hundreds

11. <u>1</u>,430   Thousands

12. <u>3</u>,101   Thousands

13. <u>4</u>52,723   Hundred-thousands

14. <u>6</u>55,878   Hundred-thousands

 15. <u>1</u>,023,676,207   Billions

16. <u>3</u>,111,901,211   Billions

17. <u>2</u>2,422   Ten-thousands

18. <u>5</u>8,106   Ten-thousands

19. 51,<u>0</u>33,201   Millions

20. 9<u>3</u>,971,224   Millions

21. The number of U.S. travelers abroad in a recent year was <u>1</u>0,677,881. **(See Example 2.)**   Ten-millions

22. The area of Lake Superior is 31,<u>8</u>20 mi^2.   Thousands

23. For a recent year, the total number of U.S. $1 bills in circulation was <u>7</u>,653,468,440.   Billions

24. For a certain flight, the cruising altitude of a commercial jet is <u>3</u>1,000 ft.   Ten-thousands

Concept 2: Standard Notation and Expanded Notation

For Exercises 25–32, convert the numbers to expanded form. **(See Example 3.)**

25. 58 5 tens + 8 ones;
5 × 10 + 8 × 1

26. 71 7 tens + 1 one;
7 × 10 + 1 × 1

27. 539 5 hundreds + 3 tens + 9 ones;
5 × 100 + 3 × 10 + 9 × 1

28. 382 3 hundreds + 8 tens + 2 ones;
3 × 100 + 8 × 10 + 2 × 1

29. 503 5 hundreds + 3 ones;
5 × 100 + 3 × 1

30. 809 8 hundreds + 9 ones;
8 × 100 + 9 × 1

31. 10,241 1 ten-thousand + 2 hundreds + 4 tens + 1 one;
1 × 10,000 + 2 × 100 + 4 × 10 + 1 × 1

32. 20,873 2 ten-thousands + 8 hundreds +
7 tens + 3 ones;
2 × 10,000 + 8 × 100 +
7 × 10 + 3 × 1

For Exercises 33–40, convert the numbers to standard form. **(See Example 4.)**

33. 5 hundreds + 2 tens + 4 ones   524

34. 3 hundreds + 1 ten + 8 ones
318

35. 1 hundred + 5 tens   150

36. 6 hundreds + 2 tens   620

37. 1 thousand + 9 hundreds + 6 ones   1,906

38. 4 thousands + 2 hundreds + 1 one   4,201

39. 8 ten-thousands + 5 thousands + 7 ones   85,007

40. 2 ten-thousands + 6 thousands + 2 ones   26,002

41. Name the first four periods of a number (from right to left).   Ones, thousands, millions, billions

42. Name the first four place values of a number (from right to left).   Ones, tens, hundreds, thousands

Concept 3: Writing Numbers in Words

For Exercises 43–50, write the number in words. **(See Example 5.)**

43. 241
Two hundred forty-one

44. 327
Three hundred twenty-seven

45. 603
Six hundred three

46. 108
One hundred eight

47. 31,530 Thirty-one
thousand, five hundred thirty

48. 52,160 Fifty-two thousand,
one hundred sixty

49. 100,234 One hundred
thousand, two hundred thirty-four

50. 400,199 Four hundred
thousand, one hundred ninety-nine

51. The Shuowen jiezi dictionary, an ancient Chinese dictionary that dates back to the year 100, contained 9,535 characters. Write 9,535 in words.
Nine thousand, five hundred thirty-five

52. Researchers calculate that about 590,712 stone blocks were used to construct the Great Pyramid. Write 590,712 in words.
Five hundred ninety thousand, seven hundred twelve

53. Mt. McKinley in Alaska is 20,320 ft high. Write 20,320 in words.
Twenty thousand, three hundred twenty

54. There are 1,800 seats in the Regal Champlain Theater in Plattsburgh, New York. Write 1,800 in words.   One thousand, eight hundred

55. Interstate I-75 is 1,377 miles (mi) long. Write the number 1,377 in words.
One thousand, three hundred seventy-seven

56. In the United States, there are approximately 60,000,000 cats living in households. Write the number 60,000,000 in words.   Sixty million

 Writing    Translating Expression   Geometry   Scientific Calculator   Video   NS E

For Exercises 57–62, convert the number to standard form. **(See Example 6.)**

57. Six thousand, five   6,005

58. Four thousand, four   4,004

59. Six hundred seventy-two thousand   672,000

60. Two hundred forty-eight thousand   248,000

61. One million, four hundred eighty-four thousand, two hundred fifty   1,484,250

62. Two million, six hundred forty-seven thousand, five hundred twenty   2,647,520

Concept 4: The Number Line and Order

For Exercises 63–64, graph the numbers on the number line.

63. a. 6   **b.** 13   **c.** 8   **d.** 1

64. a. 5   **b.** 3   **c.** 11   **d.** 9

65. On a number line, what number is 4 units to the right of 6?   10

66. On a number line, what number is 8 units to the left of 11?   3

67. On a number line, what number is 3 units to the left of 7?   4

68. On a number line, what number is 5 units to the right of 0?   5

For Exercises 69–72, translate the inequality to words.

69. $8 > 2$
8 is greater than 2, or
2 is less than 8

70. $6 < 11$
6 is less than 11, or
11 is greater than 6

71. $3 < 7$
3 is less than 7, or
7 is greater than 3

72. $14 > 12$
14 is greater than 12, or
12 is less than 14

For Exercises 73–84, fill in the blank with the inequality symbol $<$ or $>$. **(See Example 7.)**

73. 6 ☐ 11 $<$

74. 14 ☐ 13 $>$

75. 21 ☐ 18 $>$

76. 5 ☐ 7 $<$

77. 3 ☐ 7 $<$

78. 14 ☐ 24 $<$

79. 95 ☐ 89 $>$

80. 28 ☐ 30 $<$

81. 0 ☐ 3 $<$

82. 8 ☐ 0 $>$

83. 90 ☐ 91 $<$

84. 48 ☐ 47 $>$

Expanding Your Skills

85. Answer true or false. 12 is a digit.
False

86. Answer true or false. 26 is a digit.
False

87. What is the greatest two-digit number?   99

88. What is the greatest three-digit number?   999

89. What is the greatest whole number?
There is no greatest whole number.

90. What is the least whole number?   0

91. How many zeros are there in the number ten million?   7

92. How many zeros are there in the number one hundred billion?   11

93. What is the greatest three-digit number that can be formed from the digits 6, 9, and 4? Use each digit only once.   964

94. What is the greatest three-digit number that can be formed from the digits 0, 4, and 8? Use each digit only once.   840

Addition of Whole Numbers and Perimeter

Concepts

1. Addition of Whole Numbers Using the Number Line
2. Addition of Whole Numbers
3. Properties of Addition
4. Translations and Applications Involving Addition
5. Perimeter

1. Addition of Whole Numbers Using the Number Line

We use addition of whole numbers to represent an increase in quantity. For example, suppose Jonas typed 5 pages of a report before lunch. Later in the afternoon he typed 3 more pages. The total number of pages that he typed is found by adding 5 and 3.

$$5 \text{ pages} + 3 \text{ pages} = 8 \text{ pages}$$

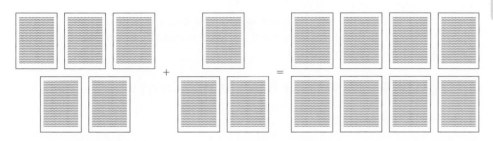

The result of an addition problem is called the **sum**, and the numbers being added are called **addends**. Thus,

$$5 + 3 = 8$$

addends   sum

The number line is a useful tool to visualize the operation of addition. To add 5 and 3 on a number line, begin at 0 and move 5 units to the right. Then move an additional 3 units to the right. The final location indicates the sum.

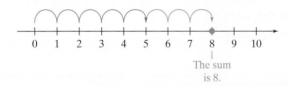

The sum is 8.

You can use a number line to find the sum of any pair of digits. The sums for all possible pairs of one-digit numbers should be memorized (see Exercise 9). Memorizing these basic addition facts will make it easier for you to add larger numbers.

Concept Connections

1. Identify the addends and the sum.
   $3 + 7 + 12 = 22$

2. Addition of Whole Numbers

To add whole numbers, line up the numbers vertically by place value. Then add the digits in the corresponding place positions.

Example 1   Adding Whole Numbers

Add. $24 + 61$

Solution:

$$
\begin{array}{rl}
24 = & 2 \text{ tens} + 4 \text{ ones} \\
+\,61 = & 6 \text{ tens} + 1 \text{ one} \\
\hline
85 = & 8 \text{ tens} + 5 \text{ ones}
\end{array}
$$

Skill Practice

2. Add. $\begin{array}{r} 47 \\ +\,32 \\ \hline \end{array}$

Classroom Example: p. 17, Exercise 16

Answers

1. Addends: 3, 7, and 12; sum: 22
2. 79

Skill Practice

Add.

3. 4135 + 210

Classroom Example: p. 17,
Exercise 26

| Example 2 | Adding Whole Numbers |

Add.   261 + 28

Solution:

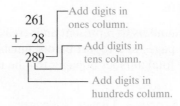

$$
\begin{array}{r}
261 \\
+\ 28 \\
\hline
289 \\
\end{array}
$$

—Add digits in ones column.

—Add digits in tens column.

—Add digits in hundreds column.

Sometimes when adding numbers, the sum of the digits in a given place position is greater than 9. If this occurs, we must do what is called *carrying* or *regrouping*. Example 3 illustrates this process.

| Example 3 | Adding Whole Numbers with Carrying |

Add.   35 + 48

Solution:

$$
\begin{array}{l}
35 = 3 \text{ tens} +\ 5 \text{ ones} \\
+\ 48 = 4 \text{ tens} +\ 8 \text{ ones} \\
\hline
\quad\quad\quad 7 \text{ tens} + 13 \text{ ones} \longleftarrow
\end{array}
$$

The sum of the digits in the ones place exceeds 9. But 13 ones is the same as 1 ten and 3 ones. We can *carry* 1 ten to the tens column while leaving the 3 ones in the ones column. Notice that we placed the carried digit above the tens column.

$$
\begin{array}{l}
\overset{1}{3}5 = 3 \text{ tens} + 5 \text{ ones} \\
+\ 48 = 4 \text{ tens} + 8 \text{ ones} \\
\hline
83 = 8 \text{ tens} + 3 \text{ ones}
\end{array}
$$

The sum is 83.

Skill Practice

Add.

4. 43 + 29

Classroom Example: p. 17,
Exercise 38

| Example 4 | Adding Whole Numbers with Carrying |

Add.   458 + 67

Solution:

$$
\begin{array}{r}
\overset{1}{4}58 \\
+\ 67 \\
\hline
5 \\
\end{array}
$$

Add the digits in the ones column: $8 + 7 = 15$. Write 5 in the ones column, and carry the 1 to the tens column.

$$
\begin{array}{r}
\overset{11}{4}58 \\
+\ 67 \\
\hline
25 \\
\end{array}
$$

Add the digits in the tens column (including the carry): $1 + 5 + 6 = 12$. Write the 2 in the tens column, and carry the 1 to the hundreds column.

$$
\begin{array}{r}
\overset{11}{4}58 \\
+\ 67 \\
\hline
525 \\
\end{array}
$$

Add the digits in the hundreds column. The sum is 525.

Skill Practice

Add.

5. 3087 + 25,686

Classroom Example: p. 17,
Exercise 48

Answers

3. 4345
4. 72
5. 28,773

Addition of numbers may include more than two addends.

Example 5   **Adding Whole Numbers**

Add. $21{,}076 + 84{,}158 + 2419$

Solution:

$$\begin{array}{r} \overset{1}{} \ \overset{1\,2}{} \\ 21{,}076 \\ 84{,}158 \\ +\ \ 2{,}419 \\ \hline 107{,}653 \end{array}$$

In this example, the sum of the digits in the ones column is 23. Therefore, we write the 3 and carry the 2.

Classroom Example: p. 17, Exercise 50

3. Properties of Addition

We present three properties of addition that you may have already discovered.

Addition Property of 0

The sum of any number and 0 is that number.

Examples: $5 + 0 = 5$

$0 + 2 = 2$

Instructor Note: Ask students if subtraction is commutative.

Commutative Property of Addition

Changing the order of two addends does not affect the sum.

Example: $5 + 7$ is equivalent to $7 + 5$

In mathematics we use parentheses ( ) as grouping symbols. To add more than two numbers, we can group them and then add. For example:

$(2 + 3) + 8$ Parentheses indicate that $2 + 3$ is added first. Then 8 is added to the result.
$= 5 + 8$
$= 13$

$2 + (3 + 8)$ Parentheses indicate that $3 + 8$ is added first. Then the result is added to 2.
$= 2 + 11$
$= 13$

Associative Property of Addition

The manner in which addends are grouped does not affect the sum.

Example: $(1 + 7) + 3$ is equivalent to $1 + (7 + 3)$

Answer

6. 71,147

Example 6   Applying the Properties of Addition

a. Rewrite 9 + 6, using the commutative property of addition.

b. Rewrite (15 + 9) + 5, using the associative property of addition.

Solution:

a. 9 + 6 = 6 + 9   Change the order of the addends.

b. (15 + 9) + 5 = 15 + (9 + 5)   Change the grouping of the addends.

4. Translations and Applications Involving Addition

In the English language, there are many different words and phrases that imply addition. A partial list is given in Table 1-1.

Table 1-1

Word/Phrase	Example	In Symbols
Sum	The sum of 6 and 2	6 + 2
Added to	3 added to 8	8 + 3
Increased by	7 increased by 2	7 + 2
More than	10 more than 6	6 + 10
Plus	8 plus 3	8 + 3
Total of	The total of 9 and 6	9 + 6

Example 7   Translating an English Phrase to a Mathematical Statement

Translate each phrase to an equivalent mathematical statement and simplify.

a. 12 added to 109

b. The sum of 1386 and 376

Solution:

a. 109 + 12
$$\begin{array}{r} \overset{1}{1}09 \\ + \ 12 \\ \hline 121 \end{array}$$

b. 1386 + 376
$$\begin{array}{r} \overset{11}{1386} \\ + \ 376 \\ \hline 1762 \end{array}$$

Addition of whole numbers is sometimes necessary to solve application problems.

Example 8 Solving an Application Problem

Carlita works as a waitress at El Pinto restaurant in Albuquerque, New Mexico. Her tips for the last five nights were $30, $18, $66, $102, and $45. Find the total amount she made in tips.

Solution:

To find the total, we add.

$$
\begin{array}{r}
\overset{1\,2}{\$\ 30} \\
18 \\
66 \\
102 \\
+\quad 45 \\
\hline
\$261
\end{array}
$$

Carlita made $261 in tips.

Skill Practice

11. Talita received test scores of 92, 100, 84, and 96 on her first four math tests. She also earned 8 points of extra credit. How many total points did she earn?

Classroom Example: p. 18, Exercise 80

Tables and graphs are often used to summarize information in an organized manner. Examples 9 and 10 demonstrate the interpretation of these tools.

Example 9 Solving an Application Problem Involving a Table

The following table gives the number of hits for five popular websites for a recent month. Find the total number of visitors.

Website	Number of Visitors
AOL Time Warner Network	97,995
MSN-Microsoft sites	89,819
Yahoo! sites	83,433
Google sites	37,460
Terra Lycos	36,173

Solution:

$$
\begin{array}{r}
\overset{3\,3 2\ 2 2}{97{,}995} \\
89{,}819 \\
83{,}433 \\
37{,}460 \\
+\quad 36{,}173 \\
\hline
344{,}880
\end{array}
$$

There were 344,880 combined visitors to these websites.

Skill Practice

12. The table gives the number of gold, silver, and bronze medals won in the 2010 Winter Olympics for selected countries. Find the total number of medals won by Canada.

	Gold	Silver	Bronze
Germany	10	13	7
USA	9	15	13
Canada	14	7	5

Classroom Example: p. 19, Exercise 86

Answers

11. 380 points **12.** 26 medals

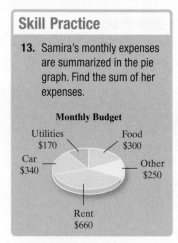
Example 10 Solving an Application Problem Involving a Graph

The bar graph in Figure 1-3 gives the estimated number of cases of HIV/AIDS in the United States for three selected years. The pink bars in the graph represent the values for the number of women. The blue bars in the graph represent the values for the number of men.

Find the total number of HIV/AIDS cases for women for the years 1–3.

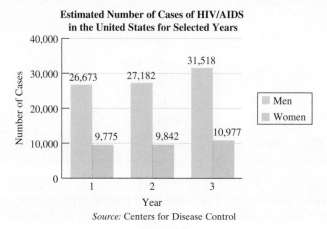

Estimated Number of Cases of HIV/AIDS in the United States for Selected Years

Source: Centers for Disease Control

Figure 1-3

Solution:

The question asks for the number of HIV/AIDS cases for women only. Therefore, add the values corresponding to the pink bars in the graph.

$$\begin{array}{r} \overset{2\ 2\ 1\ 1}{9{,}775} \\ 9{,}842 \\ +\ 10{,}977 \\ \hline 30{,}594 \end{array}$$

There were 30,594 HIV/AIDS cases among women for years 1–3.

5. Perimeter

One special application of addition is to find the perimeter of a polygon. A **polygon** is a flat closed figure formed by line segments connected at their ends. Familiar figures such as triangles, rectangles, and squares are examples of polygons. See Figure 1-4.

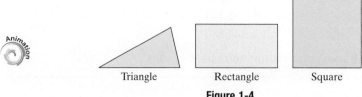

Triangle   Rectangle   Square

Figure 1-4

The **perimeter** of any polygon is the distance around the outside of the figure. To find the perimeter, add the lengths of the sides.

Example 11   **Finding Perimeter**

Find the perimeter of the triangle.

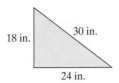

Solution:

The perimeter is the sum of the lengths of the sides.

$$
\begin{array}{r}
\overset{1}{18} \text{ in.} \\
24 \text{ in.} \\
+\ 30 \text{ in.} \\
\hline
72 \text{ in.}
\end{array}
$$

The perimeter is 72 inches.

Skill Practice

14. Find the perimeter of the rectangle.

Classroom Example: p. 20, Exercise 92

Example 12   **Finding Perimeter**

A paving company wants to edge the perimeter of a parking lot with concrete curbing. Find the perimeter of the parking lot.

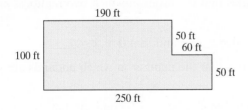

Solution:

The perimeter is the sum of the lengths of the sides.

$$
\begin{array}{r}
\overset{3}{190} \text{ ft} \\
50 \text{ ft} \\
60 \text{ ft} \\
50 \text{ ft} \\
250 \text{ ft} \\
+\ 100 \text{ ft} \\
\hline
700 \text{ ft}
\end{array}
$$

The distance around the parking lot (the perimeter) is 700 ft.

Skill Practice

15. Find the perimeter of the garden.

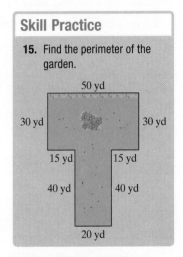

Classroom Example: p. 20, Exercise 96

Instructor Note: Tell students their answers should include the unit of measure and not just a number.

Answers
14. 22 ft
15. 240 yd

Section 1.2   Practice Exercises

Study Skills Exercise

For additional exercises, see Classroom Activities 1.2A–1.2D in the *Student's Resource Manual* at www.mhhe.com/moh.

Taking 12 credit-hours is the equivalent of a full-time job. Often students try to work too many hours while taking classes at school.

a. Write down how many hours you work per week and the number of credit-hours you are taking this term.

Number of hours worked per week _____

Number of credit-hours this term _____

b. The table gives a recommended limit on the number of hours you should work based on the number of credit-hours you are taking at school. (Keep in mind that other responsibilities in your life such as your family might also make it necessary to limit your hours at work even more.) How do your numbers from part (a) compare to those in the table? Are you working too many hours?

Number of Credit-hours	Maximum Number of Hours of Work per Week
3	40
6	30
9	20
12	10
15	0

Vocabulary and Key Concepts

1. a. The numbers being added in an addition problem are called the _____addends_____.

    b. The result of an addition problem is called the _____sum_____.

    c. The _____commutative_____ property of addition states that the order in which two numbers are added does not affect the sum.

    d. The addition property of 0 indicates that $4 + 0 = \underline{\ 4\ }$ and that $0 + 4 = \underline{\ 4\ }$.

    e. The _____associative_____ property of addition states that the manner in which addends are grouped does not affect the sum.

    f. A _____polygon_____ is a flat closed figure formed by line segments connected at their ends.

    g. The _____perimeter_____ of a polygon is the sum of the lengths of the sides.

Review Exercises

For Exercises 2–8, write the number in the form indicated.

2. Write 5,024 in expanded form.
5 thousands + 2 tens + 4 ones; $5 \times 1000 + 2 \times 10 + 4 \times 1$

3. Write 351 in expanded form.
3 hundreds + 5 tens + 1 one; $3 \times 100 + 5 \times 10 + 1 \times 1$

4. Write 351 in words.   Three hundred fifty-one

5. Write 107 in expanded form.
1 hundred + 7 ones; $1 \times 100 + 7 \times 1$

6. Write in standard form: two thousand, four   2004

7. Write in standard form: four thousand, twelve
4012

8. Write in standard form:
6 thousands + 2 hundreds + 6 ones   6206

 Writing    Translating Expression   Geometry   Scientific Calculator   Video   NS E

Concept 1: Addition of Whole Numbers Using the Number Line

9. Fill in the table. Use the number line if necessary.

+	0	1	2	3	4	5	6	7	8	9
0	0	1	2	3	4	5	6	7	8	9
1	1	2	3	4	5	6	7	8	9	10
2	2	3	4	5	6	7	8	9	10	11
3	3	4	5	6	7	8	9	10	11	12
4	4	5	6	7	8	9	10	11	12	13
5	5	6	7	8	9	10	11	12	13	14
6	6	7	8	9	10	11	12	13	14	15
7	7	8	9	10	11	12	13	14	15	16
8	8	9	10	11	12	13	14	15	16	17
9	9	10	11	12	13	14	15	16	17	18

For Exercises 10–15, identify the addends and the sum.

10. $5 + 9 = 14$
Addends: 5, 9; sum: 14

11. $2 + 8 = 10$
Addends: 2, 8; sum: 10

12. $12 + 5 = 17$
Addends: 12, 5; sum: 17

13. $11 + 10 = 21$
Addends: 11, 10; sum: 21

14. $1 + 13 + 4 = 18$
Addends: 1, 13, 4; sum: 18

15. $5 + 8 + 2 = 15$
Addends: 5, 8, 2; sum: 15

Concept 2: Addition of Whole Numbers

For Exercises 16–31, add. **(See Examples 1 and 2.)**

16.
$$\begin{array}{r} 42 \\ + 33 \\ \hline 75 \end{array}$$

17.
$$\begin{array}{r} 21 \\ + 53 \\ \hline 74 \end{array}$$

18.
$$\begin{array}{r} 39 \\ + 20 \\ \hline 59 \end{array}$$

19.
$$\begin{array}{r} 15 \\ + 43 \\ \hline 58 \end{array}$$

20.
$$\begin{array}{r} 12 \\ 15 \\ + 32 \\ \hline 59 \end{array}$$

21.
$$\begin{array}{r} 10 \\ 8 \\ + 30 \\ \hline 48 \end{array}$$

22.
$$\begin{array}{r} 7 \\ 21 \\ + 10 \\ \hline 38 \end{array}$$

23.
$$\begin{array}{r} 6 \\ 11 \\ + 2 \\ \hline 19 \end{array}$$

24. $341 + 225$ 566

25. $407 + 181$ 588

26. $890 + 107$ 997

27. $444 + 354$ 798

28. $4 + 13 + 102$ 119

29. $11 + 221 + 5$ 237

30. $31 + 7 + 430$ 468

31. $24 + 14 + 160$ 198

For Exercises 32–51, add the whole numbers with carrying. **(See Examples 3–5.)**

32.
$$\begin{array}{r} 76 \\ + 45 \\ \hline 121 \end{array}$$

33.
$$\begin{array}{r} 25 \\ + 59 \\ \hline 84 \end{array}$$

34.
$$\begin{array}{r} 87 \\ + 24 \\ \hline 111 \end{array}$$

35.
$$\begin{array}{r} 38 \\ + 77 \\ \hline 115 \end{array}$$

36.
$$\begin{array}{r} 658 \\ + 231 \\ \hline 889 \end{array}$$

37.
$$\begin{array}{r} 642 \\ + 295 \\ \hline 937 \end{array}$$

38.
$$\begin{array}{r} 152 \\ + 549 \\ \hline 701 \end{array}$$

39.
$$\begin{array}{r} 462 \\ + 388 \\ \hline 850 \end{array}$$

40. $15 + 5 + 9$ 29

41. $2 + 31 + 8$ 41

42. $14 + 9 + 17$ 40

43. $7 + 18 + 4$ 29

44. $79 + 112 + 12$
203

45. $62 + 907 + 34$
1003

46. $331 + 422 + 76$
829

47. $87 + 119 + 630$
836

48. $4980 + 10,223$
15,203

49. $23,112 + 892$
24,004

50. $10,223 + 25,782 + 4980$
40,985

51. $92,377 + 5622 + 34,659$
132,658

   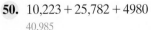 

Writing   Translating Expression   Geometry   Scientific Calculator   Video   NSE

Concept 3: Properties of Addition

For Exercises 52–55, rewrite the addition problem, using the commutative property of addition. **(See Example 6.)**

52. $12 + 6 = \square + \square$
   $6 + 12$

53. $30 + 21 = \square + \square$
   $21 + 30$

54. $101 + 44 = \square + \square$
   $44 + 101$

55. $8 + 13 = \square + \square$
   $13 + 8$

For Exercises 56–59, rewrite the addition problem using the associative property of addition, by inserting a pair of parentheses.

56. $(4 + 8) + 13 = 4 + 8 + 13$
   $4 + (8 + 13)$

57. $(23 + 9) + 10 = 23 + 9 + 10$
   $23 + (9 + 10)$

58. $7 + (12 + 8) = 7 + 12 + 8$
   $(7 + 12) + 8$

59. $41 + (3 + 22) = 41 + 3 + 22$
   $(41 + 3) + 22$

60. Explain the difference between the commutative and associative properties of addition.
   The commutative property changes the order of the addends, and the associative property changes the grouping.

61. Explain the addition property of 0. Then simplify the expressions.

   a. $423 + 0$ **b.** $0 + 25$ **c.** $\begin{array}{r} 67 \\ + \ 0 \\ \hline \end{array}$

The sum of any number and 0 is that number.
a. 423   **b.** 25   **c.** 67

Concept 4: Translations and Applications Involving Addition

For Exercises 62–70, translate the English phrase into a mathematical statement and simplify. **(See Example 7.)**

62. The sum of 13 and 7
   $13 + 7; 20$

63. The sum of 100 and 42
   $100 + 42; 142$

64. 45 added to 7
   $7 + 45; 52$

65. 81 added to 23
   $23 + 81; 104$

66. 5 more than 18
   $18 + 5; 23$

67. 2 more than 76
   $76 + 2; 78$

68. 1523 increased by 90
   $1523 + 90; 1613$

69. 1320 increased by 448
   $1320 + 448; 1768$

70. The total of 5, 39, and 81
   $5 + 39 + 81; 125$

For Exercises 71–78, write an English phrase from the mathematical statement. Answers may vary.

71. $54 + 24$
For example: The sum of 54 and 24

72. $33 + 15$
For example: The sum of 33 and 15

73. $12 + 88$
For example: 88 added to 12

74. $70 + 15$
For example: 15 added to 70

75. $4 + 23 + 77$
For example: The total of 4, 23, and 77

76. $11 + 41 + 53$
For example: The total of 11, 41, and 53

77. $10 + 8$
For example: 10 increased by 8

78. $25 + 14$
For example: 25 increased by 14

79. The attendance at a high school play during one weekend was as follows: 103 on Friday, 112 on Saturday, and 61 at the Sunday matinee. What was the total attendance? **(See Example 8.)** 276 people

80. To schedule enough drivers for an upcoming week, a local pizza shop manager recorded the number of deliveries each day from the previous week: 38, 54, 44, 61, 97, 103, 124. What was the total number of deliveries for the week? 521 deliveries

81. Three top television shows entertained the following number of viewers in one week: 21,209,000 for *Dancing with the Stars*, 20,836,000 for *American Idol (Tuesday)*, and 16,448,000 for *NCIS*. Find the total number of viewers for these shows. 58,493,000 viewers

82. To travel from Houston to Corpus Christi, a salesperson must stop in San Antonio. If it is 195 mi from Houston to San Antonio and 228 mi from San Antonio to Corpus Christi, how far will she travel on this trip? 423 mi

83. Nora earned $43,000 last year. This year her salary was increased by $2500. What is her present salary? $45,500

84. The number of participants in the Special Olympics increased by 1,205,655 since it began in 1968 with 1000 athletes. How many athletes are presently participating? 1,206,655 athletes

85. A portion of Jonathan's checking account register is shown. What is the total amount of the four checks written? **(See Example 9.)** $245

Check No.	Description	Deposit	Payment	Balance
1871	Electric bill		$60	$180
1872	Groceries		52	128
1873	Department store		75	53
	Payroll deposit	$1256		1309
1874	Restaurant		58	1251
	Deposit from savings	150		1401

86. The table gives the number of desks and chairs delivered each quarter to an office supply store. Find the total number of desks delivered for the year. 423 desks

	Chairs	Desks
March	220	115
June	185	104
September	201	93
December	198	111

87. The Student Career Experience Program is a program that places students in government jobs. The bar graph displays the number of participants in the top six agencies. Find the total number of participants in the program. **(See Example 10.)**
13,538 participants

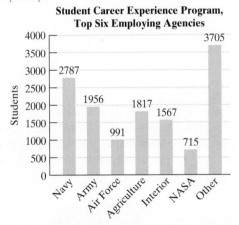

Student Career Experience Program, Top Six Employing Agencies

Source: U.S. Office of Personnel Management

88. The bar graph displays the number of public school teachers in the United States. Find the total number of elementary school, prekindergarten, and kindergarten teachers.
1691 thousand teachers

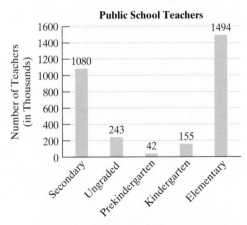

Public School Teachers

Source: National Center for Education Statistics

89. The staff for U.S. public schools is categorized in the pie graph. Determine the number of staff other than teachers. 821,024 nonteachers

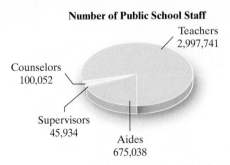

Number of Public School Staff

Teachers 2,997,741
Counselors 100,052
Supervisors 45,934
Aides 675,038

Source: National Center for Education Statistics

90. The pie graph shows the costs incurred in managing Sub-World sandwich shop for one month. From this information, determine the total cost for one month. $21,637

Sub-World Monthly Expenses

Food $7329
Labor $9560
Nonfood items $1248
Overhead $3500

 Writing   Translating Expression   Geometry   Scientific Calculator   Video   NS&E

Concept 5: Perimeter

For Exercises 91–98, find the perimeter. (See Examples 11 and 12.)

91.

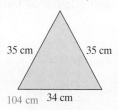

35 cm   35 cm

35 cm

104 cm   34 cm

92.

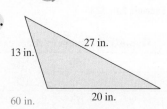

27 in.

13 in.

60 in.   20 in.

93.

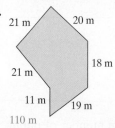

21 m   20 m

18 m

21 m

11 m   19 m

110 m

94.

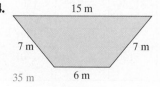

15 m

7 m   7 m

35 m   6 m

95.

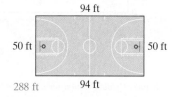

6 yd

7 yd

5 yd   10 yd

3 yd

42 yd   11 yd

96.

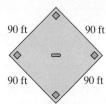

200 yd

38 yd

58 yd

98 yd   136 yd

672 yd   142 yd

97. Find the perimeter of an NBA basketball court.

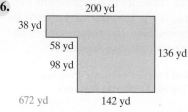

94 ft

50 ft   50 ft

288 ft   94 ft

98. A major league baseball diamond is in the shape of a square. Find the distance a batter must run if he hits a home run. (After hitting a home run, the batter must run the perimeter of the diamond.) 360 ft

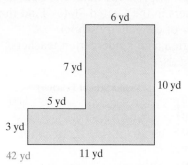

90 ft   90 ft

90 ft   90 ft

Calculator Connections

Topic: Adding Whole Numbers

The following keystrokes demonstrate the procedure to add numbers on a calculator. The **ENTER** key (or, on some calculators, the **=** key or **EXE** key) tells the calculator to complete the calculation. Notice that commas used in large numbers are not entered into the calculator.

Expression	Keystrokes	Result
$92{,}406 + 83{,}168$	92406 **+** 83168 **ENTER**	175574

Your calculator may use the **=** key or **EXE** key instead.

Calculator Exercises

For Exercises 99–106, add by using a calculator.

99. $9{,}084{,}037 + 452{,}903$ 9,536,940

100. $899{,}382 + 9406$ 908,788

101. $7{,}201{,}529 + 962{,}411$ 8,163,940

102.
$$\begin{array}{r} 45{,}418 \\ 81{,}990 \\ 9{,}063 \\ + 56{,}309 \end{array}$$ 192,780

103.
$$\begin{array}{r} 9{,}300{,}050 \\ 7{,}803{,}513 \\ 3{,}480{,}009 \\ + 907{,}822 \end{array}$$ 21,491,394

104.
$$\begin{array}{r} 3{,}421{,}019 \\ 822{,}761 \\ 1{,}003{,}721 \\ + 9{,}678 \end{array}$$ 5,257,179

105. The amount of money spent on television advertisements during the NCAA tournament is given in the table. This represents the four largest contributors. Determine the total amount spent on television advertisements for these four companies. $148,500,000

Advertisers	Amount Spent
General Motors Corporation	$64,700,000
AT&T	$36,500,000
National Collegiate Athletic Association	$24,100,000
Coca-Cola Company	$23,200,000

106. The number of votes tallied for the leading presidential candidates for the 2012 election is given in the table. (*Source:* Federal Election Commission) Find the total number of votes for these three candidates.   128,107,616 votes

Candidate	Number of Votes
Obama	65,899,660
Romney	60,932,152
Johnson	1,275,804

 Writing   Translating Expression   Geometry   Scientific Calculator   Video   NS E

Section 1.3 Subtraction of Whole Numbers

1. Introduction to Subtraction

Jeremy bought a case of 12 sodas, and on a hot afternoon he drank 3 of the sodas. We can use the operation of subtraction to find the number of sodas remaining.

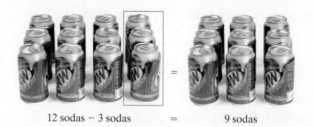

12 sodas − 3 sodas = 9 sodas

The symbol "−" between two numbers is a subtraction sign, and the result of a subtraction is called the **difference**. The number being subtracted (in this case, 3) is called the **subtrahend**. The number 12 from which 3 is subtracted is called the **minuend**.

Instructor Note: The two parts of a subtraction problem have different names (minuend and subtrahend) because subtraction is not commutative.

With addition, both parts are called addends because the order is not important. Addition is commutative.

$$12 - 3 = 9 \quad \text{is read as} \quad \text{"12 minus 3 is equal to 9"}$$

minuend subtrahend difference

Subtraction is the inverse operation of addition. To find the number of sodas that remain after Jeremy takes 3 sodas away from 12 sodas, we ask the question:

"3 added to what number equals 12?"

That is,

$$12 - 3 = ? \quad \text{is equivalent to} \quad ? + 3 = 12$$

Subtraction can also be visualized on the number line. To evaluate $7 - 4$, start from the point on the number line corresponding to the minuend (7 in this case). Then move to the *left* 4 units. The resulting position on the number line is the difference.

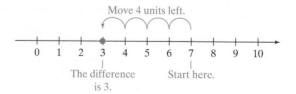

Move 4 units left.

0 1 2 3 4 5 6 7 8 9 10

The difference is 3. Start here.

To check the result, we can use addition.

$$7 - 4 = 3 \quad \text{because} \quad 3 + 4 = 7$$

Example 1 Subtracting Whole Numbers

Subtract and check the answer, using addition.

a. $8 - 2$ **b.** $10 - 6$ **c.** $5 - 0$ **d.** $3 - 3$

Solution:

a. $8 - 2 = 6$ because $6 + 2 = 8$

b. $10 - 6 = 4$ because $4 + 6 = 10$

c. $5 - 0 = 5$ because $5 + 0 = 5$

d. $3 - 3 = 0$ because $0 + 3 = 3$

Skill Practice

Subtract. Check by using addition.
1. $11 - 5$ **2.** $8 - 0$
3. $7 - 2$ **4.** $5 - 5$

Classroom Example: p. 28, Exercise 20

2. Subtraction of Whole Numbers

When subtracting large numbers, it is usually more convenient to write the numbers vertically. We write the minuend on top and the subtrahend below it. Starting from the ones column, we subtract digits having corresponding place values.

Example 2 Subtracting Whole Numbers

Subtract and check the answer by using addition.

a. $\begin{array}{r} 976 \\ -124 \end{array}$ **b.** $\begin{array}{r} 2498 \\ -197 \end{array}$

Skill Practice

Subtract. Check by using addition.
5. $\begin{array}{r} 472 \\ -261 \end{array}$ **6.** $\begin{array}{r} 3947 \\ -137 \end{array}$

Classroom Example: p. 29, Exercise 30

Solution:

a. $\begin{array}{r} 976 \\ -124 \\ \hline 852 \end{array}$ Check: $\begin{array}{r} 852 \\ +124 \\ \hline 976 \checkmark \end{array}$

└─── Subtract the ones column digits.
└─── Subtract the tens column digits.
└─── Subtract the hundreds column digits.

b. $\begin{array}{r} 2498 \\ -197 \\ \hline 2301 \end{array}$ Check: $\begin{array}{r} 2301 \\ +197 \\ \hline 2498 \checkmark \end{array}$

When a digit in the subtrahend (bottom number) is larger than the corresponding digit in the minuend (top number), we must "regroup" or borrow a value from the column to the left.

$92 = 9$ tens $+ 2$ ones
$-74 = 7$ tens $+ 4$ ones

In the ones column, we cannot take 4 away from 2. We will regroup by borrowing 1 ten from the minuend. Furthermore, 1 ten = 10 ones.

$\begin{array}{l} ^{8\,+10} \quad ^{8} \qquad ^{+10\,ones} \\ 9\,2 = 9 \text{ tens } + 2 \text{ ones} \\ -7\,4 = 7 \text{ tens } + 4 \text{ ones} \end{array}$

We now have 12 ones in the minuend.

$\begin{array}{l} ^{8\,12} \quad ^{8} \\ 9\,2 = 9 \text{ tens } + 12 \text{ ones} \\ -7\,4 = 7 \text{ tens } + 4 \text{ ones} \\ \hline 1\,8 = 1 \text{ ten } + 8 \text{ ones} \end{array}$

TIP: The process of *borrowing* in subtraction is the reverse of *carrying* in addition.

Answers
1. 6 **2.** 8 **3.** 5 **4.** 0
5. 211 **6.** 3810

Skill Practice

Subtract. Check by addition.
7.　23,126
　　 − 6,048

Classroom Example: p. 29,
Exercise 50

Example 3　Subtracting Whole Numbers with Borrowing

Subtract and check the result with addition.

$$134{,}616$$
$$- 53{,}438$$

Solution:

$$\begin{array}{r} {}^{0\ 16} \\ 1\,3\,4{,}6\,\cancel{1}\,\cancel{6} \\ -\ \ 5\,3{,}4\,3\,8 \\ \hline 8 \end{array}$$

In the ones place, 8 is greater than 6. We borrow 1 ten from the tens place.

$$\begin{array}{r} {}^{10} \\ {}^{5\ \cancel{0}\ 16} \\ 1\,3\,4{,}\cancel{6}\,\cancel{1}\,\cancel{6} \\ -\ \ 5\,3{,}4\,3\,8 \\ \hline 7\,8 \end{array}$$

In the tens place, 3 is greater than 0. We borrow 1 hundred from the hundreds place.

$$\begin{array}{r} {}^{0\ 13\ \ \ 5\ \cancel{0}\ 16}_{\ \ \ \ \ \ \ \ \ 10} \\ \cancel{1}\,\cancel{3}\,4{,}\cancel{6}\,\cancel{1}\,\cancel{6} \\ -\ \ 5\,3{,}4\,3\,8 \\ \hline 8\,1{,}1\,7\,8 \end{array}$$

In the ten-thousands place, 5 is greater than 3. We borrow 1 hundred-thousand from the hundred-thousands place.

$$\underline{\text{Check:}} \quad \begin{array}{r} {}^{1\ \ \ \ 1\,1} \\ 81{,}178 \\ +\ 53{,}438 \\ \hline 134{,}616\ \checkmark \end{array}$$

Skill Practice

Subtract. Check by addition.
8. 700 − 531

Classroom Example: p. 29,
Exercise 56

Example 4　Subtracting Whole Numbers with Borrowing

Subtract and check the result with addition.　　500 − 247

Solution:

$$\begin{array}{r} 500 \\ -\ 247 \end{array}$$

In the ones place, 7 is greater than 0. We try to borrow 1 ten from the tens place. However, the tens place digit is 0. Therefore we must first borrow from the hundreds place.

$$\begin{array}{r} {}^{4\ 10} \\ \cancel{5}\,\cancel{0}\,0 \\ -\ 2\,4\,7 \end{array}$$

$$\begin{array}{r} {}^{\ \ \ \ 9} \\ {}^{4\ \cancel{10}\ 10} \\ \cancel{5}\,\cancel{0}\,\cancel{0} \\ -\ 2\,4\,7 \\ \hline 2\,5\,3 \end{array}$$ ←—Now we can borrow 1 ten to add to the ones place.

Subtract.

$$\underline{\text{Check:}} \quad \begin{array}{r} {}^{1\ 1} \\ 253 \\ +\ 247 \\ \hline 500\ \checkmark \end{array}$$

Answers

7. 17,078　　**8.** 169

3. Translations and Applications Involving Subtraction

In applications of mathematics, several words and phrases imply subtraction. A partial list is provided in Table 1-2.

Table 1-2

Word/Phrase	Example	In Symbols
Minus	15 minus 10	$15 - 10$
Difference	The difference of 10 and 2	$10 - 2$
Decreased by	9 decreased by 1	$9 - 1$
Less than	5 less than 12	$12 - 5$
Subtract . . . from	Subtract 3 from 8	$8 - 3$

In Table 1-2, make a note of the last two entries. The phrases *less than* and *subtract . . . from* imply a specific order in which the subtraction is performed. In both cases, begin with the second number listed and subtract the first number listed.

Instructor Note: Students might confuse this use of "less than" with the phrase "is less than" used to describe $3 < 5$.

Example 5 Translating an English Phrase to a Mathematical Statement

Translate the English phrase to a mathematical statement and simplify.

a. The difference of 150 and 38

b. 30 subtracted from 82

Solution:

a. From Table 1-2, the *difference* of 150 and 38 implies $150 - 38$.

$$
\begin{array}{r}
\overset{4\ 10}{1\cancel{5}0} \\
-\ \ 38 \\
\hline
112
\end{array}
$$

b. The phrase "30 subtracted from 82" implies that 30 is taken away from 82. We have $82 - 30$.

$$
\begin{array}{r}
82 \\
-\ 30 \\
\hline
52
\end{array}
$$

Skill Practice

Translate the English phrase into a mathematical statement and simplify.

9. Twelve decreased by eight

10. Subtract three from nine.

Classroom Examples: p. 29, Exercises 66 and 70

In Section 1.2 we saw that the operation of addition is commutative. That is, the order in which two numbers are added does not affect the sum. This is *not* true for subtraction. For example, $82 - 30$ is not equal to $30 - 82$. The symbol $\neq$ means "is not equal to." Thus, $82 - 30 \neq 30 - 82$.

Most applications of subtraction generally fall into two categories.

1. The first type is phrased as a subtraction problem in which the minuend and subtrahend are given.

Answers
9. $12 - 8$; 4
10. $9 - 3$; 6

Example: Shawn has $52 and then spends $40. How much money does he have left? (In this problem, we subtract $40 from $52.)

$$\$52 - \$40 = \$12$$

2. The second type is phrased as an addition problem with a missing addend.

Example: Maria received 72 points on her last math test, but needed 90 points to receive an A. How many more points would she have needed to earn an A? (In this problem, the addition problem can be translated to subtraction.)

$$72 + ? = 90 \quad \text{is equivalent to} \quad 90 - 72 = ?$$

Because $90 - 72 = 18$, Maria would have needed 18 more points.

Skill Practice

11. The temperature at 1:00 P.M. in Denver was 47°F. Three hours later, the temperature was 34°F. By how much did the temperature drop?

Classroom Example: p. 29, Exercise 80

Example 6 Solving an Application Problem

A biology class started with 35 students. By midsemester, 7 students had dropped. How many students are still in the class?

Solution:

$$35 - 7 = 28 \quad \text{There are 28 students still in the class.}$$

Skill Practice

12. Teresa earned test scores of 98, 84, and 90 on her first three exams. How many points must she score on the fourth exam to earn a total of 360 points?

Classroom Example: p. 30, Exercise 88

Example 7 Solving an Application Problem

A surveyor knows that the perimeter of the lot shown is 620 ft. Find the missing length. See Figure 1-5.

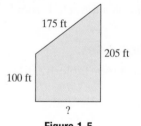

Figure 1-5

Solution:

Recall that the perimeter of a polygon is the sum of the lengths of its sides. The sum of the three known side lengths in Figure 1-5 is 480 ft:

$$
\begin{array}{r}
\overset{1}{1}00 \\
175 \\
+\ 205 \\
\hline
480
\end{array}
$$

We can subtract 480 ft from the perimeter to find the missing length:

$$620 \text{ ft} - 480 \text{ ft} = ?$$

$$
\begin{array}{r}
\overset{5\ 12}{\cancel{6}2\,0} \\
-\ 4\,8\,0 \\
\hline
1\,4\,0
\end{array}
$$

The missing length is 140 ft.

Answers

11. 13°F
12. 88 points

A third application of subtraction is to compute a change (increase or decrease) in an amount.

Example 8 **Solving an Application Problem**

A criminal justice student did a study of the number of robberies that occurred in the United State over a period of several years. The bar graph shows the results for five selected years.

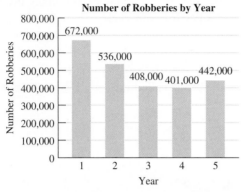

Number of Robberies by Year

Source: Federal Bureau of Investigation

a. Find the increase in the number of reported robberies from year 4 to year 5.

b. Find the decrease in the number of reported robberies from year 1 to year 2.

Solution:

For the purpose of finding an amount of increase or decrease, we will subtract the smaller number from the larger number.

a. Because the number of robberies went *up* from year 4 to year 5, there was an *increase*. To find the amount of the increase, we subtract the smaller number from the larger number.

$$\begin{array}{r} 442,000 \\ -\,401,000 \\ \hline 41,000 \end{array}$$

From year 4 to year 5, there was an increase of 41,000 reported robberies in the United States.

b. Because the number of robberies went *down* from year 1 to year 2, there was a *decrease*. To find the amount of the decrease, we subtract the smaller number from the larger number.

$$\begin{array}{r} {\scriptstyle 6\ \ 12} \\ 6\,\cancel{7}2,000 \\ -\,536,000 \\ \hline 136,000 \end{array}$$

From year 1 to year 2 there was a decrease of 136,000 reported robberies in the United States.

Skill Practice

Refer to the graph for Example 8.

13. a. Has the number of robberies increased or decreased between year 2 and year 3?
   b. Determine the amount of increase or decrease.

Classroom Examples: p. 31, Exercises 92 and 94

Answers

13. a. decreased   b. 128,000 robberies

Section 1.3 Practice Exercises

For additional exercises, see Classroom Activities 1.3A–1.3C in the *Student's Resource Manual* at www.mhhe.com/moh.

Study Skills Exercise

It is very important to attend class every day. Math is cumulative in nature, and you must master the material learned in the previous class to understand today's lesson. Because this is so important, many instructors tie attendance into the final grade. Write down the attendance policy for your class.

Vocabulary and Key Concepts

1. Given the subtraction statement $15 - 4 = 11$, the number 15 is called the ___minuend___, the number 4 is called the ___subtrahend___, and the number 11 is called the ___difference___.

Review Exercises

For Exercises 2–5, add.

2. $89 + 45$ 134

3. $330 + 821$ 1151

4.
$$\begin{array}{r} 782 \\ 21 \\ + 1046 \\ \hline 1849 \end{array}$$

5.
$$\begin{array}{r} 46 \\ 804 \\ + 49 \\ \hline 899 \end{array}$$

6. Circle the true statement:

$14 > 21, 14 < 21$ $14 < 21$

7. Circle the true statement:

$0 < 10, 0 > 10$ $0 < 10$

↤↦ **8.** Write the inequality in words:

$22 < 25$ Twenty-two is less than twenty-five.

Concept 1: Introduction to Subtraction

For Exercises 9–14, identify the minuend, subtrahend, and the difference.

9. $12 - 8 = 4$ Minuend: 12; subtrahend: 8; difference: 4

10. $6 - 1 = 5$ Minuend: 6; subtrahend: 1; difference: 5

11. $21 - 12 = 9$ Minuend: 21; subtrahend: 12; difference: 9

12. $32 - 2 = 30$ Minuend: 32; subtrahend: 2; difference: 30

13.
$$\begin{array}{r} 9 \\ - 6 \\ \hline 3 \end{array}$$
Minuend: 9; subtrahend: 6; difference: 3

14.
$$\begin{array}{r} 17 \\ - 3 \\ \hline 14 \end{array}$$
Minuend: 17; subtrahend: 3; difference: 14

For Exercises 15–18, write the subtraction problem as a related addition problem. For example, $19 - 6 = 13$ can be written as $13 + 6 = 19$.

15. $27 - 9 = 18$
$18 + 9 = 27$

16. $20 - 8 = 12$
$12 + 8 = 20$

17. $102 - 75 = 27$
$27 + 75 = 102$

18. $211 - 45 = 166$
$166 + 45 = 211$

For Exercises 19–24, subtract, then check the answer by using addition. **(See Example 1.)**

19. $8 - 3$ Check: $\boxed{} + 3 = 8$
5

20. $7 - 2$ Check: $\boxed{} + 2 = 7$
5

21. $4 - 1$ Check: $\boxed{} + 1 = 4$
3

22. $9 - 1$ Check: $\boxed{} + 1 = 9$
8

23. $6 - 0$ Check: $\boxed{} + 0 = 6$
6

24. $3 - 0$ Check: $\boxed{} + 0 = 3$
3

Concept 2: Subtraction of Whole Numbers

For Exercises 25–36, subtract and check the answer by using addition. **(See Example 2.)**

25.
$$\begin{array}{r} 68 \\ - 23 \\ \hline 45 \end{array}$$

26.
$$\begin{array}{r} 54 \\ - 31 \\ \hline 23 \end{array}$$

27.
$$\begin{array}{r} 88 \\ - 27 \\ \hline 61 \end{array}$$

28.
$$\begin{array}{r} 75 \\ - 50 \\ \hline 25 \end{array}$$

29.   1347
        − 221
        ‾‾‾‾‾
         1126

30.   4865
        − 713
        ‾‾‾‾‾
         4152

31.   1525
        − 1204
        ‾‾‾‾‾
          321

32.   8843
        − 5612
        ‾‾‾‾‾
         3231

33. 12,806 − 2802   10,004   **34.** 12,771 − 1240   11,531   **35.** 14,356 − 13,253   1103   **36.** 34,550 − 31,450   3100

For Exercises 37–60, subtract the whole numbers involving borrowing. **(See Examples 3 and 4.)**

37.   76
        − 59
        ‾‾‾‾
         17

38.   64
        − 48
        ‾‾‾‾
         16

39.   87
        − 38
        ‾‾‾‾
         49

40.   94
        − 75
        ‾‾‾‾
         19

41.   240
        − 136
        ‾‾‾‾‾
         104

42.   360
        − 225
        ‾‾‾‾‾
         135

43.   710
        − 189
        ‾‾‾‾‾
         521

44.   850
        − 303
        ‾‾‾‾‾
         547

45.   4350
        − 4327
        ‾‾‾‾‾
          23

46.   7293
        − 7255
        ‾‾‾‾‾
          38

47.   6002
        − 1238
        ‾‾‾‾‾
         4764

48.   3000
        − 2356
        ‾‾‾‾‾
          644

49.   10,425
        − 9,022
        ‾‾‾‾‾‾
          1403

50.   23,901
        − 8,064
        ‾‾‾‾‾‾
         15,837

51.   62,088
        − 59,871
        ‾‾‾‾‾‾
          2217

52.   32,112
        − 28,334
        ‾‾‾‾‾‾
          3778

53. 470 − 92   378   **54.** 674 − 89   585   **55.** 3700 − 2987   713   **56.** 8000 − 3788   4212

57. 32,439 − 1498   30,941   **58.** 21,335 − 4123   17,212   **59.** 8,007,234 − 2,345,115
                                                                        5,662,119
60. 3,045,567 − 1,871,495
        1,174,072

Concept 3: Translations and Applications Involving Subtraction

For Exercises 61–72, translate the English phrase into a mathematical statement and simplify. **(See Example 5.)**

61. 78 minus 23
     78 − 23; 55

62. 45 minus 17
     45 − 17; 28

63. 78 decreased by 6
     78 − 6; 72

64. 50 decreased by 12
     50 − 12; 38

65. Subtract 100 from 422.
     422 − 100; 322

66. Subtract 42 from 89.
     89 − 42; 47

67. 72 less than 1090
     1090 − 72; 1018

68. 60 less than 3111
     3111 − 60; 3051

69. The difference of 50 and 13
     50 − 13; 37

70. The difference of 405 and 103
     405 − 103; 302

71. Subtract 35 from 103.
     103 − 35; 68

72. Subtract 14 from 91.
     91 − 14; 77

For Exercises 73–76, write an English phrase for the mathematical statement. (Answers will vary.)

73. 93 − 27 For example:
        93 minus 27

74. 80 − 20 For example:
        80 decreased by 20

75. 165 − 85 For example:
        Subtract 85 from 165.

76. 171 − 42 For example:
        42 less than 171

77. Use the expression 7 − 4 to explain why subtraction is not commutative.
The expression 7 − 4 means 7 minus 4, yielding a difference of 3.
The expression 4 − 7 means 4 minus 7 which results in a difference
of −3. (This is a mathematical skill we have not yet learned.)

78. Is subtraction associative? Use the numbers 10, 6, 2 to explain.   Subtraction is not associative. For example,
10 − (6 − 2) = 10 − 4 = 6 and (10 − 6) − 2 = 4 − 2 = 2.
Therefore, 10 − (6 − 2) does not equal (10 − 6) − 2.

79. A $50 bill was used to purchase $17 worth of gasoline. Find the amount of change received.
(See Example 6.) $33

80. There are 55 DVDs to shelve one evening at a video rental store. If Jason puts away 39 before leaving for the day, how many are left for Patty to put away?
16 DVDs

81. The songwriting team of John Lennon and Paul McCartney had 118 chart hits while Mick Jagger and Keith Richards had 63. How many more chart hits did Lennon and McCartney have than Jagger and Richards?
55 more hits

82. Due to severe drought in the state of Alabama in 2007, a local well driller said that the minimum depth to drill for water had increased from 150 ft in 2006 to 200 ft in 2007. In 2007, the driller had to dig 505 ft to find water in one rural community. How many more feet above the 2007 minimum depth did the driller have to drill?   305 ft

83. In landscaping a yard, Lily would like 26 plants for a border. If she has 18 plants in her truck, how many more will she need to finish the job?   8 plants

84. A collection is taken to buy flowers for a co-worker who is in the hospital. If $37 has been collected and the flower arrangement costs $50, how much more needs to be collected? $13

85. A recent report indicated that the play, *The Lion King*, had been performed 5149 times on Broadway. At that time, the play, *Wicked*, had been performed 2670 times on Broadway. How many more times had *The Lion King* been performed than *Wicked*?   2479 times

86. At the time of Kurt Warner's retirement from football, his total passing yardage was 32,344 yd. Drew Brees had 30,646 yd as of 2009. How many more yards does Drew Brees need to reach Kurt Warner's total?   1698 yd

 For Exercises 87 and 88, for each figure find the missing length.

87. The perimeter of the triangle is 39 m.

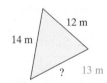

88. The perimeter of the figure is 547 cm.

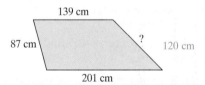

89. A homeowner knows that the perimeter of his backyard is 56 yd. Find the missing length.
(See Example 7.)   10 yd

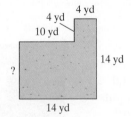

90. Barbara has 15 ft of molding to install in her bathroom, as shown in the figure. What is the missing length? *Note*: There will be no molding by the tub or door.   4 ft

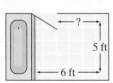

For Exercises 91–94, use the information from the bar graph. **(See Example 8.)**

91. What is the difference in the number of marriages between year 3 and year 4?   30 thousand marriages

92. Find the decrease in the number of marriages in the United States between year 4 and year 5.
89 thousand marriages

93. What is the difference in the number of marriages between the year having the greatest and the year having the least?   119 thousand marriages

94. Between which two consecutive years did the greatest increase in the number of marriages occur? What is the increase?   year 5 and year 6; 45 thousand marriages

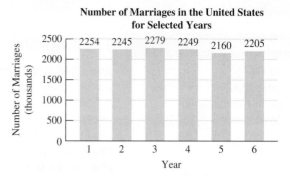

Number of Marriages in the United States for Selected Years

Source: National Center for Health Statistics
Figure for Exercises 91–94

Calculator Connections

Topic: Subtracting Whole Numbers

To subtract numbers on a calculator, use the subtraction key − . Do not confuse the subtraction key with the (−) key. The (−) is presented later to enter negative numbers.

Expression	Keystrokes	Result
$345{,}899 - 43{,}018$	345899 − 43018 **ENTER**	302881

Calculator Exercises

For Exercises 95–97, subtract by using a calculator.

95. $\begin{array}{r} 4{,}905{,}620 \\ -\ \ \ 458{,}318 \\ \hline \end{array}$
4,447,302

96. $\begin{array}{r} 953{,}400{,}415 \\ -\ \ 56{,}341{,}902 \\ \hline \end{array}$
897,058,513

97. $\begin{array}{r} 82{,}025{,}160 \\ -79{,}118{,}705 \\ \hline \end{array}$
2,906,455

For Exercises 98–101, refer to the table showing the land area for five states.

98. Find the difference in land area between Colorado and Wisconsin.   49,408 mi²

99. Find the difference in land area between Tennessee and West Virginia.   17,139 mi²

100. Find the difference in land area between the state with the greatest land area and the state with the least land area.
The difference in land area between Colorado and Rhode Island is 102,673 mi².

101. How much more land area does Wisconsin have than Tennessee?   13,093 mi²

State	Land Area (mi²)
Rhode Island	1,045
Tennessee	41,217
West Virginia	24,078
Wisconsin	54,310
Colorado	103,718

| Section 1.4 | Rounding and Estimating |

Concepts

1. Rounding
2. Estimation
3. Using Estimation in Applications

1. Rounding

Rounding a whole number is a common practice when we do not require an exact value. For example, Madagascar lost 3956 mi² of rainforest between 1990 and 2008. We might round this number to the nearest thousand and say that there was approximately 4000 mi² lost. In mathematics, we use the symbol ≈ to read "is approximately equal to." Therefore, 3956 mi² ≈ 4000 mi².

A number line is a helpful tool to understand rounding. For example, 48 is closer to 50 than it is to 40. Therefore, 48 rounded to the nearest ten is 50.

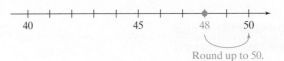

Round up to 50.

Concept Connections

1. Is 82 closer to 80 or to 90? Round 82 to the nearest ten.
2. Is 65 closer to 60 or to 70? Round the number to the nearest ten.

On the other hand, 43 is closer to 40 than to 50. Therefore, 43 rounded to the nearest ten is 40.

Round down to 40.

Note 45 is halfway between 40 and 50. In such a case, our convention will be to round *up* to the next-larger ten.

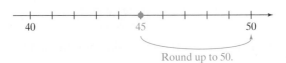

Round up to 50.

The decision to round up or down to a given place value is determined by the digit to the *right* of the given place value. The following steps outline the procedure.

Animation

> **Rounding Whole Numbers**
> **Step 1** Identify the digit one position to the right of the given place value.
> **Step 2** If the digit in step 1 is a 5 or greater, add 1 to the digit in the given place value. Then replace each digit to the right of the given place value by 0.
> **Step 3** If the digit in step 1 is less than 5, replace it and each digit to its right by 0. Note that in this case, the digit in the original given place value does not change.

Answers

1. Closer to 80; 80
2. The 65 is the same distance from 60 and 70; round up to 70.

Example 1   Rounding a Whole Number

Round 3741 to the nearest hundred.

Solution:

3 7 [4] 1 ≈ 3700

↑ hundreds place

↑ This is the digit to the right of the hundreds place. Because 4 is less than 5, the digit in the hundreds place remains the same and replace the digits to its right by zeros.

Skill Practice

3. Round 12,461 to the nearest thousand.

Classroom Example: p. 36, Exercise 18

Example 1 could also have been solved by drawing a number line. Use the part of a number line between 3700 and 3800.

Round down to 3700.

Example 2   Rounding a Whole Number

Round 1,790,641 to the nearest hundred-thousand.

Solution:

1, 7 [9] 0, 6 4 1 ≈ 1,800,000

↑ hundred-thousands place

↑ This is the digit to the right of the given place value. Because 9 is greater than 5, add 1 to the hundred-thousands place, add: 7 + 1 = 8. Replace the digits to the right of the hundred-thousands place by zeros.

Skill Practice

4. Round 147,316 to the nearest ten-thousand.

Classroom Example: p. 36, Exercise 20

Example 3   Rounding a Whole Number

Round 1503 to the nearest thousand.

Solution:

1 [5] 0 3 ≈ 2000

↑ thousands place

↑ This is the digit to the right of the thousands place. Because this digit is 5, we round up. We increase the thousands place digit by 1. That is, 1 + 1 = 2. Replace the digits to its right by zeros.

Skill Practice

5. Round 7,521,460 to the nearest million.

Classroom Example: p. 36, Exercise 22

Answers

3. 12,000   **4.** 150,000
5. 8,000,000

Skill Practice

6. Round 39,823 to the nearest thousand.

Classroom Example: p. 36, Exercise 24

Example 4 **Rounding a Whole Number**

Round the number 24,961 to the hundreds place.

Solution:

$$2\,4,9\,\boxed{6}\,1 \approx 25,000$$

This value is greater than 5. Therefore, add 1 to the hundreds place digit. Replace the digits to the right of the hundreds place with 0.

2. Estimation

We use the process of rounding to estimate the result of numerical calculations. For example, to estimate the following sum, we can round each addend to the nearest ten.

31	rounds to	→	30
12	rounds to	→	10
+ 49	rounds to	→	+ 50
			90

The estimated sum is 90 (the actual sum is 92).

Skill Practice

Estimate the sum by rounding each number to the nearest hundred.

7. 3162 + 4931 + 2206

Classroom Example: p. 37, Exercise 34

Example 5 **Estimating a Sum**

Estimate the sum by rounding to the nearest thousand.

$$6109 + 976 + 4842 + 11,619$$

Solution:

6,109	rounds to	→	6,000
976	rounds to	→	1,000
4,842	rounds to	→	5,000
+ 11,619	rounds to	→	+ 12,000
			24,000

The estimated sum is 24,000 (the actual sum is 23,546).

Skill Practice

Estimate the difference by rounding each number to the nearest million.

8. 35,264,000 − 21,906,210

Classroom Example: p. 37, Exercise 40

Example 6 **Estimating a Difference**

Estimate the difference 4817 − 2106 by rounding each number to the nearest hundred.

Solution:

4817	rounds to	→	4800
− 2106	rounds to	→	− 2100
			2700

The estimated difference is 2700 (the actual difference is 2711).

Answers
6. 40,000   7. 10,300
8. 13,000,000

3. Using Estimation in Applications

Example 7 Estimating a Sum in an Application

A driver for a delivery service must drive from Chicago, Illinois, to Dallas, Texas, and make several stops on the way. The driver follows the route given on the map. Estimate the total mileage by rounding each distance to the nearest hundred miles.

Solution:

$$
\begin{array}{rcl}
541 & \text{rounds to} \longrightarrow & 500 \\
418 & \text{rounds to} \longrightarrow & 400 \\
354 & \text{rounds to} \longrightarrow & 400 \\
+\,205 & \text{rounds to} \longrightarrow & +\,200 \\
\hline
 & & 1500
\end{array}
$$

The driver traveled approximately 1500 mi.

Example 8 Estimating a Difference in an Application

In a recent year, the U.S. Census Bureau reported that the number of males over the age of 18 was 100,994,367. The same year, the number of females over 18 was 108,133,727. Round each value to the nearest million. Estimate how many more females over 18 there were than males over 18.

Solution:

The number of males was approximately 101,000,000. The number of females was approximately 108,000,000.

$$
\begin{array}{r}
108,000,000 \\
-\,101,000,000 \\
\hline
7,000,000
\end{array}
$$

There were approximately 7 million more women over age 18 in the United States than men.

Section 1.4 Practice Exercises

Study Skills Exercise

For additional exercises, see Classroom Activities 1.4A–1.4B in the *Student's Resource Manual* at www.mhhe.com/moh.

Purchase a three-ring binder for your math notes and homework. Use section dividers to separate each chapter that you cover in the text. Keep your homework and notes in the appropriate section. What other course materials might you keep organized in your notebook?

Vocabulary and Key Concepts

1. A process called _rounding_ is common practice when the exact value of a number is not required.

Review Exercises

2. A triangle has sides of length 5 ft, 12 ft, and 13 ft. Find the perimeter.   30 ft

For Exercises 3–6, add or subtract as indicated.

3.   59
     $-$ 33
      26

4.   130
      $-$ 98
       32

5.   4009
      $+$ 998
       5007

6.   12,033
      $+$ 23,441
       35,474

7. Determine the place value of the digit 6 in the number 1,860,432.   Ten-thousands

8. Determine the place value of the digit 4 in the number 1,860,432.   Hundreds

Concept 1: Rounding

9. Explain how to round a whole number to the hundreds place.
If the digit in the tens place is 0, 1, 2, 3, or 4, then change the tens and ones digits to 0.
If the digit in the tens place is 5, 6, 7, 8, or 9, increase the digit in the hundreds place by 1 and change the tens and ones digits to 0.

10. Explain how to round a whole number to the tens place.
If the digit in the ones place is 0, 1, 2, 3, or 4, then change the ones digit to 0.
If the digit in the ones place is 5, 6, 7, 8, or 9, increase the digit in the tens place by 1 and change the ones digit to 0.

For Exercises 11–28, round each number to the given place value. **(See Examples 1–4.)**

11. 342; tens   340

12. 834; tens   830

13. 725; tens   730

14. 445; tens   450

15. 9384; hundreds   9400

16. 8363; hundreds   8400

17. 8539; hundreds   8500

18. 9817; hundreds   9800

19. 34,992; thousands   35,000

20. 76,831; thousands   77,000

21. 2578; thousands   3000

22. 3511; thousands   4000

23. 9982; hundreds   10,000

24. 7974; hundreds   8000

25. 109,337; thousands   109,000

26. 437,208; thousands   437,000

27. 489,090; ten-thousands   490,000

28. 388,725; ten-thousands   390,000

29. In the first weekend of its release, the movie *Avatar* grossed $77,025,481. Round this number to the millions place. $77,000,000

30. The average per capita personal income in the United States in a recent year was $33,050. Round this number to the nearest thousand. $33,000

31. The average center-to-center distance from the Earth to the Moon is 238,863 mi. Round this to the thousands place.   239,000 mi

32. A shopping center in Edmonton, Alberta, Canada, covers an area of 492,000 square meters (m^2). Round this number to the hundred-thousands place.   500,000 m^2

 Writing Translating Expression  Geometry   Scientific Calculator Video NS E

Concept 2: Estimation

For Exercises 33–36, estimate the sum by first rounding each number to the nearest ten. **(See Example 5.)**

 33.   57
           82
       + 21
       160

34.   33
           78
       + 41
       150

35.   41
           12
       + 129
       180

36.   29
           73
       + 113
       210

For Exercises 37–40, estimate the difference by first rounding each number to the nearest hundred. **(See Example 6.)**

37.   898
        − 422
        500

38.   731
        − 584
        100

39.   3412
        − 1252
        2100

40.   9771
        − 4544
        5300

Concept 3: Using Estimation in Applications

For Exercises 41 and 42, refer to the table.

Brand	Manufacturer	Sales ($)
M&Ms	Mars	97,404,576
Hershey's Milk Chocolate	Hershey Chocolate	81,296,784
Reese's Peanut Butter Cups	Hershey Chocolate	54,391,268
Snickers	Mars	53,695,428
KitKat	Hershey Chocolate	38,168,580

41. Round the sales to the nearest million to estimate the total sales brought in by the Mars company. **(See Example 7.)** $151,000,000

42. Round the sales to the nearest million to estimate the total sales brought in by the Hershey Chocolate Company. $173,000,000

43. Neil Diamond earned $71,339,710 in U.S. tours in one year while Paul McCartney earned $59,684,076. Round each value to the nearest million dollars to estimate how much more Neil Diamond earned. **(See Example 8.)** $11,000,000 more

44. The average annual salary for a public school teacher in Iowa is $43,130. The average salary for a public school teacher in California is $63,640. Round each value to the nearest thousand to estimate how much more a school teacher in California makes compared to one from Iowa. Approximately $21,000

For Exercises 45–48, use the given table.

45. Round the revenue to the nearest hundred-thousand to estimate the total revenue for the years 1 through 3. $10,000,000

46. Round the revenue to the nearest hundred-thousand to estimate the total revenue for the years 4 through 6. $8,700,000

47. a. Determine the year with the greatest revenue. Round this revenue to the nearest hundred-thousand. year 4; $3,500,000

b. Determine the year with the least revenue. Round this revenue to the nearest hundred-thousand. year 6; $2,000,000

48. Estimate the difference between the year with the greatest revenue and the year with the least revenue. $1,500,000

Beach Parking Revenue for Daytona Beach, Florida	
Year	Revenue
1	$3,316,897
2	3,272,028
3	3,360,289
4	3,470,295
5	3,173,050
6	1,970,380

Source: Daytona Beach News Journal
Table for Exercises 45–48

For Exercises 49–52, use the bar graph provided.

Number of Students Enrolled in Grades 6–12 for Selected States

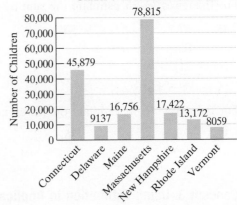

49. Determine the state with the greatest number of students enrolled in grades 6–12. Round this number to the nearest thousand. Massachusetts; 79,000 students

50. Determine the state with the least number of students enrolled in grades 6–12. Round this number to the nearest thousand. Vermont; 8000 students

51. Use the information in Exercises 49 and 50 to estimate the difference between the number of students in the state with the highest enrollment and that of the lowest enrollment. 71,000 students

Source: National Center for Education Statistics

Figure for Exercises 49–52

52. Estimate the total number of students enrolled in grades 6–12 in the selected states by first rounding the number of students to the thousands place. 189,000 students

53. If you were to estimate the following sum, what place value would you round to and why?

$$389,220 + 2988 + 12,824 + 101,333$$

Answers may vary.

54. Identify the place value that you would round to when estimating the answer to the following problem. Then round the values and estimate the answer. Thousands place; 4000

$$4208 - 932 + 1294$$

Expanding Your Skills

For Exercises 55–58, round the numbers to estimate the perimeter of each figure. (Answers may vary.)

55.
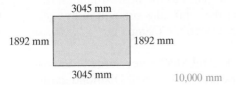
3045 mm
1892 mm   1892 mm
3045 mm   10,000 mm

56.
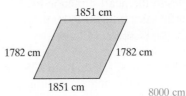
1851 cm
1782 cm   1782 cm
1851 cm   8000 cm

57.
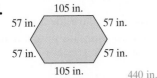
105 in.
57 in.   57 in.
57 in.   57 in.
105 in.   440 in.

58.
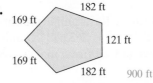
182 ft
169 ft
121 ft
169 ft
182 ft   900 ft

Multiplication of Whole Numbers and Area

1. Introduction to Multiplication

Suppose that Carmen buys three cartons of eggs to prepare a large family brunch. If there are 12 eggs per carton, then the total number of eggs can be found by adding three 12s.

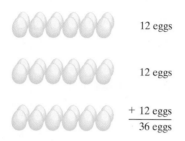

12 eggs

12 eggs

+ 12 eggs
——————
36 eggs

When each addend in a sum is the same, we have what is called *repeated* addition. Repeated addition is also called **multiplication**. We use the multiplication sign × to express repeated addition more concisely.

$$12 + 12 + 12 \quad \text{is equal to} \quad 3 \times 12$$

The expression 3×12 is read "3 times 12" to signify that the number 12 is added 3 times. The numbers 3 and 12 are called **factors**, and the result, 36, is called the **product**.

The symbol · may also be used to denote multiplication such as in the expression $3 \cdot 12 = 36$. Two factors written adjacent to each other with no other operator between them also implies multiplication. The quantity $2y$, for example, is understood to be 2 times y. If we use this notation to multiply two numbers, parentheses are used to group one or both factors. For example:

$$3(12) = 36 \quad (3)12 = 36 \quad \text{and} \quad (3)(12) = 36$$

all represent the product of 3 and 12.

> **TIP:** In the expression 3(12), the parentheses are necessary because two adjacent factors written together with no grouping symbol would look like the number 312.

The products of one-digit numbers such as $4 \times 5 = 20$ and $2 \times 7 = 14$ are basic facts. All products of one-digit numbers should be memorized (see Exercise 6).

Example 1 Identifying Factors and Products

Identify the factors and the product.

a. $6 \times 3 = 18$ **b.** $5 \cdot 2 \cdot 7 = 70$

Solution:

a. Factors: 6, 3; product: 18   **b.** Factors: 5, 2, 7; product: 70

2. Properties of Multiplication

Recall from Section 1.2 that the order in which two numbers are added does not affect the result. The same is true for multiplication. This is stated formally as the *commutative property of multiplication.*

Concepts

1. Introduction to Multiplication
2. Properties of Multiplication
3. Multiplying Many-Digit Whole Numbers
4. Estimating Products by Rounding
5. Translations and Applications Involving Multiplication
6. Area of a Rectangle

Concept Connections

1. How can multiplication be used to compute the sum $4 + 4 + 4 + 4 + 4 + 4 + 4$?

Skill Practice

Identify the factors and the product.
2. $3 \times 11 = 33$
3. $2 \cdot 5 \cdot 8 = 80$

Classroom Example: p. 48, Exercise 14

Answers

1. 7×4
2. Factors: 3 and 11; product: 33
3. Factors: 2, 5, and 8; product: 80

Commutative Property of Multiplication

Changing the order of two factors does not affect the product.

Example: 2×5 is equivalent to 5×2

Rectangular arrays help us visualize the commutative property of multiplication.

$2 \times 5 = 10$ 2 rows of 5

$5 \times 2 = 10$ 5 rows of 2

Multiplication is also an associative operation.

Associative Property of Multiplication

The manner in which factors are grouped under multiplication does not affect the product.

Example: $(3 \times 5) \times 2$ is equivalent to $3 \times (5 \times 2)$

Consider these products.

$$(3 \times 5) \times 2 \qquad 3 \times (5 \times 2)$$
$$= 15 \times 2 \qquad\quad = 3 \times 10$$
$$= 30 \qquad\qquad\quad = 30$$

Skill Practice

4. Rewrite the expression 6×5, using the commutative property of multiplication. Then find the product.
5. Rewrite the expression $3 \times (1 \times 7)$, using the associative property of multiplication. Then find the product.

Classroom Examples: p. 48, Exercises 24 and 26

Answers

4. 5×6; product is 30
5. $(3 \times 1) \times 7$; product is 21

Example 2   Applying Properties of Multiplication

a. Rewrite the expression 3×9, using the commutative property of multiplication. Then find the product.

b. Rewrite the expression $(4 \times 2) \times 3$, using the associative property of multiplication. Then find the product.

Solution:

a. $3 \times 9 = 9 \times 3$. The product is 27.

b. $(4 \times 2) \times 3 = 4 \times (2 \times 3)$.

To find the product, we have

$$4 \times (2 \times 3)$$
$$= 4 \times (6)$$
$$= 24$$

The product is 24.

Two other important properties of multiplication involve factors of 0 and 1.

Multiplication Property of 0

The product of any number and 0 is 0.

Examples: $\qquad\qquad\qquad$ $5 \times 0 = 0$

$\qquad\qquad\qquad\qquad\qquad\quad$ $0 \times 12 = 0$

The product $5 \times 0 = 0$ can easily be understood by writing the product as repeated addition.

$$\underbrace{0 + 0 + 0 + 0 + 0}_{\text{Add 0 five times.}} = 0$$

Multiplication Property of 1

The product of any number and 1 is that number.

Examples: $\qquad\qquad\qquad$ $1 \times 4 = 4$

$\qquad\qquad\qquad\qquad\qquad\quad$ $3 \times 1 = 3$

The last property of multiplication involves both addition and multiplication. First consider the expression $2(4 + 3)$. By performing the operation within parentheses first, we have

$$2(4 + 3) = 2(7) = 14$$

We get the same result by multiplying 2 times each addend within the parentheses:

$$2(4 + 3) = (2 \times 4) + (2 \times 3) = 8 + 6 = 14$$

This result illustrates the distributive property of multiplication over addition (sometimes we simply say *distributive property* for short).

Distributive Property of Multiplication over Addition

The product of a number and a sum can be found by multiplying the number by each addend.

Example: $\qquad\qquad$ $5(7 + 3) = (5 \times 7) + (5 \times 3)$

Example 3 **Applying the Distributive Property of Multiplication Over Addition**

Apply the distributive property and simplify.

a. $3(4 + 8)$ $\qquad$ **b.** $7(3 + 0)$

Solution:

a. $3(4 + 8) = (3 \times 4) + (3 \times 8) = 12 + 24 = 36$

b. $7(3 + 0) = (7 \times 3) + (7 \times 0) = 21 + 0 = 21$

Skill Practice

Apply the distributive property and simplify.

6. $2(6 + 4)$

7. $5(0 + 8)$

Classroom Example: p. 48, Exercise 28

Answers

6. $(2 \times 6) + (2 \times 4)$; 20

7. $(5 \times 0) + (5 \times 8)$; 40

3. Multiplying Many-Digit Whole Numbers

When multiplying numbers with several digits, it is sometimes necessary to carry. To see why, consider the product 3×29. By writing the factors in expanded form, we can apply the distributive property. In this way, we see that 3 is multiplied by both 20 and 9.

$$3 \times 29 = 3(20 + 9) = (3 \times 20) + (3 \times 9)$$
$$= 60 + 27$$
$$= 6 \text{ tens} + 2 \text{ tens} + 7 \text{ ones}$$
$$= 8 \text{ tens} + 7 \text{ ones}$$
$$= 87$$

Now we will multiply 29×3 in vertical form.

$$\begin{array}{r} \overset{2}{2}\,9 \\ \times\ \ 3 \\ \hline 7 \end{array}$$ Multiply $3 \times 9 = 27$. Write the 7 in the ones column and carry the 2.

$$\begin{array}{r} \overset{2}{2}\,9 \\ \times\ \ 3 \\ \hline 8\ 7 \end{array}$$ Multiply 3×2 tens $= 6$ tens. Add the carry: 6 tens $+$ 2 tens $=$ 8 tens. Write the 8 in the tens place.

Skill Practice

Multiply.
  8.   247
         $\times\ \ 3$

Classroom Example: p. 48, Exercise 34

 Example 4   Multiplying a Many-Digit Number by a One-Digit Number

Multiply.   368
            $\times\ \ 5$

Solution:

Using the distributive property, we have

$$5(300 + 60 + 8) = 1500 + 300 + 40 = 1840$$

This can be written vertically as:

$$\begin{array}{r} 368 \\ \times\ \ 5 \\ \hline 40 \\ 300 \\ +\ 1500 \\ \hline 1840 \end{array}$$ Multiply 5×8.
Multiply 5×60.
Multiply 5×300.
Add.

The numbers 40, 300, and 1500 are called *partial products*. The product of 386 and 5 is found by adding the partial products. The product is 1840.

The solution to Example 4 can also be found by using a shorter form of multiplication. We outline the procedure:

$$\begin{array}{r} \overset{4}{3}68 \\ \times\ \ \ 5 \\ \hline 0 \end{array}$$ Multiply $5 \times 8 = 40$. Write the 0 in the ones place and carry the 4.

Answer

8. 741

$$\overset{\overset{3\,4}{}}{368}$$
$$\underline{\times\quad 5}$$
$$40$$

Multiply 5×6 tens $= 300$. Add the carry. $300 + 4$ tens $= 340$.
Write the 4 in the tens place and carry the 3.

$$\overset{\overset{3\,4}{}}{368}$$
$$\underline{\times\quad 5}$$
$$1840$$

Multiply 5×3 hundreds $= 1500$. Add the carry. $1500 + 3$ hundreds $= 1800$. Write the 8 in the hundreds place and the 1 in the thousands place.

Example 5 demonstrates the process to multiply two factors with many digits.

Example 5 Multiplying a Many-Digit Number by a Many-Digit Number

Multiply.
$$72$$
$$\underline{\times\, 83}$$

Solution:

Writing the problem vertically and computing the partial products, we have

$$\overset{\overset{1}{}}{72}$$
$$\underline{\times\, 83}$$
$$216 \qquad \text{Multiply } 3 \times 72.$$
$$\underline{+\, 5760} \qquad \text{Multiply } 80 \times 72.$$
$$5976 \qquad \text{Add.}$$

The product is 5976.

Example 6 Multiplying Two Multidigit Whole Numbers

Multiply. 368×497

Solution:

$$\overset{\overset{2\,3}{\overset{6\,7}{\overset{4\,5}{}}}}{368}$$
$$\underline{\times\, 497}$$
$$2576$$
$$33120$$
$$\underline{+\, 147200}$$
$$182,896$$

4. Estimating Products by Rounding

A special pattern occurs when one or more factors in a product end in zero. Consider the following products:

$$12 \times 20 = 240 \qquad\qquad 120 \times 20 = 2400$$
$$12 \times 200 = 2400 \qquad\qquad 1200 \times 20 = 24{,}000$$
$$12 \times 2000 = 24{,}000 \qquad 12{,}000 \times 20 = 240{,}000$$

Answers
9. 1534   10. 160,564

Notice in each case the product is $12 \times 2 = 24$ followed by the total number of zeros from each factor. Consider the product 1200×20.

$$\begin{array}{r} 12\,|\,00 \\ \times\ \ 2\,|\,0 \\ \hline 24\,|\,000 \end{array}$$

Shift the numbers 1200 and 20 so that the zeros appear to the right of the multiplication process. Multiply $12 \times 2 = 24$. Write the product 24 followed by the total number of zeros from each factor.

Example 7 **Estimating a Product**

Estimate the product 795×4060 by rounding 795 to the nearest hundred and 4060 to the nearest thousand.

Solution:

$$\begin{array}{rclcr} 795 & \text{rounds to} \longrightarrow & 800 & & 8\,|\,00 \\ 4060 & \text{rounds to} \longrightarrow & 4000 & \times & 4\,|\,000 \\ & & & & \hline \\ & & & & 32\,|\,00000 \end{array}$$

The product is approximately 3,200,000.

Example 8 **Estimating a Product in an Application**

For a trip from Atlanta to Los Angeles, the average cost of a plane ticket was $495. If the plane carried 218 passengers, estimate the total revenue for the airline. (*Hint*: Round each number to the hundreds place and find the product.)

Solution:

$$\begin{array}{rclcr} \$495 & \text{rounds to} \longrightarrow & \$\ 5\,|\,00 \\ 218 & \text{rounds to} \longrightarrow & \times 2\,|\,00 \\ & & \hline \\ & & \$10\,|\,0000 \end{array}$$

The airline received approximately $100,000 in revenue.

---

5. Translations and Applications Involving Multiplication

In English there are many different words that imply multiplication. A partial list is given in Table 1-3.

Table 1-3

Word/Phrase	Example	In Symbols
Product	The product of 4 and 7	4×7
Times	8 times 4	8×4
Multiply . . . by . . .	Multiply 6 by 3	6×3

Multiplication may also be warranted in applications involving unit rates. In Example 8, we multiplied the cost per customer ($495) by the number of customers (218). The value $495 is a unit rate because it gives the cost per one customer (per one unit).

Example 9 Solving an Application Involving Multiplication

The average weekly income for production workers is $489. How much does a production worker make in 1 year (assume 52 weeks in 1 year)?

Solution:

The value $489 per week is a unit rate. The total earnings for 1 year is given by $489 × 52.

$$\begin{array}{r} \overset{4\,4}{} \\ \overset{1\,1}{} \\ 489 \\ \times\ 52 \\ \hline 978 \\ +\ 24450 \\ \hline 25,428 \end{array}$$ The yearly earnings are $25,428.

> **TIP:** This product can be estimated quickly by rounding the factors.
>
> | 489 | rounds to ⟶ | 5|00 |
> | 52 | rounds to ⟶ | × 5|0 |
> | | | 25|000 |
>
> The total yearly income is approximately $25,000. Estimating gives a quick approximation of a product. Furthermore, it also checks for the reasonableness of our exact product. In this case $25,000 is close to our exact value of $25,428.

Skill Practice

13. Ella can type 65 words per minute. How many words can she type in 45 minutes?

Classroom Example: p. 49, Exercise 78

6. Area of a Rectangle

Another application of multiplication of whole numbers lies in finding the area of a region. **Area** measures the amount of surface contained within the region. For example, a square that is 1 in. by 1 in. occupies an area of 1 square inch, denoted as 1 in.^{2}. Similarly, a square that is 1 centimeter (cm) by 1 cm occupies an area of 1 square centimeter. This is denoted by 1 cm^2.

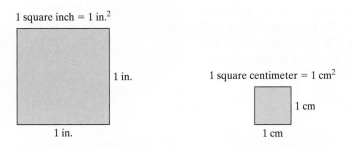

1 square inch = 1 in.^{2}

1 in.

1 in.

1 square centimeter = 1 cm^2

1 cm

1 cm

Animation

The units of square inches and square centimeters (in.^{2} and cm^2) are called *square units*. To find the area of a region, measure the number of square units occupied in that region. For example, the region in Figure 1-6 occupies 6 cm^2.

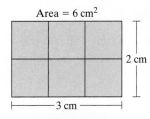

Area = 6 cm^2

2 cm

3 cm

Figure 1-6

Answer

13. 2925 words

The 3-cm by 2-cm region in Figure 1-6 suggests that to find the **area of a rectangle**, multiply the length by the width. If the area is represented by A, the length is represented by l, and the width is represented by w, then we have

$$\text{Area of rectangle} = (\text{length}) \times (\text{width})$$

$$A = l \times w$$

The letters A, l, and w are called **variables** because their values *vary* as they are replaced by different numbers.

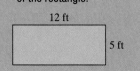
Example 10   Finding the Area of a Rectangle

Find the area and perimeter of the rectangle.

Solution:

Area:

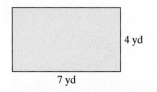

4 yd

7 yd

$$A = l \times w$$

$$A = (7 \text{ yd}) \times (4 \text{ yd})$$

$$= 28 \text{ yd}^2$$

Recall from Section 1.2 that the perimeter of a polygon is the sum of the lengths of the sides. In a rectangle the opposite sides are equal in length.

Perimeter:

$$P = 7 \text{ yd} + 4 \text{ yd} + 7 \text{ yd} + 4 \text{ yd}$$

$$= 22 \text{ yd}$$

7 yd

4 yd   4 yd

7 yd

The area is 28 yd^2 and the perimeter is 22 yd.

Example 11   Finding Area in an Application

The state of Wyoming is approximately the shape of a rectangle (Figure 1-7). Its length is 355 mi and its width is 276 mi. Approximate the total area of Wyoming by rounding the length and width to the nearest ten.

Solution:

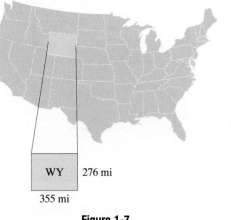

WY   276 mi

355 mi

Figure 1-7

$$
\begin{array}{lll}
355 & \text{rounds to} \longrightarrow & \overset{\overset{1}{4}}{36}\,|\,0 \\
276 & \text{rounds to} \longrightarrow & \times\, 28\,|\,0 \\
& & \overline{288} \\
& & 720 \\
& & \overline{1008\,|\,00}
\end{array}
$$

The area of Wyoming is approximately 100,800 mi^2.

Answers

14. Area: 60 ft^2; perimeter: 34 ft

15. 20,000 ft^2

Section 1.5 Practice Exercises

For additional exercises, see Classroom Activities 1.5A–1.5D in the *Student's Resource Manual* at www.mhhe.com/moh.

Study Skills Exercise

Write down your instructor's policies for the following:

a. Missing a test   **b.** Missing a class   **c.** Doing homework   **d.** Late homework policy

Vocabulary and Key Concepts

1. a. Two numbers being multiplied are called _____factors_____ and the result is called the _____product_____.

b. The _____commutative_____ property of multiplication states that the order of factors does not affect the product.

c. The _____associative_____ property of multiplication indicates that the manner in which factors are grouped under multiplication does not affect the product.

d. The multiplication property of 0 indicates that $7 \cdot 0 =$ _____0_____ and $0 \cdot 7 =$ _____0_____.

e. The multiplication property of 1 indicates that $7 \cdot 1 =$ _____7_____ and $1 \cdot 7 =$ _____7_____.

f. The statement $3(4 + 6) = 3 \cdot 4 + 3 \cdot 6$ is an example of the _____distributive_____ property of multiplication over addition.

g. The _____area_____ of a region measures the amount of surface contained within the region.

h. Given a rectangle of length l and width w, the area A is given by $A =$ _____$l \times w$_____.

Review Exercises

For Exercises 2–5, estimate the answer by rounding to the indicated place value.

2. $5981 + 7206$; thousands   13,000

3. $869,240 + 34,921 + 108,332$; ten-thousands   1,010,000

4. $907,801 - 413,560$; hundred-thousands   500,000

5. $8821 - 3401$; hundreds   5400

Concept 1: Introduction to Multiplication

6. Fill in the table of multiplication facts.

×	0	1	2	3	4	5	6	7	8	9
0	0	0	0	0	0	0	0	0	0	0
1	0	1	2	3	4	5	6	7	8	9
2	0	2	4	6	8	10	12	14	16	18
3	0	3	6	9	12	15	18	21	24	27
4	0	4	8	12	16	20	24	28	32	36
5	0	5	10	15	20	25	30	35	40	45
6	0	6	12	18	24	30	36	42	48	54
7	0	7	14	21	28	35	42	49	56	63
8	0	8	16	24	32	40	48	56	64	72
9	0	9	18	27	36	45	54	63	72	81

For Exercises 7–10, write the repeated addition as multiplication and simplify.

7. $5 + 5 + 5 + 5 + 5 + 5$ $6 \times 5; 30$

8. $2 + 2 + 2 + 2 + 2 + 2 + 2 + 2 + 2$ $9 \times 2; 18$

9. $9 + 9 + 9$ $3 \times 9; 27$

10. $7 + 7 + 7 + 7$ $4 \times 7; 28$

For Exercises 11–14, identify the factors and the product. **(See Example 1.)**

11. $13 \times 42 = 546$ Factors: 13, 42; product: 546

12. $26 \times 9 = 234$ Factors: 26, 9; product: 234

13. $3 \cdot 5 \cdot 2 = 30$ Factors: 3, 5, 2; product: 30

14. $4 \cdot 3 \cdot 8 = 96$ Factors: 4, 3, 8; product: 96

15. Write the product of 5 and 12, using three different notations. (Answers may vary.)
For example: $5 \times 12; 5 \cdot 12; 5(12)$

16. Write the product of 23 and 14, using three different notations. (Answers may vary.)
For example: $23 \times 14; 23 \cdot 14; (23)(14)$

Concept 2: Properties of Multiplication

For Exercises 17–22, match the property with the statement.

17. $8 \times 1 = 8$ d

18. $6 \cdot 13 = 13 \cdot 6$ a

19. $2(6 + 12) = 2 \cdot 6 + 2 \cdot 12$ e

20. $5 \cdot (3 \cdot 2) = (5 \cdot 3) \cdot 2$ b

21. $0 \times 4 = 0$ c

22. $7(14) = 14(7)$ a

a. Commutative property of multiplication

b. Associative property of multiplication

c. Multiplication property of 0

d. Multiplication property of 1

e. Distributive property of multiplication over addition

For Exercises 23–28, rewrite the expression, using the indicated property. **(See Examples 2 and 3.)**

23. 14×8; commutative property of multiplication
8×14

24. 3×9; commutative property of multiplication
9×3

25. $6 \times (2 \times 10)$; associative property of multiplication
$(6 \times 2) \times 10$

26. $(4 \times 15) \times 5$; associative property of multiplication
$4 \times (15 \times 5)$

27. $5(7 + 4)$; distributive property of multiplication over addition $(5 \times 7) + (5 \times 4)$

28. $3(2 + 6)$; distributive property of multiplication over addition $(3 \times 2) + (3 \times 6)$

Concept 3: Multiplying Many-Digit Whole Numbers

For Exercises 29–60, multiply. **(See Examples 4–6.)**

29. $\begin{array}{r} 24 \\ \times\ 6 \\ \hline 144 \end{array}$

30. $\begin{array}{r} 18 \\ \times\ 5 \\ \hline 90 \end{array}$

31. $\begin{array}{r} 26 \\ \times\ 2 \\ \hline 52 \end{array}$

32. $\begin{array}{r} 71 \\ \times\ 3 \\ \hline 213 \end{array}$

33. $\begin{array}{r} 131 \\ \times\ 5 \\ \hline 655 \end{array}$

34. $\begin{array}{r} 725 \\ \times\ 3 \\ \hline 2175 \end{array}$

35. $\begin{array}{r} 344 \\ \times\ 4 \\ \hline 1376 \end{array}$

36. $\begin{array}{r} 105 \\ \times\ 9 \\ \hline 945 \end{array}$

37. $\begin{array}{r} 1410 \\ \times\ 8 \\ \hline 11{,}280 \end{array}$

38. $\begin{array}{r} 2016 \\ \times\ 6 \\ \hline 12{,}096 \end{array}$

39. $\begin{array}{r} 3312 \\ \times\ 7 \\ \hline 23{,}184 \end{array}$

40. $\begin{array}{r} 4801 \\ \times\ 5 \\ \hline 24{,}005 \end{array}$

41. 42,014
    × 9
    378,126

42. 51,006
    × 8
    408,048

43. 32
    × 14
    448

44. 41
    × 21
    861

45. 68 · 24   1632

46. 55 · 41   2255

47. 72 · 12   864

48. 13 · 46   598

49. (143)(17)   2431

50. (722)(28)   20,216

51. (349)(19)   6631

52. (512)(31)   15,872

53. 151
    × 127
    19,177

54. 703
    × 146
    102,638

55. 222
    × 841
    186,702

56. 387
    × 506
    195,822

57. 3532
    × 6014
    21,241,448

58. 2810
    × 1039
    2,919,590

59. 4122
    × 982
    4,047,804

60. 7026
    × 528
    3,709,728

Concept 4: Estimating Products by Rounding

For Exercises 61–68, multiply the numbers, using the method found on page 44. **(See Example 7.)**

61. 600
    × 40
    24,000

62. 900
    × 50
    45,000

63. 3000
    × 700
    2,100,000

64. 4000
    × 400
    1,600,000

65. 8000
    × 9000
    72,000,000

66. 1000
    × 2000
    2,000,000

67. 90,000
    × 400
    36,000,000

68. 50,000
    × 6000
    300,000,000

For Exercises 69–72, estimate the product by first rounding the number to the indicated place value.

69. 11,784 × 5201; thousands place   60,000,000

70. 45,046 × 7812; thousands place   360,000,000

71. 82,941 × 29,740; ten-thousands place
    2,400,000,000

72. 630,229 × 71,907; ten-thousands place
    44,100,000,000

73. Suppose a hotel room costs $189 per night. Round this number to the nearest hundred to estimate the cost for a five-night stay. **(See Example 8.)** $1000

74. The science department of Comstock High School must purchase a set of calculators for a class. If the cost of one calculator is $129, estimate the cost of 28 calculators by rounding the numbers to the tens place. $3900

75. The average price for a ticket to see Kenny Chesney is $272. If a concert stadium seats 10,256 fans, estimate the amount of money received during that performance by rounding the number of seats to the nearest ten-thousand. $2,720,000

76. A breakfast buffet at a local restaurant serves 48 people. Estimate the maximum revenue for one week (7 days) if the price of a breakfast is $12. $3500

Concept 5: Translations and Applications Involving Multiplication

77. The 4-gigabyte (4-GB) iPod nano is advertised to store approximately 1000 songs. Assuming the average length of a song is 4 minutes, how many minutes of music can be stored on the iPod nano? **(See Example 9.)**   4000 minutes

78. One CD can hold 700 megabytes (MB) of data. How many megabytes can 15 CDs hold?   10,500 MB

79. It costs about $45 for a cat to have a medical exam. If a humane society has 37 cats, find the cost of medical exams for their cats. $1665

80. A can of Coke contains 12 fluid ounces (fl oz). Find the number of ounces in a case of Coke containing 12 cans.   144 fl oz

 Writing   Translating Expression   Geometry   Scientific Calculator   Video   NS&E

81. PaperWorld shipped 115 cases of copy paper to a business. There are 5 reams of paper in each case and 500 sheets of paper in each ream. Find the number of sheets of paper delivered to the business. 287,500 sheets

82. A dietary supplement bar has 14 grams (g) of protein. If Kathleen eats 2 bars a day for 6 days, how many grams of protein will she get from this supplement? 168 g

83. Tylee's car gets 31 miles per gallon (mpg) on the highway. How many miles can he travel if he has a full tank of gas (12 gal)? 372 mi

84. Sherica manages a small business called Pizza Express. She has 23 employees who work an average of 32 hours (hr) per week. How many hours of work does Sherica have to schedule each week? 736 hr

Concept 6: Area of a Rectangle

For Exercises 85–88, find the area. **(See Example 10.)**

85. 276 ft²

86. 62 m²

87. 5329 cm²

88. 1681 yd²

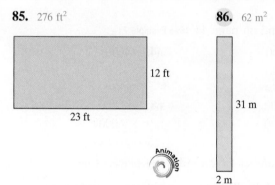

12 ft

23 ft

31 m

2 m

73 cm

73 cm

41 yd

41 yd

89. The state of Colorado is approximately the shape of a rectangle. Its length is 388 mi and its width is 269 mi. Approximate the total area of Colorado by rounding the length and width to the nearest ten. **(See Example 11.)**
105,300 mi²

90. A parcel of land has a width of 132 yd and a length of 149 yd. Approximate the total area by rounding each dimension to the nearest ten.
19,500 yd²

91. The front of a building has windows that are 44 in. by 58 in.

   a. Approximate the area of one window. 2400 in.²

   b. If the building has three floors and each floor has 14 windows, how many windows are there? 42 windows

   c. What is the approximate total area of all of the windows? 100,800 in.²

92. The length of a carport is 51 ft and its width is 29 ft. Approximate the area of the carport. 1500 ft²

93. Mr. Slackman wants to paint his garage door that is 8 ft by 16 ft. To decide how much paint to buy, he must find the area of the door. What is the area of the door?
128 ft²

94. To carpet a rectangular room, Erika must find the area of the floor. If the dimensions of the room are 10 yd by 15 yd, how much carpeting does she need? 150 yd²

 Writing   Translating Expression   Geometry   Scientific Calculator   Video   NS E

| Division of Whole Numbers | Section 1.6 |

1. Introduction to Division

Suppose 12 pieces of pizza are to be divided evenly among 4 children (Figure 1-8). The number of pieces that each child would receive is given by 12 ÷ 4, read "12 divided by 4."

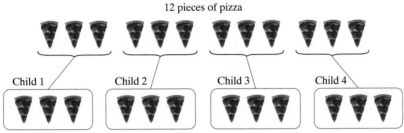

12 pieces of pizza

Child 1   Child 2   Child 3   Child 4

Figure 1-8

The process of separating 12 pieces of pizza evenly among 4 children is called **division**. The statement 12 ÷ 4 = 3 indicates that each child receives 3 pieces of pizza. The number 12 is called the **dividend**. It represents the number to be divided. The number 4 is called the **divisor**, and it represents the number of groups. The result of the division (in this case 3) is called the **quotient**. It represents the number of items in each group.

Division can be represented in several ways. For example, the following are all equivalent statements.

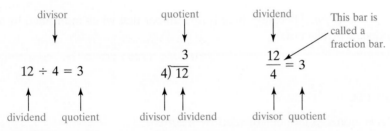

divisor   quotient   dividend   This bar is called a fraction bar.

$$12 \div 4 = 3 \qquad 4\overline{)12}^{\,3} \qquad \frac{12}{4} = 3$$

dividend quotient   divisor dividend   divisor quotient

Recall that subtraction is the inverse operation of addition. In the same way, division is the inverse operation of multiplication. For example, we say 12 ÷ 4 = 3 because 3 × 4 = 12.

Example 1 **Identifying the Dividend, Divisor, and Quotient**

Simplify each expression. Then identify the dividend, divisor, and quotient.

a. 48 ÷ 6 **b.** $9\overline{)36}$ **c.** $\frac{63}{7}$

Solution:

a. 48 ÷ 6 = 8   because 8 × 6 = 48
The dividend is 48, the divisor is 6, and the quotient is 8.

b. $9\overline{)36}^{\,4}$ because 4 × 9 = 36

The dividend is 36, the divisor is 9, and the quotient is 4.

c. $\frac{63}{7} = 9$ because 9 × 7 = 63

The dividend is 63, the divisor is 7, and the quotient is 9.

2. Properties of Division

Example 2 illustrates the important properties of division.

Example 2   Dividing Whole Numbers

Divide.

a. $8 \div 8$ **b.** $\dfrac{6}{6}$ **c.** $5 \div 1$

d. $1\overline{)7}$ **e.** $0 \div 6$ **f.** $\dfrac{0}{4}$ **g.** $6 \div 0$

Solution:

a. $8 \div 8 = 1$ because $1 \times 8 = 8$

b. $\dfrac{6}{6} = 1$ because $1 \times 6 = 6$

c. $5 \div 1 = 5$ because $5 \times 1 = 5$

d. $1\overline{)7}^{\,7}$ because $7 \times 1 = 7$

e. $0 \div 6 = 0$ because $0 \times 6 = 0$

f. $\dfrac{0}{4} = 0$ because $0 \times 4 = 0$

g. $6 \div 0$ is *undefined* because there is no number that when multiplied by 0 will produce a product of 6.

Properties of Division

  1. Any nonzero number divided by itself is 1.
       Example: $9 \div 9 = 1$

  2. Any number divided by 1 is the number itself.
       Example: $3 \div 1 = 3$

  3. Zero divided by any nonzero number is zero.
       Example: $0 \div 5 = 0$

  4. Any number divided by zero is undefined.
       Example: $9 \div 0$ is undefined

To help remember the difference between $0 \div 2$ and $2 \div 0$, consider this application:

$\$8 \div 2 = \$4$ means that if we divide \$8 between 2 people, each will receive \$4.

$\$0 \div 2 = \$0$ means that if we divide \$0 between 2 people, each will receive \$0.

But $\$2 \div 0$ means that we would like to divide \$2 between 0 people. This cannot be done. So $2 \div 0$ is undefined.

You should also note that unlike addition and multiplication, division is neither commutative nor associative. In other words, reversing the order of the dividend and divisor may produce a different quotient. Similarly, changing the manner in which numbers are grouped with division may affect the outcome. See Exercises 31 and 32.

3. Long Division

To divide larger numbers we use a process called **long division**. This process uses a series of estimates to find the quotient. We illustrate long division in Example 3.

Example 3 Using Long Division

Divide. $7\overline{)161}$

Solution:

Estimate $7\overline{)161}$ by first estimating $7\overline{)16}$ and writing the result in the tens place of the quotient. Since $7 \times 2 = 14$, there are at least 2 sevens in 16.

$$
\begin{array}{r}
2 \\
7\overline{)161} \\
-140 \\
\hline
21
\end{array}
$$

The 2 in the tens place represents 20 in the quotient.
←Multiply 7×20 and write the result under the dividend.
Subtract 140. We see that our estimate leaves 21.

Repeat the process. Now divide $7\overline{)21}$ and write the result in the ones place of the quotient.

$$
\begin{array}{r}
23 \\
7\overline{)161} \\
-140 \\
\hline
21 \\
-21 \\
\hline
0
\end{array}
$$

← Multiply 7×3.
Subtract.

The quotient is 23.

Check:
$$
\begin{array}{r}
23 \\
\times\, 7 \\
\hline
161\ ✔
\end{array}
$$

Skill Practice

Divide.

12. $8\overline{)136}$

Classroom Example: p. 59, Exercise 38

We can streamline the process of long division by "bringing down" digits of the dividend one at a time.

Example 4 Using Long Division

Divide. $6138 \div 9$

Solution:

$$
\begin{array}{r}
682 \\
9\overline{)6138} \\
-54 \\
\hline
73 \\
-72 \\
\hline
18 \\
-18 \\
\hline
0
\end{array}
$$

$9 \times 6 = 54$ and subtract.
Bring down the 3.
$9 \times 8 = 72$ and subtract.
Bring down the 8.
$9 \times 2 = 18$ and subtract.

The quotient is 682.

Check:
$$
\begin{array}{r}
\overset{7\,1}{682} \\
\times\ 9 \\
\hline
6138\ ✔
\end{array}
$$

Skill Practice

Divide.

13. $2891 \div 7$

Classroom Example: p. 59, Exercise 46

Answers

12. 17   **13.** 413

In many instances, quotients do not come out evenly. For example, suppose we had 13 pieces of pizza to distribute among 4 children (Figure 1-9).

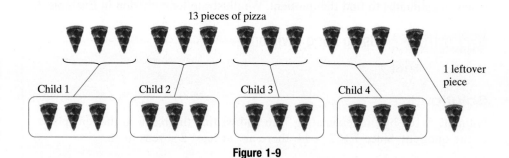

13 pieces of pizza

Child 1   Child 2   Child 3   Child 4   1 leftover piece

Figure 1-9

The mathematical term given to the "leftover" piece is called the **remainder**. The division process may be written as

$$\begin{array}{r} 3 \text{ R1} \\ 4\overline{)13} \\ -12 \\ \hline 1 \end{array}$$

The remainder is written next to the 3.

The **whole part of the quotient** is 3, and the remainder is 1. Notice that the remainder is written next to the whole part of the quotient.

We can check a division problem that has a remainder. To do so, multiply the divisor by the whole part of the quotient and then add the remainder. The result must equal the dividend. That is,

(Divisor)(whole part of quotient) + remainder = dividend

Thus,

$$(4)(3) + 1 \overset{?}{=} 13$$
$$12 + 1 \overset{?}{=} 13$$
$$13 = 13 \checkmark$$

Skill Practice

Divide.
14. $5107 \div 5$

Classroom Example: p. 59, Exercise 68

Example 5 Using Long Division

Divide. $1253 \div 6$

Solution:

$$\begin{array}{r} 208 \text{ R5} \\ 6\overline{)1253} \\ -12 \\ \hline 05 \\ -00 \\ \hline 53 \\ -48 \\ \hline 5 \end{array}$$

$6 \times 2 = 12$ and subtract.

Bring down the 5.

Note that 6 does not divide into 5, so we put a 0 in the quotient.

Bring down the 3.

$6 \times 8 = 48$ and subtract.

The remainder is 5.

To check, verify that $6 \times 208 + 5 = 1253$. ✔

Answer
14. 1021 R2

4. Dividing by a Many-Digit Divisor

When the divisor has more than one digit, we still use a series of estimations to find the quotient.

Example 6 Dividing by a Two-Digit Number

Divide. $32\overline{)1259}$

Solution:

To estimate the leading digit of the quotient, estimate the number of times 30 will go into 125. Since $30 \cdot 4 = 120$, our estimate is 4.

$$
\begin{array}{r}
4 \\
32\overline{)1259} \\
-128
\end{array}
$$

$32 \times 4 = 128$ is too big. We cannot subtract 128 from 125. Revise the estimate in the quotient to 3.

$$
\begin{array}{r}
3 \\
32\overline{)1259} \\
-96\downarrow \\
\hline
299
\end{array}
$$

$32 \times 3 = 96$ and subtract.
Bring down the 9.

Now estimate the number of times 30 will go into 299. Because $30 \times 9 = 270$, our estimate is 9.

$$
\begin{array}{r}
39 \text{ R}11 \\
32\overline{)1259} \\
-96\downarrow \\
\hline
299 \\
-288 \\
\hline
11
\end{array}
$$

$32 \times 9 = 288$ and subtract.
The remainder is 11.

To check, verify that $32 \times 39 + 11 = 1259$. ✔

Example 7 Dividing by a Many-Digit Number

Divide. $\dfrac{82,705}{602}$

Solution:

$$
\begin{array}{r}
137 \text{ R}231 \\
602\overline{)82,705} \\
-602\downarrow \\
\hline
2250 \\
-1806\downarrow \\
\hline
4445 \\
-4214 \\
\hline
231
\end{array}
$$

$602 \times 1 = 602$ and subtract.
Bring down the 0.
$602 \times 3 = 1806$ and subtract.
Bring down the 5.
$602 \times 7 = 4214$ and subtract.
The remainder is 231.

To check, verify that $602 \times 137 + 231 = 82,705$. ✔

Skill Practice

Divide.
15. $63\overline{)4516}$

Classroom Example: p. 60, Exercise 72

Skill Practice

Divide.
16. $304\overline{)62,405}$

Classroom Example: p. 60, Exercise 82

Answers
15. 71 R43 **16.** 205 R85

5. Translations and Applications Involving Division

Several words and phrases imply division. A partial list is given in Table 1-4.

Table 1-4

Word/Phrase	Example	In Symbols
Divide	Divide 12 by 3	$12 \div 3$ or $\dfrac{12}{3}$ or $3\overline{)12}$
Quotient	The quotient of 20 and 2	$20 \div 2$ or $\dfrac{20}{2}$ or $2\overline{)20}$
Per	110 mi per 2 hr	$110 \div 2$ or $\dfrac{110}{2}$ or $2\overline{)110}$
Divides into	4 divides into 28	$28 \div 4$ or $\dfrac{28}{4}$ or $4\overline{)28}$
Divided, or shared equally among	64 shared equally among 4	$64 \div 4$ or $\dfrac{64}{4}$ or $4\overline{)64}$

Skill Practice

17. Four players play Hearts with a standard 52-card deck of cards. If the cards are equally distributed, how many cards does each player get?

Classroom Example: p. 60, Exercise 90

Example 8 Solving an Application Involving Division

A painting business employs 3 painters. The business collects $1950 for painting a house. If all painters are paid equally, how much does each person make?

Solution:

This is an example where $1950 is shared equally among 3 people. Therefore, we divide.

$$
\begin{array}{r}
650 \\
3\overline{)1950} \\
\end{array}
$$

-18 $3 \times 6 = 18$ and subtract.
  15   Bring down the 5.
-15 $3 \times 5 = 15$ and subtract.
  00   Bring down the 0.
-0 $3 \times 0 = 0$ and subtract.
   0   The remainder is 0.

Each painter makes $650.

Skill Practice

18. A college has budgeted $4800 to buy graphing calculators. Each calculator costs $119. Estimate the number of calculators that the college can buy by rounding the cost to the nearest ten.

Classroom Example: p. 60, Exercise 98

Example 9 Solving an Application Involving Division with Estimation

Elaine and Max drove from South Bend, Indiana, to Bonita Springs, Florida. The total driving distance was 1089 mi, and the driving time was approximately 20 hr. Estimate the average speed by rounding the distance to the nearest hundred.

Answers

17. 13 cards   **18.** 40 calculators

Solution:

1089 mi rounds to 1100 mi. The speed is represented by 1100 mi per 20 hr, or

$$
\begin{array}{r}
55 \\
20\overline{)1100} \\
-100\!\downarrow \\
\hline
100 \\
-100 \\
\hline
0
\end{array}
$$

$20 \times 5 = 100$ and subtract.

Bring down the 0.

$20 \times 5 = 100$ and subtract.

Max and Elaine averaged approximately 55 miles per hour (mph).

Example 10   **Solving an Application Involving Division**

The bar graph in Figure 1-10 depicts the number of calories burned per hour for selected activities.

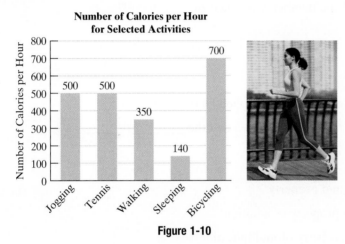

Number of Calories per Hour for Selected Activities

Figure 1-10

a. Janie wants to burn 3500 calories per week exercising. For how many hours must she jog?

b. For how many hours must Janie bicycle to burn 3500 calories?

Solution:

a. The total number of calories must be divided into 500-calorie increments. Thus, the number of hours required is given by $3500 \div 500$.

$$
\begin{array}{r}
7 \\
500\overline{)3500} \\
-3500 \\
\hline
0
\end{array}
$$

Janie requires 7 hr of jogging to burn 3500 calories.

b. 3500 calories must be divided into 700-calorie increments. The number of hours required is given by $3500 \div 700$.

$$
\begin{array}{r}
5 \\
700\overline{)3500} \\
-3500 \\
\hline
0
\end{array}
$$

Janie requires 5 hr of bicycling to burn 3500 calories.

Skill Practice

19. The cost for four different types of pastry at a French bakery is shown in the bar graph.

Cost for Selected Pastries

Melissa has $360 to spend on desserts.

a. If she spends all the money on chocolate éclairs, how many can she buy?

b. If she spends all the money on apple tarts, how many can she buy?

Classroom Example: p. 61, Exercise 100

Answers
19. a. 180 chocolate éclairs
   b. 72 apple tarts

Section 1.6   Practice Exercises

Study Skills Exercise

For additional exercises, see Classroom Activities 1.6A–1.6C in the *Student's Resource Manual* at www.mhhe.com/moh.

In your next math class, take notes by drawing a vertical line about three-fourths of the way across the paper, as shown. On the left side, write down what your instructor puts on the board or overhead. On the right side, make your own comments about important words, procedures, or questions that you have.

Vocabulary and Key Concepts

1. a. Given the division statement $15 \div 3 = 5$, the number 15 is called the _____dividend_____, the number 3 is called the _____divisor_____, and the number 5 is called the _____quotient_____.

   b. $5 \div 5 =$ _____1_____

   c. $5 \div 1 =$ _____5_____

   d. $0 \div 5 =$ _____0_____

   e. $5 \div 0$ is _____undefined_____ because no number multiplied by 0 equals 5.

   f. If 17 is divided by 5, the whole part of the quotient is 3 and the _____remainder_____ is 2.

Review Exercises

2. Rewrite each statement using the indicated property.

   a. $2 + 5 =$ _____$5 + 2$_____; commutative property of addition

   b. $2 \cdot 5 =$ _____$5 \cdot 2$_____; commutative property of multiplication

   c. $3 + (10 + 2) =$ _____$(3 + 10) + 2$_____; associative property of addition

   d. $3 \cdot (10 \cdot 2) =$ _____$(3 \cdot 10) \cdot 2$_____; associative property of multiplication

For Exercises 3–10, add, subtract, or multiply as indicated.

3. $48 \cdot 103$ 4944

4. $678 - 83$ 595

5. $1008 + 245$ 1253

6. $14(220)$ 3080

7. 5230×127 664,210

8. $789(25)$ 19,725

9. $4890 - 3988$ 902

10. $38,002 + 3902$ 41,904

Concept 1: Introduction to Division

For Exercises 11–16, simplify each expression. Then identify the dividend, divisor, and quotient. **(See Example 1.)**

11. $72 \div 8$
Dividend: 72; divisor: 8; quotient: 9

12. $32 \div 4$
Dividend: 32; divisor: 4; quotient: 8

13. $8\overline{)64}$
Dividend: 64; divisor: 8; quotient: 8

14. $5\overline{)35}$
Dividend: 35; divisor: 5; quotient: 7

15. $\dfrac{45}{9}$
Dividend: 45; divisor: 9; quotient: 5

16. $\dfrac{20}{5}$
Dividend: 20; divisor: 5; quotient: 4

Concept 2: Properties of Division

17. In your own words, explain the difference between dividing a number by zero and dividing zero by a number.
You cannot divide a number by zero (the quotient is undefined). If you divide zero by a number (other than zero), the quotient is always zero.

18. Explain what happens when a number is either divided or multiplied by 1.
A number divided or multiplied by 1 remains unchanged.

Writing   Translating Expression   Geometry   Scientific Calculator   Video   NS E

For Exercises 19–30, use the properties of division to simplify the expression, if possible. **(See Example 2.)**

19. $15 \div 1$ 15

20. $21\overline{)21}$ 1

21. $0 \div 10$ 0

22. $\dfrac{0}{3}$ 0

23. $0\overline{)9}$ Undefined

24. $4 \div 0$ Undefined

25. $\dfrac{20}{20}$ 1

26. $1\overline{)9}$ 9

27. $\dfrac{16}{0}$ Undefined

28. $\dfrac{5}{1}$ 5

29. $8\overline{)0}$ 0

30. $13 \div 13$ 1

31. Show that $6 \div 3 = 2$ but $3 \div 6 \neq 2$ by using multiplication to check. $2 \times 3 = 6, \; 2 \times 6 \neq 3$

32. Show that division is not associative, using the numbers 36, 12, and 3.
$(36 \div 12) \div 3 = 3 \div 3 = 1$ but $36 \div (12 \div 3) = 36 \div 4 = 9$

Concept 3: Long Division

33. Explain the process for checking a division problem when there is no remainder.
Multiply the quotient and the divisor to get the dividend.

34. Show how checking by multiplication can help us remember that $0 \div 5 = 0$ and that $5 \div 0$ is undefined.
In checking $0 \div 5 = 0$, we get $0 \times 5 = 0$ which is true. In trying to check $5 \div 0 = ?$, we need to find a number that when multiplied by 0, gives us 5. Since no such number exists, the answer to $5 \div 0$ is undefined.

For Exercises 35–46, divide and check by multiplying. **(See Examples 3 and 4.)**

35. $78 \div 6$ 13
Check: $6 \times \square = 78$

36. $364 \div 7$ 52
Check: $7 \times \square = 364$

37. $5\overline{)205}$ 41
Check: $5 \times \square = 205$

38. $8\overline{)152}$ 19
Check: $8 \times \square = 152$

39. $\dfrac{972}{2}$ 486

40. $\dfrac{582}{6}$ 97

41. $1227 \div 3$ 409

42. $236 \div 4$ 59

43. $5\overline{)1015}$ 203

44. $5\overline{)2035}$ 407

45. $\dfrac{4932}{6}$ 822

46. $\dfrac{3619}{7}$ 517

For Exercises 47–54, check each division problem. If it is incorrect, find the correct answer.

47. $4\overline{)224}^{\,56}$
Correct

48. $7\overline{)574}^{\,82}$
Correct

49. $761 \div 3 = 253$
Incorrect; 253 R2

50. $604 \div 5 = 120$
Incorrect; 120 R4

51. $\dfrac{1021}{9} = 113 \text{ R}4$
Correct

52. $\dfrac{1311}{6} = 218 \text{ R}3$
Correct

53. $8\overline{)203}^{\,25 \text{ R}6}$
Incorrect; 25 R3

54. $7\overline{)821}^{\,117 \text{ R}5}$
Incorrect; 117 R2

For Exercises 55–70, divide and check the answer. **(See Example 5.)**

55. $61 \div 8$ 7 R5

56. $89 \div 3$ 29 R2

57. $9\overline{)92}$ 10 R2

58. $5\overline{)74}$ 14 R4

59. $\dfrac{55}{2}$ 27 R1

60. $\dfrac{49}{3}$ 16 R1

61. $593 \div 3$ 197 R2

62. $801 \div 4$ 200 R1

63. $\dfrac{382}{9}$ 42 R4

64. $\dfrac{428}{8}$ 53 R4

65. $3115 \div 2$ 1557 R1

66. $4715 \div 6$ 785 R5

67. $6014 \div 8$ 751 R6

68. $9013 \div 7$ 1287 R4

69. $6\overline{)5012}$ 835 R2

70. $2\overline{)1101}$ 550 R1

Concept 4: Dividing by a Many-Digit Divisor

For Exercises 71–82, divide. **(See Examples 6 and 7.)**

71. $9110 \div 19$ 479 R9   **72.** $3505 \div 13$ 269 R8   **73.** $24\overline{)1051}$ 43 R19   **74.** $41\overline{)8104}$ 197 R27

75. $\dfrac{8008}{26}$ 308   **76.** $\dfrac{9180}{15}$ 612   **77.** $68{,}012 \div 54$ 1259 R26   **78.** $92{,}013 \div 35$ 2628 R33

79. $69{,}712 \div 304$ 229 R96   **80.** $51{,}107 \div 221$ 231 R56   **81.** $114\overline{)34{,}428}$ 302   **82.** $421\overline{)87{,}989}$ 209

Concept 5: Translations and Applications Involving Division

For Exercises 83–88, for each English sentence, write a mathematical expression and simplify.

83. Find the quotient of 497 and 71. $497 \div 71$; 7   **84.** Find the quotient of 1890 and 45. $1890 \div 45$; 42

85. Divide 877 by 14. $877 \div 14$; 62 R9   **86.** Divide 722 by 53. $722 \div 53$; 13 R33

87. Divide 6 into 42. $42 \div 6$; 7   **88.** Divide 9 into 108. $108 \div 9$; 12

89. There are 392 students signed up for Anatomy 101. If each classroom can hold 28 students, find the number of classrooms needed. **(See Example 8.)**   14 classrooms

90. A wedding reception is planned to take place in the fellowship hall of a church. The bride anticipates 120 guests, and each table will seat 8 people. How many tables should be set up for the reception to accommodate all the guests?   15 tables

91. A case of tomato sauce contains 32 cans. If a grocer has 168 cans, how many cases can he fill completely? How many cans will be left over?   5 cases; 8 cans left over

92. Austin has $425 to spend on dining room chairs. If each chair costs $52, does he have enough to purchase 8 chairs? If so, will he have any money left over?   Yes, $9 left over

93. Pauline drove 312 mi in 6 hr. Find Pauline's average speed (in miles per hour).   52 mph

94. A house cleaning company charges $144 to clean a 3-room apartment. At this rate, how much does it cost to clean one room? $48

95. If it takes 2200 lb of grapes to make 100 gal of white wine, how many pounds are needed for 1 gal?   22 lb

96. There are 7280 acres of ferns in Florida that are owned by 260 farmers. Find the average size of each farm.   28 acres

97. Suppose Genny can type 1234 words in 22 min. Round each number to estimate her rate in words per minute. **(See Example 9.)** $1200 \div 20 = 60$; approximately 60 words per minute

98. On a trip to California from Illinois, Lavu drove 2780 mi. The gas tank in his car allows him to travel 405 mi. Round each number to the hundreds place to estimate the number of tanks of gas needed for the trip.   Approximately 7 tanks of gas

99. A group of 18 people go to a concert. Ticket prices are given in the bar graph. If the group has $450, can they all attend the concert? If so, which type of seats can they buy? **(See Example 10.)**

Yes, they can all attend if they sit in the second balcony.

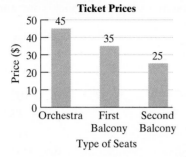

Ticket Prices

Figure for Exercise 99

100. The bar graph gives the average annual income for four professions: teacher, professor, CEO, and programmer. Find the monthly income for each of the four professions. Teacher: $3000; professor: $5000; CEO: $10,000; programmer: $4000

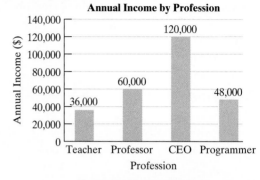

Annual Income by Profession

Figure for Exercise 100

101. The labels on most laundry detergents include the total number of wash loads that can be done. This makes comparison shopping easier since the amount of soap to be used can vary from one brand to the next. The label on Planet laundry detergent does not show the total number of loads. The instructions on the back label suggest using 4 fl oz of detergent per load of laundry.

  a. How many loads of laundry can be done with a 50-fl-oz bottle of Planet? 12 loads

  b. How many ounces of detergent are left over? 2 oz

102. At an elementary school, parents may pick up their children in a loading area in front of the school. Nine cars are allowed to pull up at one time, and it takes approximately 2 min to load and release the group of 9 cars. If it takes 26 minutes to get all cars through the line, how many cars are in line waiting to pick up children?

117 cars

Calculator Connections

Topic: Multiplying and Dividing Whole Numbers

To multiply and divide numbers on a calculator, use the ⊠ and ÷ keys, respectively.

Expression	**Keystrokes**	**Result**
38,319 × 1561	38319 ⊠ 1561 **ENTER**	59815959
2,449,216 ÷ 6248	2449216 ÷ 6248 **ENTER**	392

Calculator Exercises

For Exercises 103–106, solve the problem. Use a calculator to perform the calculations.

103. The United States consumes approximately 21,000,000 barrels (bbl) of oil per day. (*Source:* U.S. Energy Information Administration) How much does it consume in 1 year? 7,665,000,000 bbl

104. The average time to commute to work for people living in Washington State is 26 min (round trip 52 min). (*Source:* U.S. Census Bureau) How much time does a person spend commuting to and from work in 1 year if the person works 5 days a week for 50 weeks per year? 13,000 min

105. The budget for the U.S. federal government for 2010 was approximately $3552 billion. (*Source:* www.gpo.gov) How much could the government spend each quarter and still stay within its budget? $888 billion

106. At a weigh station, a truck carrying 96 crates weighs in at 34,080 lb. If the truck weighs 9600 lb when empty, how much does each crate weigh? Each crate weighs 255 lb.

Problem Recognition Exercises

Operations on Whole Numbers

For Exercises 1–18, perform the indicated operations. For each problem, estimate to check your answer.

1. a.
$$\begin{array}{r} 52 \\ + 13 \\ \hline 65 \end{array}$$

b.
$$\begin{array}{r} 52 \\ \times 13 \\ \hline 676 \end{array}$$

c.
$$\begin{array}{r} 52 \\ - 13 \\ \hline 39 \end{array}$$

d. $13\overline{)52}$ 4

2. a. $17\overline{)102}$ 6

b.
$$\begin{array}{r} 102 \\ - 17 \\ \hline 85 \end{array}$$

c.
$$\begin{array}{r} 102 \\ \times 17 \\ \hline 1734 \end{array}$$

d.
$$\begin{array}{r} 102 \\ + 17 \\ \hline 119 \end{array}$$

3. a.
$$\begin{array}{r} 5064 \\ \times 58 \\ \hline 293{,}712 \end{array}$$

b.
$$\begin{array}{r} 5064 \\ + 58 \\ \hline 5122 \end{array}$$

c. $58\overline{)5064}$ 87 R18

d.
$$\begin{array}{r} 5064 \\ - 58 \\ \hline 5006 \end{array}$$

4. a.
$$\begin{array}{r} 1226 \\ - 114 \\ \hline 1112 \end{array}$$

b. $114\overline{)1226}$ 10 R86

c.
$$\begin{array}{r} 1226 \\ + 114 \\ \hline 1340 \end{array}$$

d.
$$\begin{array}{r} 1226 \\ \times 114 \\ \hline 139{,}764 \end{array}$$

5. a.
$$\begin{array}{r} 156 \\ + 41 \\ \hline 197 \end{array}$$

b.
$$\begin{array}{r} 197 \\ - 41 \\ \hline 156 \end{array}$$

6. a.
$$\begin{array}{r} 6004 \\ + 221 \\ \hline 6225 \end{array}$$

b.
$$\begin{array}{r} 6225 \\ - 221 \\ \hline 6004 \end{array}$$

7. 418×10 4180

8. 418×100 41,800

9. 418×1000 418,000

10. $418 \times 10{,}000$ 4,180,000

11. $350{,}000 \div 10$ 35,000

12. $350{,}000 \div 100$ 3500

13. $350{,}000 \div 1000$ 350

14. $350{,}000 \div 10{,}000$ 35

15. 82×3000 246,000

16. $47 \times 60{,}000$ 2,820,000

17. $50 \cdot 400$ 20,000

18. $600 \cdot 900$ 540,000

Section 1.7 Exponents, Square Roots, and the Order of Operations

Concepts

1. Exponents
2. Square Roots
3. Order of Operations
4. Computing a Mean (Average)

1. Exponents

Thus far in the text we have learned to add, subtract, multiply, and divide whole numbers. We now present the concept of an **exponent** to represent repeated multiplication. For example, the product

$$3 \cdot 3 \cdot 3 \cdot 3 \cdot 3 \qquad \text{can be written as} \qquad 3^5$$

where the exponent is 5 and the base is 3.

The expression 3^5 is written in exponential form. The exponent, or **power**, is 5 and represents the number of times the **base**, 3, is used as a factor. The expression 3^5 is read as "three to the fifth power." Other expressions in exponential form are shown next.

5^2	is read as	"five squared" or "five to the second power"
5^3	is read as	"five cubed" or "five to the third power"
5^4	is read as	"five to the fourth power"
5^5	is read as	"five to the fifth power"

Instructor Note: Show students that 5^2 is not the same as $2 \cdot 5$.

TIP: The expression $5^1 = 5$. Any number without an exponent explicitly written has a power of 1.

Exponential form is a shortcut notation for repeated multiplication. However, to simplify an expression in exponential form, we often write out the individual factors.

Example 1   **Evaluating Exponential Expressions**

Evaluate.

a. 6^2 **b.** 5^3 **c.** 2^4

Solution:

a. $6^2 = 6 \cdot 6$ The exponent, 2, indicates the number of times the base, 6, is used as a factor.

     $= 36$

b. $5^3 = 5 \cdot 5 \cdot 5$ When three factors are multiplied, we can group the first two factors and perform the multiplication.

     $= (5 \cdot 5) \cdot 5$

     $= (25) \cdot 5$ Then multiply the product of the first two factors by the last factor.

     $= 125$

c. $2^4 = 2 \cdot 2 \cdot 2 \cdot 2$

     $= (2 \cdot 2) \cdot 2 \cdot 2$ Group the first two factors.

     $= 4 \cdot 2 \cdot 2$ Multiply the first two factors.

     $= (4 \cdot 2) \cdot 2$ Multiply the product by the next factor to the right.

     $= 8 \cdot 2$

     $= 16$

Skill Practice

Evaluate.
  1. 8^2 **2.** 4^3 **3.** 2^5

Classroom Example: p. 68, Exercise 28

One important application of exponents lies in recognizing **powers of 10**, that is, 10 raised to a whole-number power. For example, consider the following expressions.

$$10^1 = 10$$
$$10^2 = 10 \cdot 10 = 100$$
$$10^3 = 10 \cdot 10 \cdot 10 = 1000$$
$$10^4 = 10 \cdot 10 \cdot 10 \cdot 10 = 10,000$$
$$10^5 = 10 \cdot 10 \cdot 10 \cdot 10 \cdot 10 = 100,000$$
$$10^6 = 10 \cdot 10 \cdot 10 \cdot 10 \cdot 10 \cdot 10 = 1,000,000$$

From these examples, we see that a power of 10 results in a 1 followed by several zeros. The number of zeros is the same as the exponent on the base of 10.

Answers
1. 64   **2.** 64   **3.** 32

2. Square Roots

To square a number means that we multiply the base times itself. For example, $5^2 = 5 \cdot 5 = 25$.

To find a positive **square root** of a number means that we reverse the process of squaring. For example, finding the square root of 25 is equivalent to asking, "What positive number, when squared, equals 25?" The symbol $\sqrt{}$ (called a *radical sign*) is used to denote the positive square root of a number. Therefore, $\sqrt{25}$ is the positive number that, when squared, equals 25. Thus, $\sqrt{25} = 5$ because $(5)^2 = 25$.

Skill Practice

Find the square roots.

4. $\sqrt{4}$
5. $\sqrt{100}$
6. $\sqrt{400}$
7. $\sqrt{121}$

Classroom Example: p. 68, Exercise 46

Example 2 Evaluating Square Roots

Find the square roots.

  a. $\sqrt{9}$ **b.** $\sqrt{64}$ **c.** $\sqrt{1}$ **d.** $\sqrt{0}$

Solution:

  a. $\sqrt{9} = 3$ because $(3)^2 = 3 \cdot 3 = 9$

  b. $\sqrt{64} = 8$ because $(8)^2 = 8 \cdot 8 = 64$

  c. $\sqrt{1} = 1$ because $(1)^2 = 1 \cdot 1 = 1$

  d. $\sqrt{0} = 0$ because $(0)^2 = 0 \cdot 0 = 0$

TIP: To simplify square roots, it is advisable to become familiar with these squares and square roots.

$0^2 = 0 \longrightarrow \sqrt{0} = 0$	$7^2 = 49 \longrightarrow \sqrt{49} = 7$
$1^2 = 1 \longrightarrow \sqrt{1} = 1$	$8^2 = 64 \longrightarrow \sqrt{64} = 8$
$2^2 = 4 \longrightarrow \sqrt{4} = 2$	$9^2 = 81 \longrightarrow \sqrt{81} = 9$
$3^2 = 9 \longrightarrow \sqrt{9} = 3$	$10^2 = 100 \longrightarrow \sqrt{100} = 10$
$4^2 = 16 \longrightarrow \sqrt{16} = 4$	$11^2 = 121 \longrightarrow \sqrt{121} = 11$
$5^2 = 25 \longrightarrow \sqrt{25} = 5$	$12^2 = 144 \longrightarrow \sqrt{144} = 12$
$6^2 = 36 \longrightarrow \sqrt{36} = 6$	$13^2 = 169 \longrightarrow \sqrt{169} = 13$

3. Order of Operations

A numerical expression may contain more than one operation. For example, the following expression contains both multiplication and subtraction.

$$18 - 5(2)$$

The order in which the multiplication and subtraction are performed will affect the overall outcome.

Multiplying first yields	**Subtracting first yields**
$18 - 5(2) = 18 - 10$	$18 - 5(2) = 13(2)$
$= 8$ (correct)	$= 26$ (incorrect)

To avoid confusion, mathematicians have outlined the proper order of operations. In particular, multiplication is performed before addition or subtraction. The guidelines for the order of operations are given next. These rules must be followed in all cases.

Answers

4. 2
5. 10
6. 20
7. 11

Order of Operations
Step 1 Perform all operations inside parentheses first.
Step 2 Simplify any expressions containing exponents or square roots.
Step 3 Perform multiplication or division in the order that they appear from left to right.
Step 4 Perform addition or subtraction in the order that they appear from left to right.

Example 3 Using the Order of Operations

Simplify.

a. $15 - 10 \div 2 + 3$ **b.** $(5 - 2) \cdot 7 - 1$ **c.** $\sqrt{100} - 2^3$

Solution:

a. $15 - 10 \div 2 + 3$

$= 15 - 5 + 3$ Perform the division $10 \div 2$ first.

$= 10 + 3$ Perform addition and subtraction from left to right.

$= 13$ Add.

b. $(5 - 2) \cdot 7 - 1$

$= (3) \cdot 7 - 1$ Perform the operation inside parentheses first.

$= 21 - 1$ Perform multiplication before subtraction.

$= 20$ Subtract.

c. $\sqrt{100} - 2^3$

$= 10 - 8$ Simplify any expressions with exponents or square roots. Note that $\sqrt{100} = 10$, and $2^3 = 2 \cdot 2 \cdot 2 = 8$.

$= 2$ Subtract.

Skill Practice

Simplify.
8. $18 + 6 \div 2 - 4$
9. $(20 - 4) \div 2 + 1$
10. $2^3 - \sqrt{25}$

Classroom Examples: p. 68, Exercises 54, 64, and 66

Example 4 Using the Order of Operations

Simplify.

a. $300 \div (7 - 2)^2 - 2^2$ **b.** $36 + (7^2 - 3)$

Solution:

a. $300 \div (7 - 2)^2 - 2^2$

$= 300 \div (5)^2 - 2^2$ Perform the operation within parentheses first.

$= 300 \div 25 - 4$ Simplify exponents: $5^2 = 5 \cdot 5 = 25$ and $2^2 = 2 \cdot 2 = 4$.

$= 12 - 4$ Perform division before subtraction.

$= 8$ Subtract.

Skill Practice

Simplify.
11. $40(3 - 1)^2 - 5^2$
12. $42 - (50 - 6^2)$

Classroom Examples: p. 68, Exercises 80 and 88

Answers

8. 17   **9.** 9   **10.** 3
11. 135   **12.** 28

b. $\quad 36 + (7^2 - 3)$

$= 36 + (49 - 3)$ Perform the operations within parentheses first. The guidelines indicate that we simplify the expression with the exponent before we subtract: $7^2 = 49$.

$= 36 + \quad 46$ Continue simplifying within parentheses.

$= 82$ Add.

4. Computing a Mean (Average)

The order of operations must be used when we compute an average. The technical term for the average of a list of numbers is the **mean** of the numbers. To find the mean of a set of numbers, first compute the sum of the values. Then divide the sum by the number of values. This is represented by the formula

$$\text{Mean} = \frac{\text{sum of the values}}{\text{number of values}}$$

Classroom Example: p. 69, Exercise 98

Skill Practice

13. The ages (in years) of 5 students in an algebra class are given here. Find the mean age.

22, 18, 22, 32, 46

Example 5   **Computing a Mean (Average)**

Ashley took 6 tests in Chemistry. Find her mean (average) score.

$$89, 91, 72, 86, 94, 96$$

Solution:

$$\text{Average score} = \frac{89 + 91 + 72 + 86 + 94 + 96}{6}$$

$$= \frac{528}{6} \quad \text{Add the values in the list first.}$$

$$= 88 \qquad \text{Divide.} \qquad \begin{array}{r} 88 \\ 6\overline{)528} \\ -48 \\ \hline 48 \\ -48 \\ \hline 0 \end{array}$$

Ashley's mean (average) score is 88.

TIP: The division bar in $\dfrac{89 + 91 + 72 + 86 + 94 + 96}{6}$ is also a grouping symbol and implies parentheses:

$$\frac{(89 + 91 + 72 + 86 + 94 + 96)}{6}$$

Answer

13. 28 years

Section 1.7   Practice Exercises

Study Skills Exercise

For additional exercises, see Classroom Activities 1.7A–1.7B in the *Student's Resource Manual* at www.mhhe.com/moh.

Look over the notes that you took today. Do you understand what you wrote? If there were any rules, definitions, or formulas, highlight them so that they can be easily found when studying for the test. You may want to begin by highlighting the order of operations.

Vocabulary and Key Concepts

1. a. Given the expression 5^4, the _____base_____ is 5 and the exponent is _____4_____.

   b. The values 10^1, 10^2, 10^3, and so on are called _____powers_____ of 10.

   c. The positive _____square_____ _____root_____ of 81 is denoted by $\sqrt{81}$. The value of $\sqrt{81}$ is 9 because $9 \cdot 9 = $ _____81_____.

   d. The _____order_____ of _____operations_____ is a process used to simplify expressions involving more than one mathematical operation.

   e. A _____variable_____ is a letter or symbol such as x, y, or z that represents a number. Quantities that do not change such as 5 and 11 are called _____constants_____.

   f. The _____mean_____ or average of a set of numbers is found by taking the sum of the values and then dividing by the number of values.

Review Exercises

For Exercises 2–8, write true or false for each statement.

2. Subtraction is associative; for example, $10 - (3 - 2) = (10 - 3) - 2$.   False

3. Addition is commutative; for example, $5 + 3 = 3 + 5$.   True

4. Subtraction is commutative; for example, $5 - 3 = 3 - 5$.   False

5. $6 \cdot 0 = 6$ False   **6.** $0 \div 8 = 0$ True   **7.** $0 \cdot 8 = 0$ True   **8.** $5 \div 0$ is undefined.   True

Concept 1: Exponents

9. Write an exponential expression with 9 as the base and 4 as the exponent. 9^4

10. Write an exponential expression with 3 as the base and 8 as the exponent. 3^8

11. Write an exponential expression with 7 as the exponent and 2 as the base. 2^7

12. Write an exponential expression with 5 as the exponent and 6 as the base. 6^5

For Exercises 13–16, write the repeated multiplication in exponential form. Do not simplify.

13. $3 \cdot 3 \cdot 3 \cdot 3 \cdot 3 \cdot 3$ **14.** $7 \cdot 7 \cdot 7 \cdot 7$ **15.** $4 \cdot 4 \cdot 4 \cdot 4 \cdot 2 \cdot 2 \cdot 2$ **16.** $5 \cdot 5 \cdot 5 \cdot 10 \cdot 10 \cdot 10$
    3^6 7^4 $4^4 \cdot 2^3$ $5^3 \cdot 10^3$

For Exercises 17–20, expand the exponential expression as a repeated multiplication. Do not simplify.

17. 8^4 **18.** 2^6 **19.** 4^8 **20.** 6^2
    $8 \cdot 8 \cdot 8 \cdot 8$ $2 \cdot 2 \cdot 2 \cdot 2 \cdot 2 \cdot 2$ $4 \cdot 4 \cdot 4 \cdot 4 \cdot 4 \cdot 4 \cdot 4 \cdot 4$ $6 \cdot 6$

For Exercises 21–36, evaluate the exponential expressions. **(See Example 1.)**

21. 2^3 8

22. 4^2 16

23. 3^2 9

24. 5^2 25

25. 3^3 27

26. 11^2 121

27. 5^3 125

28. 4^3 64

29. 2^5 32

30. 6^3 216

31. 3^4 81

32. 5^4 625

33. 1^2 1

34. 1^3 1

35. 1^4 1

36. 1^5 1

37. Explain what happens when the number 1 is raised to any power. **(See Exercises 33–36.)**
The number 1 raised to any power equals 1.

For Exercises 38–41, evaluate the powers of 10.

38. 10^2 100

39. 10^3 1000

40. 10^4 10,000

41. 10^5 100,000

42. Explain how to get 10^9 *without* doing the repeated multiplication. **(See Exercises 38–41.)**
The term 10^9 simplifies to a number with a 1 followed by 9 zeros: 1,000,000,000

Concept 2: Square Roots

For Exercises 43–50, evaluate the square roots. **(See Example 2.)**

43. $\sqrt{4}$ 2

44. $\sqrt{9}$ 3

45. $\sqrt{36}$ 6

46. $\sqrt{81}$ 9

47. $\sqrt{100}$ 10

48. $\sqrt{49}$ 7

49. $\sqrt{0}$ 0

50. $\sqrt{16}$ 4

Concept 3: Order of Operations

51. Does the order of operations indicate that addition is always performed before subtraction? Explain.
No, addition and subtraction should be performed in the order in which they appear from left to right.

52. Does the order of operations indicate that multiplication is always performed before division? Explain.
No, multiplication and division should be performed in the order in which they appear from left to right.

For Exercises 53–93, simplify using the order of operations. **(See Examples 3 and 4.)**

53. $6 + 10 \cdot 2$ 26

54. $4 + 3 \cdot 7$ 25

55. $10 - 3^2$ 1

56. $11 - 2^2$ 7

57. $(10 - 3)^2$ 49

58. $(11 - 2)^2$ 81

59. $36 \div 2 \div 6$ 3

60. $48 \div 4 \div 2$ 6

61. $15 - (5 + 8)$ 2

62. $41 - (13 + 8)$ 20

63. $(13 - 2) \cdot 5 - 2$ 53

64. $(8 + 4) \cdot 6 + 8$ 80

65. $4 + 12 \div 3$ 8

66. $9 + 15 \div \sqrt{25}$ 12

67. $30 \div 2 \cdot \sqrt{9}$ 45

68. $55 \div 11 \cdot 5$ 25

69. $7^2 - 5^2$ 24

70. $3^3 - 2^3$ 19

71. $(7 - 5)^2$ 4

72. $(3 - 2)^3$ 1

73. $100 \div 5(2)$ 40

74. $60 \div 3(2)$ 40

75. $90 \div 3 \cdot 3$ 90

76. $80 \div 2 \cdot 2$ 80

77. $\sqrt{81} + 2(9 - 1)$ 25

78. $\sqrt{121} + 3(8 - 3)$ 26

79. $36 \div (2^2 + 5)$ 4

80. $42 \div (3^2 - 2)$ 6

81. $80 - (20 \div 4) + 6$ 81

82. $120 - (48 \div 8) - 40$ 74

83. $(43 - 26) \cdot 2 - 4^2$ 18

84. $(51 - 48) \cdot 3 + 7^2$ 58

85. $(18 - 5) - (23 - \sqrt{100})$ 0

86. $(\sqrt{36} + 11) - (31 - 16)$ 2

87. $80 \div (9^2 - 7 \cdot 11)^2$ 5

88. $108 \div (3^3 - 6 \cdot 4)^2$ 12

89. $22 - 4(\sqrt{25} - 3)^2$ 6

90. $17 + 3(7 - \sqrt{9})^2$ 65

91. $96 - 3(42 \div 7 \cdot 6 - 5)$ 3

92. $50 - 2(36 \div 12 \cdot 2 - 4)$ 46

93. $16 + 5(20 \div 4 \cdot 8 - 3)$ 201

Concept 4: Computing a Mean (Average)

For Exercises 94–96, find the mean (average) of each set of numbers. **(See Example 5.)**

94. 19, 21, 18, 21, 16   19

95. 105, 114, 123, 101, 100, 111   109

96. 1480, 1102, 1032, 1002   1154

97. Neelah took 6 quizzes and received the following scores: 19, 20, 18, 19, 18, 14. Find her quiz average.   18

 **98.** Shawn's scores on his last 4 tests were 83, 95, 87, and 91. What is his average for these tests?   89

99. Toyota Motor Company has been very competitive in building fuel-efficient autos. In a recent year, five of Toyota's most efficient cars had mileage ratings of 49 mpg, 30 mpg, 34 mpg, 31 mpg, and 26 mpg. What was the average mileage rating for these vehicles?   The average mileage rating was 34 mpg.

100. On a trip, Stephen had his car washed 4 times and paid $7, $10, $8, and $7. What was the average amount spent per wash? $8 per wash

101. The monthly rainfall for Seattle, Washington, is given in the table. All values are in millimeters (mm).

	Jan.	Feb.	Mar.	Apr.	May	Jun.	Jul.	Aug.	Sep.	Oct.	Nov.	Dec.
Rainfall	122	94	80	52	47	40	15	21	44	90	118	123

Find the average monthly rainfall for the months of November, December, and January.

121 mm per month

102. The monthly snowfall for Alpena, Michigan, is given in the table. All values are in inches.

	Jan.	Feb.	Mar.	Apr.	May	Jun.	Jul.	Aug.	Sep.	Oct.	Nov.	Dec.
Snowfall	22	16	13	5	1	0	0	0	0	1	9	20

Find the average monthly snowfall for the months of November, December, January, February, and March.   16 in. per month

Expanding Your Skills

Sometimes an expression will have parentheses within parentheses. This is called *nested parentheses.* Often different shapes such as ( ), [ ], or { } are used to make it easier to match up the pairs of parentheses, for example,

$$\{300 - 4[4 + (5 + 2)^2] + 8\} - 31$$

It is important to note that the symbols ( ), [ ], or { } all represent parentheses and are used for grouping. When nested parentheses occur, simplify the innermost set first. Then work your way out. For example, simplify

$$\{300 - 4[4 + (5 + 2)^2] + 8\} - 31$$

The solution is

$\{300 - 4[4 + (5 + 2)^2] + 8\} - 31$

$= \{300 - 4[4 + (7)^2] + 8\} - 31$ Simplify within the innermost parentheses first ( ).

$= \{300 - 4[4 + 49] + 8\} - 31$ Simplify the exponent.

$= \{300 - 4[53] + 8\} - 31$ Simplify within the next innermost parentheses [ ].

$= \{300 - 212 + 8\} - 31$ Multiply before adding.

$= \{88 + 8\} - 31$ Subtract and add in order from left to right within the parentheses { }.

$= 96 - 31$ Simplify within the parentheses { }.

$= 65$ Simplify.

For Exercises 103–106, simplify the expressions with nested parentheses.

103. $3[4 + (6 - 3)^2] - 15$ 24

104. $2[5(4 - 1) + 3] \div 6$ 6

105. $5\{21 - [3^2 - (4 - 2)]\}$ 70

 106. $4\{18 - [(10 - 8) + 2^3]\}$ 32

Calculator Connections

Topic: Evaluating Expressions with Exponents

Many calculators use the $\boxed{x^2}$ key to square a number. To raise a number to a higher power, use the $\boxed{\wedge}$ key (or on some calculators, the $\boxed{x^y}$ key or $\boxed{y^x}$ key).

Expression	Keystrokes	Result
26^2	26 $\boxed{x^2}$ $\boxed{\text{ENTER}}$	676

On some calculators, you do not need to press $\boxed{\text{ENTER}}$

| 3^4 | 3 $\boxed{\wedge}$ 4 $\boxed{\text{ENTER}}$ | 81 |
| | or   3 $\boxed{y^x}$ 4 $\boxed{=}$ | 81 |

Calculator Exercises

For Exercises 107–112, use a calculator to perform the indicated operations.

107. 156^2 24,336

108. 418^2 174,724

109. 12^5 248,832

110. 35^4 1,500,625

111. 43^3 79,507

112. 71^3 357,911

For Exercises 113–118, simplify the expressions by using the order of operations. For each step use the calculator to simplify the given operation.

113. $8126 - 54,978 \div 561$ 8028

114. $92,168 + 6954 \times 29$ 293,834

115. $(3548 - 3291)^2$ 66,049

116. $(7500 \div 625)^3$ 1728

117. $\dfrac{89,880}{384 + 2184}$ *Hint:* This expression has implied grouping symbols. $\dfrac{89,880}{(384 + 2184)}$ 35

118. $\dfrac{54,137}{3393 - 2134}$ *Hint:* This expression has implied grouping symbols. $\dfrac{54,137}{(3393 - 2134)}$ 43

Problem-Solving Strategies

1. Problem-Solving Strategies

In this section, we offer additional practice with applications of whole numbers. Keep in mind that all word problems are different and that there is no magic "trick" to solve an application problem. However, we can offer the following guidelines.

Concepts

1. Problem-Solving Strategies
2. Applications Involving One Operation
3. Applications Involving Multiple Operations

Guidelines for Problem Solving

Step 1 Read the problem carefully and familiarize yourself with the situation. If possible, draw a diagram or write down an appropriate formula. Sometimes you may be able to estimate a reasonable answer.

Step 2 Write down what information is given and what must be found.

Step 3 Form a strategy. Identify what mathematical operation applies (addition, subtraction, multiplication, or division). Sometimes a combination of operations is necessary.

Step 4 Perform the mathematical operations to solve for the unknown.

Step 5 Check the answer. If the answer is reasonable and checks, state the answer in words.

2. Applications Involving One Operation

We illustrate these guidelines with a variety of examples. To assist with step 3 where we must identify an appropriate mathematical operation, we summarize some of our key words and phrases. See Table 1-5.

Table 1-5

Operation	Key Word or Phrase
Addition	Sum, added to, increased by, more than, plus, total of
Subtraction	Difference, minus, decreased by, less, subtract
Multiplication	Product, times, multiply
Division	Quotient, divide, per, shared equally

Example 1 Solving a Travel Application

Kent travels from Columbus, Ohio, to Indianapolis, Indiana, and then on to Springfield, Illinois. The total distance he drives is 351 mi. The distance between Columbus and Indianapolis is 168 mi. Find the distance between Indianapolis and Springfield.

Solution:

Familiarize and draw a picture.

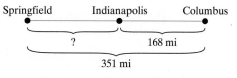

Springfield   Indianapolis   Columbus

? 168 mi

351 mi

Given: In this case, we know the total distance and one of the parts.

Find: Find the second distance (between Indianapolis and Springfield).

Skill Practice

1. The odometer of a car read 24,316 mi last year. This year the reading is 37,134. How many miles was the car driven during the year?

Classroom Example: p. 76, Exercise 20

Answer

1. 12,818 mi

Operation: This problem can be phrased as an addition problem with a missing addend or as an equivalent subtraction problem.

$$? + 168 = 351 \quad \text{or} \quad 351 - 168 = ?$$

Subtracting yields

$$351 - 168 = 183$$

The distance between Indianapolis and Springfield is 183 mi.

Example 2 Solving a Sales Application

A used car business keeps records of vehicle sales by type of vehicle and month.

	July	August	September
Cars	23	28	32
Trucks	13	8	10
SUVs	15	18	21

Find the total number of vehicles sold in July.

Solution:

Given: The number of each type of vehicle sold in July (highlighted in red)

Find: The total number of vehicles sold in July

Operation: The word *total* indicates addition.

$$\text{Total} = 23 + 13 + 15$$
$$= 51$$

There were 51 vehicles sold in July.

Example 3 Solving a Business Application

A ream of paper holds 500 sheets. Kim purchases 24 reams for her office. How many sheets of paper is this?

Solution:

Familiarize and draw a picture.

500 sheets + 500 sheets + 500 sheets + ··· + 500 sheets

Given: There are 24 reams (packages) of paper with 500 sheets per package.

Find: The total number of sheets of paper

Operation: This situation calls for repeated addition. Therefore, we will multiply.

$$24(500) = 12,000$$

There are 12,000 sheets of paper.

Example 4 Solving a Consumer Application

A 5-speed Jeep Cherokee gets 23 mpg (miles per gallon) on the highway. How many gallons of gas would be required for a 667-mi drive from El Paso to Dallas?

Solution:

Familiarize and draw a picture.

Given: The total distance, 667 mi, and the gas mileage, 23 mpg

Find: How many increments of 23 mi would be required for the trip?

Operation: This is a situation where 667 mi must be divided into 23-mi increments. Use the operation of division.

$$
\begin{array}{r}
29 \\
23\overline{)667} \\
-46 \\
\hline
207 \\
-207 \\
\hline
0
\end{array}
$$

> **TIP:** The solution to Example 4 can be checked by multiplication. Twenty-nine gallons of gas at 23 mpg produces
>
> (29 gal)(23 mi/gal) = 667 mi

The drive from El Paso to Dallas in a Jeep Cherokee will require 29 gal of gas.

3. Applications Involving Multiple Operations

Sometimes more than one operation is needed to solve an application problem.

Example 5 Solving a Consumer Application

Jorge bought a car for $18,340. He paid $2500 down and then paid the rest in equal monthly payments over a 4-year period. Find the amount of Jorge's monthly payment (not including interest).

Solution:

Familiarize and draw a picture.

Given: Total price: $18,340
Down payment: $2500

Payment plan: 4 years (48 months)

Find: Monthly payment

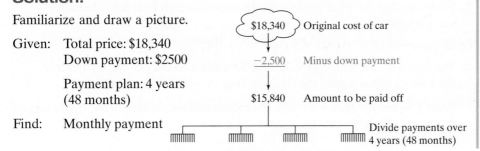

Operations:

1. The amount of the loan to be paid off is equal to the original cost of the car minus the down payment. We use subtraction:

$$\begin{array}{r} \$18,340 \\ -\ 2,500 \\ \hline \$15,840 \end{array}$$

2. This money is distributed in equal payments over a 4-year period. Because there are 12 months in 1 year, there are $4 \times 12 = 48$ months in a 4-year period. To distribute $15,840 among 48 equal payments, we divide.

$$\begin{array}{r} 330 \\ 48\overline{)15,840} \\ -144 \\ \hline 144 \\ -144 \\ \hline 00 \end{array}$$

> **TIP:** The solution to Example 5 can be checked by multiplication. Forty-eight payments of $330 each amount to 48($330) = $15,840. This added to the down payment totals $18,340 as desired.

Jorge's monthly payments will be $330.

Skill Practice

6. Taylor makes $18 per hour for the first 40 hr worked each week. His overtime rate is $27 per hour for hours exceeding the normal 40-hr workweek. If his total salary for one week is $963, determine the number of hours of overtime worked.

Classroom Example: p. 78, Exercise 38

Example 6   Solving a Travel Application

Linda must drive from Clayton to Oakley. She can travel directly from Clayton to Oakley on a mountain road, but will only average 40 mph. On the route through Pearson, she travels on highways and can average 60 mph. Which route will take less time?

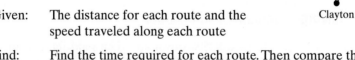

Solution:

Read and familiarize: A map is presented in the problem.

Given: The distance for each route and the speed traveled along each route

Find: Find the time required for each route. Then compare the times to determine which will take less time.

Operations:

1. First note that the total distance of the route through Pearson is found by using addition.

$$85 \text{ mi} + 95 \text{ mi} = 180 \text{ mi}$$

2. The speed of the vehicle gives us the distance traveled per hour. Therefore, the time of travel equals the total distance divided by the speed.

From Clayton to Oakley through the mountains, we divide 120 mi by 40-mph increments to determine the number of hours.

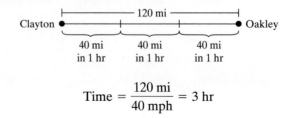

$$\text{Time} = \frac{120 \text{ mi}}{40 \text{ mph}} = 3 \text{ hr}$$

From Clayton to Oakley through Pearson, we divide 180 mi by 60-mph increments to determine the number of hours.

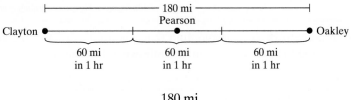

$$\text{Time} = \frac{180 \text{ mi}}{60 \text{ mph}} = 3 \text{ hr}$$

Therefore, each route takes the same amount of time, 3 hr.

Example 7 **Solving a Construction Application**

A rancher must fence the corral shown in Figure 1-11. However, no fencing is required on the side adjacent to the barn. If fencing costs $4 per foot, what is the total cost?

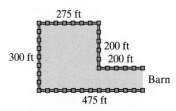

Figure 1-11

Solution:

Read and familiarize: A figure is provided.

Strategy

With some application problems, it helps to work backward from your final goal. In this case, our final goal is to find the total cost. However, to find the total cost, we must first find the total distance to be fenced. To find the total distance, we add the lengths of the sides that are being fenced.

$$
\begin{array}{r}
\overset{1\ 1}{275} \text{ ft} \\
200 \text{ ft} \\
200 \text{ ft} \\
475 \text{ ft} \\
+\ 300 \text{ ft} \\
\hline
1450 \text{ ft}
\end{array}
$$

Therefore,

$$\begin{pmatrix}\text{Total cost} \\ \text{of fencing}\end{pmatrix} = \begin{pmatrix} \text{total} \\ \text{distance} \\ \text{in feet} \end{pmatrix}\begin{pmatrix} \text{cost} \\ \text{per foot} \end{pmatrix}$$

$$= (1450 \text{ ft})(\$4 \text{ per ft})$$

$$= \$5800$$

The total cost of fencing is $5800.

Skill Practice

7. Alain wants to put molding around the base of the room shown in the figure. No molding is needed where the door, closet, and bathroom are located. Find the total cost if molding is $2 per foot.

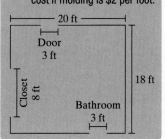

Classroom Example: p. 78, Exercise 42

Answer

7. $124

Section 1.8 Practice Exercises

Study Skills Exercise

For additional exercises, see Classroom Activities 1.8A–1.8B in the *Student's Resource Manual* at www.mhhe.com/moh.

Sometimes you may run into a problem with homework, or you may find that you are having trouble keeping up with the pace of the class. A tutor can be a good resource. Answer the following questions.

a. Does your college offer tutoring? **b.** Is it free? **c.** Where would you go to sign up for a tutor?

Vocabulary and Key Concepts

1. Which expression is undefined, $4 \div 0$ or $0 \div 4$? $\quad 4 \div 0$

Review Exercises

For Exercises 2–13, translate the English phrase into a mathematical statement and simplify.

2. 89 decreased by 66
$89 - 66; 23$

3. 71 increased by 14
$71 + 14; 85$

4. 16 more than 42
$42 + 16; 58$

5. Twice 14
$2 \cdot 14; 28$

6. The difference of 93 and 79
$93 - 79; 14$

7. Subtract 32 from 102
$102 - 32; 70$

8. Divide 12 into 60
$60 \div 12; 5$

9. The product of 10 and 13
$10 \cdot 13; 130$

10. The total of 12, 14, and 15
$12 + 14 + 15; 41$

11. The quotient of 24 and 6
$24 \div 6; 4$

12. 41 less than 78
$78 - 41; 37$

13. The sum of 5, 13, and 25
$5 + 13 + 25; 43$

Concept 1: Problem-Solving Strategies

14. In your own words, list the guidelines or strategy that you would use to solve an application problem.
See page 71.

For Exercises 15–18, write two or more key words or phrases that represent the given operation. Answers may vary.

15. Addition
For example: sum, added to, increased by, more than, plus, total of

16. Multiplication
For example: product, times, multiply

17. Subtraction
For example: difference, minus, decreased by, less, subtract

18. Division
For example: quotient, divide, per, distributed equally, shared equally

Concept 2: Applications Involving One Operation

19. White Mountain Peak in California is 14,246 ft high. Denali in Alaska is 20,320 ft high. How much higher is Denali than White Mountain Peak? **(See Example 1.)** Denali is 6074 ft higher than White Mountain Peak.

20. In a recent year, *Reader's Digest* was the best-selling U.S. magazine with 12,212,000 yearly subscriptions. *Sports Illustrated* was 15th overall and had 3,252,900 yearly subscriptions. How many more subscriptions did *Reader's Digest* have than *Sports Illustrated*? Reader's Digest had 8,959,100 more subscriptions.

For Exercises 21–22, refer to the table.

Country	Crude Oil Consumption in 2009 (in barrels per day)	Population 2009
United States	18,690,000	307,000,000
China	8,220,000	1,339,000,000
Japan	4,360,000	127,000,000
Russia	4,210,000	140,000,000
Canada	2,170,000	33,000,000

Source: U.S. Energy Information Administration

21. Find the total amount of oil (in barrels per day) consumed by China, Japan, Russia, and Canada. **(See Example 2.)**   18,960,000 barrels per day

22. Find the total population of China, Japan, Russia, and Canada.   1,639,000,000 people

23. A graphing calculator screen consists of an array of rectangular dots called *pixels*. If the screen has 96 rows of pixels and 126 pixels in each row, how many pixels are in the whole screen? **(See Example 3.)**   The whole screen has 12,096 pixels.

24. The floor of a rectangular room has 62 rows of tile with 38 tiles in each row. How many total tiles are there?   There are 2356 tiles.

25. At one time, San Antonio College had 3000 students who registered for Prealgebra. If the average class size is 25 students, how many Prealgebra classes will the college have to offer?   There will be 120 classes of Prealgebra.

26. Eight people are to share equally in an inheritance of $84,480. How much money will each person receive?   Each person will receive $10,560.

27. The Honda Insight gets 45 miles per gallon (mpg) in stop-and-go traffic. How many gallons will it use in 405 mi of stop-and-go driving? **(See Example 4.)**   There will be 9 gal used.

28. A couple traveled at an average speed of 52 mph for a cross-country trip. If the couple drove 1352 mi, how many hours was the trip?   The couple traveled for 26 hr.

29. Jeannette has two children who each attended college in Boston. Her son Ricardo attended Bunker Hill Community College where the yearly tuition and fees came to $3024. Her daughter Ricki attended M.I.T. where the yearly tuition and fees totaled $39,212. If Jeannette paid the full amount for both children to go to school, what was her total expense for tuition and fees for 1 year?   Jeannette will pay $42,236 for 1 year.

30. Clyde and Mason each leave a rest area on the Florida Turnpike. Clyde travels north and Mason travels south. After 2 hr, Clyde has gone 138 mi and Mason, who ran into heavy traffic, traveled only 96 mi. How far apart are they?   They are 234 mi apart.

31. The Toyota Prius gets 55 mpg on the highway. How many miles can it go on 20 gal?   The Prius can go 1100 mi.

32. A 3 credit-hour class at a certain college meets 3 hr per week. If a semester is 16 weeks long, how many hours will the class meet during the semester?   The class will meet for 48 hr during the semester.

33. A movie theater has 70 rows and 45 seats in a row. What is the maximum seating capacity?   The maximum capacity is 3150 seats.

34. A square checkerboard has 8 boxes per row and 8 rows. What is the total number of boxes?   There are 64 boxes in a checkerboard.

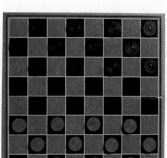

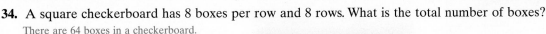

Concept 3: Applications Involving Multiple Operations

35. Jackson purchased a used car for $16,540. He paid $2500 down and paid the rest in equal monthly payments over a 36-month period. How much were his monthly payments? **(See Example 5.)**
Jackson's monthly payment was $390.

36. Lucio purchased a refrigerator for $1170. He paid $150 at the time of purchase and then paid off the rest in equal monthly payments over 1 year. How much was his monthly payment?   Lucio's monthly payment was $85.

37. Monika must drive from Watertown to Utica. She can travel directly from Watertown to Utica on a small county road, but will only average 40 mph. On the route through Syracuse, she travels on highways and can average 60 mph. Which route will take less time? **(See Example 6.)**   Each trip will take 2 hr.

Watertown

70 mi   80 mi

Syracuse   50 mi   Utica

Figure for Exercise 37

38. Rex has a choice of two routes to drive from Oklahoma City to Fort Smith. On the interstate, the distance is 220 mi and he can drive 55 mph. If he takes the back roads, he can only travel 40 mph, but the distance is 200 mi. Which route will take less time?   The interstate will take 4 hr, and the back roads will take 5 hr. The interstate will take less time.

39. If you wanted to line the outside of a garden with a decorative border, would you need to know the area of the garden or the perimeter of the garden?   Perimeter

40. If you wanted to know how much sod to lay down within a rectangular backyard, would you need to know the area of the yard or the perimeter of the yard?   Area

41. Arisu wants to buy molding for a room that is 12 ft by 11 ft. No molding is needed for the doorway, which measures 3 ft. See the figure. If molding costs $2 per foot, how much money will it cost? **(See Example 7.)**   The cost will be $86.

11 ft

3 ft

12 ft   12 ft

11 ft

42. A homeowner wants to fence her rectangular backyard. The yard is 75 ft by 90 ft. If fencing costs $5 per foot, how much will it cost to fence the yard?
It will cost $1650.

43. What is the cost to carpet the room whose dimensions are shown in the figure? Assume that carpeting costs $34 per square yard and that there is no waste.   The cost is $1020.

6 yd

5 yd

44. What is the cost to tile the room whose dimensions are shown in the figure? Assume that tile costs $3 per square foot.   The cost is $720.

12 ft

20 ft

45. The balance in Gina's checking account is $278. If she writes checks for $82, $59, and $101, how much will be left over?
There will be $36 left in Gina's account.

46. The balance in Jose's checking account is $3455. If he write checks for $587, $36, and $156, how much will be left over?   There will be $2676 left in Jose's account.

47. A community college bought 72 new computers and 6 new printers for a computer lab. If computers were purchased for $2118 each and the printers for $256 each, what was the total bill (not including tax)?
The total bill was $154,032.

48. Tickets to the San Diego Zoo in California cost $27 for children aged 3–11 and $37 for adults. How much money is required to buy tickets for a class of 33 children and 6 adult chaperones? The amount of money required is $1113.

49. A discount music store buys used CDs from its customers for $3. Furthermore, a customer can buy any used CD in the store for $8. Latayne sells 16 CDs.

   a. How much money does she receive by selling the 16 CDs?
   Latayne will receive $48.
   b. How many CDs can she then purchase with the money?
   She can buy 6 CDs.

50. Shevona earns $12 per hour and works a 40-hr workweek. At the end of the week, she cashes her paycheck and then buys two tickets to an Alicia Keys concert.

   a. How much money is her paycheck worth?
   Shevona's paycheck is worth $480.
   b. If the concert tickets cost $89 each, how much money does she have left over from her paycheck after buying the tickets? She will have $302 left.

51. During his 13-year career with the Chicago Bulls, Michael Jordan scored 12,192 field goals (worth 2 points each). He scored 581 three-point shots and 7327 free-throws (worth 1 point each). How many total points did he score during his career with the Bulls?
Michael Jordan scored 33,454 points with the Bulls.

52. A matte is to be cut and placed over five small square pictures before framing. Each picture is 5 in. wide, and the matte frame is 37 in. wide, as shown in the figure. If the pictures are to be equally spaced (including the space on the left and right edges), how wide is the matte between them?
There will be 2 in. of matte between the pictures.

53. Mortimer the cat was prescribed a suspension of methimazole for hyperthyroidism. This suspension comes in a 60-milliliter bottle with instructions to give 1 milliliter twice a day. The label also shows there is one refill, but it must be called in 2 days ahead. Mortimer had his first two doses on September 1.

   a. For how many days will one bottle last?
   One bottle will last for 30 days.
   b. On what day, at the latest, should his owner order a refill to avoid running out of medicine?
   The owner should order a refill no later than September 28.

54. Recently, the American Medical Association reported that there were 630,300 male doctors and 205,900 female doctors in the United States.

   a. What is the difference between the number of male doctors and the number of female doctors?
   The difference between male and female doctors is 424,400.
   b. What is the total number of doctors?
   The total number of doctors is 836,200.

55. On a map, each inch represents 60 mi.

   a. If Las Vegas and Salt Lake City are approximately 6 in. apart on the map, what is the actual distance between the cities?
   The distance is 360 mi.
   b. If Madison, Wisconsin, and Dallas, Texas, are approximately 840 mi apart, how many inches would this represent on the map?
   14 in. represents 840 mi.

56. On a map, each inch represents 40 mi.

   a. If Wichita, Kansas, and Des Moines, Iowa, are approximately 8 in. apart on the map, what is the actual distance between the cities?
   The distance is 320 mi.
   b. If Seattle, Washington, and Sacramento, California, are approximately 600 mi apart, how many inches would this represent on the map? 15 in. represents 600 mi.

57. A textbook company ships books in boxes containing a maximum of 12 books. If a bookstore orders 1250 books, how many boxes can be filled completely? How many books will be left over?
104 boxes will be filled completely with 2 books left over.

58. A farmer sells eggs in containers holding a dozen eggs. If he has 4257 eggs, how many containers will be filled completely? How many eggs will be left over?

354 containers will be filled completely with 9 eggs left over.

 59. Marc pays for an $84 dinner with $20 bills.

    a. How many bills must he use?   Marc needs five $20 bills.

    b. How much change will he receive?   He will receive $16 in change.

60. Shawn buys 3 CDs for a total of $54 and pays with $10 bills.

    a. How many bills must he use?   Shawn needs six $10 bills.

    b. How much change will he receive?   He will receive $6 in change.

61. Ling has three jobs. He works for a lawn maintenance service 4 days a week. He also tutors math and works as a waiter on weekends. His hourly wage and the number of hours for each job are given for a 1-week period. How much money did Ling earn for the week?

He earned $520.

	Hourly Wage	Number of Hours
Tutor	$30/hr	4
Waiter	10/hr	16
Lawn maintenance	8/hr	30

62. An electrician, a plumber, a mason, and a carpenter work at a certain construction site. The hourly wage and the number of hours each person worked are summarized in the table. What was the total amount paid for all four workers?   The total amount paid was $2748.

	Hourly Wage	Number of Hours
Electrician	$36/hr	18
Plumber	28/hr	15
Mason	26/hr	24
Carpenter	22/hr	48

Group Activity

Becoming a Successful Student

Materials: Computer with Internet access

Estimated time: 15 minutes

Group Size: 4

Good time management, good study skills, and good organization will help you be successful in this course. Answer the following questions and compare your answers with your group members.

Answers will vary.

1. For the week, write down the times each day that you plan to study math.

Monday	Tuesday	Wednesday	Thursday	Friday	Saturday	Sunday

2. Write down the date of your next math test. _____

Writing   Translating Expression   Geometry   Scientific Calculator   Video   NS E

3. How many hours do you work each week outside of your school work? _____
Do you think that this is impacting your success in this class?

4. Look through the book in Chapter 1 and find the page number corresponding to each feature in the book. Discuss with your group members how you might use each feature.

Problem Recognition Exercises: page _____

Chapter Summary: page _____

Chapter Review Exercises: page _____

Chapter Test: page _____

5. Look at the Skill Practice exercises in the margin (for example, find Skill Practice exercises 8–10 in Section 1.7). Where are the answers to these exercises located? Discuss with your group members how you might use the Skill Practice exercises.

6. Discuss with your group members places where you can go for extra help in math. Then write down three of the suggestions.

_____

_____

_____

7. Do you keep an organized notebook for this class? _____ Can you think of any suggestions that you can share with your group members to help them keep their materials organized?

8. Some students favor different methods of learning over others. For example, you might prefer:

- Learning through listening and hearing.
- Learning through seeing images, watching demonstrations, and visualizing diagrams and charts.
- Learning by experience through a hands-on approach by doing things.
- Learning through reading and writing.

Most experts believe that the most effective learning comes when a student engages in *all* of these activities. However, each individual is different and may benefit from one activity more than another. You can visit a number of different websites to determine your "learning style." Try doing a search on the Internet with the key words "*learning styles assessment.*" Once you have found a suitable website, answer the questionnaire and the site will give you feedback on what method of learning works best for you.

9. As you read through Chapter 1, try to become familiar with the features of this textbook. Then match the feature in column B with its description in column A.

<u>Column A</u>	<u>Column B</u>
1. Allows you to check your work as you do your homework e	**a.** Tips
2. Shows you how to avoid common errors g	**b.** ConnectMath
3. Provides an online tutorial and exercise supplement b	**c.** Skill Practice Exercises
4. Outlines key concepts for each section in the chapter f	**d.** Problem Recognition Exercises
5. Provides exercises that allow you to distinguish between different types of problems d	**e.** Answers to odd exercises
6. Offers helpful hints and insight a	**f.** Chapter Summary
7. Offers practice exercises that go along with each example c	**g.** Avoiding Mistakes

Chapter 1   Summary

Section 1.1   Introduction to Whole Numbers

Key Concepts

The place value for each **digit** of a number is shown in the chart.

Billions Period			Millions Period			Thousands Period			Ones Period		
Hundred-billions	Ten-billions	Billions	Hundred-millions	Ten-millions	Millions	Hundred-thousands	Ten-thousands	Thousands	Hundreds	Tens	Ones
		3,	4	0	9,	1	1	2			

Numbers can be written in different forms, for example:

Standard Form

3,409,112

Expanded Form

3 millions + 4 hundred-thousands +
9 thousands + 1 hundred + 1 ten + 2 ones
$= 3 \times 1,000,000 + 4 \times 100,000 + 9 \times 1,000 + 1 \times 100$
$\quad + 1 \times 10 + 2 \times 1$
$= 3,000,000 + 400,000 + 9,000 + 100 + 10 + 2$

Words

three million, four hundred nine thousand, one hundred twelve

The order of whole numbers can be visualized by placement on a number line.

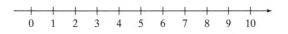

Examples

Example 1

The digit 9 in the number 24,891,321 is in the ten-thousands place.

Example 2

The standard form of the number forty-one million, three thousand, fifty-six is 41,003,056.

Example 3

The expanded form of the number 76,903 is
7 ten-thousands + 6 thousands + 9 hundreds + 3 ones
$= 7 \times 10,000 + 6 \times 1,000 + 9 \times 100 + 3 \times 1$
$= 70,000 + 6,000 + 900 + 3$

Example 4

In words the number 2504 is two thousand, five hundred four.

Example 5

To show that $8 > 4$, note the placement on the number line: 8 is to the right of 4.

Section 1.2   Addition of Whole Numbers and Perimeter

Key Concepts

The **sum** is the result of adding numbers called **addends**.

Addition is performed with and without carrying.

Addition Property of Zero

The sum of any number and zero is that number.

Commutative Property of Addition

Changing the order of the addends does not affect the sum.

Associative Property of Addition

The manner in which the addends are grouped does not affect the sum.

There are several words and phrases that indicate addition, such as *sum, added to, increased by, more than, plus,* and *total of.*

The **perimeter** of a **polygon** is the distance around the outside of the figure. To find perimeter, take the sum of the lengths of all sides of the figure.

Examples

Example 1

For $2 + 7 = 9$, the addends are 2 and 7, and the sum is 9.

Example 2

$$\begin{array}{r} 23 \\ +\,41 \\ \hline 64 \end{array} \qquad \begin{array}{r} {}^{1\,1}189 \\ +\,76 \\ \hline 265 \end{array}$$

Example 3

$16 + 0 = 16$

Example 4

$3 + 12 = 12 + 3$

Example 5

$2 + (19 + 3) = (2 + 19) + 3$

Example 6

The sum of 6 and 18 translates to $6 + 18$.

Example 7

The expression $5 + 4$ can be translated as 5 increased by 4, or 4 more than 5.

Example 8

The perimeter is found by adding the lengths of all sides.

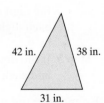

42 in.   38 in.

31 in.

Perimeter $= 42$ in. $+ 38$ in. $+ 31$ in. $= 111$ in.

Section 1.3 Subtraction of Whole Numbers

Key Concepts

The **difference** is the result of subtracting the **subtrahend** from the **minuend**.

Subtract numbers with and without borrowing.

There are several words and phrases that indicate subtraction, such as *minus, difference, decreased by, less than,* and *subtract from.*

Examples

Example 1

For $19 - 13 = 6$, the minuend is 19, the subtrahend is 13, and the difference is 6.

Example 2

$$\begin{array}{r} 398 \\ -\,227 \\ \hline 171 \end{array} \qquad \begin{array}{r} {}^{9}\\[-4pt] {}^{1\,\cancel{10}\,14}\\[-2pt] 2\cancel{0}4 \\ -\,\,88 \\ \hline 116 \end{array}$$

Example 3

The difference of 15 and 7 translates to $15 - 7$.

Example 4

The expression $31 - 20$ can be translated to 31 decreased by 20, or subtract 20 from 31.

Section 1.4 Rounding and Estimating

Key Concepts

To **round a number**, follow these steps.

Step 1 Identify the digit one position to the right of the given place value.

Step 2 If the digit in step 1 is a 5 or greater, add 1 to the digit in the given place value. Then replace each digit to the right of the given place value by 0.

Step 3 If the digit in step 1 is less than 5, replace it and each digit to its right by 0.

Round to estimate sums and differences.

Examples

Example 1

Round each number to the indicated place.

a. 4942; hundreds place $\longrightarrow$ 4900
b. 3712; thousands place $\longrightarrow$ 4000
c. 135; tens place $\longrightarrow$ 140

Example 2

Round to the thousands place to estimate the sum: $3929 + 2528 + 5452$.

Solution $\quad 4000 + 3000 + 5000 = 12,000$

The sum is approximately 12,000.

Section 1.5   Multiplication of Whole Numbers and Area

Key Concepts

Multiplication is repeated addition.

The **product** is the result of multiplying **factors**.

Properties of Multiplication

1. Commutative Property of Multiplication: Changing the order of the factors does not affect the product.
2. Associative Property of Multiplication: The manner in which the factors are grouped does not affect the product.
3. Multiplication Property of 0: The product of any number and 0 is 0.
4. Multiplication Property of 1: The product of any number and 1 is that number.
5. Distributive Property of Multiplication over Addition

Multiply whole numbers.

The **area of a rectangle** with length l and width w is given by $A = l \cdot w$.

Examples

Example 1

$16 + 16 + 16 + 16 = 4 \times 16 = 64$

Example 2

For $3 \times 13 \times 2 = 78$ the factors are 3, 13, and 2, and the product is 78.

Example 3

1. $4 \times 7 = 7 \times 4$

2. $6 \times (5 \times 7) = (6 \times 5) \times 7$

3. $43 \times 0 = 0$

4. $290 \times 1 = 290$

5. $5 \times (4 + 8) = (5 \times 4) + (5 \times 8)$

Example 4

$3 \times 14 = 42 \qquad 7(4) = 28$

$$
\begin{array}{r}
312 \\
\times\ 23 \\
\hline
936 \\
6240 \\
\hline
7176
\end{array}
$$

Example 5

Find the area of the rectangle.

23 cm □

70 cm

$A = (23\ \text{cm}) \cdot (70\ \text{cm}) = 1610\ \text{cm}^2$

Section 1.6   Division of Whole Numbers

Key Concepts

A **quotient** is the result of dividing the **dividend** by the **divisor**.

Properties of Division

1. Any number divided by itself is 1.
2. Any number divided by 1 is the number itself.
3. Zero divided by any nonzero number is zero.
4. A number divided by zero is undefined.

Long division, with and without a **remainder**

Examples

Example 1

For $36 \div 4 = 9$, the dividend is 36, the divisor is 4, and the quotient is 9.

Example 2

1. $13 \div 13 = 1$

2.
$$\begin{array}{r} 37 \\ 1\overline{)37} \end{array}$$

3. $\dfrac{0}{2} = 0$

Example 3

$\dfrac{2}{0}$ is undefined.

Example 4

$$\begin{array}{r} 263 \\ 3\overline{)789} \\ -6 \\ \hline 18 \\ -18 \\ \hline 09 \\ -9 \\ \hline 0 \end{array}$$

$$\begin{array}{r} 41 \text{ R}12 \\ 21\overline{)873} \\ -84 \\ \hline 33 \\ -21 \\ \hline 12 \end{array}$$

Section 1.7   Exponents, Square Roots, and the Order of Operations

Key Concepts

A number raised to an **exponent** represents repeated multiplication.

For 6^3, 6 is the **base** and 3 is the exponent or **power**.

The **square root** of 16 is 4 because $4^2 = 16$. That is, $\sqrt{16} = 4$.

Order of Operations

1. Perform all operations inside parentheses first.
2. Simplify any expressions containing exponents or square roots.
3. Perform multiplication or division in the order that they appear from left to right.
4. Perform addition or subtraction in the order that they appear from left to right.

Powers of 10

$10^1 = 10$

$10^2 = 100$

$10^3 = 1000$ and so on.

The **mean** is the average of a set of numbers. To find the mean, add all the values and divide by the number of values.

Examples

Example 1

$9^4 = 9 \cdot 9 \cdot 9 \cdot 9 = 6561$

Example 2

$\sqrt{49} = 7$

Example 3

$$32 \div \sqrt{16} + (9 - 6)^2$$
$$= 32 \div \sqrt{16} + (3)^2$$
$$= 32 \div 4 + 9$$
$$= 8 + 9$$
$$= 17$$

Example 4

$10^5 = 100{,}000$ 1 followed by 5 zeros

Example 5

Find the mean (average) of Michael's scores from his homework assignments.

40, 41, 48, 38, 42, 43

Solution

$$\frac{40 + 41 + 48 + 38 + 42 + 43}{6} = \frac{252}{6} = 42$$

The average is 42.

Section 1.8   Problem-Solving Strategies

Key Concepts

Guidelines for Problem Solving

Step 1 Read the problem carefully. Draw a diagram or write an appropriate formula. Estimate a reasonable answer.

Step 2 Write down what information is given and what must be found.

Step 3 Form a strategy. Identify what mathematical operation or operations apply.

Step 4 Perform the mathematical operations to solve for the unknown.

Step 5 Check the answer.

Examples

Example 1

Nolan received a doctor's bill for $984. His insurance will pay $200, and the balance can be paid in 4 equal monthly payments. How much will each payment be?

Solution

To find the amount not paid by insurance, subtract $200 from the total bill.

$$984 - 200 = 784$$

To find Nolan's 4 equal payments, divide the amount not covered by insurance by 4.

$$784 \div 4 = 196$$

Nolan must make 4 payments of $196 each.

Chapter 1   Review Exercises

Section 1.1

For Exercises 1 and 2, determine the place value for each underlined digit.

1. 10,024
   Ten-thousands

2. 821,811
   Hundred-thousands

For Exercises 3 and 4, convert the numbers to standard form.

3. 9 ten-thousands + 2 thousands + 4 tens + 6 ones   92,046

4. 5 hundred-thousands + 3 thousands + 1 hundred + 6 tens   503,160

For Exercises 5 and 6, convert the numbers to expanded form.

5. 3,400,820   3 millions + 4 hundred-thousands + 8 hundreds + 2 tens;
$3 \times 1,000,000 + 4 \times 100,000 + 8 \times 100 + 2 \times 10$

6. 30,554   3 ten-thousands + 5 hundreds + 5 tens + 4 ones;
$3 \times 10,000 + 5 \times 100 + 5 \times 10 + 4 \times 1$

For Exercises 7 and 8, write the numbers in words.

7. 245   Two hundred forty-five

8. 30,861   Thirty thousand, eight hundred sixty-one

For Exercises 9 and 10, write the numbers in standard form.

9. Three thousand, six-hundred two   3602

10. Eight hundred thousand, thirty-nine   800,039

For Exercises 11 and 12, place the numbers on the number line.

11. 2

12. 7

For Exercises 13 and 14, determine if the inequality is true or false.

13. $3 < 10$ True

14. $10 > 12$ False

Section 1.2

For Exercises 15 and 16, identify the addends and the sum.

15. $105 + 119 = 224$
   Addends: 105, 119; sum: 224

16.
$$\begin{array}{r} 53 \\ + 21 \\ \hline 74 \end{array}$$
   Addends: 53, 21; sum: 74

For Exercises 17–20, add.

17. $18 + 24 + 29$ 71

18. $27 + 9 + 18$ 54

19.
$$\begin{array}{r} 8403 \\ + 9007 \\ \hline \end{array}$$ 17,410

20.
$$\begin{array}{r} 68,421 \\ + 2,221 \\ \hline \end{array}$$ 70,642

 Writing   Translating Expression   Geometry   Scientific Calculator   Video   NS&E

21. For each of the mathematical statements, identify the property used. Choose from the commutative property or the associative property.

a. $6 + (8 + 2) = (8 + 2) + 6$ Commutative property

b. $6 + (8 + 2) = (6 + 8) + 2$ Associative property

c. $6 + (8 + 2) = 6 + (2 + 8)$ Commutative property

For Exercises 22–25, translate the English phrase to a mathematical statement and simplify.

22. The sum of 403 and 79
403 + 79; 482

23. 92 added to 44
44 + 92; 136

24. 7 more than 36
36 + 7; 43

25. 23 increased by 6
23 + 6; 29

26. The table gives the number of cars sold by three dealerships during one week.

	Honda	**Ford**	**Toyota**
Bob's Discount Auto	23	21	34
AA Auto	31	25	40
Car World	33	20	22

a. What is the total number of cars sold by AA Auto?   96 cars

b. What is the total number of Fords sold by these three dealerships?   66 Fords

27. The bar graph represents the distribution of the U.S. population by age group for a recent year. Determine the number of seniors (aged 60 and over).   45,797 thousand seniors

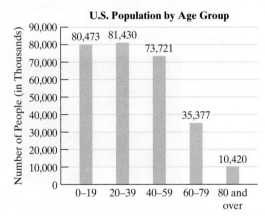

U.S. Population by Age Group

Source: U.S. Census Bureau

28. Find the perimeter of the figure.   177 m

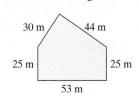

30 m, 44 m, 25 m, 25 m, 53 m

Section 1.3

For Exercises 29 and 30, identify the minuend, subtrahend, and difference.

29. $14 - 8 = 6$
Minuend: 14; subtrahend: 8; difference: 6

30.   102
      $- 78$
      ‾‾‾‾
       24
Minuend: 102; subtrahend: 78; difference: 24

For Exercises 31 and 32, subtract and check your answer by addition.

31.   37
       $- 11$
       ‾‾‾‾
        26
Check: ☐ $+ 11 = 37$

32.   61
       $- 41$
       ‾‾‾‾
        20
Check: ☐ $+ 41 = 61$

For Exercises 33–36, subtract.

33.   2005
        $- 1884$
        ‾‾‾‾‾
         121

34. $1389 - 299$
1090

35. $86,000 - 54,981$
31,019

36. $67,000 - 32,812$
34,188

For Exercises 37–40, translate the English phrase into a mathematical statement and simplify.

37. 38 minus 31   38−31; 7

38. 111 decreased by 15   111−15; 96

39. Subtract 42 from 251   251−42; 209

40. The difference of 90 and 52   90−52; 38

41. There were 95,191,761 tons of watermelons and 23,299,323 tons of cantaloupes produced in a recent year. What is the difference between the weight of the watermelons and the weight of the cantaloupes?   71,892,438 tons

42. For a recent year, Phil Mickelson earned $25,800,000 from both the golf tours and endorsements. Vijay Singh earned $18,600,000. Find the difference in their earnings. $7,200,000

 Writing   Translating Expression   Geometry   Scientific Calculator   Video   NS&E

43. The graph gives the estimated number of overseas visitors (in thousands) for five cities in the United States for a recent year. What is the difference between the number of visitors to New York and the number of visitors to Orlando?

2336 thousand visitors

Overseas Visitors in U.S. Cities

Source: U.S. Department of Commerce

Section 1.4

For Exercises 44 and 45, round each number to the given place value.

44. 5,234,446; millions   5,000,000

45. 9,332,945; ten-thousands   9,330,000

For Exercises 46 and 47, estimate the sum by rounding to the indicated place value.

46. 894,004 − 123,883; hundred-thousands   800,000

47. 330 + 489 + 123 + 571; hundreds   1500

48. In a recent year, the population of Russia was 140,041,247, and the population of Japan was 127,078,679. Estimate the difference in their populations by rounding to the nearest million.
13,000,000 people

49. The state of Missouri has two dams: Fort Peck with a volume of 96,050 cubic meters (m^3) and Oahe with a volume of 66,517 m^3. Round the numbers to the nearest thousand to estimate the total volume of these two dams.   163,000 m^3

Section 1.5

For Exercises 50 and 51, identify the factors and the product.

50. $32 \cdot 12 = 384$ Factors: 32, 12; product: 384

51. $33 \times 40 = 1320$ Factors: 33, 40; product: 1320

52. Indicate whether the statement is equal to the product of 8 and 13.

a. 8(13)   Yes

b. $(8) \cdot 13$ Yes

c. $(8) + (13)$ No

For Exercises 53–57, for each property listed, choose an expression from the right column that demonstrates the property.

53. Associative property of multiplication   c   **a.** $3(4) = 4(3)$

54. Distributive property of multiplication over addition   e   **b.** $19 \times 1 = 19$

55. Multiplication property of 0   d   **c.** $(1 \cdot 8) \cdot 3 = 1 \cdot (8 \cdot 3)$

56. Commutative property of multiplication   a   **d.** $0 \cdot 29 = 0$

57. Multiplication property of 1   b   **e.** $4(3 + 1) = 4 \cdot 3 + 4 \cdot 1$

For Exercises 58–60, multiply.

58. 142
 $\times\ 43$
 ———
 6106

59. (1024)(51)
 52,224

60. 6000
 $\times\ 500$
 ————
 3,000,000

61. A discussion group needs to purchase books that are accompanied by a workbook. The price of the book is $26, and the workbook costs an additional $13. If there are 11 members in the group, how much will it cost the group to purchase both the text and workbook for each student? $429

62. Orcas, or killer whales, eat 551 pounds (lb) of food a day. If Sea World has two adult killer whales, how much food will they eat in 1 week?   7714 lb

Section 1.6

For Exercises 63 and 64, perform the division. Then identify the divisor, dividend, and quotient.

63. $42 \div 6$
7; divisor: 6, dividend: 42, quotient: 7

64. $4\overline{)52}$
13; divisor: 4, dividend: 52, quotient: 13

 Writing   Translating Expression   Geometry   Scientific Calculator   Video   NS E

For Exercises 65–68, use the properties of division to simplify the expression, if possible.

65. $3 \div 1$ 3

66. $3 \div 3$ 1

67. $3 \div 0$ Undefined

68. $0 \div 3$ 0

69. Explain how you check a division problem if there is no remainder.
Multiply the quotient and the divisor to get the dividend.

70. Explain how you check a division problem if there is a remainder.
Multiply the whole number part of the quotient and the divisor, and then add the remainder to get the dividend.

For Exercises 71–73, divide and check the answer.

71. $348 \div 6$
58

72. $11\overline{)458}$
41 R7

73. $\dfrac{1043}{20}$
52 R3

→ For Exercises 74 and 75, write the English phrase as a mathematical expression and simplify.

74. The quotient of 72 and 4 $\dfrac{72}{4}$; 18

75. 108 divided by 9 $9\overline{)108}$; 12

76. Quinita has 105 photographs that she wants to divide equally among herself and three siblings. How many photos will each person receive? How many photos will be left over?
26 photos with 1 left over

77. Ashley has $60 to spend on souvenirs at a surf shop. The prices of several souvenirs are given in the graph.

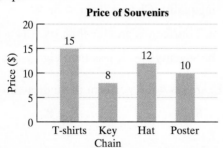

Price of Souvenirs

a. How many souvenirs can Ashley buy if she chooses all T-shirts?   4 T-shirts

b. How many souvenirs can Ashley buy if she chooses all hats?   5 hats

Section 1.7

For Exercises 78 and 79, write the repeated multiplication in exponential form. Do not simplify.

78. $8 \cdot 8 \cdot 8 \cdot 8 \cdot 8$ 8^5

79. $2 \cdot 2 \cdot 2 \cdot 2 \cdot 5 \cdot 5 \cdot 5$ $2^4 \cdot 5^3$

For Exercises 80–83, evaluate the exponential expressions.

80. 5^3 125

81. 4^4 256

82. 1^7 1

83. 10^6 1,000,000

For Exercises 84 and 85, evaluate the square roots.

84. $\sqrt{64}$ 8

85. $\sqrt{144}$ 12

For Exercises 86–91, evaluate the expression using the order of operations.

86. $14 \div 7 \cdot 4 - 1$ 7

87. $10^2 - 5^2$ 75

88. $90 - 4 + 6 \div 3 \cdot 2$ 90

89. $2 + 3 \cdot 12 \div 2 - \sqrt{25}$ 15

90. $6^2 - 4^2 + (9 - 7)^3$ 28

91. $26 - 2(10 - 1) + (3 + 4 \cdot 11)$ 55

92. Find the mean (average) for the set of numbers $7, 6, 12, 5, 7, 6, 13$.   8

93. Carolyn's electric bills for the past 5 months have been $80, $78, $101, $92, and $94. Find her average monthly charge. $89

94. The table shows the number of homes sold by a realty company in the last 6 months. Determine the average number of houses sold per month for these 6 months.   8 houses per month

Month	Number of Houses
May	6
June	9
July	11
August	13
September	5
October	4

Section 1.8

95. The Cincinnati Zoo houses about 17,000 animals that represent 750 species. The San Diego Zoo has 4000 animals representing 800 species.

    a. Which zoo has the most animals? How many more animals does it have? The Cincinnati Zoo has 13,000 more animals than the San Diego Zoo.

    b. Which zoo has the most species? How many more species does it have? The San Diego Zoo has 50 more species than the Cincinnati Zoo.

96. Doris drives her son to extracurricular activities each week. She drives 5 mi round-trip to baseball practice 3 times a week and 6 mi round-trip to piano lessons once a week.

    a. How many miles does she drive in 1 week to get her child to his activities? 21 mi

    b. Approximately how many miles does she travel during a school year consisting of 10 months (there are approximately 4 weeks per month)? 840 mi

97. At one point in his baseball career, Alex Rodriquez signed a contract for $252,000,000 for a 9-year period between 2001 and 2010. Suppose federal taxes amount to $75,600,000 for the contract. After taxes, how much will Alex receive per year? He will receive $19,600,000 per year.

98. Aletha wants to buy plants for a rectangular garden in her backyard that measures 12 ft by 8 ft. She wants to divide the garden into 2-square-foot (2 ft^2) areas, one for each plant.

    a. How many plants should Aletha buy? She should purchase 48 plants.

    b. If the plants cost $3 each, how much will it cost Aletha for the plants? The plants will cost $144.

    c. If she puts a fence around the perimeter of the garden that costs $2 per foot, how much will it cost for the fence? The fence will cost $80.

    d. What will be Aletha's total cost for this garden? Aletha's total cost will be $224.

Chapter 1   Test

1. Determine the place value for the underlined digit.

    a. <u>4</u>92   **b.** 2<u>3</u>,441   **c.** <u>2</u>,340,711   **d.** 3<u>4</u>0,592

      Hundreds   Thousands   Millions   Ten-thousands

2. Fill in the table with either the word name for the number or the number in standard form.

	Population	
State / Province	**Standard Form**	**Word Name**
a. Kentucky		Four million, sixty-five thousand
b. Texas	21,325,000	
c. Pennsylvania	12,287,000	
d. New Brunswick, Canada		Seven hundred twenty-nine thousand
e. Ontario, Canada	11,410,000	

a. 4,065,000
b. Twenty-one million, three hundred twenty-five thousand
c. Twelve million, two hundred eighty-seven thousand
d. 729,000   e. Eleven million, four hundred ten thousand

3. Translate the phrase by writing the numbers in standard form and inserting the appropriate inequality. Choose from $<$ or $>$.

    a. Fourteen is greater than six. $14 > 6$

    b. Seventy-two is less than eighty-one. $72 < 81$

For Exercises 4–17, perform the indicated operation.

4. $\begin{array}{r} 51 \\ + 78 \\ \hline 129 \end{array}$ **5.** $\begin{array}{r} 82 \\ \times 4 \\ \hline 328 \end{array}$

6. $\begin{array}{r} 154 \\ - 41 \\ \hline 113 \end{array}$ **7.** $4\overline{)908}$ 227

8. $58 \cdot 49$ 2842   **9.** $149 + 298$ 447

10. $324 \div 15$ 21 R9   **11.** $3002 - 2456$ 546

12. $10,984 - 2881$ 8103   **13.** $\dfrac{840}{42}$ 20

14. $(500,000)(3000)$ 1,500,000,000   **15.** $34 + 89 + 191 + 22$ 336

16. $403(0)$ 0   **17.** $0\overline{)16}$ Undefined

 Writing    Translating Expression   Geometry   Scientific Calculator   Video   NS&E

18. For each of the mathematical statements, identify the property used. Choose from the commutative property of multiplication and the associative property of multiplication. Explain your answer.

a. $(11 \cdot 6) \cdot 3 = 11 \cdot (6 \cdot 3)$
The associative property of multiplication; the expression shows a change in grouping.

b. $(11 \cdot 6) \cdot 3 = 3 \cdot (11 \cdot 6)$
The commutative property of multiplication; the expression shows a change in the order of the factors.

19. Round each number to the indicated place value.

a. 4850; hundreds
4900
b. 12,493; thousands
12,000
c. 7,963,126; hundred-thousands
8,000,000

20. The attendance to the Van Gogh and Gauguin exhibit in Chicago was 690,951. The exhibit moved to Amsterdam, and the attendance was 739,117. Round the numbers to the ten-thousands place to estimate the total attendance for this exhibit.
There were approximately 1,430,000 people.

For Exercises 21–24, simplify, using the order of operations.

21. $8^2 \div 2^4$
4

22. $26 \cdot \sqrt{4} - 4(8 - 1)$
24

23. $36 \div 3(14 - 10)$
48

24. $65 - 2(5 \cdot 3 - 11)^2$
33

25. Brittany and Jennifer are taking an online course in business management. Brittany has taken 6 quizzes worth 30 points each and received the following scores: 29, 28, 24, 27, 30, and 30. Jennifer has only taken 5 quizzes so far, and her scores are 30, 30, 29, 28, and 28. At this point in the course, which student has a higher average?
Jennifer has a higher average of 29. Brittany has an average of 28.

26. The number of people who used the Internet during four recent years in the United States is shown in the graph.

a. Find the change in the number of Internet users from year 2 to year 3.
442 thousand users

b. Of the years presented in the graph, between which two consecutive years was the increase the greatest? How much was the increase?
The largest increase was between year 3 and year 4. The increase was 15,430 thousand.

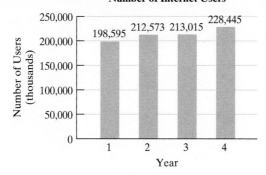

Number of Internet Users

27. The table gives the number of calls to three fire departments during a selected number of weeks. Find the number of calls per week of each department to determine which department is the busiest. The North Side Fire Department is the busiest with an average of five calls per week.

	Number of Calls	Time Period (Number of Weeks)
North Side Fire Department	80	16
South Side Fire Department	72	18
East Side Fire Department	84	28

28. Find the perimeter of the figure. 156 mm

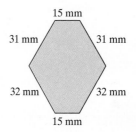

29. Find the perimeter and the area of the rectangle.

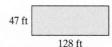

Perimeter: 350 ft; area: 6016 ft²

30. Round to the nearest hundred to estimate the area of the rectangle. 4,560,000 m²

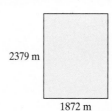

 Writing   Translating Expression   Geometry Scientific Calculator   Video NS E

Fractions and Mixed Numbers: Multiplication and Division

2

Chapter 2

In Chapter 2 we study the concept of a fraction and a mixed number. We learn how to simplify fractions to lowest terms.

Review Your Skills

To prepare for work with fractions, practice performing the order of operations with whole numbers. To complete the puzzle, first answer the questions and fill in the appropriate box. Then fill in the grid so that every row, every column, and every 3 × 2 box contains the digits 1 through 6.

A. $5 - 2(10 - 8)$

B. $3^2 - 2^3$

C. $3 \cdot 4 - 16 \div 2$

D. $1142 - (121 + 1015)$

E. $(10 + 11 + 14) \div 7$

F. $6^2 - 8 \cdot 4$

G. $80 \cdot 2 - 155$

H. $\sqrt{25} - 3$

I. $\sqrt{100} - 12 \div 3$

J. $16 \div 8 \cdot 2 - 3$

3	5	6	**A** 1	2	4
B 1	2	3	**C** 4	**D** 6	**E** 5
6	4	2	5	3	1
2	1	**F** 4	6	5	3
G 5	3	1	**H** 2	4	**I** 6
4	6	5	3	**J** 1	2

Section 2.1 Introduction to Fractions and Mixed Numbers

Concepts

1. Definition of a Fraction
2. Proper and Improper Fractions
3. Mixed Numbers
4. Fractions and the Number Line

Instructor Note: Students may not realize that a fraction such as $\frac{1}{5}$ represents one of five *equal* parts. For example, they may incorrectly think that each part of the following figure is represented by $\frac{1}{2}$.

1. Definition of a Fraction

In Chapter 1, we studied operations on whole numbers. In this chapter, we work with numbers that represent part of a whole. When a whole unit is divided into equal parts, we call the parts **fractions** of a whole. For example, the pizza in Figure 2-1 is divided into 5 equal parts. One-fifth ($\frac{1}{5}$) of the pizza has been eaten, and four-fifths ($\frac{4}{5}$) of the pizza remains.

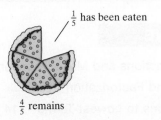

$\frac{1}{5}$ has been eaten

$\frac{4}{5}$ remains

Figure 2-1

A fraction is written in the form $\frac{a}{b}$, where a and b are whole numbers and $b \neq 0$. In the fraction $\frac{5}{8}$, the "top" number, 5, is called the **numerator**. The "bottom" number, 8, is called the **denominator**.

$$\text{numerator} \longrightarrow \frac{5}{8} \longleftarrow \text{denominator}$$

Skill Practice

Identify the numerator and denominator.

1. $\frac{4}{11}$ **2.** $\frac{0}{5}$ **3.** $\frac{6}{1}$

Classroom Example: p. 103, Exercise 4

Avoiding Mistakes

The fraction $\frac{a}{b}$ can also be written as a/b. However, we discourage the use of the "slanted" fraction bar. In later applications of algebra, the slanted fraction bar can cause confusion.

Example 1 Identifying the Numerator and Denominator of a Fraction

For each fraction, identify the numerator and denominator.

a. $\frac{3}{5}$ **b.** $\frac{1}{8}$ **c.** $\frac{8}{1}$

Solution:

a. $\frac{3}{5}$ The numerator is 3. The denominator is 5.

b. $\frac{1}{8}$ The numerator is 1. The denominator is 8.

c. $\frac{8}{1}$ The numerator is 8. The denominator is 1.

The *denominator* of a fraction denotes the number of equal pieces into which a whole unit is divided. The *numerator* denotes the number of pieces being considered.

Answers

1. Numerator: 4, denominator: 11
2. Numerator: 0, denominator: 5
3. Numerator: 6, denominator: 1

For example, the garden in Figure 2-2 is divided into 10 equal parts. Three sections contain tomato plants. Therefore, $\frac{3}{10}$ of the garden contains tomato plants.

$\frac{3}{10}$ tomato plants

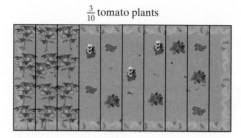

Figure 2-2

Example 2 **Writing Fractions**

Write a fraction for the shaded portion and a fraction for the unshaded portion of the figure.

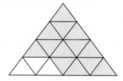

Solution:

Shaded portion: $\dfrac{13}{16}$ ←—— 13 pieces are shaded.
                              ←—— The triangle is divided into 16 equal pieces.

Unshaded portion: $\dfrac{3}{16}$ ←—— 3 pieces are not shaded.
                                 ←—— The triangle is divided into 16 equal pieces.

Skill Practice

4. Write a fraction for the shaded portion and a fraction for the unshaded portion.

Classroom Example: p. 103, Exercise 20

Example 3 **Writing Fractions**

What portion of the group of doctors shown below is female?

Solution:

The group consists of 5 members. Therefore, the denominator is 5. There are 2 women being considered. Thus, $\frac{2}{5}$ of the group is female.

Skill Practice

5. Refer to Example 3. What portion of the group of doctors is male?

Classroom Example: p. 103, Exercise 26

In Section 1.6 we learned that fractions represent division. For example, note that the fraction $\frac{5}{1} = 5 \div 1 = 5$. In general, a fraction of the form $\frac{n}{1} = n$. This implies that any whole number can be written as a fraction by writing the whole number over 1.

Answers

4. Shaded portion: $\frac{3}{8}$; unshaded portion: $\frac{5}{8}$
5. $\frac{3}{5}$

Further recall that for $a \neq 0$, $0 \div a = 0$ and $a \div 0$ is undefined. Therefore,

$$\frac{0}{a} = 0 \text{ and } \frac{a}{0} \text{ is undefined. Example: } \frac{0}{5} = 0 \text{ and } \frac{5}{0} \text{ is undefined.}$$

2. Proper and Improper Fractions

If the numerator is less than the denominator in a fraction, then the fraction is called a **proper fraction**. Furthermore, a proper fraction represents a number less than 1 whole unit. The following are proper fractions.

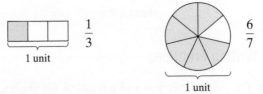

An **improper fraction** is a fraction in which the numerator is greater than or equal to the denominator. For example:

$$\text{numerator greater} \longrightarrow \frac{4}{3} \text{ and } \frac{7}{7} \longleftarrow \text{numerator equal}$$
$$\text{than denominator} \qquad\qquad\qquad\qquad\qquad\qquad \text{to denominator}$$

An improper fraction represents a quantity greater than 1 whole unit or equal to 1 whole unit.

Instructor Note: Tell students that the word "improper" does not imply that anything is wrong with the fraction.

Classroom Examples: p. 104, Exercises 32, 34, and 36

Example 4 **Categorizing Fractions**

Identify each fraction as proper or improper.

a. $\frac{12}{5}$ **b.** $\frac{5}{12}$ **c.** $\frac{12}{12}$

Solution:

a. $\frac{12}{5}$ Improper fraction (numerator is greater than denominator)

b. $\frac{5}{12}$ Proper fraction (numerator is less than denominator)

c. $\frac{12}{12}$ Improper fraction (numerator is equal to denominator)

Answers

6. $\frac{0}{3} = 0$ means $0 \cdot 3 = 0$

7. $\frac{6}{0}$ is undefined because there is no number that when multiplied by 0 equals 6.

8. Improper   **9.** Proper

10. Improper

Example 5 **Writing Improper Fractions**

a. Write an improper fraction representing the length of the screw shown in the figure.

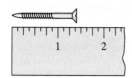

Avoiding Mistakes

Each whole unit is divided into 8 pieces. Therefore the screw is $\frac{11}{8}$ in., not $\frac{11}{16}$ in.

b. Write an improper fraction representing the shaded area.

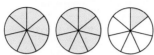

Solution:

a. Each 1-in. unit is divided into 8 parts, and the screw extends for 11 parts. Therefore, the screw is $\frac{11}{8}$ in.

b. Each circle is divided into 7 sections of equal size. Of these sections, 15 are shaded. Therefore, the shaded area can be represented by $\frac{15}{7}$.

Skill Practice

11. Write an improper fraction representing the length of the nail shown in the figure.

Classroom Example: p. 104, Exercise 44

3. Mixed Numbers

Sometimes a mixed number is used instead of an improper fraction to denote a quantity greater than one whole. For example, suppose a typist typed $\frac{9}{4}$ pages of a report. We would be more likely to say that the typist typed $2\frac{1}{4}$ pages (read as "two and one-fourth pages"). The number $2\frac{1}{4}$ is called a *mixed number* and represents 2 wholes plus $\frac{1}{4}$ of a whole.

$$\frac{9}{4} = 2\frac{1}{4}$$

In general, a **mixed number** is a sum of a whole number and a fractional part of a whole. However, by convention the plus sign is left out.

$$3\frac{1}{2} \quad \text{means} \quad 3 + \frac{1}{2}$$

Suppose we want to change a mixed number to an improper fraction. From Figure 2-3, we see that the mixed number $3\frac{1}{2}$ is the same as $\frac{7}{2}$.

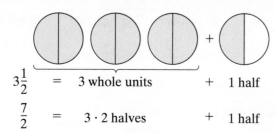

$3\frac{1}{2}$ = 3 whole units + 1 half

$\frac{7}{2}$ = 3 · 2 halves + 1 half

Figure 2-3

Answer

11. $\frac{9}{4}$

This process to convert a mixed number to an improper fraction can be summarized as follows.

> **Changing a Mixed Number to an Improper Fraction**
> **Step 1** Multiply the whole number by the denominator.
> **Step 2** Add the result to the numerator.
> **Step 3** Write the result from step 2 over the denominator.

For example,

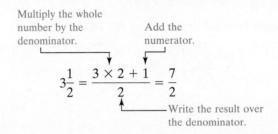

$$3\frac{1}{2} = \frac{3 \times 2 + 1}{2} = \frac{7}{2}$$

Skill Practice

Convert the mixed number to an improper fraction.
12. $10\frac{5}{8}$ **13.** $15\frac{1}{2}$

Classroom Example: p. 105, Exercise 54

> **Example 6**   **Converting Mixed Numbers to Improper Fractions**
>
> Convert the mixed number to an improper fraction.
>
> **a.** $7\frac{1}{4}$ **b.** $8\frac{2}{5}$
>
> **Solution:**
>
> **a.** $7\frac{1}{4} = \frac{7 \times 4 + 1}{4}$ **b.** $8\frac{2}{5} = \frac{8 \times 5 + 2}{5}$
>
> $\qquad = \frac{28 + 1}{4}$ $\qquad = \frac{40 + 2}{5}$
>
> $\qquad = \frac{29}{4}$ $\qquad = \frac{42}{5}$

Now suppose we want to convert an improper fraction to a mixed number. In Figure 2-4, the improper fraction $\frac{13}{5}$ represents 13 slices of pizza where each slice is $\frac{1}{5}$ of a whole pizza. If we divide the 13 pieces into groups of 5, we make 2 whole pizzas with 3 pieces left over. Thus,

$$\frac{13}{5} = 2\frac{3}{5} \qquad 13 \text{ pieces} = \quad 2 \text{ groups of 5} \quad + \ 3 \text{ left over}$$

$$\frac{13}{5} = \qquad\qquad 2 \qquad\qquad + \qquad \frac{3}{5}$$

Figure 2-4

Avoiding Mistakes

When writing a mixed number, the + sign between the whole number and fraction should not be written.

This process can be accomplished by division.

$$\frac{13}{5} \longrightarrow \begin{array}{r} 2 \\ 5\overline{)13} \\ -10 \\ \hline 3 \end{array} \quad 2\frac{3}{5}$$

remainder

divisor

Answers
12. $\frac{85}{8}$ **13.** $\frac{31}{2}$

Changing an Improper Fraction to a Mixed Number

Step 1 Divide the numerator by the denominator to obtain the quotient and remainder.

Step 2 The mixed number is then given by

$$\text{Quotient} + \frac{\text{remainder}}{\text{divisor}}$$

Note: The + sign is not written but understood.

Instructor Note: Remind students that the word "quotient" in this context means the whole number part of the quotient.

Example 7 **Converting Improper Fractions to Mixed Numbers**

Convert to a mixed number.

a. $\dfrac{25}{6}$ **b.** $\dfrac{162}{41}$

Solution:

a. $\dfrac{25}{6} \longrightarrow$
$\begin{array}{r} 4 \\ 6\overline{)25} \\ -24 \\ \hline 1 \end{array}$ $4\dfrac{1}{6}$ remainder ⟶ divisor

b. $\dfrac{162}{41} \longrightarrow$
$\begin{array}{r} 3 \\ 41\overline{)162} \\ -123 \\ \hline 39 \end{array}$ $3\dfrac{39}{41}$ remainder ⟶ divisor

Skill Practice

Convert the improper fraction to a mixed number.

14. $\dfrac{14}{5}$ **15.** $\dfrac{95}{22}$

Classroom Example: p. 105, Exercise 68

The process to convert an improper fraction to a mixed number indicates that the result of a division problem can be written as a mixed number.

Example 8 **Writing a Quotient as a Mixed Number**

Divide. Write the quotient as a mixed number.

$$28\overline{)4217}$$

Solution:

$\begin{array}{r} 150 \\ 28\overline{)4217} \\ -28 \\ \hline 141 \\ -140 \\ \hline 17 \\ -0 \\ \hline 17 \end{array}$ $150\dfrac{17}{28}$ remainder ⟶ divisor

Skill Practice

Divide and write the quotient as a mixed number.

16. $5967 \div 41$

Classroom Example: p. 105, Exercise 84

Instructor Note: In Example 8, remind students that when they learned division in Chapter 1, they wrote 150 R17, which is equivalent to $150\frac{17}{28}$.

Answers

14. $2\dfrac{4}{5}$ **15.** $4\dfrac{7}{22}$ **16.** $145\dfrac{22}{41}$

4. Fractions and the Number Line

Fractions can be visualized on a number line. For example, to graph the fraction $\frac{3}{4}$, divide the distance between 0 and 1 into 4 equal parts. To plot the number $\frac{3}{4}$, start at 0 and count over 3 parts.

Example 9   **Plotting Fractions on a Number Line**

Plot the point on the number line corresponding to each fraction.

a. $\frac{1}{2}$ **b.** $\frac{5}{6}$ **c.** $\frac{21}{5}$

Solution:

a. $\frac{1}{2}$ Divide the distance between 0 and 1 into 2 equal parts.

b. $\frac{5}{6}$ Divide the distance between 0 and 1 into 6 equal parts.

c. $\frac{21}{5} = 4\frac{1}{5}$ Write $\frac{21}{5}$ as a mixed number.

Thus, $\frac{21}{5} = 4\frac{1}{5}$ is located one-fifth of the way between 4 and 5 on the number line.

Section 2.1   **Practice Exercises**

Study Skills Exercise

After doing a section of homework, check the odd-numbered answers in the back of the text. Choose a method to identify the exercises that gave you trouble (i.e., circle the number or put a star by the number). List some reasons why it is important to label these problems.

Vocabulary and Key Concepts

1. a. When a whole unit is divided into parts, we call the parts _____fractions_____ of the whole unit.

b. Given a fraction $\frac{a}{b}$, where a and b are whole numbers and $b \neq 0$, the top number a is called the _____numerator_____ and the bottom number b is called the _____denominator_____.

c. A fraction whose numerator is less than its denominator is called a _____proper_____ fraction.

d. A fraction whose numerator is greater than or equal to its denominator is called an _____improper_____ fraction.

e. A _____mixed_____ number is the sum of a whole number and a fractional part of a whole.

Concept 1: Definition of a Fraction

2. Write a fraction representing 2 parts of a whole unit divided into 7 parts of equal size. $\frac{2}{7}$

For Exercises 3–6, identify the numerator and the denominator for each fraction. **(See Example 1.)**

3. $\frac{2}{3}$ Numerator: 2; denominator: 3

4. $\frac{8}{9}$ Numerator: 8; denominator: 9

5. $\frac{12}{11}$ Numerator: 12; denominator: 11

6. $\frac{1}{2}$ Numerator: 1; denominator: 2

For Exercises 7–14, write the fraction as a division problem and simplify, if possible.

7. $\frac{6}{1}$ $6 \div 1;\ 6$

8. $\frac{9}{1}$ $9 \div 1;\ 9$

9. $\frac{2}{2}$ $2 \div 2;\ 1$

10. $\frac{8}{8}$ $8 \div 8;\ 1$

11. $\frac{0}{3}$ $0 \div 3;\ 0$

12. $\frac{0}{7}$ $0 \div 7;\ 0$

13. $\frac{2}{0}$ $2 \div 0;$ undefined

14. $\frac{11}{0}$ $11 \div 0;$ undefined

For Exercises 15–22, write a fraction that represents the shaded area. **(See Example 2.)**

15. $\frac{3}{4}$

16. $\frac{1}{2}$

17. $\frac{5}{9}$

18. $\frac{3}{5}$

19.  $\frac{1}{6}$

20. $\frac{4}{7}$

21. $\frac{3}{8}$

22.  $\frac{2}{3}$

23. Write a fraction to represent the portion of gas in a gas tank represented by the gauge. $\frac{3}{4}$

24. Write a fraction that represents the portion of medicine left in the bottle. $\frac{1}{4}$

25. The scoreboard for a recent men's championship swim meet in Melbourne, Australia, shows the final standings in the men's 100-m freestyle event. What portion of the finalists are from the USA? **(See Example 3.)** $\frac{1}{8}$

26. Refer to the scoreboard from Exercise 25. What portion of the finalists are from the Republic of South Africa (RSA)? $\frac{2}{8}$ or $\frac{1}{4}$

27. The graph categorizes a sample of people by blood type. What portion of the sample represents people with type O blood? $\frac{41}{103}$

28. Refer to the graph from Exercise 27. What portion of the sample represents people with type A blood? $\frac{43}{103}$

Name	Country	Time
Maginni, Filippo	ITA	48.43
Hayden, Brent	CAN	48.43
Sullivan, Eamon	AUS	48.47
Cielo Filho, Cesar	BRA	48.51
Lezak, Jason	USA	48.52
Van Den Hoogenband, Pieter	NED	48.63
Schoeman, Roland Mark	RSA	48.72
Neethling, Ryk	RSA	48.81

Sample by Blood Type

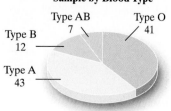

Type AB 7
Type O 41
Type B 12
Type A 43

 Writing   Translating Expression   Geometry Scientific Calculator   Video   NS&E

29. A class has 21 children—11 girls and 10 boys. What fraction of the class is made up of boys? $\frac{10}{21}$

30. In a neighborhood in Ft. Lauderdale, 10 houses are for sale and 53 are not for sale. Write a fraction representing the portion of houses that are for sale.

$\frac{10}{63}$

Concept 2: Proper and Improper Fractions

For Exercises 31–38, label the fraction as proper or improper. **(See Example 4.)**

31. $\frac{7}{8}$ Proper

32. $\frac{2}{3}$ Proper

33. $\frac{10}{10}$ Improper

34. $\frac{3}{3}$ Improper

35. $\frac{7}{2}$ Improper

36. $\frac{21}{20}$ Improper

37. $\frac{15}{17}$ Proper

38. $\frac{13}{21}$ Proper

For Exercises 39–42, write an improper fraction for the shaded portion of each group of figures. **(See Example 5.)**

39.

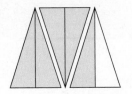

$\frac{5}{2}$

40.

$\frac{4}{3}$

41.

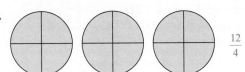

$\frac{12}{4}$

42.

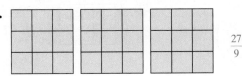

$\frac{27}{9}$

For Exercises 43 and 44, write an improper fraction representing the lengths of the objects shown in the figure.
(See Example 5.)

43.

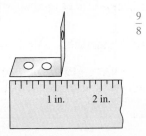

$\frac{9}{8}$

44.

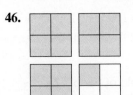

$\frac{7}{4}$

Concept 3: Mixed Numbers

For Exercises 45 and 46, write an improper fraction and a mixed number for the shaded portion of each group of figures.

45.

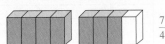

$\frac{7}{4}; 1\frac{3}{4}$

46.

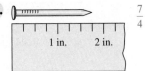

$\frac{13}{4}; 3\frac{1}{4}$

47. Write an improper fraction and a mixed number to represent the length of the screw.

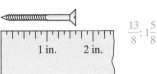

$\frac{13}{8}; 1\frac{5}{8}$

48. Write an improper fraction and a mixed number for the number of cups indicated in the figure.

$\frac{5}{2}; 2\frac{1}{2}$

$\frac{1}{2}$

For Exercises 49–60, convert the mixed number to an improper fraction. **(See Example 6.)**

49. $1\frac{3}{4}$ $\frac{7}{4}$

50. $6\frac{1}{3}$ $\frac{19}{3}$

51. $4\frac{2}{9}$ $\frac{38}{9}$

52. $3\frac{1}{5}$ $\frac{16}{5}$

53. $3\frac{3}{7}$ $\frac{24}{7}$

54. $8\frac{2}{3}$ $\frac{26}{3}$

55. $7\frac{1}{4}$ $\frac{29}{4}$

56. $10\frac{3}{5}$ $\frac{53}{5}$

57. $11\frac{5}{12}$ $\frac{137}{12}$

58. $12\frac{1}{6}$ $\frac{73}{6}$

59. $21\frac{3}{8}$ $\frac{171}{8}$

60. $15\frac{1}{2}$ $\frac{31}{2}$

61. How many eighths are in $2\frac{3}{8}$?   19

62. How many fifths are in $2\frac{3}{5}$?   13

63. How many fourths are in $1\frac{3}{4}$?   7

64. How many thirds are in $5\frac{2}{3}$?   17

For Exercises 65–76, convert the improper fraction to a mixed number. **(See Example 7.)**

65. $\frac{37}{8}$ $4\frac{5}{8}$

66. $\frac{13}{7}$ $1\frac{6}{7}$

67. $\frac{39}{5}$ $7\frac{4}{5}$

68. $\frac{19}{4}$ $4\frac{3}{4}$

69. $\frac{27}{10}$ $2\frac{7}{10}$

70. $\frac{43}{18}$ $2\frac{7}{18}$

71. $\frac{52}{9}$ $5\frac{7}{9}$

72. $\frac{67}{12}$ $5\frac{7}{12}$

73. $\frac{133}{11}$ $12\frac{1}{11}$

74. $\frac{51}{10}$ $5\frac{1}{10}$

75. $\frac{23}{6}$ $3\frac{5}{6}$

76. $\frac{115}{7}$ $16\frac{3}{7}$

For Exercises 77–84, divide. Write the quotient as a mixed number. **(See Example 8.)**

77. $7\overline{)309}$ $44\frac{1}{7}$

78. $4\overline{)921}$ $230\frac{1}{4}$

79. $5281 \div 5$ $1056\frac{1}{5}$

80. $7213 \div 8$ $901\frac{5}{8}$

81. $8913 \div 11$ $810\frac{3}{11}$

82. $4257 \div 23$ $185\frac{2}{23}$

83. $15\overline{)187}$ $12\frac{7}{15}$

84. $34\overline{)695}$ $20\frac{15}{34}$

Concept 4: Fractions and the Number Line

For Exercises 85–94, plot the fraction on the number line. **(See Example 9.)**

85. $\frac{3}{4}$

86. $\frac{1}{2}$

87. $\frac{1}{3}$

88. $\frac{1}{5}$

89. $\dfrac{2}{3}$

90. $\dfrac{5}{6}$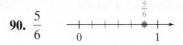

91. $\dfrac{7}{6}$

92. $\dfrac{7}{5}$

93. $\dfrac{5}{3}$

94. $\dfrac{3}{2}$

Expanding Your Skills

95. True or false? Whole numbers can be written both as proper and improper fractions. False

96. True or false? Suppose m and n are nonzero numbers, where $m > n$. Then $\dfrac{m}{n}$ is an improper fraction. True

97. True or false? Suppose m and n are nonzero numbers, where $m > n$. Then $\dfrac{n}{m}$ is a proper fraction. True

98. True or false? Suppose m and n are nonzero numbers, where $m > n$. Then $\dfrac{n}{3m}$ is a proper fraction. True

---

Section 2.2 Prime Numbers and Factorization

Concepts

1. Factors and Factorizations
2. Divisibility Rules
3. Prime and Composite Numbers
4. Prime Factorization
5. Identifying All Factors of a Whole Number

1. Factors and Factorizations

Recall from Section 1.5 that two numbers multiplied to form a product are called factors. For example, $2 \cdot 3 = 6$ indicates that 2 and 3 are factors of 6. Likewise, because $1 \cdot 6 = 6$, the numbers 1 and 6 are factors of 6. In general, a **factor** of a number n is a nonzero whole number that divides evenly into n.

The products $2 \cdot 3$ and $1 \cdot 6$ are called factorizations of 6. In general, a **factorization** of a number n is a product of factors that equals n.

Skill Practice

1. Find four different factorizations of 18.

Classroom Example: p. 111, Exercise 12

> **Example 1** Finding Factorizations of a Number
>
> Find four different factorizations of 12.
>
> **Solution:**
>
> $$12 = \begin{cases} 1 \cdot 12 \\ 2 \cdot 6 \\ 3 \cdot 4 \\ 2 \cdot 2 \cdot 3 \end{cases}$$
>
> **TIP:** Notice that a factorization may include more than two factors.

Answer

1. For example: $1 \cdot 18$
$\quad\quad\quad 2 \cdot 9$
$\quad\quad\quad 3 \cdot 6$
$\quad\quad\quad 2 \cdot 3 \cdot 3$

Writing   Translating Expression   Geometry   Scientific Calculator   Video   NS&E

2. Divisibility Rules

The number 20 is said to be divisible by 5 because 5 divides evenly into 20. To determine whether one number is divisible by another, we can perform the division and note whether the remainder is zero. However, there are several rules by which we can quickly determine whether a number is divisible by 2, 3, 5, or 10. These are called divisibility rules.

Divisibility Rules for 2, 3, 5, and 10

- *Divisibility by 2.* A whole number is divisible by 2 if it is an even number. That is, the ones-place digit is 0, 2, 4, 6, or 8.
  Examples: 26 and 384
- *Divisibility by 3.* A whole number is divisible by 3 if the sum of its digits is divisible by 3.
  Example: 312   (sum of digits is $3 + 1 + 2 = 6$ which is divisible by 3)
- *Divisibility by 5.* A whole number is divisible by 5 if its ones-place digit is 5 or 0.
  Examples: 45 and 260
- *Divisibility by 10.* A whole number is divisible by 10 if its ones-place digit is 0.
  Examples: 30 and 170

Example 2 **Applying the Divisibility Rules**

Determine whether the given number is divisible by 2, 3, 5, or 10.

a. 624   **b.** 82   **c.** 720

Solution:

		Test for Divisibility	
a. 624	By 2:	Yes.	The number 624 is even.
	By 3:	Yes.	The sum $6 + 2 + 4 = 12$ is divisible by 3.
	By 5:	No.	The ones-place digit is not 5 or 0.
	By 10:	No.	The ones-place digit is not 0.
b. 82	By 2:	Yes.	The number 82 is even.
	By 3:	No.	The sum $8 + 2 = 10$ is not divisible by 3.
	By 5:	No.	The ones-place digit is not 5 or 0.
	By 10:	No.	The ones-place digit is not 0.
c. 720	By 2:	Yes.	The number 720 is even.
	By 3:	Yes.	The sum $7 + 2 + 0 = 9$ is divisible by 3.
	By 5:	Yes.	The ones-place digit is 0.
	By 10:	Yes.	The ones-place digit is 0.

Skill Practice

Determine whether the given number is divisible by 2, 3, 5, or 10.

2. 428   **3.** 75   **4.** 2100

Classroom Example: p. 112, Exercise 22

TIP: When in doubt about divisibility, you can check by division. When we divide 624 by 3, the remainder is zero.

Divisibility rules for the numbers 2, 3, 5, and 10 are used most often for factoring a number into prime factorization. We address other divisibility rules for 4, 6, 8, and 9 in the Expanding Your Skills portion of the exercises.

Answers
2. Divisible by 2
3. Divisible by 3 and 5
4. Divisible by 2, 3, 5, and 10

3. Prime and Composite Numbers

Two important classifications of whole numbers are prime numbers and composite numbers.

Definition of Prime and Composite Numbers

- A **prime number** is a whole number greater than 1 that has only two factors (itself and 1).
- A **composite number** is a whole number greater than 1 that is not prime. That is, a composite number will have at least one factor other than 1 and the number itself.

Note: The whole numbers 0 and 1 are neither prime nor composite.

Skill Practice

Determine whether the number is prime, composite, or neither.
5. 39   **6.** 0   **7.** 41

Classroom Examples: p. 112, Exercises 34 and 38

Example 3 Identifying Prime and Composite Numbers

Determine whether the number is prime, composite, or neither.

a. 19   **b.** 51   **c.** 1

Solution:

a. The number 19 is prime because its only factors are 1 and 19.

b. The number 51 is composite because $3 \cdot 17 = 51$. That is, 51 has factors other than 1 and 51.

c. The number 1 is neither prime nor composite by definition.

TIP: The number 2 is the only even prime number.

Prime numbers are used in a variety of ways in mathematics. We advise you to become familiar with the first several prime numbers: 2, 3, 5, 7, 11, 13, 17, 19, 23, 29, ...

4. Prime Factorization

In Example 1 we found four factorizations of 12.

$$1 \cdot 12$$
$$2 \cdot 6$$
$$3 \cdot 4$$
$$2 \cdot 2 \cdot 3$$

The last factorization $2 \cdot 2 \cdot 3$ consists of only prime-number factors. Therefore, we say $2 \cdot 2 \cdot 3$ is the prime factorization of 12.

Concept Connections

8. Is the product $2 \cdot 3 \cdot 10$ the prime factorization of 60? Explain.

Prime Factorization

The **prime factorization** of a number is the factorization in which every factor is a prime number.

Note: The order in which the factors are written does not affect the product.

Answers

5. Composite **6.** Neither **7.** Prime
8. No. The factor 10 is not a prime number. The prime factorization of 60 is $2 \cdot 2 \cdot 3 \cdot 5$.

Prime factorizations of numbers will be particularly helpful when we add, subtract, multiply, divide, and simplify fractions.

Example 4 **Determining the Prime Factorization of a Number**

Find the prime factorization of 220.

Solution:

One method to factor a whole number is to make a factor tree. Begin by determining any two numbers that when multiplied equal 220. Then continue factoring each factor until the branches "end" in prime numbers.

←Branches end in prime numbers.

Therefore, the prime factorization of 220 is $2 \cdot 2 \cdot 5 \cdot 11$.

In Example 4, note that the result of a prime factorization does not depend on the original two-number factorization. Similarly, the order in which the factors are written does not affect the product, for example,

$$220 = 2 \cdot 2 \cdot 5 \cdot 11 \qquad 220 = 2 \cdot 2 \cdot 5 \cdot 11 \qquad 220 = 11 \cdot 2 \cdot 2 \cdot 5$$

TIP: You can check the prime factorization of any number by multiplying the factors.

Another technique to find the prime factorization of a number is to divide the number by the smallest known prime factor. Then divide the quotient by its smallest known prime factor. Continue dividing in this fashion until the quotient is a prime number. The prime factorization is the product of divisors and the final quotient. For example,

> 2 is the smallest prime factor of 220 ⟶ 2)220
> 2 is the smallest prime factor of 110 ⟶ 2)110
> 5 is the smallest prime factor of 55 ⟶ 5)55
> the last quotient is prime ⟶ 11

Therefore, the prime factorization of 220 is $2 \cdot 2 \cdot 5 \cdot 11$ or $2^2 \cdot 5 \cdot 11$.

Example 5 Determining Prime Factorizations

Find the prime factorization.

a. 198   **b.** 153

Solution:

a. Since 198 is even, we → 2)198
know it is divisible by 2.   3)99 ← The sum of the digits $9 + 9 = 18$ is
                              3)33   divisible by 3.
                                11

The prime factorization of 198 is $2 \cdot 3 \cdot 3 \cdot 11$ or $2 \cdot 3^2 \cdot 11$.

b.   3)153
                     3)51
                       17

The prime factorization of 153 is $3 \cdot 3 \cdot 17$ or $3^2 \cdot 17$.

5. Identifying All Factors of a Whole Number

Sometimes it is necessary to identify all factors (both prime and other) of a number. Take the number 30, for example. A list of all factors of 30 is a list of all whole numbers that divide evenly into 30.

Factors of 30:   1, 2, 3, 5, 6, 10, 15, and 30

Example 6 Listing All Factors of a Number

List all factors of 36.

Solution:

Begin by listing all the two-number factorizations of 36. This can be accomplished by systematically dividing 36 by 1, 2, 3, and so on. Notice, however, that after the product $6 \cdot 6$, the two-number factorizations are repetitious, and we can stop the process.

$1 \cdot 36$

$2 \cdot 18$

$3 \cdot 12$

$4 \cdot 9$

$6 \cdot 6$

$9 \cdot 4$ ⎫
$12 \cdot 3$ ⎪ These products repeat the factorizations
$18 \cdot 2$ ⎪ above. Therefore, we can stop at $6 \cdot 6$.
$36 \cdot 1$ ⎭

TIP: When listing a set of factors, it is not necessary to write the numbers in any specified order. However, in general we list the factors in order from smallest to largest.

The list of all factors of 36 consists of the individual factors in the products. The factors are 1, 2, 3, 4, 6, 9, 12, 18, 36.

Answers

10. $2 \cdot 2 \cdot 2 \cdot 3 \cdot 7$ or $2^3 \cdot 3 \cdot 7$
11. $2 \cdot 3 \cdot 3 \cdot 5 \cdot 11$ or $2 \cdot 3^2 \cdot 5 \cdot 11$
12. 1, 3, 5, 9, 15, 45

Section 2.2 Practice Exercises

Study Skills Exercise

For additional exercises, see Classroom Activities 2.2A–2.2C in the *Student's Resource Manual* at www.mhhe.com/moh.

In general, 2 to 3 hours of study time per week is needed for each 1 hour per week of class time. Based on the number of hours you are in class this semester, how many hours per week should you be studying?

Vocabulary and Key Concepts

1. **a.** A _____factor_____ of a number n is a nonzero whole number that divides evenly into n.

   b. A _____prime_____ number is a whole number greater than 1 that has only two factors, itself and 1.

   c. A _____composite_____ number is a whole number greater than 1 that is not prime.

   d. The _____prime_____ factorization of a number n is the product of prime numbers that equals n.

Review Exercises

2. Where would the number $\frac{7}{3}$ be located on a number line?

   a. Between 0 and 1   **b.** Between 1 and 2   **c.** Between 2 and 3   c

For Exercises 3–5, write two fractions, one representing the shaded area and one representing the unshaded area.

3.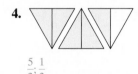
   $\frac{8}{12}, \frac{4}{12}$

4. $\frac{5}{2}, \frac{1}{2}$

5.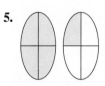
   $\frac{5}{4}, \frac{3}{4}$

6. Write a fraction with numerator 6 and denominator 5. Is this fraction proper or improper? $\frac{6}{5}$; improper

7. Write a fraction with denominator 12 and numerator 7. Is this fraction proper or improper? $\frac{7}{12}$; proper

8. Write a fraction with denominator 6 and numerator 6. Is this fraction proper or improper? $\frac{6}{6}$; improper

9. Write the improper fraction $\frac{23}{5}$ as a mixed number.
   $4\frac{3}{5}$

10. Write the mixed number $6\frac{2}{7}$ as an improper fraction.
    $\frac{44}{7}$

Concept 1: Factors and Factorizations

For Exercises 11–14, find two different factorizations of each number. (Answers may vary.) **(See Example 1.)**

11. 8   For example: $2 \cdot 4$ and $1 \cdot 8$

12. 20   For example: $2 \cdot 10$ and $4 \cdot 5$

13. 24   For example: $4 \cdot 6$ and $2 \cdot 2 \cdot 2 \cdot 3$

14. 14   For example: $1 \cdot 14$ and $2 \cdot 7$

15. Find two factors whose product is the number in the top row and whose sum is the number in the bottom row. The first column is done for you as an example.

Product	36	42	30	15	81
Factor	12	7	30	15	27
Factor	3	6	1	1	3
Sum	15	13	31	16	30

16. Find two factors whose product is the number in the top row and whose difference is the number in the bottom row. The first column is done for you as an example.

Product	36	42	45	72	24
Factor	9	7	15	18	8
Factor	4	6	3	4	3
Difference	5	1	12	14	5

Concept 2: Divisibility Rules

17. State the divisibility rule for dividing by 2.
A whole number is divisible by 2 if it is an even number.

18. State the divisibility rule for dividing by 10.
A whole number is divisible by 10 if its ones-place digit is 0.

19. State the divisibility rule for dividing by 3.
A whole number is divisible by 3 if the sum of its digits is divisible by 3.

20. State the divisibility rule for dividing by 5.
A whole number is divisible by 5 if its ones-place digit is 5 or 0.

For Exercises 21–28, determine if the number is divisible by   **a.** 2   **b.** 3   **c.** 5   **d.** 10
(See Example 2.)

21. 45   a. No   b. Yes   c. Yes   d. No

22. 100   a. Yes   b. No   c. Yes   d. Yes

23. 137   a. No   b. No   c. No   d. No

24. 241   a. No   b. No   c. No   d. No

25. 108   a. Yes   b. Yes   c. No   d. No

26. 1040   a. Yes   b. No   c. Yes   d. Yes

27. 3140   a. Yes   b. No   c. Yes   d. Yes

28. 2115   a. No   b. Yes   c. Yes   d. No

29. Ms. Berglund has 28 students in her class. Can she distribute a package of 84 candies evenly to her students?   Yes

30. Mr. Blankenship has 22 students in an algebra class. He has 110 sheets of graph paper. Can he distribute the graph paper evenly among his students?   Yes

Concept 3: Prime and Composite Numbers

For Exercises 31–46, determine whether the number is prime, composite, or neither. (See Example 3.)

31. 7   Prime

32. 17   Prime

33. 10   Composite

34. 21   Composite

35. 51   Composite

36. 57   Composite

37. 23   Prime

38. 31   Prime

39. 1   Neither

40. 0   Neither

41. 121   Composite

42. 69   Composite

43. 19   Prime

44. 29   Prime

45. 39   Composite

46. 49   Composite

47. Are there any whole numbers that are not prime or composite? If so, list them.
There are two whole numbers that are neither prime nor composite, 0 and 1.

48. True or false? The square of any prime number is also a prime number.   False

49. True or false? All odd numbers are prime.   False

50. True or false? All even numbers are composite.   False

51. One method for finding prime numbers is the *sieve of Eratosthenes*. The natural numbers from 2 to 50 are shown in the table. Start at the number 2 (the smallest prime number). Leave the number 2 and cross out every second number after the number 2. This will eliminate all numbers that are multiples of 2. Then go back to the beginning of the chart and leave the number 3, but cross out every third number after the number 3 (thus eliminating the multiples of 3). Begin at the next open number and continue this process. The numbers that remain are prime numbers. Use this process to find the prime numbers less than 50.

	2	3	4	5	6	7	8	9	10
11	12	13	14	15	16	17	18	19	20
21	22	23	24	25	26	27	28	29	30
31	32	33	34	35	36	37	38	39	40
41	42	43	44	45	46	47	48	49	50

2, 3, 5, 7, 11, 13, 17, 19, 23, 29, 31, 37, 41, 43, 47

52. Use the sieve of Eratosthenes to find the prime numbers less than 80.

	2	3	4	5	6	7	8	9	10
11	12	13	14	15	16	17	18	19	20
21	22	23	24	25	26	27	28	29	30
31	32	33	34	35	36	37	38	39	40
41	42	43	44	45	46	47	48	49	50
51	52	53	54	55	56	57	58	59	60
61	62	63	64	65	66	67	68	69	70
71	72	73	74	75	76	77	78	79	80

2, 3, 5, 7, 11, 13, 17, 19, 23, 29, 31, 37, 41, 43, 47, 53, 59, 61, 67, 71, 73, 79

Concept 4: Prime Factorization

For Exercises 53–56, determine whether or not the factorization represents the prime factorization. If not, explain why.

53. $36 = 2 \cdot 2 \cdot 9$
No, 9 is not a prime number.

54. $48 = 2 \cdot 3 \cdot 8$
No, 8 is not a prime number.

55. $210 = 5 \cdot 2 \cdot 7 \cdot 3$
Yes

56. $126 = 3 \cdot 7 \cdot 3 \cdot 2$
Yes

For Exercises 57–68, find the prime factorization. **(See Examples 4 and 5.)**

57. 70 $2 \cdot 5 \cdot 7$

58. 495 $3 \cdot 3 \cdot 5 \cdot 11$ or $3^2 \cdot 5 \cdot 11$

59. 260 $2 \cdot 2 \cdot 5 \cdot 13$ or $2^2 \cdot 5 \cdot 13$

60. 175 $5 \cdot 5 \cdot 7$ or $5^2 \cdot 7$

61. 147 $3 \cdot 7 \cdot 7$ or $3 \cdot 7^2$

62. 102 $2 \cdot 3 \cdot 17$

63. 138 $2 \cdot 3 \cdot 23$

64. 231 $3 \cdot 7 \cdot 11$

65. 616 $2 \cdot 2 \cdot 2 \cdot 7 \cdot 11$ or $2^3 \cdot 7 \cdot 11$

66. 364 $2 \cdot 2 \cdot 7 \cdot 13$ or $2^2 \cdot 7 \cdot 13$

67. 47   Prime

68. 41   Prime

Concept 5: Identifying All Factors of a Whole Number

For Exercises 69–76, list all the factors of the number. **(See Example 6.)**

69. 12   1, 2, 3, 4, 6, 12

70. 18   1, 2, 3, 6, 9, 18

71. 32   1, 2, 4, 8, 16, 32

72. 55   1, 5, 11, 55

73. 81   1, 3, 9, 27, 81

74. 60   1, 2, 3, 4, 5, 6, 10, 12, 15, 20, 30, 60

75. 48   1, 2, 3, 4, 6, 8, 12, 16, 24, 48

76. 72   1, 2, 3, 4, 6, 8, 9, 12, 18, 24, 36, 72

Expanding Your Skills

For Exercises 77–80, determine whether the number is divisible by 4. Use the following divisibility rule: A whole number is divisible by 4 if the number formed by its last two digits is divisible by 4.

77. 230   No

78. 1046   No

79. 4616   Yes

80. 10,264   Yes

For Exercises 81–84, determine whether the number is divisible by 8. Use the following divisibility rule: A whole number is divisible by 8 if the number formed by its last three digits is divisible by 8.

81. 1032   Yes

82. 2520   Yes

83. 17,126   No

84. 25,058   No

For Exercises 85–88, determine whether the number is divisible by 9. Use the following divisibility rule: A whole number is divisible by 9 if the sum of its digits is divisible by 9.

85. 396   Yes

86. 414   Yes

87. 8453   No

88. 1587   No

For Exercises 89–92, determine whether the number is divisible by 6. Use the following divisibility rule: A whole number is divisible by 6 if it is divisible by both 2 and 3 (use the divisibility rules for 2 and 3 together).

89. 522   Yes

90. 546   Yes

91. 5917   No

92. 6394   No

| **Section 2.3** | **Simplifying Fractions to Lowest Terms** |

Concepts

1. Equivalent Fractions
2. Simplifying Fractions to Lowest Terms
3. Applications of Simplifying Fractions

1. Equivalent Fractions

The fractions $\frac{3}{6}$, $\frac{2}{4}$, and $\frac{1}{2}$ all represent the same portion of a whole. See Figure 2-5. Therefore, we say that the fractions are *equivalent*.

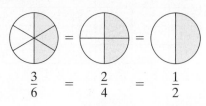

$$\frac{3}{6} \quad = \quad \frac{2}{4} \quad = \quad \frac{1}{2}$$

Figure 2-5

Avoiding Mistakes

The test to determine whether two fractions are equivalent is not the same process as multiplying fractions. Multiplying of fractions is covered in Section 2.4.

One method to show that two fractions are equivalent is to calculate their cross products. If their cross products are equal, then the fractions are equivalent. For example, to show that $\frac{3}{6} = \frac{2}{4}$, we have

$$\frac{3}{6} \diagdown\!\!\!\!\!\diagup \frac{2}{4}$$

$$3 \times 4 \overset{?}{=} 6 \times 2$$

$$12 = 12 \qquad \text{Yes. The fractions are equivalent.}$$

Skill Practice

Fill in the blank ☐ with = or ≠.

1. $\dfrac{13}{24}$ ☐ $\dfrac{6}{11}$

2. $\dfrac{9}{4}$ ☐ $\dfrac{54}{24}$

Classroom Examples: p. 119, Exercises 18 and 22

| **Example 1** | **Determining Whether Two Fractions Are Equivalent** |

Fill in the blank ☐ with = or ≠.

a. $\dfrac{18}{39}$ ☐ $\dfrac{6}{13}$ **b.** $\dfrac{5}{7}$ ☐ $\dfrac{7}{9}$

Solution:

a. $\dfrac{18}{39} \diagdown\!\!\!\!\!\diagup \dfrac{6}{13}$

$18 \times 13 \overset{?}{=} 39 \times 6$

$234 = 234$

Therefore, $\dfrac{18}{39} \boxed{=} \dfrac{6}{13}$.

b. $\dfrac{5}{7} \diagdown\!\!\!\!\!\diagup \dfrac{7}{9}$

$5 \times 9 \overset{?}{=} 7 \times 7$

$45 \neq 49$

Therefore, $\dfrac{5}{7} \boxed{\neq} \dfrac{7}{9}$.

2. Simplifying Fractions to Lowest Terms

In Figure 2-5, we see that $\frac{3}{6}$, $\frac{2}{4}$, and $\frac{1}{2}$ all represent equal quantities. However, the fraction $\frac{1}{2}$ is said to be in **lowest terms** because the numerator and denominator share no common factors other than 1.

To simplify a fraction to lowest terms, we apply the following important principle.

Answers

1. ≠   2. =

Fundamental Principle of Fractions

Suppose that a number, c, is a common factor in the numerator and denominator of a fraction. Then

$$\frac{a \cdot c}{b \cdot c} = \frac{a}{b} \cdot \frac{c}{c} = \frac{a}{b} \cdot 1 = \frac{a}{b}$$

To simplify a fraction, we begin by factoring the numerator and denominator into prime factors. This will help identify the common factors.

Example 2 Simplifying a Fraction to Lowest Terms

Simplify to lowest terms.

a. $\dfrac{6}{10}$ b. $\dfrac{170}{102}$ c. $\dfrac{20}{24}$

Solution:

a. $\dfrac{6}{10} = \dfrac{3 \cdot 2}{5 \cdot 2}$ Factor the numerator and denominator. Notice that 2 is a common factor.

$= \dfrac{3}{5} \cdot \dfrac{2}{2}$ Apply the fundamental principle of fractions.

$= \dfrac{3}{5} \cdot 1$ Any nonzero number divided by itself is 1.

$= \dfrac{3}{5}$

b. $\dfrac{170}{102} = \dfrac{5 \cdot 2 \cdot 17}{3 \cdot 2 \cdot 17}$ Factor the numerator and denominator.

$= \dfrac{5}{3} \cdot \dfrac{2}{2} \cdot \dfrac{17}{17}$ Apply the fundamental principle of fractions.

$= \dfrac{5}{3} \cdot 1 \cdot 1$ Any nonzero number divided by itself is 1.

$= \dfrac{5}{3}$

c. $\dfrac{20}{24} = \dfrac{5 \cdot 2 \cdot 2}{3 \cdot 2 \cdot 2 \cdot 2}$ Factor the numerator and denominator.

$= \dfrac{5}{3 \cdot 2} \cdot \dfrac{2}{2} \cdot \dfrac{2}{2}$ Apply the fundamental principle of fractions.

$= \dfrac{5}{6} \cdot 1 \cdot 1$

$= \dfrac{5}{6}$

Skill Practice

Simplify to lowest terms.

3. $\dfrac{15}{35}$ 4. $\dfrac{26}{195}$ 5. $\dfrac{150}{105}$

Classroom Example: p. 119, Exercise 28

Instructor Note: Students will probably not be able to simplify the fraction $\frac{170}{102}$ mentally, because they may not recognize the common factor of 17. In situations like this, the prime factor method to simplify a fraction is particularly good.

Answers

3. $\dfrac{3}{7}$ 4. $\dfrac{2}{15}$ 5. $\dfrac{10}{7}$

In Example 2, we show numerous steps to simplify fractions to lowest terms. However, the process is often made easier. For instance, we sometimes "divide out" common factors, and replace them with the new common factor of 1.

$$\frac{20}{24} = \frac{5 \cdot \overset{1}{2} \cdot \overset{1}{2}}{3 \cdot 2 \cdot \underset{1}{2} \cdot \underset{1}{2}} = \frac{5}{6}$$

The largest number that divides evenly into two natural numbers is called the **greatest common factor** or **(GCF)**. To find the greatest common factor of 20 and 24, factor each number into its prime factors. Then identify the common factors.

$20 = \boxed{2 \cdot 2} \cdot 5$ The factors that are common to both lists are circled.
$24 = \boxed{2 \cdot 2} \cdot 2 \cdot 3$ The GCF is $2 \cdot 2 = 4$

By identifying the greatest common factor between the numerator and denominator of a fraction we can easily simplify the fraction.

$$\frac{20}{24} = \frac{5 \cdot \overset{1}{4}}{6 \cdot \underset{1}{4}} = \frac{5}{6}$$

Notice that "dividing out" the common factor of 4 has the same effect as dividing the numerator and denominator by 4. This is often done mentally.

$$\frac{\overset{5}{20}}{\underset{6}{24}} = \frac{5}{6} \longleftarrow \text{20 divided by 4 equals 5.} \atop \longleftarrow \text{24 divided by 4 equals 6.}$$

TIP: Simplifying a fraction is also called reducing a fraction to lowest terms. For example, the simplified (or reduced) form of $\frac{20}{24}$ is $\frac{5}{6}$.

Skill Practice

Simplify the fraction. Write the answer as a fraction or whole number.

6. $\frac{28}{36}$ **7.** $\frac{39}{3}$ **8.** $\frac{15}{90}$

Classroom Example: p. 119, Exercise 46

Avoiding Mistakes

Do not forget to write the "1" in the numerator of the fraction $\frac{1}{5}$.

Answers

6. $\frac{7}{9}$ **7.** 13 **8.** $\frac{1}{6}$

Example 3 **Simplifying Fractions to Lowest Terms**

Simplify the fraction. Write the answer as a fraction or whole number.

a. $\frac{110}{99}$ **b.** $\frac{75}{25}$ **c.** $\frac{12}{60}$

Solution:

a. $\frac{110}{99} = \frac{10 \cdot \overset{1}{11}}{9 \cdot \underset{1}{11}} = \frac{10}{9}$ The greatest common factor in the numerator and denominator is 11.

Or alternatively: $\frac{\overset{10}{110}}{\underset{9}{99}} = \frac{10}{9} \longleftarrow \text{110 divided by 11 equals 10.} \atop \longleftarrow \text{99 divided by 11 equals 9.}$

b. $\frac{75}{25} = \frac{3 \cdot \overset{1}{25}}{1 \cdot \underset{1}{25}} = \frac{3}{1} = 3$ The greatest common factor in the numerator and denominator is 25.

Or alternatively: $\frac{\overset{3}{75}}{\underset{1}{25}} = \frac{3}{1} \longleftarrow \text{75 divided by 25 equals 3.} \atop \longleftarrow \text{25 divided by 25 equals 1.}$

$= 3$

TIP: Recall that any fraction of the form $\frac{n}{1} = n$. Therefore, $\frac{3}{1} = 3$.

c. $\frac{12}{60} = \frac{1 \cdot \overset{1}{12}}{5 \cdot \underset{1}{12}} = \frac{1}{5}$ The greatest common factor in the numerator and denominator is 12.

Or alternatively: $= \frac{\overset{1}{12}}{\underset{5}{60}} = \frac{1}{5} \longleftarrow \text{12 divided by 12 equals 1.} \atop \longleftarrow \text{60 divided by 12 equals 5.}$

Avoiding Mistakes

Suppose that you do not recognize the *greatest* common factor in the numerator and denominator. You can still divide by *any* common factor. However, you will have to repeat this process more than once to simplify the fraction completely. For instance, consider the fraction from Example 3(c).

$$\frac{\overset{2}{\cancel{12}}}{\underset{10}{\cancel{60}}} = \frac{2}{10}$$ Dividing by the common factor of 6 leaves a fraction that can be simplified further.

$$= \frac{\overset{1}{\cancel{2}}}{\underset{5}{\cancel{10}}} = \frac{1}{5}$$ Divide again, this time by 2. The fraction is now simplified completely because there are no other common factors in the numerator and denominator.

Example 4 Simplifying Fractions by 10, 100, and 1000

Simplify each fraction to lowest terms by first reducing by 10, 100, or 1000. Write the answer as a fraction.

a. $\dfrac{170}{30}$ b. $\dfrac{2500}{7500}$ c. $\dfrac{5000}{130,000}$

Solution:

a. $\dfrac{170}{30} = \dfrac{17 \cdot \overset{1}{\cancel{10}}}{3 \cdot \underset{1}{\cancel{10}}} = \dfrac{17}{3}$ Notice that dividing numerator and denominator by 10 has the effect of eliminating the 0 in the ones place from each number: $\dfrac{17\cancel{0}}{3\cancel{0}}$

b. $\dfrac{2500}{7500} = \dfrac{25\cancel{00}}{75\cancel{00}}$ Both 2500 and 7500 are divisible by 100. "Strike through" two zeros.

$= \dfrac{\overset{1}{\cancel{25}}}{\underset{3}{\cancel{75}}}$ Simplify further.

$= \dfrac{1}{3}$

c. $\dfrac{5000}{130,000} = \dfrac{5\cancel{000}}{130,\cancel{000}}$ The numbers 5000 and 130,000 are both divisible by 1000. "Strike through" three zeros.

$= \dfrac{\overset{1}{\cancel{5}}}{\underset{26}{\cancel{130}}}$ Simplify further.

$= \dfrac{1}{26}$

Skill Practice

Simplify to lowest terms by first reducing by 10, 100, or 1000.

9. $\dfrac{630}{190}$ 10. $\dfrac{1300}{52,000}$

11. $\dfrac{21,000}{35,000}$

Classroom Examples: p. 120, Exercises 62 and 68

Avoiding Mistakes

The "strike through" method only works for the digit 0 at the *end* of the numerator and denominator.

Instructor Note: To reinforce understanding, remind students that $\frac{2500}{7500}$ and $\frac{1}{3}$ represent the *same* portion of a whole unit.

Concept Connections

12. How many zeros may be eliminated from the numerator and denominator of the fraction $\frac{430,000}{154,000,000}$?

Answers

9. $\dfrac{63}{19}$ 10. $\dfrac{1}{40}$ 11. $\dfrac{3}{5}$

12. Four zeros; the numerator and denominator are both divisible by 10,000.

3. Applications of Simplifying Fractions

Example 5   Simplifying Fractions in an Application

Madeleine got 28 out of 35 problems correct on an algebra exam. David got 27 out of 45 questions correct on a different algebra exam.

a. What fractional part of the exam did each student answer correctly?

b. Which student performed better?

Solution:

a. Fractional part correct for Madeleine:

$$\frac{28}{35} \quad \text{or equivalently} \quad \frac{\overset{4}{28}}{\underset{5}{35}} = \frac{4}{5}$$

Fractional part correct for David:

$$\frac{27}{45} \quad \text{or equivalently} \quad \frac{\overset{3}{27}}{\underset{5}{45}} = \frac{3}{5}$$

b. From the simplified form of each fraction, we see that Madeleine performed better because $\frac{4}{5} > \frac{3}{5}$. That is, 4 parts out of 5 is greater than 3 parts out of 5. This is also easily verified on a number line.

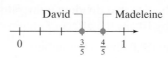

Skill Practice

13. Joanne planted 77 seeds in her garden and 55 sprouted. Geoff planted 140 seeds and 80 sprouted.
  a. What fractional part of the seeds sprouted for Joanne and what part sprouted for Geoff?
  b. For which person did a greater portion of seeds sprout?

Classroom Example: p. 120, Exercise 74

Instructor Note: Remind students that the expression $\frac{4}{5} > \frac{3}{5}$ means the same thing as $\frac{3}{5} < \frac{4}{5}$.

Answers

13. a. Joanne: $\frac{5}{7}$; Geoff: $\frac{4}{7}$
  b. Joanne had a greater portion of seeds sprout.

Section 2.3   Practice Exercises

For additional exercises, see Classroom Activities 2.3A–2.3C in the *Student's Resource Manual* at www.mhhe.com/moh.

Study Skills Exercise

Sometimes, test anxiety can be greatly reduced by adequate preparation and practice. List some places in the text where you can find extra problems for practice.

Vocabulary and Key Concepts

1. A fraction is said to be in _____lowest_____ terms if the numerator and denominator share no common factors other than 1.

Review Exercises

2. Determine whether 405 is divisible by

  a. 2   No   **b.** 3   Yes   **c.** 5   Yes   **d.** 10   No

For Exercises 3–10, write the prime factorization for each number.

3. 145 $5 \cdot 29$ **4.** 114 $2 \cdot 3 \cdot 19$ **5.** 92 $2 \cdot 2 \cdot 23$ or $2^2 \cdot 23$ **6.** 153 $3 \cdot 3 \cdot 17$ or $3^2 \cdot 17$

7. 85 $5 \cdot 17$ **8.** 120 $2 \cdot 2 \cdot 2 \cdot 3 \cdot 5$ or $2^3 \cdot 3 \cdot 5$ **9.** 195 $3 \cdot 5 \cdot 13$ **10.** 180 $2 \cdot 2 \cdot 3 \cdot 3 \cdot 5$ or $2^2 \cdot 3^2 \cdot 5$

 Writing   Translating Expression   Geometry    Scientific Calculator   Video   NS E

Concept 1: Equivalent Fractions

For Exercises 11–14, shade the second figure so that it expresses a fraction equivalent to the first figure

11. 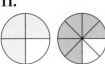   **12.** 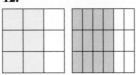   **13.** 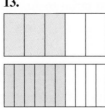   **14.**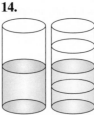

15. True or false? The fractions $\frac{4}{5}$ and $\frac{5}{4}$ are equivalent.   False

16. In your own words, explain the concept of equivalent fractions.

Two fractions are equivalent if they both represent the same part of a whole.

For Exercises 17–24, determine if the fractions are equivalent. Then fill in the blank with either = or ≠.
(See Example 1.)

17. $\frac{2}{3} \square \frac{3}{5}$ ≠   **18.** $\frac{1}{4} \square \frac{2}{9}$ ≠   **19.** $\frac{1}{2} \square \frac{3}{6}$ =   **20.** $\frac{6}{16} \square \frac{3}{8}$ =

21. $\frac{12}{16} \square \frac{3}{4}$ =   **22.** $\frac{4}{5} \square \frac{12}{15}$ =   **23.** $\frac{8}{9} \square \frac{20}{27}$ ≠   **24.** $\frac{5}{6} \square \frac{12}{18}$ ≠

Concept 2: Simplifying Fractions to Lowest Terms

For Exercises 25–52, simplify the fraction to lowest terms. Write the answer as a fraction or a whole number.
(See Examples 2 and 3.)

25. $\frac{12}{24}$ $\frac{1}{2}$ **26.** $\frac{15}{18}$ $\frac{5}{6}$ **27.** $\frac{6}{18}$ $\frac{1}{3}$ **28.** $\frac{21}{24}$ $\frac{7}{8}$

29. $\frac{36}{20}$ $\frac{9}{5}$ **30.** $\frac{49}{42}$ $\frac{7}{6}$ **31.** $\frac{15}{12}$ $\frac{5}{4}$ **32.** $\frac{30}{25}$ $\frac{6}{5}$

33. $\frac{20}{25}$ $\frac{4}{5}$ **34.** $\frac{8}{16}$ $\frac{1}{2}$ **35.** $\frac{14}{14}$ 1 **36.** $\frac{8}{8}$ 1

37. $\frac{50}{25}$ 2 **38.** $\frac{24}{6}$ 4 **39.** $\frac{9}{9}$ 1 **40.** $\frac{2}{2}$ 1

41. $\frac{105}{140}$ $\frac{3}{4}$ **42.** $\frac{84}{126}$ $\frac{2}{3}$ **43.** $\frac{33}{11}$ 3 **44.** $\frac{65}{5}$ 13

45. $\frac{77}{110}$ $\frac{7}{10}$ **46.** $\frac{85}{153}$ $\frac{5}{9}$ **47.** $\frac{130}{150}$ $\frac{13}{15}$ **48.** $\frac{70}{120}$ $\frac{7}{12}$

49. $\frac{385}{195}$ $\frac{77}{39}$ **50.** $\frac{39}{130}$ $\frac{3}{10}$ **51.** $\frac{34}{85}$ $\frac{2}{5}$ **52.** $\frac{69}{92}$ $\frac{3}{4}$

For Exercises 53–60, simplify the fractions. Operations in the numerator and denominator must be performed first.

53. $\frac{6-2}{10+4}$ $\frac{2}{7}$ **54.** $\frac{9-1}{15+3}$ $\frac{4}{9}$ **55.** $\frac{5-5}{7-2}$ 0 **56.** $\frac{11-11}{4+7}$ 0

57. $\dfrac{7-2}{5-5}$ Undefined **58.** $\dfrac{4+7}{11-11}$ Undefined **59.** $\dfrac{8-2}{8+2}$ $\dfrac{3}{5}$ **60.** $\dfrac{15+3}{15-3}$ $\dfrac{3}{2}$

For Exercises 61–68, simplify to lowest terms by first reducing the powers of 10. **(See Example 4.)**

61. $\dfrac{120}{160}$ $\dfrac{3}{4}$ **62.** $\dfrac{720}{800}$ $\dfrac{9}{10}$ **63.** $\dfrac{3000}{1800}$ $\dfrac{5}{3}$ **64.** $\dfrac{2000}{1500}$ $\dfrac{4}{3}$

65. $\dfrac{42{,}000}{22{,}000}$ $\dfrac{21}{11}$ **66.** $\dfrac{50{,}000}{65{,}000}$ $\dfrac{10}{13}$ **67.** $\dfrac{5100}{30{,}000}$ $\dfrac{17}{100}$ **68.** $\dfrac{9800}{28{,}000}$ $\dfrac{7}{20}$

Concept 3: Applications of Simplifying Fractions

69. André tossed a coin 48 times and heads came up 20 times. What fractional part of the tosses came up heads? What fractional part came up tails?

70. At Pizza Company, Lee made 70 pizzas one day. There were 105 pizzas sold that day. What fraction of the pizzas did Lee make? $\dfrac{2}{3}$

Heads: $\dfrac{5}{12}$; tails: $\dfrac{7}{12}$

71. a. What fraction of the alphabet is made up of vowels? (Include the letter y as a vowel, not a consonant.) $\dfrac{3}{13}$

b. What fraction of the alphabet is made up of consonants? $\dfrac{10}{13}$

72. Of the 88 constellations that can be seen in the night sky, 12 are associated with astrological horoscopes. The names of as many as 36 constellations are associated with animals or mythical creatures.

a. Of the 88 constellations, what fraction is associated with horoscopes? $\dfrac{3}{22}$

b. What fraction of the constellations have names associated with animals or mythical creatures? $\dfrac{9}{22}$

73. Jonathan and Jared both sold candy bars for a fund-raiser. Jonathan sold 25 of his 35 candy bars, and Jared sold 24 of his 28 candy bars. **(See Example 5.)**

a. What fractional part of his total number of candy bars did each boy sell? Jonathan: $\dfrac{5}{7}$; Jared: $\dfrac{6}{7}$

b. Which boy sold the greater fractional part? Jared sold the greater fractional part.

74. Lisa and Lynette are taking online courses. Lisa has completed 14 out of 16 assignments in her course while Lynette has completed 15 out of 24 assignments.

a. What fractional part of her total number of assignments did each woman complete? Lisa: $\dfrac{7}{8}$; Lynette: $\dfrac{5}{8}$

b. Which woman has completed more of her course? Lisa has completed more of her course.

75. Raymond read 720 pages of a 792-page book. His roommate, Travis, read 540 pages from a 660-page book.

a. What fractional part of the book did each person read? Raymond: $\dfrac{10}{11}$; Travis: $\dfrac{9}{11}$

b. Which of the roommates read a greater fraction of his book? Raymond read the greater fractional part.

76. Mr. Zahnen and Ms. Waymire both gave exams today. By mid-afternoon, Mr. Zahnen had finished grading 16 out of 36 exams, and Ms. Waymire had finished grading 15 out of 27 exams.

a. What fractional part of her total has Ms. Waymire completed? $\dfrac{5}{9}$

b. What fractional part of his total has Mr. Zahnen completed? $\dfrac{4}{9}$

 Writing   Translating Expression   Geometry Scientific Calculator   Video NS E

77. For a recent year, the population of the United States was reported to be 296,000,000. During the same year, the population of California was 36,458,000.

   a. Round the U.S. population to the nearest hundred million.   300,000,000

   b. Round the population of California to the nearest million.   36,000,000

   c. Using the results from parts (a) and (b), write a simplified fraction showing the portion of the U.S. population represented by California. $\frac{3}{25}$

78. For a recent year, the population of the United States was reported to be 296,000,000. During the same year, the population of Ethiopia was 75,067,000.

   a. Round the U.S. population to the nearest hundred million.   300,000,000

   b. Round the population of Ethiopia to the nearest million.   75,000,000

   c. Using the results from parts (a) and (b), write a simplified fraction comparing the population of the United States to the population of Ethiopia. $\frac{4}{1}$

   d. Based on the result from part (c), how many times greater is the U.S. population than the population of Ethiopia?   4 times greater

Expanding Your Skills

79. Write three fractions equivalent to $\frac{3}{4}$.   For example: $\frac{6}{8}, \frac{9}{12}, \frac{12}{16}$

80. Write three fractions equivalent to $\frac{1}{3}$.   For example: $\frac{2}{6}, \frac{3}{9}, \frac{4}{12}$

81. Write three fractions equivalent to $\frac{12}{18}$.   For example: $\frac{6}{9}, \frac{4}{6}, \frac{2}{3}$

82. Write three fractions equivalent to $\frac{80}{100}$.   For example: $\frac{40}{50}, \frac{8}{10}, \frac{4}{5}$

Calculator Connections

Topic: Simplifying Fractions on a Calculator

Some calculators have a fraction key, $\boxed{a^{b/c}}$. To enter a fraction, follow this example.

Expression	Keystrokes	Result
$\dfrac{3}{4}$	3 $\boxed{a^{b/c}}$ 4 $\boxed{=}$	$\boxed{\quad 3\rfloor 4 \quad}$

                                                numerator   denominator

To simplify a fraction to lowest terms, follow this example.

Expression	Keystrokes	Result
$\dfrac{22}{10}$	22 $\boxed{a^{b/c}}$ 10 $\boxed{=}$	$\boxed{\quad 2\_1\rfloor 5 \quad} = 2\dfrac{1}{5}$

                                      whole number   fraction

To convert to an improper fraction, follow this example.

Expression	Keystrokes	Result
$2\dfrac{1}{5}$	2 $\boxed{a^{b/c}}$ 1 $\boxed{a^{b/c}}$ 5 $\boxed{2^{nd}}$ $\boxed{d/e}$	$\boxed{\quad 11\rfloor 5 \quad} = \dfrac{11}{5}$

Calculator Exercises

For Exercises 83–90, use a calculator to simplify the fractions. Write the answer as a proper or improper fraction.

83. $\dfrac{792}{891}$ $\frac{8}{9}$ **84.** $\dfrac{728}{784}$ $\frac{13}{14}$ **85.** $\dfrac{779}{969}$ $\frac{41}{51}$ **86.** $\dfrac{462}{220}$ $\frac{21}{10}$

87. $\dfrac{493}{510}$ $\frac{29}{30}$ **88.** $\dfrac{871}{469}$ $\frac{13}{7}$ **89.** $\dfrac{969}{646}$ $\frac{3}{2}$ **90.** $\dfrac{713}{437}$ $\frac{31}{19}$

Section 2.4   Multiplication of Fractions and Applications

Concepts

1. Multiplication of Fractions
2. Fractions and the Order of Operations
3. Area of a Triangle
4. Applications of Multiplying Fractions

1. Multiplication of Fractions

Suppose Elija takes $\frac{1}{3}$ of a cake and then gives $\frac{1}{2}$ of this portion to his friend Max. Max gets $\frac{1}{2}$ of $\frac{1}{3}$ of the cake. This is equivalent to the expression $\frac{1}{2} \cdot \frac{1}{3}$. See Figure 2-6.

Elija takes $\frac{1}{3}$

Max gets
$\frac{1}{2}$ of $\frac{1}{3} = \frac{1}{6}$

Figure 2-6

From the illustration, the product $\frac{1}{2} \cdot \frac{1}{3} = \frac{1}{6}$. Notice that the product $\frac{1}{6}$ is found by multiplying the numerators and multiplying the denominators. This is true in general to multiply fractions.

Concept Connections

1. What fraction is $\frac{1}{2}$ of $\frac{1}{4}$ of a whole?

Multiplying Fractions

To multiply fractions, write the product of the numerators over the product of the denominators. Then simplify the resulting fraction, if possible.

$$\frac{a}{b} \cdot \frac{c}{d} = \frac{a \cdot c}{b \cdot d} \qquad \text{provided } b \text{ and } d \text{ are not equal to } 0$$

Skill Practice

Multiply. Write the answer as a fraction.

2. $\frac{2}{3} \cdot \frac{5}{9}$ **3.** $\frac{7}{12} \times 11$

Classroom Example: p. 129, Exercise 20

Example 1   Multiplying Fractions

Multiply and write the answer as a fraction.

a. $\frac{2}{5} \cdot \frac{4}{7}$ **b.** $\frac{8}{3} \times 5$

Solution:

a. $\frac{2}{5} \cdot \frac{4}{7} = \frac{2 \cdot 4}{5 \cdot 7} = \frac{8}{35}$ ⟵ Multiply the numerators.
⟵ Multiply the denominators.

Notice that the product $\frac{8}{35}$ is simplified completely because there are no common factors shared by 8 and 35.

b. $\frac{8}{3} \times 5 = \frac{8}{3} \times \frac{5}{1}$ First write the whole number as a fraction.

$= \frac{8 \times 5}{3 \times 1}$ Multiply the numerators. Multiply the denominators.

$= \frac{40}{3}$ The product is not reducible because there are no common factors shared by 40 and 3.

Answers

1. $\frac{1}{8}$ **2.** $\frac{10}{27}$ **3.** $\frac{77}{12}$

Example 2 illustrates a case where the product of fractions must be simplified.

Example 2   Multiplying and Simplifying Fractions

Multiply the fraction and simplify if possible.

$$\frac{4}{30} \cdot \frac{5}{14}$$

Solution:

$$\frac{4}{30} \cdot \frac{5}{14} = \frac{4 \cdot 5}{30 \cdot 14}$$ Multiply the numerators. Multiply the denominators.

$$= \frac{20}{420}$$ Simplify by first dividing 20 and 420 by 10.

$$= \frac{\overset{1}{2}}{\underset{21}{42}}$$ Simplify further by dividing 2 and 42 by 2.

$$= \frac{1}{21}$$

Skill Practice

Multiply and simplify.

4. $\frac{7}{20} \cdot \frac{4}{3}$

Classroom Example: p. 129, Exercise 32

It is often easier to simplify *before* multiplying. Consider the product from Example 2.

$$\frac{4}{30} \cdot \frac{5}{14} = \frac{\overset{2}{4}}{\underset{6}{30}} \cdot \frac{\overset{1}{5}}{\underset{7}{14}}$$ 4 and 14 share a common factor of 2.
30 and 5 share a common factor of 5.

$$= \frac{\overset{\overset{1}{2}}{4}}{\underset{\underset{3}{6}}{30}} \cdot \frac{\overset{1}{5}}{\underset{7}{14}}$$ 2 and 6 share a common factor of 2.

$$= \frac{1}{21}$$

As a general rule, this method is used most often in the text.

Example 3   Multiplying and Simplifying Fractions

Multiply and simplify.

$$\frac{10}{18} \times \frac{21}{55}$$

Solution:

$$\frac{10}{18} \times \frac{21}{55} = \frac{\overset{2}{10}}{\underset{6}{18}} \times \frac{\overset{7}{21}}{\underset{11}{55}}$$ 10 and 55 share a common factor of 5.
18 and 21 share a common factor of 3.

$$= \frac{\overset{\overset{1}{2}}{10}}{\underset{\underset{3}{6}}{18}} \times \frac{\overset{7}{21}}{\underset{11}{55}}$$ We can simplify further because 2 and 6 share a common factor of 2.

$$= \frac{7}{33}$$

Skill Practice

Multiply and simplify.

5. $\frac{6}{25} \times \frac{15}{18}$

Classroom Example: p. 129, Exercise 34

Answers

4. $\frac{7}{15}$ 5. $\frac{1}{5}$

Example 4 **Multiplying and Simplifying Fractions**

Multiply and simplify. Write the answers as fractions.

a. $6\left(\dfrac{3}{8}\right)$ **b.** $\dfrac{21}{25} \cdot \dfrac{15}{39} \cdot \dfrac{65}{24}$

Solution:

a. $6\left(\dfrac{3}{8}\right) = \dfrac{6}{1} \cdot \dfrac{3}{8}$ Write the whole number as a fraction.

$= \dfrac{\overset{3}{6}}{1} \cdot \dfrac{3}{\underset{4}{8}}$ Reduce before multiplying.

$= \dfrac{9}{4}$ Multiply.

b. $\dfrac{21}{25} \cdot \dfrac{15}{39} \cdot \dfrac{65}{24} = \left(\dfrac{\overset{7}{21}}{\underset{5}{25}} \cdot \dfrac{\overset{3}{15}}{\underset{13}{39}}\right) \cdot \dfrac{65}{24}$ Multiply the first two fractions.

$= \left(\dfrac{21}{65}\right) \cdot \dfrac{65}{24}$

$= \dfrac{\overset{7}{21}}{\underset{1}{65}} \cdot \dfrac{\overset{1}{65}}{\underset{8}{24}}$ Simplify.

$= \dfrac{7}{8}$

2. Fractions and the Order of Operations

For problems with more than one operation, recall the order of operations.

Order of Operations

Step 1 Perform all operations inside parentheses first.

Step 2 Simplify any expressions containing exponents or square roots.

Step 3 Perform multiplication or division in the order that they appear from left to right.

Step 4 Perform addition or subtraction in the order that they appear from left to right.

Example 5 Simplifying Expressions

Simplify.

a. $\left(\dfrac{2}{5}\right)^3$ b. $\left(\dfrac{2}{15} \cdot \dfrac{3}{4}\right)^2$

Solution:

a. $\left(\dfrac{2}{5}\right)^3 = \dfrac{2}{5} \cdot \dfrac{2}{5} \cdot \dfrac{2}{5}$ With an exponent of 3, multiply 3 factors of the base.

$= \dfrac{2 \cdot 2 \cdot 2}{5 \cdot 5 \cdot 5}$ Multiply the numerators. Multiply the denominators.

$= \dfrac{8}{125}$

b. $\left(\dfrac{2}{15} \cdot \dfrac{3}{4}\right)^2 = \left(\dfrac{\overset{1}{2}}{\underset{5}{15}} \cdot \dfrac{\overset{1}{3}}{\underset{2}{4}}\right)^2$ Perform the multiplication within the parentheses. Simplify.

$= \left(\dfrac{1}{10}\right)^2$ Multiply fractions within parentheses.

$= \dfrac{1}{10} \cdot \dfrac{1}{10}$ Square $\dfrac{1}{10}$ by multiplying $\dfrac{1}{10}$ times itself.

$= \dfrac{1}{100}$

Skill Practice

Simplify. Write the answer as a fraction.

8. $\left(\dfrac{4}{3}\right)^2$ 9. $\left(\dfrac{6}{5} \cdot \dfrac{1}{12}\right)^3$

Classroom Examples: p. 130, Exercises 58 and 62

In Section 1.7 we learned to recognize powers of 10. These are $10^1 = 10$, $10^2 = 100$, and so on. In this section, we learn to recognize the **powers of one-tenth**, that is, $\frac{1}{10}$ raised to a whole-number power. For example, consider the following expressions.

$$\left(\dfrac{1}{10}\right)^1 = \dfrac{1}{10}$$

$$\left(\dfrac{1}{10}\right)^2 = \dfrac{1}{10} \cdot \dfrac{1}{10} = \dfrac{1}{100}$$

$$\left(\dfrac{1}{10}\right)^3 = \dfrac{1}{10} \cdot \dfrac{1}{10} \cdot \dfrac{1}{10} = \dfrac{1}{1000}$$

$$\left(\dfrac{1}{10}\right)^4 = \dfrac{1}{10} \cdot \dfrac{1}{10} \cdot \dfrac{1}{10} \cdot \dfrac{1}{10} = \dfrac{1}{10,000}$$

From these examples, we see that a power of one-tenth results in a fraction with a 1 in the numerator. The denominator has a 1 followed by the same number of zeros as the exponent on the base of $\frac{1}{10}$.

3. Area of a Triangle

Recall that the area of a rectangle with length l and width w is given by

$$A = l \times w$$

w

l

Answers

8. $\dfrac{16}{9}$ 9. $\dfrac{1}{1000}$

Instructor Note: Sometimes students remember the formula for area of a triangle better if you compute the area of rectangle

$$A = l \cdot w$$

then draw a diagonal and explain that area of triangle is

$$A = \frac{1}{2}(l \cdot w)$$

Area of a Triangle

The formula for the area of a triangle is given by $A = \frac{1}{2}bh$, read "one-half times base times height."

The value of b is the measure of the base of the triangle. The value of h is the measure of the height of the triangle. The base b can be chosen as the length of any of the sides of the triangle. However, once you have chosen the base, the height must be measured as the shortest distance from the base to the opposite vertex (or point) of the triangle.

Figure 2-7 shows the same triangle with different choices for the base. Figure 2-8 shows a situation in which the height must be drawn "outside" the triangle. In such a case, notice that the height is drawn down to an imaginary extension of the base line.

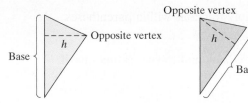

Figure 2-7 **Figure 2-8**

Classroom Example: p. 130, Exercise 72

Skill Practice

Find the area of the triangle.

10.

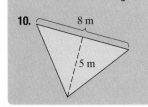

8 m

5 m

Answer

10. 20 m²

Example 6 **Finding the Area of a Triangle**

Find the area of the triangle.

4 m

6 m

Solution:

$b = 6 \, \text{m}$ and $h = 4 \, \text{m}$ Identify the measure of the base and the height.

$A = \frac{1}{2}bh$

$= \frac{1}{2}(6 \, \text{m})(4 \, \text{m})$ Apply the formula for the area of a triangle.

$= \frac{1}{2}\left(\frac{6}{1}\,\text{m}\right)\left(\frac{4}{1}\,\text{m}\right)$ Write the whole numbers as fractions.

$= \frac{1}{2}\left(\frac{\overset{3}{6}}{1}\,\text{m}\right)\left(\frac{4}{1}\,\text{m}\right)$ Simplify.

$= \frac{12}{1}\,\text{m}^2$ Multiply numerators. Multiply denominators.

$= 12 \, \text{m}^2$ The area of the triangle is 12 square meters (m²).

Example 7 **Finding the Area of a Triangle**

Find the area of the triangle.

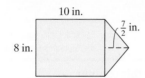
$\frac{3}{4}$ ft

$\frac{5}{3}$ ft

Solution:

$b = \dfrac{5}{3}$ ft   and $h = \dfrac{3}{4}$ ft   Identify the measure of the base and the height.

$A = \dfrac{1}{2}bh$

$= \dfrac{1}{2}\left(\dfrac{5}{3}\text{ ft}\right)\left(\dfrac{3}{4}\text{ ft}\right)$ Apply the formula for the area of a triangle.

$= \dfrac{1}{2}\left(\dfrac{5}{\underset{1}{3}}\text{ ft}\right)\left(\dfrac{\overset{1}{3}}{4}\text{ ft}\right)$ Simplify.

$= \dfrac{5}{8}\text{ ft}^2$ The area of the triangle is $\dfrac{5}{8}$ square feet (ft²).

Skill Practice

11. Find the area of the triangle.

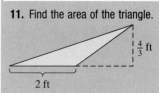
$\frac{4}{3}$ ft

2 ft

Classroom Example: p. 130,
Exercise 74

Example 8 **Find the Area of a Composite Geometric Figure**

Find the area.

10 in.

8 in.

$\frac{7}{2}$ in.

Solution:

The total area is the sum of the areas of the rectangular region and the triangular region. That is,

Total area =

10 in.

8 in.

$+$ 8 in. $\left\{\right.$

$\frac{7}{2}$ in.

(area of rectangle) + (area of triangle)
     $(l \times w)$ $\left(\frac{1}{2}bh\right)$

Area of rectangle: (10 in.)(8 in.)   Area of triangle: $\dfrac{1}{2}(8\text{ in.})\left(\dfrac{7}{2}\text{ in.}\right)$

$= 80\text{ in.}^2$ $= \dfrac{1}{2}\left(\dfrac{\overset{4}{8}}{\underset{1}{1}}\text{ in.}\right)\left(\dfrac{7}{2}\text{ in.}\right)$

$= \left(\dfrac{\overset{2}{4}}{1}\text{ in.}\right)\left(\dfrac{7}{\underset{1}{2}}\text{ in.}\right)$

$= \dfrac{14}{1}\text{ in.}^2$

$= 14\text{ in.}^2$

Total area = 80 in.² + 14 in.² = 94 in.²

The total area of the region is 94 square inches (in.²).

Skill Practice

12. Find the area of the kite.

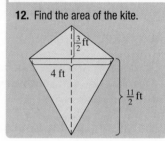
$\frac{3}{2}$ ft

4 ft

$\frac{11}{2}$ ft

Classroom Example: p. 131,
Exercise 82

Answers

11. $\dfrac{4}{3}$ or $1\dfrac{1}{3}$ ft²   **12.** 14 ft²

4. Applications of Multiplying Fractions

> **Example 9**   **Multiplying Fractions in an Application**
>
> The population of Texas comprises roughly $\frac{2}{25}$ of the population of the United States. If the U.S. population is approximately 306,000,000, approximate the population of Texas.
>
> **Solution:**
>
> We must find $\frac{2}{25}$ of the U.S. population. This translates to
>
> $$\frac{2}{25} \cdot 306{,}000{,}000 = \frac{2}{25} \cdot \frac{306{,}000{,}000}{1}$$
>
> $$= \frac{612{,}000{,}000}{25} \quad \longleftarrow \quad \text{Multiply the numerators.}$$
> $$\phantom{= \frac{612{,}000{,}000}{25}} \quad \longleftarrow \quad \text{Multiply the denominators.}$$
>
> $$= 24{,}480{,}000 \qquad \text{Divide by 25.}$$
>
> The population of Texas is approximately 24,480,000.

Skill Practice

13. Find the population of Tennessee if it is approximately $\frac{1}{50}$ of the U.S. population. Assume that the U.S. population is approximately 306,000,000.

Classroom Example: p. 131, Exercise 86

TIP: To take a fractional portion of a quantity, *multiply* the quantity by the fraction.

Answer

13. 6,120,000

Section 2.4   Practice Exercises

For additional exercises, see Classroom Activities 2.4A–2.4C in the *Student's Resource Manual* at www.mhhe.com/moh.

Study Skills Exercise

Write down the page number(s) for the Chapter Summary for this chapter. _____
Describe one way in which you can use the summary found at the end of each chapter.

Vocabulary and Key Concepts

1. a. The values $\left(\frac{1}{10}\right)^1, \left(\frac{1}{10}\right)^2, \left(\frac{1}{10}\right)^3$, and so on are called powers of __one-tenth__. In simplified form these equal $\frac{1}{10}, \frac{1}{100}, \frac{1}{1000}$, and so on.

b. Given a triangle with base b and height h, the area of the triangle is given by $A = \underline{\frac{1}{2}bh}$.

Review Exercises

2. a. Convert the fraction $\frac{17}{5}$ to a mixed number. $3\frac{2}{5}$

b. Convert the mixed number $4\frac{1}{8}$ to an improper fraction. $\frac{33}{8}$

For Exercises 3–6, identify the numerator and the denominator. Then simplify the fraction to lowest terms.

3. $\frac{10}{14}$ Numerator: 10; denominator: 14; $\frac{5}{7}$

4. $\frac{32}{36}$ Numerator: 32; denominator: 36; $\frac{8}{9}$

5. $\frac{25}{15}$ Numerator: 25; denominator: 15; $\frac{5}{3}$

6. $\frac{2100}{7000}$ Numerator: 2100; denominator: 7000; $\frac{3}{10}$

Concept 1: Multiplication of Fractions

7. Shade the portion of the figure that represents $\frac{1}{6}$ of $\frac{1}{2}$.

8. Shade the portion of the figure that represents $\frac{1}{2}$ of $\frac{1}{2}$.

 9. Shade the portion of the figure that represents $\frac{1}{4}$ of $\frac{1}{4}$.

10. Shade the portion of the figure that represents $\frac{1}{3}$ of $\frac{1}{4}$.

11. Find $\frac{1}{2}$ of $\frac{1}{4}$. $\frac{1}{8}$

12. Find $\frac{2}{3}$ of $\frac{1}{5}$. $\frac{2}{15}$

13. Find $\frac{3}{4}$ of 8. 6

14. Find $\frac{2}{5}$ of 20. 8

For Exercises 15–26, multiply the fractions. Write the answer as a fraction. **(See Example 1.)**

15. $\frac{1}{2} \times \frac{3}{8}$ $\frac{3}{16}$

16. $\frac{2}{3} \times \frac{1}{3}$ $\frac{2}{9}$

17. $\frac{14}{9} \cdot \frac{1}{9}$ $\frac{14}{81}$

18. $\frac{1}{8} \cdot \frac{9}{8}$ $\frac{9}{64}$

19. $\left(\frac{12}{7}\right)\left(\frac{2}{5}\right)$ $\frac{24}{35}$

20. $\left(\frac{9}{10}\right)\left(\frac{7}{4}\right)$ $\frac{63}{40}$

21. $8 \cdot \left(\frac{1}{11}\right)$ $\frac{8}{11}$

22. $3 \cdot \left(\frac{2}{7}\right)$ $\frac{6}{7}$

23. $\frac{4}{5} \cdot 6$ $\frac{24}{5}$

24. $\frac{5}{8} \cdot 5$ $\frac{25}{8}$

25. $\frac{13}{9} \times \frac{5}{4}$ $\frac{65}{36}$

26. $\frac{6}{5} \times \frac{7}{5}$ $\frac{42}{25}$

For Exercises 27–50, multiply the fractions and simplify to lowest terms. Write the answer as a fraction or whole number. **(See Examples 2–4.)**

27. $\frac{2}{9} \times \frac{3}{5}$ $\frac{2}{15}$

28. $\frac{1}{8} \times \frac{4}{7}$ $\frac{1}{14}$

29. $\frac{5}{6} \times \frac{3}{4}$ $\frac{5}{8}$

30. $\frac{7}{12} \times \frac{18}{5}$ $\frac{21}{10}$

31. $\frac{21}{5} \cdot \frac{25}{12}$ $\frac{35}{4}$

32. $\frac{16}{25} \cdot \frac{15}{32}$ $\frac{3}{10}$

33. $\frac{24}{15} \cdot \frac{5}{3}$ $\frac{8}{3}$

34. $\frac{49}{24} \cdot \frac{6}{7}$ $\frac{7}{4}$

35. $\left(\frac{6}{11}\right)\left(\frac{22}{15}\right)$ $\frac{4}{5}$

36. $\left(\frac{12}{45}\right)\left(\frac{5}{4}\right)$ $\frac{1}{3}$

37. $\left(\frac{17}{9}\right)\left(\frac{72}{17}\right)$ 8

38. $\left(\frac{39}{11}\right)\left(\frac{11}{13}\right)$ 3

39. $\frac{21}{4} \cdot \frac{16}{7}$ 12

40. $\frac{85}{6} \cdot \frac{12}{10}$ 17

41. $12 \times \frac{15}{42}$ $\frac{30}{7}$

42. $4 \times \frac{8}{92}$ $\frac{8}{23}$

43. $\frac{9}{15} \times \frac{16}{3} \times \frac{25}{8}$ 10

44. $\frac{49}{8} \times \frac{4}{5} \times \frac{20}{7}$ 14

45. $\frac{5}{2} \times \frac{10}{21} \times \frac{7}{5}$ $\frac{5}{3}$

46. $\frac{55}{9} \times \frac{18}{32} \times \frac{24}{11}$ $\frac{15}{2}$

47. $\frac{7}{10} \cdot \frac{3}{28} \cdot 5$ $\frac{3}{8}$

48. $\frac{11}{18} \cdot \frac{2}{20} \cdot 15$ $\frac{11}{12}$

49. $\frac{100}{49} \times 21 \times \frac{14}{25}$ 24

50. $\frac{38}{22} \times 11 \times \frac{5}{19}$ 5

Concept 2: Fractions and the Order of Operations

For Exercises 51–54, simplify the powers of $\frac{1}{10}$.

51. $\left(\frac{1}{10}\right)^3$ $\frac{1}{1000}$

52. $\left(\frac{1}{10}\right)^4$ $\frac{1}{10,000}$

53. $\left(\frac{1}{10}\right)^6$ $\frac{1}{1,000,000}$

54. $\left(\frac{1}{10}\right)^9$ $\frac{1}{1,000,000,000}$

 Writing   Translating Expression   Geometry   Scientific Calculator   Video NS E

For Exercises 55–66, simplify. Write the answer as a fraction or whole number. **(See Example 5.)**

55. $\left(\dfrac{1}{9}\right)^2$ $\dfrac{1}{81}$

56. $\left(\dfrac{1}{4}\right)^2$ $\dfrac{1}{16}$

57. $\left(\dfrac{3}{2}\right)^3$ $\dfrac{27}{8}$

58. $\left(\dfrac{4}{3}\right)^3$ $\dfrac{64}{27}$

59. $\left(4\cdot\dfrac{3}{4}\right)^3$ 27

60. $\left(5\cdot\dfrac{2}{5}\right)^3$ 8

61. $\left(\dfrac{1}{9}\cdot\dfrac{3}{5}\right)^2$ $\dfrac{1}{225}$

62. $\left(\dfrac{10}{3}\cdot\dfrac{1}{100}\right)^2$ $\dfrac{1}{900}$

63. $\dfrac{1}{3}\cdot\left(\dfrac{21}{4}\cdot\dfrac{8}{7}\right)$ 2

64. $\dfrac{1}{6}\cdot\left(\dfrac{24}{5}\cdot\dfrac{30}{8}\right)$ 3

 65. $\dfrac{16}{9}\cdot\left(\dfrac{1}{2}\right)^3$ $\dfrac{2}{9}$

66. $\dfrac{28}{6}\cdot\left(\dfrac{3}{2}\right)^2$ $\dfrac{21}{2}$

Concept 3: Area of a Triangle

 For Exercises 67–70, label the height with h and the base with b, as shown in the figure.

67.

68.

69.

70.

 For Exercises 71–80, find the area of the figure. **(See Examples 6 and 7.)**

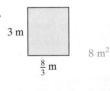

71.

8 cm
11 cm
44 cm²

72.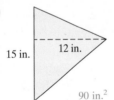

15 in.
12 in.
90 in.²

73.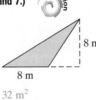

8 m
8 m
32 m²

 74.

1 ft
$\dfrac{7}{4}$ ft
$\dfrac{7}{8}$ ft²

75.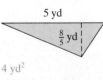

5 yd
$\dfrac{8}{5}$ yd
4 yd²

76.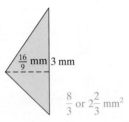

$\dfrac{16}{9}$ mm 3 mm
$\dfrac{8}{3}$ or $2\dfrac{2}{3}$ mm²

77.

$\dfrac{3}{4}$ cm
$\dfrac{1}{3}$ cm
$\dfrac{1}{4}$ cm²

78.

3 m
$\dfrac{8}{3}$ m
8 m²

79.

$\dfrac{15}{16}$ in.
$\dfrac{13}{16}$ in.

$\dfrac{195}{256}$ in.²

80.

$\dfrac{3}{4}$ ft
$\dfrac{23}{24}$ ft

$\dfrac{23}{32}$ ft²

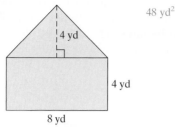 For Exercises 81–84, find the area of the shaded region. **(See Example 8.)**

81.

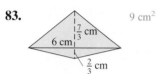

48 yd²

4 yd

4 yd

8 yd

82.

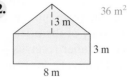

36 m²

3 m

3 m

8 m

83.

9 cm²

$\frac{7}{3}$ cm

6 cm

$\frac{2}{3}$ cm

84.

24 m²

$\frac{9}{4}$ m

8 m

$\frac{15}{4}$ m

Concept 4: Applications of Multiplying Fractions

85. Ms. Robbin's car holds 16 gallons (gal) of gas. If the fuel gauge indicates that there is $\frac{5}{8}$ of a tank left, how many gallons of gas are left in the tank? **(See Example 9.)** The amount left is 10 gal.

86. Land in a rural part of Bowie County, Texas, sells for $11,000 per acre. If Ms. Anderson purchased $\frac{3}{4}$ acre, how much did it cost? The cost is $8250.

87. Trey has half a pizza left over from dinner. If he eats $\frac{1}{4}$ of this for breakfast, how much pizza did he eat for breakfast?

Trey ate $\frac{1}{8}$ of the pizza for breakfast.

88. In a certain sample of individuals, $\frac{2}{5}$ are known to have blood type O. Of the individuals with blood type O, $\frac{1}{4}$ are Rh-negative. What fraction of the individuals in the sample have O-negative blood? $\frac{1}{10}$ of the sample has O-negative blood.

89. In planning a luncheon for a conference, Corrine finds out that only $\frac{3}{4}$ of the people invited will be attending. She originally ordered $5\frac{1}{2}$ lb of cold cuts. If she prepares only $\frac{3}{4}$ of the order, how many pounds will she prepare?

Corrine will prepare $4\frac{1}{8}$ lb.

90. A grocery store received a shipment of $140\frac{2}{3}$ lb of beef. It was later determined that $\frac{3}{8}$ of the shipment was contaminated and had to be destroyed. How much of the beef had to be destroyed? $52\frac{3}{4}$ lb must be destroyed.

91. One week, Nielsen Media Research reported that a new episode of *C.S.I.* was the second most watched television show with 9,825,000 total viewers. The following week, a repeat episode was shown, but only $\frac{2}{3}$ of the viewers returned to watch. How many viewers tuned in to watch the repeat episode? 6,550,000

92. Nancy spends $\frac{3}{4}$ hr 3 times a day walking and playing with her dog. What is the total time she spends walking and playing with the dog each day? Nancy spends $\frac{9}{4}$ or $2\frac{1}{4}$ hr a day.

93. The Bishop Gaming Center hosts a football pool. There is $1200 in prize money. The first-place winner receives $\frac{2}{3}$ of the prize money. The second-place winner receives $\frac{1}{4}$ of the prize money, and the third-place winner receives $\frac{1}{12}$ of the prize money. How much money does each person get?

First place: $800; second place: $300; third place: $100

 94. Frankie's lawn measures 40 yd by 36 yd. In the morning he mowed $\frac{2}{3}$ of the lawn. How many square yards of lawn did he already mow? How much is left to be mowed?

Frankie mowed 960 yd². He has 480 yd² left to mow.

Expanding Your Skills

95. Evaluate.

a. $\left(\frac{1}{6}\right)^2$ $\frac{1}{36}$ **b.** $\sqrt{\frac{1}{36}}$ $\frac{1}{6}$

96. Evaluate.

a. $\left(\frac{2}{7}\right)^2$ $\frac{4}{49}$ **b.** $\sqrt{\frac{4}{49}}$ $\frac{2}{7}$

For Exercises 97–100, evaluate the square roots.

97. $\sqrt{\frac{1}{25}}$ $\frac{1}{5}$

98. $\sqrt{\frac{1}{100}}$ $\frac{1}{10}$

99. $\sqrt{\frac{64}{81}}$ $\frac{8}{9}$

100. $\sqrt{\frac{9}{4}}$ $\frac{3}{2}$

101. Find the next number in the sequence: $\frac{1}{2}, \frac{1}{4}, \frac{1}{8}, \frac{1}{16},$ _____ $\frac{1}{32}$

102. Find the next number in the sequence: $\frac{2}{3}, \frac{2}{9}, \frac{2}{27},$ _____ $\frac{2}{81}$

103. Which is greater, $\frac{1}{2}$ of $\frac{1}{8}$ or $\frac{1}{8}$ of $\frac{1}{2}$?
They are the same.

104. Which is greater, $\frac{2}{3}$ of $\frac{1}{4}$ or $\frac{1}{4}$ of $\frac{2}{3}$?
They are the same.

Section 2.5   Division of Fractions and Applications

Concepts

1. Reciprocal of a Fraction
2. Division of Fractions
3. Order of Operations
4. Applications of Multiplication and Division of Fractions

Concept Connections

Fill in the blank.

1. The product of a number and its reciprocal is _____.

1. Reciprocal of a Fraction

Two numbers whose product is 1 are said to be *reciprocals* of each other. For example, consider the product of $\frac{3}{8}$ and $\frac{8}{3}$.

$$\frac{3}{8} \cdot \frac{8}{3} = \frac{\overset{1}{\cancel{3}}}{\cancel{8}} \cdot \frac{\overset{1}{\cancel{8}}}{\cancel{3}} = 1$$

Because the product equals 1, we say that $\frac{3}{8}$ is the reciprocal of $\frac{8}{3}$ and vice versa.

To divide fractions, first we need to learn how to find the reciprocal of a fraction.

> **Finding the Reciprocal of a Fraction**
> To find the **reciprocal** of a nonzero fraction, interchange the numerator and denominator of the fraction. Thus, the reciprocal of $\frac{a}{b}$ is $\frac{b}{a}$ (provided $a \neq 0$ and $b \neq 0$).

Answer

1. 1

Example 1   **Finding Reciprocals**

Find the reciprocal.

a. $\dfrac{2}{5}$ **b.** $\dfrac{1}{9}$ **c.** 5   **d.** 0

Solution:

a. The reciprocal of $\frac{2}{5}$ is $\frac{5}{2}$.

b. The reciprocal of $\frac{1}{9}$ is $\frac{9}{1}$, or 9.

c. First write the whole number 5 as the improper fraction $\frac{5}{1}$.
The reciprocal of $\frac{5}{1}$ is $\frac{1}{5}$.

d. The number 0 has no reciprocal because $\frac{1}{0}$ is undefined.

2. Division of Fractions

To understand the division of fractions, consider $6 \div \frac{1}{2}$. This statement asks, "How many halves $\left(\frac{1}{2}\right)$ can be found in 6 wholes?" The answer is 12.

$$6 \div \frac{1}{2} = 12$$

This result can also be found by multiplying.

$$6 \cdot \frac{2}{1} = 6 \cdot 2 = 12$$

That is, dividing by $\frac{1}{2}$ is equivalent to multiplying by the reciprocal, $\frac{2}{1} = 2$.

In general, to divide two nonzero numbers we can multiply the dividend by the reciprocal of the divisor. This is how we divide by a fraction.

Instructor Note: Remind students that division is not commutative. Thus, $\frac{2}{3} \div \frac{1}{6} \neq \frac{1}{6} \div \frac{2}{3}$.

Dividing Fractions

To divide two fractions, multiply the dividend (the "first" fraction) by the reciprocal of the divisor (the "second" fraction).

The process to divide fractions can be written symbolically as

Change division to multiplication.

$$\frac{a}{b} \div \frac{c}{d} = \frac{a}{b} \cdot \frac{d}{c}$$ provided b, c, and d are not 0.

Take the reciprocal of the divisor.

Example 2 **Dividing Fractions**

Divide and simplify, if possible. Write the answer as a fraction.

a. $\dfrac{2}{5} \div \dfrac{7}{4}$ **b.** $\dfrac{2}{27} \div \dfrac{8}{15}$ **c.** $\dfrac{35}{14} \div 7$ **d.** $12 \div \dfrac{8}{3}$

Solution:

a. $\dfrac{2}{5} \div \dfrac{7}{4} = \dfrac{2}{5} \cdot \dfrac{4}{7}$ Multiply by the reciprocal of the divisor ("second" fraction).

$= \dfrac{2 \cdot 4}{5 \cdot 7}$ Multiply numerators. Multiply denominators.

$= \dfrac{8}{35}$

b. $\dfrac{2}{27} \div \dfrac{8}{15} = \dfrac{2}{27} \cdot \dfrac{15}{8}$ Multiply by the reciprocal of the divisor ("second" fraction).

$= \dfrac{\overset{1}{2}}{\underset{9}{27}} \cdot \dfrac{\overset{5}{15}}{\underset{4}{8}}$ Simplify.

$= \dfrac{5}{36}$ Multiply.

c. $\dfrac{35}{14} \div 7 = \dfrac{35}{14} \div \dfrac{7}{1}$ Write the whole number 7 as an improper fraction *before* multiplying by the reciprocal.

$= \dfrac{35}{14} \cdot \dfrac{1}{7}$ Multiply by the reciprocal of the divisor.

$= \dfrac{\overset{5}{35}}{14} \cdot \dfrac{1}{\underset{1}{7}}$ Simplify.

$= \dfrac{5}{14}$ Multiply.

d. $12 \div \dfrac{8}{3} = \dfrac{12}{1} \div \dfrac{8}{3}$ Write the whole number 12 as an improper fraction.

$= \dfrac{12}{1} \cdot \dfrac{3}{8}$ Multiply by the reciprocal of the divisor.

$= \dfrac{\overset{3}{12}}{1} \cdot \dfrac{3}{\underset{2}{8}}$ Simplify.

$= \dfrac{9}{2}$ Multiply.

3. Order of Operations

When simplifying fractional expressions with more than one operation, be sure to follow the order of operations. Simplify within parentheses first. Then simplify expressions with exponents, followed by multiplication or division in the order of appearance from left to right.

Example 3   Applying the Order of Operations

Simplify. Write the answer as a fraction.

$$\frac{2}{3} \div \frac{4}{9} \div 6$$

Solution:

$\frac{2}{3} \div \frac{4}{9} \div 6$

$= \left(\frac{2}{3} \div \frac{4}{9}\right) \div 6$ We will divide from left to right. To emphasize this order, we can insert parentheses around the first two fractions.

$= \left(\frac{\overset{1}{2}}{3} \cdot \frac{\overset{3}{9}}{\underset{1}{4}}\right) \div 6$ Simplify within parentheses.

$= \left(\frac{3}{2}\right) \div \frac{6}{1}$ Simplify within parentheses and write the whole number as an improper fraction.

$= \left(\frac{3}{2}\right) \cdot \frac{1}{6}$ Multiply by the reciprocal of the divisor.

$= \frac{\overset{1}{3}}{2} \cdot \frac{1}{\underset{2}{6}}$ Simplify.

$= \frac{1}{4}$ Multiply.

Skill Practice

Simplify.

12. $\dfrac{6}{7} \div 2 \div \dfrac{3}{14}$

Classroom Example: p. 140, Exercise 68

TIP: In Example 3 we could also have written each division as multiplication of the reciprocal right from the start.

$\frac{2}{3} \div \frac{4}{9} \div 6 = \left(\frac{2}{3} \cdot \frac{9}{4}\right) \cdot \frac{1}{6}$

$= \frac{\overset{1}{2}}{3} \cdot \frac{\overset{3}{9}}{\underset{2}{4}} \cdot \frac{1}{\underset{2}{6}}$ Simplify.

$= \frac{1}{4}$ Multiply.

Example 4   Applying the Order of Operations

Simplify. Write the answer as a fraction.

$$\left(\frac{3}{5} \div \frac{2}{15}\right)^2$$

Solution:

$\left(\frac{3}{5} \div \frac{2}{15}\right)^2$ Perform operations within parentheses first.

$= \left(\frac{3}{5} \cdot \frac{15}{2}\right)^2$ Multiply by the reciprocal of the divisor.

$= \left(\frac{3}{\underset{1}{5}} \cdot \frac{\overset{3}{15}}{2}\right)^2$ Simplify.

$= \left(\frac{9}{2}\right)^2$ Multiply within parentheses.

$= \frac{9}{2} \cdot \frac{9}{2}$ With an exponent of 2, multiply 2 factors of the base.

$= \frac{81}{4}$ Multiply.

Skill Practice

Simplify.

13. $\left(\dfrac{4}{7} \div \dfrac{8}{3}\right)^2$

Classroom Example: p. 140, Exercise 74

Answers

12. 2   **13.** $\dfrac{9}{196}$

4. Applications of Multiplication and Division of Fractions

Sometimes it is difficult to determine whether multiplication or division is appropriate to solve an application problem. Division is generally used for a problem that requires you to separate or "split up" a quantity into pieces. Multiplication is generally used if it is necessary to take a fractional part of a quantity.

Example 5 Using Division in an Application

A road crew must mow the grassy median along a stretch of highway I-95. If they can mow $\frac{5}{8}$ mile (mi) in 1 hr, how long will it take them to mow a 15-mi stretch?

Solution:

Read and familiarize.
Strategy/operation: From the figure, we must separate or "split up" a 15-mi stretch of highway into pieces that are $\frac{5}{8}$ mi in length. Therefore, we must divide 15 by $\frac{5}{8}$.

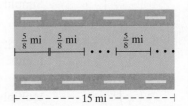

$$15 \div \frac{5}{8} = \frac{15}{1} \cdot \frac{8}{5}$$ Write the whole number as a fraction. Multiply by the reciprocal of the divisor.

$$= \frac{\overset{3}{\cancel{15}}}{1} \cdot \frac{8}{\underset{1}{\cancel{5}}}$$

$$= 24$$

The 15-mi stretch of highway will take 24 hr to mow.

Example 6 Using Division in an Application

A $\frac{9}{4}$-ft length of wire must be cut into pieces of equal length that are $\frac{3}{8}$ ft long. How many pieces can be cut?

Solution:

Read and familiarize.
Operation: Here we divide the total length of wire into pieces of equal length.

$$\frac{9}{4} \div \frac{3}{8} = \frac{9}{4} \cdot \frac{8}{3}$$ Multiply by the reciprocal of the divisor.

$$= \frac{\overset{3}{\cancel{9}}}{\underset{1}{\cancel{4}}} \cdot \frac{\overset{2}{\cancel{8}}}{\underset{1}{\cancel{3}}}$$ Simplify.

$$= 6$$

Six pieces of wire can be cut.

Example 7 Using Multiplication in an Application

Carson estimates that his total cost for college for 1 year is $12,600. He has financial aid to pay $\frac{2}{3}$ of the cost.

a. How much money is the financial aid worth?

b. How much money will Carson have to pay?

c. If Carson's parents help him by paying $\frac{1}{3}$ of the amount not paid by financial aid, how much money will be paid by Carson's parents?

Solution:

a. Carson's financial aid will pay $\frac{2}{3}$ of $12,600. Because we are looking for a fraction of a quantity, we multiply.

$$\frac{2}{3} \cdot 12{,}600 = \frac{2}{3} \cdot \frac{12{,}600}{1}$$

$$= \frac{2}{\underset{1}{3}} \cdot \frac{\overset{4200}{\cancel{12{,}600}}}{1}$$

$$= 8400$$

Financial aid will pay $8400.

b. Carson will have to pay the remaining portion of the cost. This can be found by subtraction.

$$\$12{,}600 - \$8400 = \$4200$$

Carson will have to pay $4200.

TIP: The answer to Example 7(b) could also have been found by noting that financial aid paid $\frac{2}{3}$ of the cost. This means that Carson must pay $\frac{1}{3}$ of the cost, or

$$\frac{1}{3} \cdot \frac{\$12{,}600}{1} = \frac{1}{\underset{1}{3}} \cdot \frac{\overset{4200}{\cancel{\$12{,}600}}}{1}$$

$$= \$4200$$

c. Carson's parents will pay $\frac{1}{3}$ of $4200.

$$\frac{1}{\underset{1}{3}} \cdot \frac{\overset{1400}{\cancel{4200}}}{1}$$

Carson's parents will pay $1400.

Skill Practice

16. A new school will cost $20,000,000 to build, and the state will pay $\frac{3}{5}$ of the cost.
  a. How much will the state pay?
  b. How much will the state not pay?
  c. The county school district issues bonds to pay $\frac{4}{5}$ of the money not covered by the state. How much money will be covered by bonds?

Classroom Example: p. 141, Exercise 90

Answers

16. a. $12,000,000
  b. $8,000,000
  c. $6,400,000

17. In a certain police department, $\frac{1}{3}$ of the department is female. Of the female officers, $\frac{2}{7}$ were promoted within the last year. What fraction of the police department consists of females who were promoted within the last year?

Classroom Example: p. 141, Exercise 92

Example 8 **Using Multiplication in an Application**

Three-fifths of the students in the freshman class are female. Of these students, $\frac{5}{9}$ are over the age of 25. What fraction of the freshman class is female over the age of 25?

Solution:

Read and familiarize.
Strategy/operation: We must find $\frac{5}{9}$ of $\frac{3}{5}$ of one whole freshman class. This implies multiplication.

$$\frac{5}{9} \times \frac{3}{5} = \frac{\overset{1}{5}}{\underset{3}{9}} \times \frac{\overset{1}{3}}{\underset{1}{5}} = \frac{1}{3}$$

Answer

17. $\frac{2}{21}$

One-third of the freshman class consists of female students over the age of 25.

Section 2.5 Practice Exercises

For additional exercises see Classroom Activities 2.5A–2.5D in the *Student's Resource Manual* at www.mhhe.com/moh.

Study Skills Exercise

Write down the page number(s) for the Problem Recognition Exercises for this chapter. How do you think that the Problem Recognition Exercises can help you?

Vocabulary and Key Concepts

1. Two numbers whose product is 1 are called _____reciprocals_____ of each other.

Review Exercises

2. Write the prime factorization of 108. $2^2 \cdot 3^3$

For Exercises 3–11, multiply and simplify to lowest terms. Write the answer as a fraction or whole number.

3. $\frac{9}{11} \times \frac{22}{5}$ $\frac{18}{5}$

4. $\frac{24}{7} \cdot \frac{7}{8}$ 3

5. $\frac{34}{5} \cdot \frac{5}{17}$ 2

6. $3 \cdot \left(\frac{7}{6}\right)$ $\frac{7}{2}$

7. $8 \cdot \left(\frac{5}{24}\right)$ $\frac{5}{3}$

8. $\left(\frac{2}{7}\right)\left(\frac{7}{2}\right)$ 1

9. $\left(\frac{9}{5}\right)\left(\frac{5}{9}\right)$ 1

10. $\frac{1}{10} \times 10$ 1

11. $\frac{1}{3} \times 3$ 1

Concept 1: Reciprocal of a Fraction

12. For each number, determine whether the number has a reciprocal.

  a. $\dfrac{1}{2}$ Yes   **b.** $\dfrac{5}{3}$ Yes   **c.** 6 Yes   **d.** 0 No

For Exercises 13–20, find the reciprocal of the number, if it exists. **(See Example 1.)**

13. $\dfrac{7}{8}$ $\frac{8}{7}$ **14.** $\dfrac{5}{6}$ $\frac{6}{5}$ **15.** $\dfrac{10}{9}$ $\frac{9}{10}$ **16.** $\dfrac{14}{5}$ $\frac{5}{14}$

17. 4 $\frac{1}{4}$ **18.** 9 $\frac{1}{9}$ **19.** 0 No reciprocal exists. **20.** $\dfrac{0}{4}$ No reciprocal exists.

Concept 2: Division of Fractions

For Exercises 21–24, fill in the blank.

21. Dividing by 3 is the same as multiplying by _____.
$\frac{1}{3}$

22. Dividing by 5 is the same as multiplying by _____.
$\frac{1}{5}$

23. Dividing by 8 is the same as _____ by $\dfrac{1}{8}$.
multiplying

24. Dividing by 12 is the same as _____ by $\dfrac{1}{12}$.
multiplying

For Exercises 25–40, divide and simplify the answer to lowest terms. Write the answer as a fraction or whole number. **(See Example 2.)**

25. $\dfrac{2}{15} \div \dfrac{5}{12}$ $\frac{8}{25}$ **26.** $\dfrac{11}{3} \div \dfrac{6}{5}$ $\frac{55}{18}$ **27.** $\dfrac{7}{13} \div \dfrac{2}{5}$ $\frac{35}{26}$ **28.** $\dfrac{8}{7} \div \dfrac{3}{10}$ $\frac{80}{21}$

29. $\dfrac{14}{3} \div \dfrac{6}{5}$ $\frac{35}{9}$ **30.** $\dfrac{11}{2} \div \dfrac{3}{4}$ $\frac{22}{3}$ **31.** $\dfrac{15}{2} \div \dfrac{3}{2}$ 5 **32.** $\dfrac{9}{10} \div \dfrac{9}{2}$ $\frac{1}{5}$

33. $\dfrac{3}{4} \div \dfrac{3}{4}$ 1 **34.** $\dfrac{6}{5} \div \dfrac{6}{5}$ 1 **35.** $7 \div \dfrac{2}{3}$ $\frac{21}{2}$ **36.** $4 \div \dfrac{3}{5}$ $\frac{20}{3}$

37. $\dfrac{12}{5} \div 4$ $\frac{3}{5}$ **38.** $\dfrac{20}{6} \div 5$ $\frac{2}{3}$ **39.** $\dfrac{9}{50} \div \dfrac{18}{25}$ $\frac{1}{4}$ **40.** $\dfrac{30}{40} \div \dfrac{15}{8}$ $\frac{2}{5}$

Mixed Exercises

For Exercises 41–64, multiply or divide as indicated. Write the answer as a fraction or whole number.

41. $\dfrac{10}{9} \div \dfrac{1}{18}$ 20 **42.** $\dfrac{4}{3} \div \dfrac{1}{3}$ 4 **43.** $12 \cdot \dfrac{4}{3}$ 16 **44.** $24 \cdot \dfrac{5}{8}$ 15

45. $\dfrac{9}{100} \div \dfrac{13}{1000}$ $\frac{90}{13}$ **46.** $\dfrac{1000}{17} \div \dfrac{10}{3}$ $\frac{300}{17}$ **47.** $\dfrac{36}{5} \cdot \dfrac{25}{9}$ 20 **48.** $\dfrac{13}{5} \cdot \dfrac{10}{17}$ $\frac{26}{17}$

49. $\dfrac{7}{8} \div \dfrac{1}{4}$ $\frac{7}{2}$ **50.** $\dfrac{7}{12} \div \dfrac{5}{3}$ $\frac{7}{20}$ **51.** $\dfrac{5}{8} \cdot \dfrac{2}{9}$ $\frac{5}{36}$ **52.** $\dfrac{1}{16} \cdot \dfrac{4}{3}$ $\frac{1}{12}$

53. $6 \cdot \dfrac{4}{3}$ 8 **54.** $12 \cdot \dfrac{5}{6}$ 10 **55.** $\dfrac{16}{5} \div 8$ $\frac{2}{5}$ **56.** $\dfrac{42}{11} \div 7$ $\frac{6}{11}$

57. $\dfrac{16}{3} \div \dfrac{2}{5}$ $\dfrac{40}{3}$

58. $\dfrac{17}{8} \div \dfrac{1}{4}$ $\dfrac{17}{2}$

59. $\dfrac{1}{8} \cdot 16$ 2

60. $\dfrac{2}{3} \cdot 9$ 6

61. $\dfrac{22}{7} \cdot \dfrac{5}{16}$ $\dfrac{55}{56}$

62. $\dfrac{40}{21} \cdot \dfrac{18}{25}$ $\dfrac{48}{35}$

63. $8 \div \dfrac{16}{3}$ $\dfrac{3}{2}$

64. $5 \div \dfrac{15}{4}$ $\dfrac{4}{3}$

Concept 3: Order of Operations

$\dfrac{2}{3} \cdot 6$ multiplies $\dfrac{2}{3}$ by $\dfrac{6}{1}$, and $\dfrac{2}{3} \div 6$ multiplies $\dfrac{2}{3}$ by $\dfrac{1}{6}$. So $\dfrac{2}{3} \cdot 6 = 4$ and $\dfrac{2}{3} \div 6 = \dfrac{1}{9}$.

65. Explain the difference in the process to evaluate $\dfrac{2}{3} \cdot 6$ versus $\dfrac{2}{3} \div 6$. Then evaluate each expression.

66. Explain the difference in the process to evaluate $8 \cdot \dfrac{2}{3}$ versus $8 \div \dfrac{2}{3}$. Then evaluate each expression.

$8 \cdot \dfrac{2}{3}$ multiplies 8 by $\dfrac{2}{3}$, and $8 \div \dfrac{2}{3}$ multiplies 8 by $\dfrac{3}{2}$. So $8 \cdot \dfrac{2}{3} = \dfrac{16}{3}$ and $8 \div \dfrac{2}{3} = 12$.

For Exercises 67–78, simplify by using the order of operations. Write the answer as a fraction or whole number.
(See Examples 3 and 4.)

67. $\dfrac{54}{21} \div \dfrac{2}{3} \div 9$ $\dfrac{3}{7}$

68. $\dfrac{48}{56} \div \dfrac{3}{8} \div 8$ $\dfrac{2}{7}$

69. $\dfrac{3}{5} \div \dfrac{6}{7} \cdot \dfrac{5}{3}$ $\dfrac{7}{6}$

70. $\dfrac{5}{8} \div \dfrac{35}{16} \cdot \dfrac{1}{4}$ $\dfrac{1}{14}$

71. $\left(\dfrac{3}{8}\right)^2 \div \dfrac{9}{14}$ $\dfrac{7}{32}$

72. $\dfrac{7}{8} \div \left(\dfrac{1}{2}\right)^2$ $\dfrac{7}{2}$

73. $\left(\dfrac{2}{5} \div \dfrac{8}{3}\right)^2$ $\dfrac{9}{400}$

74. $\left(\dfrac{5}{12} \div \dfrac{2}{3}\right)^2$ $\dfrac{25}{64}$

75. $\left(\dfrac{63}{8} \div \dfrac{9}{4}\right)^2 \cdot 4$ 49

76. $\left(\dfrac{25}{3} \div \dfrac{50}{9}\right)^2 \cdot 8$ 18

77. $\dfrac{15}{16} \cdot \left(\dfrac{2}{3}\right)^2 \div \dfrac{20}{21}$ $\dfrac{7}{16}$

78. $\dfrac{8}{27} \cdot \left(\dfrac{3}{4}\right)^2 \div \dfrac{13}{18}$ $\dfrac{3}{13}$

Concept 4: Applications of Multiplication and Division of Fractions

79. How many eighths are in $\dfrac{9}{4}$? 18

80. How many sixths are in $\dfrac{4}{3}$? 8

81. During the month of December, a department store wraps packages free of charge. Each package requires $\dfrac{2}{3}$ yd of ribbon. If Li used up a 36-yd roll of ribbon, how many packages were wrapped? **(See Example 5.)** Li wrapped 54 packages.

82. A developer sells lots of land in increments of $\dfrac{3}{4}$ acre. If the developer has 60 acres, how many lots can be sold?
She can sell 80 parcels of land.

83. If one cup is $\dfrac{1}{16}$ gal, how many cups of orange juice can be filled from three half gallons ($\dfrac{3}{2}$ gal)? **(See Example 6.)** 24 cups of juice

84. If 1 centimeter (cm) is $\dfrac{1}{100}$ meter (m), how many centimeters are in a $\dfrac{5}{4}$-m piece of rope? 125 cm

85. Dorci buys 16 sheets of plywood, each $\dfrac{3}{4}$ in. thick, to cover her windows in the event of a hurricane. She stacks the wood in the garage. How high will the stack be? The stack will be 12 in. high.

86. Davey built a bookshelf 36 in. long. Can the shelf hold a set of encyclopedias if there are 24 books and each book averages $\frac{5}{4}$ in. thick? **Explain your answer.** Yes, the books will take up only 30 in.

87. A radio station allows 18 minutes (min) of advertising each hour. How many 40-second ($\frac{2}{3}$-min) commercials can be run in

    a. 1 hr   **b.** 1 day   648 commercials in 1 day
    27 commercials in 1 hr

88. A television station has 20 min of advertising each hour. How many 30-second ($\frac{1}{2}$-min) commercials can be run in

    a. 1 hr   **b.** 1 day   960 commercials in 1 day
    40 commercials in 1 hr

89. Ricardo wants to buy a new house for $240,000. The bank requires $\frac{1}{10}$ of the cost of the house as a down payment. As a gift, Ricardo's mother will pay $\frac{2}{3}$ of the down payment. **(See Example 7.)**

    a. How much money will Ricardo's mother pay toward the down payment?   Ricardo's mother will pay $16,000.

    b. How much money will Ricardo have to pay toward the down payment?   Ricardo will have to pay $8000.

    c. How much is left over for Ricardo to finance?   He will have to finance $216,000.

90. Althea wants to buy a Toyota Camry for a total cost of $19,560. The dealer requires $\frac{1}{12}$ of the money as a down payment. Althea's parents have agreed to pay one-half of the down payment for her.

    a. How much is the down payment?   The down payment is $1630.

    b. How much will Althea pay toward the down payment?   Althea will have to pay $815.

    c. How much will Althea have to finance?   She will have to finance $17,930.

91. A landowner has $\frac{9}{4}$ acres of land. She plans to sell $\frac{1}{3}$ of the land. **(See Example 8.)**

    a. How much land will she sell?   **b.** How much land will she retain?
    She plans to sell $\frac{3}{4}$ acre.   She will keep $\frac{3}{2}$ or $1\frac{1}{2}$ acres.

92. Josh must read 24 pages for his English class and 18 pages for psychology. He has read $\frac{1}{6}$ of the pages.

    a. How many pages has he read?   Josh has read 7 pages.

    b. How many pages does he still have to read?   He still must read 35 pages.

93. A lab technician has $\frac{7}{4}$ liters (L) of alcohol. If she needs samples of $\frac{1}{8}$ L, how many samples can she prepare?   She can prepare 14 samples.

94. Troy has a $\frac{7}{8}$-in. nail that he must hammer into a board. Each strike of the hammer moves the nail $\frac{1}{16}$ in. into the board. How many strikes of the hammer must he make?   Tony must make 14 strikes.

Expanding Your Skills

95. The rectangle shown here has an area of 30 ft². Find the length.

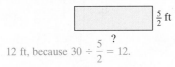

$\frac{5}{2}$ ft

12 ft, because $30 \div \frac{5}{2} = 12$.

96. The rectangle shown here has an area of 8 m². Find the width.

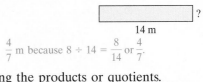

14 m

$\frac{4}{7}$ m because $8 \div 14 = \frac{8}{14}$ or $\frac{4}{7}$.

For Exercises 97–100, answer the questions without calculating the products or quotients.

97. Will the product $47 \times \frac{3}{5}$ be more or less than 47?
Less

98. Will the product $81 \times \frac{4}{7}$ be more or less than 81?
Less

99. Will the quotient $25 \div \frac{2}{3}$ be more or less than 25?
More

100. Will the quotient $41 \div \frac{2}{11}$ be more or less than 41?
More

Writing   Translating Expression   Geometry   Scientific Calculator   Video   NS E

Problem Recognition Exercises

Multiplication and Division of Fractions

Multiply or divide as indicated. Write the answer as a whole number or a fraction.

1. a. $\dfrac{8}{3} \cdot \dfrac{6}{5}$ $\dfrac{16}{5}$ **b.** $\dfrac{6}{5} \cdot \dfrac{8}{3}$ $\dfrac{16}{5}$ **2. a.** $\dfrac{10}{3} \cdot \dfrac{12}{7}$ $\dfrac{40}{7}$ **b.** $\dfrac{12}{7} \cdot \dfrac{10}{3}$ $\dfrac{40}{7}$

c. $\dfrac{8}{3} \div \dfrac{6}{5}$ $\dfrac{20}{9}$ **d.** $\dfrac{6}{5} \div \dfrac{8}{3}$ $\dfrac{9}{20}$ **c.** $\dfrac{10}{3} \div \dfrac{12}{7}$ $\dfrac{35}{18}$ **d.** $\dfrac{12}{7} \div \dfrac{10}{3}$ $\dfrac{18}{35}$

3. a. $12 \cdot \dfrac{9}{8}$ $\dfrac{27}{2}$ **b.** $\dfrac{9}{8} \cdot 12$ $\dfrac{27}{2}$ **4. a.** $15 \cdot \dfrac{3}{5}$ 9 **b.** $\dfrac{3}{5} \cdot 15$ 9

c. $12 \div \dfrac{9}{8}$ $\dfrac{32}{3}$ **d.** $\dfrac{9}{8} \div 12$ $\dfrac{3}{32}$ **c.** $15 \div \dfrac{3}{5}$ 25 **d.** $\dfrac{3}{5} \div 15$ $\dfrac{1}{25}$

5. a. $\dfrac{5}{6} \cdot \dfrac{5}{6}$ $\dfrac{25}{36}$ **b.** $\dfrac{5}{6} \cdot \dfrac{6}{5}$ 1 **6. a.** $\dfrac{9}{8} \cdot 0$ 0 **b.** $0 \cdot \dfrac{9}{8}$ 0

c. $\dfrac{5}{6} \div \dfrac{5}{6}$ 1 **d.** $\dfrac{5}{6} \div \dfrac{6}{5}$ $\dfrac{25}{36}$ **c.** $\dfrac{9}{8} \div 0$ Undefined **d.** $0 \div \dfrac{9}{8}$ 0

7. a. $\dfrac{1}{12} \cdot \dfrac{2}{3} \cdot \dfrac{16}{21}$ $\dfrac{8}{189}$ **b.** $\dfrac{1}{12} \cdot \dfrac{2}{3} \div \dfrac{16}{21}$ $\dfrac{7}{96}$ **8. a.** $\dfrac{1}{2} \cdot \dfrac{7}{9} \cdot \dfrac{2}{3}$ $\dfrac{7}{27}$ **b.** $\dfrac{1}{2} \cdot \dfrac{7}{9} \div \dfrac{2}{3}$ $\dfrac{7}{12}$

c. $\dfrac{1}{12} \div \dfrac{2}{3} \cdot \dfrac{16}{21}$ $\dfrac{2}{21}$ **d.** $\dfrac{1}{12} \div \dfrac{2}{3} \div \dfrac{16}{21}$ $\dfrac{21}{128}$ **c.** $\dfrac{1}{2} \div \dfrac{7}{9} \cdot \dfrac{2}{3}$ $\dfrac{3}{7}$ **d.** $\dfrac{1}{2} \div \dfrac{7}{9} \div \dfrac{2}{3}$ $\dfrac{27}{28}$

9. a. $\dfrac{9}{10} \cdot 6 \cdot \dfrac{1}{4}$ $\dfrac{27}{20}$ **b.** $\dfrac{9}{10} \cdot 6 \div \dfrac{1}{4}$ $\dfrac{108}{5}$ **10. a.** $\dfrac{4}{5} \cdot \dfrac{1}{20} \cdot 10$ $\dfrac{2}{5}$ **b.** $\dfrac{4}{5} \cdot \dfrac{1}{20} \div 10$ $\dfrac{1}{250}$

c. $\dfrac{9}{10} \div 6 \cdot \dfrac{1}{4}$ $\dfrac{3}{80}$ **d.** $\dfrac{9}{10} \div 6 \div \dfrac{1}{4}$ $\dfrac{3}{5}$ **c.** $\dfrac{4}{5} \div \dfrac{1}{20} \cdot 10$ 160 **d.** $\dfrac{4}{5} \div \dfrac{1}{20} \div 10$ $\dfrac{8}{5}$

11. a. $\dfrac{2}{3} \cdot 1$ $\dfrac{2}{3}$ **b.** $1 \cdot \dfrac{2}{3}$ $\dfrac{2}{3}$ **12. a.** $6 \div 10$ $\dfrac{3}{5}$ **b.** $10 \div 6$ $\dfrac{5}{3}$

c. $\dfrac{2}{3} \div 1$ $\dfrac{2}{3}$ **d.** $1 \div \dfrac{2}{3}$ $\dfrac{3}{2}$ **c.** $6 \cdot 10$ 60 **d.** $10 \cdot 6$ 60

13. a. $8 \div \dfrac{1}{4}$ 32 **b.** $8 \cdot \dfrac{1}{4}$ 2 **14. a.** $\dfrac{1}{7} \div 2$ $\dfrac{1}{14}$ **b.** $\dfrac{1}{7} \cdot 2$ $\dfrac{2}{7}$

c. $8 \div 4$ 2 **d.** $8 \cdot 4$ 32 **c.** $\dfrac{1}{7} \cdot \dfrac{1}{2}$ $\dfrac{1}{14}$ **d.** $\dfrac{1}{7} \div \dfrac{1}{2}$ $\dfrac{2}{7}$

15. a. $4^2 \cdot \dfrac{1}{6}$ $\dfrac{8}{3}$ **b.** $4^2 \div \dfrac{1}{6}$ 96 **16. a.** $\left(\dfrac{1}{2}\right)^2 \cdot \dfrac{2}{3}$ $\dfrac{1}{6}$ **b.** $\left(\dfrac{1}{2}\right)^2 \div \dfrac{2}{3}$ $\dfrac{3}{8}$

c. $4 \cdot \left(\dfrac{1}{6}\right)^2$ $\dfrac{1}{9}$ **d.** $4 \div \left(\dfrac{1}{6}\right)^2$ 144 **c.** $\dfrac{1}{2} \cdot \left(\dfrac{2}{3}\right)^2$ $\dfrac{2}{9}$ **d.** $\dfrac{1}{2} \div \left(\dfrac{2}{3}\right)^2$ $\dfrac{9}{8}$

| **Multiplication and Division of Mixed Numbers** | **Section 2.6** |

1. Multiplication of Mixed Numbers

Recall that to multiply fractions, we write the product of the numerators over the product of the denominators and then simplify. In order to apply this procedure to mixed numbers, we must first convert the mixed numbers to fractions. We have outlined the steps below.

> **Multiplying Mixed Numbers**
>
> **Step 1** Change each mixed number to an improper fraction.
>
> **Step 2** Multiply the improper fractions and simplify to lowest terms, if possible (see Section 2.4).
>
> Answers greater than 1 may be written as an improper fraction or as a mixed number, depending on the directions of the problem.

Example 1 demonstrates this process.

Concepts

1. Multiplication of Mixed Numbers
2. Division of Mixed Numbers
3. Applications of Multiplication and Division of Mixed Numbers

---

Example 1 **Multiplying Mixed Numbers**

Multiply and write the answer as a mixed number.

$$\left(3\frac{1}{5}\right)\left(4\frac{3}{4}\right)$$

Solution:

$$\left(3\frac{1}{5}\right)\left(4\frac{3}{4}\right) = \frac{16}{5} \cdot \frac{19}{4} \qquad \text{Write each mixed number as an improper fraction.}$$

$$= \frac{\overset{4}{16}}{5} \cdot \frac{19}{\underset{1}{4}} \qquad \text{Simplify.}$$

$$= \frac{76}{5} \qquad \text{Multiply.}$$

$$= 15\frac{1}{5} \qquad \text{Write the improper fraction as a mixed number.}$$

$$\begin{array}{r} 15 \\ 5\overline{)76} \\ \underline{-5} \\ 26 \\ \underline{-25} \\ 1 \end{array}$$

Skill Practice

Multiply and write the answer as a mixed number.

1. $\left(4\frac{3}{5}\right)\left(5\frac{5}{6}\right)$

Classroom Example: p. 147, Exercise 18

TIP: To check whether the answer from Example 1 is reasonable, we can round each factor and estimate the product.

$\left.\begin{array}{l} 3\frac{1}{5} \text{ rounds to 3.} \\[2mm] 4\frac{3}{4} \text{ rounds to 5.} \end{array}\right\}$ Thus, $\left(3\frac{1}{5}\right)\left(4\frac{3}{4}\right) \approx (3)(5) = 15$, which is close to $15\frac{1}{5}$.

Answer

1. $26\frac{5}{6}$

Skill Practice

Multiply and write the answer as a mixed number or whole number.

2. $16\frac{1}{2} \cdot 3\frac{7}{11}$

3. $7\frac{1}{6} \cdot 10$

Classroom Example: p. 148, Exercise 26

Example 2 **Multiplying Mixed Numbers**

Multiply and write the answer as a mixed number or whole number.

a. $25\frac{1}{2} \cdot 4\frac{2}{3}$ **b.** $12 \cdot \left(8\frac{7}{9}\right)$

Solution:

a. $25\frac{1}{2} \cdot 4\frac{2}{3} = \frac{51}{2} \cdot \frac{14}{3}$ Write each mixed number as an improper fraction.

$= \frac{\overset{17}{\cancel{51}}}{\underset{1}{\cancel{2}}} \cdot \frac{\overset{7}{\cancel{14}}}{\underset{1}{\cancel{3}}}$ Simplify.

$= \frac{119}{1}$ Multiply.

$= 119$

Avoiding Mistakes

Do not try to multiply mixed numbers by multiplying the whole-number parts and multiplying the fractional parts. You will not get the correct answer.

For the expression $25\frac{1}{2} \cdot 4\frac{2}{3}$, it would be incorrect to multiply $(25)(4)$ and $\frac{1}{2} \cdot \frac{2}{3}$. Notice that these values do not equal 119.

b. $12 \cdot \left(8\frac{7}{9}\right) = \frac{12}{1} \cdot \frac{79}{9}$ Write the whole number and mixed number as improper fractions.

$= \frac{\overset{4}{\cancel{12}}}{1} \cdot \frac{79}{\underset{3}{\cancel{9}}}$ Simplify.

$= \frac{316}{3}$ Multiply.

$= 105\frac{1}{3}$ Write the improper fraction as a mixed number.

$$\begin{array}{r} 105 \\ 3{\overline{\smash{\big)}\,316}} \\ \underline{-3} \\ 16 \\ \underline{-15} \\ 1 \end{array}$$

2. Division of Mixed Numbers

To divide mixed numbers, we use the following steps.

Dividing Mixed Numbers

Step 1 Change each mixed number to an improper fraction.

Step 2 Divide the improper fractions and simplify to lowest terms, if possible. Recall that to divide fractions, we multiply the dividend by the reciprocal of the divisor (see Section 2.5).

Answers greater than 1 may be written as an improper fraction or as a mixed number, depending on the directions of the problem.

Answers

2. 60   **3.** $71\frac{2}{3}$

Example 3 Dividing Mixed Numbers

Divide and write the answer as a mixed number or whole number.

a. $7\frac{1}{2} \div 4\frac{2}{3}$　　**b.** $6 \div 5\frac{1}{7}$　　**c.** $13\frac{5}{6} \div 7$

Solution:

a. $7\frac{1}{2} \div 4\frac{2}{3} = \frac{15}{2} \div \frac{14}{3}$　　Write the mixed numbers as improper fractions.

$\qquad = \frac{15}{2} \cdot \frac{3}{14}$　　Multiply by the reciprocal of the divisor.

$\qquad = \frac{45}{28}$　　Multiply.

$\qquad = 1\frac{17}{28}$　　Write the improper fraction as a mixed number.

b. $6 \div 5\frac{1}{7} = \frac{6}{1} \div \frac{36}{7}$　　Write the whole number and mixed number as improper fractions.

$\qquad = \frac{6}{1} \cdot \frac{7}{36}$　　Multiply by the reciprocal of the divisor.

$\qquad = \frac{\overset{1}{6}}{1} \cdot \frac{7}{\underset{6}{36}}$　　Simplify.

$\qquad = \frac{7}{6}$　　Multiply.

$\qquad = 1\frac{1}{6}$　　Write the improper fraction as a mixed number.

c. $13\frac{5}{6} \div 7 = \frac{83}{6} \div \frac{7}{1}$　　Write the whole number and mixed number as improper fractions.

$\qquad = \frac{83}{6} \cdot \frac{1}{7}$　　Multiply by the reciprocal of the divisor.

$\qquad = \frac{83}{42}$　　Multiply.

$\qquad = 1\frac{41}{42}$　　Write the improper fraction as a mixed number.

3. Applications of Multiplication and Division of Mixed Numbers

Examples 4 and 5 demonstrate multiplication and division of mixed numbers in day-to-day applications.

7. A recipe calls for $2\frac{3}{4}$ cups of flour. How much flour is required for $2\frac{1}{2}$ times the recipe?

Classroom Example: p. 148, Exercise 50

TIP: The solution to Example 4 can be estimated. In this case, the painter might want to over-estimate the answer to be sure that he doesn't run short of paint. We can round $5\frac{4}{5}$ gal to 6 gal. Therefore, our estimate is 8(6 gal) = 48 gal of paint.

Example 4 Applying Multiplication of Mixed Numbers

Antonio has a painting company, and he recently won a contract to paint eight large classrooms at a university. Each room requires $5\frac{4}{5}$ gal of paint. How much paint is needed for this job?

Solution:

Amount needed for 8 rooms:

$$8 \cdot \left(5\frac{4}{5}\text{ gal}\right) = 8 \cdot \left(\frac{29}{5}\right)\text{ gal}$$

$$= \frac{8}{1} \cdot \frac{29}{5}\text{ gal}$$

$$= \frac{232}{5}\text{ gal}$$

$$= 46\frac{2}{5}\text{ gal}$$

The painter requires $46\frac{2}{5}$ gal of paint.

Skill Practice

8. A department store wraps packages for $2 each. Ribbon $2\frac{5}{8}$ ft long is used to wrap each package. How many packages can be wrapped from a roll of ribbon 168 ft long?

Classroom Example: p. 148, Exercise 54

Example 5 Applying Division of Mixed Numbers

A construction site brings in $6\frac{2}{3}$ tons of soil. Each truck holds $\frac{2}{3}$ ton. How many truckloads are necessary?

Solution:

The $6\frac{2}{3}$ tons of soil must be distributed in $\frac{2}{3}$-ton increments. This will require division.

$6\frac{2}{3}$ tons

$\frac{2}{3}$ ton $\frac{2}{3}$ ton $\cdots$ $\frac{2}{3}$ ton

$$6\frac{2}{3} \div \frac{2}{3} = \frac{20}{3} \div \frac{2}{3} \qquad \text{Write the mixed number as an improper fraction.}$$

$$= \frac{20}{3} \cdot \frac{3}{2} \qquad \text{Multiply by the reciprocal of the divisor.}$$

$$= \frac{\overset{10}{\cancel{20}}}{\underset{1}{\cancel{3}}} \cdot \frac{\overset{1}{\cancel{3}}}{\underset{1}{\cancel{2}}} \qquad \text{Simplify.}$$

$$= \frac{10}{1} \qquad \text{Multiply.}$$

$$= 10$$

A total of 10 truckloads of soil will be required.

Answers

7. $6\frac{7}{8}$ cups of flour   8. 64 packages

Section 2.6   Practice Exercises

For additional exercises, see Classroom Activities 2.6A–2.6C in the *Student's Resource Manual* at www.mhhe.com/moh.

Study Skills Exercise

Find the page numbers for the Chapter Review Exercises, Chapter Test, and Cumulative Review Exercises for this chapter.

Chapter Review Exercises, page(s) _____ .

Chapter Test, page(s) _____ .

Cumulative Review Exercises, page(s) _____ .

Compare these features and state the advantages of each.

Vocabulary and Key Concepts

1. To multiply or divide mixed numbers, they must first be changed to _____improper_____ fractions.

Review Exercises

For Exercises 2–7, multiply or divide the fractions. Write your answers as fractions.

2. $\dfrac{5}{6} \cdot \dfrac{2}{9}$ $\frac{5}{27}$

3. $\dfrac{13}{5} \cdot \dfrac{10}{9}$ $\frac{26}{9}$

4. $\dfrac{20}{9} \div \dfrac{10}{3}$ $\frac{2}{3}$

5. $\dfrac{42}{11} \div \dfrac{7}{2}$ $\frac{12}{11}$

6. $\dfrac{32}{15} \div 4$ $\frac{8}{15}$

7. $\dfrac{52}{18} \div 13$ $\frac{2}{9}$

8. Explain the process to change a mixed number to an improper fraction.
    1. Multiply the whole number by the denominator. 2. Add the result to the numerator.
    3. Write the result from step 2 over the denominator.

For Exercises 9–12, write the mixed number as an improper fraction.

9. $3\dfrac{2}{5}$ $\frac{17}{5}$

10. $2\dfrac{7}{10}$ $\frac{27}{10}$

11. $1\dfrac{4}{7}$ $\frac{11}{7}$

12. $4\dfrac{1}{8}$ $\frac{33}{8}$

For Exercises 13–16, write the improper fraction as a mixed number.

13. $\dfrac{77}{6}$ $12\frac{5}{6}$

14. $\dfrac{57}{11}$ $5\frac{2}{11}$

15. $\dfrac{39}{4}$ $9\frac{3}{4}$

16. $\dfrac{31}{2}$ $15\frac{1}{2}$

Concept 1: Multiplication of Mixed Numbers

For Exercises 17–32, multiply the mixed numbers. Write the answer as a mixed number or whole number.
(See Examples 1 and 2.)

17. $\left(2\dfrac{2}{5}\right)\left(3\dfrac{1}{12}\right)$ $7\frac{2}{5}$

18. $\left(5\dfrac{1}{5}\right)\left(3\dfrac{3}{4}\right)$ $19\frac{1}{2}$

19. $2\dfrac{1}{3} \cdot \dfrac{5}{7}$ $1\frac{2}{3}$

20. $6\dfrac{1}{8} \cdot \dfrac{4}{7}$ $3\frac{1}{2}$

21. $4\dfrac{2}{9} \cdot 9$ 38

22. $3\dfrac{1}{3} \cdot 6$ 20

23. $\left(5\dfrac{3}{16}\right)\left(5\dfrac{1}{3}\right)$ $27\frac{2}{3}$

24. $\left(8\dfrac{2}{3}\right)\left(2\dfrac{1}{13}\right)$ 18

25. $\left(7\frac{1}{4}\right) \cdot 10$ $72\frac{1}{2}$ **26.** $\left(2\frac{2}{3}\right) \cdot 3$ 8   **27.** $4\frac{5}{8} \cdot 0$ 0   **28.** $0 \cdot 6\frac{1}{10}$ 0

29. $\left(3\frac{1}{2}\right)\left(2\frac{1}{7}\right)$ $7\frac{1}{2}$ **30.** $\left(1\frac{3}{10}\right)\left(1\frac{1}{4}\right)$ $1\frac{5}{8}$ **31.** $\left(5\frac{2}{5}\right)\left(\frac{2}{9}\right)\left(1\frac{4}{5}\right)$ $2\frac{4}{25}$ **32.** $\left(6\frac{1}{8}\right)\left(2\frac{3}{4}\right)\left(\frac{8}{7}\right)$ $19\frac{1}{4}$

Concept 2: Division of Mixed Numbers

For Exercises 33–48, divide the mixed numbers. Write the answer as a mixed number, proper fraction, or whole number. **(See Example 3.)**

33. $1\frac{7}{10} \div 2\frac{3}{4}$ $\frac{34}{55}$ **34.** $5\frac{1}{10} \div \frac{3}{4}$ $6\frac{4}{5}$ **35.** $5\frac{8}{9} \div 1\frac{1}{3}$ $4\frac{5}{12}$ **36.** $12\frac{4}{5} \div 2\frac{3}{5}$ $4\frac{12}{13}$

37. $2\frac{1}{2} \div 1\frac{1}{16}$ $2\frac{6}{17}$ **38.** $7\frac{3}{5} \div 1\frac{7}{12}$ $4\frac{4}{5}$  **39.** $4\frac{1}{2} \div 2\frac{1}{4}$ 2   **40.** $5\frac{5}{6} \div 2\frac{1}{3}$ $2\frac{1}{2}$

41. $0 \div 6\frac{7}{12}$ 0   **42.** $0 \div 1\frac{9}{11}$ 0   **43.** $2\frac{5}{6} \div \frac{1}{6}$ 17   **44.** $6\frac{1}{2} \div \frac{1}{2}$ 13

45. $1\frac{1}{3} \div \frac{2}{7}$ $4\frac{2}{3}$ **46.** $2\frac{1}{7} \div \frac{5}{13}$ $5\frac{4}{7}$ **47.** $3\frac{1}{2} \div 2$ $1\frac{3}{4}$ **48.** $4\frac{2}{3} \div 3$ $1\frac{5}{9}$

Concept 3: Applications of Multiplication and Division of Mixed Numbers

49. Tabitha charges $8 per hour for baby sitting. If she works for $4\frac{3}{4}$ hr, how much should she be paid? **(See Example 4.)**   Tabitha earned $38.

50. Kurt bought $2\frac{2}{3}$ acres of land. If land costs $10,500 per acre, how much will the land cost him?   The land will cost Kurt $28,000.

51. According to the U.S. Census Bureau's Valentine's Day Press Release, the average American consumes $25\frac{7}{10}$ lb of chocolate in a year. Over the course of 25 years, how many pounds of chocolate would the average American consume? $642\frac{1}{2}$ lb

52. Kayla has a bottle that contains 12 oz of medicine. She is instructed to take a dose of $\frac{3}{4}$ oz twice a day. How many doses will the 12-oz bottle provide?   Kayla will have 16 doses.

53. The age of a small kitten can be approximated by the following rule. The kitten's age is given as 1 week for every quarter pound of weight. **(See Example 5.)**

   a. Approximately how old is a $1\frac{3}{4}$-lb kitten?   7 weeks old

   b. Approximately how old is a $2\frac{1}{8}$-lb kitten? $8\frac{1}{2}$ weeks old

54. Richard's estate is to be split equally among his three children. If his estate is worth $1\frac{3}{4}$ million, how much will each child inherit?   Each child will inherit $\frac{7}{12}$ million.

55. Lucy earns $14 per hour and Ricky earns $10 per hour. Suppose Lucy worked $35\frac{1}{2}$ hr last week and Ricky worked $42\frac{1}{2}$ hr.

   a. Who earned more money and by how much?   Lucy earned $72 more than Ricky.

   b. How much did they earn altogether?   Together they earned $922.

56. A roll of wallpaper covers an area of 28 ft². If the roll is $1\frac{17}{24}$ ft wide, how long is the roll?   The roll is $16\frac{16}{41}$ ft long.

Writing   Translating Expression   Geometry   Scientific Calculator   Video   NS

Mixed Exercises

For Exercises 57–76, perform the indicated operation. Write the answer as a mixed number, proper fraction, or whole number.

57. $2\frac{1}{5} \div 1\frac{1}{10}$ 2

58. $3\frac{3}{4} \cdot 1\frac{5}{6}$ $6\frac{7}{8}$

59. $6 \div 1\frac{1}{8}$ $5\frac{1}{3}$

60. $8 \div 2\frac{1}{3}$ $3\frac{3}{7}$

61. $\frac{2}{3} \cdot 2\frac{7}{10}$ $1\frac{4}{5}$

62. $\frac{4}{3} \cdot 5\frac{1}{8}$ $6\frac{5}{6}$

63. $4\frac{1}{12} \cdot 0$ 0

64. $5\frac{1}{3} \cdot 6$ 32

65. $10\frac{1}{2} \div 9$ $1\frac{1}{6}$

66. $\frac{2}{7} \cdot 1\frac{8}{9}$ $\frac{34}{63}$

67. $0 \div 9\frac{2}{3}$ 0

68. $\frac{3}{8} \div 2\frac{1}{2}$ $\frac{3}{20}$

69. $12 \cdot \frac{1}{8}$ $1\frac{1}{2}$

70. $20 \cdot \frac{2}{15}$ $2\frac{2}{3}$

71. $6\frac{8}{9} \div 0$ Undefined

72. $0 \cdot 2\frac{1}{8}$ 0

73. $\left(3\frac{2}{5}\right)\left(\frac{7}{34}\right)\left(3\frac{3}{4}\right)$ $2\frac{5}{8}$

74. $\left(5\frac{1}{6}\right)\left(1\frac{4}{7}\right)\left(\frac{14}{33}\right)$ $3\frac{4}{9}$

75. $7\frac{1}{8} \div 1\frac{1}{3} \div 2\frac{1}{4}$ $2\frac{3}{8}$

76. $3\frac{1}{8} \div 5\frac{5}{7} \div 1\frac{5}{16}$ $\frac{5}{12}$

Expanding Your Skills

77. A landscaper will use a decorative concrete border around the cactus garden shown. Each concrete brick is $1\frac{1}{4}$ ft long and costs $3. What is the total cost of the border?   The total cost is $168.

78. Sara drives a total of $64\frac{1}{2}$ mi to work and back. Her car gets $21\frac{1}{2}$ miles per gallon (mpg) of gas. If gas is $5 per gallon, how much does it cost her to commute to and from work each day?   It costs Sara $15 each day.

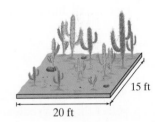

15 ft

20 ft

Calculator Connections

Topic: Multiplying and Dividing Fractions and Mixed Numbers

Expression	Keystrokes	Result
$\dfrac{8}{15} \cdot \dfrac{25}{28}$	8 $a^{b/c}$ 15 × 25 $a^{b/c}$ 28 =	$\boxed{10 \lrcorner 21} = \dfrac{10}{21}$
$2\dfrac{3}{4} \div \dfrac{1}{6}$	2 $a^{b/c}$ 3 $a^{b/c}$ 4 ÷ 1 $a^{b/c}$ 6 = This is how you enter the mixed number $2\frac{3}{4}$.	$\boxed{16 \_ 1 \lrcorner 2} = 16\dfrac{1}{2}$

To convert the result to an improper fraction, press 2nd d/c

$\boxed{33 \lrcorner 2} = \dfrac{33}{2}$

Calculator Exercises

For Exercises 79–86, use a calculator to multiply or divide and simplify. Write the answer as a mixed number.

79. $12\frac{2}{3} \cdot 25\frac{1}{8}$ $318\frac{1}{4}$

80. $38\frac{1}{3} \div 12\frac{1}{2}$ $3\frac{1}{15}$

81. $56\frac{5}{6} \div 3\frac{1}{6}$ $17\frac{18}{19}$

82. $25\frac{1}{5} \cdot 18\frac{1}{2}$ $466\frac{1}{5}$

83. $32\frac{7}{12} \div 12\frac{1}{6}$ $2\frac{99}{146}$

84. $106\frac{1}{9} \div 41\frac{5}{6}$ $2\frac{404}{753}$

85. $11\frac{1}{2} \cdot 41\frac{3}{4}$ $480\frac{1}{8}$

86. $9\frac{8}{9} \cdot 28\frac{1}{3}$ $280\frac{5}{27}$

 Writing   Translating Expression   Geometry   Scientific Calculator   Video   NS&E

Group Activity

Cooking for Company

Estimated time: 15 minutes

Group Size: 3

This recipe for chili serves 6 people.

CHILI

1 tablespoon oil

2 onions

$1\frac{1}{2}$ lb ground beef

14 oz canned tomatoes

$1\frac{1}{4}$ teaspoons chili powder

$\frac{3}{8}$ teaspoon cumin

$1\frac{1}{8}$ teaspoons salt

$\frac{1}{8}$ teaspoon cayenne pepper

4 whole cloves

15 oz canned kidney beans

1. Suppose that this recipe is to be used for a Super Bowl party for 30 people. Each student in the group should select three ingredients and determine the amount needed to make this recipe for 30 people. Then write the revised recipe.

   5 T oil; 10 onions; $7\frac{1}{2}$ lb beef; 70 oz tomatoes; $6\frac{1}{4}$ teaspoons chili power; $1\frac{7}{8}$ teaspoons cumin; $5\frac{5}{8}$ teaspoons salt; $\frac{5}{8}$ teaspoon cayenne pepper; 20 whole cloves, 75 oz beans

2. Suppose that this recipe is to be used for a dinner party for 9 guests. Each student in the group should select three ingredients and determine the amount needed to make this recipe for 9 people. Then write the revised recipe.

   $1\frac{1}{2}$ T oil; 3 onions; $2\frac{1}{4}$ lb beef; 21 oz tomatoes; $1\frac{7}{8}$ teaspoons chili power; $\frac{9}{16}$ teaspoon cumin; $1\frac{11}{16}$ teaspoons salt; $\frac{3}{16}$ teaspoon cayenne pepper; 6 whole cloves, $22\frac{1}{2}$ oz beans

Chapter 2   Summary

Section 2.1   Introduction to Fractions and Mixed Numbers

Key Concepts

A **fraction** represents a part of a whole unit. For example, $\frac{1}{3}$ represents one part of a whole unit that is divided into 3 equal pieces.

In the fraction $\frac{1}{3}$, the "top" number, 1, is the **numerator**, and the "bottom" number, 3, is the **denominator**.

A fraction in which the numerator is less than the denominator is called a **proper fraction**. A fraction in which the numerator is greater than or equal to the denominator is called an **improper fraction**.

An improper fraction can be written as a **mixed number** by dividing the numerator by the denominator. Write the quotient as a whole number, and write the remainder over the divisor.

A mixed number can be written as an improper fraction by multiplying the whole number by the denominator and adding the numerator. Then write that total over the denominator.

Fractions can be represented on a number line. For example,

$$\text{represents } \frac{3}{10}.$$

Examples

Example 1

$\frac{1}{3}$ of the pie is shaded.

Example 2

For the fraction $\frac{7}{9}$, the numerator is 7 and the denominator is 9.

Example 3

$\frac{5}{3}$ is an improper fraction, $\frac{3}{5}$ is a proper fraction, and $\frac{3}{3}$ is an improper fraction.

Example 4

$\frac{10}{3}$ can be written as $3\frac{1}{3}$ because

$$\begin{array}{r} 3 \\ 3\overline{)10} \\ -9 \\ \hline 1 \end{array}$$

Example 5

$2\frac{4}{5}$ can be written as $\frac{14}{5}$ because $\dfrac{2 \cdot 5 + 4}{5} = \dfrac{14}{5}$.

Example 6

Section 2.2   Prime Numbers and Factorization

Key Concepts

A **factorization** of a number is a product of factors that equals the number.

A number is divisible by another if their quotient leaves no remainder.

Divisibility Rules for 2, 3, 5, and 10

- A whole number is divisible by 2 if the ones-place digit is 0, 2, 4, 6, or 8.
- A whole number is divisible by 3 if the sum of the digits of the number is divisible by 3.
- A whole number is divisible by 5 if the ones-place digit is 0 or 5.
- A whole number is divisible by 10 if the ones-place digit is 0.

A **prime number** is a whole number greater than 1 that has exactly two factors, 1 and itself.

Composite numbers are whole numbers that have more than two factors. The numbers 0 and 1 are neither prime nor composite.

Prime Factorization

The **prime factorization** of a number is the factorization in which every factor is a prime number.

A factor of a number n is any number that divides evenly into n. For example, the factors of 80 are 1, 2, 4, 5, 8, 10, 16, 20, 40, and 80.

Examples

Example 1

$4 \cdot 4$ and $8 \cdot 2$ are two factorizations of 16.

Example 2

15 is divisible by 5 because 5 divides into 15 evenly.

Example 3

382 is divisible by 2.
640 is divisible by 2, 5, and 10.
735 is divisible by 3 and 5.

Example 4

9 is a composite number.
2 is a prime number.
1 is neither prime nor composite.

Example 5

$$2)\overline{756}$$
$$2)\overline{378}$$
$$3)\overline{189}$$
$$3)\overline{63}$$
$$3)\overline{21}$$
$$7$$

The prime factorization of 756 is

$$2 \cdot 2 \cdot 3 \cdot 3 \cdot 3 \cdot 7 \qquad \text{or} \qquad 2^2 \cdot 3^3 \cdot 7$$

Section 2.3   Simplifying Fractions to Lowest Terms

Key Concepts

Equivalent fractions are fractions that represent the same portion of a whole unit.

To determine if two fractions are equivalent, calculate the cross products. If the cross products are equal, then the fractions are equivalent.

To simplify fractions to **lowest terms**, use the fundamental principle of fractions:

Consider the fraction $\frac{a}{b}$ and the nonzero number c. Then

$$\frac{a \cdot c}{b \cdot c} = \frac{a}{b} \cdot \frac{c}{c} = \frac{a}{b} \cdot 1 = \frac{a}{b}$$

To simplify fractions with common powers of 10, "strike through" the common zeros first.

Examples

Example 1

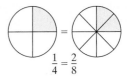

$$\frac{1}{4} = \frac{2}{8}$$

Example 2

a. Compare $\frac{5}{3}$ and $\frac{6}{4}$.

$$\frac{5}{3} \overset{?}{\times} \frac{6}{4}$$

$$20 \neq 18$$

Fractions are
not equivalent

b. Compare $\frac{4}{5}$ and $\frac{8}{10}$.

$$\frac{4}{5} \overset{?}{\times} \frac{8}{10}$$

$$40 = 40$$

Fractions are
equivalent

Example 3

$$\frac{25}{15} = \frac{5 \cdot 5}{3 \cdot 5} = \frac{5}{3} \cdot \frac{5}{5} = \frac{5}{3} \cdot 1 = \frac{5}{3}$$

Example 4

$$\frac{3{,}\cancel{000}}{12{,}\cancel{000}} = \frac{3}{12} = \frac{\overset{1}{\cancel{3}}}{\underset{4}{\cancel{12}}} = \frac{1}{4}$$

Section 2.4 · Multiplication of Fractions and Applications

Key Concepts

Multiplication of Fractions

To multiply fractions, write the product of the numerators over the product of the denominators. Then simplify the resulting fraction, if possible.

When multiplying a whole number and a fraction, first write the whole number as a fraction by writing the whole number over 1.

Powers of one-tenth can be expressed as a fraction with 1 in the numerator. The denominator has a 1 followed by the same number of zeros as the exponent of the base of $\frac{1}{10}$.

The formula for the area of a triangle is given by $A = \frac{1}{2}bh$.

Area is expressed in square units such as ft^2, in.^{2}, yd^2, m^2, and cm^2.

Examples

Example 1

$$\frac{4}{7} \times \frac{6}{5} = \frac{24}{35}$$

$$\left(\frac{15}{16}\right)\left(\frac{4}{5}\right) = \frac{\overset{3}{\cancel{15}}}{\underset{4}{\cancel{16}}} \cdot \frac{\overset{1}{\cancel{4}}}{\underset{1}{\cancel{5}}} = \frac{3}{4}$$

Example 2

$$8 \cdot \left(\frac{5}{6}\right) = \frac{8}{1} \cdot \frac{5}{\underset{3}{\cancel{6}}} = \frac{20}{3}$$

Example 3

$$\left(\frac{1}{10}\right)^4 = \frac{1}{10,000}$$

Example 4

The area of the triangle is

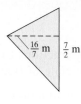

$$A = \frac{1}{2}bh$$

$$= \left(\frac{1}{2} \cdot \frac{7}{2}\right) \cdot \frac{16}{7}$$

$$= \frac{\overset{1}{\cancel{7}}}{\underset{1}{4}} \cdot \frac{\overset{4}{\cancel{16}}}{\underset{1}{\cancel{7}}}$$

$$= 4$$

The area is 4 m^2.

Section 2.5   Division of Fractions and Applications

Key Concepts

The **reciprocal** of $\frac{a}{b}$ is $\frac{b}{a}$ for $a, b \neq 0$. The product of a fraction and its reciprocal is 1. For example, $\frac{6}{11} \cdot \frac{11}{6} = 1$.

Dividing Fractions

To divide two fractions, multiply the dividend (the "first" fraction) by the reciprocal of the divisor (the "second" fraction).

When dividing by a whole number, first write the whole number as a fraction by writing the whole number over 1. Then multiply by its reciprocal.

When simplifying expressions with more than one operation, follow the order of operations.

Examples

Example 1

The reciprocal of $\frac{5}{8}$ is $\frac{8}{5}$.
The reciprocal of 4 is $\frac{1}{4}$.

The number 0 does not have a reciprocal because $\frac{1}{0}$ is undefined.

Example 2

$$\frac{18}{25} \div \frac{30}{35} = \frac{\overset{3}{\cancel{18}}}{\underset{5}{\cancel{25}}} \cdot \frac{\overset{7}{\cancel{35}}}{\underset{5}{\cancel{30}}} = \frac{21}{25}$$

Example 3

$$\frac{9}{8} \div 4 = \frac{9}{8} \div \frac{4}{1} = \frac{9}{8} \cdot \frac{1}{4} = \frac{9}{32}$$

Example 4

$$\frac{5}{42} \div \left[\left(\frac{1}{6}\right)^2 \cdot 3 \right] = \frac{5}{42} \div \left[\frac{1}{\underset{12}{\cancel{36}}} \cdot \frac{\overset{1}{\cancel{3}}}{1} \right]$$

$$= \frac{5}{42} \div \frac{1}{12}$$

$$= \frac{5}{\underset{7}{\cancel{42}}} \cdot \frac{\overset{2}{\cancel{12}}}{1} = \frac{10}{7}$$

Section 2.6   Multiplication and Division of Mixed Numbers

Key Concepts

Multiply Mixed Numbers

Step 1 Change each mixed number to an improper fraction.
Step 2 Multiply the improper fractions and reduce to lowest terms, if possible.

Divide Mixed Numbers

Step 1 Change each mixed number to an improper fraction.
Step 2 Divide the improper fractions and reduce to lowest terms, if possible. Recall that to divide fractions, multiply the dividend by the reciprocal of the divisor.

Examples

Example 1

$$4\frac{4}{5} \cdot 2\frac{1}{2} = \frac{\overset{12}{\cancel{24}}}{\underset{1}{\cancel{5}}} \cdot \frac{\overset{1}{\cancel{5}}}{\underset{1}{\cancel{2}}} = \frac{12}{1} = 12$$

Example 2

$$6\frac{2}{3} \div 2\frac{7}{9} = \frac{20}{3} \div \frac{25}{9} = \frac{\overset{4}{\cancel{20}}}{\underset{1}{\cancel{3}}} \cdot \frac{\overset{3}{\cancel{9}}}{\underset{5}{\cancel{25}}} = \frac{12}{5} = 2\frac{2}{5}$$

Chapter 2 Review Exercises

Section 2.1

For Exercises 1 and 2, write a fraction that represents the shaded area.

1.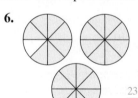
$\frac{1}{2}$

2.
$\frac{4}{7}$

3. **a.** Write a fraction that has denominator 3 and numerator 5. $\frac{5}{3}$

   b. Label this fraction as proper or improper.
   Improper

4. **a.** Write a fraction that has numerator 1 and denominator 6. $\frac{1}{6}$

   b. Label this fraction as proper or improper.
   Proper

5. In an office supply store, 7 of the 15 computers displayed are laptops. Write a fraction representing the computers that are laptops. $\frac{7}{15}$

For Exercises 6 and 7, write a fraction and a mixed number that represent the shaded area.

6.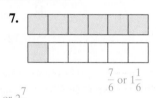

$\frac{23}{8}$ or $2\frac{7}{8}$

7.
$\frac{7}{6}$ or $1\frac{1}{6}$

For Exercises 8 and 9, convert the mixed number to a fraction.

8. $6\frac{1}{7}$ $\frac{43}{7}$

9. $11\frac{2}{5}$ $\frac{57}{5}$

10. How many fourths are in $4\frac{1}{4}$?   17

For Exercises 11 and 12, convert the improper fraction to a mixed number.

11. $\frac{47}{9}$ $5\frac{2}{9}$

12. $\frac{23}{21}$ $1\frac{2}{21}$

For Exercises 13–15, locate the numbers on the number line.

13. $\frac{10}{5}$ 14. $\frac{7}{8}$ 15. $\frac{13}{8}$

For Exercises 16 and 17, divide. Write the answer as a mixed number.

16. $7\overline{)941}$ $134\frac{3}{7}$

17. $26\overline{)1582}$ $60\frac{11}{13}$

Section 2.2

For Exercises 18–20, refer to this list of numbers:
21, 43, 51, 55, 58, 124, 140, 260, 1200.

18. List all the numbers that are divisible by 3.
21, 51, 1200

19. List all the numbers that are divisible by 5.
55, 140, 260, 1200

20. List all the numbers that are divisible by 2.
58, 124, 140, 260, 1200

For Exercises 21–24, determine if the number is prime, composite, or neither.

21. 61   Prime

22. 44   Composite

23. 1   Neither

24. 0   Neither

For Exercises 25–27, find the prime factorization.

25. 64
$2 \cdot 2 \cdot 2 \cdot 2 \cdot 2 \cdot 2$ or 2^6

26. 330 $2 \cdot 3 \cdot 5 \cdot 11$

27. 900 $2 \cdot 2 \cdot 3 \cdot 3 \cdot 5 \cdot 5$ or $2^2 \cdot 3^2 \cdot 5^2$

For Exercises 28 and 29, list all the factors of the number.

28. 48   1, 2, 3, 4, 6, 8, 12, 16, 24, 48

29. 80   1, 2, 4, 5, 8, 10, 16, 20, 40, 80

Section 2.3

For Exercises 30 and 31, determine if the fractions are equivalent. Fill in the blank with = or ≠ .

30. $\frac{3}{6} \square \frac{5}{9}$ ≠

31. $\frac{15}{21} \square \frac{10}{14}$ =

 Writing    Translating Expression   Geometry   Scientific Calculator   Video   NS E

For Exercises 32–39, simplify the fraction to lowest terms. Write the answer as a fraction.

32. $\dfrac{5}{20}$ $\quad \frac{1}{4}$ **33.** $\dfrac{14}{49}$ $\quad \frac{2}{7}$ **34.** $\dfrac{24}{16}$ $\quad \frac{3}{2}$

35. $\dfrac{63}{27}$ $\quad \frac{7}{3}$ **36.** $\dfrac{17}{17}$ $\quad 1$ **37.** $\dfrac{42}{21}$ $\quad 2$

38. $\dfrac{120}{150}$ $\quad \frac{4}{5}$ **39.** $\dfrac{1400}{2000}$ $\quad \frac{7}{10}$

40. On his final exam, Gareth got 42 out of 45 questions correct. What fraction of the test represents correct answers? What fraction represents incorrect answers? $\frac{14}{15}; \frac{1}{15}$

41. Isaac proofread 6 pages of his 10-page term paper. Yulisa proofread 6 pages of her 15-page term paper.

    a. What fraction of his paper did Isaac proofread? $\frac{3}{5}$

    b. What fraction of her paper did Yulisa proofread? $\frac{2}{5}$

Section 2.4

For Exercises 42–47, multiply the fractions and simplify to lowest terms. Write the answer as a fraction or whole number.

42. $\dfrac{3}{5} \times \dfrac{2}{7}$ $\quad \frac{6}{35}$ **43.** $\dfrac{4}{3} \times \dfrac{8}{3}$ $\quad \frac{32}{9}$ **44.** $14 \cdot \dfrac{9}{2}$ $\quad 63$

45. $33 \cdot \dfrac{5}{11}$ $\quad 15$ **46.** $\dfrac{2}{9} \cdot \dfrac{5}{8} \cdot \dfrac{36}{25}$ $\quad \frac{1}{5}$ **47.** $\dfrac{45}{7} \cdot \dfrac{6}{10} \cdot \dfrac{28}{63}$ $\quad \frac{12}{7}$

For Exercises 48–51, evaluate by using the order of operations.

48. $\left(\dfrac{1}{10}\right)^4$ $\quad \frac{1}{10,000}$ **49.** $\left(\dfrac{2}{5}\right)^2 \cdot \left(\dfrac{1}{10}\right)^2$ $\quad \frac{1}{625}$

50. $\left(\dfrac{3}{20} \cdot \dfrac{2}{3}\right)^3$ $\quad \frac{1}{1000}$ **51.** $\left(\dfrac{1}{10}\right)^3 \left(\dfrac{1000}{17}\right)$ $\quad \frac{1}{17}$

52. Write the formula for the area of a triangle.

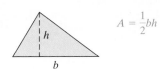

$A = \dfrac{1}{2}bh$

53. Write the formula for the area of a rectangle.

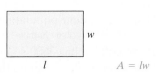

$A = lw$

For Exercises 54–56, find the area of the shaded region.

54.

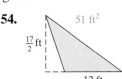

51 ft^2
$\frac{17}{2}$ ft
12 ft

55.

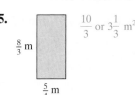

$\frac{8}{3}$ m
$\frac{5}{4}$ m
$\frac{10}{3}$ or $3\frac{1}{3}$ m^2

56.

6 yd
3 yd
$\frac{20}{3}$ yd   40 yd^2

57. Maximus wants to build a workbench for his garage. He needs four boards, each cut into $\frac{7}{8}$-yd pieces. How many yards of lumber does he require?
Maximus requires $\frac{7}{2}$ or $3\frac{1}{2}$ yd of lumber.

For Exercises 58–61, refer to the graph. The graph represents the distribution of the students at a college by race/ethnicity.

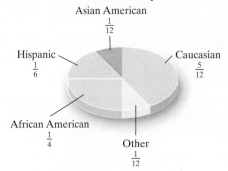

Distribution of Student Body by Race/Ethnicity

Asian American $\frac{1}{12}$
Caucasian $\frac{5}{12}$
Other $\frac{1}{12}$
African American $\frac{1}{4}$
Hispanic $\frac{1}{6}$

58. If the college has 3600 students, how many are African American?
There are 900 African American students.

59. If the college has 3600 students, how many are Asian American?
There are 300 Asian American students.

60. If the college has 3600 students, how many are Hispanic females (assume that one-half of the Hispanic student population is female)?
There are 300 Hispanic female students.

61. If the college has 3600 students, how many are Caucasian males (assume that one-half of the Caucasian student population is male)?
There are 750 Caucasian male students.

Section 2.5

For Exercises 62 and 63, multiply.

62. $\frac{3}{4} \cdot \frac{4}{3}$ 1

63. $\frac{1}{12} \cdot 12$ 1

For Exercises 64–67, find the reciprocal of the number, if it exists.

64. $\frac{7}{2}$ $\frac{2}{7}$

65. 7 $\frac{1}{7}$

66. 0
Reciprocal does not exist.

67. $\frac{1}{6}$ 6

68. Dividing by 5 is the same as multiplying by _____. $\frac{1}{5}$

69. Dividing by $\frac{2}{9}$ is the same as __multiplying__ by $\frac{9}{2}$.

For Exercises 70–75, divide and simplify the answer to lowest terms. Write the answer as a fraction or whole number.

70. $\frac{28}{15} \div \frac{21}{20}$ $\frac{16}{9}$

71. $\frac{7}{9} \div \frac{35}{63}$ $\frac{7}{5}$

72. $\frac{6}{7} \div 18$ $\frac{1}{21}$

73. $\frac{3}{10} \div \frac{9}{5}$ $\frac{1}{6}$

74. $\frac{200}{51} \div \frac{25}{17}$ $\frac{8}{3}$

75. $12 \div \frac{6}{7}$ 14

For Exercises 76–79, simplify by using the order of operations. Write the answer as a fraction.

76. $\left(\frac{2}{19} \div \frac{8}{19}\right)^3$ $\frac{1}{64}$

77. $\left(\frac{12}{5}\right)^2 \div \frac{36}{5}$ $\frac{4}{5}$

78. $\frac{81}{55} \div \frac{3}{11} \div \frac{3}{2}$ $\frac{18}{5}$

79. $\frac{4}{13} \cdot \left(\frac{1}{2}\right)^3 \div 2$ $\frac{1}{52}$

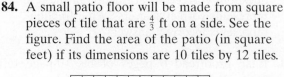 For Exercises 80 and 81, translate to a mathematical statement. Then simplify.

80. How much is $\frac{4}{5}$ of 20? $\frac{4}{5} \times 20$; 16

81. How many $\frac{2}{3}$'s are in 18? $18 \div \frac{2}{3}$; 27

82. How many $\frac{2}{3}$-lb bags of candy can be filled from a 24-lb sack of candy?   36 bags of candy

83. Amelia worked only $\frac{4}{5}$ of her normal 40-hr workweek. If she makes $18 per hour, how much money did she earn for the week?
Amelia earned $576.

84. A small patio floor will be made from square pieces of tile that are $\frac{4}{3}$ ft on a side. See the figure. Find the area of the patio (in square feet) if its dimensions are 10 tiles by 12 tiles.

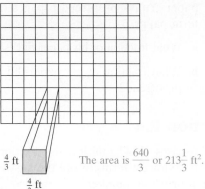

$\frac{4}{3}$ ft   The area is $\frac{640}{3}$ or $213\frac{1}{3}$ ft^2.

$\frac{4}{3}$ ft

85. Chuck is an elementary school teacher and needs 22 pieces of wood, $\frac{3}{8}$ ft long, for a class project. If he has a 9-ft board from which to cut the pieces, will he have enough $\frac{3}{8}$-ft pieces for his class? Explain.   Yes. $9 \div \frac{3}{8} = 24$ so he will have 24 pieces, which is more than enough for his class.

Section 2.6

For Exercises 86–96, multiply or divide as indicated.

86. $\left(3\frac{2}{3}\right)\left(6\frac{2}{5}\right)$ $23\frac{7}{15}$

87. $\left(11\frac{1}{3}\right)\left(2\frac{3}{34}\right)$ $23\frac{2}{3}$

88. $6\frac{1}{2} \cdot 1\frac{3}{13}$ 8

89. $4 \cdot \left(5\frac{5}{8}\right)$ $22\frac{1}{2}$

90. $45\frac{5}{13} \cdot 0$ 0

91. $4\frac{5}{16} \div 2\frac{7}{8}$ $1\frac{1}{2}$

92. $3\frac{5}{11} \div 3\frac{4}{5}$ $\frac{10}{11}$

93. $7 \div 1\frac{5}{9}$ $4\frac{1}{2}$

94. $4\frac{6}{11} \div 2$ $2\frac{3}{11}$

95. $10\frac{1}{5} \div 17$ $\frac{3}{5}$

96. $0 \div 3\frac{5}{12}$ 0

97. It takes $1\frac{1}{4}$ gal of paint for Neva to paint her living room. If her great room (including the dining area) is $2\frac{1}{2}$ times larger than the living room, how many gallons will it take to paint that room? It will take $3\frac{1}{8}$ gal.

98. A roll of ribbon contains $12\frac{1}{2}$ yd. How many pieces of length $1\frac{1}{4}$ yd can be cut from this roll? There will be 10 pieces.

Chapter 2   Test

1. a. Write a fraction that represents the shaded portion of the figure.

 $\frac{5}{8}$

b. Is the fraction proper or improper?   Proper

2. a. Write a fraction that represents the total shaded portion of the three figures.

 $\frac{7}{3}$

b. Is the fraction proper or improper?   Improper

3. Write an improper fraction and a mixed number that represent the shaded region.

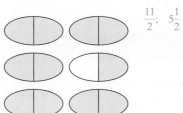 $\frac{11}{2}$; $5\frac{1}{2}$

4. Is the fraction $\frac{7}{7}$ a proper or an improper fraction? Explain. $\frac{7}{7}$ is an improper fraction because the numerator is greater than or equal to the denominator.

5. a. Write $\frac{11}{3}$ as a mixed number. $3\frac{2}{3}$

b. Write $3\frac{7}{9}$ as an improper fraction. $\frac{34}{9}$

For Exercises 6–9, plot the fraction on the number line.

6. $\frac{1}{2}$

7. $\frac{3}{4}$

8. $\frac{7}{12}$

9. $\frac{13}{5}$

10. Label the following numbers as prime, composite, or neither.

a. 15   Composite
b. 0   Neither
c. 53   Prime
d. 1   Neither
e. 29   Prime
f. 39   Composite

11. a. List all the factors of 45.   1, 3, 5, 9, 15, 45

b. Write the prime factorization of 45.
$3 \cdot 3 \cdot 5$ or $3^2 \cdot 5$

12. a. What is the divisibility rule for 3?
Add the digits of the number. If the sum is divisible by 3, then
b. Is 1,981,011 divisible by 3?   the original number
Yes.   is divisible by 3.

13. Determine whether 1155 is divisible by

a. 2   No
b. 3   Yes
c. 5   Yes
d. 10   No

For Exercises 14 and 15, determine if the fractions are equivalent. Then fill in the blank with either = or ≠ .

14. $\frac{15}{12}$ ☐ $\frac{5}{4}$ $=$ **15.** $\frac{2}{5}$ ☐ $\frac{4}{25}$ ≠

For Exercises 16 and 17, simplify the fractions to lowest terms.

16. $\frac{150}{105}$ $\frac{10}{7}$ or $1\frac{3}{7}$ **17.** $\frac{1,200,000}{1,400,000}$ $\frac{6}{7}$

18. Christine and Brad are putting their photographs in scrapbooks. Christine has placed 15 of her 25 photos and Brad has placed 16 of his 20 photos.

    a. What fractional part of the total photos has each person placed? Christine: $\frac{3}{5}$; Brad: $\frac{4}{5}$

    b. Which person has a greater fractional part completed? Brad has the greater fractional part completed.

For Exercises 19–26, multiply or divide as indicated. Simplify the fraction to lowest terms.

19. $\frac{2}{9} \times \frac{57}{46}$ $\frac{19}{69}$ **20.** $\left(\frac{75}{24}\right) \cdot 4$ $\frac{25}{2}$ or $12\frac{1}{2}$

21. $\frac{28}{24} \div \frac{21}{8}$ $\frac{4}{9}$ **22.** $\frac{105}{42} \div 5$ $\frac{1}{2}$

23. $\frac{2}{18} \times \frac{9}{25} \times \frac{40}{6}$ $\frac{4}{15}$ **24.** $\frac{600}{1200} \div \frac{50}{65} \div \frac{13}{15}$ $\frac{3}{4}$

25. $\frac{10}{21} \div 4\frac{1}{6}$ $\frac{4}{35}$ **26.** $4\frac{4}{17} \cdot 2\frac{4}{15}$ $9\frac{3}{5}$

27. Perform the order of operations. Simplify the fraction to lowest terms.

$$\frac{52}{72} \div \left[\left(\frac{1}{2}\right)^2 \cdot \frac{8}{3}\right] \quad \frac{13}{12}$$

 28. Find the area of the triangle.

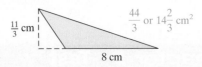

29. Which is greater, $20 \cdot \frac{1}{4}$ or $20 \div \frac{1}{4}$? $20 \div \frac{1}{4}$

30. How many "quarter-pounders" can be made from 12 lb of ground beef? 48 quarter-pounders

31. The Humane Society has 120 dogs. Of the 120, $\frac{5}{8}$ are female. Among the female dogs, $\frac{1}{15}$ are pure breeds. How many of the dogs are female and pure breeds? 5 dogs are female pure breeds.

32. A zoning requirement indicates that a house built on less than 1 acre of land may take up no more than one-half of the land. If Liz and George purchased a $\frac{4}{5}$-acre lot of land, what is the maximum land area that they can use to build the house?

They can build on a maximum of $\frac{2}{5}$ acre.

Chapters 1–2   Cumulative Review Exercises

1. Fill in the table with either the word name for the number or the number in standard form.

Mountain	Height (ft)	
	Standard Form	**Words**
Mt. Foraker (Alaska)	17,400	Seventeen thousand, four hundred
Mt. Kilimanjaro (Tanzania)	19,340	Nineteen thousand, three hundred forty
El Libertador (Argentina)	22,047	Twenty-two thousand, forty-seven
Mont Blanc (France-Italy)	15,771	Fifteen thousand, seven hundred seventy-one

For Exercises 2–13, perform the indicated operation.

2. $432 + 998$ 1430   **3.** $572 - 433$ 139

4. 4122×52 214,344   **5.** $384 \div 16$ 24

6. $23(81)$ 1863   **7.** $4\overline{)74}$ 18 R2

8. $\begin{array}{r} 3,000,000 \\ \times\ 40,000 \\ \hline 120,000,000,000 \end{array}$ **9.** $\begin{array}{r} 1007 \\ -\ 823 \\ \hline 184 \end{array}$

10. $\frac{48}{8}$ 6   **11.** $6 + 2 \cdot 8$ 22

12. $5^2 - 3^2$ 16   **13.** $(5 - 3)^2$ 4

For Exercises 14–18, match the algebraic statement with the property that it demonstrates.

14. $5 \cdot 8 = 8 \cdot 5$ d   **a.** Commutative property of addition

15. $4(3 + 2)$ c   **b.** Associative property of addition
    $= 4 \cdot 3 + 4 \cdot 2$

16. $(12 + 3) + 5$ b   **c.** Distributive property of multiplication over addition
    $= 12 + (3 + 5)$

17. $8 \cdot (7 \cdot 2)$ e   **d.** Commutative property of multiplication
    $= (8 \cdot 7) \cdot 2$

18. $32 + 9$ a   **e.** Associative property of multiplication
    $= 9 + 32$

19. Write a fraction that represents the shaded area.

a.
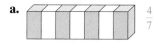
$\frac{4}{7}$

b.

$\frac{7}{3}$ or $2\frac{1}{3}$

20. Identify each fraction as proper or improper.

a. $\frac{7}{8}$ **b.** $\frac{8}{7}$ **c.** $\frac{8}{8}$ Improper
    Proper   Improper

21. a. List all the factors of 30.   1, 2, 3, 5, 6, 10, 15, 30

b. Write the prime factorization of 30. $2 \cdot 3 \cdot 5$

22. Simplify the fractions to lowest terms.

a. $\frac{144}{84}$ $\frac{12}{7}$ **b.** $\frac{60,000}{150,000}$ $\frac{2}{5}$

23. Multiply and simplify to lowest terms. $\frac{35}{27} \cdot \frac{51}{95}$ $\frac{119}{171}$

24. Divide and simplify to lowest terms. $5\frac{2}{3} \div 6\frac{4}{5}$ $\frac{5}{6}$

25. Is multiplication of fractions a commutative operation? Explain, using the fractions $\frac{8}{13}$ and $\frac{5}{16}$.
Yes. $\frac{8}{13} \cdot \frac{5}{16} = \frac{5}{26}$ and $\frac{5}{16} \cdot \frac{8}{13} = \frac{5}{26}$

26. Is multiplication of fractions an associative operation? Explain, using the fractions $\frac{1}{2}, \frac{2}{9}$, and $\frac{5}{3}$.
Yes. $\left(\frac{1}{2} \cdot \frac{2}{9}\right) \cdot \frac{5}{3} = \frac{1}{9} \cdot \frac{5}{3} = \frac{5}{27}$ and $\frac{1}{2} \cdot \left(\frac{2}{9} \cdot \frac{5}{3}\right) = \frac{1}{2} \cdot \frac{10}{27} = \frac{5}{27}$

27. Simplify. $\left(\frac{5}{6} \cdot \frac{12}{25}\right)^2 \div \frac{2}{3}$ $\frac{6}{25}$

28. Find the area of the rectangle. $\frac{11}{9}$ or $1\frac{2}{9}$ m²
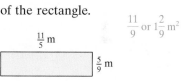
$\frac{11}{5}$ m
$\frac{5}{9}$ m

29. Find the area of the triangle.   50 ft²
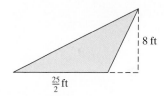
8 ft
$\frac{25}{2}$ ft

30. At one college $\frac{3}{4}$ of the students are male, and of the males, $\frac{1}{10}$ are from out of state. What fraction of the students are males who are from out of state?
$\frac{3}{40}$ of the students are males from out of state.

 Writing   ← → Translating Expression   Geometry   Scientific Calculator   Video   NS/E

Fractions and Mixed Numbers: Addition and Subtraction

3

Chapter 3

In this chapter, we learn about addition and subtraction of fractions as well as operations on mixed numbers.

Review Your Skills

As we continue in this chapter, review some of the terms and vocabulary relating to fractions. Try the crossword puzzle to review these terms.

Across

2. The top number of a fraction.

5. To divide two fractions, multiply the first fraction by the _____ of the second fraction.

6. A fraction whose numerator is greater than or equal to the denominator is called _____.

Down

1. The bottom number of a fraction.

3. A fraction whose numerator is less than the denominator is called _____.

4. A whole number greater than 1 that has only 1 and itself as factors is called a _____ number.

Section 3.1   Addition and Subtraction of Like Fractions

1. Addition and Subtraction of Like Fractions

In Chapter 2 we learned how to multiply and divide fractions. The main focus of this chapter is to add and subtract fractions. The operation of addition can be thought of as combining like groups of objects. For example:

$$3 \text{ apples} + 1 \text{ apple} = 4 \text{ apples}$$

$$\text{three-fifths} + \text{one-fifth} = \text{four-fifths}$$

$$\frac{3}{5} + \frac{1}{5} = \frac{4}{5}$$

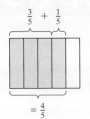

The fractions $\frac{3}{5}$ and $\frac{1}{5}$ are said to be **like fractions** because their denominators are the same. That is, the fractions have a **common denominator**. In general, two or more like fractions may be added or subtracted according to the following procedure.

> **Adding and Subtracting Like Fractions**
>
> **Step 1**   Add or subtract the numerators.
> **Step 2**   Write the sum or difference over the common denominator.
> **Step 3**   Simplify the fraction to lowest terms, if possible.

Skill Practice

Add. Write the answer as a fraction or whole number.

1. $\dfrac{2}{9} + \dfrac{4}{9}$

2. $\dfrac{7}{12} + \dfrac{5}{12} + \dfrac{11}{12}$

Classroom Examples: p. 168, Exercises 12 and 22

Instructor Note: Show students that when adding whole numbers we also follow this procedure. $7 + 8 = \frac{7}{1} + \frac{8}{1} = \frac{15}{1}$, *not* $\frac{15}{2}$.

Example 1   **Adding Like Fractions**

Add. Write the answer as a fraction or whole number.

a. $\dfrac{1}{4} + \dfrac{5}{4}$ **b.** $\dfrac{7}{15} + \dfrac{2}{15} + \dfrac{1}{15}$

Solution:

a. $\dfrac{1}{4} + \dfrac{5}{4} = \dfrac{1+5}{4}$ Add the numerators.

$= \dfrac{6}{4}$ Write the sum over the common denominator.

$= \dfrac{\overset{3}{\cancel{6}}}{\underset{2}{\cancel{4}}}$ Simplify to lowest terms.

$= \dfrac{3}{2}$

> **Avoiding Mistakes**
>
> Notice that when adding fractions, we do not add the denominators. We add *only* the numerators.

b. $\dfrac{7}{15} + \dfrac{2}{15} + \dfrac{1}{15} = \dfrac{7+2+1}{15}$ Add the numerators.

$= \dfrac{10}{15}$ Write the sum over the common denominator.

$= \dfrac{\overset{2}{\cancel{10}}}{\underset{3}{\cancel{15}}}$ Simplify to lowest terms.

$= \dfrac{2}{3}$

Answers

1. $\dfrac{2}{3}$ 2. $\dfrac{23}{12}$

Avoiding Mistakes

Note that the process to add fractions is different from the process to multiply fractions.

$$\frac{2}{7} \times \frac{3}{7} = \frac{6}{49} \quad \text{but} \quad \frac{2}{7} + \frac{3}{7} = \frac{5}{7}$$

Concept Connections

3. Which is the correct sum for $\frac{2}{3} + \frac{5}{3}$?

$$\frac{7}{6} \quad \text{or} \quad \frac{7}{3}$$

4. Which is the correct product for $\frac{2}{3} \cdot \frac{5}{3}$?

$$\frac{10}{9} \quad \text{or} \quad \frac{10}{3}$$

Example 2   **Subtracting Like Fractions**

Subtract. Write the answer as a fraction or whole number.

a. $\dfrac{13}{9} - \dfrac{2}{9}$ **b.** $\dfrac{4}{3} - \dfrac{1}{3}$

Solution:

a. $\dfrac{13}{9} - \dfrac{2}{9} = \dfrac{13 - 2}{9}$ Subtract the numerators.

$= \dfrac{11}{9}$ Write the difference over the common denominator. The fraction is already in lowest terms because 11 and 9 share no common factors.

b. $\dfrac{4}{3} - \dfrac{1}{3} = \dfrac{4 - 1}{3}$ Subtract the numerators.

$= \dfrac{3}{3}$ Write the difference over the common denominator.

$= 1$ Simplify.

Skill Practice

Subtract. Write the answer as a fraction or whole number.

5. $\dfrac{14}{11} - \dfrac{8}{11}$ **6.** $\dfrac{5}{2} - \dfrac{3}{2}$

Classroom Example: p. 168, Exercise 38

2. Order of Operations

Example 3 reviews the order of operations.

Order of Operations

Step 1   Perform all operations inside parentheses first.

Step 2   Simplify expressions containing exponents or square roots.

Step 3   Perform multiplication or division in the order that they appear from left to right.

Step 4   Perform addition or subtraction in the order that they appear from left to right.

Answers

3. $\dfrac{7}{3}$ **4.** $\dfrac{10}{9}$ **5.** $\dfrac{6}{11}$ **6.** 1

Example 3 **Applying the Order of Operations**

Simplify.

a. $\left(\dfrac{2}{7} + \dfrac{1}{7}\right)^2$ **b.** $\dfrac{2}{11} + \dfrac{1}{2} \div \dfrac{11}{6}$

Solution:

a. $\left(\dfrac{2}{7} + \dfrac{1}{7}\right)^2 = \left(\dfrac{2 + 1}{7}\right)^2$ Add fractions within parentheses first.

$\qquad = \left(\dfrac{3}{7}\right)^2$

$\qquad = \dfrac{3}{7} \cdot \dfrac{3}{7}$ Square the fraction $\dfrac{3}{7}$.

$\qquad = \dfrac{9}{49}$ The fraction is in lowest terms.

b. $\dfrac{2}{11} + \dfrac{1}{2} \div \dfrac{11}{6} = \dfrac{2}{11} + \left(\dfrac{1}{2} \div \dfrac{11}{6}\right)$ We can insert parentheses to emphasize that division is performed before addition.

$\qquad = \dfrac{2}{11} + \left(\dfrac{1}{2} \cdot \dfrac{6}{11}\right)$ Multiply by the reciprocal of the divisor.

$\qquad = \dfrac{2}{11} + \left(\dfrac{1}{2} \cdot \dfrac{\overset{3}{\cancel{6}}}{11}\right)$ Multiply the fractions.

$\qquad = \dfrac{2}{11} + \left(\dfrac{3}{11}\right)$

$\qquad = \dfrac{5}{11}$ Add the fractions.

3. Applications of Addition and Subtraction of Fractions

Recall that the perimeter of a polygon is found by adding the lengths of the sides.

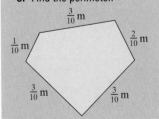
Example 4 **Finding Perimeter**

Find the perimeter.

Solution:

$$\text{Perimeter} = \dfrac{10}{12} + \dfrac{5}{12} + \dfrac{15}{12} + \dfrac{5}{12}$$

$$= \dfrac{10 + 5 + 15 + 5}{12}$$

$$= \dfrac{35}{12} \quad \text{or} \quad 2\dfrac{11}{12}$$

The perimeter is $2\frac{11}{12}$ yd.

| **Example 5** | **Applying Addition and Subtraction of Fractions** |

Pam mixed $\frac{12}{16}$ gal of water with $\frac{1}{16}$ gal of liquid fertilizer. Then she used $\frac{7}{16}$ gal of the mixture to fertilize her blueberry bushes. How much mixture was left over?

Solution:

The net amount of liquid remaining is given by

$$\overbrace{\frac{12}{16}}^{\text{Amount Pam started with}} + \frac{1}{16} - \overbrace{\frac{7}{16}}^{\text{Amount Pam used}}$$

$$= \frac{12 + 1 - 7}{16}$$

$$= \frac{\overset{3}{\cancel{6}}}{\underset{8}{\cancel{16}}}$$

$$= \frac{3}{8}$$

There was $\frac{3}{8}$ gal of mixture left over.

Skill Practice

10. Jamie mixed $\frac{5}{8}$ gal of green paint with $\frac{7}{8}$ gal of white paint. Then she used $\frac{3}{8}$ gal of the mixture to paint a mural. How much paint is left over?

Classroom Example: p. 169, Exercise 74

Answer

10. $\frac{9}{8}$ or $1\frac{1}{8}$ gal

Section 3.1   Practice Exercises

Study Skills Exercise

For additional exercises, see Classroom Activities 3.1A–3.1C in the *Student's Resource Manual* at www.mhhe.com/moh.

> How can you use the Concept Connections and Skill Practice exercises in the margins of this text?

Vocabulary and Key Concepts

1. Two fractions are _____ fractions if they have the same denominator.   like

Concept 1: Addition and Subtraction of Like Fractions

For Exercises 2–7, add the like units.

2. 9 cm + 11 cm   20 cm

3. 3 ft + 5 ft   8 ft

4. 7 chairs + 2 chairs   9 chairs

5. 7 m + 13 m   20 m

6. 8 thirds + 2 thirds   10 thirds

7. 1 fourth + 6 fourths   7 fourths

For Exercises 8 and 9, shade in the portion of the third figure that represents the addition of the first two figures.

8.
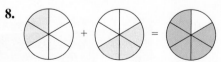

9.

10. Explain the difference between evaluating the two expressions $\frac{2}{5} \times \frac{7}{5}$ and $\frac{2}{5} + \frac{7}{5}$. $\frac{2}{5} \times \frac{7}{5}$ is multiplication in which we multiply the numerators and multiply the denominators: $\frac{2}{5} \times \frac{7}{5} = \frac{14}{25}$. The expression $\frac{2}{5} + \frac{7}{5}$ is addition in which we add the numerators but keep the denominator: $\frac{2}{5} + \frac{7}{5} = \frac{9}{5}$.

Writing   ←→ Translating Expression   Geometry   Scientific Calculator   Video   NS E

For Exercises 11–22, add the like fractions. Write the answer as a fraction in lowest terms or as a whole number. **(See Example 1.)**

11. $\dfrac{6}{11} + \dfrac{7}{11}$ $\quad \frac{13}{11}$

12. $\dfrac{5}{3} + \dfrac{2}{3}$ $\quad \frac{7}{3}$

13. $\dfrac{6}{5} + \dfrac{3}{5}$ $\quad \frac{9}{5}$

14. $\dfrac{3}{10} + \dfrac{4}{10}$ $\quad \frac{7}{10}$

15. $\dfrac{1}{4} + \dfrac{3}{4}$ $\quad 1$

16. $\dfrac{1}{8} + \dfrac{3}{8}$ $\quad \frac{1}{2}$

17. $\dfrac{2}{9} + \dfrac{4}{9}$ $\quad \frac{2}{3}$

18. $\dfrac{3}{2} + \dfrac{5}{2}$ $\quad 4$

19. $\dfrac{3}{20} + \dfrac{8}{20} + \dfrac{15}{20}$ $\quad \frac{13}{10}$

20. $\dfrac{5}{8} + \dfrac{4}{8} + \dfrac{9}{8}$ $\quad \frac{9}{4}$

 21. $\dfrac{18}{14} + \dfrac{11}{14} + \dfrac{6}{14}$ $\quad \frac{5}{2}$

22. $\dfrac{7}{18} + \dfrac{22}{18} + \dfrac{10}{18}$ $\quad \frac{13}{6}$

23. Bethany pours $\frac{1}{4}$ cup of bleach into a container and then adds $\frac{9}{4}$ cups of water. How many cups of bleach and water mixture does she have?
Bethany has $\frac{5}{2}$ or $2\frac{1}{2}$ cups of bleach and water mixture.

24. Austin rode his bike $\frac{7}{6}$ mi before he got a flat tire. He then had to walk another $\frac{1}{6}$ mi. How far did Austin travel?
Austin traveled a total of $\frac{4}{3}$ or $1\frac{1}{3}$ mi.

For Exercises 25–28, subtract the like units.

25. 15 baskets − 4 baskets
11 baskets

26. 52 cards − 13 cards
39 cards

27. 7 fifths − 1 fifth
6 fifths

28. 18 tenths − 11 tenths
7 tenths

For Exercises 29 and 30, shade in the portion of the third figure that represents the subtraction of the first two figures.

29.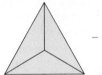

30.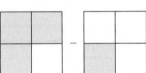

For Exercises 31–42, subtract the like fractions. Write the answer as a fraction or whole number. **(See Example 2.)**

31. $\dfrac{9}{8} - \dfrac{6}{8}$ $\quad \frac{3}{8}$

32. $\dfrac{7}{9} - \dfrac{6}{9}$ $\quad \frac{1}{9}$

33. $\dfrac{9}{2} - \dfrac{6}{2}$ $\quad \frac{3}{2}$

34. $\dfrac{10}{4} - \dfrac{5}{4}$ $\quad \frac{5}{4}$

35. $\dfrac{13}{3} - \dfrac{7}{3}$ $\quad 2$

36. $\dfrac{13}{10} - \dfrac{3}{10}$ $\quad 1$

 37. $\dfrac{23}{12} - \dfrac{15}{12}$ $\quad \frac{2}{3}$

38. $\dfrac{13}{6} - \dfrac{5}{6}$ $\quad \frac{4}{3}$

39. $\dfrac{28}{25} - \dfrac{14}{25} - \dfrac{4}{25}$ $\quad \frac{2}{5}$

40. $\dfrac{34}{15} - \dfrac{6}{15} - \dfrac{3}{15}$ $\quad \frac{5}{3}$

41. $\dfrac{10}{16} - \dfrac{1}{16} - \dfrac{5}{16}$ $\quad \frac{1}{4}$

42. $\dfrac{31}{40} - \dfrac{14}{40} - \dfrac{12}{40}$ $\quad \frac{1}{8}$

43. A chemist has $\frac{5}{8}$ gram (g) of NaCl (salt). If he uses $\frac{3}{8}$ g, how much is left?
$\frac{1}{4}$ g is left.

44. Jason bought $\frac{11}{4}$ acres of land and then sold $\frac{3}{4}$ acre. How much land does he have left? Jason has 2 acres left.

Mixed Exercises

For Exercises 45–56, add or subtract as indicated. Write the answer as a fraction or whole number.

45. $\dfrac{7}{8} + \dfrac{5}{8}$ $\dfrac{3}{2}$

46. $\dfrac{1}{21} + \dfrac{13}{21}$ $\dfrac{2}{3}$

47. $\dfrac{14}{5} - \dfrac{2}{5}$ $\dfrac{12}{5}$

48. $\dfrac{5}{3} - \dfrac{2}{3}$ 1

49. $\dfrac{6}{13} + \dfrac{7}{13}$ 1

50. $\dfrac{20}{35} + \dfrac{12}{35}$ $\dfrac{32}{35}$

51. $\dfrac{14}{15} + \dfrac{2}{15} - \dfrac{4}{15}$ $\dfrac{4}{5}$

52. $\dfrac{19}{6} - \dfrac{11}{6} + \dfrac{5}{6}$ $\dfrac{13}{6}$

53. $\dfrac{7}{2} - \dfrac{3}{2} + \dfrac{1}{2}$ $\dfrac{5}{2}$

54. $\dfrac{8}{3} + \dfrac{2}{3} - \dfrac{1}{3}$ 3

55. $\dfrac{19}{12} - \dfrac{5}{12} + \dfrac{7}{12}$ $\dfrac{7}{4}$

56. $\dfrac{7}{18} + \dfrac{13}{18} - \dfrac{5}{18}$ $\dfrac{5}{6}$

Concept 2: Order of Operations

For Exercises 57–68, simplify the expression by using the order of operations. Write the answer as a fraction or whole number. **(See Example 3.)**

57. $\left(\dfrac{11}{10} - \dfrac{2}{10}\right)^2$ $\dfrac{81}{100}$

58. $\left(\dfrac{7}{3} - \dfrac{5}{3}\right)^3$ $\dfrac{8}{27}$

 59. $\dfrac{5}{4} \div \dfrac{3}{2} + \dfrac{5}{6}$ $\dfrac{5}{3}$

60. $\dfrac{1}{7} \div \dfrac{2}{21} + \dfrac{5}{2}$ 4

61. $\dfrac{6}{5} + \dfrac{7}{5} - \dfrac{4}{5}$ $\dfrac{9}{5}$

62. $\dfrac{10}{3} - \dfrac{2}{3} + \dfrac{5}{3}$ $\dfrac{13}{3}$

63. $\dfrac{3}{7} + \dfrac{13}{14} \cdot 2$ $\dfrac{16}{7}$

64. $\dfrac{13}{6} - \dfrac{5}{18} \cdot 3$ $\dfrac{4}{3}$

65. $\left(\dfrac{2}{21} + \dfrac{11}{21}\right) \div \dfrac{1}{7}$ $\dfrac{13}{3}$

66. $\left(\dfrac{17}{30} - \dfrac{12}{30}\right) \div \dfrac{5}{6}$ $\dfrac{1}{5}$

67. $\dfrac{17}{30} - \dfrac{1}{2} \cdot \dfrac{7}{15}$ $\dfrac{1}{3}$

68. $\dfrac{5}{12} - \dfrac{1}{2} \cdot \dfrac{1}{6}$ $\dfrac{1}{3}$

Concept 3: Applications of Addition and Subtraction of Fractions

For Exercises 69 and 70, find the perimeter. **(See Example 4.)**

69.

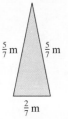

$\dfrac{12}{7}$ or $1\dfrac{5}{7}$ m

$\dfrac{5}{7}$ m $\dfrac{5}{7}$ m

$\dfrac{2}{7}$ m

70.

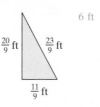

6 ft

$\dfrac{20}{9}$ ft $\dfrac{23}{9}$ ft

$\dfrac{11}{9}$ ft

71. Find the perimeter of the stamp.

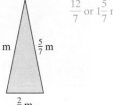

$\dfrac{13}{16}$ in.

$\dfrac{7}{2}$ or $3\dfrac{1}{2}$ in.

$\dfrac{15}{16}$ in.

72. Find the perimeter of the top of the table.

$\dfrac{8}{3}$ yd   8 yd

$\dfrac{4}{3}$ yd

73. Gabby mixed $\frac{1}{10}$ gal of red paint with $\frac{7}{10}$ gal of white paint. She used $\frac{3}{10}$ gal of the mixture to paint a room. How much mixture was left over? **(See Example 5.)**

There was $\dfrac{1}{2}$ gal left over.

74. Emeril mixed $\frac{3}{8}$ cup balsamic vinegar with $\frac{4}{8}$ cup oil to make a salad dressing. Then he used $\frac{1}{8}$ cup of the mixture for a large salad. How much oil and vinegar mixture was left over?

$\dfrac{3}{4}$ cup was left over.

Writing   Translating Expression   Geometry   Scientific Calculator   Video   NS E

75. A chemist mixed $\frac{5}{8}$ liter (L) of water with $\frac{7}{8}$ L of alcohol. Then he used one-quarter of the mixture in an experiment. How much mixture did he use?
He used $\frac{3}{8}$ L.

76. Malcom planted tomatoes in $\frac{2}{7}$ of his garden. He planted cucumbers in $\frac{3}{7}$ of the garden and cabbage in $\frac{1}{7}$. The remaining $\frac{1}{7}$ still has not yet been planted. A deer came in one night and ate $\frac{1}{3}$ of the plants in the planted area. What fraction of the garden did the deer eat? The deer ate $\frac{2}{7}$ of the garden.

 77. Thilan has taken up a new exercise program. He walks 6 days per week. One week he walked the distances given in the table.

   a. Find the total distance he walked for the week.
   Thilan walked $5\frac{1}{2}$ mi total.

   b. Find the average distance walked per day.
   He walked an average of $\frac{11}{12}$ mi per day.

Day	Distance
Monday	$\frac{4}{10}$ mi
Tuesday	$\frac{7}{10}$ mi
Wednesday	$\frac{9}{10}$ mi
Thursday	$\frac{5}{10}$ mi
Friday	$\frac{13}{10}$ mi
Saturday	$\frac{17}{10}$ mi

78. Denzel recorded the weekly rainfall for his town for 4 weeks of summer.

   a. Find the total amount of rainfall for this 4-week period.
   The total amount of rain is $3\frac{1}{2}$ in.

   b. Find the average rainfall per week.
   The average is $\frac{7}{8}$ in. per week.

Week	Amount of Rainfall
1	$\frac{2}{10}$ in.
2	$\frac{7}{10}$ in.
3	$\frac{9}{10}$ in.
4	$\frac{17}{10}$ in.

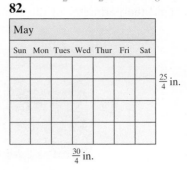

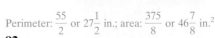 For Exercises 79–82, find the perimeter and the area.
Perimeter: $\frac{55}{2}$ or $27\frac{1}{2}$ in.; area: $\frac{375}{8}$ or $46\frac{7}{8}$ in.^{2}

79.

$\frac{3}{8}$ ft
$\frac{5}{8}$ ft

Perimeter: 2 ft; area: $\frac{15}{64}$ ft^2

80.
$\frac{7}{8}$ m
$\frac{15}{8}$ m

Perimeter: $\frac{11}{2}$ or $5\frac{1}{2}$ m;
area: $\frac{105}{64}$ or $1\frac{41}{64}$ m^2

81.

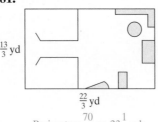
$\frac{13}{3}$ yd
$\frac{22}{3}$ yd

Perimeter: $\frac{70}{3}$ or $23\frac{1}{3}$ yd;
area: $\frac{286}{9}$ or $31\frac{7}{9}$ yd^2

82.

May						
Sun	Mon	Tues	Wed	Thur	Fri	Sat

$\frac{25}{4}$ in.
$\frac{30}{4}$ in.

 For Exercises 83–86, translate the phrase to a mathematical expression, then simplify.

83. The sum of three-fifths and two-fifths $\frac{3}{5} + \frac{2}{5}; 1$

84. Seven-ninths more than five-ninths $\frac{5}{9} + \frac{7}{9}; \frac{4}{3}$

85. The difference of eleven-fifteenths and eight-fifteenths $\frac{11}{15} - \frac{8}{15}; \frac{1}{5}$

86. Two-sevenths subtracted from five-sevenths
$\frac{5}{7} - \frac{2}{7}; \frac{3}{7}$

Writing   Translating Expression   Geometry   Scientific Calculator   Video   NSE

Least Common Multiple

1. Least Common Multiple

In Section 3.1 we learned how to add and subtract like fractions. To add or subtract fractions with different denominators, we must learn how to convert unlike fractions into like fractions. An essential concept in this process is the idea of a least common multiple of two or more numbers.

When we multiply a number by the whole numbers 1, 2, 3, and so on, we form the **multiples** of the number. For example, some of the multiples of 6 and 9 are shown below.

Multiples of 6	Multiples of 9
$6 \times 1 = 6$	$9 \times 1 = 9$
$6 \times 2 = 12$	$9 \times 2 = 18$
$6 \times 3 = 18$	$9 \times 3 = 27$
$6 \times 4 = 24$	$9 \times 4 = 36$
$6 \times 5 = 30$	$9 \times 5 = 45$
$6 \times 6 = 36$	$9 \times 6 = 54$
$6 \times 7 = 42$	$9 \times 7 = 63$
$6 \times 8 = 48$	$9 \times 8 = 72$
$6 \times 9 = 54$	$9 \times 9 = 81$

In red, we have indicated several multiples that are common to both 6 and 9.

Concepts

1. Least Common Multiple
2. Finding the LCM by Using Prime Factors
3. Finding the LCM by Using Division by Primes
4. Applications of the LCM
5. Equivalent Fractions and Ordering Fractions

The **least common multiple (LCM)** of two given numbers is the smallest whole number that is a multiple of each given number. For example, the LCM of 6 and 9 is 18.

Multiples of 6: 6, 12, 18, 24, 30, 36, 42, . . .

Multiples of 9: 9, 18, 27, 36, 45, 54, 63, . . .

Concept Connections

1. Explain the difference between a multiple of a number and a factor of a number.

TIP: There are infinitely many numbers that are common multiples of both 6 and 9. These include 18, 36, 54, 72, and so on. However, 18 is the smallest, and is therefore the *least* common multiple.

If one number is a multiple of another number, then the LCM is the larger of the two numbers. For example, the LCM of 4 and 8 is 8.

Multiples of 4: 4, 8, 12, 16, . . .

Multiples of 8: 8, 16, 24, 32, . . .

Example 1 Finding the LCM by Listing Multiples

Find the LCM of the given numbers by listing several multiples of each number.

a. 15 and 12   **b.** 10, 15, and 8

Solution:

a. Multiples of 15:   15, 30, 45, 60
Multiples of 12:   12, 24, 36, 48, 60

The LCM of 15 and 12 is 60.

b. Multiples of 10:   10, 20, 30, 40, 50, 60, 70, 80, 90, 100, 110, 120
Multiples of 15:   15, 30, 45, 60, 75, 90, 105, 120
Multiples of 8:   8, 16, 24, 32, 40, 48, 56, 64, 72, 80, 88, 96, 104, 112, 120

The LCM of 10, 15, and 8 is 120.

Skill Practice

Find the LCM by listing several multiples of each number.
  2. 15 and 25 **3.** 4, 6, and 10

Classroom Example: p. 177, Exercise 16

Answers

1. A multiple of a number is the product of the number and a whole number 1 or greater. A factor of a number is a value that divides evenly into the number.
2. 75 **3.** 60

Instructor Note: Mention to students that, although the first method for finding an LCM always works, it can be tedious and inefficient. Therefore, other methods will be presented.

2. Finding the LCM by Using Prime Factors

In Example 1 we used the method of listing multiples to find the LCM of two or more numbers. As you can see, the solution to Example 1(b) required several long lists of multiples. Here we offer another method to find the LCM of two given numbers by using their prime factors.

> **Using Prime Factors to Find the LCM of Two Numbers**
> **Step 1** Write each number as a product of prime factors.
> **Step 2** The LCM is the product of unique prime factors from both numbers. Use repeated factors the maximum number of times they appear in either factorization.

This process is demonstrated in Example 2.

Skill Practice

Find the LCM by using prime factors.
4. 9 and 24
5. 16 and 9
6. 36, 42, and 30

Classroom Example: p. 177, Exercise 22

TIP: The product $2 \cdot 2 \cdot 3 \cdot 7$ can also be written as $2^2 \cdot 3 \cdot 7$.

Example 2 Finding the LCM by Using Prime Factors

Find the LCM.

a. 14 and 12   **b.** 50 and 24   **c.** 45, 54, and 50

Solution:

a. Find the prime factorization for 14 and 12.

	2's	3's	7's
14 =	2 ·		⑦
12 =	②·2·	③	

LCM = $2 \cdot 2 \cdot 3 \cdot 7 = 84$

For the factors of 2, 3, and 7, we circle the greatest number of times each occurs. The LCM is the product.

b. Find the prime factorization for 50 and 24.

	2's	3's	5's
50 =	2 ·		⑤·5
24 =	②·2·2·	③	

LCM = $2 \cdot 2 \cdot 2 \cdot 3 \cdot 5 \cdot 5 = 600$

The factor 5 is repeated twice. The factor 2 is repeated 3 times. The factor 3 is used only once.

(The LCM can also be written as $2^3 \cdot 3 \cdot 5^2$.)

c. Find the prime factorization for 45, 54, and 50.

	2's	3's	5's
45 =		3 · 3 ·	5
54 =	2 ·	③·3·3	
50 =	② ·		⑤·5

LCM = $2 \cdot 3 \cdot 3 \cdot 3 \cdot 5 \cdot 5 = 1350$

(The LCM can also be written as $2 \cdot 3^3 \cdot 5^2$.)

Answers
4. 72 **5.** 144 **6.** 1260

3. Finding the LCM by Using Division by Primes

We present a third method for finding least common multiples. We systematically divide by prime numbers to determine which will be a factor of the LCM. This method is particularly helpful if three or more numbers are involved.

| Example 3 | Finding the LCM by Using Division by Primes

Find the LCM of 32, 48, and 30 by using division of prime factors.

Solution:

To begin this process, find any prime number that divides evenly into any of the numbers. Then divide and write the quotient as shown. We begin by dividing by the smallest prime number, 2.

$$
\begin{array}{r}
2)\overline{32\ \ 48\ \ 30} \\
16\ \ 24\ \ 15
\end{array}
$$

Repeat this process and bring down any number that is not divisible by the chosen prime.

$$
\begin{array}{r}
2)\overline{32\ \ 48\ \ 30} \\
2)\overline{16\ \ 24\ \ 15} \\
)\overline{8\ \ 12\ \ 15}
\end{array}
$$
Bring down the 15.

Continue until all quotients are 1. The LCM is the product of the prime factors at the left.

$$
\begin{array}{r}
2)\overline{32\ \ 48\ \ 30} \\
2)\overline{16\ \ 24\ \ 15} \\
2)\overline{8\ \ 12\ \ 15} \\
2)\overline{4\ \ 6\ \ 15} \\
2)\overline{2\ \ 3\ \ 15} \\
3)\overline{1\ \ 3\ \ 15} \\
5)\overline{1\ \ 1\ \ 5} \\
1\ \ 1\ \ 1
\end{array}
$$

At this point, the prime number 2 does not divide evenly into any of the quotients. We try the next-greater prime number, 3.

The LCM is $2 \cdot 2 \cdot 2 \cdot 2 \cdot 2 \cdot 3 \cdot 5 = 480$.

Skill Practice

Find the LCM by using division by prime factors.

7. 20, 36, and 15

Classroom Example: p. 177, Exercise 34

TIP: We have presented three methods to find the LCM. Try each method. Then you and your instructor can decide which methods work best for you.

4. Applications of the LCM

| Example 4 | Using the LCM in an Application

A tile wall is to be made from 6-in., 8-in., and 12-in. square tiles. A design is made by alternating rows with different-size tiles. The first row uses only 6-in. tiles, the second row uses only 8-in. tiles, and the third row uses only 12-in. tiles. Neglecting the grout seams, what is the shortest length of wall space that can be covered using only whole tiles?

Answer

7. 180

Solution:

The length of the first row must be a multiple of 6 in., the length of the second row must be a multiple of 8 in., and the length of the third row must be a multiple of 12 in. Therefore, the shortest-length wall that can be covered is given by the LCM of 6, 8, and 12.

$$6 = 2 \cdot \boxed{3}$$
$$8 = \boxed{2 \cdot 2 \cdot 2}$$
$$12 = 2 \cdot 2 \cdot 3$$

The LCM is $2 \cdot 2 \cdot 2 \cdot 3 = 24$. The shortest-length wall is 24 in.

This means that four 6-in. tiles can be placed on the first row, three 8-in. tiles can be placed on the second row, and two 12-in. tiles can be placed in the third row. See Figure 3-1.

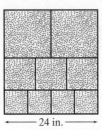

← 24 in. →

Figure 3-1

5. Equivalent Fractions and Ordering Fractions

Suppose we want to determine which of two fractions is larger. Comparing fractions with the same denominator, such as $\frac{3}{5}$ and $\frac{2}{5}$ is relatively easy. Clearly 3 parts out of 5 is greater than 2 parts out of 5.

$\frac{3}{5}$

$\frac{2}{5}$

Thus, $\frac{3}{5} > \frac{2}{5}$.

So how would we compare the relative size of two fractions with *different* denominators such as $\frac{3}{5}$ and $\frac{4}{7}$? Our first step is to write the fractions as equivalent fractions with the same denominator, called a common denominator. The **least common denominator (LCD)** of two fractions is the LCM of the denominators of the fractions. The LCD of $\frac{3}{5}$ and $\frac{4}{7}$ is 35, because this is the least common multiple of 5 and 7. In Example 5, we convert the fractions $\frac{3}{5}$ and $\frac{4}{7}$ to equivalent fractions having 35 as the denominator.

Example 5 **Writing Equivalent Fractions**

Write the fractions with the indicated denominator.

a. $\frac{3}{5} = \frac{}{35}$ b. $\frac{4}{7} = \frac{}{35}$

Solution:

a. $\frac{3}{5} = \frac{}{35}$ $\frac{3 \cdot 7}{5 \cdot 7} = \frac{21}{35}$ So, $\frac{21}{35}$ is equivalent to $\frac{3}{5}$.

What number must we multiply 5 by to get 35?   Multiply numerator and denominator by 7.

b. $\dfrac{4}{7} = \dfrac{}{35}$

What number must we
multiply 7 by to get 35?

$\dfrac{4 \cdot 5}{7 \cdot 5} = \dfrac{20}{35}$

Multiply numerator and
denominator by 5.

So, $\dfrac{20}{35}$ is equivalent to $\dfrac{4}{7}$.

TIP: Writing a fraction as an equivalent fraction is simply an application of the fundamental principle of fractions (see Section 2.3). In Example 5(a), we multiplied numerator and denominator of the fraction by 7. This is the same as multiplying the fraction by a convenient form of 1.

$$\frac{3}{5} = \frac{3}{5} \cdot 1 = \frac{3}{5} \cdot \frac{7}{7} = \frac{3 \cdot 7}{5 \cdot 7} = \frac{21}{35}$$

This is the same as multiplying
numerator and denominator by 7.

From Example 5, we know that $\dfrac{3}{5} = \dfrac{21}{35}$ and $\dfrac{4}{7} = \dfrac{20}{35}$.

Furthermore, because $\dfrac{21}{35} > \dfrac{20}{35}$, then we know that $\dfrac{3}{5} > \dfrac{4}{7}$.

Instructor Note: To motivate this discussion start by asking students if they would rather receive $\frac{3}{5}$ of $70 or $\frac{4}{7}$ of $70.

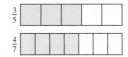

Animation

Example 6 **Comparing Two Fractions**

Fill in the blank with $<$, $>$, or $=$.

$$\frac{9}{8} \;\square\; \frac{7}{6}$$

Skill Practice

Fill in the blank with $<$, $>$, or $=$.

11. $\dfrac{4}{7} \;\square\; \dfrac{5}{9}$

Solution:

The fractions have different denominators and cannot be compared by inspection. The LCD is 24. We need to convert each fraction to an equivalent fraction with a denominator of 24.

$\dfrac{9}{8} = \dfrac{9 \cdot 3}{8 \cdot 3} = \dfrac{27}{24}$ Multiply numerator and denominator by 3, because $8 \cdot 3 = 24$.

$\dfrac{7}{6} = \dfrac{7 \cdot 4}{6 \cdot 4} = \dfrac{28}{24}$ Multiply numerator and denominator by 4, because $6 \cdot 4 = 24$.

Because $\dfrac{27}{24} < \dfrac{28}{24}$, then $\dfrac{9}{8} \;\boxed{<}\; \dfrac{7}{6}$.

The relationship between $\frac{9}{8}$ and $\frac{7}{6}$ is shown on the number line in Figure 3-2.

Classroom Example: p. 178, Exercise 62

Figure 3-2

Answer

11. $>$

Example 7 **Ranking Fractions in Order from Least to Greatest**

Rank the fractions from least to greatest.

$$\frac{9}{20}, \frac{7}{15}, \frac{4}{9}$$

Solution:

We want to convert each fraction to an equivalent fraction with a common denominator. The least common denominator is the LCM of 20, 15, and 9.

$$\left.\begin{array}{l} 20 = 2 \cdot 2 \cdot 5 \\ 15 = 3 \cdot 5 \\ 9 = 3 \cdot 3 \end{array}\right\}$$ The least common denominator is $2 \cdot 2 \cdot 3 \cdot 3 \cdot 5 = 180$.

Now convert each fraction to an equivalent fraction with a denominator of 180.

$\dfrac{9}{20} = \dfrac{9 \cdot 9}{20 \cdot 9} = \dfrac{81}{180}$ Multiply numerator and denominator by 9 because $20 \cdot 9 = 180$.

$\dfrac{7}{15} = \dfrac{7 \cdot 12}{15 \cdot 12} = \dfrac{84}{180}$ Multiply numerator and denominator by 12 because $15 \cdot 12 = 180$.

$\dfrac{4}{9} = \dfrac{4 \cdot 20}{9 \cdot 20} = \dfrac{80}{180}$ Multiply numerator and denominator by 20 because $9 \cdot 20 = 180$.

Ranking the fractions from least to greatest, we have $\frac{80}{180}, \frac{81}{180}, \frac{84}{180}$. This is equivalent to $\frac{4}{9}, \frac{9}{20}, \frac{7}{15}$.

Section 3.2 Practice Exercises

Study Skills Exercise

For additional exercises, see Classroom Activities 3.2A–3.2B in the *Student's Resource Manual* at www.mhhe.com/moh.

Where do you usually do your homework? Is this the best place for you to concentrate? Explain.

Vocabulary and Key Concepts

1. a. A ___multiple___ of a number is the product of the number and a nonzero whole number.

b. The ___least___ ___common___ ___multiple___ (LCM) of two numbers is the smallest whole number that is a multiple of each given number.

c. The ___least___ ___common___ ___denominator___ (LCD) of two fractions is the LCM of the denominators of the fractions.

Review Exercises

2. Simplify.

a. $\dfrac{2}{5} + \dfrac{1}{5}$ $\frac{3}{5}$ **b.** $\dfrac{2}{5} \cdot \dfrac{1}{5}$ $\frac{2}{25}$ **c.** $\dfrac{2}{5} \div \dfrac{1}{5}$ 2 **d.** $\dfrac{2}{5} - \dfrac{1}{5}$ $\frac{1}{5}$

For Exercises 3–8, add and subtract as indicated. Write the answer as a whole number or fraction simplified to lowest terms.

3. $\frac{19}{6} - \frac{16}{6}$ $\frac{1}{2}$

4. $\frac{28}{4} - \frac{22}{4}$ $\frac{3}{2}$

5. $\frac{31}{15} + \frac{2}{15} - \frac{8}{15}$ $\frac{5}{3}$

6. $\frac{8}{5} + \frac{12}{5}$ 4

7. $\frac{11}{3} + \frac{7}{3}$ 6

8. $\frac{5}{19} - \frac{2}{19}$ $\frac{3}{19}$

Concept 1: Least Common Multiple

9. a. Circle the multiples of 24: 4, 8, 48, 72, 12, 240 48, 72, 240

  b. Circle the factors of 24: 4, 8, 48, 72, 12, 240 4, 8, 12

10. a. Circle the multiples of 30: 15, 90, 120, 3, 5, 60 90, 120, 60

  b. Circle the factors of 30: 15, 90, 120, 3, 5, 60 15, 3, 5

11. a. Circle the multiples of 36: 72, 6, 360, 12, 9, 108 72, 360, 108

  b. Circle the factors of 36: 72, 6, 360, 12, 9, 108 6, 12, 9

12. a. Circle the multiples of 28: 7, 4, 2, 56, 140, 280 56, 140, 280

  b. Circle the factors of 28: 7, 4, 2, 56, 140, 280 7, 4, 2

For Exercises 13–18, find the LCM by listing several multiples of each number. **(See Example 1.)**

13. 10 and 25 50

14. 21 and 14 42

15. 16 and 12 48

16. 20 and 12 60

17. 8, 10, and 12 120

18. 4, 6, and 14 84

Concepts 2 and 3: Finding the LCM

For Exercises 19–38, find the LCM. **(See Examples 2 and 3.)**

19. 18 and 24 72

20. 9 and 30 90

21. 12 and 15 60

22. 27 and 45 135

23. 15 and 25 75

24. 16 and 24 48

25. 24 and 30 120

26. 14 and 35 70

27. 42 and 70 210

28. 6 and 21 42

29. 20, 18, and 27 540

30. 9, 15, and 42 630

31. 12, 15, and 20 60

32. 20, 30, and 40 120

33. 16, 24, and 30 240

34. 20, 42, and 35 420

35. 6, 12, 18, and 20 180

36. 21, 35, 50, and 75 1050

37. 5, 15, 18, and 20 180

38. 28, 10, 21, and 35 420

Concept 4: Applications of the LCM

39. A tile floor is to be made from 10-in., 12-in., and 15-in. square tiles. A design is made by alternating rows with different-size tiles. The first row uses only 10-in. tiles, the second row uses only 12-in. tiles, and the third row uses only 15-in. tiles. Neglecting the grout seams, what is the shortest length of floor space that can be covered evenly by each row? **(See Example 4.)**

The shortest length of floor space is 60 in. (5 ft).

40. A patient admitted to the hospital was prescribed a pain medication to be given every 4 hr and an antibiotic to be given every 5 hr. Bandages applied to the patient's external injuries needed changing every 12 hr. The nurse changed the bandages and gave the patient both medications at 6:00 A.M. Monday morning.

  a. How many hours will pass before the patient is given both medications and has his bandages changed at the same time? 60 hr

  b. What day and time will this be?

Wednesday at 6:00 P.M.

Writing Translating Expression Geometry Scientific Calculator Video NS E

41. Four satellites revolve around the earth once every 6, 8, 10, and 15 hr, respectively. If the satellites are initially "lined up," how many hours must pass before they will again be lined up?

It will take 120 hr (5 days) for the satellites to be lined up again.

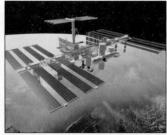

42. Mercury, Venus, and Earth revolve around the Sun approximately once every 3 months, 7 months, and 12 months, respectively (see the figure). If the planets begin "lined up," what is the minimum number of months required for them to be aligned again? (Assume that the planets lie roughly in the same plane.)

It will take 84 months (7 years) for the planets to be aligned again.

Earth
Venus
Mercury

Concept 5: Equivalent Fractions and Ordering Fractions

For Exercises 43–60, rewrite each fraction with the indicated denominators. **(See Example 5.)**

43. $\dfrac{2}{3} = \dfrac{}{21}$ $\quad \frac{14}{21}$

44. $\dfrac{7}{4} = \dfrac{}{32}$ $\quad \frac{56}{32}$

45. $\dfrac{5}{8} = \dfrac{}{16}$ $\quad \frac{10}{16}$

46. $\dfrac{2}{9} = \dfrac{}{27}$ $\quad \frac{6}{27}$

47. $\dfrac{3}{4} = \dfrac{}{16}$ $\quad \frac{12}{16}$

48. $\dfrac{3}{10} = \dfrac{}{50}$ $\quad \frac{15}{50}$

49. $\dfrac{4}{5} = \dfrac{}{15}$ $\quad \frac{12}{15}$

50. $\dfrac{3}{7} = \dfrac{}{70}$ $\quad \frac{30}{70}$

51. $\dfrac{7}{6} = \dfrac{}{42}$ $\quad \frac{49}{42}$

52. $\dfrac{10}{3} = \dfrac{}{18}$ $\quad \frac{60}{18}$

53. $\dfrac{11}{9} = \dfrac{}{99}$ $\quad \frac{121}{99}$

54. $\dfrac{7}{5} = \dfrac{}{35}$ $\quad \frac{49}{35}$

55. $\dfrac{5}{13} = \dfrac{}{39}$ $\quad \frac{15}{39}$

56. $\dfrac{6}{17} = \dfrac{}{34}$ $\quad \frac{12}{34}$

57. $\dfrac{11}{4} = \dfrac{}{4000}$ $\quad \frac{11,000}{4000}$

58. $\dfrac{18}{7} = \dfrac{}{700}$ $\quad \frac{1800}{700}$

59. $\dfrac{3}{14} = \dfrac{}{70}$ $\quad \frac{15}{70}$

60. $\dfrac{5}{66} = \dfrac{}{198}$ $\quad \frac{15}{198}$

For Exercises 61–68, fill in the blanks with $<$, $>$, or $=$. **(See Example 6.)**

61. $\dfrac{7}{8} \ \square \ \dfrac{3}{4}$ $\quad >$

62. $\dfrac{7}{15} \ \square \ \dfrac{11}{20}$ $\quad <$

63. $\dfrac{13}{10} \ \square \ \dfrac{22}{15}$ $\quad <$

64. $\dfrac{15}{4} \ \square \ \dfrac{21}{6}$ $\quad >$

65. $\dfrac{3}{12} \ \square \ \dfrac{2}{8}$ $\quad =$

66. $\dfrac{5}{20} \ \square \ \dfrac{4}{16}$ $\quad =$

67. $\dfrac{5}{18} \ \square \ \dfrac{8}{27}$ $\quad <$

68. $\dfrac{9}{24} \ \square \ \dfrac{8}{21}$ $\quad <$

NS E **69.** Which of the following fractions has the greatest value? $\dfrac{2}{3}, \dfrac{7}{8}, \dfrac{5}{6}, \dfrac{1}{2}$ $\dfrac{7}{8}$

NS E **70.** Which of the following fractions has the least value? $\dfrac{1}{6}, \dfrac{1}{4}, \dfrac{2}{15}, \dfrac{2}{9}$ $\dfrac{2}{15}$

For Exercises 71–76, rank the fractions from least to greatest. **(See Example 7.)**

71. $\dfrac{7}{8}, \dfrac{2}{3}, \dfrac{3}{4}$ $\dfrac{2}{3} \; \dfrac{3}{4} \; \dfrac{7}{8}$

72. $\dfrac{5}{12}, \dfrac{3}{8}, \dfrac{2}{3}$ $\dfrac{3}{8} \; \dfrac{5}{12} \; \dfrac{2}{3}$

73. $\dfrac{5}{16}, \dfrac{3}{8}, \dfrac{1}{4}$ $\dfrac{1}{4} \; \dfrac{5}{16} \; \dfrac{3}{8}$

74. $\dfrac{2}{5}, \dfrac{3}{10}, \dfrac{5}{6}$ $\dfrac{3}{10} \; \dfrac{2}{5} \; \dfrac{5}{6}$

75. $\dfrac{4}{3}, \dfrac{13}{12}, \dfrac{17}{15}$ $\dfrac{13}{12} \; \dfrac{17}{15} \; \dfrac{4}{3}$

76. $\dfrac{5}{7}, \dfrac{11}{21}, \dfrac{18}{35}$ $\dfrac{18}{35} \; \dfrac{11}{21} \; \dfrac{5}{7}$

77. A patient had three cuts that needed stitches. A nurse recorded the lengths of the cuts. Where did the patient have the longest cut? Where did the patient have the shortest cut?

upper right arm ¾ in.
Right hand $\frac{11}{16}$ in.
above left eye $\frac{7}{8}$ in.

The longest cut is above the left eye. The shortest cut is on the right hand.

78. Three screws have lengths equal to $\frac{3}{4}$ in., $\frac{5}{8}$ in., and $\frac{11}{16}$ in. Which screw is the longest? Which is the shortest?
The $\frac{3}{4}$-in. screw is the longest, and the $\frac{5}{8}$-in. screw is the shortest.

NS E **79.** Susan buys $\frac{2}{3}$ lb of smoked turkey, $\frac{3}{5}$ lb of ham, and $\frac{5}{8}$ lb of roast beef. Which type of meat did she buy in the greatest amount? Which type did she buy in the least amount? The greatest amount is $\frac{2}{3}$ lb of turkey. The least amount is $\frac{3}{5}$ lb of ham.

NS E **80.** For a party, Aman had $\frac{3}{4}$ lb of cheddar cheese, $\frac{7}{8}$ lb of Swiss cheese, and $\frac{4}{5}$ lb of pepper jack cheese. Which type of cheese is in the least amount? Which type is in the greatest amount? The least amount is $\frac{3}{4}$ lb of cheddar, and the greatest amount is $\frac{7}{8}$ lb of Swiss.

Expanding Your Skills

NS E **81.** Which of the following fractions is between $\frac{1}{4}$ and $\frac{5}{6}$? Identify all that apply.

a. $\dfrac{5}{12}$ **b.** $\dfrac{2}{3}$ **c.** $\dfrac{1}{8}$ a and b

NS E **82.** Which of the following fractions is between $\frac{1}{3}$ and $\frac{11}{15}$? Identify all that apply.

a. $\dfrac{2}{3}$ **b.** $\dfrac{4}{5}$ **c.** $\dfrac{2}{5}$ a and c

---

Addition and Subtraction of Unlike Fractions | Section 3.3

1. Addition and Subtraction of Unlike Fractions

Two fractions are **unlike fractions** if they have different denominators. In this section, we will use the least common denominator of two or more fractions to help us add and subtract unlike fractions. The first step in adding or subtracting unlike fractions is to identify the LCD. Then we change the unlike fractions to like fractions having the LCD as the denominator.

Concepts

1. Addition and Subtraction of Unlike Fractions
2. Order of Operations
3. Applications Involving Unlike Fractions

 Writing   Translating Expression   Geometry   Scientific Calculator   Video   NS E

Skill Practice

Add.

1. $\dfrac{1}{10} + \dfrac{1}{15}$

Classroom Example: p. 186, Exercise 18

Example 1 Adding Unlike Fractions

Add. $\dfrac{1}{6} + \dfrac{3}{4}$

Solution:

We cannot add $\frac{1}{6}$ and $\frac{3}{4}$ as they are because they have different denominators. Using the LCD of 12, we can convert the individual fractions to be like fractions.

$$\frac{1}{6} = \frac{1 \cdot 2}{6 \cdot 2} = \frac{2}{12} \qquad \text{Multiply numerator and denominator by 2 because } 6 \cdot 2 = 12.$$

$$\frac{3}{4} = \frac{3 \cdot 3}{4 \cdot 3} = \frac{9}{12} \qquad \text{Multiply numerator and denominator by 3 because } 4 \cdot 3 = 12.$$

Thus, $\dfrac{1}{6} + \dfrac{3}{4}$ becomes $\dfrac{2}{12} + \dfrac{9}{12} = \dfrac{11}{12}$.

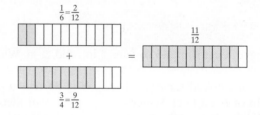

Concept Connections

2. Use the figure to add the fractions.

$$\frac{1}{3} + \frac{2}{5}$$

TIP: In Example 1, we multiplied the fraction $\frac{1}{6}$ by $\frac{2}{2}$. This is equivalent to multiplying $\frac{1}{6}$ by 1 and does not change the value.

$$\frac{1}{6} = \frac{1}{6} \cdot 1 = \frac{1}{6} \cdot \frac{2}{2} = \frac{2}{12}$$

The fraction $\frac{2}{12}$ is equivalent to $\frac{1}{6}$.

Adding fractions can be visualized by using a diagram. For example, the sum $\frac{1}{2} + \frac{1}{3}$ is illustrated in Figure 3-3.

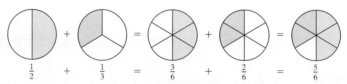

Figure 3-3

The general procedure to add or subtract unlike fractions is outlined as follows.

Answers

1. $\dfrac{1}{6}$

2.

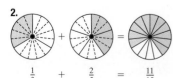

Adding and Subtracting Unlike Fractions

Step 1 Identify the LCD.

Step 2 Write each individual fraction as an equivalent fraction with the LCD.

Step 3 Add or subtract the resulting fractions as indicated.

Step 4 Simplify to lowest terms, if possible.

Example 2	Adding Unlike Fractions

Add. $\dfrac{3}{10} + \dfrac{1}{5}$

Solution:

$\dfrac{3}{10} + \dfrac{1}{5}$ The LCD is 10. We must convert $\frac{1}{5}$ to an equivalent fraction with 10 as the denominator.

$= \dfrac{3}{10} + \dfrac{1 \cdot 2}{5 \cdot 2}$ Multiply numerator and denominator by 2 because $5 \cdot 2 = 10$.

$= \dfrac{3}{10} + \dfrac{2}{10}$ The fractions are now like.

$= \dfrac{3 + 2}{10}$ Add the like fractions.

$= \dfrac{5}{10}$

$= \dfrac{\overset{1}{\cancel{5}}}{\underset{2}{\cancel{10}}}$ Simplify to lowest terms.

$= \dfrac{1}{2}$

Skill Practice

Add.

3. $\dfrac{9}{5} + \dfrac{7}{10}$

Classroom Example: p. 186, Exercise 20

Avoiding Mistakes

Do not confuse addition and subtraction of fractions with multiplication of fractions. In multiplication, we multiply denominators. In addition we do not add denominators. We get a common denominator and then add only the numerators.

Example 3	Subtracting Unlike Fractions

Subtract. $\dfrac{4}{15} - \dfrac{1}{10}$

Solution:

$\dfrac{4}{15} - \dfrac{1}{10}$ The LCD is 30.

$= \dfrac{4 \cdot 2}{15 \cdot 2} - \dfrac{1 \cdot 3}{10 \cdot 3}$ Write the fractions as equivalent fractions with the LCD.

$= \dfrac{8}{30} - \dfrac{3}{30}$

$= \dfrac{8 - 3}{30}$

$= \dfrac{5}{30}$ Subtract.

$= \dfrac{\overset{1}{\cancel{5}}}{\underset{6}{\cancel{30}}}$ Simplify to lowest terms.

$= \dfrac{1}{6}$

Skill Practice

Subtract.

4. $\dfrac{7}{12} - \dfrac{1}{8}$

Classroom Example: p. 186, Exercise 28

Answers

3. $\dfrac{5}{2}$ 4. $\dfrac{11}{24}$

Sometimes when denominators are large, it is helpful to write the denominators as a product of prime factors. This is demonstrated in Example 4.

Skill Practice

Add or subtract as indicated.

5. $\dfrac{7}{18} + \dfrac{4}{15} - \dfrac{17}{30}$

Classroom Example: p. 187, Exercise 48

Example 4 Adding and Subtracting Unlike Fractions

Add or subtract as indicated.

$$\frac{7}{12} - \frac{2}{15} + \frac{5}{48}$$

Solution:

$$\frac{7}{12} - \frac{2}{15} + \frac{5}{48} \quad \text{To find the LCD, factor each denominator.}$$

$$= \frac{7}{2 \cdot 2 \cdot 3} - \frac{2}{3 \cdot 5} + \frac{5}{2 \cdot 2 \cdot 2 \cdot 2 \cdot 3}$$

$$\left.\begin{array}{l} 12 = 2 \cdot 2 \cdot \boxed{3} \\ 15 = 3 \cdot \boxed{5} \\ 48 = \boxed{2 \cdot 2 \cdot 2 \cdot 2} \cdot 3 \end{array}\right\} \quad \text{The LCD is } 2 \cdot 2 \cdot 2 \cdot 2 \cdot 3 \cdot 5 = 240.$$

We want to convert each fraction to an equivalent fraction having a denominator of $2 \cdot 2 \cdot 2 \cdot 2 \cdot 3 \cdot 5 = 240$. Multiply numerator and denominator of each original fraction by the factors missing from the denominator.

$$= \frac{7 \cdot (2 \cdot 2 \cdot 5)}{2 \cdot 2 \cdot 3 \cdot (2 \cdot 2 \cdot 5)} - \frac{2 \cdot (2 \cdot 2 \cdot 2 \cdot 2)}{3 \cdot 5 \cdot (2 \cdot 2 \cdot 2 \cdot 2)} + \frac{5 \cdot (5)}{2 \cdot 2 \cdot 2 \cdot 2 \cdot 3 \cdot (5)}$$

$$= \frac{140}{240} - \frac{32}{240} + \frac{25}{240} \quad \text{The fractions are now like fractions.}$$

$$= \frac{140 - 32 + 25}{240} \quad \text{Add and subtract as indicated.}$$

$$= \frac{133}{240} \quad \begin{array}{l} \text{The prime factors of 240 are 2, 3, and 5.} \\ \text{Using divisibility rules we can see that 133 is} \\ \text{not divisible by 2, 3, or 5. Therefore, } \frac{133}{240} \text{ is in} \\ \text{lowest terms.} \end{array}$$

2. Order of Operations

In Examples 5 and 6, we must apply the order of operations to simplify the expressions.

Skill Practice

Simplify.

6. $\left(\dfrac{2}{3} - \dfrac{1}{7}\right)^2$

Classroom Example: p. 187, Exercise 52

Example 5 Applying the Order of Operations

Simplify. $\left(\dfrac{1}{4} + \dfrac{2}{3}\right)^2$

Answers

5. $\dfrac{4}{45}$ **6.** $\dfrac{121}{441}$

Solution:

$$\left(\frac{1}{4} + \frac{2}{3}\right)^2$$

Perform the operation within parentheses first.

$$= \left(\frac{1 \cdot 3}{4 \cdot 3} + \frac{2 \cdot 4}{3 \cdot 4}\right)^2$$

The common denominator is 12.

$$= \left(\frac{3}{12} + \frac{8}{12}\right)^2$$

The fractions within parentheses are now like fractions.

$$= \left(\frac{11}{12}\right)^2$$

Add fractions within parentheses.

$$= \frac{11}{12} \cdot \frac{11}{12}$$

To square a number, multiply the number by itself.

$$= \frac{121}{144}$$

The fraction is in lowest terms.

Instructor Note: Remind students that

$$\left(\frac{1}{4} + \frac{2}{3}\right)^2 \neq \left(\frac{1}{4}\right)^2 + \left(\frac{2}{3}\right)^2$$

Example 6 **Applying the Order of Operations**

Simplify. $\dfrac{5}{12} - \dfrac{1}{4} \div \dfrac{3}{2}$

Solution:

$$\frac{5}{12} - \frac{1}{4} \div \frac{3}{2}$$

Perform the division before the subtraction.

$$= \frac{5}{12} - \frac{1}{4} \cdot \frac{\overset{1}{2}}{\underset{2}{3}}$$

To divide fractions, multiply by the reciprocal of the divisor.

$$= \frac{5}{12} - \frac{1}{6}$$

To subtract, we need the LCD of 12.

$$= \frac{5}{12} - \frac{1 \cdot 2}{6 \cdot 2}$$

Multiply numerator and denominator by 2 because $6 \cdot 2 = 12$.

$$= \frac{5}{12} - \frac{2}{12}$$

The fractions are now like fractions.

$$= \frac{3}{12}$$

Subtract.

$$= \frac{\overset{1}{3}}{\underset{4}{12}}$$

Simplify to lowest terms.

$$= \frac{1}{4}$$

Skill Practice

Simplify.

7. $\dfrac{4}{15} \div \dfrac{2}{5} - \dfrac{1}{6}$

Classroom Example: p. 187, Exercise 56

Answer

7. $\dfrac{1}{2}$

3. Applications Involving Unlike Fractions

Example 7 **Applying Operations on Unlike Fractions**

A new Kelly Safari SUV tire has $\frac{7}{16}$-in. tread. After being driven 50,000 mi, the tread depth has worn down to $\frac{7}{32}$ in. By how much has the tread depth worn away?

Tread

Solution:

In this case, we are looking for the difference in the tread depth.

$$\begin{array}{l}\text{Difference in}\\ \text{tread depth}\end{array} = \left(\begin{array}{c}\text{original}\\ \text{tread depth}\end{array}\right) - \left(\begin{array}{c}\text{final}\\ \text{tread depth}\end{array}\right)$$

$$= \frac{7}{16} - \frac{7}{32} \qquad \text{The LCD is 32.}$$

$$= \frac{7 \cdot 2}{16 \cdot 2} - \frac{7}{32} \qquad \begin{array}{l}\text{Multiply numerator}\\ \text{and denominator by 2 because } 16 \cdot 2 = 32.\end{array}$$

$$= \frac{14}{32} - \frac{7}{32} \qquad \text{The fractions are now like.}$$

$$= \frac{7}{32} \qquad \text{Subtract.}$$

The tire lost $\frac{7}{32}$ in. in tread depth after 50,000 mi of driving.

> **TIP:** You can check your result by adding the final tread depth to the difference in tread depth to get the original tread depth.
>
> $$\frac{7}{32} + \frac{7}{32} = \frac{14}{32} = \frac{7}{16}$$

Example 8 **Applying Operations on Unlike Fractions**

An oil tank contains 2 liters (L) of oil. A slow leak has occurred, and oil leaks out at a rate of $\frac{1}{16}$ L per day. After 7 days, a mechanic fixed the leak and poured in $\frac{3}{8}$ L of oil back into the tank. How much oil is now in the tank?

Solution:

To find the current amount in the tank, we must determine the amount lost and the amount added.

The amount lost is given by the amount lost per day times 7 days.

$$\left(\frac{1}{16}\right)(7) = \frac{1}{16} \cdot \frac{7}{1} = \frac{7}{16} \qquad \text{The amount lost in 7 days is } \frac{7}{16} \text{ L.}$$

Therefore, the current amount in the tank is given by

$$\begin{array}{l}\text{Current}\\ \text{amount}\end{array} = \begin{pmatrix}\text{original}\\ \text{amount}\end{pmatrix} - \begin{pmatrix}\text{amount}\\ \text{lost}\end{pmatrix} + \begin{pmatrix}\text{amount}\\ \text{replaced}\end{pmatrix}$$

$$= 2 - \frac{7}{16} + \frac{3}{8}$$

$$= \frac{2}{1} - \frac{7}{16} + \frac{3}{8} \qquad \text{Write the whole number as a fraction.}$$

$$= \frac{2 \cdot 16}{1 \cdot 16} - \frac{7}{16} + \frac{3 \cdot 2}{8 \cdot 2} \qquad \text{The LCD is 16.}$$

$$= \frac{32}{16} - \frac{7}{16} + \frac{6}{16} \qquad \text{The fractions are now like fractions.}$$

$$= \frac{32 - 7 + 6}{16} \qquad \text{Add and subtract as indicated.}$$

$$= \frac{31}{16} \text{ or } 1\frac{15}{16} \qquad \text{The fraction is in lowest terms.}$$

The tank now contains $1\frac{15}{16}$ L.

> **TIP:** For Example 8, both $\frac{31}{16}$ L and $1\frac{15}{16}$ L are correct answers but, in general, we will give the answer to an application as a mixed number as opposed to an improper fraction.

Example 9 **Finding Perimeter**

A parcel of land has the following dimensions. Find the perimeter.

Solution:

To find the perimeter, add the lengths of the sides.

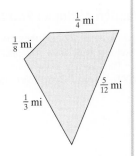

$$\frac{1}{8} + \frac{1}{4} + \frac{5}{12} + \frac{1}{3} \qquad \text{The LCD is 24.}$$

$$= \frac{1 \cdot 3}{8 \cdot 3} + \frac{1 \cdot 6}{4 \cdot 6} + \frac{5 \cdot 2}{12 \cdot 2} + \frac{1 \cdot 8}{3 \cdot 8} \qquad \text{Convert to like fractions.}$$

$$= \frac{3}{24} + \frac{6}{24} + \frac{10}{24} + \frac{8}{24}$$

$$= \frac{27}{24} \qquad \text{Add the fractions.}$$

$$= \frac{9}{8} \text{ or } 1\frac{1}{8} \qquad \text{Simplify to lowest terms.}$$

The perimeter is $1\frac{1}{8}$ mi.

Skill Practice

10. Twelve members of a college hiking club hiked the perimeter of a canyon. How far did they hike?

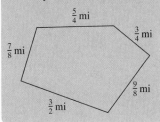

Classroom Example: p. 188, Exercise 74

Answer
10. They hiked $\frac{11}{2}$ mi or equivalently $5\frac{1}{2}$ mi.

Section 3.3 Practice Exercises

Study Skills Exercise

For additional exercises, see Classroom Activities 3.3A–3.3C in the *Student's Resource Manual* at www.mhhe.com/moh.

Do you need complete silence, or do you listen to music while you do your homework?
Try something different today so that you can compare and choose the best working environment for you.

Vocabulary and Key Concepts

1. **a.** To add or subtract fractions, a common denominator (is/is not) needed.   is

   b. To multiply or divide fractions, a common denominator (is/is not) needed.   is not

Review Exercises

For Exercises 2–13, rewrite the fraction with the given denominator.

2. $\dfrac{3}{5} = \dfrac{}{15}$ $\frac{9}{15}$

3. $\dfrac{6}{7} = \dfrac{}{14}$ $\frac{12}{14}$

4. $\dfrac{4}{9} = \dfrac{}{36}$ $\frac{16}{36}$

5. $\dfrac{2}{3} = \dfrac{}{21}$ $\frac{14}{21}$

6. $\dfrac{3}{1} = \dfrac{}{10}$ $\frac{30}{10}$

7. $\dfrac{5}{1} = \dfrac{}{5}$ $\frac{25}{5}$

8. $\dfrac{4}{1} = \dfrac{}{12}$ $\frac{48}{12}$

9. $\dfrac{2}{1} = \dfrac{}{4}$ $\frac{8}{4}$

10. $\dfrac{3}{4} = \dfrac{}{12}$ $\frac{9}{12}$

11. $\dfrac{4}{5} = \dfrac{}{100}$ $\frac{80}{100}$

12. $\dfrac{3}{2} = \dfrac{}{18}$ $\frac{27}{18}$

13. $\dfrac{1}{8} = \dfrac{}{40}$ $\frac{5}{40}$

Concept 1: Addition and Subtraction of Unlike Fractions

14. Explain the difference between the procedures to add fractions and to multiply fractions.   To multiply two fractions, multiply their numerators and multiply their denominators. Then simplify to lowest terms. To add two fractions, rewrite the fractions so that they have a common denominator. Then add the numerators and keep the common denominator. Simplify the answer to lowest terms.

For Exercises 15–50, add or subtract. Write the answer as a fraction simplified to lowest terms. **(See Examples 1–4.)**

15. $\dfrac{2}{3} + \dfrac{1}{8}$ $\frac{19}{24}$

16. $\dfrac{3}{4} + \dfrac{2}{5}$ $\frac{23}{20}$

17. $\dfrac{1}{15} + \dfrac{1}{10}$ $\frac{1}{6}$

18. $\dfrac{5}{6} + \dfrac{3}{8}$ $\frac{29}{24}$

19. $\dfrac{1}{10} + \dfrac{3}{20}$ $\frac{1}{4}$

20. $\dfrac{4}{15} + \dfrac{2}{5}$ $\frac{2}{3}$

21. $\dfrac{5}{6} + \dfrac{8}{7}$ $\frac{83}{42}$

22. $\dfrac{2}{11} + \dfrac{4}{5}$ $\frac{54}{55}$

23. $\dfrac{7}{8} - \dfrac{1}{5}$ $\frac{27}{40}$

24. $\dfrac{9}{10} - \dfrac{1}{3}$ $\frac{17}{30}$

25. $\dfrac{13}{12} - \dfrac{3}{4}$ $\frac{1}{3}$

26. $\dfrac{29}{30} - \dfrac{7}{10}$ $\frac{4}{15}$

27. $\dfrac{10}{9} - \dfrac{5}{12}$ $\frac{25}{36}$

28. $\dfrac{7}{6} - \dfrac{1}{15}$ $\frac{11}{10}$

29. $\dfrac{5}{8} - \dfrac{0}{11}$ $\frac{5}{8}$

30. $\dfrac{7}{12} - \dfrac{0}{5}$ $\frac{7}{12}$

31. $2 + \dfrac{9}{8}$ $\frac{25}{8}$

32. $3 + \dfrac{11}{9}$ $\frac{38}{9}$

33. $4 - \dfrac{4}{3}$ $\frac{8}{3}$

34. $2 - \dfrac{3}{8}$ $\frac{13}{8}$

35. $\dfrac{14}{3} + 1$ $\frac{17}{3}$

36. $\dfrac{12}{5} + 2$ $\frac{22}{5}$

37. $\dfrac{16}{7} - 2$ $\frac{2}{7}$

38. $\dfrac{15}{4} - 3$ $\frac{3}{4}$

39. $\dfrac{7}{10} + \dfrac{19}{100}$ $\frac{89}{100}$

40. $\dfrac{3}{10} + \dfrac{27}{100}$ $\frac{57}{100}$

41. $\dfrac{1}{10} - \dfrac{9}{100}$ $\frac{1}{100}$

42. $\dfrac{3}{100} - \dfrac{21}{1000}$ $\frac{9}{1000}$

43. $\dfrac{3}{10} + \dfrac{9}{100} + \dfrac{1}{1000}$ $\frac{391}{1000}$

44. $\dfrac{1}{10} + \dfrac{3}{100} + \dfrac{7}{1000}$ $\frac{137}{1000}$

45. $\dfrac{5}{3} - \dfrac{7}{6} + \dfrac{5}{8}$ $\frac{9}{8}$

46. $\dfrac{7}{12} - \dfrac{2}{15} + \dfrac{5}{18}$ $\frac{131}{180}$

47. $\dfrac{1}{20} + \dfrac{5}{8} - \dfrac{7}{24}$ $\dfrac{23}{60}$ **48.** $\dfrac{5}{8} + \dfrac{3}{10} - \dfrac{1}{12}$ $\dfrac{101}{120}$ **49.** $\dfrac{1}{2} + \dfrac{1}{4} - \dfrac{1}{8} - \dfrac{1}{16}$ $\dfrac{9}{16}$ **50.** $\dfrac{1}{3} - \dfrac{1}{9} + \dfrac{1}{27} - \dfrac{1}{81}$ $\dfrac{20}{81}$

Concept 2: Order of Operations

For Exercises 51–64, simplify by applying the order of operations. Write the answer as a fraction. **(See Examples 5 and 6.)**

51. $\left(\dfrac{1}{2} - \dfrac{1}{3}\right)^2$ $\dfrac{1}{36}$ **52.** $\left(\dfrac{2}{3} + \dfrac{1}{6}\right)^2$ $\dfrac{25}{36}$ **53.** $\dfrac{2}{3} \div \dfrac{1}{2} - \dfrac{3}{4}$ $\dfrac{7}{12}$ **54.** $\dfrac{3}{5} \div \dfrac{6}{7} - \dfrac{2}{5}$ $\dfrac{3}{10}$

55. $\dfrac{5}{6} + \dfrac{3}{8} \div \dfrac{1}{4}$ $\dfrac{7}{3}$ **56.** $\dfrac{11}{12} + \dfrac{1}{9} \div \dfrac{7}{9}$ $\dfrac{89}{84}$ **57.** $\left(\dfrac{7}{10} - \dfrac{1}{5}\right) \cdot \dfrac{8}{3}$ $\dfrac{4}{3}$ **58.** $\left(\dfrac{2}{5} + \dfrac{9}{10}\right) \cdot \dfrac{5}{6}$ $\dfrac{13}{12}$

59. $\dfrac{4}{5} + \dfrac{5}{8} \cdot \dfrac{16}{35}$ $\dfrac{38}{35}$ **60.** $\dfrac{1}{6} + \dfrac{3}{7} \cdot \dfrac{14}{15}$ $\dfrac{17}{30}$ **61.** $\left(\dfrac{2}{5}\right)^3 + \dfrac{1}{25}$ $\dfrac{13}{125}$ **62.** $\left(\dfrac{3}{2}\right)^3 - \dfrac{5}{4}$ $\dfrac{17}{8}$

63. $\left(\dfrac{1}{4}\right)^2 \div \left(\dfrac{5}{6} - \dfrac{2}{3}\right) + \dfrac{7}{12}$ $\dfrac{23}{24}$ **64.** $\left(\dfrac{1}{2} + \dfrac{1}{3}\right) \cdot \left(\dfrac{2}{5}\right)^2 + \dfrac{3}{10}$ $\dfrac{13}{30}$

Concept 3: Applications Involving Unlike Fractions

65. When doing her laundry, Inez added $\frac{3}{4}$ cup of bleach to $\frac{3}{8}$ cup of liquid detergent. How much total liquid is added to her wash? Inez added $1\frac{1}{8}$ cups.

66. What is the smallest possible length of screw needed to pass through two pieces of wood, one that is $\frac{7}{8}$ in. thick and one that is $\frac{1}{2}$ in. thick? A screw that is at least $1\frac{3}{8}$ in. long is needed.

67. Before a storm, a rain gauge has $\frac{1}{8}$ in. of water. After the storm, the gauge has $\frac{9}{32}$ in. How many inches of rain did the storm deliver? **(See Example 7.)** The storm delivered $\frac{5}{32}$ in. of rain.

68. In one week it rained $\frac{5}{16}$ in. If a garden needs $\frac{9}{8}$ in. of water per week, how much more water does it need? The garden needs $\frac{13}{16}$ in. more water.

69. A watering trough holds 5 gal of water. In the summer, water evaporates at a rate of approximately $\frac{3}{8}$ gal per day. After 4 days, Jeff pours in another gallon and a half of water ($\frac{3}{2}$ gal). How much water is in the trough? **(See Example 8.)** The trough now holds the original amount of 5 gal.

70. A contractor hired two electricians to do a job. One did $\frac{3}{5}$ of the job and the other did $\frac{3}{8}$ of the job. Did the job get completed? If not, what fraction of the job is left? The job did not get completed. There is still $\frac{1}{40}$ of the job left.

71. The information in the graph shows the distribution of a college student body by class.

    a. What fraction of the student body consists of upper classmen (juniors and seniors)? $\dfrac{13}{36}$

    b. What fraction of the student body consists of freshmen and sophomores? $\dfrac{23}{36}$

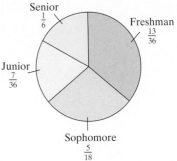

Distribution of Student Body by Class

Senior $\frac{1}{6}$

Freshman $\frac{13}{36}$

Junior $\frac{7}{36}$

Sophomore $\frac{5}{18}$

72. A group of college students took part in a survey. One of the survey questions read:

"Do you think the government should spend more money on research to produce alternative forms of fuel?"

The results of the survey are shown in the figure.

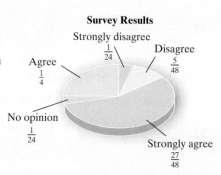

Survey Results

a. What fraction of the survey participants chose to strongly agree or agree? $\frac{13}{16}$

b. What fraction of the survey participants chose to strongly disagree or disagree? $\frac{7}{48}$

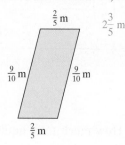 For Exercises 73 and 74, find the perimeter. **(See Example 9.)**

73.

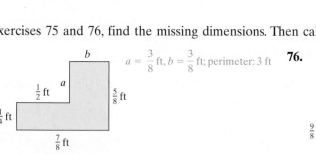

$2\frac{3}{5}$ m

74.

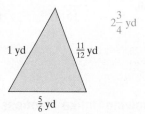

$2\frac{3}{4}$ yd

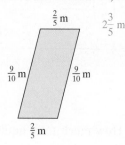 For Exercises 75 and 76, find the missing dimensions. Then calculate the perimeter.

75.

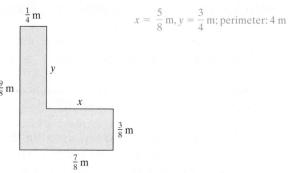

$a = \frac{3}{8}$ ft, $b = \frac{3}{8}$ ft; perimeter: 3 ft

76.

$x = \frac{5}{8}$ m, $y = \frac{3}{4}$ m; perimeter: 4 m

Expanding Your Skills

77. Which fraction is closest to $\frac{1}{2}$?

    a. $\frac{3}{4}$ **b.** $\frac{7}{10}$ **c.** $\frac{5}{6}$ b

78. Which fraction is closest to $\frac{3}{4}$?

    a. $\frac{5}{8}$ **b.** $\frac{7}{12}$ **c.** $\frac{5}{6}$ c

Writing   Translating Expression   Geometry   Scientific Calculator   Video   NS&E

Addition and Subtraction of Mixed Numbers

1. Addition of Mixed Numbers

In this section, we learn to add and subtract mixed numbers. To find the sum of two or more mixed numbers, add the whole-number parts and add the fractional parts.

Example 1 Adding Mixed Numbers

Add. $1\frac{5}{9} + 2\frac{1}{9}$

Solution:

$$1\frac{5}{9}$$
$$+2\frac{1}{9}$$
$$\overline{3\frac{6}{9}} = 3\frac{2}{3}$$

Add the whole ⟶ numbers.   ⟵ Add the fractional parts.

The sum is $3\frac{2}{3}$.

TIP: To understand why mixed numbers can be added in this way, recall that $1\frac{5}{9} = 1 + \frac{5}{9}$ and $2\frac{1}{9} = 2 + \frac{1}{9}$. Therefore,

$$1\frac{5}{9} + 2\frac{1}{9} = 1 + \frac{5}{9} + 2 + \frac{1}{9}$$
$$= 3 + \frac{6}{9}$$
$$= 3\frac{6}{9}$$
$$= 3\frac{2}{3}$$

Skill Practice

Add.

1. $7\frac{2}{15} + 2\frac{1}{15}$

Classroom Example: p. 195, Exercise 12

When we perform operations on mixed numbers, it is often desirable to estimate the answer first. When rounding a mixed number, we offer the following convention.

Rounding a Mixed Number

Step 1 If the fractional part of a mixed number is greater than or equal to $\frac{1}{2}$ (that is, if the numerator is half the denominator or greater), round to the next-greater whole number. For example, $6\frac{9}{16}$ and $6\frac{1}{2}$ both round to 7.

Step 2 If the fractional part of a mixed number is less than $\frac{1}{2}$ (that is, if the numerator is less than half the denominator), the mixed number rounds down to the whole number. For example $6\frac{7}{16}$ rounds to 6.

Example 2 Adding Mixed Numbers

Estimate the sum and then find the actual sum.

$$42\frac{1}{12} + 17\frac{7}{8}$$

Solution:

To estimate the sum, first round the addends.

$$42\frac{1}{12} \qquad \text{rounds to} \qquad 42$$
$$\underline{+\, 17\frac{7}{8}} \qquad \text{rounds to} \qquad \underline{+\, 18}$$
$$\hspace{7.5cm} 60 \qquad \text{The estimated value is 60.}$$

Skill Practice

Estimate the sum and then find the actual sum.

2. $6\frac{1}{11} + 3\frac{1}{2}$

Classroom Example: p. 195, Exercise 16

Answers

1. $9\frac{1}{5}$
2. Estimate: 10; actual sum: $9\frac{13}{22}$

To find the actual sum, we must first write the fractional parts as like fractions. The LCD is 24.

$$42\frac{1}{12} = \quad 42\frac{1 \cdot 2}{12 \cdot 2} = \quad 42\frac{2}{24}$$

$$+ \ 17\frac{7}{8} = \ + \ 17\frac{7 \cdot 3}{8 \cdot 3} = \ + \ 17\frac{21}{24}$$

$$\overline{\phantom{+ \ 17\frac{7}{8}}} \qquad \overline{\phantom{+ \ 17\frac{7 \cdot 3}{8 \cdot 3}}} \qquad 59\frac{23}{24}$$

The actual sum is $59\frac{23}{24}$. This is close to our estimate of 60.

Skill Practice

Estimate the sum and then find the actual sum.

3. $5\frac{2}{5} + 7\frac{8}{9}$

Classroom Example: p. 195, Exercise 30

| **Example 3** | **Adding Mixed Numbers with Carrying** |

Estimate the sum and then find the actual sum.

$$7\frac{5}{6} + 3\frac{3}{5}$$

Solution:

$$7\frac{5}{6} \qquad \text{rounds to} \qquad 8$$

$$+ \ 3\frac{3}{5} \qquad \text{rounds to} \qquad \frac{+ \ 4}{12} \qquad \text{The estimated value is 12.}$$

To find the actual sum, first write the fractional parts as like fractions. The LCD is 30.

$$7\frac{5}{6} = \quad 7\frac{5 \cdot 5}{6 \cdot 5} = \quad 7\frac{25}{30}$$

$$+ \ 3\frac{3}{5} = \ + \ 3\frac{3 \cdot 6}{5 \cdot 6} = \ + \ 3\frac{18}{30}$$

$$\overline{\phantom{+ \ 3\frac{3}{5}}} \qquad \overline{\phantom{+ \ 3\frac{3 \cdot 6}{5 \cdot 6}}} \qquad 10\frac{43}{30}$$

Concept Connections

4. Explain how you would rewrite $2\frac{9}{8}$ as a mixed number containing a proper fraction.

Notice that the number $\frac{43}{30}$ is an improper fraction. By convention, a mixed number is written as a whole number and a *proper* fraction. We have $\frac{43}{30} = 1\frac{13}{30}$. Therefore,

$$10\frac{43}{30} = 10 + 1\frac{13}{30} = 11\frac{13}{30}$$

The sum is $11\frac{13}{30}$. This is close to our estimate of 12.

We have shown how to add mixed numbers by writing the numbers in columns. Another approach to add or subtract mixed numbers is to write the numbers first as improper fractions. Then add or subtract the fractions, as you learned in Section 3.3. To demonstrate this process, we add the mixed numbers from Example 3.

Answers

3. Estimate: 13; actual sum: $13\frac{13}{45}$

4. Write the improper fraction $\frac{9}{8}$ as $1\frac{1}{8}$, and add the result to 2. The result is $3\frac{1}{8}$.

Example 4 Adding Mixed Numbers by Using Improper Fractions

Add. $7\frac{5}{6} + 3\frac{3}{5}$

Solution:

$7\frac{5}{6} + 3\frac{3}{5}$ Write each mixed number as an improper fraction.

$\qquad\qquad 7\frac{5}{6} = \dfrac{7 \cdot 6 + 5}{6} = \dfrac{47}{6}$ and $3\frac{3}{5} = \dfrac{3 \cdot 5 + 3}{5} = \dfrac{18}{5}$

$= \dfrac{47}{6} + \dfrac{18}{5}$

$= \dfrac{47 \cdot 5}{6 \cdot 5} + \dfrac{18 \cdot 6}{5 \cdot 6}$ Convert the fractions to like fractions.
The LCD is 30.

$= \dfrac{235}{30} + \dfrac{108}{30}$ The fractions are now like fractions.

$= \dfrac{343}{30}$ Add the like fractions.

$= 11\dfrac{13}{30}$ Convert the improper fraction to a
mixed number.

$$\begin{array}{r} 11 \\ 30\overline{)343} \quad 11\frac{13}{30} \\ -30 \\ \hline 43 \\ -30 \\ \hline 13 \end{array}$$

The mixed number $11\frac{13}{30}$ is the same as the value obtained in Example 3.

Skill Practice

Add the mixed numbers by first
converting the addends to
improper fractions. Write the
answer as a mixed number.

5. $12\frac{1}{3} + 4\frac{3}{4}$

Classroom Example: p. 196,
Exercise 64

2. Subtraction of Mixed Numbers

To subtract mixed numbers, we subtract the fractional parts and subtract the whole-number parts.

Example 5 Subtracting Mixed Numbers

Estimate the difference and then find the actual difference.

$$15\frac{2}{3} - 4\frac{1}{6}$$

Solution:

To estimate the difference, first round to the nearest whole number.

$$\begin{array}{lll} 15\frac{2}{3} & \text{rounds to} & 16 \\[2mm] -\ 4\frac{1}{6} & \text{rounds to} & \underline{-\ 4} \\[2mm] & & 12 \qquad \text{The estimated value is 12.} \end{array}$$

Skill Practice

Subtract.

6. $6\frac{3}{4} - 2\frac{1}{3}$

Classroom Example: p. 196,
Exercise 40

Answers

5. $17\frac{1}{12}$ **6.** $4\frac{5}{12}$

To subtract the fractional parts, we need a common denominator. The LCD is 6.

$$15\frac{2}{3} = 15\frac{2 \cdot 2}{3 \cdot 2} = 15\frac{4}{6}$$

$$-4\frac{1}{6} = -4\frac{1}{6} \quad = -4\frac{1}{6}$$

$$11\frac{3}{6} = 11\frac{1}{2}$$

Subtract the whole numbers. ⎺⎹ ⎹⎺ Subtract the fractional parts.

The difference is $11\frac{1}{2}$. This is close to the estimate of 12.

Borrowing is sometimes necessary when subtracting mixed numbers.

Example 6   **Subtracting Mixed Numbers with Borrowing**

Subtract.

a. $17\frac{2}{7} - 11\frac{5}{7}$ **b.** $14\frac{2}{9} - 9\frac{3}{5}$

Solution:

a. We cannot subtract $\frac{5}{7}$ from $\frac{2}{7}$. Therefore, borrow 1 from 17. The borrowed 1 is written as $\frac{7}{7}$ because the common denominator is 7.

$$17\frac{2}{7} = \overset{16}{\cancel{17}}\frac{2}{7} + \frac{7}{7} = 16\frac{9}{7}$$

$$-11\frac{5}{7} = -11\frac{5}{7} \quad = -11\frac{5}{7}$$

$$5\frac{4}{7}$$

The difference is $5\frac{4}{7}$.

b. To subtract the fractional parts, we need a common denominator. The LCD is 45.

$$14\frac{2}{9} = 14\frac{2 \cdot 5}{9 \cdot 5} = 14\frac{10}{45}$$
$$-9\frac{3}{5} = -9\frac{3 \cdot 9}{5 \cdot 9} = -9\frac{27}{45}$$

We cannot subtract $\frac{27}{45}$ from $\frac{10}{45}$. Therefore, borrow 1 (or equivalently $\frac{45}{45}$) from 14.

$$= \overset{13}{\cancel{14}}\frac{10}{45} + \frac{45}{45} = 13\frac{55}{45}$$

$$= -9\frac{27}{45} \quad = -9\frac{27}{45}$$

$$4\frac{28}{45}$$

The difference is $4\frac{28}{45}$.

Classroom Examples: p. 196, Exercises 48 and 52

Skill Practice

Subtract.

7. $24\frac{2}{7} - 8\frac{5}{7}$

8. $9\frac{2}{3} - 8\frac{3}{4}$

Answers

7. $15\frac{4}{7}$ **8.** $\frac{11}{12}$

| **Example 7** | **Subtracting Mixed Numbers with Borrowing** |

Subtract. $4 - 2\frac{5}{8}$

Solution:

$$
\begin{array}{r}
4 \\
- 2\frac{5}{8} \\
\hline
\end{array}
$$

In this case, we have no fractional part from which to subtract.

$$
\begin{array}{r}
4\frac{8}{8} \\
\end{array}
$$

We can borrow 1 or equivalently $\frac{8}{8}$ from the whole number 4.

$$
\begin{array}{r}
4\frac{8}{8} \\
- 2\frac{5}{8} \\
\hline
1\frac{3}{8}
\end{array}
$$

TIP: The borrowed 1 is written as $\frac{8}{8}$ because the common denominator is 8.

The difference is $1\frac{3}{8}$.

TIP: The subtraction problem $4 - 2\frac{5}{8} = 1\frac{3}{8}$ can be checked by adding:

$$1\frac{3}{8} + 2\frac{5}{8} = 3\frac{8}{8} = 3 + 1 = 4 \checkmark$$

Skill Practice

Subtract.

9. $10 - 3\frac{1}{6}$

Classroom Example: p. 196, Exercise 54

Instructor Note: First ask students to estimate the difference.

| **Example 8** | **Subtracting Mixed Numbers by Using Improper Fractions** |

Subtract by first converting to improper fractions. Write the answer as a mixed number.

$$10\frac{2}{5} - 4\frac{3}{4}$$

Solution:

$$10\frac{2}{5} - 4\frac{3}{4} = \frac{52}{5} - \frac{19}{4}$$

Write each mixed number as an improper fraction.

$$= \frac{52 \cdot 4}{5 \cdot 4} - \frac{19 \cdot 5}{4 \cdot 5}$$

Convert the fractions to like fractions. The LCD is 20.

$$= \frac{208}{20} - \frac{95}{20}$$

$$= \frac{113}{20}$$

Subtract the like fractions.

$$= 5\frac{13}{20}$$

Write the result as a mixed number.

$$
\begin{array}{r}
5 \\
20\overline{)113} \\
-100 \\
\hline
13
\end{array}
$$

Skill Practice

Subtract by first converting to improper fractions. Write the answer as a mixed number.

10. $8\frac{2}{9} - 3\frac{5}{6}$

Classroom Example: p. 196, Exercise 70

Answers

9. $6\frac{5}{6}$ **10.** $4\frac{7}{18}$

As you can see from Examples 4 and 8, when we convert mixed numbers to improper fractions, the numerators of the fractions become larger numbers. Thus, we must add (or subtract) larger numerators than if we had used the method involving columns. This is one drawback. However, an advantage to converting to improper fractions first is that there is no need for carrying or borrowing.

3. Applications of Mixed Numbers

Example 9 **Subtracting Mixed Numbers in an Application**

The average height of a 3-year-old girl is $38\frac{1}{3}$ in. The average height of a 4-year-old girl is $41\frac{3}{4}$ in. On average, by how much does a girl grow between the ages of 3 and 4?

Solution:

We use subtraction to find the difference in heights.

$$41\frac{3}{4} = 41\frac{3 \cdot 3}{4 \cdot 3} = 41\frac{9}{12}$$
$$-38\frac{1}{3} = -38\frac{1 \cdot 4}{3 \cdot 4} = -38\frac{4}{12}$$
$$\overline{ \quad\quad\quad\quad\quad 3\frac{5}{12}}$$

The average amount of growth is $3\frac{5}{12}$ in.

Skill Practice

11. On December 1, the snow base at the Bear Mountain Ski Resort was $4\frac{1}{3}$ ft. By January 1, the base was $6\frac{1}{2}$ ft. By how much did the base amount of snow increase?

Classroom Example: p. 197, Exercise 78

Answer

11. $2\frac{1}{6}$ ft

Section 3.4 Practice Exercises

For additional exercises, see Classroom Activities 3.4A–3.4C in the *Student's Resource Manual* at www.mhhe.com/moh.

Study Skills Exercise

Write the page number(s) for the Problem Recognition Exercises for this chapter. _____
Explain the purpose of this set of exercises.

Review Exercises

For Exercises 1–8, add or subtract as indicated. Write the answer as a fraction or whole number.

1. $\dfrac{8}{15} + \dfrac{11}{15}$ $\frac{19}{15}$

2. $\dfrac{3}{16} + \dfrac{7}{12}$ $\frac{37}{48}$

3. $\dfrac{25}{8} - \dfrac{23}{24}$ $\frac{13}{6}$

4. $4 - \dfrac{15}{7}$ $\frac{13}{7}$

5. $\dfrac{9}{5} + 3$ $\frac{24}{5}$

6. $\dfrac{23}{6} + \dfrac{5}{6} - \dfrac{2}{3}$ 4

7. $\dfrac{125}{32} - \dfrac{51}{32} - \dfrac{58}{32}$ $\frac{1}{2}$

8. $\dfrac{17}{10} - \dfrac{23}{100} + \dfrac{321}{1000}$ $\frac{1791}{1000}$

 Writing ←→ Translating Expression Geometry Scientific Calculator Video NS E

Concept 1: Addition of Mixed Numbers

For Exercises 9–16, add the mixed numbers. **(See Examples 1 and 2.)**

9. $2\frac{1}{11}$
   $+ 5\frac{3}{11}$
   $7\frac{4}{11}$

10. $5\frac{2}{7}$
    $+ 4\frac{3}{7}$
    $9\frac{5}{7}$

11. $12\frac{1}{14}$
    $+ 3\frac{5}{14}$
    $15\frac{3}{7}$

12. $1\frac{3}{20}$
    $+ 17\frac{7}{20}$
    $18\frac{1}{2}$

13. $4\frac{5}{16}$
    $+ 11\frac{1}{4}$
    $15\frac{9}{16}$

14. $21\frac{2}{9}$
    $+ 10\frac{1}{3}$
    $31\frac{5}{9}$

15. $6\frac{2}{3}$
    $+ 4\frac{1}{5}$
    $10\frac{13}{15}$

16. $7\frac{1}{6}$
    $+ 3\frac{5}{8}$
    $10\frac{19}{24}$

For Exercises 17–20, round the mixed number to the nearest whole number.

17. $5\frac{1}{3}$ 5

18. $2\frac{7}{8}$ 3

19. $1\frac{3}{5}$ 2

20. $6\frac{3}{7}$ 6

For Exercises 21–24, write the mixed number in proper form (that is, as a whole number with a proper fraction that is simplified to lowest terms).

21. $2\frac{6}{5}$ $3\frac{1}{5}$

22. $4\frac{8}{7}$ $5\frac{1}{7}$

23. $7\frac{5}{3}$ $8\frac{2}{3}$

24. $1\frac{9}{5}$ $2\frac{4}{5}$

For Exercises 25–34, round the numbers to estimate the answer. Then find the exact sum. In Exercise 25, the estimate is done for you. **(See Example 3.)**

	Estimate	Exact		Estimate	Exact		Estimate	Exact
25.	7	$6\frac{3}{4}$	26.		$8\frac{3}{5}$	27.		$14\frac{7}{8}$
	$+ 8$	$+ 7\frac{3}{4}$		$+$	$+ 13\frac{4}{5}$		$+$	$+ 8\frac{1}{4}$
	15	$14\frac{1}{2}$			$22\frac{2}{5}$			$23\frac{1}{8}$

	Estimate	Exact		Estimate	Exact		Estimate	Exact
28.		$21\frac{3}{5}$	29.		$3\frac{7}{16}$	30.		$7\frac{7}{9}$
	$+$	$+ 24\frac{9}{10}$		$+$	$+ 15\frac{11}{12}$		$+$	$+ 8\frac{5}{6}$
		$46\frac{1}{2}$			$19\frac{17}{48}$			$16\frac{11}{18}$

31. $3 + 6\frac{7}{8}$ $9\frac{7}{8}$

32. $5 + 11\frac{1}{13}$ $16\frac{1}{13}$

33. $32\frac{2}{7} + 10$ $42\frac{2}{7}$

34. $2\frac{18}{37} + 16$ $18\frac{18}{37}$

Concept 2: Subtraction of Mixed Numbers

For Exercises 35–42, subtract the mixed numbers. **(See Example 5.)**

35. $21\frac{9}{10}$
    $- 10\frac{3}{10}$
    $11\frac{3}{5}$

36. $19\frac{2}{3}$
    $- 4\frac{1}{3}$
    $15\frac{1}{3}$

37. $5\frac{9}{15}$
    $- 3\frac{7}{15}$
    $2\frac{2}{15}$

38. $33\frac{11}{12}$
    $- 14\frac{5}{12}$
    $19\frac{1}{2}$

39. $18\dfrac{5}{6}$

$-\ 6\dfrac{2}{3}$

$12\dfrac{1}{6}$

40. $21\dfrac{17}{20}$

$-\ 20\dfrac{1}{10}$

$1\dfrac{3}{4}$

41. $11\dfrac{5}{7}$

$-\ 9\dfrac{5}{14}$

$2\dfrac{5}{14}$

42. $5\dfrac{9}{11}$

$-\ 2\dfrac{13}{22}$

$3\dfrac{5}{22}$

For Exercises 43–46, rewrite the number 1 as a fraction having the given denominator.

43. $3\quad\dfrac{3}{3}$

44. $5\quad\dfrac{5}{5}$

45. $12\quad\dfrac{12}{12}$

46. $6\quad\dfrac{6}{6}$

NS&E For Exercises 47–60, round the numbers to estimate the answer. Then find the exact difference. In Exercise 47, the estimate is done for you. **(See Examples 6 and 7.)**

Estimate   Exact

47. $\quad 25 \qquad 25\dfrac{1}{4}$

$\underline{-\ 14} \qquad \underline{-\ 13\dfrac{3}{4}}$

$\quad 11 \qquad\qquad 11\dfrac{1}{2}$

Estimate   Exact

48. $\qquad\qquad 36\dfrac{1}{5}$

$\underline{-\qquad} \qquad \underline{-\ 12\dfrac{3}{5}}$

$\qquad\qquad 23\dfrac{3}{5}$

Estimate   Exact

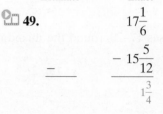 **49.** $\qquad\qquad 17\dfrac{1}{6}$

$\underline{-\qquad} \qquad \underline{-\ 15\dfrac{5}{12}}$

$\qquad\qquad 1\dfrac{3}{4}$

Estimate   Exact

50. $\qquad\qquad 22\dfrac{5}{18}$

$\underline{-\qquad} \qquad \underline{-\ 10\dfrac{7}{9}}$

$\qquad\qquad 11\dfrac{1}{2}$

Estimate   Exact

51. $\qquad\qquad 46\dfrac{3}{7}$

$\underline{-\qquad} \qquad \underline{-\ 38\dfrac{1}{2}}$

$\qquad\qquad 7\dfrac{13}{14}$

Estimate   Exact

52. $\qquad\qquad 23\dfrac{1}{2}$

$\underline{-\qquad} \qquad \underline{-\ 18\dfrac{10}{13}}$

$\qquad\qquad 4\dfrac{19}{26}$

53. $6 - 2\dfrac{5}{6}\quad 3\dfrac{1}{6}$

54. $9 - 4\dfrac{1}{2}\quad 4\dfrac{1}{2}$

55. $12 - 9\dfrac{2}{9}\quad 2\dfrac{7}{9}$

56. $10 - 9\dfrac{1}{3}\quad \dfrac{2}{3}$

57. $5\dfrac{3}{17} - 3\quad 2\dfrac{3}{17}$

58. $16\dfrac{4}{11} - 5\quad 11\dfrac{4}{11}$

59. $23\dfrac{5}{14} - 17\quad 6\dfrac{5}{14}$

60. $21\dfrac{3}{4} - 10\quad 11\dfrac{3}{4}$

Mixed Exercises

For Exercises 61–76, add or subtract the mixed numbers. Write the answers as mixed numbers, if possible. **(See Examples 4 and 8.)**

61. $2\dfrac{2}{3} + 4\dfrac{5}{8}\quad 7\dfrac{7}{24}$

62. $5\dfrac{1}{4} - 3\dfrac{1}{2}\quad 1\dfrac{3}{4}$

63. $1\dfrac{11}{15} + 4\dfrac{2}{5}\quad 6\dfrac{2}{15}$

64. $2\dfrac{10}{11} + 2\dfrac{1}{2}\quad 5\dfrac{9}{22}$

65. $3\dfrac{7}{8} - 3\dfrac{3}{16}\quad \dfrac{11}{16}$

66. $3\dfrac{1}{6} - 1\dfrac{23}{24}\quad 1\dfrac{5}{24}$

67. $4\dfrac{1}{12} + 5\dfrac{1}{9}\quad 9\dfrac{7}{36}$

68. $10\dfrac{2}{25} - 7\dfrac{13}{20}\quad 2\dfrac{43}{100}$

69. $9\dfrac{5}{32} - 8\dfrac{1}{4}\quad \dfrac{29}{32}$

70. $4\dfrac{3}{40} - 2\dfrac{7}{8}\quad 1\dfrac{1}{5}$

71. $6\dfrac{11}{14} + 4\dfrac{1}{6}\quad 10\dfrac{20}{21}$

72. $8\dfrac{3}{22} + 4\dfrac{1}{4}\quad 12\dfrac{17}{44}$

73. $12\dfrac{1}{5} - 11\dfrac{2}{7}\quad \dfrac{32}{35}$

74. $5\dfrac{11}{30} + 5\dfrac{3}{4}\quad 11\dfrac{7}{60}$

75. $10\dfrac{1}{8} - 2\dfrac{17}{18}\quad 7\dfrac{13}{72}$

76. $3\dfrac{8}{21} + 6\dfrac{8}{9}\quad 10\dfrac{17}{63}$

Concept 3: Applications of Mixed Numbers

For Exercises 77–80, use the table to find the lengths of several common birds.

Bird	Length
Cuban Bee Hummingbird	$2\frac{1}{4}$ in.
Sedge Wren	$3\frac{1}{2}$ in.
Great Carolina Wren	$5\frac{1}{2}$ in.
Belted Kingfisher	$11\frac{1}{4}$ in.

77. How much longer is the Belted Kingfisher than the Sedge Wren? **(See Example 9.)** $7\frac{3}{4}$ in.

78. How much longer is the Great Carolina Wren than the Cuban Bee Hummingbird? $3\frac{1}{4}$ in.

79. Estimate or measure the length of your index finger. Which is longer, your index finger or a Cuban Bee Hummingbird?
The index finger is longer.

80. For a recent year, the smallest living dog in the United States was Brandy, a female Chihuahua, who measures 6 in. in length. How much longer is a Belted Kingfisher than Brandy?
A Belted Kingfisher is $5\frac{1}{4}$ in. longer.

81. A student has three part-time jobs. She tutors, delivers newspapers, and takes notes for a blind student. During a typical week she works $8\frac{2}{3}$ hr delivering newspapers, $4\frac{1}{2}$ hr tutoring, and $3\frac{3}{4}$ hr note-taking. What is the total number of hours worked in a typical week?
The total is $16\frac{11}{12}$ hr.

82. A contractor ordered three loads of gravel. The orders were for $2\frac{1}{2}$ tons, $3\frac{1}{8}$ tons, and $4\frac{1}{3}$ tons. What is the total amount of gravel ordered?
The total amount is $9\frac{23}{24}$ tons.

83. A plumber fits together two pipes. Find the length of the larger piece. $3\frac{5}{12}$ ft

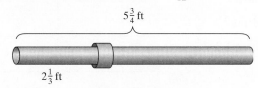

$5\frac{3}{4}$ ft

$2\frac{1}{3}$ ft

84. Find the thickness of the carpeting and pad. $\frac{13}{16}$ in.

$\frac{3}{8}$ in.

$\frac{7}{16}$ in.

85. A pipe has an outer diameter of $1\frac{3}{16}$ in. and an inner diameter of $1\frac{1}{8}$ in. Find the thickness of the pipe. $\frac{1}{32}$ in.

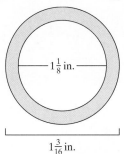

$1\frac{1}{8}$ in.

$1\frac{3}{16}$ in.

86. Find the inside diameter of the washer. $\frac{7}{16}$ in.

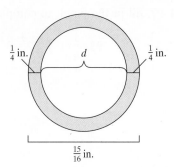

$\frac{1}{4}$ in. $\quad d \quad$ $\frac{1}{4}$ in.

$\frac{15}{16}$ in.

87. When using the word processor, Microsoft Word, the default margins are $1\frac{1}{4}$ in. for the left and right margins. If an $8\frac{1}{2}$-in. by 11-in. piece of paper is used, what is the width of the printing area?

The printing area width is 6 in.

88. A water gauge in a pond measured $25\frac{7}{8}$ in. on Monday. After 2 days of rain and runoff, the gauge read $32\frac{1}{2}$ in. By how much did the water level rise?

The water rose $6\frac{5}{8}$ in.

89. A flight from Atlanta to San Diego takes $5\frac{1}{3}$ hr. After $2\frac{1}{2}$ hr, how much time remains?   There is $2\frac{5}{6}$ hr remaining.

90. In a triathlon, an athlete must swim $\frac{1}{4}$ mi, bike $10\frac{1}{2}$ mi, and run $3\frac{1}{5}$ mi. What is the total distance?

The total distance is $13\frac{19}{20}$ mi.

91. Vertical blinds were purchased for a window that is $3\frac{5}{12}$ ft high. The blinds are $3\frac{3}{4}$ ft in length. Find the distance that the blinds will hang below the window.

The blinds will hang $\frac{1}{3}$ ft below the window.

92. The number of hours worked per day for a plumber is given in the table. How many more hours did he work on Monday than on Saturday?

Monday	Tuesday	Wednesday	Thursday	Friday	Saturday	Sunday
$9\frac{1}{6}$ hr	$7\frac{3}{4}$ hr	$8\frac{1}{3}$ hr	$8\frac{1}{2}$ hr	$4\frac{1}{2}$ hr	$3\frac{3}{4}$ hr	0

He worked $5\frac{5}{12}$ hr more on Monday.

93. A patient admitted to the hospital was dehydrated. In addition to intravenous (IV) fluids, the doctor told the patient that she must drink at least 4 L of an electrolyte solution within the next 12 hr. A nurse recorded the amounts the patient drank in the patient's chart.

   a. How many liters of electrolyte solution did the patient drink? $3\frac{3}{8}$ L

   b. How much more would the patient need to drink to reach 4 L? $\frac{5}{8}$ L

Time	Amount
7 A.M.–10 A.M.	$1\frac{1}{4}$ L
10 A.M.–1 P.M.	$\frac{7}{8}$ L
1 P.M.–4 P.M.	$\frac{3}{4}$ L
4 P.M.–7 P.M.	$\frac{1}{2}$ L

Expanding Your Skills

For Exercises 94–97, fill in the blank to complete the pattern.

94. $1, 1\frac{1}{3}, 1\frac{2}{3}, 2, 2\frac{1}{3}, \Box$

$2\frac{2}{3}$

95. $\frac{1}{4}, 1, 1\frac{3}{4}, 2\frac{1}{2}, 3\frac{1}{4}, \Box$

4

96. $\frac{5}{6}, 1\frac{1}{6}, 1\frac{1}{2}, 1\frac{5}{6}, \Box$

$2\frac{1}{6}$

97. $\frac{1}{2}, 1\frac{1}{4}, 2, 2\frac{3}{4}, 3\frac{1}{2}, \Box$

$4\frac{1}{4}$

Writing   Translating Expression   Geometry   Scientific Calculator   Video   NSE

Calculator Connections

Topic: Adding and Subtracting Fractions and Mixed Numbers on a Calculator

Expression	Keystrokes	Result
$\dfrac{7}{18} + \dfrac{1}{3}$	7 [a%] 18 + 1 [a%] 3 [=]	$\boxed{13 \lrcorner 18} = \dfrac{13}{18}$

$7\dfrac{5}{8} - 4\dfrac{2}{3}$ 7 [a%] 5 [a%] 8 [−] 4 [a%] 2 [a%] 3 [=] $\boxed{2\_23 \lrcorner 24} = 2\dfrac{23}{24}$

$\underbrace{\qquad\qquad}_{7\frac{5}{8}}$ $\underbrace{\qquad\qquad}_{4\frac{2}{3}}$

To convert the result to an improper fraction, press [2nd] [d/c] $\boxed{71 \lrcorner 24} = \dfrac{71}{24}$

Calculator Exercises

Add or subtract as indicated.

98. $\dfrac{23}{42} + \dfrac{17}{24}$ $\frac{211}{168}$ or $1\frac{43}{168}$ **99.** $\dfrac{14}{75} + \dfrac{9}{50}$ $\frac{11}{30}$ **100.** $\dfrac{31}{44} - \dfrac{14}{33}$ $\frac{37}{132}$ **101.** $\dfrac{29}{68} - \dfrac{7}{92}$ $\frac{137}{391}$

102. $32\dfrac{7}{18} + 14\dfrac{2}{27}$ **103.** $21\dfrac{3}{28} + 4\dfrac{31}{42}$ **104.** $7\dfrac{11}{21} - 2\dfrac{10}{33}$ **105.** $5\dfrac{14}{17} - 2\dfrac{47}{68}$

$\frac{2509}{54}$ or $46\frac{25}{54}$ $\frac{2171}{84}$ or $25\frac{71}{84}$ $\frac{402}{77}$ or $5\frac{17}{77}$ $\frac{213}{68}$ or $3\frac{9}{68}$

Problem Recognition Exercises

Operations on Fractions and Mixed Numbers

Perform the indicated operations. Check the reasonableness of your answers by estimating.

1. a. $\dfrac{7}{5} + \dfrac{2}{5}$ $\frac{9}{5}$ or $1\frac{4}{5}$ **b.** $\dfrac{7}{5} \times \dfrac{2}{5}$ $\frac{14}{25}$ **c.** $\dfrac{7}{5} \div \dfrac{2}{5}$ $\frac{7}{2}$ or $3\frac{1}{2}$ **d.** $\dfrac{7}{5} - \dfrac{2}{5}$ 1

2. a. $\dfrac{4}{3} \times \dfrac{5}{6}$ $\frac{10}{9}$ or $1\frac{1}{9}$ **b.** $\dfrac{4}{3} \div \dfrac{5}{6}$ $\frac{8}{5}$ or $1\frac{3}{5}$ **c.** $\dfrac{4}{3} + \dfrac{5}{6}$ $\frac{13}{6}$ or $2\frac{1}{6}$ **d.** $\dfrac{4}{3} - \dfrac{5}{6}$ $\frac{1}{2}$

3. a. $2\dfrac{3}{4} + 1\dfrac{1}{2}$ $\frac{17}{4}$ or $4\frac{1}{4}$ **b.** $2\dfrac{3}{4} - 1\dfrac{1}{2}$ $\frac{5}{4}$ or $1\frac{1}{4}$ **c.** $2\dfrac{3}{4} \div 1\dfrac{1}{2}$ $\frac{11}{6}$ or $1\frac{5}{6}$ **d.** $2\dfrac{3}{4} \times 1\dfrac{1}{2}$ $\frac{33}{8}$ or $4\frac{1}{8}$

4. a. $\left(4\dfrac{1}{3}\right) \times \left(2\dfrac{5}{6}\right)$ $\frac{221}{18}$ or $12\frac{5}{18}$ **b.** $\left(4\dfrac{1}{3}\right) \div \left(2\dfrac{5}{6}\right)$ $\frac{26}{17}$ or $1\frac{9}{17}$ **c.** $\left(4\dfrac{1}{3}\right) - \left(2\dfrac{5}{6}\right)$ $\frac{3}{2}$ or $1\frac{1}{2}$ **d.** $\left(4\dfrac{1}{3}\right) + \left(2\dfrac{5}{6}\right)$ $\frac{43}{6}$ or $7\frac{1}{6}$

5. a. $4 - \dfrac{3}{8}$ $\frac{29}{8}$ or $3\frac{5}{8}$ **b.** $4 \times \dfrac{3}{8}$ $\frac{3}{2}$ or $1\frac{1}{2}$ **c.** $4 \div \dfrac{3}{8}$ $\frac{32}{3}$ or $10\frac{2}{3}$ **d.** $4 + \dfrac{3}{8}$ $\frac{35}{8}$ or $4\frac{3}{8}$

6. a. $3\dfrac{2}{3} \div 2$ $\frac{11}{6}$ or $1\frac{5}{6}$ **b.** $3\dfrac{2}{3} - 2$ $\frac{5}{3}$ or $1\frac{2}{3}$ **c.** $3\dfrac{2}{3} + 2$ $\frac{17}{3}$ or $5\frac{2}{3}$ **d.** $3\dfrac{2}{3} \cdot 2$ $\frac{22}{3}$ or $7\frac{1}{3}$

 Writing    Translating Expression   Geometry   Scientific Calculator   Video   NSE

7. a. $4\frac{1}{5} - \frac{2}{3}$ $\frac{53}{15}$ or $3\frac{8}{15}$ **b.** $4\frac{1}{5} + \frac{2}{3}$ $\frac{73}{13}$ or $4\frac{13}{15}$ **c.** $\left(4\frac{1}{5}\right) \cdot \frac{2}{3}$ $\frac{14}{5}$ or $2\frac{4}{5}$ **d.** $\left(4\frac{1}{5}\right) \div \frac{2}{3}$ $\frac{63}{10}$ or $6\frac{3}{10}$

8. a. $\frac{25}{9} \div 2$ $\frac{25}{18}$ or $1\frac{7}{18}$ **b.** $\frac{25}{9} \cdot 2$ $\frac{50}{9}$ or $5\frac{5}{9}$ **c.** $\frac{25}{9} - 2$ $\frac{7}{9}$ **d.** $\frac{25}{9} + 2$ $\frac{43}{9}$ or $4\frac{7}{9}$

9. a. $1\frac{4}{5} \cdot \frac{5}{9}$ 1 **b.** $1\frac{4}{5} + \frac{5}{9}$ $\frac{106}{45}$ or $2\frac{16}{45}$ **c.** $1\frac{4}{5} \div \frac{5}{9}$ $\frac{81}{25}$ or $3\frac{6}{25}$ **d.** $1\frac{4}{5} - \frac{5}{9}$ $\frac{56}{45}$ or $1\frac{11}{45}$

10. a. $\frac{7}{3} \cdot \frac{3}{7}$ 1 **b.** $\frac{7}{3} + \frac{3}{7}$ $\frac{58}{21}$ or $2\frac{16}{21}$ **c.** $\frac{7}{3} - \frac{3}{7}$ $\frac{40}{21}$ or $1\frac{19}{21}$ **d.** $\frac{7}{3} \div \frac{3}{7}$ $\frac{49}{9}$ or $5\frac{4}{9}$

Section 3.5   Order of Operations and Applications of Fractions and Mixed Numbers

Concepts

1. Order of Operations
2. Applications of Fractions and Mixed Numbers
3. Applications of Geometry

1. Order of Operations

At this point in the text, we have learned how to add, subtract, multiply, and divide whole numbers, fractions, and mixed numbers. In this section we practice putting all these skills to use by applying the order of operations.

To review the steps in the order of operations, refer to Section 3.1, page 165.

Skill Practice

Simplify.

1. $\left(4 - \frac{5}{2}\right)^2$

Classroom Example: p. 205, Exercise 18

Example 1   Applying the Order of Operations

Simplify. $\left(3 - \frac{3}{4}\right)^2$

Solution:

$$\left(3 - \frac{3}{4}\right)^2 = \left(\frac{3}{1} - \frac{3}{4}\right)^2$$ First subtract the numbers within parentheses. Write the whole number as an improper fraction.

$$= \left(\frac{3 \cdot 4}{1 \cdot 4} - \frac{3}{4}\right)^2$$ Convert the fractions to like fractions. The LCD is 4.

$$= \left(\frac{12}{4} - \frac{3}{4}\right)^2$$ The fractions are now like.

$$= \left(\frac{9}{4}\right)^2$$ Subtract.

$$= \frac{9}{4} \cdot \frac{9}{4}$$ Square the quantity $\frac{9}{4}$.

$$= \frac{81}{16} \text{ or } 5\frac{1}{16}$$

Answer

1. $\frac{9}{4}$

Example 2 Applying the Order of Operations

Simplify. $1\dfrac{2}{3} \div 6 \cdot \left(\dfrac{3}{10}\right)$

Solution:

$1\dfrac{2}{3} \div 6 \cdot \left(\dfrac{3}{10}\right) = \dfrac{5}{3} \div \dfrac{6}{1} \cdot \dfrac{3}{10}$
Write the mixed number and whole number as improper fractions. Multiply and divide in order from left to right.

$= \dfrac{5}{3} \cdot \dfrac{1}{6} \cdot \dfrac{3}{10}$
Multiply by the reciprocal of the second fraction.

$= \dfrac{\overset{1}{5}}{3} \cdot \dfrac{1}{6} \cdot \dfrac{\overset{1}{3}}{\underset{2}{10}}$
Simplify.

$= \dfrac{1}{12}$
Multiply.

Skill Practice

Simplify.

2. $3\dfrac{1}{4} \div 4 \cdot \left(\dfrac{2}{5}\right)$

Classroom Example: p. 205, Exercise 20

Example 3 Applying the Order of Operations

Simplify. $\left(\dfrac{2}{5}\right)^2 + \left(2\dfrac{5}{8}\right) \cdot \dfrac{3}{7}$

Solution:

$\left(\dfrac{2}{5}\right)^2 + \left(2\dfrac{5}{8}\right) \cdot \dfrac{3}{7}$
Perform the exponent operation first.

$= \dfrac{2}{5} \cdot \dfrac{2}{5} + \left(2\dfrac{5}{8}\right) \cdot \dfrac{3}{7}$
Square the quantity $\frac{2}{5}$.

$= \dfrac{4}{25} + \dfrac{21}{8} \cdot \dfrac{3}{7}$
Write the mixed number as an improper fraction.

$= \dfrac{4}{25} + \dfrac{\overset{3}{21}}{8} \cdot \dfrac{3}{\underset{1}{7}}$
Multiply before adding. Simplify common factors within the second two fractions.

$= \dfrac{4}{25} + \dfrac{9}{8}$

$= \dfrac{4 \cdot 8}{25 \cdot 8} + \dfrac{9 \cdot 25}{8 \cdot 25}$
Add the fractions. The LCD is $25 \cdot 8 = 200$.

$= \dfrac{32}{200} + \dfrac{225}{200}$
The fractions are now like.

$= \dfrac{257}{200} \text{ or } 1\dfrac{57}{200}$
The answer can be written as either an improper fraction or a mixed number.

Skill Practice

Simplify.

3. $3\dfrac{2}{5} - \left(\dfrac{1}{3}\right)^2 \cdot 6$

Classroom Example: p. 206, Exercise 32

Avoiding Mistakes

Do not try to "cancel" the 4 and the 8. The fraction $\frac{4}{25}$ is being *added*, not multiplied.

Answers

2. $\dfrac{13}{40}$ **3.** $\dfrac{41}{15}$ or $2\dfrac{11}{15}$

2. Applications of Fractions and Mixed Numbers

Examples 4 and 5 use operations on fractions and mixed numbers in real-world applications.

Example 4 Using Mixed Numbers in a Sports Application

The graph in Figure 3-4 gives the winning height for the men's high jump for selected Olympic games.

a. What is the difference between the winning high jump in 1992 versus 1948?

b. What is the average height from the 1948, 1960, 1968, and 1992 results?

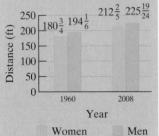
Classroom Example: p. 206, Exercise 36

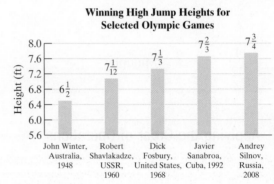

Winning High Jump Heights for Selected Olympic Games

Source: International Association of Athletics Federations

Figure 3-4

Solution:

a. The word "difference" implies subtraction. We subtract the 1948 height from the 1992 height.

$$1992 \text{ height} \longrightarrow \quad 7\frac{2}{3} = \quad 7\frac{2 \cdot 2}{3 \cdot 2} = \quad 7\frac{4}{6}$$

$$1948 \text{ height} \longrightarrow \quad -6\frac{1}{2} = \quad -6\frac{1 \cdot 3}{2 \cdot 3} = \quad -6\frac{3}{6}$$

$$\underline{\phantom{1948 \text{ height} \longrightarrow \quad -6\frac{1}{2} = \quad -6\frac{1 \cdot 3}{2 \cdot 3} = \quad}} 1\frac{1}{6}$$

The difference between the winning heights in 1992 and 1948 is $1\frac{1}{6}$ ft.

b. The average height is found by taking the sum of the four heights and dividing by 4. The sum of the heights is given by

$$\left.\begin{array}{rcl}
6\frac{1}{2} & = & 6\frac{6}{12} \\[6pt]
7\frac{1}{12} & = & 7\frac{1}{12} \\[6pt]
7\frac{1}{3} & = & 7\frac{4}{12} \\[6pt]
+7\frac{2}{3} & = & +7\frac{8}{12}
\end{array}\right\} \quad \text{The LCD of all four fractions is 12.}$$

$$27\frac{19}{12} = 27 + 1\frac{7}{12} = 28\frac{7}{12}$$

Now divide by 4 to find the average:

$$28\frac{7}{12} \div 4 = \frac{343}{12} \div \frac{4}{1}$$ Write the mixed number and whole number as improper fractions.

$$= \frac{343}{12} \cdot \frac{1}{4}$$ Multiply by the reciprocal of the divisor.

$$= \frac{343}{48}$$ Multiply.

$$= 7\frac{7}{48}$$ Write the result as a mixed number. $$\begin{array}{r} 7 \\ 48\overline{)343} \\ -336 \\ \hline 7 \end{array}$$

The average winning height for the men's high jump for the selected years is $7\frac{7}{48}$ ft.

Example 5 **Using Multiplication of Fractions in a Finance Application**

Sheila makes $75,000 as a self-employed graphics artist. However, she pays approximately $\frac{2}{5}$ of her income in income tax, Medicare tax, and social security. If she has business deductions of $15,000 per year, how much total tax does she pay?

Solution:

Sheila will be taxed on her income, minus her business deductions. Therefore, her total tax is given by

$$\text{Total tax} = \frac{2}{5}(75,000 - 15,000)$$

$$= \frac{2}{5}(60,000)$$

$$= \frac{2}{\overset{}{\underset{1}{5}}}\left(\frac{\overset{12,000}{\cancel{60,000}}}{1}\right)$$

$$= 24,000$$

Sheila will have to pay $24,000 in taxes.

3. Applications of Geometry

In Example 6, we review the concepts of perimeter and area in applications.

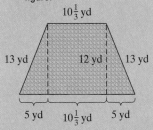
Example 6 **Using Fractions and Mixed Numbers in a Geometry Application**

Jason and Sara plan to paint a side of their house (Figure 3-5).

a. How much area will they have to paint?

b. They want to string Christmas lights around the triangular portion of the house. What length is required for the string of lights?

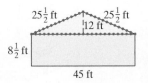

Figure 3-5

Solution:

a. The area of the side of the house is given by the sum of the rectangular area and the triangular area.

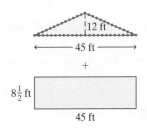

$$\text{Area of triangle} = \frac{1}{2}bh$$

$$= \frac{1}{2}(45 \text{ ft})(12 \text{ ft})$$

$$= \frac{1}{2}\left(\frac{45}{1}\text{ ft}\right)\left(\frac{\overset{6}{12}}{1}\text{ ft}\right)$$

$$= 270 \text{ ft}^2$$

$$\text{Area of rectangle} = l \cdot w$$

$$= (45 \text{ ft})\left(8\frac{1}{2}\text{ ft}\right)$$

$$= \left(\frac{45}{1}\text{ ft}\right)\left(\frac{17}{2}\text{ ft}\right)$$

$$= \frac{765}{2}\text{ ft}^2 \quad \text{or} \quad 382\frac{1}{2}\text{ ft}^2$$

Avoiding Mistakes

In reality, Jason and Sara might want to overestimate the amount of paint needed so that they don't run out of supplies during the job. For example:

Triangle area $= 270 \text{ ft}^2$

Rectangle area $\approx (45 \text{ ft})(10 \text{ ft})$

$\approx 450 \text{ ft}^2$

Total area $\approx 270 \text{ ft}^2 + 450 \text{ ft}^2$

$\approx 720 \text{ ft}^2$

The total area is given by $270 \text{ ft}^2 + 382\frac{1}{2}\text{ ft}^2 = 652\frac{1}{2}\text{ ft}^2$.

The total area to paint is $652\frac{1}{2}\text{ ft}^2$.

b. The perimeter of the triangle is found by adding the lengths of the sides.

$$\begin{array}{r}
25\frac{1}{2}\text{ ft} \\
25\frac{1}{2}\text{ ft} \\
+\ 45\ \ \text{ ft} \\
\hline
95\frac{2}{2}\text{ ft} = 96 \text{ ft}
\end{array}$$

Jason and Sara will need 96 ft of lights.

Section 3.5 Practice Exercises

Study Skills Exercise

For additional exercises, see Classroom Activities 3.5A–3.5B in the *Student's Resource Manual* at www.mhhe.com/moh.

When you take a test, go through the test, doing all the problems that you know first. Then go back and work on the problems that were more difficult. Give yourself a time limit for how much time you spend on each problem (maybe 3 to 5 min the first time through). Circle the importance of each statement.

	Not important	Somewhat important	Very important
a. Read through the entire test first.	1	2	3
b. If time allows, go back and check each problem.	1	2	3
c. Write out all steps instead of doing the work in your head.	1	2	3

Review Exercises

For Exercises 1–8, perform the indicated operation.

1. $4\frac{3}{7} - 1\frac{4}{7}$ $\quad 2\frac{6}{7}$

2. $7\frac{3}{10} + 2\frac{14}{15}$ $\quad 10\frac{7}{30}$

3. $16 - 3\frac{7}{9}$ $\quad 12\frac{2}{9}$

4. $5\frac{5}{8} \cdot 2\frac{1}{9}$ $\quad 11\frac{7}{8}$

5. $7\frac{1}{9} \div 2\frac{2}{3}$ $\quad 2\frac{2}{3}$

6. $24\frac{3}{5} - 14\frac{3}{4}$ $\quad 9\frac{17}{20}$

7. $\left(1\frac{5}{6}\right)^2$ $\quad 3\frac{13}{36}$

8. $13\frac{1}{14} + 4\frac{5}{7}$ $\quad 17\frac{11}{14}$

For Exercises 9–12, convert the mixed number to an improper fraction.

9. $5\frac{2}{13}$ $\quad \frac{67}{13}$

10. $2\frac{7}{11}$ $\quad \frac{29}{11}$

11. $3\frac{9}{10}$ $\quad \frac{39}{10}$

12. $1\frac{15}{16}$ $\quad \frac{31}{16}$

For Exercises 13–16, convert the improper fraction to a mixed number.

13. $\frac{29}{5}$ $\quad 5\frac{4}{5}$

14. $\frac{50}{7}$ $\quad 7\frac{1}{7}$

15. $\frac{30}{19}$ $\quad 1\frac{11}{19}$

16. $\frac{25}{8}$ $\quad 3\frac{1}{8}$

Concept 1: Order of Operations

For Exercises 17–34, simplify, using the order of operations. Write the answer as a mixed number, if possible. **(See Examples 1–3.)**

17. $\left(2 - \frac{1}{2}\right)^2$ $\quad 2\frac{1}{4}$

18. $\left(3 - \frac{2}{5}\right)^2$ $\quad 6\frac{19}{25}$

19. $1\frac{5}{6} \cdot 2\frac{1}{2} \div 1\frac{1}{4}$ $\quad 3\frac{2}{3}$

20. $2\frac{1}{7} \div 1\frac{1}{3} \cdot \frac{7}{10}$ $\quad 1\frac{1}{8}$

21. $6\frac{1}{6} + 2\frac{1}{3} \div 1\frac{3}{4}$ $\quad 7\frac{1}{2}$

22. $8\frac{7}{9} + 2\frac{1}{6} \cdot 3\frac{1}{3}$ $\quad 16$

23. $6 - 5\frac{1}{7} \cdot \frac{1}{3}$ $\quad 4\frac{2}{7}$

24. $11 - 6\frac{1}{3} \div 1\frac{1}{6}$ $\quad 5\frac{4}{7}$

25. $\left(3\frac{1}{4} + 1\frac{5}{8}\right) \cdot 2\frac{2}{3}$ $\quad 13$

26. $\left(1\frac{3}{5} + 2\frac{4}{7}\right) \cdot 5\frac{5}{6}$ $\quad 24\frac{1}{3}$

27. $\left(1\frac{1}{5}\right)^2 \cdot \left(1\frac{7}{9} - 1\frac{5}{12}\right)$ $\quad \frac{13}{25}$

28. $\left(1\frac{1}{3}\right)^3 \div \left(2\frac{7}{9} + 1\frac{2}{3}\right)$ $\quad \frac{8}{15}$

 Writing   Translating Expression    Geometry   Scientific Calculator   Video   NS E

29. $\left(6\frac{3}{4} - 2\frac{1}{8}\right) \div \left(1\frac{1}{2}\right)^3$ $1\frac{10}{27}$

30. $\left(2\frac{1}{2} + 1\frac{7}{8}\right) \cdot \left(1\frac{1}{7}\right)^2$ $5\frac{5}{7}$

31. $\left(\frac{1}{2}\right)^2 + \left(2\frac{1}{4}\right) \cdot \frac{1}{3}$ 1

32. $\left(\frac{2}{3}\right)^2 + \left(2\frac{1}{2}\right) \cdot \frac{2}{9}$ 1

33. $\left(5 - 1\frac{7}{8}\right) \div \left(3 - \frac{13}{16}\right)$ $1\frac{3}{7}$

34. $\left(4 + 2\frac{1}{9}\right) \div \left(2 - 1\frac{11}{36}\right)$ $8\frac{4}{5}$

Concept 2: Applications of Fractions and Mixed Numbers

35. Acceleration on cars is often compared by the time it takes to go from 0 to 60 miles per hour (mph). The graph gives the times for four cars. **(See Example 4.)**

a. What is the difference between the times for the Lamborghini and the Caterham? The difference is $\frac{3}{10}$ sec.

b. Find the average time for these four cars.

The average is $3\frac{3}{5}$ sec.

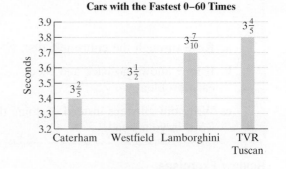
Cars with the Fastest 0–60 Times

36. Luis keeps a portfolio and tracks the number of hours he spends working on math each day.

a. How much time did Luis spend last week working on math? The total hours is $19\frac{1}{4}$ hr.

b. Find the average amount of time spent per day working on math. The average per day is $2\frac{3}{4}$ hr.

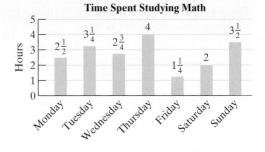
Time Spent Studying Math

37. Six women went on a weight loss program for 4 weeks. The amount of weight lost for each woman is given in the graph.

a. Find the total weight loss.
The total weight loss is 51 lb.

b. Find the average weight loss per person. The average is $8\frac{1}{2}$ lb.

c. Find the difference between the maximum weight loss and the minimum weight loss. The difference is $6\frac{1}{2}$ lb.

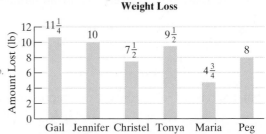
Weight Loss

38. Geoff ran $4\frac{1}{4}$ mi on Monday, $2\frac{1}{2}$ mi on Tuesday, 3 mi on Wednesday, and $8\frac{3}{4}$ mi on Thursday.

a. What is the total distance he ran? The total distance is $18\frac{1}{2}$ mi.

b. What is the average distance he ran per day? The average is $4\frac{5}{8}$ mi.

39. A financial analyst followed a certain stock. On Monday, the stock was at $\$15\frac{5}{8}$. By Friday, it had fallen to $\$11\frac{3}{4}$. By how much did the stock drop?
The stock dropped $\$3\frac{7}{8}$.

40. On Monday a stock closed at $\$12\frac{3}{4}$. On Tuesday, the stock rose $\$1\frac{1}{2}$. What was the closing price of the stock on Tuesday? The closing price was $\$14\frac{1}{4}$.

41. George will get $\frac{1}{3}$ of an $80,250 inheritance. How much money will he receive? **(See Example 5.)**
George will receive $26,750.

42. Aaron pays about $\frac{7}{25}$ of his annual salary (after deductions) in federal income tax. If he makes $60,000 per year and claims deductions of $7250, how much tax must he pay?
He must pay $14,770 in federal income tax.

43. A $15\frac{1}{4}$-ft cable is cut into 4 pieces of equal length. How long is each piece?
Each piece is $3\frac{13}{16}$ ft.

44. Twenty-seven pounds of candy is distributed in $\frac{3}{4}$-lb bags. How many bags can be filled?
36 bags can be filled.

 Writing  Translating Expression Geometry Scientific Calculator   Video   NS&E

45. A cheese plate advertises a total of 3 lb of assorted cheeses. At the end of a party, there was $\frac{1}{4}$ lb of Swiss cheese, $\frac{1}{3}$ lb of cheddar, and $\frac{1}{6}$ lb of Jack cheese left over. How many pounds of cheese were eaten? $2\frac{1}{4}$ lb of cheese was eaten.

46. Meade gave $\frac{1}{4}$ of a candy bar to Max and then ate $\frac{1}{3}$ of the candy bar himself. What fraction of the candy bar is left? There is $\frac{5}{12}$ of the candy bar left.

47. A bread recipe calls for $3\frac{1}{4}$ cups of flour. If the Daily Bread bakery has large bags of flour containing 65 cups each, how many loaves of bread can be made?
    20 loaves can be made.

48. A carpenter worked $37\frac{1}{4}$ hr last week and earned $894. What is his hourly rate?
    His hourly rate is $24 per hour.

49. If interest rates average $6\frac{1}{2}$ points and go up $\frac{3}{4}$ point, what is the new rate?
    The new rate is $7\frac{1}{4}$ points.

50. The annual consumption of tea in Hong Kong is $1\frac{9}{25}$ kg per capita. The per capita tea consumption in the United Kingdom is $\frac{21}{25}$ kg more than that in Hong Kong. What is the per capita amount of tea consumed in the United Kingdom? United Kingdom consumes $2\frac{1}{5}$ kg per capita.

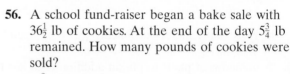

51. Stephanie is planning to sew her bridesmaids' dresses. The pattern calls for $2\frac{1}{2}$ yd of material for one dress with an additional $1\frac{1}{4}$ yd for the matching jacket. If she has three bridesmaids, how many yards of material will she need to buy? Stephanie will need $11\frac{1}{4}$ yd for the dresses.

52. Grace travels $1\frac{1}{4}$ hr to work each day. If she works 5 days a week, how much time is spent traveling to and from work?

    Grace spends $12\frac{1}{2}$ hr each week traveling to and from work.

53. Wilma has $6\frac{1}{2}$ lb of mixed nuts. If she gives Fred one-half of the mixture and Barney one-third of the mixture, how many pounds of nuts does she have left? Wilma has $1\frac{1}{12}$ lb left.

54. Jeremy mowed $\frac{2}{3}$ of his front lawn in the morning and then $\frac{1}{4}$ of the lawn in the evening. What portion of the lawn still needs to be mowed? Jeremy has $\frac{1}{12}$ of the lawn left to mow.

55. Joan waters her plants each day with $22\frac{3}{4}$ gal of water. With a new irrigation system in place, she uses only $17\frac{2}{3}$ gal of water. How many gallons of water does she save in a 30-day period with the new irrigation system?
    Joan saves $152\frac{1}{2}$ gal.

56. A school fund-raiser began a bake sale with $36\frac{1}{2}$ lb of cookies. At the end of the day $5\frac{3}{4}$ lb remained. How many pounds of cookies were sold?

    $30\frac{3}{4}$ lb of cookies was sold.

Concept 3: Applications of Geometry

57. A decorator has $14\frac{2}{3}$ ft of wallpaper border. How much more does she need to place wallpaper border around the walls of the bathroom shown in the figure? (*Note:* No border is needed on the door.) She needs $15\frac{1}{3}$ ft more.

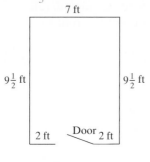

7 ft

$9\frac{1}{2}$ ft $9\frac{1}{2}$ ft

2 ft   Door   2 ft

58. Leonie wants to put a border of shrubs around her garden, as seen in the figure. Find the distance that must be lined with shrubs.
    The distance is $35\frac{1}{4}$ yd.

$8\frac{3}{4}$ yd

$7\frac{5}{8}$ yd

$3\frac{3}{8}$ yd

$2\frac{1}{4}$ yd

59. A stop sign is in the shape of an 8-sided polygon in which each of the sides is $12\frac{1}{2}$ in. Find the perimeter.   The perimeter is 100 in.

$12\frac{1}{2}$ in.

60. The shutters on the front of the house need to be painted. Determine the area of the four shutters.
The total area of the four shutters is $26\frac{8}{9}$ ft².

$3\frac{2}{3}$ ft

$1\frac{5}{6}$ ft

61. Matt needs to replace gutters on his townhouse. If the front and back of the townhouse are each $20\frac{1}{2}$ ft long and the side of the house is $35\frac{1}{3}$ ft long, how much gutter should he buy to go around the three sides of the house?

Matt needs $76\frac{1}{3}$ ft of gutter.

62. Find the area of the triangular portion of the roof.
The area is $212\frac{1}{2}$ ft².

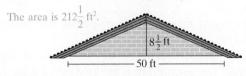

$8\frac{1}{2}$ ft

50 ft

63. The cost of a new roof depends on the area of the roof. Find the area of the roof of the house in the figure.   The area of the whole roof is $1022\frac{7}{16}$ ft².

$14\frac{1}{4}$ ft

$35\frac{7}{8}$ ft

64. Find the area of the calculator screen.
The area is $22\frac{47}{50}$ cm².

$6\frac{1}{5}$ cm

$3\frac{7}{10}$ cm

65. The Krajewskis want to improve the backyard by fertilizing the grass and putting up a decorative fence. To know how much fertilizer to use, they must know the area of the yard. **(See Example 6.)**

a. Using the figure, determine the area of the yard.   The area is $247\frac{1}{2}$ m².

b. How many meters of fencing will they need?

They will need 65 m.

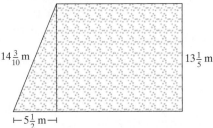

16 m

$14\frac{3}{10}$ m

$13\frac{1}{5}$ m

$5\frac{1}{2}$ m

66. A homeowner plans to put an addition onto her house. The dimensions of the front face of the addition are shown in the figure.

a. If the homeowner wants to paint the front face of the house, how many square feet of paint will be needed?   135 ft²

b. If the homeowner strings lights along the slanted edge and left edge, how long a string of lights will be needed? $22\frac{1}{4}$ ft

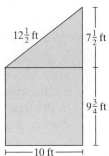

$12\frac{1}{2}$ ft

$7\frac{1}{2}$ ft

$9\frac{3}{4}$ ft

10 ft

Expanding Your Skills

67. Richard wants to paint his garage. Determine the area that needs to be painted. (He will paint the garage door, front, back, and sides of the garage, but not the roof.)

$152\frac{3}{4}$ m²

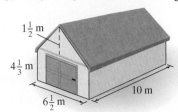

$1\frac{1}{2}$ m

$4\frac{1}{3}$ m

$6\frac{1}{2}$ m

10 m

 Writing   Translating Expression   Geometry   Scientific Calculator   Video   NS E

Group Activity

Card Games with Fractions

Materials: A deck of fraction cards for each group. These can be made from index cards where one side of the card is blank, and the other side has a fraction written on it. The deck should consist of several cards of each of the following fractions.

$$\frac{1}{4}, \frac{1}{2}, \frac{3}{4}, \frac{1}{6}, \frac{1}{3}, \frac{2}{3}, \frac{5}{6}, \frac{1}{8}, \frac{3}{8}, \frac{5}{8}, \frac{7}{8}, \frac{1}{5}, \frac{2}{5}, \frac{3}{5}, \frac{4}{5}, \frac{1}{10}, \frac{3}{10}, \frac{4}{9}, \frac{2}{9}, \frac{3}{7}$$

Estimated time: Instructor discretion

In this activity, we outline three different games for students to play in their groups as a fun way to reinforce skills of adding fractions, recognizing equivalent fractions, and ordering fractions.

Game 1 "Blackjack"

Group Size: 3

1. In this game, one student in the group will be the dealer, and the other two will be players. The dealer will deal each player one card face down and one card face up. Then the players individually may elect to have more cards given to them (face up). The goal is to have the sum of the fractions get as close to "2" without going over.

2. Once the players have taken all the cards that they want, they will display their cards face up for the group to see. The player who has a sum closest to "2" without going over wins. The dealer will resolve any "disputes."

3. The members of the group should rotate after several games so that each person has the opportunity to be a player and to be the dealer.

Game 2 "War"

Group Size: 2

1. In this game, each player should start with half of the deck of cards. The players should shuffle the cards and then stack them neatly face down on the table. Then each player will select the top card from the deck, turn it over, and place it on the table. The player who has the fraction with the greatest value "wins" that round and takes both cards.

2. Continue overturning cards and deciding who "wins" each round until all of the cards have been overturned. Then the players will count the number of cards they each collected. The player with the most cards wins.

Game 3 "Bingo"

Group Size: The whole class

1. Each student gets five fraction cards. The instructor will call out fractions that are not in lowest terms. The students must identify whether the fraction that was called is the same as one of the fractions on their cards. For example, if the instructor calls out "three-ninths," then students with the fraction card $\frac{1}{3}$ would have a match.

2. The student who first matches all five cards, wins.

Chapter 3   Summary

Section 3.1   Addition and Subtraction of Like Fractions

Key Concepts

Adding Like Fractions

1. Add the numerators.
2. Write the sum over the common denominator.
3. Simplify the fraction to lowest terms, if possible.

Subtracting Like Fractions

1. Subtract the numerators.
2. Write the difference over the common denominator.
3. Simplify the fraction to lowest terms, if possible.

Example 3 is an application involving subtraction of **like fractions.**

Examples

Example 1

$$\frac{5}{8} + \frac{7}{8} = \frac{12}{8} = \frac{\overset{3}{\cancel{12}}}{\underset{2}{\cancel{8}}} = \frac{3}{2}$$

Example 2

$$\frac{25}{10} - \frac{7}{10} = \frac{18}{10} = \frac{\overset{9}{\cancel{18}}}{\underset{5}{\cancel{10}}} = \frac{9}{5}$$

Example 3

A nail that is $\frac{13}{8}$ in. long is driven through a board that is $\frac{11}{8}$ in. thick. How much of the nail extends beyond the board?

$$\frac{13}{8} - \frac{11}{8} = \frac{2}{8} = \frac{1}{4}$$

The nail will extend $\frac{1}{4}$ in.

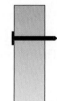

Section 3.2 Least Common Multiple

Key Concepts

The numbers obtained by multiplying a number by the whole numbers 1, 2, 3, and so on are called the **multiples** of the number.

The **least common multiple (LCM)** of two given numbers is the smallest whole number that is a multiple of each given number.

Using Prime Factors to Find the LCM of Two Numbers

1. Write each number as a product of prime factors.
2. The LCM is the product of unique prime factors from both numbers. Use repeated factors the maximum number of times they appear in either factorization.

Writing Equivalent Fractions

Use the fundamental principle of fractions to convert a fraction to an equivalent fraction with a given denominator.

Ordering Fractions

Write the fractions with a common denominator. Then compare the numerators.

Examples

Example 1

The numbers 5, 10, 15, 20, 25, 30, 35, and 40 are several multiples of 5.

Example 2

Find the LCM of 8 and 10.
Some multiples of 8 are 8, 16, 24, 32, 40.
Some multiples of 10 are 10, 20, 30, 40.

40 is the least common multiple.

Example 3

Find the LCM for the numbers 24 and 16.

$24 = 2 \cdot 2 \cdot 2 \cdot 3$
$16 = 2 \cdot 2 \cdot 2 \cdot 2$
$\text{LCM} = 2 \cdot 2 \cdot 2 \cdot 2 \cdot 3 = 48$

Example 4

Write the fraction with the indicated denominator.

$$\frac{3}{4} = \frac{}{36}$$

$$\frac{3 \cdot 9}{4 \cdot 9} = \frac{27}{36}$$

The fraction $\frac{27}{36}$ is equivalent to $\frac{3}{4}$.

Example 5

Fill in the blank with the appropriate symbol, < or >.

$$\frac{5}{9} \ \square \ \frac{7}{12} \qquad \text{The LCD is 36.}$$

$$\frac{5 \cdot 4}{9 \cdot 4} \ \square \ \frac{7 \cdot 3}{12 \cdot 3}$$

$$\frac{20}{36} \ \boxed{<} \ \frac{21}{36}$$

Section 3.3   Addition and Subtraction of Unlike Fractions

Key Concepts

To add or subtract **unlike fractions,** first we must write each fraction as an equivalent fraction with a common denominator.

Steps to Add or Subtract Unlike Fractions

1. Identify the LCD.
2. Write each individual fraction as an equivalent fraction with the LCD.
3. Add or subtract the resulting fractions as indicated.
4. Simplify to lowest terms, if possible.

Examples

Example 1

Simplify. $\dfrac{7}{5} - \dfrac{3}{10} + \dfrac{13}{15}$

$\dfrac{7 \cdot 6}{5 \cdot 6} - \dfrac{3 \cdot 3}{10 \cdot 3} + \dfrac{13 \cdot 2}{15 \cdot 2}$ The LCD = 30.

$= \dfrac{42}{30} - \dfrac{9}{30} + \dfrac{26}{30}$

$= \dfrac{42 - 9 + 26}{30}$

$= \dfrac{59}{30}$

Section 3.4   Addition and Subtraction of Mixed Numbers

Key Concepts

Addition of Mixed Numbers

To find the sum of two or more mixed numbers, add the whole-number parts and add the fractional parts.

Examples

Example 1

$$3\dfrac{5}{8} = 3\dfrac{10}{16}$$
$$+\ 1\dfrac{1}{16} = 1\dfrac{1}{16}$$
$$\overline{\hphantom{+\ 1\dfrac{1}{16} = }4\dfrac{11}{16}}$$

Example 2

$$2\dfrac{9}{10} = 2\dfrac{27}{30}$$
$$+\ 6\dfrac{5}{6} = 6\dfrac{25}{30}$$
$$\overline{\hphantom{+\ 6\dfrac{5}{6} = }8\dfrac{52}{30}} = 8 + 1\dfrac{22}{30}$$
$$= 9\dfrac{11}{15}$$

Subtraction of Mixed Numbers

To subtract mixed numbers, subtract the fractional parts and subtract the whole-number parts.

When the fractional part in the subtrahend is larger than the fractional part in the minuend, we borrow from the whole number part of the minuend.

Example 3

$$5\dfrac{3}{4} = 5\dfrac{9}{12}$$
$$-\ 2\dfrac{2}{3} = 2\dfrac{8}{12}$$
$$\overline{\hphantom{-\ 2\dfrac{2}{3} = }3\dfrac{1}{12}}$$

Example 4

$$7\dfrac{1}{2} = \overset{6}{\cancel{7}}\dfrac{\overset{5+10}{5}}{10} = 6\dfrac{15}{10}$$
$$-\ 3\dfrac{4}{5} = 3\dfrac{8}{10} = 3\dfrac{8}{10}$$
$$\overline{\hphantom{-\ 3\dfrac{4}{5} = 3\dfrac{8}{10} = }3\dfrac{7}{10}}$$

We can also add or subtract mixed numbers by writing the numbers as improper fractions. Then add or subtract the fractions.

Example 5

$$4\dfrac{7}{8} + 2\dfrac{1}{16} - 3\dfrac{1}{4} = \dfrac{39}{8} + \dfrac{33}{16} - \dfrac{13}{4}$$
$$= \dfrac{78}{16} + \dfrac{33}{16} - \dfrac{52}{16} = \dfrac{59}{16} = 3\dfrac{11}{16}$$

Section 3.5   Order of Operations and Applications of Fractions and Mixed Numbers

Key Concepts

Order of Operations

1. Perform all operations inside parentheses first.
2. Simplify expressions containing exponents or square roots.
3. Perform multiplication or division in the order that they appear from left to right.
4. Perform addition or subtraction in the order that they appear from left to right.

Example 2 is an example of an application involving mixed numbers and fractions.

Examples

Example 1

$$4\frac{2}{9} + 2\frac{1}{12} \cdot 5\frac{1}{3} = 4\frac{2}{9} + \frac{25}{\underset{3}{12}} \cdot \frac{\overset{4}{16}}{3}$$

$$= 4\frac{2}{9} + \frac{100}{9}$$

$$= \frac{38}{9} + \frac{100}{9}$$

$$= \frac{138}{9} = 15\frac{3}{9} = 15\frac{1}{3}$$

Example 2

Wallace has a budget of $750 for putting a curb around an area in his front yard. Curbing costs $3\frac{1}{2}$ per foot. To stay within his budget, how many feet of curbing can he afford?

Solution:

$$750 \div 3\frac{1}{2} = \frac{750}{1} \div \frac{7}{2}$$

$$= \frac{750}{1} \cdot \frac{2}{7}$$

$$= \frac{1500}{7} = 214\frac{2}{7}$$

Wallace can purchase up to $214\frac{2}{7}$ ft of curbing.

Chapter 3   Review Exercises

Section 3.1

For Exercises 1–4, add or subtract the like units.

1. 5 books + 3 books
   8 books
2. 12 cm + 6 cm
   18 cm
3. 25 mi – 13 mi
   12 mi
4. 13 CDs – 2 CDs
   11 CDs
5. Explain what is meant by the term like fractions.
   Fractions with the same denominators are considered like fractions.
6. Give an example of two like fractions and two unlike fractions. Answers may vary.
   For example: like fractions: $\frac{4}{7}, \frac{2}{7}$; unlike fractions: $\frac{1}{9}, \frac{3}{16}$.

For Exercises 7–12, add or subtract the like fractions. Simplify the answer to lowest terms.

7. $\frac{5}{6} + \frac{4}{6}$ $\frac{3}{2}$
8. $\frac{4}{15} + \frac{6}{15}$ $\frac{2}{3}$
9. $\frac{5}{12} + \frac{1}{12}$ $\frac{1}{2}$
10. $\frac{2}{9} + \frac{7}{9}$ 1
11. $\frac{15}{7} - \frac{6}{7}$ $\frac{9}{7}$
12. $\frac{21}{5} - \frac{6}{5}$ 3

 Writing   Translating Expression   Geometry    Scientific Calculator   Video   NS E

For Exercises 13–16, simplify the expression by using the order of operations.

13. $\frac{3}{8} \cdot \frac{3}{2} + \frac{3}{16}$ $\frac{3}{4}$

14. $\frac{4}{9} + \left(\frac{4}{3}\right)^2$ $\frac{20}{9}$

15. $\frac{21}{13} - \frac{5}{2} \div \frac{13}{4}$ $\frac{11}{13}$

16. $\left(\frac{7}{10} - \frac{2}{10}\right)^3 \cdot \frac{8}{7}$ $\frac{1}{7}$

17. Find the perimeter of the picture. 12 in. or 1 ft

$\frac{13}{4}$ in.

$\frac{11}{4}$ in.

18. Refer to the table that gives the average snowfall for selected cities.

	January	February	March
Spartanburg, SC	$\frac{25}{10}$ in.	$\frac{19}{10}$ in.	$\frac{12}{10}$ in.
Amarillo, TX	$\frac{39}{10}$ in.	$\frac{36}{10}$ in.	$\frac{28}{10}$ in.
Portland, OR	$\frac{33}{10}$ in.	$\frac{10}{10}$ in.	$\frac{4}{10}$ in.

a. What was the total snowfall for the months of January, February, and March for Spartanburg? $\frac{28}{5}$ in.

b. How much more snow did Amarillo get than Portland in the month of March? $\frac{12}{5}$ in.

Section 3.2

19. List the first four multiples for each number.

a. 7   **b.** 13   **c.** 22
7, 14, 21, 28   13, 26, 39, 52   22, 44, 66, 88

20. Explain the difference between a common multiple and the *least* common multiple of a pair of numbers. Use the numbers 6 and 8 in your explanation. 6 and 8 have many common multiples including 24, 48, and 72. Of all the common multiples, 24 is the least.

21. List all factors for each number.

a. 100   **b.** 65   **c.** 70
1, 2, 4, 5, 10,   1, 5, 13, 65   1, 2, 5, 7, 10,
20, 25, 50, 100   14, 35, 70

22. Find the prime factorization.

a. 100   **b.** 65   **c.** 70
2 · 2 · 5 · 5   5 · 13   2 · 5 · 7

For Exercises 23–26, find the LCM by using any method.

23. 30 and 25 150

24. 22 and 144 1584

25. 105 and 28 420

26. 16, 24, and 32 96

27. Sharon and Tonya signed up at a gym on the same day. Sharon will be able to go to the gym every third day and Tonya will go to the gym every fourth day. In how many days will they meet again at the gym? They will meet on the 12th day.

For Exercises 28–31, rewrite each fraction with the indicated denominator.

28. $\frac{5}{16} = \frac{}{48}$ $\frac{15}{48}$

29. $\frac{9}{5} = \frac{}{35}$ $\frac{63}{35}$

30. $\frac{7}{12} = \frac{}{60}$ $\frac{35}{60}$

31. $\frac{17}{15} = \frac{}{150}$ $\frac{170}{150}$

For Exercises 32–34, fill in the blanks with $<$, $>$, or $=$.

32. $\frac{11}{24} \square \frac{7}{12}$ $<$

33. $\frac{5}{6} \square \frac{7}{9}$ $>$

34. $\frac{5}{6} \square \frac{15}{18}$ $=$

35. Rank the following numbers from least to greatest: $\frac{7}{10}, \frac{72}{105}, \frac{8}{15}, \frac{27}{35}$ $\frac{8}{15}, \frac{72}{105}, \frac{7}{10}, \frac{27}{35}$

Section 3.3

For Exercises 36–46, add or subtract. Write the answer as a fraction simplified to lowest terms.

36. $\frac{1}{8} + \frac{7}{12}$ $\frac{17}{24}$

37. $\frac{9}{10} - \frac{61}{100}$ $\frac{29}{100}$

38. $\frac{11}{25} - \frac{2}{5}$ $\frac{1}{25}$

39. $\frac{3}{26} + \frac{5}{13}$ $\frac{1}{2}$

40. $\frac{25}{11} + 2$ $\frac{47}{11}$

41. $4 - \frac{37}{20}$ $\frac{43}{20}$

42. $\frac{4}{15} - \frac{0}{3}$ $\frac{4}{15}$

43. $\frac{0}{17} + \frac{1}{34}$ $\frac{1}{34}$

44. $\frac{7}{100} - \frac{33}{1000}$ $\frac{37}{1000}$

45. $\frac{2}{15} + \frac{5}{8} - \frac{1}{3}$ $\frac{17}{40}$

46. $\frac{11}{14} - \frac{4}{7} + \frac{3}{2}$ $\frac{12}{7}$

For Exercises 47 and 48, simplify by applying the order of operations. 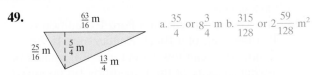 NS/E

47. $\left(\dfrac{2}{5} + \dfrac{1}{40}\right) \div \dfrac{15}{8} - \dfrac{4}{25}$ $\dfrac{1}{15}$

48. $\dfrac{20}{7} \cdot \left(\dfrac{11}{15} - \dfrac{1}{3}\right)^2 + \dfrac{1}{7}$ $\dfrac{3}{5}$

For Exercises 49 and 50, find (a) the perimeter and (b) the area.

49.

$\frac{63}{16}$ m

$\frac{25}{16}$ m $\frac{5}{4}$ m

$\frac{13}{4}$ m

a. $\dfrac{35}{4}$ or $8\dfrac{3}{4}$ m   b. $\dfrac{315}{128}$ or $2\dfrac{59}{128}$ m²

50.

$\frac{3}{2}$ yd

$\frac{7}{3}$ yd

a. $\dfrac{23}{3}$ or $7\dfrac{2}{3}$ yd   b. $\dfrac{7}{2}$ or $3\dfrac{1}{2}$ yd²

Section 3.4

For Exercises 51–62, add or subtract the mixed numbers.

51. $9\dfrac{8}{9}$ $11\dfrac{11}{63}$
$+\ 1\dfrac{2}{7}$

52. $10\dfrac{1}{2}$ $14\dfrac{7}{16}$
$+\ 3\dfrac{15}{16}$

53. $7\dfrac{5}{24}$ $2\dfrac{5}{8}$
$-\ 4\dfrac{7}{12}$

54. $5\dfrac{1}{6}$ $1\dfrac{11}{12}$
$-\ 3\dfrac{1}{4}$

55. $5\dfrac{3}{8}$ $3\dfrac{1}{24}$
$-\ 2\dfrac{1}{3}$

56. $3\dfrac{4}{5}$ $2\dfrac{8}{15}$
$-\ 1\dfrac{4}{15}$

57. $6\dfrac{4}{7}$ $12\dfrac{5}{14}$
$+\ 5\dfrac{11}{14}$

58. $3\dfrac{3}{8}$ $6\dfrac{3}{16}$
$+\ 2\dfrac{13}{16}$

59. 6 $3\dfrac{2}{5}$
$-\ 2\dfrac{3}{5}$

60. 8 $3\dfrac{3}{14}$
$-\ 4\dfrac{11}{14}$

61. $42\dfrac{1}{8}$ $63\dfrac{15}{16}$
$+\ 21\dfrac{13}{16}$

62. $38\dfrac{9}{10}$ $50\dfrac{1}{2}$
$+\ 11\dfrac{3}{5}$

For Exercises 63–66, round the numbers to estimate the answer. Then find the exact sum or difference.

63. $2\dfrac{1}{4} + 4\dfrac{2}{9} + 1\dfrac{29}{36}$

Estimate: _____ 8

Exact: _____ $8\dfrac{5}{18}$

64. $5\dfrac{2}{5} + 1\dfrac{9}{10} + 3\dfrac{19}{30}$

Estimate: _____ 11

Exact: _____ $10\dfrac{14}{15}$

65. $65\dfrac{1}{8} - 14\dfrac{9}{10}$

Estimate: _____ 50

Exact: _____ $50\dfrac{9}{40}$

66. $43\dfrac{13}{15} - 20\dfrac{23}{25}$

Estimate: _____ 23

Exact: _____ $22\dfrac{71}{75}$

67. Corry drove for $4\dfrac{1}{2}$ hr in the morning and $3\dfrac{2}{3}$ hr in the afternoon. Find the total number of hours he drove.
Corry drove a total of $8\dfrac{1}{6}$ hr.

68. Denise owned $2\dfrac{1}{8}$ acres of land. If she sells $1\dfrac{1}{4}$ acres, how much will she have left?
Denise will have $\dfrac{7}{8}$ acre left.

Section 3.5

For Exercises 69–74, simplify by using the order of operations. Write the answer as a mixed number, if possible.

69. $1\dfrac{1}{5} + 4\dfrac{9}{10} \cdot 2\dfrac{2}{7}$ $12\dfrac{2}{5}$

70. $5\dfrac{3}{4} - 23\dfrac{1}{2} \div 5\dfrac{2}{9}$ $1\dfrac{1}{4}$

71. $\left(8\dfrac{1}{9} - 6\dfrac{2}{3}\right) \div 9\dfrac{3}{4}$ $\dfrac{4}{27}$

72. $\left(5\dfrac{1}{8} + 1\dfrac{1}{16}\right) \cdot 2\dfrac{10}{11}$ 18

73. $\left(1\dfrac{1}{5}\right)^2 \cdot \left(4\dfrac{1}{2} + 3\dfrac{5}{6}\right)$ 12

74. $\left(1\dfrac{5}{16}\right) \div \left(11\dfrac{1}{8} - 10\dfrac{3}{4}\right)^2$ $9\dfrac{1}{3}$

75. In a certain region, the appraised value of a house is $\dfrac{9}{10}$ of its market value. If the market value of Owen's house is $160,000, what is the appraised value? The appraised value is $144,000.

76. Nuts 'N Things makes a nut mixture from $2\dfrac{1}{4}$ lb of cashews, $7\dfrac{3}{4}$ lb of peanuts, and $2\dfrac{1}{2}$ lb of pecans. The mixture is then divided into 10 bags. How many pounds are in each bag?

There are $1\dfrac{1}{4}$ lb of nuts in each bag.

Chapter 3   Test

For Exercises 1 and 2, add or subtract the like fractions.

1. $\dfrac{4}{5} + \dfrac{3}{5}$ $\dfrac{7}{5}$

2. $\dfrac{23}{16} - \dfrac{15}{16}$ $\dfrac{1}{2}$

3. Explain the difference between evaluating these two expressions:

$\dfrac{5}{11} - \dfrac{3}{11}$ and $\dfrac{5}{11} \times \dfrac{3}{11}$

When subtracting like fractions, keep the same denominator and subtract the numerators. When multiplying fractions, multiply the denominators as well as the numerators.

4. **a.** List the first four multiples of 24.
24, 48, 72, 96

  b. List all factors of 24.
1, 2, 3, 4, 6, 8, 12, 24

  c. Write the prime factorization of 24.
$2 \cdot 2 \cdot 2 \cdot 3$ or $2^3 \cdot 3$

5. Find the LCM for the numbers 16, 24, and 30.
240

For Exercises 6–8, write each fraction with the indicated denominator.

6. $\dfrac{5}{9} = \dfrac{}{63}$ $\dfrac{35}{63}$
7. $\dfrac{11}{21} = \dfrac{}{63}$ $\dfrac{33}{63}$
8. $\dfrac{4}{7} = \dfrac{}{63}$ $\dfrac{36}{63}$

9. Rank the fractions in Exercises 6–8 from least to greatest. $\dfrac{11}{21}, \dfrac{5}{9}, \dfrac{4}{7}$

For Exercises 10–13, add or subtract as indicated. Write the answer as a fraction.

10. $\dfrac{3}{8} + \dfrac{3}{16}$ $\dfrac{9}{16}$

11. $\dfrac{7}{3} - 2$ $\dfrac{1}{3}$

12. $\dfrac{7}{12} - \dfrac{1}{4}$ $\dfrac{1}{3}$

13. $\dfrac{3}{5} + \dfrac{1}{15}$ $\dfrac{2}{3}$

For Exercises 14–17, add or subtract the mixed numbers. Write the answer as a mixed number.

14. $6\dfrac{3}{4} + 10\dfrac{5}{8}$
$17\dfrac{3}{8}$

15. $12 - 9\dfrac{10}{11}$
$2\dfrac{1}{11}$

16. $22\dfrac{1}{4} + 35\dfrac{1}{2} + 2\dfrac{2}{3}$
$60\dfrac{5}{12}$

17. $15\dfrac{1}{6} - 12\dfrac{3}{8} - 1\dfrac{7}{24}$
$1\dfrac{1}{2}$

For Exercises 18–21, perform the indicated operations.

18. $4\dfrac{5}{8} \cdot 2\dfrac{2}{3} - 8\dfrac{1}{6}$
$\dfrac{25}{6}$ or $4\dfrac{1}{6}$

19. $3\dfrac{1}{3} \div 2\dfrac{1}{2} + 5\dfrac{2}{3}$ 7

20. $\left(\dfrac{2}{5}\right)^2 \div \left(1\dfrac{1}{10} + 2\dfrac{5}{6}\right)$ $\dfrac{12}{295}$

21. $\left(7\dfrac{1}{4} - 5\dfrac{1}{6}\right) \cdot 1\dfrac{3}{5}$ $\dfrac{10}{3}$ or $3\dfrac{1}{3}$

22. A fudge recipe calls for $1\dfrac{1}{2}$ lb of chocolate. How many pounds are required for $\dfrac{2}{3}$ of the recipe?
1 lb is needed.

23. The towing capacity of a Ford Expedition is $4\dfrac{19}{40}$ times that of a Buick Rendezvous. If the Rendezvous can tow 1 ton (2000 lb), what is the towing capacity of the Expedition (in pounds)?
The Ford Expedition can tow 8950 lb.

24. Find the area and perimeter of this parking area.
Area: $25\dfrac{2}{25}$ m²; perimeter: $20\dfrac{1}{5}$ m

$5\dfrac{7}{10}$ m

$4\dfrac{2}{5}$ m

25. Justin has a budget of $14,000 to redecorate his kitchen. If he spends $\dfrac{3}{28}$ of the money on a stove and $\dfrac{1}{7}$ on a refrigerator, how much is left for new cabinets?   Justin has $10,500 for cabinets.

26. The figure gives the national records for long jump and high jump for men's and women's indoor track and field for a recent year. What is the difference for the record long jump between men and women?   The difference is $4\dfrac{2}{3}$ ft.

Track and Field Records

$28\dfrac{41}{48}$

$24\dfrac{3}{16}$

$7\dfrac{23}{24}$

$6\dfrac{19}{24}$

Feet

Men's high jump   Women's high jump   Men's long jump   Women's long jump

Source: International Association of Athletics Federations

Chapters 1-3 Cumulative Review Exercises

1. Write the number in words: 23,400,806
Twenty-three million, four hundred thousand, eight hundred six

2. Find the sum of 72 and 24.   96

3. Find the difference of 72 and 24.   48

4. Find the product of 72 and 24.   1728

5. Find the quotient of 72 and 24.   3

6. Round the numbers to the ten-thousands place to estimate the product: 54,923 × 28,543.
1,500,000,000

7. Write the expression by using exponents:
$4 \cdot 4 \cdot 5 \cdot 5 \cdot 5 \cdot 5 \cdot 8 \cdot 8$ $4^2 \cdot 5^4 \cdot 8^2$

8. Simplify. $72 \div (4^2 - 10) \cdot 3$ 36

9. List all the prime numbers between 15 and 35.
17, 19, 23, 29, 31

10. Write the prime factorization of 70.
$2 \cdot 5 \cdot 7$

11. Label the numerator and denominator of the fraction $\frac{21}{17}$.   Numerator: 21; denominator: 17

12. What fraction is represented by the figure?

$\frac{4}{16}$ or $\frac{1}{4}$

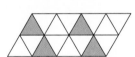

13. Kevin delivered 22 pizzas one evening. Of the 22 pizzas, 17 had pepperoni. What fraction of the pizzas had pepperoni? What fraction did not have pepperoni?
$\frac{17}{22}$ had pepperoni and $\frac{5}{22}$ did not have pepperoni.

14. Label the fractions as proper or improper.

a. $\frac{13}{5}$ **b.** $\frac{5}{13}$ **c.** $\frac{13}{13}$

Improper   Proper   Improper

15. Which of the numbers is divisible by 3 and 5?   b

a. 2390   **b.** 1245   **c.** 9321

16. Label the numbers as prime, composite, or neither.

a. 51   **b.** 52   **c.** 53

Composite   Composite   Prime

17. Find the prime factorization of 360.
$2 \cdot 2 \cdot 2 \cdot 3 \cdot 3 \cdot 5$ or $2^3 \cdot 3^2 \cdot 5$

18. Simplify the fraction to lowest terms: $\frac{180}{900}$ $\frac{1}{5}$

19. Multiply: $\frac{15}{16} \cdot \frac{2}{5}$ $\frac{3}{8}$

20. Divide: $\frac{20}{63} \div \frac{5}{9}$ $\frac{4}{7}$

21. Subtract: $\frac{13}{8} - \frac{7}{8}$ $\frac{3}{4}$

22. Add: $\frac{3}{16} + \frac{5}{16} + \frac{25}{16}$
$\frac{33}{16}$

23. Subtract: $4 - \frac{18}{5}$ $\frac{2}{5}$

24. Multiply: $6\frac{1}{10} \cdot 2\frac{3}{11}$
$\frac{305}{22}$ or $13\frac{19}{22}$

25. Divide: $2\frac{3}{5} \div 1\frac{7}{10}$ $\frac{26}{17}$ or $1\frac{9}{17}$

26. Simplify the expression.

$\left(8\frac{1}{4} \div 2\frac{3}{4}\right)^2 \cdot \frac{5}{18} + \frac{5}{6}$ $\frac{10}{3}$ or $3\frac{1}{3}$

27. To approximate the distance around a circle, multiply the diameter by the fraction $\frac{22}{7}$. Find the distance around a circle with diameter 28 cm.
The distance around is approximately 88 cm.

28. Find the perimeter of the triangle. $4\frac{1}{3}$ yd

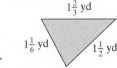

29. Find the area of the triangle.
$7\frac{7}{8}$ m²

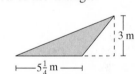

30. The figure gives the earthquake intensity on the Richter scale for four earthquakes in 2010.

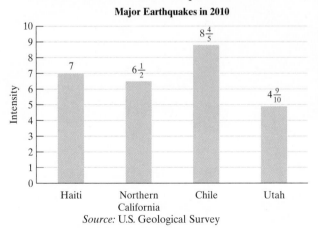

Major Earthquakes in 2010

Source: U.S. Geological Survey

a. What was the difference in Richter scale values between the earthquake in Northern California and in Chile? $2\frac{3}{10}$

b. What is the average of the Richter scale values for these earthquakes? $6\frac{4}{5}$

 Writing   Translating Expression   Geometry   Scientific Calculator   Video   NS E

Decimals

Chapter 4

This chapter is devoted to the study of decimal numbers. We begin with a discussion of place values and then perform addition, subtraction, multiplication, and division. The applications of decimal numbers are far-reaching. Almost every day, we use decimals in transactions involving money.

Review Your Skills

In the first section, we show the relationship between fractions and decimals. Use what you have learned with fractions to prepare for operations on decimals. Match statements 1–8 with the appropriate letter. Then record the letter in the space below to complete the sentence.

1. Given $\dfrac{3}{25}$, select an equivalent fraction.

2. The LCD of $\dfrac{7}{10}$ and $\dfrac{81}{100}$

3. Sum of $\dfrac{41}{100}$ and $\dfrac{9}{10}$

4. Difference of $1\dfrac{39}{50}$ and $\dfrac{27}{100}$

5. Given $\dfrac{2}{5}$, select an equivalent fraction.

6. The LCD of $\dfrac{1}{10}, \dfrac{1}{100},$ and $\dfrac{1}{1000}$

7. Sum of $\dfrac{3}{10}, \dfrac{7}{100},$ and $\dfrac{9}{1000}$

8. Difference of $\dfrac{9}{10}$ and $\dfrac{1}{5}$

i. $1\dfrac{51}{100}$ **d.** $\dfrac{12}{100}$

l. $\dfrac{7}{10}$ **c.** $\dfrac{131}{100}$

a. 1000   **m.** $\dfrac{40}{100}$

e. 100

l. $\dfrac{379}{1000}$

Mathematicians shop at the $\underset{1}{\underline{d}}\ \underset{2}{\underline{e}}\ \underset{3}{\underline{c}}\ \underset{4}{\underline{i}}\ \underset{5}{\underline{m}}\ \underset{6}{\underline{a}}\ \underset{7}{\underline{l}}\ \underset{8}{\underline{l}}$.

Section 4.1 Decimal Notation and Rounding

1. Decimal Notation

In Chapters 2 and 3, we studied fraction notation to denote equal parts of a whole. In this chapter, we introduce decimal notation to denote equal parts of a whole. We first introduce the concept of a decimal fraction. A **decimal fraction** is a fraction whose denominator is a power of 10. The following are examples of decimal fractions.

$$\frac{3}{10} \text{ is read as "three-tenths"}$$

$$\frac{7}{100} \text{ is read as "seven-hundredths"}$$

$$\frac{9}{1000} \text{ is read as "nine-thousandths"}$$

We now want to write these fractions in **decimal notation**. This means that we will write the numbers by using place values, as we did with whole numbers. The place value chart from Section 1.1 can be extended as shown in Figure 4-1.

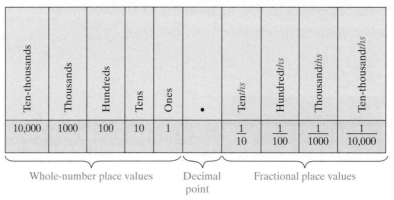

Figure 4-1

From Figure 4-1, we see that the decimal point separates the whole-number part from the fractional part. The place values for decimal fractions are located to the right of the decimal point. Their place value names are similar to those for whole numbers, but end in *ths*. Notice the correspondence between the tens place and the ten*ths* place. Similarly notice the hundreds place and the hundred*ths* place. Each place value on the left has a corresponding place value on the right, with the exception of the ones place. There is no "one*ths*" place.

Example 1   Identify the Place Values

Identify the place value of each underlined digit.

   a. 30,804.0<u>9</u>   **b.** 0.8469<u>2</u>0   **c.** 2<u>9</u>3.604

Solution:

   a. 30,804.0<u>9</u>   The digit 9 is in the hundredths place.

   b. 0.8469<u>2</u>0   The digit 2 is in the hundred-thousandths place.

   c. 2<u>9</u>3.604   The digit 9 is in the tens place.

For a whole number, the decimal point is understood to be after the ones place, and is usually not written. For example:

$$42. = 42$$

Using Figure 4-1, we can write the numbers $\frac{3}{10}$, $\frac{7}{100}$, and $\frac{9}{1000}$ in decimal notation.

Fraction	Word name	Decimal notation
$\frac{3}{10}$	Three-tenths	0.3 ↑ tenths place
$\frac{7}{100}$	Seven-hundredths	0.07 ↑ hundredths place
$\frac{9}{1000}$	Nine-thousandths	0.009 ↑ thousandths place

Now consider the number $15\frac{7}{10}$. This value represents 1 ten + 5 ones + 7 tenths. In decimal form we have 15.7.

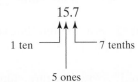

The decimal point is interpreted as the word *and*. Thus, 15.7 is read as "fifteen *and* seven tenths." The number 356.29 can be represented as

$$356 + 2 \text{ tenths} + 9 \text{ hundredths} = 356 + \frac{2}{10} + \frac{9}{100}$$

$$= 356 + \frac{20}{100} + \frac{9}{100} \qquad \text{We can use the LCD of } 100 \text{ to add the fractions.}$$

$$= 356\frac{29}{100}$$

We can read the number 356.29 as "three hundred fifty-six *and* twenty-nine hundredths."

    This discussion leads to a quicker method to read decimal numbers.

> **Reading a Decimal Number**
>
> **Step 1** The part of the number to the left of the decimal point is read as a whole number. *Note:* If there is no whole-number part, skip to step 3.
>
> **Step 2** The decimal point is read *and*.
>
> **Step 3** The part of the number to the right of the decimal point is read as a whole number but is followed by the name of the place position of the digit farthest to the right.

Example 2   **Reading Decimal Numbers**

Write the word name for each number.

a. 1028.4   **b.** 2.0736   **c.** 0.478

Solution:

a. 1028.4   is written as "one thousand, twenty-eight and four-tenths."

b. 2.0736   is written as "two and seven hundred thirty-six ten-thousandths."

c. 0.478   is written as "four hundred seventy-eight thousandths."

Example 3   **Writing a Numeral from a Word Name**

Write the word name as a numeral.

a. Four hundred eight and fifteen ten-thousandths

b. Five thousand eight hundred and twenty-three hundredths

Solution:

a. Four hundred eight and fifteen ten-thousandths: 408.0015

Place "15" so that the digit "5" is in the ten-thousandths place. Then place zeros as placeholders in the tenths place and hundredths place.

b. Five thousand eight hundred and twenty-three hundredths: 5800.23

2. Writing Decimals as Mixed Numbers or Fractions

A fractional part of a whole may be written as a fraction or as a decimal. To convert a decimal to an equivalent fraction, it is helpful to think of the decimal in words. For example:

Decimal	Word Name	Fraction	
0.3	Three tenths	$\dfrac{3}{10}$	
0.67	Sixty-seven hundredths	$\dfrac{67}{100}$	
0.048	Forty-eight thousandths	$\dfrac{48}{1000} = \dfrac{6}{125}$	(simplified)
6.8	Six and eight-tenths	$6\dfrac{8}{10} = 6\dfrac{4}{5}$	(simplified)

From the list, we notice several patterns that can be summarized as follows.

Converting a Decimal to a Mixed Number or Proper Fraction

Step 1 The digits to the right of the decimal point are written as the numerator of the fraction.

Step 2 The place value of the digit farthest to the right of the decimal point determines the denominator.

Step 3 The whole-number part of the number is left unchanged.

Step 4 Once the number is converted to a fraction or mixed number, simplify the fraction to lowest terms, if possible.

Example 4 **Writing Decimals as Proper Fractions or Mixed Numbers**

Write the decimals as proper fractions or mixed numbers and simplify.

a. 0.847   **b.** 0.0025   **c.** 4.16

Solution:

a. $0.847 = \dfrac{847}{1000}$

thousandths place

b. $0.0025 = \dfrac{25}{10{,}000} = \dfrac{\overset{1}{25}}{\underset{400}{10{,}000}} = \dfrac{1}{400}$

ten-thousandths place

c. $4.16 = 4\dfrac{16}{100} = 4\dfrac{\overset{4}{16}}{\underset{25}{100}} = 4\dfrac{4}{25}$

hundredths place

Skill Practice

Write the decimals as proper fractions or mixed numbers.

10. 0.034   **11.** 0.00086

12. 3.184

Classroom Example: p. 227, Exercise 54

A decimal number greater than 1 can be written as a mixed number or as an improper fraction. The number 4.16 from Example 4(c) can be expressed as follows.

$$4.16 = 4\frac{16}{100} = 4\frac{4}{25} \quad \text{or} \quad \frac{104}{25}$$

A quick way to obtain an improper fraction for a decimal number greater than 1 is outlined here.

Writing a Decimal Number Greater Than 1 as an Improper Fraction

Step 1 The denominator is determined by the place position of the last digit to the right of the decimal point.

Step 2 The numerator is obtained by removing the decimal point of the original number. The resulting whole number is then written over the denominator.

Step 3 Simplify the improper fraction to lowest terms, if possible.

Concept Connections

13. Which is a correct representation of 3.17?

$3\dfrac{17}{100}$ or $\dfrac{317}{100}$

Instructor Note: Explain to students that this works because

$$4.16 = 4 + \frac{16}{100}$$

$$= \frac{400}{100} + \frac{16}{100}$$

$$= \frac{416}{100}$$

Answers

10. $\dfrac{17}{500}$ **11.** $\dfrac{43}{50{,}000}$ **12.** $3\dfrac{23}{125}$

13. They are both correct representations.

For example:

Remove decimal point.

$$\overbrace{4.16} = \frac{416}{100} = \frac{104}{25} \quad \text{(simplified)}$$

hundredths
place

> **Example 5**   **Writing Decimals as Improper Fractions**
>
> Write the decimals as improper fractions and simplify.
>
> **a.** 40.2   **b.** 2.113
>
> **Solution:**
>
> **a.** $40.2 = \dfrac{402}{10} = \dfrac{\overset{201}{\cancel{402}}}{\underset{5}{\cancel{10}}} = \dfrac{201}{5}$
>
> **b.** $2.113 = \dfrac{2113}{1000}$ Note that the fraction is already in lowest terms.

3. Ordering Decimal Numbers

It is often necessary to compare the values of two decimal numbers. One way of doing this is to compare the numbers in fractional form. First note that adding 0 after the last digit in a decimal number does not change its value. For example,

$$0.7 = 0.70 \quad \text{because} \quad \frac{7}{10} = \frac{70}{100}$$

> **Example 6**   **Comparing Decimal Numbers**
>
> Write the numbers from least to greatest.
>
> $$2.1, \quad 2.09, \quad 2.15$$
>
> **Solution:**
>
> First write each number with the same number of digits to the right of the decimal point.
>
> | 2.10, | 2.09, | 2.15 | We can now write each number as a decimal |
> |---|---|---|---|
> | $\dfrac{210}{100}$, | $\dfrac{209}{100}$, | $\dfrac{215}{100}$ | fraction with the same denominator. The value 209 hundredths is less than 210 hundredths, which is less than 215 hundredths. |
>
> Writing the numbers from least to greatest, we have 2.09, 2.1, and 2.15.

A quicker way to compare two decimals is outlined next.

> **Comparing Two Decimal Numbers**
>
> **Step 1**   Starting at the left (and moving toward the right), compare the digits in each corresponding place position.
>
> **Step 2**   As we move from left to right, the first instance in which the digits differ determines the order of the numbers. The number having the greater digit is greater overall.

Ordering Decimals

Fill in the blank with $<$ or $>$.

   a. 0.68 ☐ 0.7   **b.** 3.462 ☐ 3.4619

Solution:

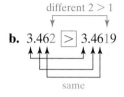

Skill Practice

Fill in the blank with $<$ or $>$.

17. 4.163 ☐ 4.159

18. 218.38 ☐ 218.41

Classroom Example: p. 228, Exercise 72

4. Rounding Decimals

The process to round the decimal part of a number is nearly the same as rounding whole numbers (see Section 1.4). The main difference is that the digits to the right of the rounding place position are dropped instead of being replaced by zeros.

> **Rounding Decimals to a Place Value to the Right of the Decimal Point**
>
> **Step 1** Identify the digit one position to the right of the given place value.
> **Step 2** If the digit in step 1 is 5 or greater, add 1 to the given digit. If the digit in step 1 is less than 5, leave the given digit unchanged.
> **Step 3** Discard all digits to the right of the given digit.

 Animation

Example 8 **Rounding Decimal Numbers**

   a. Round 4.81542 to the thousandths place.

   b. Round 52.9999 to the hundredths place.

Solution:

remaining digits discarded

a. $4.81542 \approx 4.815$

thousandths place — This digit is less than 5. Discard it and all digits to the right.

discard remaining digits

b. $5\,2.\overset{1}{9}\,\overset{1}{9}\,\overset{+1}{9}\,9\,9$

hundredths place — This digit is greater than 5. Add 1 to the hundredths place digit.

   • Since the hundredths place digit is 9, adding 1 requires us to carry 1 to the tenths place digit.
   • Since the tenths place digit is 9, adding 1 requires us to carry 1 to the ones place digit.

≈ 53.00

Skill Practice

19. Round 45.372 to the hundredths place.

20. Round 134.9996 to the thousandths place.

Classroom Examples: p. 229, Exercises 90 and 94

Answers

17. $>$ **18.** $<$
19. 45.37   **20.** 135.000

Classroom Example: p. 229, Exercise 88

Instructor Note: Discuss with students the way a newborn baby's weight is recorded (where a high degree of accuracy is essential) versus the way their own weights are recorded (where such a high degree of accuracy is *not* essential).

Answers

21. 187.26

22. 187.2650

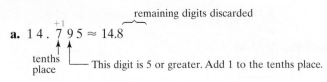

Example 9 **Rounding Decimal Numbers**

Round 14.795 to the indicated place value.

a. Tenths   **b.** Hundredths

Solution:

remaining digits discarded

a. $\overset{+1}{1\,4\,.\,7}\,9\,5 \approx 14.8$

tenths place — This digit is 5 or greater. Add 1 to the tenths place.

remaining digit discarded

b. $1\,4\,.\,7\,\overset{+1}{9}\,5 \approx 14.80$

hundredths place — This digit is 5 or greater. Add 1 to the hundredths place.

• Since the hundredths place digit is 9, adding 1 requires us to carry 1 to the tenths place digit.

In Example 9(b) the 0 in 14.80 indicates that the number was rounded to the hundredths place. It would be incorrect to drop the zero. Even though 14.8 has the same numerical value as 14.80, it implies a different level of accuracy. For example, when measurements are taken using some instrument such as a ruler or scale, the measured values are not exact. The place position to which a number is rounded reflects the accuracy of the measuring device. Thus, the value 14.8 lb indicates that the scale is accurate to the nearest tenth of a pound. The value 14.80 lb indicates that the scale is accurate to the nearest hundredth of a pound.

Section 4.1 Practice Exercises

Study Skills Exercise

For additional exercises, see Classroom Activities 4.1A–4.1C in the *Student's Resource Manual* at www.mhhe.com/moh.

After you get a test back, it is a good idea to correct the test so that you do not make the same errors again. One recommended approach is to use a clean sheet of paper and divide the paper down the middle vertically, as shown. For each problem that you missed on the test, rework the problem correctly on the left-hand side of the paper. Then write a written explanation on the right-hand side of the paper.

Take the time this week to make corrections from your last test.

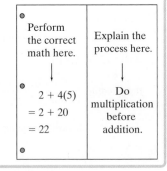

Perform the correct math here.	Explain the process here.
↓	↓
$2 + 4(5)$	Do multiplication before addition.
$= 2 + 20$	
$= 22$	

Vocabulary and Key Concepts

1. a. A ___decimal___ fraction is a fraction whose denominator is a power of 10.

   b. The first three place values to the right of the decimal point are the ___tenths___ place, the ___hundredths___ place, and the ___thousandths___ place.

Concept 1: Decimal Notation

2. Write the word name for the number 3005.   Three thousand, five

Writing   Translating Expression   Geometry   Scientific Calculator   Video   NSE

For Exercises 3–6, expand the powers of 10.

3. 10^2 100

4. 10^3 1000

5. 10^4 10,000

6. 10^5 100,000

For Exercises 7–10, expand the powers of $\frac{1}{10}$.

7. $\left(\frac{1}{10}\right)^2$ $\frac{1}{100}$

8. $\left(\frac{1}{10}\right)^3$ $\frac{1}{1000}$

9. $\left(\frac{1}{10}\right)^4$ $\frac{1}{10,000}$

10. $\left(\frac{1}{10}\right)^5$ $\frac{1}{100,000}$

For Exercises 11–22, identify the place value of each underlined digit. **(See Example 1.)**

11. 3.<u>9</u>83   Tenths

12. 34.8<u>2</u>   Hundredths

13. 440.3<u>9</u>   Hundredths

14. 2<u>4</u>8.94   Tens

15. 4<u>8</u>9.02   Tens

16. 4.092<u>8</u>4   Ten-thousandths

17. 9.283<u>4</u>5   Ten-thousandths

18. 0.32<u>1</u>   Thousandths

 19. 0.48<u>9</u>   Thousandths

20. 5<u>8</u>.211   Ones

21. 9<u>3</u>.834   Ones

22. 5.00000<u>1</u>   Millionths

For Exercises 23–30, write the word name for each decimal fraction.

23. $\frac{9}{10}$ Nine-tenths

24. $\frac{7}{10}$ Seven-tenths

25. $\frac{23}{100}$ Twenty-three hundredths

26. $\frac{19}{100}$ Nineteen hundredths

27. $\frac{33}{1000}$ Thirty-three thousandths

28. $\frac{51}{1000}$ Fifty-one thousandths

29. $\frac{407}{10,000}$ Four hundred seven ten-thousandths

30. $\frac{20}{10,000}$ Twenty ten-thousandths

For Exercises 31–38, write the word name for the decimal. **(See Example 2.)**

31. 3.24   Three and twenty-four hundredths

32. 4.26   Four and twenty-six hundredths

33. 5.9   Five and nine-tenths

34. 3.4   Three and four-tenths

35. 52.3   Fifty-two and three-tenths

36. 21.5   Twenty-one and five-tenths

37. 6.219   Six and two hundred nineteen thousandths

 38. 7.338   Seven and three hundred thirty-eight thousandths

For Exercises 39–44, write the word name as a numeral. **(See Example 3.)**

39. Eight thousand, four hundred seventy-two and fourteen thousandths   8472.014

40. Sixty thousand, twenty-five and four hundred one ten-thousandths   60,025.0401

 41. Seven hundred and seven hundredths   700.07

42. Nine thousand and nine thousandths   9000.009

43. Two million, four hundred sixty-nine thousand and five hundred six thousandths   2,469,000.506

44. Eighty-two million, six hundred fourteen and ninety-seven ten-thousandths   82,000,614.0097

Concept 2: Writing Decimals as Mixed Numbers or Fractions

For Exercises 45–56, write the decimal as a proper fraction or as a mixed number and simplify. **(See Example 4.)**

45. 3.7 $3\frac{7}{10}$

46. 1.9 $1\frac{9}{10}$

 47. 2.8 $2\frac{4}{5}$

48. 4.2 $4\frac{1}{5}$

49. 0.25 $\frac{1}{4}$

50. 0.75 $\frac{3}{4}$

51. 0.55 $\frac{11}{20}$

52. 0.45 $\frac{9}{20}$

53. 20.812 $20\frac{203}{250}$

54. 32.905 $32\frac{181}{200}$

55. 15.0005 $15\frac{1}{2000}$

56. 4.0015 $4\frac{3}{2000}$

For Exercises 57–64, write the decimal as an improper fraction and simplify. **(See Example 5.)**

57. 8.4 $\frac{42}{5}$ **58.** 2.5 $\frac{5}{2}$ **59.** 3.14 $\frac{157}{50}$ **60.** 5.65 $\frac{113}{20}$

61. 23.5 $\frac{47}{2}$ **62.** 14.6 $\frac{73}{5}$ **63.** 11.91 $\frac{1191}{100}$ **64.** 21.33 $\frac{2133}{100}$

Concept 3: Ordering Decimal Numbers

For Exercises 65–68, arrange the numbers from least to greatest. **(See Example 6.)**

65. 34.25, 34.2, 34.3, 34.29   34.2, 34.25, 34.29, 34.3   **66.** 12.46, 12.4, 12.5, 12.49   12.4, 12.46, 12.49, 12.5

67. 0.42, 0.043, $\frac{4}{10}$, 0.042, 0.43   0.042, 0.043, $\frac{4}{10}$, 0.42, 0.43   **68.** 0.04999, 0.0499, $\frac{5}{10}$, 0.4999, 0.05001

                                                                                     0.0499, 0.04999, 0.05001, 0.4999, $\frac{5}{10}$

For Exercises 69–76, fill in the blank with < or >. **(See Example 7.)**

69. 6.312 [<] 6.321   **70.** 8.503 [<] 8.530   **71.** 11.21 [>] 11.2099   **72.** 10.51 [>] 10.5098

73. 0.762 [>] 0.76   **74.** 0.1291 [>] 0.129   **75.** 51.72 [<] 51.721   **76.** 49.06 [<] 49.062

77. Which number is between 3.12 and 3.13? Circle all that apply.   a, b

    a. 3.127   **b.** 3.129   **c.** 3.134   **d.** 3.139

78. Which number is between 42.73 and 42.86? Circle all that apply.   a, c, d

    a. 42.81   **b.** 42.64   **c.** 42.79   **d.** 42.85

79. The batting averages for five legends are given in the table. Rank the players' batting averages from lowest to highest. (*Source:* Baseball Almanac)

Player	Average
Joe Jackson	0.3558
Ty Cobb	0.3664
Lefty O'Doul	0.3493
Ted Williams	0.3444
Rogers Hornsby	0.3585

0.3444, 0.3493, 0.3558, 0.3585, 0.3664

80. The average speed, in miles per hour (mph), of the Daytona 500 for selected years is given in the table. Rank the speeds from slowest to fastest.

Year	Driver	Speed (mph)
1989	Darrell Waltrip	148.466
1991	Ernie Irvan	148.148
1997	Jeff Gordon	148.295
2007	Kevin Harvick	149.333

Source: NASCAR

148.148, 148.295, 148.466, 149.333

Concept 4: Rounding Decimals

81. The numbers given all have equivalent value. However, suppose they represent measured values from a scale. Explain the difference in the interpretation of these numbers.

$$0.25, \quad 0.250, \quad 0.2500, \quad 0.25000$$

These numbers are equivalent, but they represent different levels of accuracy.

82. Which number properly represents 3.499999 rounded to the thousandths place?   a

   a. 3.500   **b.** 3.5   **c.** 3.500000   **d.** 3.499

83. Which value is rounded to the nearest tenth, 7.1 or 7.10?   7.1

84. Which value is rounded to the nearest hundredth, 34.50 or 34.5?   34.50

For Exercises 85–96, round the decimals to the indicated place values. **(See Examples 8 and 9.)**

85. 49.943; tenths   49.9

86. 12.7483; tenths   12.7

87. 33.416; hundredths   33.42

88. 4.359; hundredths   4.36

89. 9.0955; thousandths   9.096

90. 2.9592; thousandths   2.959

91. 21.0239; tenths   21.0

92. 16.804; hundredths   16.80

93. 6.9995; thousandths   7.000

94. 21.9997; thousandths   22.000

95. 0.0079499; ten-thousandths   0.0079

96. 0.00084985; ten-thousandths   0.0008

97. A snail moves at a rate of about 0.00362005 miles per hour. Round the decimal value to the ten-thousandths place.   0.0036 mph

For Exercises 98–101, round the number to the indicated place value.

	Number	Hundreds	Tens	Tenths	Hundredths	Thousandths
98.	349.2395	300	350	349.2	349.24	349.240
99.	971.0948	1000	970	971.1	971.09	971.095
100.	79.0046	100	80	79.0	79.00	79.005
101.	21.9754	0	20	22.0	21.98	21.975

Expanding Your Skills

102. What is the least number with three places to the right of the decimal that can be created with the digits 2, 9, and 7? Assume that the digits cannot be repeated.   0.279

103. What is the greatest number with three places to the right of the decimal that can be created from the digits 2, 9, and 7? Assume that the digits cannot be repeated.   0.972

Concepts

1. Addition and Subtraction of Decimals
2. Applications of Addition and Subtraction of Decimals

Instructor Note: To prepare students, ask them to add the following amounts of money: $5.67 + $3.12. Students are usually quite adept at calculating sums of money.

1. Addition and Subtraction of Decimals

In this section, we learn to add and subtract decimals. To begin, consider the sum $5.67 + 3.12$.

$$5.67 = \quad 5 + \frac{6}{10} + \frac{7}{100}$$

$$\underline{+\ 3.12} = \underline{+\ 3 + \frac{1}{10} + \frac{2}{100}}$$

$$8 + \frac{7}{10} + \frac{9}{100} = 8.79$$

Notice that the decimal points and place positions are lined up to add the numbers. In this way, we can add digits with the same place values because we are effectively adding decimal fractions with like denominators. The intermediate step of using fraction notation is often skipped. We can get the same result more quickly by adding digits in like place positions.

> **Adding and Subtracting Decimals**
>
> **Step 1** Write the numbers in a column with the decimal points and corresponding place values lined up. (You may insert additional zeros as placeholders after the last digit to the right of the decimal point.)
>
> **Step 2** Add or subtract the digits in columns from right to left, as you would whole numbers. The decimal point in the answer should be lined up with the decimal points from the original numbers.

Skill Practice

Add.
1. 184.218 + 14.12

Classroom Example: p. 235, Exercise 16

Example 1   **Adding Decimals**

Add. $27.486 + 6.37$

Solution:

$$\begin{array}{r} 27.486 \\ +\ 6.370 \\ \end{array}$$ Line up the decimal points.
Insert an extra zero as a placeholder.

$$\begin{array}{r} \overset{1\ \ 1}{27.486} \\ +\ 6.370 \\ \hline 33.856 \end{array}$$ Add digits with common place values.

Line up the decimal point in the answer.

Concept Connections

2. Check your answer to Skill Practice exercise 1 by estimation.

Answers

1. 198.338
2. $\approx 184 + 14 = 198$, which is close to 198.338.

With operations on decimals it is important to locate the correct position of the decimal point. A quick estimate can help you determine whether your answer is reasonable. From Example 1, we have

$$\begin{array}{lll} 27.486 & \text{rounds to} & 27 \\ 6.370 & \text{rounds to} & \underline{+6} \\ & & 33 \end{array}$$

The estimated value, 33, is close to the actual value of 33.856.

Example 2   Adding Decimals

Add.   3.7026 + 43 + 816.3

Solution:

$$
\begin{array}{r}
3.7026 \\
43.0000 \\
+\ 816.3000 \\
\end{array}
$$

Line up the decimal points.
Insert a decimal point and four zeros after it.
Insert three zeros.

$$
\begin{array}{r}
{}^{1\ 1}\\
3.7026 \\
43.0000 \\
+\ 816.3000 \\
\hline
863.0026
\end{array}
$$

Add the digits with common place values.

Line up the decimal point in the answer.

The sum is 863.0026.

> **TIP:** To check that the answer is reasonable, round each addend.
>
> | | | |
> |---|---|---|
> | 3.7026 | rounds to | 4 |
> | 43 | rounds to | 43 |
> | 816.3 | rounds to | + 816 |
> | | | 863 |
>
> (with a small ^{1} above 43)
>
> which is close to the actual sum, 863.0026.

Skill Practice

Add.
3. 2.90741 + 15.13 + 3

Classroom Example: p. 235, Exercise 22

Example 3   Subtracting Decimals

Subtract.

a. 0.2868 − 0.056   **b.** 139 − 28.63   **c.** 192.4 − 89.387

Solution:

a.
$$
\begin{array}{r}
0.2868 \\
-\ 0.0560 \\
\hline
0.2308
\end{array}
$$

Line up the decimal points.
Insert an extra zero as a placeholder.
Subtract digits with common place values.
Decimal point in the answer is lined up.

b.
$$
\begin{array}{r}
139.00 \\
-\ 28.63 \\
\end{array}
$$

Line up the decimal points.
Insert extra zeros as placeholders.

$$
\begin{array}{r}
{}^{9}\\
{}^{8\ \cancel{10}\ 10}\\
13\cancel{9}.\cancel{0}\,\cancel{0} \\
-\ 28.63 \\
\hline
110.37
\end{array}
$$

Subtract digits with common place values.
Borrow where necessary.
Line up the decimal point in the answer.

c.
$$
\begin{array}{r}
192.400 \\
-\ 89.387 \\
\end{array}
$$

Line up the decimal points.
Insert extra zeros as placeholders.

$$
\begin{array}{r}
{}^{9}\\
{}^{8\ 12\ 3\ \cancel{10}\ 10}\\
19\cancel{2}.\cancel{4}\,\cancel{0}\,\cancel{0} \\
-\ 89.3\,8\,7 \\
\hline
103.013
\end{array}
$$

Subtract digits with common place values.
Borrow where necessary.
Line up the decimal point in the answer.

Skill Practice

Subtract.
4. 3.194 − 0.512
5. 0.397 − 0.1584
6. 566.4 − 414.231

Classroom Examples: p. 236, Exercises 36 and 38

Instructor Note: Students seem to have difficulty subtracting decimal numbers from whole numbers. Provide lots of examples, such as: 4 − 2.16 and 5 − 4.73.

Answers
3. 21.03741   **4.** 2.682
5. 0.2386   **6.** 152.169

Skill Practice

Simplify.

7. $416.04 + 67.2 - 291.76$

Classroom Example: p. 236,
Exercise 50

Example 4 Adding and Subtracting Decimals

Simplify. $27.819 - 13.78 + 9.6$

Solution:

$\underbrace{27.819 - 13.78} + 9.6$ We must apply the order of operations.

First subtract: $27.819 - 13.78$

$$\begin{array}{r} 27.\overset{7}{\cancel{8}}\overset{11}{\cancel{1}}9 \\ -\ 13.780 \\ \hline 14.039 \end{array}$$

$= 14.039 + 9.6$ Now add 9.6 to the result.

$$\begin{array}{r} \overset{1}{1}4.039 \\ +\ 9.600 \\ \hline 23.639 \end{array}$$

$= 23.639$

2. Applications of Addition and Subtraction of Decimals

Decimals are used often in measurements and in day-to-day applications.

Example 5 Subtracting Decimals in an Application

A graph of the U.S. population (in millions) is given for selected years (Figure 4-2).

a. What is the difference in population between the years 1970 and 1960?

b. What is the difference in population between the years 2000 and 1990?

Skill Practice

8. The following graph represents the average height (in inches) for girls for selected ages.
   a. What is the difference in height between a 9-year-old and a 3-year-old?
   b. What is the difference in height between an 11-year-old and a 9-year-old?

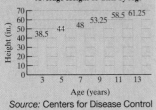

Average Height of Girls by Age

Source: Centers for Disease Control

Classroom Example: p. 236,
Exercise 60

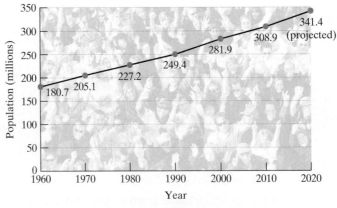

U.S. Population for Selected Years

Source: U.S. Census Bureau

Figure 4-2

Answers

7. 191.48
8. a. 14.75 in.   **b.** 5.25 in.

Solution:

a. The difference in population between the years 1970 and 1960 is given by

$$
\begin{array}{r}
\overset{1}{2}\,\overset{10}{\cancel{0}}\,\overset{4}{\cancel{5}}.\overset{11}{\cancel{1}} \\
-1\,8\,0.7 \\
\hline
2\,4.4
\end{array}
$$ The difference in population is 24.4 million.

b. The difference in population between the years 2000 and 1990 is given by

$$
\begin{array}{r}
2\,8\,\overset{7}{\cancel{1}}.\overset{11}{9} \\
-2\,4\,9.4 \\
\hline
3\,2.5
\end{array}
$$ The difference in population is 32.5 million.

Comparing the values from parts (a) and (b), we see that the U.S. population increased during both 10-year periods. However, there was a greater increase between 1990 and 2000. This indicates that the rate of increase in population is increasing.

Example 6 **Applying Addition and Subtraction of Decimals in a Checkbook**

Fill in the balance for each line in the checkbook register, shown in Figure 4-3. What is the ending balance?

Check No.	Description	Payment	Deposit	Balance
				$684.60
2409	Doctor	$ 75.50		
2410	Mechanic	215.19		
2411	Home Depot	94.56		
	Paycheck		$981.46	
2412	Veterinarian	49.90		

Figure 4-3

Solution:

We begin with $684.60 in the checking account. For each payment, we subtract. For each deposit, we add.

Check No.	Description	Payment	Deposit	Balance	
				$ 684.60	
2409	Doctor	$ 75.50		609.10	= $684.60 − $75.50
2410	Mechanic	215.19		393.91	= $609.10 − $215.19
2411	Home Depot	94.56		299.35	= $393.91 − $94.56
	Paycheck		$981.46	1280.81	= $299.35 + $981.46
2412	Veterinarian	49.90		1230.91	= $1280.81 − $49.90

The ending balance is $1230.91.

Skill Practice

9. Fill in the balance for each line in the checkbook register.

Check	Payment	Deposit	Balance
			$437.80
1426	$82.50		
Pay		$514.02	
1427	26.04		

Classroom Example: p. 237, Exercise 64

Answer

9.

Check	Payment	Deposit	Balance
			$ 437.80
1426	$82.50		355.30
Pay		$514.02	869.32
1427	26.04		843.28

Skill Practice

10. Consider the figure.

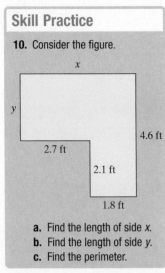

a. Find the length of side *x*.
b. Find the length of side *y*.
c. Find the perimeter.

Classroom Example: p. 238,
Exercise 70

Example 7 **Applying Decimals to Perimeter**

a. Find the length of the side labeled *x*.

b. Find the length of the side labeled *y*.

c. Find the perimeter of the figure.

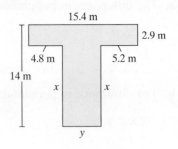

Solution:

a. If we extend the line segment labeled *x* with the dashed line as shown below, we see that the sum of side *x* and the dashed line must equal 14 m. Therefore, subtract 14 − 2.9 to find the length of side *x*.

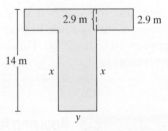

$$\text{Length of side } x: \quad \begin{array}{r} {}^{3\ 10} \\ 1\cancel{4}.\cancel{0} \\ -\ 2.9 \\ \hline 11.1 \end{array}$$

Side *x* is 11.1 m long.

b. The dashed line in the figure below has the same length as side *y*. We also know that $4.8 + 5.2 + y$ must equal 15.4. Since $4.8 + 5.2 = 10.0$,

$$y = 15.4 - 10.0$$
$$= 5.4$$

The length of side *y* is 5.4 m.

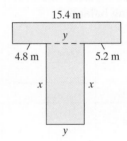

c. Now that we have the lengths of all sides, add them to get the perimeter.

$$\begin{array}{r} {}^{2\ 3} \\ 15.4 \\ 2.9 \\ 5.2 \\ 11.1 \\ 5.4 \\ 11.1 \\ 4.8 \\ +\ 2.9 \\ \hline 58.8 \end{array}$$

The perimeter is 58.8 m.

Section 4.2   Practice Exercises

Study Skills Exercise

For additional exercises, see Classroom Activities 4.2A–4.2C in the *Student's Resource Manual* at www.mhhe.com/moh.

Go to the online services that accompany this text. List two options that this online service offers that could help you in this course.

a. _____   b. _____

Review Exercises

1. Which number is equal to 2.007? Circle all that apply.

    a. 2.070   **b.** 2.0070   **c.** 2.00700   **d.** 2.7   b, c

NS E **2.** Which number is equal to 5.03? Circle all that apply.

    a. 5.030   **b.** 5.30   **c.** 5.0300   **d.** 5.3   a, c

NS E **3.** Which number is equal to $\frac{7}{100}$? Circle all that apply.

    a. 0.7   **b.** 0.07   **c.** 0.070   **d.** 0.007   b, c

NS E **4.** Which number is equal to $\frac{9}{10}$? Circle all that apply.

    a. 0.09   **b.** 0.090   **c.** 0.90   **d.** 0.900   c, d

For Exercises 5–10, round the decimals to the indicated place values.

5. 23.489; tenths   23.5   **6.** 42.314; hundredths   42.31   **7.** 8.6025; thousandths   8.603

8. 0.981; tenths   1.0   **9.** 2.82998; ten-thousandths   2.8300   **10.** 2.78999; thousandths   2.790

Concept 1: Addition and Subtraction of Decimals

NS E For Exercises 11–16, add the decimal numbers. Then round the numbers and find the sum to determine if your answer is reasonable. The first estimate is done for you. **(See Examples 1 and 2.)**

Expression	Estimate	Expression	Estimate
11. 44.6 + 18.6 63.2	45 + 19 = 64	**12.** 28.2 + 23.2 51.4	
13. 5.306 + 3.645 8.951		**14.** 3.451 + 7.339 10.79	
15. 12.9 + 3.091 15.991		**16.** 4.125 + 5.9 10.025	

For Exercises 17–28, add the decimals. **(See Examples 1 and 2.)**

17. 78.9 + 0.9005
79.8005

18. 44.2 + 0.7802
44.9802

19. 23 + 8.0148
31.0148

20. 7.9302 + 34
41.9302

21. 34 + 23.0032 + 5.6
62.6032

22. 23 + 8.01 + 1.0067
32.0167

23. 68.394 + 32.02
100.414

24. 2.904 + 34.229
37.133

25. 103.94 + 24.5
128.44

26. 93.2 + 43.336
136.536

27. 54.2 + 23.993 + 3.87
82.063

28. 13.9001 + 72.4 + 34.13
120.4301

Writing   Translating Expression   Geometry   Scientific Calculator   Video   NS E

NS&E For Exercises 29–34, subtract the decimal numbers. Then round the numbers and find the difference to determine if your answer is reasonable. The first estimate is done for you. **(See Example 3.)**

Expression	Estimate		Expression	Estimate

29. 35.36 − 21.12   14.24   35 − 21 = 14   **30.** 53.9 − 22.4   31.5

31. 7.24 − 3.56   3.68   **32.** 23.3 − 20.8   2.5

33. 45.02 − 32.7   12.32   **34.** 66.15 − 42.9   23.25

For Exercises 35–46, subtract the decimals. **(See Example 3.)**

35. 14.5 − 8.823   5.677   **36.** 33.2 − 21.932   11.268   **37.** 2 − 0.123   1.877

38. 4 − 0.42   3.58   **39.** 103.4 − 45.05 − 0.982   57.368   **40.** 98.5 − 23.21 − 0.144   75.146

41. 55.9 − 34.2354   21.6646   **42.** 49.1 − 24.481   24.619   **43.** 18.003 − 3.238   14.765

44. 21.03 − 16.446   4.584   **45.** 183.01 − 23.452   159.558   **46.** 164.23 − 44.3893   119.8407

Mixed Exercises

For Exercises 47–58, add and subtract as indicated. **(See Example 4.)**

47. 6.007 + 12.74 − 3.4   15.347   **48.** 3.005 + 25.127 − 13.7   14.432   **49.** 23.37 − 21.9 + 5.111   6.581

50. 0.78 − 0.028 + 6.1   6.852   **51.** 8.962 + 51 − 40.05   19.912   **52.** 11.957 + 45 − 3.55   53.407

53. 5.3 + 5.03 + 5.003 − 5.0003
10.3327
54. 2.6 + 2.06 + 2.006 − 2.0006
4.6654
55. 5.84 + 5.084 − 5.0084
5.9156

56. 85.3 − 47.0092 + 4.06
42.3508
57. 10 − 0.9 − 0.09 − 0.009
9.001
58. 5 − 0.9 − 0.99 − 0.999
2.111

Concept 2: Applications of Addition and Subtraction of Decimals

59. The amount of time that it takes Mercury, Venus, Earth, and Mars to revolve about the Sun is given in the graph. **(See Example 5.)**

a. How much longer does it take Mars to complete a revolution around the Sun than the Earth?   321.724 days

b. How much longer does it take Venus than Mercury to revolve around the Sun?   156.73 days

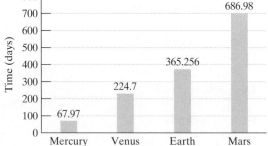

Period of Revolution about the Sun

Time (days): Mercury 67.97, Venus 224.7, Earth 365.256, Mars 686.98

Source: National Aeronautics and Space Administration

60. The birth weights of the Dilley sextuplets are given in the graph.

a. What is the difference between the weights of Julian and Quinn?   The difference is 0.6875 lb.

b. What is the total weight of all six babies?
The total weight is 15.25 lb.

Birth Weights of the Dilley Sextuplets

Weight (lb): Brenna 2.375, Ian 2.6875, Julian 2.8125, Quinn 2.125, Claire 2.4375, Adrian 2.8125

61. Water flows into a pool at a constant rate. The water level is recorded at several 1-hr intervals.

Time	Water Level
9:00 A.M.	4.2 in.
10:00 A.M.	5.9 in.
11:00 A.M.	7.6 in.
12:00 P.M.	9.3 in.

   a. From the table, how many inches is the water level rising each hour? The water is rising 1.7 in./hr.

   b. At this rate, what will the water level be at 1:00 P.M.?
    At 1:00 P.M. the level will be 11 in.

   c. At this rate, what will the water level be at 3:00 P.M.?
    At 3:00 P.M. the level will be 14.4 in.

62. The total gross earnings in the United States for four top animated films are given in the table.

Movie	Earnings ($ millions)
Shrek 2	441.2
Finding Nemo	339.7
The Lion King	328.5
Shrek the Third	322.7

   a. What was the difference between the earnings for *Shrek 2* and *The Lion King*? $112.7 million

   b. What was the difference between the earnings for *Finding Nemo* and *Shrek the Third*? $17 million

   c. What was the total earnings for all four films? $1432.1 million (or $1.4321 billion)

63. Fill in the balance for each line in the checkbook register shown in the figure. What was the ending balance?
(See Example 6.)

Check No.	Description	Payment	Deposit	Balance
				$ 245.62
2409	Electric bill	$ 52.48		193.14
2410	Groceries	72.44		120.70
2411	Department store	108.34		12.36
	Paycheck		$1084.90	1097.26
2412	Restaurant	23.87		1073.39
	Transfer from savings		200	1273.39

64. A section of a bank statement is shown in the figure. Find the mistake that was made by the bank.
The bank subtracted $1500.00 on January 6 instead of $150.00.

Date	Action	Payment	Deposit	Balance
				$1124.35
Jan. 2	Check #4214	$749.32		375.03
Jan. 3	Check #4215	37.29		337.74
Jan. 4	Transfer from savings		$ 400.00	737.74
Jan. 5	Paycheck		1451.21	2188.95
Jan. 6	Cash withdrawal	150.00		688.95

65. A normal human red blood cell count is between 4.2 and 6.9 million cells per microliter (μL). A cancer patient undergoing chemotherapy has a red blood cell count of 2.85 million cells per microliter. How far below the lower normal limit is this? 1.35 million cells per microliter

66. A laptop computer was originally priced at $1299.99 and was discounted to $998.95. By how much was it marked down? It was marked down by $301.04.

 Writing    Translating Expression   Geometry   Scientific Calculator   Video   NS E

67. The table shows the thickness of four U.S. coins. If you stacked three quarters and a dime in one pile and two nickels and two pennies in another pile, which pile would be higher?
The pile containing the two nickels and two pennies is higher.

68. How much thicker is a nickel than a quarter?
0.2 mm

Coin	Thickness
Quarter	1.75 mm
Dime	1.35 mm
Nickel	1.95 mm
Penny	1.55 mm

Source: U.S. Department of the Treasury

For Exercises 69–72, find the lengths of the sides labeled x and y. Then find the perimeter. **(See Example 7.)**

69.

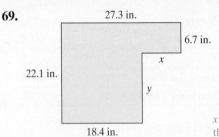

$x = 8.9$ in.; $y = 15.4$ in.; the perimeter is 98.8 in.

70.

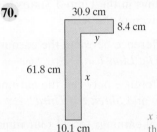

$x = 53.4$ cm; $y = 20.8$ cm; the perimeter is 185.4 cm.

71.

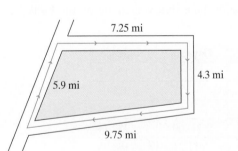

$x = 2.075$ ft; $y = 2.59$ ft; the perimeter is 22.17 ft.

72.

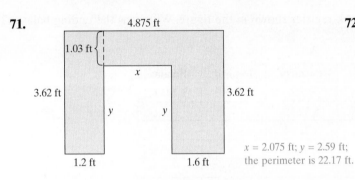

$x = 26.7$ yd; $y = 37.13$ yd; the perimeter is 144.36 yd.

73. A city bus follows the route shown in the map. How far does it travel in one circuit?   27.2 mi

74. Santos built a new deck and needs to put a railing around the sides. He does not need railing where the deck is against the house. How much railing should he purchase?   32.5 m

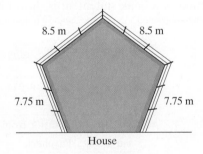

Expanding Your Skills

In a circle, the length of a line segment connecting two points on the circle and passing through the center is called a *diameter*.

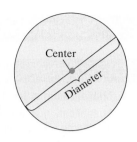

Use the definition of a diameter for Exercises 75 and 76.

75. The wire in a cable has a diameter of 6 mm. The insulation is 0.5 mm. What is the diameter of the cable with the insulation included?   7 mm

76. Find the inner diameter of the cable if the total diameter is 1.65 cm and the insulation is 0.15 cm thick.   1.35 cm

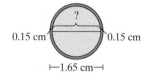

Calculator Connections

Topic: Entering Decimals on a Calculator

To enter decimals on a calculator, use the [.] key.

Expression	Keystrokes	Result
984.126 + 37.11	984 [.] 126 [+] 37 [.] 11 **ENTER**	1021.236

Calculator Exercises

For Exercises 77–82, refer to the table. The table gives the closing stock prices (in dollars per share) for the first day of trading for the given month.

Stock	Jan.	Feb.	March	April	May
IBM	132.45	125.53	128.57	128.25	130.46
FedEx	83.45	80.67	85.81	92.17	90.01

77. By how much did the IBM stock decrease between January and May?
   IBM decreased by $1.99 per share.

78. By how much did the FedEx stock increase between January and May?
   FedEx increased by $6.56 per share.

79. Between which two consecutive months did the FedEx stock increase the most? What was the amount of increase?   Between March and April, FedEx increased the most, by $6.36 per share.

80. Between which two consecutive months did the IBM stock increase the most? What was the amount of increase?   Between February and March, IBM increased the most, by $3.04 per share.

81. Between which two consecutive months did the FedEx stock decrease the most? What was the amount of decrease?   Between January and February, FedEx decreased the most, by $2.78 per share.

82. Between which two consecutive months did the IBM stock decrease the most? What was the amount of decrease?   Between January and February, IBM decreased the most, by $6.92 per share.

Section 4.3   Multiplication of Decimals

Concepts

1. Multiplication of Decimals
2. Multiplication by a Power of 10 and by a Power of 0.1
3. Applications Involving Multiplication of Decimals

1. Multiplication of Decimals

Multiplication of decimals is much like multiplication of whole numbers. However, we need to know where to place the decimal point in the product. Consider the product 0.3×0.41. One way to multiply these numbers is to write them first as decimal fractions.

$$0.3 \times 0.41 = \frac{3}{10} \times \frac{41}{100} = \frac{123}{1000} \text{ or } 0.123$$

Another method multiplies the factors vertically. First we multiply the numbers as though they were whole numbers. We temporarily disregard the decimal point in the product because it will be placed later.

$$\begin{array}{r} 0.41 \\ \times\ 0.3 \\ \hline 123 \end{array} \longleftarrow \text{ decimal point not yet placed}$$

From the first method, we know that the correct answer to this problem is 0.123. Notice that 0.123 contains the same number of decimal places as the two factors combined. That is,

$$\begin{array}{r} 0.41 \\ \times\ 0.3 \\ \hline .123 \end{array} \quad \begin{array}{l} \longleftarrow \quad 2 \text{ decimal places} \\ \longleftarrow \quad \underline{1 \text{ decimal place}} \\ \longleftarrow \quad 3 \text{ decimal places} \end{array}$$

The process to multiply decimals is summarized as follows.

Concept Connections

1. How many decimal places will be in the product 2.72×1.4?
2. Explain the difference between the process to multiply 123×51 and the process to multiply 1.23×5.1.

TIP: When multiplying decimals, it is *not* necessary to line up the decimal points as we do when we add or subtract decimals. Instead, we write the factors "right-justified."

> **Multiplying Two Decimals**
>
> **Step 1** Multiply as you would whole numbers.
>
> **Step 2** Place the decimal point in the product so that the number of decimal places equals the combined number of decimal places of both factors.
>
> *Note:* You may need to insert zeros to the left of the whole-number product to get the correct number of decimal places in the answer.

In Example 1, we multiply decimals by using this process.

Skill Practice

Multiply.
3. 19.7×4.1

Classroom Example: p. 245, Exercise 16

Answers

1. 3
2. The actual process of vertical multiplication is the same for both cases. However, for the product 1.23×5.1, the decimal point must be placed so that the product has the same number of decimal places as both factors combined (in this case, 3).
3. 80.77

Example 1   Multiplying Decimals

Multiply. $\begin{array}{r} 11.914 \\ \times\ 0.8 \\ \hline \end{array}$

Solution:

$$\begin{array}{r} \overset{17\ 13}{11.914} \\ \times\ 0.8 \\ \hline 9.5312 \end{array} \qquad \begin{array}{l} 3 \text{ decimal places} \\ +\ 1 \text{ decimal place} \\ \hline 4 \text{ decimal places} \end{array}$$

The product is 9.5312.

| Example 2 | **Multiplying Decimals** |

Multiply. Then use estimation to check the location of the decimal point.

$$29.3 \times 2.8$$

Solution:

Actual product:

$$
\begin{array}{r}
\overset{\scriptstyle 1}{} \\
\overset{\scriptstyle 7\,2}{29.3} \\
\times\ 2.8 \\
\hline
2344 \\
5860 \\
\hline
82.04
\end{array}
$$

29.3   1 decimal place
× 2.8   + 1 decimal place

82.04   2 decimal places   The product is 82.04.

To check the answer, we can round the factors and estimate the product. The purpose of the estimate is primarily to determine whether we have placed the decimal point correctly. Therefore, it is usually sufficient to round each factor to the left-most nonzero digit. This is called **front-end rounding.** Thus,

$$
\begin{array}{rcl}
29.3 & \text{rounds to} & 30 \\
\times\ 2.8 & \text{rounds to} & \times\ 3 \\
\hline
& & 90
\end{array}
$$

The first digit for the actual product 8̲2.04 and the first digit for the estimate 9̲0 is the tens place. Therefore, we are reasonably sure that we have located the decimal point correctly. The estimate 90 is close to 82.04.

Skill Practice

4. Multiply 1.9×29.1, and check your answer using estimation.

Classroom Example: p. 245, Exercise 36

| Example 3 | **Multiplying Decimals** |

Multiply. Then use estimation to check the location of the decimal point.

$$2.79 \times 0.0003$$

Solution:

Actual product: Estimate:

$$
\begin{array}{r}
\overset{\scriptstyle 2\ 2}{2.79} \\
\times\ 0.0003 \\
\hline
.000837
\end{array}
$$

2.79   2 decimal places
× 0.0003   + 4 decimal places

.000837   6 decimal places
(insert 3 zeros to the left)

$$
\begin{array}{rcl}
2.79 & \text{rounds to} & 3 \\
\times\ 0.0003 & \text{rounds to} & \times\ 0.0003 \\
\hline
& & 0.0009
\end{array}
$$

The first digit for both the actual product and the estimate is in the ten-thousandths place. We are reasonably sure the decimal point is positioned correctly.

The product is 0.000837.

Skill Practice

5. Multiply 4.6×0.00008, and check your answer using estimation.

Classroom Example: p. 245, Exercise 40

Answers

4. 55.29; $\approx 2 \times 30 = 60$ which is close to 55.29.
5. 0.000368; using front-end rounding, we have $\approx 5 \times 0.00008 = 0.0004$ which is close to 0.000368.

Example 4 Squaring a Decimal

Simplify. $(0.05)^2$

Solution:

$(0.05)^2 = 0.05 \times 0.05$

$$
\begin{array}{r r}
0.05 & \text{2 decimal places} \\
\times\ 0.05 & \underline{\text{2 decimal places}} \\
\hline
0.0025 & \text{4 decimal places}
\end{array}
$$

When squaring a decimal, remember to count the decimal places to the right of the decimal points in both factors.

$$
\text{Consider:} \quad
\begin{aligned}
(0.2)^2 &= 0.04 \\
(0.02)^2 &= 0.0004 \\
(0.002)^2 &= 0.000004
\end{aligned}
$$

2. Multiplication by a Power of 10 and by a Power of 0.1

Consider the number 2.7 multiplied by the powers of 10; that is, 10, 100, 1000 . . .

$$
\begin{array}{r}
10 \\
\times\ 2.7 \\
\hline
70 \\
\underline{200} \\
27.0
\end{array}
\qquad
\begin{array}{r}
100 \\
\times\ 2.7 \\
\hline
700 \\
\underline{2000} \\
270.0
\end{array}
\qquad
\begin{array}{r}
1000 \\
\times\ 2.7 \\
\hline
7000 \\
\underline{20000} \\
2700.0
\end{array}
$$

Multiplying 2.7 by 10 moves the decimal point 1 place to the right.
Multiplying 2.7 by 100 moves the decimal point 2 places to the right.
Multiplying 2.7 by 1000 moves the decimal point 3 places to the right.

This leads us to the following generalization.

> **Multiplying a Decimal by a Power of 10**
> Move the decimal point to the right the same number of decimal places as the number of zeros in the power of 10.

Example 5 Multiplying by Powers of 10

Multiply.

a. $14.78 \times 10,000$ **b.** 0.0064×100 **c.** $8.271 \times 1,000,000$

Solution:

a. $14.78 \times 10,000 = 147,800$ Move the decimal point 4 places to the right.

b. $0.0064 \times 100 = 0.64$ Move the decimal point 2 places to the right.

c. $8.271 \times 1,000,000 = 8,271,000$ Move the decimal point 6 places to the right.

Multiplying a decimal by 10, 100, 1000, and so on increases its value. Therefore, it makes sense to move the decimal point to the *right*. Now suppose we multiply a decimal by 0.1, 0.01, and 0.001. These numbers represent the decimal fractions $\frac{1}{10}$, $\frac{1}{100}$, and $\frac{1}{1000}$, respectively, and are easily recognized as powers of 0.1 (see Section 2.4). Taking one-tenth of a number or one-hundredth of a number makes the number smaller. To multiply by 0.1, 0.01, 0.001, and so on (powers of 0.1), move the decimal point to the *left*.

$$
\begin{array}{ccc}
3.6 & 3.6 & 3.6 \\
\times\ 0.1 & \times\ 0.01 & \times\ 0.001 \\
\hline
.36 & .036 & .0036
\end{array}
$$

> **Multiplying a Decimal by Powers of 0.1**
>
> Move the decimal point to the left the same number of places as there are decimal places in the power of 0.1.

Example 6 **Multiplying by Powers of 0.1**

Multiply.

a. 62.074×0.0001 **b.** 7965.3×0.1 **c.** 0.0057×0.00001

Solution:

a. $62.074 \times 0.0001 = 0.0062074$ Move the decimal point 4 places to the left. Insert extra zeros.

b. $7965.3 \times 0.1 = 796.53$ Move the decimal point 1 place to the left.

c. $0.0057 \times 0.00001 = 0.000000057$ Move the decimal point 5 places to the left.

Sometimes people prefer to use number names to express very large numbers. For example, we might say that the U.S. population in a recent year was approximately 280 million. To write this in decimal form, we note that 1 million = 1,000,000. In this case, we have 280 of this quantity. Thus,

$$280 \text{ million} = 280 \times 1,000,000 \text{ or } 280,000,000$$

Example 7 **Naming Large Numbers**

Write the decimal number representing each word name.

a. The distance between the Earth and Sun is approximately 92.9 million miles.

b. The number of deaths in the United States due to heart disease in 2010 was projected to be 8 hundred thousand.

c. A recent estimate claimed that collectively Americans throw away 472 billion pounds of garbage each year.

Solution:

a. 92.9 million = 92.9 × 1,000,000 = 92,900,000

b. 8 hundred thousand = 8 × 100,000 = 800,000

c. 472 billion = 472 × 1,000,000,000 = 472,000,000,000

3. Applications Involving Multiplication of Decimals

Skill Practice

17. A book club ordered 12 books on www.amazon .com for $8.99 each. The shipping cost was $4.95. What was the total bill?

Classroom Example: p. 247, Exercise 90

Example 8   Applying Decimal Multiplication

Jane Marie bought 8 cans of tennis balls for $1.98 each. She paid $1.03 in tax. What was the total bill?

Solution:

The cost of the tennis balls before tax is

$$8(\$1.98) = \$15.84 \qquad \begin{array}{r} {}^{7\ 6} \\ 1.98 \\ \times\ \ 8 \\ \hline 15.84 \end{array}$$

Adding the tax to this value, we have

$$\begin{pmatrix} \text{Total} \\ \text{cost} \end{pmatrix} = \begin{pmatrix} \text{Cost of} \\ \text{tennis balls} \end{pmatrix} + (\text{Tax})$$

$$\begin{array}{r} = \$15.84 \\ + 1.03 \\ \hline \$16.87 \end{array} \qquad \text{The total cost is \$16.87.}$$

Skill Practice

18. The IMAX movie screen at the Museum of Science and Discovery in Ft. Lauderdale, Florida, is 18 m by 24.4 m. What is the area of the screen?

Classroom Example: p. 247, Exercise 98

Example 9   Finding the Area of a Rectangle

The *Mona Lisa* is perhaps the most famous painting in the world. It was painted by Leonardo da Vinci somewhere between 1503 and 1506 and now hangs in the Louvre in Paris, France. The dimensions of the painting are 30 in. by 20.875 in. What is the total area?

Solution:

Recall that the area of a rectangle is given by

$$A = \ell \cdot w$$

$$A = (30 \text{ in.})(20.875 \text{ in.}) \qquad \begin{array}{r} {}^{2\ \ 2\ 1} \\ 20.875 \\ \times\ 30 \\ \hline 0 \\ 626250 \\ \hline 626.250 \end{array}$$

$$= 626.25 \text{ in.}^2$$

The area of the *Mona Lisa* is 626.25 in.^{2}.

Answers

17. The total bill was $112.83.
18. The screen area is 439.2 m^2.

Section 4.3 Practice Exercises

Study Skills Exercise

For additional exercises, see Classroom Activities 4.3A–4.3B in the *Student's Resource Manual* at www.mhhe.com/moh.

Look through this chapter and write down page numbers in which you can find the following features.

a. Avoiding Mistakes box _____

b. TIP box _____

c. Key term (shown in bold) _____

d. Skill Practice exercises _____

Vocabulary and Key Concepts

1. Rounding a number to the left-most nonzero digit is called _____ -end rounding.   front

Review Exercises

2. Fill in the blank with $<$ or $>$. 51.4382 ☐ 51.4389 $<$

For Exercises 3–6, expand the powers of 10 and 0.1.

3. 10^3 1000

4. 0.1^3 0.001

5. 0.1^2 0.01

6. 10^2 100

Concept 1: Multiplication of Decimals

For Exercises 7–18, multiply the decimals. **(See Examples 1–3.)**

7. $\begin{array}{r} 0.8 \\ \times\ 0.5 \\ \hline 0.4 \end{array}$

8. $\begin{array}{r} 0.6 \\ \times\ 0.5 \\ \hline 0.3 \end{array}$

9. $(0.9)(4)$ 3.6

10. $(0.2)(9)$ 1.8

11. $\begin{array}{r} 0.4 \\ \times\ 20 \\ \hline 8 \end{array}$

12. $\begin{array}{r} 0.9 \\ \times\ 30 \\ \hline 27 \end{array}$

13. $(60)(0.003)$ 0.18

14. $(40)(0.005)$ 0.2

15. $\begin{array}{r} 22.38 \\ \times\ 0.8 \\ \hline 17.904 \end{array}$

16. $\begin{array}{r} 31.67 \\ \times\ 0.4 \\ \hline 12.668 \end{array}$

17. $\begin{array}{r} 14 \\ \times\ 0.002 \\ \hline 0.028 \end{array}$

18. $\begin{array}{r} 0.25 \\ \times\ 40 \\ \hline 10 \end{array}$

For Exercises 19–26, round each number by using front-end rounding.

19. 135   100

20. 481   500

21. 28   30

22. 52   50

23. 0.0672   0.07

24. 0.0807   0.08

25. 0.241   0.2

26. 0.339   0.3

For Exercises 27–40, multiply the decimals. Then estimate the answer by rounding. The first estimate is done for you.

	Exact	Estimate		Exact	Estimate		Exact	Estimate

27. $\begin{array}{r} 8.3 \\ \times\ 4.5 \\ \hline 37.35 \end{array}$ $\begin{array}{r} 8 \\ \times\ 5 \\ \hline 40 \end{array}$

28. $\begin{array}{r} 4.3 \\ \times\ 9.2 \\ \hline 39.56 \end{array}$

29. $\begin{array}{r} 0.58 \\ \times\ 7.2 \\ \hline 4.176 \end{array}$

30. $\begin{array}{r} 0.83 \\ \times\ 6.5 \\ \hline 5.395 \end{array}$

31. 5.92×0.8 4.736

32. 9.14×0.6 5.484

33. $(0.413)(7)$ 2.891

34. $(0.321)(6)$ 1.926

35. 35.9×3.2 114.88

36. 41.7×6.1 254.37

37. 562×0.004 2.248

38. 984×0.009 8.856

39. 0.0004×3.6 0.00144

40. 0.0008×6.5 0.0052

NS/E **41.** Compare the quantities $(0.3)^2$ and 0.9. Are they equal? $(0.3)^2 = 0.09$, which is not equal to 0.9.

NS/E **42.** Compare the quantities $(0.8)^2$ and 6.4. Are they equal? $(0.8)^2 = 0.64$ which is not equal to 6.4.

For Exercises 43–54, simplify the expressions. **(See Example 4.)**

43. $(0.06)^2$ 0.0036   **44.** $(0.16)^2$ 0.0256   **45.** $(2.5)^2$ 6.25   **46.** $(1.1)^2$ 1.21

47. $(0.4)^2$ 0.16   **48.** $(0.7)^2$ 0.49   **49.** $(1.3)^2$ 1.69   **50.** $(2.4)^2$ 5.76

51. $(0.1)^3$ 0.001   **52.** $(0.2)^3$ 0.008   **53.** $(0.2)^4$ 0.0016   **54.** $(0.3)^3$ 0.027

Concept 2: Multiplication by a Power of 10 and by a Power of 0.1

55. If 417.43 is multiplied by 100, will the decimal point move to the left or to the right? By how many places?
The decimal point will move to the right two places.

56. If 2498.613 is multiplied by 10,000, will the decimal point move to the left or to the right? By how many places?   The decimal point will move to the right four places.

57. Multiply the numbers.
   a. 5.1×10 51   **b.** 5.1×100 510   **c.** 5.1×1000 5100   **d.** $5.1 \times 10,000$ 51,000

58. If 256.8 is multiplied by 0.001, will the decimal point move to the left or to the right? By how many places?
The decimal point will move to the left three places.

59. If 0.45 is multiplied by 0.1, will the decimal point move to the left or to the right? By how many places?
The decimal point will move to the left one place.

60. Multiply the numbers.
   a. 5.1×0.1 0.51   **b.** 5.1×0.01 0.051   **c.** 5.1×0.001 0.0051   **d.** 5.1×0.0001 0.00051

For Exercises 61–72, multiply the numbers by the powers of 10 and 0.1. **(See Examples 5 and 6.)**

61. 34.9×100 3490   **62.** 2.163×100 216.3   **63.** 96.59×1000 96,590   **64.** 18.22×1000 18,220

65. 93.3×0.01 0.933   **66.** 80.2×0.01 0.802   **67.** 54.03×0.001 0.05403   **68.** 23.11×0.001 0.02311

69. 2.001×10 20.01   **70.** 5.932×10 59.32   **71.** 0.5×0.0001 0.00005   **72.** 0.8×0.0001 0.00008

For Exercises 73–76, write the amount in terms of cents.

73. $3.24 324¢   **74.** $21.56 2156¢   **75.** $0.37 37¢   **76.** $0.75 75¢

For Exercises 77–80, write the amount in terms of dollars.

77. 347¢ $3.47   **78.** 512¢ $5.12   **79.** 2041¢ $20.41   **80.** 5712¢ $57.12

NS/E **81. a.** Round $1.499 to the nearest dollar. $1   NS/E **82. a.** Round $20.599 to the nearest dollar. $21

   b. Round $1.499 to the nearest cent. $1.50   **b.** Round $20.599 to the nearest cent. $20.60

For Exercises 83–88, write the decimal number representing each word name. **(See Example 7.)**

83. The number of beehives in the United States is 2.6 million. (*Source:* U.S. Department of Agriculture) 2,600,000

84. The people of France collectively consume 34.7 million gallons of champagne per year. (*Source*: Food and Agriculture Organization of the United Nations) 34,700,000 gal

85. The most stolen make of car worldwide is Toyota. For a recent year, there were four hundred-thousand Toyota's stolen. (*Source:* Interpol) 400,000

86. The musical *Miss Saigon* ran for about 4 thousand performances in a 10-year period. 4000

87. The people in the United States have spent over $20.549 billion on DVDs.

$20,549,000,000

88. Coca-Cola Classic was the greatest selling brand of soft-drinks. For a recent year, over 4.8 billion gallons were sold in the United States. (*Source:* Beverage Marketing Corporation)   4,800,000,000 gal

Concept 3: Applications Involving Multiplication of Decimals

89. One gallon of gasoline weighs about 6.3 lb. However, when burned, it produces 20 lb of carbon dioxide (CO_2). This is because most of the weight of the CO_2 comes from the oxygen in the air.

   a. How many pounds of gasoline does a Hummer H2 carry when its tank is full (the tank holds 32 gal).   201.6 lb of gasoline

   b. How many pounds of CO_2 does a Hummer H2 produce after burning an entire tankful of gasoline?   640 lb of CO_2

90. Corrugated boxes for shipping cost $2.27 each. How much will 10 boxes cost including tax of $1.59?   The total cost is $24.29.

91. The Athletic Department at Broward College bought 20 pizzas for $12.95 each, 10 Greek salads for $5.95 each, and 60 soft drinks for $1.29 each. What was the total bill including a sales tax of $27.71? **(See Example 8.)**   The bill was $423.61.

92. A hotel gift shop ordered 40 T-shirts at $8.69 each, 10 hats at $3.95 each, and 20 beach towels at $4.99 each. What was the total cost of the merchandise, including the $29.21 sales tax?   The total cost was $516.11.

93. At General Tires, one tire costs $70.20. A set of four Firestone tires costs $231.99. How much can a person save by buying the set of four Firestone tires compared to four General tires? $48.81 can be saved.

94. Certain DVD titles are on sale for 2 for $36. If they regularly sell for $24.99, how much can a person save by buying 4 DVDs?

$27.96 can be saved.

For Exercises 95 and 96, find the area. **(See Example 9.)**

95.

0.05 km

0.023 km

0.00115 km^2

96.   4.5 yd   30.15 yd^2

6.7 yd

97. Blake plans to build a rectangular patio that is 15 yd by 22.2 yd. What is the total area of the patio?   The area is 333 yd^2.

98. The front page of a newspaper is 56 cm by 31.5 cm. Find the area of the page.

The area is 1764 cm^2.

Expanding Your Skills

99. Evaluate.

   a. $(0.3)^2$ 0.09   **b.** $\sqrt{0.09}$ 0.3

100. Evaluate.

   a. $(0.5)^2$ 0.25   **b.** $\sqrt{0.25}$ 0.5

For Exercises 101–104, evaluate the square roots.

101. $\sqrt{0.01}$ 0.1   **102.** $\sqrt{0.04}$ 0.2   **103.** $\sqrt{0.36}$ 0.6   **104.** $\sqrt{0.49}$ 0.7

 Writing   Translating Expression   Geometry   Scientific Calculator   Video   NS E

Section 4.4   Division of Decimals

1. Division of Decimals

Dividing decimals is much the same as dividing whole numbers. However, we must determine where to place the decimal point in the quotient.

First consider the quotient $3.5 \div 7$. We can write the numbers in fractional form and then divide.

$$3.5 \div 7 = \frac{35}{10} \div \frac{7}{1} = \frac{35}{10} \cdot \frac{1}{7} = \frac{35}{70} = \frac{5}{10} = 0.5$$

Now consider the same problem by using the efficient method of long division: $7\overline{)3.5}$.

When the divisor is a whole number, we place the decimal point directly above the decimal point in the dividend. Then we divide as we would whole numbers.

Decimal point placed above
the decimal point in the dividend.

$$7\overline{)3.5} \quad .5$$

Dividing a Decimal by a Whole Number

To divide by a whole number:

Step 1 Place the decimal point in the quotient directly above the decimal point in the dividend.

Step 2 Divide as you would whole numbers.

Example 1   Dividing by a Whole Number

Divide and check the answer by multiplying.

$$30.55 \div 13$$

Solution:

Locate the decimal point in the quotient.

$$13\overline{)30.55}$$

$$
\begin{array}{r}
2.35 \\
13\overline{)30.55} \\
-26 \\
\hline
45 \\
-39 \\
\hline
65 \\
-65 \\
\hline
0
\end{array}
$$

Divide as you would whole numbers.

Check by multiplying:

$$
\begin{array}{r}
2.35 \\
\times 13 \\
\hline
705 \\
2350 \\
\hline
30.55 \checkmark
\end{array}
$$

When dividing decimals, we do not use a remainder. Instead we insert zeros to the right of the dividend and continue dividing. This is demonstrated in Example 2.

Example 2 **Dividing by a Whole Number**

Divide and check the answer by multiplying.

$$3.5 \div 4$$

Solution:

Locate the decimal point in the quotient.

$$4\overline{)3.5}$$

$$\begin{array}{r} .8 \\ 4\overline{)3.5} \\ -32 \\ \hline 3 \end{array}$$ ◄—— Rather than using a remainder, we insert zeros in the dividend and continue dividing.

$$\begin{array}{r} .875 \\ 4\overline{)3.500} \\ -32\downarrow\,| \\ \hline 30\,| \\ -28\downarrow \\ \hline 20 \\ -20 \\ \hline 0 \end{array}$$

Check by multiplying:

$$\begin{array}{r} {\scriptstyle 3\ 2} \\ 0.875 \\ \times\ 4 \\ \hline 3.500\ \checkmark \end{array}$$

The quotient is 0.875.

Skill Practice

Divide.

2. $6.8 \div 5$

Classroom Example: p. 256, Exercise 20

Example 3 **Dividing by a Whole Number**

Divide and check the answer by multiplying. $40\overline{)5}$

Solution:

$$40\overline{)5.}$$ ◄—— The dividend is a whole number, and the decimal point is understood to be to its right. Insert the decimal point above it in the quotient.

$$\begin{array}{r} .125 \\ 40\overline{)5.000} \\ -40\downarrow\,| \\ \hline 100\,| \\ -80\downarrow \\ \hline 200 \\ -200 \\ \hline 0 \end{array}$$

Since 40 is greater than 5, we need to insert zeros to the right of the dividend.

Check by multiplying.

$$\begin{array}{r} {\scriptstyle 1\ 2} \\ 0.125 \\ \times\ 40 \\ \hline 000 \\ 5000 \\ \hline 5.000\ \checkmark \end{array}$$

The quotient is 0.125.

Skill Practice

Divide.

3. $20\overline{)3}$

Classroom Example: p. 256, Exercise 24

Answers

2. 1.36 **3.** 0.15

Sometimes when dividing decimals, the quotient follows a repeated pattern. The result is called a **repeating decimal**.

Skill Practice

Divide.

4. $2.4 \div 9$

Classroom Example: p. 256, Exercise 30

Example 4 Dividing Where the Quotient Is a Repeating Decimal

Divide. $1.7 \div 30$

Solution:

$$
\begin{array}{r}
.05666\ldots \\
30\overline{)1.70000} \\
-150 \\
\hline
200 \\
-180 \\
\hline
200 \\
-180 \\
\hline
200
\end{array}
$$

Notice that as we continue to divide, we get the same values for each successive step. This causes a pattern of repeated digits in the quotient. Therefore, the quotient is *repeating decimal*.

The quotient is $0.05666\ldots$. To denote the repeated pattern, we often use a bar over the first occurrence of the repeat cycle to the right of the decimal point. That is,

$0.05666\ldots = 0.05\overline{6}$ ← repeat bar

Avoiding Mistakes

In Example 4, notice that the repeat bar goes over only the 6. The 5 is not being repeated.

Skill Practice

Divide.

5. $11\overline{)57}$

Classroom Example: p. 256, Exercise 32

Example 5 Dividing Where the Quotient Is a Repeating Decimal

Divide. $11\overline{)68}$

Solution:

$$
\begin{array}{r}
6.1818\ldots \\
11\overline{)68.0000} \\
-66 \\
\hline
20 \\
-11 \\
\hline
90 \\
-88 \\
\hline
20 \\
-11 \\
\hline
90 \\
-88 \\
\hline
20
\end{array}
$$

Could have stopped here →

Once again, we see a repeated pattern. The quotient is a repeating decimal. Notice that we could have stopped dividing when we obtained the second value of 20.

Avoiding Mistakes

Be sure to put the repeating bar over the entire block of numbers that is being repeated. In Example 5, the bar extends over both the 1 and the 8. We have $6.\overline{18}$.

The quotient is $6.\overline{18}$.

The numbers $0.05\overline{6}$ and $6.\overline{18}$ are examples of repeating decimals. A decimal that "stops" is called a **terminating decimal**. For example, 6.18 is a terminating decimal, whereas $6.\overline{18}$ is a repeating decimal.

Answers

4. $0.2\overline{6}$ **5.** $5.\overline{18}$

In Examples 1–5, we performed division where the divisor was a whole number. Suppose now that we have a divisor that is *not* a whole number, for example, 0.56 ÷ 0.7. Because division can also be expressed in fraction notation, we have

$$0.56 \div 0.7 = \frac{0.56}{0.7}$$

If we multiply the numerator and denominator by 10, the denominator (divisor) becomes the whole number 7.

$$\frac{0.56}{0.7} = \frac{0.56 \times 10}{0.7 \times 10} = \frac{5.6}{7} \xrightarrow{\hspace{2cm}} 7\overline{)5.6}$$

Recall that multiplying decimal numbers by 10 (or any power of 10, such as 100, 1000, etc.) moves the decimal point to the right. We use this idea to divide decimal numbers when the divisor is not a whole number.

Dividing When the Divisor Is Not a Whole Number

Step 1 Move the decimal point in the divisor to the right to make it a whole number.

Step 2 Move the decimal point in the dividend to the right the same number of places as in step 1.

Step 3 Place the decimal point in the quotient directly above the decimal point in the dividend.

Step 4 Divide as you would whole numbers.

Example 6 **Dividing Decimals**

Divide.

a. 0.56 ÷ 0.7   **b.** 0.005$\overline{)3.1}$

Solution:

a. .7$\overline{).56}$ Move the decimal point in the divisor and dividend one place to the right.

$7\overline{)5.6}$ ⟵ Line up the decimal point in the quotient.

$$\begin{array}{r} .8 \\ 7\overline{)5.6} \\ \underline{-5\,6} \\ 0 \end{array}$$

The quotient is 0.8.

b. .005$\overline{)3.100}$ Move the decimal point in the divisor and dividend three places to the right. Insert additional zeros in the dividend if necessary. Line up the decimal point in the quotient.

$$\begin{array}{r} 620. \\ 5\overline{)3100.} \\ \underline{-30} \\ 10 \\ \underline{-10} \\ 00 \end{array}$$

The quotient is 620.

Skill Practice

Divide.

8. $70 \div 0.6$

Classroom Example: p. 256, Exercise 42

Calculator Connections

Repeating decimals displayed on a calculator are rounded. This is because the display cannot show an infinite number of digits.

For example, on a scientific calculator, the repeating decimal $2.\overline{66}$ may appear as

$$2.66666667$$

| **Example 7** | **Dividing Decimals** |

Divide. $50 \div 1.1$

Solution:

$$1.1\overline{)50.0}$$

Move the decimal point in the divisor and dividend one place to the right. Insert an additional zero in the dividend. Line up the decimal point in the quotient.

$$
\begin{array}{r}
45.45\ldots \\
11\overline{)500.000} \\
-44 \\
\hline
60 \\
-55 \\
\hline
50 \\
-44 \\
\hline
60 \\
-55 \\
\hline
50
\end{array}
$$

The quotient is the repeating decimal, $45.\overline{45}$.

Avoiding Mistakes

The quotient in Example 7 is a repeating decimal. The repeat cycle actually begins to the left of the decimal point. However, the repeat bar is placed on the first repeated block of digits to the *right* of the decimal point. Therefore, we write the quotient as $45.\overline{45}$.

The quotient is $45.\overline{45}$.

When we multiply a number by 10, 100, 1000, and so on, we move the decimal point to the right. However, dividing a number by 10, 100, or 1000 decreases its value. Therefore, we move the decimal point to the *left*.

For example, suppose 3.6 is divided by 10, 100, and 1000.

$$
\begin{array}{r}
.36 \\
10\overline{)3.60} \\
-30 \\
\hline
60 \\
-60 \\
\hline
0
\end{array}
\qquad
\begin{array}{r}
.036 \\
100\overline{)3.600} \\
-300 \\
\hline
600 \\
-600 \\
\hline
0
\end{array}
\qquad
\begin{array}{r}
.0036 \\
1000\overline{)3.6000} \\
-3000 \\
\hline
6000 \\
-6000 \\
\hline
0
\end{array}
$$

Dividing by a Power of 10

To divide a number by a power of 10, move the decimal point to the *left* the same number of places as there are zeros in the power of 10.

Skill Practice

Divide.

9. $162.8 \div 1000$
10. $0.0039 \div 10$

Classroom Example: p. 256, Exercise 52

| **Example 8** | **Dividing by a Power of 10** |

Divide.

a. $214.3 \div 10,000$ **b.** $0.03 \div 100$

Solution:

a. $214.3 \div 10,000 = 0.02143$ Move the decimal point four places to the left. Insert an additional zero.

b. $0.03 \div 100 = 0.0003$ Move the decimal point two places to the left. Insert two additional zeros.

Answers

8. $116.\overline{6}$ **9.** 0.1628 **10.** 0.00039

2. Rounding a Quotient

In Example 7, we found that $50 \div 1.1 = 45.\overline{45}$. To check this result, we could multiply $45.\overline{45} \times 1.1$ and show that the product equals 50. However, at this point we do not have the tools to multiply repeating decimals. What we can do is round the quotient and then multiply to see if the product is *close* to 50.

Example 9 Rounding a Repeating Decimal

Round $45.\overline{45}$ to the hundredths place. Then use the rounded value to estimate whether the product $45.\overline{45} \times 1.1$ is close to 50. (This will serve as a check to the division problem in Example 7.)

Solution:

To round the number $45.\overline{45}$, we must write out enough of the repeated pattern so that we can view the digit to the right of the rounding place. In this case, we must write out the number to the thousandths place.

$$45.\overline{45} = 45.454 \cdots \approx 45.45$$

hundredths place

This digit is less than 5. Discard it and all others to its right.

Now multiply the rounded value by 1.1.

$$
\begin{array}{r}
45.45 \\
\times\ 1.1 \\
\hline
4545 \\
45450 \\
\hline
49.995
\end{array}
$$

This value is close to 50. We are reasonably sure that we divided correctly in Example 7.

Sometimes we may want to round a quotient to a given place value. To do so, divide until you get a digit in the quotient one place value to the right of the rounding place. At this point, you can stop dividing and round the quotient.

Example 10 Rounding a Quotient

Round the quotient to the tenths place.

$$47.3 \div 5.4$$

Solution:

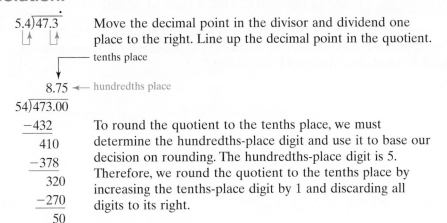

Move the decimal point in the divisor and dividend one place to the right. Line up the decimal point in the quotient.

tenths place

8.75 ← hundredths place

$$
\begin{array}{r}
54\,\overline{)473.00} \\
-432 \\
\hline
410 \\
-378 \\
\hline
320 \\
-270 \\
\hline
50
\end{array}
$$

To round the quotient to the tenths place, we must determine the hundredths-place digit and use it to base our decision on rounding. The hundredths-place digit is 5. Therefore, we round the quotient to the tenths place by increasing the tenths-place digit by 1 and discarding all digits to its right.

The quotient is approximately 8.8.

3. Applications of Decimal Division

Examples 11 and 12 show how decimal division can be used in applications. Remember that division is used when we need to distribute a quantity into equal parts.

Example 11 Applying Division of Decimals

A lunch costs $45.80, and the bill is to be split equally among 5 people. How much must each person pay?

Solution:

We want to distribute $45.80 equally among 5 people, so we must divide $45.80 ÷ 5.

$$
\begin{array}{r}
9.16 \\
5\overline{)45.80} \\
-45 \\
\hline
08 \\
-5 \\
\hline
30 \\
-30 \\
\hline
0
\end{array}
$$

Each person must pay $9.16.

TIP: To check if your answer is reasonable, divide $45 among 5 people.

$$
\begin{array}{r}
9 \\
5\overline{)45}
\end{array}
$$

The estimated value of $9 suggests that the answer to Example 11 is correct.

Division is also used in practical applications to express rates. In Example 12, we find the rate of speed in meters per second (m/sec) for the world record time in the men's 400-m run.

Example 12 Using Division to Find a Rate of Speed

In a recent year, the world-record time in the men's 400-m run was 43.2 sec. What is the speed in meters per second? Round to one decimal place. (*Source:* International Association of Athletics Federations)

Answers
13. Each piece is 7.75 ft.
14. The speed was 9.4 m/sec.

Solution:

To find the rate of speed in meters per second, we must divide the distance in meters by the time in seconds.

$$43.2\overline{)400.0}$$

— tenths place

9.25 ←—hundredths place

$$432\overline{)4000.00}$$
$$\underline{-3888}$$
$$1120$$
$$\underline{-864}$$
$$2560$$
$$\underline{-2160}$$
$$400$$

TIP: In Example 12, we had to find speed in meters per second. The units of measurement required in the answer give a hint as to the order of the division. The word *per* implies division. So to obtain meters *per* second implies 400 m ÷ 43.2 sec.

To round the quotient to the tenths place, determine the hundredths-place digit and use it to make the decision on rounding. The hundredths-place digit is 5, which is 5 or greater. Therefore, add 1 to the tenths-place digit and discard all digits to its right.

The speed is approximately 9.3 m/sec.

Section 4.4   Practice Exercises

Study Skills Exercise

For additional exercises, see Classroom Activities 4.4A–4.4C in the *Student's Resource Manual* at www.mhhe.com/moh.

Meet some of the other students in your class. They can be good resources for asking questions and discussing the material that was covered in class. Write the names of two fellow students.

Vocabulary and Key Concepts

1. a. If a decimal number has an infinite number of digits with a repeated pattern, then the number is called a _____ decimal.   repeating

   b. If a decimal number has a finite number of digits after the decimal point, then the number is called a _____ decimal.   terminating

Review Exercises

2. Round 34.99991 to the hundredths place.   35.00

For Exercises 3–8, perform the indicated operation.

3. 5.28×1000 5280

4. $8.003 - 2.2$ 5.803

5. 11.8×0.32 3.776

6. $102.4 + 1.239$ 103.639

7. $16.82 - 14.8$ 2.02

8. 5.28×0.001 0.00528

Concept 1: Division of Decimals

For Exercises 9–16, divide. Check the answer by using multiplication. **(See Example 1.)**

9. $8.1 \div 9$ Check: _____ × 9 = 8.1
0.9

10. $4.8 \div 6$ Check: _____ × 6 = 4.8
0.8

11. $6\overline{)1.08}$ Check: _____ × 6 = 1.08
0.18

12. $4\overline{)2.08}$ Check: _____ × 4 = 2.08
0.52

13. $4.24 \div 8$ 0.53

14. $5.75 \div 25$ 0.23

15. $5\overline{)105.5}$ 21.1

16. $7\overline{)221.2}$ 31.6

Writing   ←→ Translating Expression   Geometry   Scientific Calculator   Video   NSE

For Exercises 17–48, divide. **(See Examples 2–7.)**

17. $5\overline{)9.8}$ 1.96   **18.** $3\overline{)2.07}$ 0.69   **19.** $0.28 \div 8$ 0.035   **20.** $0.54 \div 8$ 0.0675

21. $5\overline{)84.2}$ 16.84   **22.** $2\overline{)89.1}$ 44.55   **23.** $50\overline{)6}$ 0.12   **24.** $80\overline{)6}$ 0.075

25. $4 \div 25$ 0.16   **26.** $12 \div 60$ 0.2   **27.** $16 \div 3$ $5.\overline{3}$ **28.** $52 \div 9$ $5.\overline{7}$

29. $19 \div 6$ $3.1\overline{6}$ **30.** $9.1 \div 3$ $3.0\overline{3}$ **31.** $33\overline{)71}$ $2.1\overline{5}$ **32.** $11\overline{)42}$ $3.8\overline{1}$

33. $5.03 \div 0.01$ 503   **34.** $3.2 \div 0.001$ 3200   **35.** $0.992 \div 0.1$ 9.92   **36.** $123.4 \div 0.01$ 12,340

37. $57.12 \div 1.02$ 56   **38.** $95.89 \div 2.23$ 43   **39.** $2.38 \div 0.8$ 2.975   **40.** $5.51 \div 0.2$ 27.55

41. $0.3\overline{)62.5}$ $208.\overline{3}$ **42.** $1.05\overline{)22.4}$ $21.\overline{3}$ **43.** $6.305 \div 0.13$ 48.5   **44.** $42.9 \div 0.25$ 171.6

45. $1.1 \div 0.001$ 1100   **46.** $4.44 \div 0.01$ 444   **47.** $420.6 \div 0.01$ 42,060   **48.** $0.31 \div 0.1$ 3.1

49. If 45.62 is divided by 100, will the decimal point move to the right or to the left? By how many places? The decimal point will move to the left two places.

50. If 5689.233 is divided by 100,000, will the decimal point move to the right or to the left? By how many places? The decimal point will move to the left five places.

For Exercises 51–58, divide by the powers of 10. **(See Example 8.)**

51. $3.923 \div 100$ 0.03923   **52.** $5.32 \div 100$ 0.0532   **53.** $98.02 \div 10$ 9.802   **54.** $11.033 \div 10$ 1.1033

55. $0.027 \div 100$ 0.00027   **56.** $0.665 \div 100$ 0.00665   **57.** $1.02 \div 1000$ 0.00102   **58.** $8.1 \div 1000$ 0.0081

Concept 2: Rounding a Quotient

59. Round $2.\overline{4}$ to the
   a. Tenths place 2.4
   b. Hundredths place 2.44
   c. Thousandths place 2.444
   (See Example 9.)

60. Round $5.\overline{2}$ to the
   a. Tenths place 5.2
   b. Hundredths place 5.22
   c. Thousandths place 5.222

61. Round $1.\overline{8}$ to the
   a. Tenths place 1.9
   b. Hundredths place 1.89
   c. Thousandths place 1.889

62. Round $4.\overline{7}$ to the
   a. Tenths place 4.8
   b. Hundredths place 4.78
   c. Thousandths place 4.778

63. Round $3.6\overline{2}$ to the
   a. Tenths place 3.6
   b. Hundredths place 3.63
   c. Thousandths place 3.626

64. Round $9.3\overline{8}$ to the
   a. Tenths place 9.4
   b. Hundredths place 9.38
   c. Thousandths place 9.384

For Exercises 65–73, divide. Round the answer to the indicated place value. Use the rounded quotient to check. **(See Example 10.)**

65. $7\overline{)1.8}$ hundredths 0.26   **66.** $2.1\overline{)75.3}$ hundredths 35.86   **67.** $54.9 \div 3.7$ tenths 14.8

68. $94.3 \div 21$ tenths 4.5   **69.** $0.24\overline{)4.96}$ thousandths 20.667   **70.** $2.46\overline{)27.88}$ thousandths 11.333

71. $0.9\overline{)32.1}$ hundredths 35.67   **72.** $0.6\overline{)81.4}$ hundredths 135.67   **73.** $2.13\overline{)237.1}$ tenths 111.3

Concept 3: Applications of Decimal Division

NS_E When multiplying or dividing decimals, it is important to place the decimal point correctly. For Exercises 74–77, determine whether you think the number is reasonable or unreasonable. If the number is unreasonable, move the decimal point to a position that makes more sense.

74. Steve computed the gas mileage for his Honda Civic to be 3.2 miles per gallon.
Unreasonable; 32 miles per gallon.

75. The sale price of a new kitchen refrigerator is $96.0. Unreasonable; $960

76. Mickey makes $8.50 per hour. He estimates his weekly paycheck to be $3400. Unreasonable; $340.00

77. Jason works in a legal office. He computes the average annual income for the attorneys in his office to be $1400 per year. Unreasonable; $140,000

For Exercises 78–86, solve the application. Check to see if your answers are reasonable.

78. The amount that Brooke owes on her mortgage including interest is $40,540.08. If her monthly payment is $965.24, how many months does she still need to pay? How many years is this?
Brooke needs to pay for 42 months, which equals 3.5 years.

79. A membership at a health club costs $560 per year. The club has a payment plan in which a member can pay $50 down and the rest in 12 equal payments. How much is each payment? **(See Example 11.)**
The monthly payment is $42.50.

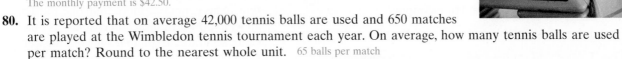

80. It is reported that on average 42,000 tennis balls are used and 650 matches are played at the Wimbledon tennis tournament each year. On average, how many tennis balls are used per match? Round to the nearest whole unit. 65 balls per match

81. A standard 75-watt lightbulb costs $0.75 and lasts about 800 hr. An energy efficient fluorescent bulb that gives off the same amount of light costs $5.00 and lasts about 10,000 hr.

   a. How many standard lightbulbs would be needed to provide 10,000 hr of light?
   13 bulbs would be needed (rounded up to the nearest whole unit).

   b. How much would it cost using standard lightbulbs to provide 10,000 hr of light?
   $9.75

   c. Which is more cost effective long term?
   The energy efficient fluorescent bulb would be more cost effective.

82. Refer to Exercise 81.

   a. If a standard 75-watt lightbulb were left on 24 hr per day, after how many days would the lightbulb have to be changed? Round to the nearest day. After 33 days

   b. If an energy efficient fluorescent bulb were left on 24 hr per day, after how many days would the lightbulb have to be changed? Round to the nearest day. After 417 days (or 1 year, 52 days)

83. In baseball, the batting average is found by dividing the number of hits by the number of times a batter was at bat. Babe Ruth had 2873 hits in 8399 times at bat. What was his batting average? Round to the thousandths place. Babe Ruth's batting average was 0.342.

84. Baseball legend, Ty Cobb, was at bat 11,434 times and had 4189 hits, giving him the all time best batting average. Find his average. Round to the thousandths place. (Refer to Exercise 83.)
Ty Cobb's batting average was 0.366.

85. Manny hikes 12 mi in 5.5 hr. What is his speed in miles per hour? Round to one decimal place. **(See Example 12.)**
2.2 mph

86. Alicia rides her bike 33.2 mi in 2.5 hr. What is her speed in miles per hour? Round to one decimal place.
13.3 mph

Expanding Your Skills

NS_E **87.** What number is halfway between 47.26 and 47.27? 47.265

NS_E **88.** What number is halfway between 22.4 and 22.5? 22.45

89. Which numbers when divided by 8.6 will produce a quotient less than 12.4? Circle all that apply.

    a. 111.8   **b.** 103.2   **c.** 107.5   **d.** 105.78 b, d

90. Which numbers when divided by 5.3 will produce a quotient greater than 15.8? Circle all that apply.

    a. 84.8   **b.** 84.27   **c.** 83.21   **d.** 79.5 a, b

Calculator Connections

Topic: Multiplying and Dividing Decimals on a Calculator

In some applications, the arithmetic on decimal numbers can be very tedious, and it is practical to use a calculator. To multiply or divide on a calculator, use the $\boxed{\times}$ and $\boxed{\div}$ keys, respectively. However, be aware that for repeating decimals, the calculator cannot give an exact value. For example, the quotient of $17 \div 3$ is the repeating decimal $5.\overline{6}$. The calculator returns the rounded value 5.666666667. This is *not* the exact value. Also, when performing division, be careful to enter the dividend and divisor into the calculator in the correct order. For example:

Expression	Keystrokes	Result
$17 \div 3$	17 $\boxed{\div}$ 3 $\boxed{=}$	5.666666667
$0.024\overline{)56.87}$	56.87 $\boxed{\div}$ 0.024 $\boxed{=}$	2369.583333
$\dfrac{82.9}{3.1}$	82.9 $\boxed{\div}$ 3.1 $\boxed{=}$	26.74193548

Calculator Exercises

For Exercises 91–96, multiply or divide as indicated.

91. $(2749.13)(418.2)$
  1149686.166

92. $(139.241)(24.5)$
  3411.4045

93. $(43.75)^2$
  1914.0625

94. $(9.3)^5$
  69,568.83693

95. $21.5\overline{)2056.75}$
  95.6627907

96. $14.2\overline{)4167.8}$
  293.5070423

97. A Hummer H3 SUV uses 1260 gal of gas to travel 12,000 mi per year. A Honda Accord uses 375 gal of gas to go the same distance. Use the current cost of gasoline in your area, to determine the amount saved per year by driving a Honda Accord rather than a Hummer. Answers will vary.

98. A Chevy Blazer gets 16.5 mpg and a Toyota Corolla averages 32 mpg. Suppose a driver drives 15,000 mi per year. Use the current cost of gasoline in your area to determine the amount saved per year by driving the Toyota rather than the Chevy. Answers will vary.

99. Recently, the U.S. capacity to generate wind power was 25,369 megawatts (MW). Texas generates approximately 9410 MW of this power. (*Source:* American Wind Energy Association)

    a. What fraction of the U.S. wind power is generated in Texas? Express this fraction as a decimal number rounded to the nearest hundredth of a megawatt. 0.37

    b. Suppose there is a claim in a news article that Texas generates more than one-third of all wind power in the United States. Is this claim accurate? Explain using your answer from part (a). Yes the claim is accurate. The decimal, 0.37 is greater than $0.\overline{3}$, which is equal to $\frac{1}{3}$.

100. Population density is defined to be the number of people per square mile of land area. If California has 42,475,000 people with a land area of 155,959 square miles, what is the population density? Round to the nearest whole unit. (*Source:* U.S. Census Bureau)
  272 people per square mile

101. The Earth travels approximately 584,000,000 mi around the Sun each year.

    a. How many miles does the Earth travel in one day? 1,600,000 mi per day

    b. Find the speed of the Earth in miles per hour. $66,666.\overline{6}$ mph

102. Although we say the time for the Earth to revolve about the Sun is 365 days, the actual time is 365.256 days. Multiply the fractional amount (0.256) by 4 to explain why we have a leap year every 4 years. (A leap year is a year in which February has an extra day, February 29.)
When we say that 1 year is 365 days, we are ignoring the 0.256 day each year. In 4 years, that amount is $4 \times 0.256 = 1.024$, which is another whole day. This is why we add one more day to the calendar every 4 years.

Problem Recognition Exercises

Operations on Decimals

For Exercises 1–20, perform the indicated operations.

1. a. $123.04 + 100$ 223.04
   b. 123.04×100 12,304
   c. $123.04 - 100$ 23.04
   d. $123.04 \div 100$ 1.2304
   e. $123.04 + 0.01$ 123.05
   f. 123.04×0.01 1.2304
   g. $123.04 \div 0.01$ 12,304
   h. $123.04 - 0.01$ 123.03

2. a. $5078.3 + 1000$ 6078.3
   b. 5078.3×1000 5,078,300
   c. $5078.3 - 1000$ 4078.3
   d. $5078.3 \div 1000$ 5.0783
   e. $5078.3 + 0.001$ 5078.301
   f. 5078.3×0.001 5.0783
   g. $5078.3 \div 0.001$ 5,078,300
   h. $5078.3 - 0.001$ 5078.299

3. a. $4.8 + 2.391$ 7.191
   b. $2.391 + 4.8$ 7.191

4. a. $632.46 + 98.0034$ 730.4634
   b. $98.0034 + 632.46$ 730.4634

5. a. 32.9×1.6 52.64
   b. 1.6×32.9 52.64

6. a. 74.23×0.8 59.384
   b. 0.8×74.23 59.384

7. a. $4(21.6)$ 86.4
   b. $0.25(21.6)$ 5.4

8. a. $2(92.5)$ 185
   b. $0.5(92.5)$ 46.25

9. a. $448 \div 5.6$ 80
   b. 5.6×80 448

10. a. $496.8 \div 9.2$ 54
   b. 54×9.2 496.8

11. 8×0.125 1

12. 20×0.05 1

13. $280 \div 0.07$ 4000

14. $6400 \div 0.001$ 6,400,000

15. $490\overline{)98{,}000{,}000}$ 200,000

16. $2000\overline{)5{,}400{,}000}$ 2700

17. $(4500)(300{,}000)$ 1,350,000,000

18. $(340)(5000)$ 1,700,000

19. $\begin{array}{r} 83.4 \\ -78.9999 \\ \hline 4.4001 \end{array}$

20. $\begin{array}{r} 124.7 \\ -47.9999 \\ \hline 76.7001 \end{array}$

Section 4.5   Fractions as Decimals

1. Writing Fractions as Decimals

Sometimes it is possible to convert a fraction to its equivalent decimal form by rewriting the fraction as a decimal fraction. That is, try to multiply the numerator and denominator by a number that will make the denominator a power of 10.

For example, the fraction $\frac{3}{5}$ can easily be written as an equivalent fraction with a denominator of 10.

$$\frac{3}{5} = \frac{3 \cdot 2}{5 \cdot 2} = \frac{6}{10} = 0.6$$

The fraction $\frac{3}{25}$ can easily be converted to a fraction with a denominator of 100.

$$\frac{3}{25} = \frac{3 \cdot 4}{25 \cdot 4} = \frac{12}{100} = 0.12$$

This technique is useful in some cases. However, some fractions such as $\frac{1}{3}$ cannot be converted to a fraction with a denominator that is a power of 10. This is because 3 is not a factor of any power of 10. For this reason, we recommend dividing the numerator by the denominator.

Skill Practice

Write each fraction or mixed number as a decimal.

1. $\frac{3}{8}$

2. $\frac{43}{20}$

3. $12\frac{5}{16}$

Classroom Examples: p. 265, Exercises 16 and 24

Example 1   Writing Fractions as Decimals

Write each fraction or mixed number as a decimal.

a. $\frac{3}{5}$ b. $\frac{68}{25}$ c. $3\frac{5}{8}$

Solution:

a. $\frac{3}{5}$ means $3 \div 5$.

$$\frac{3}{5} = 0.6$$

$$\begin{array}{r} .6 \\ 5\overline{)3.0} \\ -30 \\ \hline 0 \end{array}$$

Divide the numerator by the denominator.

b. $\frac{68}{25}$ means $68 \div 25$.

$$\frac{68}{25} = 2.72$$

$$\begin{array}{r} 2.72 \\ 25\overline{)68.00} \\ -50 \\ \hline 180 \\ -175 \\ \hline 50 \\ -50 \\ \hline 0 \end{array}$$

Divide the numerator by the denominator.

c. $3\frac{5}{8} = 3 + 5 \div 8$

$$3\frac{5}{8} = 3 + 0.625$$

$$= 3.625$$

$$\begin{array}{r} .625 \\ 8\overline{)5.000} \\ -48 \\ \hline 20 \\ -16 \\ \hline 40 \\ -40 \\ \hline 0 \end{array}$$

Divide the numerator by the denominator.

Answers

1. 0.375   **2.** 2.15   **3.** 12.3125

Writing Fractions as Decimals

Step 1   Divide the numerator by the denominator.

Step 2   Continue the division process until the quotient is a terminating decimal or a repeating pattern is recognized.

Example 2   **Converting Fractions to Repeating Decimals**

Write each fraction as a decimal.

a. $\dfrac{4}{9}$ **b.** $\dfrac{5}{6}$ **c.** $\dfrac{4}{7}$

Solution:

a. $\dfrac{4}{9}$ means $4 \div 9$.

$$
\begin{array}{r}
.44\ldots \\
9\overline{)4.00} \\
-36 \\
\hline
40 \\
-36 \\
\hline
40
\end{array}
$$

The quotient is a repeating decimal.

$\dfrac{4}{9} = 0.\overline{4}$

b. $\dfrac{5}{6}$ means $5 \div 6$.

$$
\begin{array}{r}
.833\ldots \\
6\overline{)5.000} \\
-48 \\
\hline
20 \\
-18 \\
\hline
20
\end{array}
$$

The quotient is a repeating decimal.

$\dfrac{5}{6} = 0.8\overline{3}$

c. $\dfrac{4}{7}$ means $4 \div 7$.

$$
\begin{array}{r}
.571428\ldots \\
7\overline{)4.000000} \\
-35 \\
\hline
50 \\
-49 \\
\hline
10 \\
-7 \\
\hline
30 \\
-28 \\
\hline
20 \\
-14 \\
\hline
60 \\
-56 \\
\hline
40
\end{array}
$$

The cycle will repeat.

TIP: Be sure to carry out the division far enough to see the repeating digits.

$\dfrac{4}{7} = 0.571428571428571428571428\ldots = 0.\overline{571428}$

Skill Practice

Write each fraction as a decimal.

4. $\dfrac{8}{9}$

5. $\dfrac{1}{12}$

6. $\dfrac{3}{7}$

Classroom Example: p. 266, Exercise 38

Answers

4. $0.\overline{8}$ **5.** $0.08\overline{3}$ **6.** $0.\overline{428571}$

Several fractions are used quite often. Their decimal forms are worth memorizing and are presented in Table 4-1.

Table 4-1

$\frac{1}{4} = 0.25$	$\frac{2}{4} = \frac{1}{2} = 0.5$	$\frac{3}{4} = 0.75$	
$\frac{1}{9} = 0.\overline{1}$	$\frac{2}{9} = 0.\overline{2}$	$\frac{3}{9} = \frac{1}{3} = 0.\overline{3}$	$\frac{4}{9} = 0.\overline{4}$
$\frac{5}{9} = 0.\overline{5}$	$\frac{6}{9} = \frac{2}{3} = 0.\overline{6}$	$\frac{7}{9} = 0.\overline{7}$	$\frac{8}{9} = 0.\overline{8}$

Example 3   **Converting Fractions to Decimals with Rounding**

Convert the fraction to a decimal rounded to the indicated place value.

a. $\frac{162}{7}$; tenths place   **b.** $\frac{21}{31}$; hundredths place

Solution:

a. $\frac{162}{7}$

$$
\begin{array}{r}
23.14 \\
7\overline{)162.00} \\
-14 \\
\hline
22 \\
-21 \\
\hline
10 \\
-7 \\
\hline
30 \\
-28 \\
\hline
2
\end{array}
$$

To round to the tenths place, we must determine the hundredths-place digit and use it to base our decision on rounding.

$23.14 \approx 23.1$

The fraction $\frac{162}{7}$ is approximately 23.1.

b. $\frac{21}{31}$

$$
\begin{array}{r}
.677 \\
31\overline{)21.000} \\
-186 \\
\hline
240 \\
-217 \\
\hline
230 \\
-217 \\
\hline
13
\end{array}
$$

To round to the hundredths place, we must determine the thousandths-place digit and use it to base our decision on rounding.

$0.6\overset{1}{7}7 \approx 0.68$

The fraction $\frac{21}{31}$ is approximately 0.68.

2. Writing Decimals as Fractions

In Section 4.1 we converted terminating decimals to fractions. We did this by writing the decimal as a decimal fraction and then reducing the fraction to lowest terms. For example:

$$0.46 = \frac{46}{100} = \frac{\overset{23}{\cancel{46}}}{\underset{50}{\cancel{100}}} = \frac{23}{50}$$

We do not yet have the tools to convert a repeating decimal to its equivalent fraction form. However, we can make use of our knowledge of the common fractions and their repeating decimal forms from Table 4-1.

Example 4 Writing Decimals as Fractions and Fractions as Decimals

Complete the table.

	Decimal Form	Fractional Form
a.	0.475	
b.		$\frac{3}{16}$
c.		$2\frac{4}{5}$
d.	$0.\overline{6}$	
e.		$\frac{19}{11}$

Classroom Example: p. 266, Exercise 60

Skill Practice

Complete the table.

	Decimal Form	Fractional Form
10.	0.875	
11.		$\frac{7}{20}$
12.		$2\frac{1}{3}$
13.	$0.\overline{7}$	

Solution:

a. $0.475 = \dfrac{475}{1000} = \dfrac{19 \cdot \overset{1}{\cancel{25}}}{40 \cdot \underset{1}{\cancel{25}}} = \dfrac{19}{40}$

b. $\dfrac{3}{16} = 3 \div 16$

$$16\overline{)3.0000} \qquad \text{Therefore, } \dfrac{3}{16} = 0.1875.$$

with quotient $.1875$
$$-16$$
$$140$$
$$-128$$
$$120$$
$$-112$$
$$80$$
$$-80$$
$$0$$

c. To convert $2\frac{4}{5}$ to decimal form, we need to convert $\frac{4}{5}$ to decimal form. This can be done by dividing. Or we can easily convert $\frac{4}{5}$ to a decimal fraction with a denominator of 10.

$$\frac{4}{5} = \frac{4 \cdot 2}{5 \cdot 2} = \frac{8}{10} = 0.8$$

Therefore, $2\dfrac{4}{5} = 2.8$.

d. From Table 4-1, the decimal $0.\overline{6} = \dfrac{2}{3}$.

e. $\dfrac{19}{11}$ means $19 \div 11$.

$$11\overline{)19.0000}$$ with quotient $1.7272\ldots$
$$-11$$
$$80$$
$$-77$$
$$30$$
$$-22$$
$$80 \qquad \text{The cycle will repeat.}$$

$\dfrac{19}{11} = 1.\overline{72}$

Answers

10. $\frac{7}{8}$ **11.** 0.35

12. $2.\overline{3}$ **13.** $\frac{7}{9}$

We can now complete the table.

	Decimal Form	Fractional Form
a.	0.475	$\dfrac{19}{40}$
b.	0.1875	$\dfrac{3}{16}$
c.	2.8	$2\dfrac{4}{5}$
d.	$0.\overline{6}$	$\dfrac{2}{3}$
e.	$1.\overline{72}$	$\dfrac{19}{11}$

Instructor Note: To prepare students for Concept 3, ask them to rank the five numbers in the table they just completed: {0.475, 0.1875, 2.8, $0.\overline{6}$, $1.\overline{72}$}.

3. Decimals and the Number Line

In Example 5, we rank the numbers from least to greatest and visualize the position of the numbers on the number line.

Example 5   Ordering Decimals and Fractions

Rank the numbers from least to greatest. Then approximate the position of the points on the number line.

$$0.\overline{45}, \quad 0.45, \quad \frac{1}{2}$$

Solution:

First note that $\frac{1}{2} = 0.5$ and that $0.\overline{45} = 0.454545\ldots$. By writing each number in decimal form, we can compare the decimals as we did in Section 4.1.

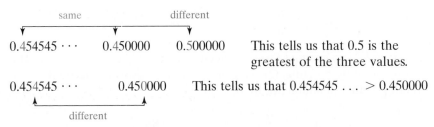

Ranking the numbers from least to greatest we have: $0.45, \quad 0.\overline{45}, \quad 0.5$

The position of these numbers can be seen on the number line. Note that we have expanded the segment of the number line between 0.4 and 0.5 to see more place values to the right of the decimal point.

Recall that numbers that lie to the left on the number line have lesser value than numbers that lie to the right.

Answer

14. $0.16, 0.161, \frac{1}{6}$

0.16 0.161 $\frac{1}{6}$

0.16   0.17

Section 4.5   Practice Exercises

Study Skills Exercise

For additional exercises see Classroom Activities 4.5A–4.5B in the *Student's Resource Manual* at www.mhhe.com/moh.

In a study group, check which activities you might try to help you learn and understand the material.

☐ Quiz one another by asking each other questions.

☐ Practice teaching one another.

☐ Share and compare class notes.

☐ Support and encourage one another.

☐ Work together on exercises and sample problems.

Review Exercises

For Exercises 1–4, write the decimal fraction in decimal form.

1. $\frac{9}{10}$ 0.9

2. $\frac{39}{100}$ 0.39

3. $\frac{141}{1000}$ 0.141

4. $\frac{71}{10,000}$ 0.0071

For Exercises 5–8, write the decimals as fractions.

5. 0.6 $\frac{3}{5}$

6. 0.0016 $\frac{1}{625}$

7. 0.35 $\frac{7}{20}$

8. 0.125 $\frac{1}{8}$

9. Round $4.\overline{25}$ to the hundredths place.   4.25

10. Round $0.\overline{37}$ to the thousandths place.   0.374

Concept 1: Writing Fractions as Decimals

For Exercises 11–14, write each fraction as a decimal fraction, that is, a fraction whose denominator is a power of 10. Then write the number in decimal form.

11. $\frac{2}{5}$ $\frac{4}{10}$; 0.4

12. $\frac{4}{5}$ $\frac{8}{10}$; 0.8

13. $\frac{49}{50}$ $\frac{98}{100}$; 0.98

14. $\frac{3}{50}$ $\frac{6}{100}$; 0.06

For Exercises 15–34, write each fraction or mixed number as a decimal. **(See Example 1.)**

15. $\frac{7}{25}$ 0.28

16. $\frac{4}{25}$ 0.16

17. $\frac{316}{500}$ 0.632

18. $\frac{19}{500}$ 0.038

19. $\frac{7}{8}$ 0.875

20. $\frac{16}{64}$ 0.25

21. $\frac{16}{5}$ 3.2

22. $\frac{68}{25}$ 2.72

23. $5\frac{3}{12}$ 5.25

24. $4\frac{1}{16}$ 4.0625

25. $1\frac{1}{5}$ 1.2

26. $6\frac{5}{8}$ 6.625

27. $\frac{18}{24}$ 0.75

28. $\frac{24}{40}$ 0.6

29. $\frac{53}{16}$ 3.3125

30. $\frac{105}{56}$ 1.875

31. $7\frac{9}{20}$ 7.45

32. $3\frac{11}{25}$ 3.44

33. $\frac{22}{25}$ 0.88

34. $\frac{11}{20}$ 0.55

Writing   ←→ Translating Expression   Geometry   Scientific Calculator   Video   NS E

For Exercises 35–46, write each fraction or mixed number as a repeating decimal. **(See Example 2.)**

35. $3\frac{8}{9}$ $3.\overline{8}$

36. $4\frac{7}{9}$ $4.\overline{7}$

37. $\frac{7}{15}$ $0.4\overline{6}$

38. $\frac{5}{18}$ $0.2\overline{7}$

39. $\frac{19}{36}$ $0.527\overline{7}$

40. $\frac{7}{12}$ $0.583\overline{3}$

41. $\frac{6}{11}$ $0.\overline{54}$

42. $\frac{8}{33}$ $0.\overline{24}$

43. $\frac{14}{111}$ $0.\overline{126}$

44. $\frac{58}{111}$ $0.\overline{522}$

45. $\frac{25}{22}$ $1.1\overline{36}$

46. $\frac{45}{22}$ $2.0\overline{45}$

For Exercises 47–56, convert the fraction to a decimal and round to the indicated place value. **(See Example 3.)**

47. $\frac{1}{7}$; thousandths 0.143

48. $\frac{2}{7}$; thousandths 0.286

49. $\frac{1}{13}$; hundredths 0.08

50. $\frac{9}{13}$; hundredths 0.69

51. $\frac{15}{16}$; tenths 0.9

52. $\frac{3}{11}$; tenths 0.3

53. $\frac{5}{7}$; hundredths 0.71

54. $\frac{1}{8}$; hundredths 0.13

55. $\frac{25}{21}$; tenths 1.2

56. $\frac{18}{13}$; tenths 1.4

57. Write the fractions as decimals. Explain how to memorize the decimal form for these fractions with a denominator of 9.

   a. $\frac{1}{9}$ $0.\overline{1}$ **b.** $\frac{2}{9}$ $0.\overline{2}$ **c.** $\frac{4}{9}$ $0.\overline{4}$ **d.** $\frac{5}{9}$ $0.\overline{5}$

   If we memorize that $\frac{1}{9} = 0.\overline{1}$, then $\frac{2}{9} = 2 \cdot \frac{1}{9} = 2 \cdot 0.\overline{1} = 0.\overline{2}$, and so on.

58. Write the fractions as decimals. Explain how to memorize the decimal forms for these fractions with a denominator of 3.

   a. $\frac{1}{3}$ $0.\overline{3}$ **b.** $\frac{2}{3}$ $0.\overline{6}$ If we memorize that $\frac{1}{3} = 0.\overline{3}$, then $\frac{2}{3} = 2 \cdot \frac{1}{3} = 2 \cdot 0.\overline{3} = 0.\overline{6}$.

Concept 2: Writing Decimals as Fractions

For Exercises 59–62, complete the table. **(See Example 4.)**

59.

	Decimal Form	Fraction Form
a.	0.45	$\frac{9}{20}$
b.	1.625	$\frac{13}{8}$ or $1\frac{5}{8}$
c.	$0.\overline{7}$	$\frac{7}{9}$
d.	$0.\overline{45}$	$\frac{5}{11}$

60.

	Decimal Form	Fraction Form
a.	$0.\overline{6}$	$\frac{2}{3}$
b.	1.6	$\frac{8}{5}$ or $1\frac{3}{5}$
c.	6.08	$\frac{152}{25}$
d.	$0.\overline{2}$	$\frac{2}{9}$

61.

	Decimal Form	Fraction Form
a.	$0.\overline{3}$	$\dfrac{1}{3}$
b.	2.125	$\dfrac{17}{8}$ or $2\dfrac{1}{8}$
c.	$0.8\overline{63}$	$\dfrac{19}{22}$
d.	1.68	$\dfrac{42}{25}$

62.

	Decimal Form	Fraction Form
a.	0.75	$\dfrac{3}{4}$
b.	$0.\overline{63}$	$\dfrac{7}{11}$
c.	$1.\overline{8}$	$\dfrac{17}{9}$ or $1\dfrac{8}{9}$
d.	2.96	$\dfrac{74}{25}$

Historically stock prices were given as fractions or mixed numbers, but are now given as decimals. For Exercises 63 and 64, complete the table that gives recent stock prices taken from the *Wall Street Journal*.

63.

Stock	Closing Price ($) (Decimal)	Closing Price ($) (Fraction)
McGraw-Hill	69.25	$69\dfrac{1}{4}$
Walgreens	44.95	$44\dfrac{19}{20}$
Home Depot	38.50	$38\dfrac{1}{2}$
General Electric	37.44	$37\dfrac{11}{25}$

64.

Stock	Closing Price ($) (Decimal)	Closing Price ($) (Fraction)
Dell	26.3	$26\dfrac{3}{10}$
StrideRite	15.72	$15\dfrac{18}{25}$
Intel	28.10	$28\dfrac{1}{10}$
Burger King	24.15	$24\dfrac{3}{20}$

Concept 3: Decimals and the Number Line

For Exercises 65–76, insert the appropriate symbol. Choose from $<$, $>$, or $=$.

65. $0.2\ \square\ \dfrac{1}{5}$ $=$

66. $1.5\ \square\ \dfrac{3}{2}$ $=$

67. $0.2\ \square\ 0.\overline{2}$ $<$

68. $\dfrac{3}{5}\ \square\ 0.\overline{6}$ $<$

69. $\dfrac{1}{3}\ \square\ 0.3$ $>$

70. $\dfrac{2}{3}\ \square\ 0.66$ $>$

71. $4\dfrac{1}{4}\ \square\ 4.\overline{25}$ $<$

72. $2.12\ \square\ 2.\overline{12}$ $<$

73. $0.\overline{5}\ \square\ \dfrac{5}{9}$ $=$

74. $\dfrac{7}{4}\ \square\ 1.75$ $=$

75. $0.27\ \square\ \dfrac{3}{11}$ $<$

76. $6.4\overline{3}\ \square\ 6.43$ $>$

For Exercises 77–80, rank the numbers from least to greatest. Then approximate the position of the points on the number line. **(See Example 5.)**

77. $0.\overline{1}, \dfrac{1}{10}, \dfrac{1}{5}$

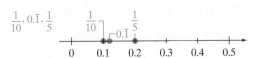

78. $3\dfrac{1}{4}, 3\dfrac{1}{3}, 3.3$

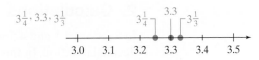

 Writing   Translating Expression Geometry Scientific Calculator Video   NS&E

79. 1.8, 1.75, 1.$\overline{7}$

1.75, 1.$\overline{7}$, 1.8

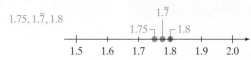

80. $5\frac{1}{6}$, 5.$\overline{6}$, 5.0$\overline{6}$

5.0$\overline{6}$, $5\frac{1}{6}$, 5.$\overline{6}$

Expanding Your Skills

81. If $0.\overline{8} = \frac{8}{9}$, then what is the fraction form of $0.\overline{9}$? $\frac{9}{9} = 1$

For Exercises 82–84, simplify.

82. 1.$\overline{9}$ $_2$ **83.** 6.$\overline{9}$ $_7$ **84.** 15.$\overline{9}$ $_{16}$

Section 4.6 Order of Operations and Applications of Decimals

Concepts

1. Order of Operations Involving Decimals
2. Calculations with Decimals and Fractions
3. Calculations with Round-off Error
4. Applications of Decimals and Fractions

Skill Practice

Simplify.
1. $(5.8 - 4.3)^2 - 2$

Classroom Example: p. 275, Exercise 14

1. Order of Operations Involving Decimals

In Example 1, we perform the order of operations with an expression involving decimal numbers. For a review of the order of operations, see page 65.

Example 1 Applying the Order of Operations by Using Decimal Numbers

Simplify. $16.4 - (6.7 - 3.5)^2$

Solution:

$16.4 - (6.7 - 3.5)^2$

$= 16.4 - (3.2)^2$ Perform the subtraction within parentheses first.

$\begin{array}{r} 6.7 \\ -\ 3.5 \\ \hline 3.2 \end{array}$

$= 16.4 - 10.24$ Perform the operation involving the exponent.

$\begin{array}{r} 3.2 \\ \times\ 3.2 \\ \hline 64 \\ 960 \\ \hline 10.24 \end{array}$

$= 6.16$ Subtract.

$\begin{array}{r} \overset{3\ 10}{1\ 6.\ 4\ \cancel{0}} \\ -\ 1\ 0.\ 2\ 4 \\ \hline 6.\ 1\ 6 \end{array}$

2. Calculations with Decimals and Fractions

In Sections 4.1 and 4.5 we learned how to convert between fraction notation and decimal notation. In this section we apply the order of operations on fractions and decimals combined.

Answer

1. 0.25

Example 2   Dividing a Decimal and a Mixed Number

Divide. $1.52 \div 1\frac{3}{5}$

Solution:

For this example we will show two approaches. In the first approach, we convert both numbers to fractional form and then divide. In the second approach, we convert the mixed number to decimal form and then divide.

Approach 1
Convert both numbers to fractional form.

$1.52 \div 1\frac{3}{5} = \frac{152}{100} \div \frac{8}{5}$ Convert the decimal and mixed number to fractional form.

$= \frac{152}{100} \cdot \frac{5}{8}$ Multiply by the reciprocal of the divisor.

$= \frac{\overset{19}{\cancel{152}}}{\underset{20}{\cancel{100}}} \cdot \frac{\overset{1}{\cancel{5}}}{\underset{1}{\cancel{8}}}$ Simplify common factors.

$= \frac{19}{20}$ Multiply fractions.

We can write the quotient as $\frac{19}{20}$ or in its equivalent decimal form as 0.95.

Approach 1 (Modified)
A modification to approach 1 is to write the decimal 1.52 as $\frac{1.52}{1}$. (Recall that any number divided by 1 equals the number.)

$1.52 \div 1\frac{3}{5} = \frac{1.52}{1} \div \frac{8}{5}$

$= \frac{1.52}{1} \cdot \frac{5}{8}$ Multiply by the reciprocal of the divisor.

$= \frac{7.6}{8}$ Multiply fractions. Note that $1.52 \times 5 = 7.6$.

$= 0.95$ Divide.
$$\begin{array}{r} .95 \\ 8\overline{)7.60} \\ -72 \\ \hline 40 \\ -40 \\ \hline 0 \end{array}$$

Approach 2
Convert the numbers to decimal form.

$1.52 \div 1\frac{3}{5} = 1.52 \div 1.6$ Convert the mixed number to a decimal.

$= 0.95$ Divide.
$$1.6\overline{)1.52} \qquad \begin{array}{r} .95 \\ 16\overline{)15.20} \\ -144 \\ \hline 80 \\ -80 \\ \hline 0 \end{array}$$

Skill Practice

Divide.

2. $4.2 \div \dfrac{4}{3}$

Classroom Example: p. 275,
Exercise 28

Answer

2. $\dfrac{63}{20}$ or 3.15

Example 3 **Applying the Order of Operations**

Simplify. $6.4 \times 2\dfrac{5}{8} \div \left(\dfrac{3}{5}\right)^2$

Solution:

Approach 1

Convert all numbers to fractional form.

$$6.4 \times 2\dfrac{5}{8} \div \left(\dfrac{3}{5}\right)^2 = \dfrac{64}{10} \times \dfrac{21}{8} \div \left(\dfrac{3}{5}\right)^2 \qquad \text{Convert the decimal and mixed number to fractions.}$$

$$= \dfrac{64}{10} \times \dfrac{21}{8} \div \dfrac{9}{25} \qquad \text{Square the quantity } \tfrac{3}{5}.$$

$$= \dfrac{64}{10} \times \dfrac{21}{8} \times \dfrac{25}{9} \qquad \text{Multiply by the reciprocal of } \tfrac{9}{25}.$$

$$= \dfrac{\overset{4}{\cancel{\overset{8}{\cancel{64}}}}}{\underset{1}{\underset{2}{\cancel{10}}}} \times \dfrac{\overset{7}{\cancel{21}}}{\underset{1}{\cancel{8}}} \times \dfrac{\overset{5}{\cancel{25}}}{\underset{3}{\cancel{9}}} \qquad \text{Simplify common factors.}$$

$$= \dfrac{140}{3} \text{ or } 46\dfrac{2}{3} \text{ or } 46.\overline{6} \qquad \text{Multiply.}$$

Approach 2

Convert all numbers to decimal form.

$$6.4 \times 2\dfrac{5}{8} \div \left(\dfrac{3}{5}\right)^2 = 6.4 \times 2.625 \div (0.6)^2 \qquad \text{The fraction } \tfrac{5}{8} = 0.625 \text{ and } \tfrac{3}{5} = 0.6.$$

$$= 6.4 \times 2.625 \div 0.36 \qquad \text{Square the quantity } 0.6. \text{ That is, } (0.6)(0.6) = 0.36.$$

$$\text{Multiply. } 6.4 \times 2.625 = 16.8$$

$$= 16.8 \div 0.36 \qquad \text{Divide } 16.8 \div 0.36. \qquad .36\overline{)16.80}$$

$$= 46.\overline{6}$$

$$\begin{array}{r} 46.6\ldots \\ 36\overline{)1680} \\ -144 \\ \hline 240 \\ -216 \\ \hline 240 \end{array}$$

3. Calculations with Round-off Error

Example 4 **Multiplying a Fraction and a Decimal**

Multiply. $2.52 \cdot \left(\dfrac{5}{6}\right)$

Solution:

Approach 1
Convert 2.52 to fractional form and then multiply fractions.

$$\frac{252}{100} \cdot \frac{5}{6} = \frac{\overset{126}{\cancel{252}}}{\underset{20}{\cancel{100}}} \cdot \frac{\overset{1}{\cancel{5}}}{\underset{3}{\cancel{6}}} = \frac{\overset{21}{\cancel{126}}}{\underset{10}{60}} = \frac{21}{10} \qquad \text{Simplify common factors and multiply fractions.}$$

$$= 2.1 \qquad \text{Divide: } 21 \div 10 = 2.1.$$

Approach 2
Convert $\frac{5}{6}$ to decimal form and then multiply the decimals. However, $\frac{5}{6} = 0.8\overline{3} = 0.8333\ldots$ and we do not know how to multiply repeating decimals. We can approximate the product by rounding the value $0.8\overline{3}$ to some desired level of accuracy. Suppose we round $0.8\overline{3}$ to the thousandths place. Then $0.8\overline{3} \approx 0.833$.

$$2.52 \cdot (0.8\overline{3}) \approx 2.52 \cdot (0.833)$$

$$= 2.09916 \qquad \text{Multiply decimals.}$$

The approximated value 2.09916 is close to 2.1.

$$
\begin{array}{r}
2.52 \\
\times\ .833 \\
\hline
756 \\
7560 \\
201600 \\
\hline
2.09916
\end{array}
$$

Notice that the second approach in Example 4 was not as accurate as the first method. This is so because we used "intermediate rounding." Intermediate rounding refers to rounding a number before it is used in a calculation. We rounded the number $0.8\overline{3}$ *before* multiplying. Keep in mind that a rounded number is not exact. Any calculation performed on a rounded number compounds the error.

To minimize the effects of round-off error, keep the fraction notation as long as possible in the expression. In this way, if you do choose to convert to decimal form, you perform the division in the last step. Any rounding is done at the end.

Example 5 **Dividing a Fraction and Decimal**

Divide $\dfrac{4}{7} \div 3.6$. Round the answer to the nearest hundredth.

Solution:

If we attempt to write $\frac{4}{7}$ as a decimal, we find that it is the repeating decimal $0.\overline{571428}$. Therefore, we choose to change 3.6 to fractional form: $3.6 = \frac{36}{10}$.

$$\frac{4}{7} \div 3.6 = \frac{4}{7} \div \frac{36}{10} \qquad \text{Write 3.6 as a fraction.}$$

$$= \frac{4}{7} \cdot \frac{\overset{1}{\cancel{10}}}{\underset{9}{\cancel{36}}} \qquad \text{Multiply by the reciprocal of the divisor.}$$

$$= \frac{10}{63} \qquad \text{Multiply and reduce to lowest terms.}$$

We must write the answer in decimal form, rounded to the nearest hundredth.

$$
\begin{array}{r}
.158 \\
63\overline{)10.00} \\
\underline{-63} \\
370 \\
\underline{-315} \\
550 \\
\underline{-504} \\
46
\end{array}
$$

Divide. To round to the hundredths place, divide until we find the thousandths-place digit in the quotient. Use that digit to make a decision for rounding.

≈ 0.16 Round to the nearest hundredth.

4. Applications of Decimals and Fractions

Example 6 Using Decimals and Fractions in a Consumer Application

Joanne filled the gas tank in her car and noted that the odometer read 22,341.9 mi. Ten days later she filled the tank again with $11\frac{1}{2}$ gal of gas. Her odometer reading at that time was 22,622.5 mi.

a. How many miles had she driven between fill-ups?

b. How many miles per gallon did she get?

Solution:

a. To find the number of miles driven, we need to subtract the initial odometer reading from the final reading.

$$
\begin{array}{r}
{\scriptstyle 5\ 12\ 1\ 15} \\
2\,2,6\,2\,2.5 \\
-\ 2\,2,3\,4\,1.9 \\
\hline
2\,8\,0.6
\end{array}
$$

Recall that to add or subtract decimals, line up the decimal points.

Joanne had driven 280.6 mi between fill-ups.

b. To find the number of miles per gallon (mi/gal), we divide the number of miles driven by the number of gallons.

$280.6 \div 11\frac{1}{2} = 280.6 \div 11.5$ We convert to decimal form because the fraction $11\frac{1}{2}$ is recognized as 11.5.

$= 24.4$

$$
11.5\overline{)280.6}
$$

$$
\begin{array}{r}
24.4 \\
115\overline{)2806.0} \\
\underline{-230} \\
506 \\
\underline{-460} \\
460 \\
\underline{-460} \\
0
\end{array}
$$

Joanne got 24.4 mi/gal.

Example 7 Using Decimals and Fractions to Compute a Lawsuit Settlement

Althea won a legal settlement for $4105.20. Her lawyer received $\frac{1}{3}$ of the settlement.

a. How much money did the lawyer get?

b. How much money did Althea get?

Solution:

a. The lawyer got $\frac{1}{3}$ of $4105.20. This implies multiplication.

$\frac{1}{3} \cdot (4105.20)$ The fraction $\frac{1}{3}$ cannot be written as a terminating decimal.

$= \frac{1}{3} \cdot \frac{4105.20}{1}$ We can write $4105.20 as $\frac{4105.20}{1}$ and multiply fractions.

$= \frac{4105.20}{3}$ Multiply numerators. Multiply denominators.

$= 1368.4$ Divide: $4105.20 \div 3 = 1368.4$.

The lawyer got $1368.40.

b. Althea got the remaining portion of the money.

$$\begin{array}{r} \$4105.20 \\ -\ \ 1368.40 \\ \hline \$2736.80 \end{array}$$

Althea received $2736.80.

TIP: We can check the answer to Example 7 by realizing that Althea will receive $\frac{2}{3}$ of the settlement.

$\frac{2}{3}(\$4105.20) = \$2736.80$

Example 8 Finding an Average

Table 4-2 represents the average snowfall for 6 winter months in Syracuse, New York. What is the mean (average) amount of snowfall per winter month? Round to the nearest tenth of an inch. (*Source:* National Climate Data Center)

Table 4-2

Month	Nov.	Dec.	Jan.	Feb.	March	April
Snowfall (in.)	9.3	26.8	29.6	26.2	17.3	4.0

Answers

8. a. He ran 15.72 mi.
  b. 10.48 mi was left.
9. The average is $484.41.

Solution:

To find the average, we must add the values. Then divide by the number of values (in this case, 6).

To find the sum, line up the addends:

$$
\begin{array}{r}
^{1\,4\,2}9.3 \\
26.8 \\
29.6 \\
26.2 \\
17.3 \\
+\ \ 4.0 \\
\hline
113.2
\end{array}
$$

The total snowfall for these 6 months is 113.2 in. To find the average, divide by 6.

$$
\begin{array}{r}
18.86 \\
6)\overline{113.20} \\
-6 \\
\hline
53 \\
-48 \\
\hline
52 \\
-48 \\
\hline
40 \\
-36 \\
\hline
4
\end{array}
$$

To round to the tenths place, divide until we find the hundredths-place digit. Use that digit to make a decision for rounding.

The number 18.86 rounds to 18.9.

Syracuse averages about 18.9 in. of snow per month during these 6 months.

Section 4.6 Practice Exercises

Study Skills Exercise

For additional exercises, see Classroom Activities 4.6A–4.6B in the *Student's Resource Manual* at www.mhhe.com/moh.

In addition to studying the material for a test, here are some other activities that people use when preparing for a test. Circle the importance of each statement.

	Not important	Somewhat important	Very important
a. Get a good night's sleep the night before the test.	1	2	3
b. Eat a good breakfast on the day of the test.	1	2	3
c. Wear comfortable clothes on the day of the test.	1	2	3
d. Arrive early to class on the day of the test.	1	2	3

Review Exercises

1. a. Write the fraction form for 0.85. $\frac{17}{20}$

 b. Write the decimal form for $\frac{23}{5}$.   4.6

For Exercises 2–9, perform the indicated operation.

2. $\left(\frac{24}{7}\right)\left(\frac{35}{36}\right)$ $\frac{10}{3}$ **3.** 34.1×9.2 313.72   **4.** $790.9 + 23.91$ 814.81   **5.** $\frac{34}{9} + \frac{5}{27}$ $\frac{107}{27}$

6. $56.7 \div 1.2$ 47.25   **7.** $\frac{55}{16} \div \frac{11}{4}$ $\frac{5}{4}$ **8.** $\frac{9}{4} - \frac{7}{8}$ $\frac{11}{8}$ **9.** $13 - 6.04$ 6.96

Concept 1: Order of Operations Involving Decimals

10. List the order of operations.   1. Perform all operations inside parentheses first.   2. Simplify expressions containing exponents.
3. Perform multiplication or division in the order that they appear from left to right.
4. Perform addition or subtraction in the order that they appear from left to right.

For Exercises 11–26, simplify by using the order of operations. **(See Example 1.)**

11. $(3.7 - 1.2)^2$ 6.25   **12.** $(6.8 - 4.7)^2$ 4.41   **13.** $16.25 - (18.2 - 15.7)^2$ 10   **14.** $11.38 - (10.42 - 7.52)^2$ 2.97

 15. $12.46 - 3.05 - 0.8^2$ 8.77   **16.** $15.06 - 1.92 - 0.4^2$ 12.98   **17.** $63.75 - 9.5(4)$ 25.75   **18.** $6.84 + (3.6)(9)$ 39.24

19. $6.8 \div 2 \div 1.7$ 2   **20.** $8.4 \div 2 \div 2.1$ 2   **21.** $2.2 + [9.34 + (1.2)^2]$ 12.98   **22.** $(3.1)^2 - (4.2 \div 2.1)$ 7.61

23. $16.04 \div [(2.2)^2 - 0.83]$ 4   **24.** $6[(3.1)(4) - 8.1]$ 25.8   **25.** $42.82 - 3(4.8 - 1.6)^2$ 12.1   **26.** $14.28 \div [(1.1)^2 + 5.79]$ 2.04

Concept 2: Calculations with Decimals and Fractions

For Exercises 27–32, simplify by using the order of operations. Express the answer in decimal form. **(See Examples 2 and 3.)**

27. $89.8 \div 1\frac{1}{3}$ 67.35   **28.** $30.12 \div 1\frac{3}{5}$ 18.825   **29.** $20.04 \div \frac{4}{5}$ 25.05

30. $(78.2 - 60.2) \div \frac{9}{13}$ 26   **31.** $14.4 \times \left(\frac{7}{4} - \frac{1}{8}\right)$ 23.4   **32.** $6.5 + \frac{1}{8} \times \left(\frac{1}{5}\right)^2$ 6.505

Concept 3: Calculations with Round-off Error

For Exercises 33–38, perform the indicated operations. Round the answer to the nearest hundredth when necessary. **(See Examples 4 and 5.)**

33. $2.3 \times \frac{5}{9}$ 1.28   **34.** $4.6 \times \frac{7}{6}$ 5.37   **35.** $6.5 \div \frac{3}{5}$ 10.83

36. $\frac{1}{12} \times 6.24 \div 2.1$ 0.25   **37.** $(42.81 - 30.01) \div \frac{9}{2}$ 2.84   **38.** $\frac{2}{7} \times 5.1 \times \frac{1}{10}$ 0.15

For Exercises 39–42, perform the indicated operation. Write the answer as a repeating decimal.

39. $\frac{2}{9} \times 4.21$ $0.93\overline{5}$ **40.** $6.02 \div \frac{22}{23}$ $6.29\overline{36}$ **41.** $5.32 \div \frac{6}{5}$ $4.4\overline{3}$ **42.** $\frac{34}{11} \times 2.5$ $7.\overline{72}$

Concept 4: Applications of Decimals and Fractions

43. Mitch and Jody traveled for $7\frac{1}{2}$ hr between Atlanta, Georgia, and Orlando, Florida. At the beginning of the trip, the car's odometer read 21,345.6 mi. When they arrived in Orlando, the odometer read 21,816.6 mi. **(See Example 6.)**

    a. How many miles had they driven on the trip?   471 mi

    b. Find the average speed in miles per hour (mph).   62.8 mph

44. Jennifer rented a car while on vacation. The odometer read 43,725.1 mi. After a week she filled the tank with 9.8 gal of gas and noticed the odometer read 43,984.8 mi.

    a. How many miles had she driven on vacation?   259.7 mi

    b. How many miles per gallon did she get?   26.5 mi/gal

45. A cell phone plan has a $39.95 monthly fee and includes 450 min. For time on the phone over 450 min, the charge is $0.40 per minute. How much is Jorge charged for a month in which he talks for 597 min? Jorge will be charged $98.75.

46. A night at a hotel in Dallas costs $185.95 with a nightly room tax of $24.17. If Radcliff stays for 5 nights, how much is his total bill? Radcliff's bill is $1050.60.

47. Susan's diet allows her 60 grams (g) of fat per day. If she has $\frac{1}{4}$ of her total fat grams for breakfast and a McDonald's Quarter Pounder (20.7 g of fat) for lunch, how many grams does she have left for dinner? **(See Example 7.)**   She has 24.3 g left for dinner.

48. Todd is establishing his beneficiaries for his life insurance policy. The policy is for $150,000.00 and $\frac{1}{2}$ will go to his daughter, $\frac{3}{8}$ will go to his stepson, and the rest will go to his grandson. What dollar amount will go to the grandson?   The grandson will receive $18,750.00.

49. Caren bought three packages of printer paper for $4.79 each. The sales tax for the merchandise was $0.86. If Caren paid with a $20 bill, how much change should she receive?   Caren should get $4.77 in change.

50. Mr. Timpson bought dinner for $28.42. He left a tip of $6.00. He paid the bill with two $20 bills. How much change should he receive?   Mr. Timpson should receive $5.58 in change.

51. Duncan earned test grades of 92, 84, 77, and 62. What is his test average?   Duncan's average is 78.75.

52. Owen earned quiz grades of 19, 14, 16, and 20. What is his quiz average?   Owen's average is 17.25.

53. The average snowfall amounts for 5 winter months in Burlington, Vermont, are given in the table. Find the average snowfall per month. **(See Example 8.)** The average snowfall per month is 14.54 in.

Month	Snowfall (in.)
November	6.6
December	18.1
January	18.8
February	16.8
March	12.4

54. The average rainfall for 4 summer months for Houston, Texas, is given in the table. Find the average rainfall per month.   The average rainfall per month is 4.45 in.

Month	Rainfall (in.)
June	5.6
July	5.2
August	3.3
September	3.7

Expanding Your Skills

Use the following formula and table for Exercises 55–58. Round values to the tenths place if necessary.

In the spring of 1998, the National Institutes of Health (NIH) issued new guidelines for determining whether an individual is overweight. The standard of measure is called the *body mass index* (BMI). Body mass index is a measure of an individual's weight in relation to the person's height. The formula for body mass index is

$$\text{BMI} = \frac{703 \cdot W}{h^2}$$

where W is weight in pounds and h is height in inches. The NIH categorizes body mass indices as shown in the table.

BMI	Weight Status
18.5–24.9	Considered ideal
25.0–29.9	Considered overweight
30.0 or above	Considered obese

55. Find your own body mass index.
Answers will vary.

56. a. Find the body mass index for a 150-lb person who is 5 ft 10 in. (70 in.) tall.   21.5

   b. Is this person's weight considered ideal, overweight, or obese?   Ideal

57. a. Find the body mass index for a 220-lb person who is 6 ft 0 in. (72 in.) tall.   29.8

   b. Is this person's weight considered ideal, overweight, or obese?   Overweight

58. a. Find the body mass index for a 141-lb person who is 5 ft 3 in. (63 in.) tall.   25.0

   b. Is this person's weight considered ideal, overweight, or obese?   Overweight

For Exercises 59–62, perform the indicated operations.

59. $0.\overline{3} \times 0.3 + 3.375$ 3.475

60. $0.\overline{5} \div 0.\overline{2} - 0.75$ 1.75

61. $(0.\overline{8} + 0.\overline{4}) \times 0.39$ 0.52

62. $(0.\overline{7} - 0.\overline{6}) \times 5.4$ 0.6

Calculator Connections

Topic: Applications Using the Order of Operations with Decimals

Calculator Exercises

63. Suppose that Deanna owns 50 shares of stock in Company A, valued at $132.05 per share. She decides to sell these shares and use the money to buy stock in Company B, valued at $27.80 per share. Assume there are no fees for either transaction.

   a. How many full shares of Company B stock can she buy?   237 shares

   b. How much money will be left over after she buys the Company B stock?
   $13.90 will be left over.

64. One megawatt (MW) of wind power produces enough electricity to supply approximately 275 homes. As of June 2007, the state of Texas was producing 3352 MW of wind power. (*Source:* American Wind Energy Association)

   a. About how many homes can be supplied with electricity using wind power produced in Texas?
   Approximately 921,800 homes could be powered.

   b. The given table outlines new proposed wind power projects in Texas. If these projects are completed, approximately how many additional homes could be supplied with electricity?

   Approximately 342,678 additional homes could be powered.

Project	MW
JD Wind IV	79.8
Buffalo Gap, Phase II	232.5
Lone Star I (3Q)	128
Sand Bluff	90
Roscoe	209
Barton Chapel	120
Stanton Wind Energy Center	120
Whirlwind Energy Center	59.8
Sweetwater V	80.5
Champion	126.5

65. Marty bought a home for $145,000. He paid $25,000 as a down payment and then financed the rest with a 30-yr mortgage. His monthly payments are $798.36 each and go toward paying off the loan and interest on the loan.

   a. How much money does Marty have to finance?   Marty will have to finance $120,000.

   b. How many months are in a 30-yr period?
   There are 360 months in 30 yr.

   c. How much money will Marty pay over a 30-yr period to pay off the loan?
   He will pay $287,409.60.

   d. How much money did Marty pay in interest over the 30-yr period?
   He will pay $167,409.60 in interest.

66. Gwen bought a home for $109,000. She paid $15,000 as a down payment and then financed the rest with a 15-yr mortgage. Her monthly payments are $849.00 each and go toward paying off the loan and toward interest on the loan.

   a. How much money does Gwen have to finance?   Gwen needs to finance $94,000.

   b. How many months are in a 15-yr period?
   There are 180 months in 15 yr.

   c. How much money will Gwen pay over a 15-yr period to pay off the loan?
   Gwen will pay $152,820.00.

   d. How much money did Gwen pay in interest over the 15-yr period?
   She will pay $58,820.00 in interest.

67. An inheritance for $80,460.60 is to be divided equally among four heirs. However, before the money can be distributed, approximately one-third of the money must go to the government for taxes. How much does each person get after the taxes have been subtracted?
   Each person will get approximately $13,410.10.

68. For the fall semester, Sylvia bought 4 textbooks for $106.97, $90.75, $133.25, and $110.15. What is the average price per textbook?
   The average price is $110.28.

Group Activity

Purchasing from a Catalog

Materials: Catalog (such as Amazon, Sears, JC Penney, or the like) or a computer to access an online catalog

Estimated time: 15 min

Group Size: 4

1. Each person in the group will choose an item to add to the list of purchases. All members of the group will keep a list of all items ordered and the prices. Answers will vary.

ORDER SHEET

Item	Price Each	Quantity	Total
		SUBTOTAL	
		SHIPPING	
		TOTAL	

2. When the order list is complete, each member will find the total cost of the items ordered (subtotal) and compare the answer with the other members of the group. When the correct subtotal has been determined, find the cost of shipping from the catalog. (This is usually found on the order page of the catalog.) Now find the total cost of this order. **Instructor Note:** If you want to include sales tax, you can have students do this activity after completing Chapter 6.

Chapter 4 Summary

Section 4.1 Decimal Notation and Rounding

Key Concepts

A **decimal fraction** is a fraction whose denominator is a power of 10.

Identify the place values of a decimal number.

1 2 3 4 . 5 6 7 8

thousands
hundreds
tens
ones
decimal point
tenths
hundredths
thousandths
ten-thousandths

Reading a Decimal Number

1. The part of the number to the left of the decimal point is read as a whole number. *Note:* If there is not a whole-number part, skip to step 3.
2. The decimal point is read "and."
3. The part of the number to the right of the decimal point is read as a whole number but is followed by the name of the place position of the digit farthest to the right.

Converting a Decimal to a Mixed Number or Proper Fraction

1. The digits to the right of the decimal point are written as the numerator of the fraction.
2. The place value of the digit farthest to the right of the decimal point determines the denominator.
3. The whole-number part of the number is left unchanged.
4. Once the number is converted to a fraction or mixed number, simplify the fraction to lowest terms, if possible.

Writing a Decimal Number Greater Than 1 as an Improper Fraction

1. The denominator is determined by the place position of the last digit to the right of the decimal point.
2. The numerator is obtained by removing the decimal point of the original number. The resulting whole number is then written over the denominator.
3. Simplify the improper fraction to lowest terms, if possible.

Examples

Example 1

$\frac{7}{10}$, $\frac{31}{100}$, and $\frac{191}{1000}$ are decimal fractions.

Example 2

In the number 34.914, the 1 is in the hundredths place.

Example 3

23.089 reads "twenty-three and eighty-nine thousandths."

Example 4

$$4.2 = 4\frac{\overset{1}{\cancel{2}}}{\underset{5}{\cancel{10}}} = 4\frac{1}{5}$$

Example 5

$$5.24 = \frac{\overset{131}{\cancel{524}}}{\underset{25}{\cancel{100}}} = \frac{131}{25}$$

Comparing Two Decimal Numbers

1. Starting at the left (and moving toward the right), compare the digits in each corresponding place position.
2. As we move from left to right, the first instance in which the digits differ determines the order of the numbers. The number having the greater digit is greater overall.

Rounding Decimals to a Place
Value to the Right of the Decimal Point

1. Identify the digit one position to the right of the given place value.
2. If the digit in step 1 is 5 or greater, add 1 to the given digit. If the digit in step 1 is less than 5, leave the given digit unchanged.
3. Discard all digits to the right of the given digit.

Example 6

$3.024 > 3.019$ because

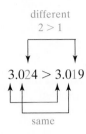

Example 7

Round 4.8935 to the nearest hundredth.

$4.8935 \approx 4.89$

This digit is less than 5.

hundredths place

Section 4.2 — Addition and Subtraction of Decimals

Key Concepts

Adding Decimals

1. Write the addends in a column with the decimal points and corresponding place values lined up.
2. Add the digits in columns from right to left as you would whole numbers. The decimal point in the answer should be lined up with the decimal points from the addends.

Examples

Example 1

Add $6.92 + 12 + 0.001$.

$$
\begin{array}{r}
6.920 \\
12.000 \\
+\ 0.001 \\
\hline
18.921
\end{array}
$$

Add zeros to the right of the decimal point as placeholders.

Check by estimating:

6.92 rounds to 7 and 0.001 rounds to 0.

$7 + 12 + 0 = 19$, which is close to 18.921.

Example 2

Subtract $41.03 - 32.4$.

$$
\begin{array}{r}
4\,1.03 \\
-\ 3\,2.40 \\
\hline
8.63
\end{array}
$$

Check by estimating:

41.03 rounds to 41 and 32.40 rounds to 32.

$41 - 32 = 9$, which is close to 8.63.

Subtracting Decimals

1. Write the numbers in a column with the decimal points and corresponding place values lined up.
2. Subtract the digits in columns from right to left as you would whole numbers. The decimal point in the answer should be lined up with the other decimal points.

Section 4.3   Multiplication of Decimals

Key Concepts

Multiplying Two Decimals

1. Multiply as you would whole numbers.
2. Place the decimal point in the product so that the number of decimal places equals the combined number of decimal places of both factors.

Multiplying a Decimal by Powers of 10

Move the decimal point to the right the same number of decimal places as the number of zeros in the power of 10.

Multiplying a Decimal by Powers of 0.1

Move the decimal point to the left the same number of places as there are decimal places in the power of 0.1.

Examples

Example 1

Multiply. 5.02×2.8

$$
\begin{array}{ll}
\overset{1}{5.02} & \text{2 decimal places} \\
\underline{\times\ 2.8} & \underline{+\ 1\ \text{decimal place}} \\
4016 & \\
\underline{10040} & \\
14.056 & \text{3 decimal places}
\end{array}
$$

Example 2

$83.251 \times 100 = 8325.1$

Move 2 places
to the right.

Example 3

$149.02 \times 0.001 = 0.14902$

Move 3 places
to the left.

Section 4.4   Division of Decimals

Key Concepts

Dividing a Decimal by a Whole Number

1. Place the decimal point in the quotient directly above the decimal point in the dividend.
2. Divide as you would whole numbers.

Dividing When the Divisor Is Not a Whole Number

1. Move the decimal point in the divisor to the right to make it a whole number.
2. Move the decimal point in the dividend to the right the same number of places as in step 1.
3. Place the decimal point in the quotient directly above the decimal point in the dividend.
4. Divide as you would whole numbers.

To round a repeating decimal, be sure to expand the repeating digits to one digit beyond the indicated rounding place.

Examples

Example 1

$62.6 \div 4$

$$
\begin{array}{r}
15.65 \\
4\overline{)62.60} \\
-4 \\
\hline
22 \\
-20 \\
\hline
26 \\
-24 \\
\hline
20 \\
-20 \\
\hline
0
\end{array}
$$

Example 2

$81.1 \div 0.9$ $.9\overline{)81.1}$

$$
\begin{array}{r}
90.11\ldots \\
9\overline{)811.00} \\
-81 \\
\hline
01 \\
\underline{00} \\
10 \\
\underline{-9} \\
10
\end{array}
$$

The pattern repeats.

The answer is the repeating decimal $90.\overline{1}$.

Example 3

Round $6.\overline{56}$ to the thousandths place.

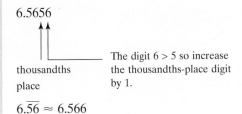

The digit 6 > 5 so increase the thousandths-place digit by 1.

$6.\overline{56} \approx 6.566$

Section 4.5   Fractions as Decimals

Key Concepts

To write a fraction as a decimal, divide the numerator by the denominator. See Examples 1 and 2.

These are some common fractions represented by decimals.

$\dfrac{1}{4} = 0.25$ $\dfrac{1}{2} = 0.5$ $\dfrac{3}{4} = 0.75$

$\dfrac{1}{9} = 0.\overline{1}$ $\dfrac{2}{9} = 0.\overline{2}$ $\dfrac{1}{3} = 0.\overline{3}$

$\dfrac{4}{9} = 0.\overline{4}$ $\dfrac{5}{9} = 0.\overline{5}$ $\dfrac{2}{3} = 0.\overline{6}$

$\dfrac{7}{9} = 0.\overline{7}$ $\dfrac{8}{9} = 0.\overline{8}$

To write a decimal as a fraction, first write the number as a decimal fraction and reduce. See Example 3.

To rank decimals from least to greatest, compare corresponding digits from left to right.

$$0.3\,4\,5\,0\,1 < 0.3\,4\,5\,1\,0$$

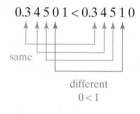

same

different
$0 < 1$

Examples

Example 1

$\dfrac{17}{20} = 0.85$
$\begin{array}{r} .85 \\ 20\overline{)17.00} \\ -160 \\ \hline 100 \\ -100 \\ \hline 0 \end{array}$

Example 2

$\dfrac{14}{3} = 4.\overline{6}$
$\begin{array}{r} 4.66... \\ 3\overline{)14.00} \\ -12 \\ \hline 20 \\ -18 \\ \hline 20 \end{array}$

The pattern repeats.

Example 3

$6.84 = \dfrac{\overset{171}{\cancel{684}}}{\underset{25}{\cancel{100}}} = \dfrac{171}{25}$ or $6\dfrac{21}{25}$

Example 4

Plot the decimals 2.6, $2.\overline{6}$, and 2.58 on a number line.

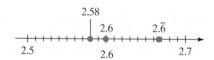

Section 4.6 Order of Operations and Applications of Decimals

Key Concepts

Examples 1 and 2 apply the order of operations with fractions and decimals. There are two approaches for simplifying.

Option 1: Write the expressions as decimals.

Option 2: Write the expressions as fractions.

Example 3 illustrates an application of the order of operations involving decimals.

Examples

Example 1

$$\left(1.6 - \frac{13}{25}\right) \div 8 = (1.6 - 0.52) \div 8$$
$$= (1.08) \div 8$$
$$= 0.135$$

Example 2

$$\frac{2}{3}\left(2.2 + \frac{7}{5}\right) = \frac{2}{3}\left(\frac{22}{10} + \frac{7}{5}\right)$$
$$= \frac{2}{3}\left(\frac{11}{5} + \frac{7}{5}\right)$$
$$= \frac{2}{3}\left(\frac{\overset{6}{18}}{\underset{1}{5}}\right) = \frac{12}{5} = 2.4$$

Example 3

For one month, Dan brought home paychecks worth $824.25, $840.05, $915.21, and $880.89. What was his average weekly pay?

$$(824.25 + 840.05 + 915.21 + 880.89) \div 4$$
$$= (3460.40) \div 4 = 865.1$$

Dan brings home an average of $865.10 a week.

Chapter 4 Review Exercises

Section 4.1

1. Identify the place value for each digit in the number 32.16.
The 3 is in the tens place, 2 is in the ones place, 1 is in the tenths place, and 6 is in the hundredths place.

2. Identify the place value for each digit in the number 2.079.
The 2 is in the ones place, 0 is in the tenths place, 7 is in the hundredths place, and 9 is in the thousandths place.

For Exercises 3–6, write the word name for the decimal.

3. 5.7
Five and seven-tenths

4. 10.21 Ten and twenty-one hundredths

5. 51.008 Fifty-one and eight thousandths

6. 109.01 One hundred nine and one-hundredth

For Exercises 7 and 8, write the word name as a numeral.

7. Thirty-three thousand, fifteen and forty-seven thousandths 33,015.047

8. One hundred and one hundredth 100.01

For Exercises 9 and 10, write the decimal as a proper fraction or mixed number.

9. 4.8 $4\frac{4}{5}$

10. 0.025 $\frac{1}{40}$

For Exercises 11 and 12, write the decimal as an improper fraction.

11. 1.3 $\frac{13}{10}$

12. 6.75 $\frac{27}{4}$

For Exercises 13 and 14, fill in the blank with either
< or >.

13. 15.032 ☐ 15.03 >   **14.** 7.209 ☐ 7.22 <

15. The earned run average (ERA) for five
members of the American League for a recent
season is given in the table. Rank the averages
from least to greatest.

Player	ERA
Jon Garland	4.5142
Cliff Lee	4.3953
Gil Meche	4.4839
Jamie Moyer	4.3875
Vicente Padilla	4.5000

4.3875, 4.3953, 4.4839, 4.5000, 4.5142

For Exercises 16 and 17, round the decimal to the
indicated place value.

16. 89.9245; hundredths 89.92

17. 34.8895; thousandths 34.890

18. A quality control manager tests the amount of
cereal in several brands of breakfast cereal
against the amount advertised on the box. She
selects one box at random. She measures the
contents of one 12.5-oz box and finds that the
box has 12.46 oz.

  a. Is the amount in the box less than or greater
  than the advertised amount?
  The amount in the box is less than the advertised
  amount.

  b. If the quality control manager rounds the
  measured value to the tenths place, what is
  the value?
  The amount rounds to 12.5 oz.

NS⊗E 19. Which number is equivalent to 571.24? Circle all
that apply.

  a. 571.240   **b.** 571.2400

  c. 571.024   **d.** 571.0024 a, b

NS⊗E 20. Which number is equivalent to 3.709? Circle all
that apply.

  a. 3.7   **b.** 3.7090

  c. 3.709000   **d.** 3.907 b, c

Section 4.2

For Exercises 21–28, add or subtract as indicated.

21. 45.03 + 4.713   **22.** 239.3 + 33.92
49.743   273.22

23. 34.89 − 29.44   **24.** 5.002 − 3.1
5.45   1.902

25. 221 − 23.04   **26.** 34 + 4.993
197.96   38.993

27. 17.3 + 3.109 − 12.6
7.809

28. 189.22 + 13.1 − 120.055
82.265

29. Find the values of x and y. Then find the
perimeter of the figure.
$x = 4.5$ in., $y = 5.07$ in.; the perimeter is 201 in.

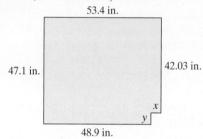

30. The closing prices for a mutual fund are given in
the graph for a 5-day period.

  a. Determine the difference in price between
  the two consecutive days for which the price
  increased the most.
  Between days 1 and 2, the increase was $0.194.

  b. Determine the difference in price between
  the two consecutive days for which the price
  decreased the most.

  Between days 3 and 4, the decrease was $0.209.

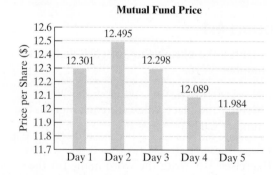

31. During a tropical storm, Dana emptied her rain
gauge four times during the day. She measured
1.4 in., 1 in., 0.09 in., and 1.25 in. What was the
total rainfall for that day?   3.74 in.

Section 4.3

For Exercises 32–39, multiply the decimals.

32. 3.9 × 2.1
8.19

33. 57.01 × 1.3
74.113

34. 60.1 × 4.4
264.44

35. 7.7 × 45
346.5

36. 85.49 × 1000
85,490

37. 1.0034 × 100
100.34

38. 92.01 × 0.01
0.9201

39. 104.22 × 0.01
1.0422

For Exercises 40 and 41, write the decimal number representing each word name.

40. In its Season 9 premier, *American Idol* had 28.1 million viewers.   28,100,000

41. The population of Guadeloupe is approximately 4.32 hundred-thousand.   432,000

42. A store advertises a package of two 9-volt batteries on sale at $3.99.

  a. What is the cost of buying 8 batteries?
  Eight batteries cost $15.96 on sale.

  b. If another store has an 8-pack for the regular price of $17.99, how much can a customer save by buying batteries at the sale price?
  A customer can save $2.03.

43. If long-distance phone calls cost $0.25 per minute, how much will a 23-min long-distance call cost?
The call will cost $5.75.

44. Find the area and perimeter of the rectangle.
Area = 940 ft²; perimeter = 127 ft

23.5 ft

40 ft

45. Population density gives the approximate number of people per square mile. The population density for Texas is given in the graph for selected years.

  a. Approximately how many people would have been located in a 200-mi² area in 1960?
  7280 people

  b. Approximately how many people would have been located in a 200-mi² area in 2008?
  18,580 people

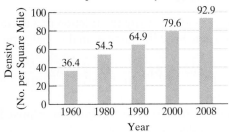

Population Density for Texas

Source: U.S. Census Bureau

Figure for Exercise 45

Section 4.4

For Exercises 46–55, divide. Write the answer in decimal form.

46. 8.55 ÷ 0.5
17.1

47. 64.2 ÷ 1.5
42.8

48. $0.06\overline{)0.248}$
4.1$\overline{3}$

49. $0.3\overline{)2.63}$
8.7$\overline{6}$

50. 18.9 ÷ 0.7
27

51. 0.036 ÷ 1.2
0.03

52. 493.93 ÷ 100
4.9393

53. 90.234 ÷ 10
9.0234

54. 553.8 ÷ 0.001
553,800

55. 2.6 ÷ 0.01
260

56. For each number, round to the indicated place.

	8.$\overline{6}$	52.$\overline{52}$	0.$\overline{409}$
Tenths	8.7	52.5	0.4
Hundredths	8.67	52.53	0.41
Thousandths	8.667	52.525	0.409
Ten-thousandths	8.6667	52.5253	0.4094

For Exercises 57 and 58, divide and round the answer to the nearest hundredth.

57. 104.6 ÷ 9
11.62

58. 71.8 ÷ 6
11.97

59. **a.** A generic package of toilet paper costs $5.99 for 12 rolls. What is the cost per roll? (Round the answer to the nearest cent, that is, the nearest hundredth of a dollar.) $0.50 per roll

  b. A package of four rolls costs $2.29. What is the cost per roll? $0.57 per roll

  c. Which of the two packages offers the better buy?   The 12-pack is better.

 Writing   Translating Expression   Geometry   Scientific Calculator   Video   NS E

Section 4.5

For Exercises 60–62, write the fraction as a decimal fraction. Then write the number in decimal form.

60. $\dfrac{3}{5}$ $\dfrac{6}{10}$; 0.6   **61.** $\dfrac{7}{20}$ $\dfrac{35}{100}$; 0.35 **62.** $\dfrac{27}{500}$ $\dfrac{54}{1000}$; 0.054

For Exercises 63–66, write the fraction or mixed number as a decimal.

63. $2\dfrac{2}{5}$ 2.4

64. $3\dfrac{13}{25}$ 3.52

65. $\dfrac{24}{125}$ 0.192

66. $\dfrac{7}{16}$ 0.4375

For Exercises 67–70, write the fraction as a repeating decimal.

67. $\dfrac{7}{12}$ $0.58\overline{3}$

68. $\dfrac{55}{36}$ $1.52\overline{7}$

69. $4\dfrac{7}{22}$ $4.3\overline{18}$

70. $\dfrac{2}{13}$ $0.\overline{153846}$

For Exercises 71–74, write the fraction as a decimal rounded to the nearest hundredth.

71. $\dfrac{5}{17}$ 0.29

72. $\dfrac{20}{23}$ 0.87

73. $\dfrac{11}{3}$ 3.67

74. $\dfrac{17}{6}$ 2.83

For Exercises 75–78, write a fraction or mixed number for the repeating decimal.

75. $0.\overline{2}$ $\dfrac{2}{9}$

76. $1.\overline{6}$ $1\dfrac{2}{3}$

77. $3.\overline{3}$ $3\dfrac{1}{3}$

78. $5.\overline{7}$ $5\dfrac{7}{9}$

79. Complete the table, giving the closing value of stocks as reported in the *Wall Street Journal*.

Stock	Closing Price ($) (Decimal)	Closing Price ($) (Fraction)
Ford	13.02	$13\dfrac{1}{50}$
Microsoft	30.50	$30\dfrac{1}{2}$
Citibank	4.37	$4\dfrac{37}{100}$

For Exercises 80–82, insert the appropriate symbol. Choose from <, >, or =.

80. $1\dfrac{1}{3}$ ☐ 1.33   >   **81.** 2.25 ☐ $\dfrac{9}{4}$ =

82. 0.14 ☐ $\dfrac{1}{7}$ <

For Exercises 83 and 84, determine what decimal number is represented by the point on the number line.

83.
0.25   0.30
0.28

84.
0.71   0.72
0.713

Section 4.6

For Exercises 85–88, perform the indicated operations. Write the answers in decimal form.

85. $7.5 \div \dfrac{3}{2}$ 5

86. $2(3.14)(20)$ 125.6

87. $3.14(5)^2$ 78.5

88. $\dfrac{1}{3}(3.14)(2)^2(6)$ 25.12

89. An audio Spanish course is available online at one website for the following prices. How much money is saved by buying the combo package versus the three levels individually?
$89.90 will be saved by buying the combo package.

Level	Price
Spanish I	$189.95
Spanish II	199.95
Spanish III	219.95
Combo (Spanish I, II, III combined)	519.95

90. Marvin drives the route shown in the figure each day, making deliveries. He completes one-third of the route before lunch. How many more miles does he still have to drive after lunch?
Marvin must drive 34 mi more.

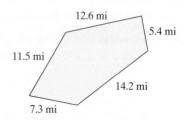
12.6 mi
5.4 mi
11.5 mi
14.2 mi
7.3 mi

Chapter 4 Test

1. Identify the place value of the underlined digit.

a. 2<u>3</u>4.17
Tens place

b. 234.1<u>7</u>
Hundredths place

2. Write the word name for 509.024.
Five hundred nine and twenty-four thousandths

3. Write the decimal 1.26 as a mixed number and as a fraction. $1\frac{13}{50}; \frac{63}{50}$

4. The field goal percentages for a recent basketball season are given in the table for four NBA teams. Rank the percentages from least to greatest. 0.4419, 0.4484, 0.4489, 0.4495

Team	Percentage
LA Lakers	0.4419
Cleveland	0.4489
San Antonio	0.4495
Utah	0.4484

5. Which statement is correct? b is correct.

a. $0.043 > 0.430$ **b.** $0.692 < 0.926$

c. $0.078 < 0.0780$

For Exercises 6–17, perform the indicated operation.

6. $49.002 + 3.83$
52.832

7. $34.09 - 12.8$
21.29

8. 28.1×4.5
126.45

9. $25.4 \div 5$
5.08

10. $4 - 2.78$
1.22

11. $12.03 + 0.1943$
12.2243

12. $39.82 \div 0.33$
120.6

13. 42.7×10.3
439.81

14. 45.92×0.1
4.592

15. 579.23×100
57,923

16. $80.12 \div 0.01$
8012

17. $2.931 \div 1000$
0.002931

18. The temperature of a cake is recorded in 10-min intervals after it comes out of the oven. See the graph.

a. What was the difference in temperature between 10 and 20 min?
61.4°F

b. What was the difference in temperature between 40 and 50 min?
1.4°F

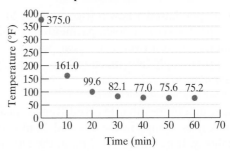

19. In the 2000 U.S. Presidential election, George Bush received approximately 50.5 million votes. In the same election, Al Gore received approximately 51.0 million votes.

a. Write a decimal number representing the number of votes received by George Bush.
50,500,000 votes

b. Write a decimal number representing the number of votes received by Al Gore.
51,000,000 votes

c. What is the approximate difference between the number of votes received by Al Gore and the number of votes received by George Bush? The difference is approximately 500,000 in favor of Al Gore.

20. A picture is framed and matted as shown in the figure.

a. Find the area of the picture itself.
67.5 in.²

b. Find the area of the matting *only*.
75.5 in.²

c. Find the area of the frame *only*.
157.3 in.²

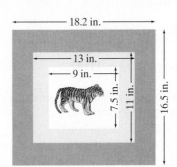

21. Jonas bought 200 shares of a stock for $36.625 per share, and had to pay a commission of $4.25. Two years later he sold all of his shares at a price of $52.16 per share. If he paid $8 for selling the stock, how much money did he make overall?

He made $3094.75.

22. Dalia purchased a new refrigerator for $1099.99. She paid $200 as a down payment and will finance the rest over a two-year period. Approximately how much will her monthly payment be?

She will pay approximately $37.50 per month.

23. Kent determines that his Ford Ranger pickup truck gets 23 mpg in the city and 26 mpg on the highway. If he drives 110.4 mi in the city and 135.2 mi on the highway how much gas will he use?

He will use 10 gal of gas.

24. The table shows the winning times in seconds for a women's speed skating event for several years. Complete the table.

Year	Time in Seconds (Decimal)	Time in Seconds (Fraction)
1998	38.24	$38\frac{6}{25}$
2002	38.23	$38\frac{23}{100}$
2006	37.30	$37\frac{3}{10}$
2010	38.21	$38\frac{21}{100}$

25. Rank the numbers and plot them on a number line. $3.2, 3\frac{1}{2}, 3.\overline{5}$

$3\frac{1}{2}, 3.\overline{5}, 3.2$

For Exercises 26 and 27, simplify by using the order of operations.

26. $(8.7)\left(1.6 - \frac{1}{2}\right)$

9.57

27. $\frac{7}{3}\left(5.25 - \frac{3}{4}\right)^2$

47.25

28. David runs almost every day. His distances for one week are given in the table.

a. Find the total distance that he ran. 38.8 mi

b. Find the average distance for the 7-day period. Round to the nearest tenth of a mile. 5.5 mi/day

Day	Distance
Monday	4.6 mi
Tuesday	5.9
Wednesday	0
Thursday	8.4
Friday	2.5
Saturday	12.8
Sunday	4.6

Chapters 1–4 Cumulative Review Exercises

1. Simplify. $(17 + 12) - (8 - 3) \cdot 3$

14

2. Convert the number to standard form.
4 thousands + 3 tens + 9 ones

4039

3. Add. $3902 + 34 + 904$ 4840

4. Subtract. $4990 - 1118$

3872

5. Multiply and round the answer to the thousands place. $23,444 \times 103$

2,415,000

6. Divide 4530 by 225. Then identify the dividend, divisor, whole-number part of the quotient, and remainder.   Dividend: 4530; divisor: 225; whole-number part of the quotient: 20; remainder: 30

7. Explain how to check the division problem in Exercise 6.   To check a division problem, multiply the whole-number part of the quotient and the divisor. Then add the remainder to get the dividend. That is, $20 \times 225 + 30 = 4530$.

8. The chart shows the retail sales for several companies. Find the difference between the highest and lowest sales in the chart.

The difference between sales for Wal-Mart and Sears is $181,956 million.

Retail Sales

Wal-Mart	Target	Kmart	Sears
217,799	39,455	36,151	35,843

(Millions of Dollars)

For Exercises 9–14, multiply or divide as indicated. Write the answer as a fraction.

9. $\frac{1}{5} \cdot \frac{6}{11}$ $\frac{6}{55}$

10. $\left(\frac{6}{15}\right)\left(\frac{10}{7}\right)$ $\frac{4}{7}$

11. $\left(\dfrac{7}{10}\right)^2$ $\dfrac{49}{100}$

12. $\left(\dfrac{32}{22}\right) \div \left(\dfrac{8}{11}\right)$ 2

13. $\dfrac{8}{3} \div 4$ $\dfrac{2}{3}$

14. $\left(\dfrac{0}{5}\right) \div \left(\dfrac{3}{5}\right)$ 0

15. A settlement for a lawsuit is made for $15,000. The attorney gets $\frac{2}{5}$ of the settlement. How much is left?
There is $9000 left.

16. Simplify.

$$\dfrac{8}{25} + \dfrac{1}{5} \div \dfrac{5}{6} - \left(\dfrac{2}{5}\right)^2 \quad \dfrac{2}{5}$$

For Exercises 17–20, add or subtract as indicated. Write the answers as fractions.

17. $\dfrac{7}{10} + \dfrac{27}{100}$ $\dfrac{97}{100}$

18. $\dfrac{5}{11} + 3$ $\dfrac{38}{11}$

19. $5 - \dfrac{2}{7}$ $\dfrac{33}{7}$

20. $\dfrac{12}{5} - \dfrac{9}{10}$ $\dfrac{3}{2}$

 21. Find the area and perimeter.
Area: $\dfrac{15}{64}$ ft²; perimeter: 2 ft

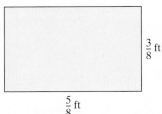

$\dfrac{3}{8}$ ft

$\dfrac{5}{8}$ ft

22. Cliff walks in the mornings for exercise. For the last 4 days he has walked for $\frac{9}{10}$, $1\frac{2}{5}$, $1\frac{1}{4}$, and $1\frac{1}{5}$ km. What is the average distance?
The average is $1\dfrac{3}{16}$ km.

For Exercises 23–27, perform the indicated operations.

23. $50.9 + 123.23$
174.13

24. $700.8 - 32.01$
668.79

25. 301.1×0.25
75.275

26. $51.2 \div 3.2$
16

27. $\dfrac{4}{3}(3.14)(9)^2$ 339.12

28. Divide $79.02 \div 1.7$ and round the answer to the nearest hundredth.   46.48

29. a. Multiply. 0.004×938.12
3.75248
  b. Multiply. 938.12×0.004
3.75248
  c. Identify the property that has been demonstrated in parts (a) and (b).
Commutative property of multiplication

30. The table gives the average length of several bones in the human body. Complete the table by writing the mixed numbers as decimals and the decimals as mixed numbers.

Bone	Length (in.) (Decimal)	Length (in.) (Mixed Number)
Femur	19.875	$19\frac{7}{8}$
Fibula	15.9375	$15\frac{15}{16}$
Humerus	14.375	$14\frac{3}{8}$
Innominate bone (hip)	7.5	$7\frac{1}{2}$

Ratio and Proportion

5

CHAPTER OUTLINE

Chapter 5

This chapter is devoted to the study of ratios, rates, and proportions. Unit rates are important when comparing two or more items such as comparing prices at the grocery store.

Review Your Skills

As we cover ratios and rates, we will have the opportunity to work with decimals again. Review decimals by performing the operations on decimals shown in hints a–d. Then use the hints to help you complete the puzzle. Each row and each column in the grid must use the numbers 1, 2, 3, 4, and 5 exactly once.

Hints

a. Simplify $0.8 + 12.12 + 7.33$. Place the tenths place digit in box a.   20.25

b. Simplify $60.75 \div 0.3$. Place the number of digits to the left of the decimal point in box b.   202.5

c. Simplify $1.7625 - 1.56$. Place the number of digits to the right of the decimal point in box c.   0.2025

d. Simplify 8.1×0.25. Place the thousandths place digit in box d.   2.025

4	1	2	5	3
2	3	4	1	5
1	ᵃ 2	5	3	ᶜ 4
3	5	1	4	2
ᵈ 5	4	ᵇ 3	2	1

Section 5.1 Ratios

1. Writing a Ratio

Thus far we have seen two interpretations of fractions.

- The fraction $\frac{5}{8}$ represents 5 parts of a whole that has been divided evenly into 8 pieces.
- The fraction $\frac{5}{8}$ represents $5 \div 8$.

Now we consider a third interpretation.

- The fraction $\frac{5}{8}$ represents the ratio of 5 to 8.

A **ratio** is a comparison of two quantities. There are three different ways to write a ratio.

Concept Connections

1. When forming the ratio $\frac{a}{b}$, why must b not equal zero?

> **Writing a Ratio**
>
> The ratio of a to b can be written as follows, provided $b \neq 0$.
>
> **1.** a to b **2.** $a : b$ **3.** $\dfrac{a}{b}$
>
> The colon means "to."   The fraction bar means "to."

Although there are three ways to write a ratio, we primarily use the fraction form.

Skill Practice

2. For a recent flight from Atlanta to San Diego, 291 seats were occupied and 29 were unoccupied. Write the ratio of
   a. The number of occupied seats to unoccupied seats
   b. The number of unoccupied seats to occupied seats
   c. The number of occupied seats to the total number of seats

Classroom Example: p. 298, Exercise 10

> **Example 1** Writing a Ratio
>
> In an algebra class there are 15 women and 17 men.
>
> **a.** Write the ratio of women to men.
>
> **b.** Write the ratio of men to women.
>
> **c.** Write the ratio of women to the total number of people in the class.
>
> **Solution:**
>
> It is important to observe the *order* of the quantities mentioned in a ratio. The first quantity mentioned is the numerator. The second quantity is the denominator.
>
> **a.** The ratio of women to men is   **b.** The ratio of men to women is
>
>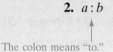
>
> $\dfrac{15}{17}$ $\dfrac{17}{15}$
>
> **c.** First find the total number of people in the class.
>
> $$\text{Total} = \text{number of women} + \text{number of men}$$
> $$= 15 + 17$$
> $$= 32$$
>
> Therefore the ratio of women to the total number of people in the class is
>
> $$\frac{15}{32}$$

Answers

1. The value b must not be zero because division by zero is undefined.
2. **a.** $\dfrac{291}{29}$ **b.** $\dfrac{29}{291}$ **c.** $\dfrac{291}{320}$

2. Simplifying a Ratio to Lowest Terms

It is often desirable to write a ratio in lowest terms. The process is similar to simplifying fractions to lowest terms.

Example 2   Writing Ratios in Lowest Terms

Write each ratio in lowest terms.

a. 15 ft to 10 ft   **b.** $20 to $10

Solution:

In part (a) we are comparing feet to feet. In part (b) we are comparing dollars to dollars. We can divide out the like units in the numerator and denominator as we would common factors.

a. $\dfrac{15 \text{ ft}}{10 \text{ ft}} = \dfrac{3 \cdot 5 \text{ ft}}{2 \cdot 5 \text{ ft}}$

$= \dfrac{3 \cdot \overset{1}{\cancel{5}} \text{ ft}}{2 \cdot \underset{1}{\cancel{5}} \text{ ft}}$ Simplify common factors.

$= \dfrac{3}{2}$

> **TIP:** *Note:* Even though the number $\frac{3}{2}$ is equivalent to $1\frac{1}{2}$, we do not write the ratio as a mixed number. Remember that a ratio is a comparison of *two* quantities. If you did convert $\frac{3}{2}$ to the mixed number $1\frac{1}{2}$, you would write the ratio as $\dfrac{1\frac{1}{2}}{1}$. This would imply that the numerator is one and one-half times as large as the denominator.

b. $\dfrac{\$20}{\$10} = \dfrac{\$\overset{2}{\cancel{20}}}{\$\underset{1}{\cancel{10}}}$ Simplify common factors.

$= \dfrac{2}{1}$

Although the fraction $\frac{2}{1}$ is equivalent to 2, we do not generally write ratios as whole numbers. Again, a ratio compares *two* quantities. In this case, we say that there is a 2-to-1 ratio between the original dollar amounts.

3. Writing Ratios of Mixed Numbers and Decimals

It is often desirable to express a ratio in lowest terms by using whole numbers in the numerator and denominator. This is demonstrated in Examples 3 and 4.

Example 3   Writing a Ratio as a Ratio of Whole Numbers

The length of a rectangular picture frame is 10.8 in., and the width is 8.64 in. Express the ratio of the length to the width. Then rewrite the ratio as a ratio of whole numbers simplified to lowest terms.

Answers

3. $\dfrac{9}{2}$ **4.** $\dfrac{6}{1}$ **5.** b

Solution:

The ratio of length to width is $\frac{10.8}{8.64}$. We now want to rewrite the ratio, using whole numbers in the numerator and denominator. If we multiply 10.8 by 10, the decimal point will move to the right one place, resulting in a whole number. If we multiply 8.64 by 100, the decimal point will move to the right two places, resulting in a whole number. Because we want to multiply the numerator and denominator by the *same* number, we choose the greater number, 100.

$$\frac{10.8}{8.64} = \frac{10.8 \times 100}{8.64 \times 100} \quad \text{Multiply numerator and denominator by 100.}$$

$$= \frac{1080}{864} \quad \begin{array}{l}\text{Because the numerator and denominator are large} \\ \text{numbers, we write the prime factorization of each.} \\ \text{The common factors are now easy to identify.}\end{array}$$

$$= \frac{\overset{1}{2} \cdot \overset{1}{2} \cdot \overset{1}{2} \cdot \overset{1}{3} \cdot \overset{1}{3} \cdot \overset{1}{3} \cdot 5}{\underset{1}{2} \cdot \underset{1}{2} \cdot \underset{1}{2} \cdot 2 \cdot 2 \cdot \underset{1}{3} \cdot \underset{1}{3} \cdot \underset{1}{3}} \quad \text{Simplify common factors to lowest terms.}$$

$$= \frac{5}{4} \quad \text{The ratio of length to width is } \tfrac{5}{4}.$$

In Example 3, we multiplied by 100 to move the decimal point *two* places to the right. Multiplying by 10 would not have been sufficient, because $8.64 \times 10 = 86.4$, which is not a whole number.

Example 4 **Writing a Ratio as a Ratio of Whole Numbers**

Ling walked $2\frac{1}{4}$ mi on Monday and $3\frac{1}{2}$ mi on Tuesday. Write the ratio of miles walked Monday to miles walked Tuesday. Then rewrite the ratio as a ratio of whole numbers reduced to lowest terms.

Solution:

The ratio of miles walked on Monday to miles walked on Tuesday is $\dfrac{2\frac{1}{4}}{3\frac{1}{2}}$.

To convert this to a ratio of whole numbers, first we rewrite each mixed number as an improper fraction. Then we can divide the fractions and simplify.

$$\frac{2\frac{1}{4}}{3\frac{1}{2}} = \frac{\frac{9}{4}}{\frac{7}{2}} \quad \longleftarrow \quad \begin{array}{l}\text{Write the mixed numbers as improper fractions.} \\ \text{Recall that a fraction bar also implies division.}\end{array}$$

$$= \frac{9}{4} \div \frac{7}{2}$$

$$= \frac{9}{4} \cdot \frac{2}{7} \quad \text{Multiply by the reciprocal of the divisor.}$$

$$= \frac{9}{\underset{2}{4}} \cdot \frac{\overset{1}{2}}{7} \quad \text{Simplify common factors to lowest terms.}$$

$$= \frac{9}{14} \quad \text{This is a ratio of whole numbers in lowest terms.}$$

Answers

6. $\dfrac{35}{24}$ **7.** $\dfrac{2\frac{1}{2}}{\frac{3}{4}}; \dfrac{10}{3}$

4. Applications of Ratios

Ratios are used in a number of applications.

Example 5   Using Ratios to Express Population Increase

The town of Roxbury, Connecticut, had 1825 people in the year 1990. By the year 2010, its population was 2514. Write a ratio depicting the increase in population to the number of people in the town in 1990.

Solution:

To write this ratio, we need to know the increase in population.

$$\text{Increase in population} = 2514 - 1825$$

$$= 689$$

The ratio of the increase in population to the number of people in 1990 is

Increase in population ⟶ $\dfrac{689}{1825}$ ⟵ number of people in 1990

Example 6   Applying Ratios to Unit Conversion

A fence is 12 yd long and $1\frac{1}{2}$ ft high.

a. Write the ratio of length to height with all units measured in yards.

b. Write the ratio of length to height with all units measured in feet.

$1\frac{1}{2}$ ft = $\frac{1}{2}$ yd

12 yd

Solution:

a. Since 3 ft = 1 yd. Therefore, $1\frac{1}{2}$ ft = $\frac{1}{2}$ yd.

Measuring in yards, we see that the ratio of length to height is
$$\dfrac{12 \text{ yd}}{\frac{1}{2}\text{ yd}} = \dfrac{12}{1} \cdot \dfrac{2}{1} = \dfrac{24}{1}.$$

b. The length is 12 yd = 36 ft. (Since 1 yd = 3 ft, then 12 yd = 12 · 3 ft = 36 ft.)

Measuring in feet, we see that the ratio of length to height is
$$\dfrac{36 \text{ ft}}{1\frac{1}{2}\text{ ft}} = \dfrac{36}{\frac{3}{2}} = \dfrac{\overset{12}{36}}{1} \cdot \dfrac{2}{\underset{1}{3}} = \dfrac{24}{1}.$$

Notice that regardless of the units used, the ratio is the same, 24 to 1. This means that the length is 24 times the height.

Section 5.1 Practice Exercises

Study Skills Exercise

For additional exercises, see Classroom Activities 5.1A–5.1C in the *Student's Resource Manual* at www.mhhe.com/moh.

Does your school have a learning resource center or a tutoring center? If so, do you remember the location and hours of operation? Write them here.

Location of learning resource center or tutoring center:

Hours of operation:

Vocabulary and Key Concepts

1. The statement "*a* to *b*" or "*a* : *b*" or $\frac{a}{b}$ represents the _____ratio_____ of *a* to *b*.

Concept 1: Writing a Ratio

2. The flour-to-sugar ratio for a recipe is 2 to 1. Explain what this means.
   The recipe calls for twice as much flour as sugar. For example, if 3 cups of sugar is used, then 6 cups of flour must be used.

For Exercises 3–8, write the ratio in two other ways.

3. 5 to 6 $5:6$ and $\frac{5}{6}$

4. 3 to 7 $3:7$ and $\frac{3}{7}$

5. 11 : 4   11 to 4 and $\frac{11}{4}$

6. 8 : 13   8 to 13 and $\frac{8}{13}$

7. $\frac{1}{2}$ $1:2$ and 1 to 2

8. $\frac{1}{8}$ $1:8$ and 1 to 8

For Exercises 9–12, write the ratios in fraction form. **(See Example 1.)**

9. Nancy has 3 cats and 2 dogs.

   a. Write a ratio of cats to dogs. $\frac{3}{2}$

   b. Write a ratio of dogs to cats. $\frac{2}{3}$

   c. Write a ratio of cats to the total number of pets. $\frac{3}{5}$

10. In a kindergarten classroom, there are 11 boys and 9 girls.

   a. Write a ratio of girls to boys. $\frac{9}{11}$

   b. Write a ratio of boys to girls. $\frac{11}{9}$

   c. Write a ratio of boys to the total number of children in class. $\frac{11}{20}$

11. There are 52 cars in the parking lot, of which 21 are silver.

   a. Write a ratio of silver cars to the total number of cars. $\frac{21}{52}$

   b. Write a ratio of silver cars to cars that are not silver. $\frac{21}{31}$

12. On one city block, there are 21 houses and 10 of them have pools in the backyard.

   a. Write a ratio of the number of houses with a pool to the total number of houses. $\frac{10}{21}$

   b. Write a ratio of the number of houses with a pool to the number of houses without a pool. $\frac{10}{11}$

Concept 2: Simplifying a Ratio to Lowest Terms

For Exercises 13–24, write the ratio in lowest terms. **(See Example 2.)**

13. 4 yr to 6 yr $\frac{2}{3}$

14. 10 lb to 14 lb $\frac{5}{7}$

15. 5 mi to 25 mi $\frac{1}{5}$

16. 20 ft to 12 ft $\frac{5}{3}$

17. 8 m to 2 m $\frac{4}{1}$

18. 14 oz to 7 oz $\frac{2}{1}$

19. 33 cm to 15 cm $\frac{11}{5}$

20. 21 days to 30 days $\frac{7}{10}$

21. \$60 to \$50 $\frac{6}{5}$

22. 75¢ to 100¢ $\frac{3}{4}$

23. 18 in. to 36 in. $\frac{1}{2}$

24. 3 cups to 9 cups $\frac{1}{3}$

Concept 3: Writing Ratios of Mixed Numbers and Decimals

For Exercises 25–36, write the ratio in lowest terms with whole numbers in the numerator and denominator.
(See Examples 3 and 4.)

25. 3.6 ft to 2.4 ft $\frac{3}{2}$ **26.** 10.15 hr to 8.12 hr $\frac{5}{4}$ **27.** 8 gal to $9\frac{1}{3}$ gal $\frac{6}{7}$ **28.** 24 yd to $13\frac{1}{3}$ yd $\frac{9}{5}$

29. $16\frac{4}{5}$ m to $18\frac{9}{10}$ m $\frac{8}{9}$ **30.** $1\frac{1}{4}$ in. to $1\frac{3}{8}$ in. $\frac{10}{11}$ **31.** $16.80 to $2.40 $\frac{7}{1}$ **32.** $18.50 to $3.70 $\frac{5}{1}$

33. $\frac{1}{2}$ day to 4 days $\frac{1}{8}$ **34.** $\frac{1}{4}$ mi to $1\frac{1}{2}$ mi $\frac{1}{6}$ **35.** 10.25 L to 8.2 L $\frac{5}{4}$ **36.** 11.55 km to 6.6 km $\frac{7}{4}$

Concept 4: Applications of Ratios

37. The temperature at 8:00 A.M. in Los Angeles was 66°F. By 2:00 P.M., the temperature had risen to 90°F. Write a ratio representing the increase in temperature to the temperature at 8:00 A.M. **(See Example 5.)** $\frac{4}{11}$

38. The city of Goddard, Kansas, had approximately 2100 people in the year 2000. In 2010, the U.S. Census Bureau indicated the population was 4344. Write a ratio representing the increase in population to the number of people in the year 2000. $\frac{187}{175}$

39. A plaque is 6 in. wide and $1\frac{1}{3}$ ft long ($\frac{1}{3}$ ft is 4 in.). **(See Example 6.)**

    a. Find the ratio of width to length with all units in inches. $\frac{6}{16} = \frac{3}{8}$

    b. Find the ratio of width to length with all units in feet. $\frac{\frac{1}{2}}{1\frac{1}{3}} = \frac{3}{8}$

40. A construction company needs 2 weeks to construct a family room and 3 days to add a porch.

    a. Find the ratio of the time it takes for constructing the porch to the time constructing the family room, with all units in weeks. $\frac{\frac{3}{7}}{2} = \frac{3}{14}$

    b. Find the ratio of the time it takes for constructing the porch to the time constructing the family room, with all units in days. $\frac{3}{14}$

For Exercises 41–44, refer to the table showing Alex Rodriguez's salary (rounded to the nearest $100,000) for selected years during his career. Write each ratio in lowest terms.

Year	Team	Salary	Position
1994	Seattle Mariners	$400,000	Shortstop
2000	Seattle Mariners	$4,400,000	Shortstop
2003	Texas Rangers	$22,000,000	Shortstop
2009	New York Yankees	$33,000,000	Third baseman

Source: USA TODAY

41. Write the ratio of Alex's salary for the year 1994 to the year 2000. $\frac{1}{11}$

42. Write a ratio of Alex's salary for the year 2009 to the year 2003. $\frac{3}{2}$

43. Write a ratio of the increase in Alex's salary between the years 1994 and the year 2000 to his salary in 1994. $\frac{10}{1}$

44. Write a ratio of the increase in Alex's salary between the years 2003 and 2009 to his salary in 2003. $\frac{1}{2}$

For Exercises 45–48, refer to the table that shows the average spending per person for reading (books, newspapers, magazines, etc.) by age group. Write each ratio in lowest terms.

45. Find the ratio of spending for the group under 25 years old to the spending for the group 75 years and over. $\frac{15}{32}$

46. Find the ratio of spending for the group 25 to 34 years old to the spending for the group of 65 to 74 years old. $\frac{37}{53}$

47. Find the ratio of spending for the group under 25 years old to the spending for the group of 55 to 64 years old. $\frac{20}{61}$

48. Find the ratio of spending for the group 35 to 44 years old to the spending for the group 45 to 54 years old. $\frac{34}{43}$

Age Group	Annual Average ($)
Under 25 years	60
25 to 34 years	111
35 to 44 years	136
45 to 54 years	172
55 to 64 years	183
65 to 74 years	159
75 years and over	128

Source: Mediamark Research Inc.

 For Exercises 49–52, find the ratio of the shortest side to the longest side. Write each ratio in lowest terms with whole numbers in the numerator and denominator.

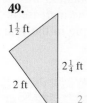 **49.**
$1\frac{1}{2}$ ft, $2\frac{1}{4}$ ft, 2 ft $\quad \frac{2}{3}$

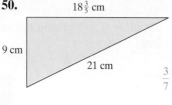 **50.**
$18\frac{3}{5}$ cm, 9 cm, 21 cm $\quad \frac{3}{7}$

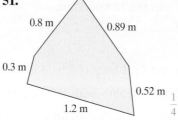 **51.**
0.8 m, 0.89 m, 0.3 m, 0.52 m, 1.2 m $\quad \frac{1}{4}$

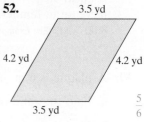 **52.**
3.5 yd, 4.2 yd, 4.2 yd, 3.5 yd $\quad \frac{5}{6}$

Expanding Your Skills

For Exercises 53–55, refer to the figure. The lengths of the sides for squares A, B, C, and D are given.

53. What are the lengths of the sides of square E? 13 units

54. Find the ratio of the lengths of the sides for the given pairs of squares.

a. Square B to square A $\quad \frac{3}{2}$

b. Square C to square B $\quad \frac{5}{3}$

c. Square D to square C $\quad \frac{8}{5}$

d. Square E to square D $\quad \frac{13}{8}$

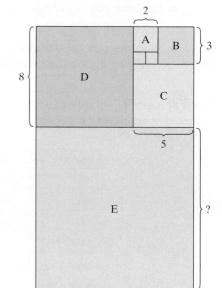

55. Write the decimal equivalents for each ratio in Exercise 54. Do these values seem to be approaching a number close to 1.618 (this is an approximation for the *golden ratio*, which is equal to $\frac{1+\sqrt{5}}{2}$)? Applications of the golden ratio are found throughout nature. In particular, as a result of the geometrically pleasing pattern, artists and architects have proportioned their work to approximate the golden ratio. a. 1.5 b. 1.6̄ c. 1.6 d. 1.625; yes

56. The ratio of a person's height to the length of the person's lower arm (from elbow to wrist) is approximately 6.5 to 1. Measure your own height and lower arm length. Is the ratio you get close to the average of 6.5 to 1? Answers will vary.

57. The ratio of a person's height to the person's shoulder width (measured from outside shoulder to outside shoulder) is approximately 4 to 1. Measure your own height and shoulder width. Is the ratio you get close to the average of 4 to 1? Answers will vary.

Rates and Unit Cost

1. Definition of a Rate

A **rate** is a type of ratio used to compare different types of quantities, for example:

$$\frac{270 \text{ mi}}{13 \text{ gal}} \quad \text{and} \quad \frac{\$8.55}{1 \text{ hr}}$$

Several key words imply rates. These are given in Table 5-1.

Table 5-1

Key Word	Example	Rate
Per	117 miles per 2 hours	$\dfrac{117 \text{ mi}}{2 \text{ hr}}$
For	$12 for 3 pounds	$\dfrac{\$12}{3 \text{ lb}}$
In	400 meters in 43.5 seconds	$\dfrac{400 \text{ m}}{43.5 \text{ sec}}$
On	270 miles on 12 gallons of gas	$\dfrac{270 \text{ mi}}{12 \text{ gal}}$

Because a rate compares two different quantities it is important to include the units in both the numerator and the denominator. It is also desirable to write rates in lowest terms.

Example 1 Writing Rates in Lowest Terms

Write each rate in lowest terms.

a. In one region, there are approximately 640 trees on 12 acres.

b. Latonya drove 138 mi on 6 gal of gas.

c. Jounne can type 625 words in 10 min.

Solution:

a. The rate of 640 trees on 12 acres can be expressed as $\dfrac{640 \text{ trees}}{12 \text{ acres}}$.

Now write this rate in lowest terms. $\dfrac{\overset{160}{\cancel{640}} \text{ trees}}{\underset{3}{\cancel{12}} \text{ acres}} = \dfrac{160 \text{ trees}}{3 \text{ acres}}$

b. The rate of 138 mi on 6 gal of gas can be expressed as $\dfrac{138 \text{ mi}}{6 \text{ gal}}$.

Now write this rate in lowest terms. $\dfrac{\overset{23}{\cancel{138}} \text{ mi}}{\underset{1}{\cancel{6}} \text{ gal}} = \dfrac{23 \text{ mi}}{1 \text{ gal}}$

Concepts

1. Definition of a Rate
2. Unit Rates
3. Unit Cost
4. Applications of Rates

Instructor Note: To prepare students, ask them the *cost per daytime minute* that they are charged on their cell phone plan. This will open a discussion on rates, then ask students what other rates they might know about (miles per gallon, cost per pound, miles per hour, etc.).

Concept Connections

1. Why is it important to include units when you are expressing a rate?

Skill Practice

Write each rate in lowest terms.

2. Maria reads 15 pages in 10 min.
3. A Chevrolet Corvette Z06 gets 206.4 mi on 8.6 gal of gas.
4. At Nash Community College in Rocky Mount, North Carolina, there are 2200 students to 120 instructors.

Classroom Example: p. 305, Exercise 14

Answers

1. The units are different in the numerator and denominator and will not "cancel."
2. $\dfrac{3 \text{ pages}}{2 \text{ min}}$
3. $\dfrac{24 \text{ mi}}{1 \text{ gal}}$ or 24 mi/gal
4. $\dfrac{55 \text{ students}}{3 \text{ instructors}}$

c. The rate of 625 words in 10 min can be represented as $\dfrac{625 \text{ words}}{10 \text{ min}}$.

Now write this rate in lowest terms. $\dfrac{\overset{125}{\cancel{625}} \text{ words}}{\underset{2}{\cancel{10}} \text{ min}} = \dfrac{125 \text{ words}}{2 \text{ min}}$

This rate indicates that 625 words in 10 min is the same speed as 125 words in 2 min.

2. Unit Rates

A rate having a denominator of 1 unit is called a **unit rate**. Furthermore the number 1 is often omitted in the denominator.

$\dfrac{23 \text{ mi}}{1 \text{ gal}} = 23 \text{ mi/gal}$ is read as "twenty-three miles per gallon."

$\dfrac{52 \text{ ft}}{1 \text{ sec}} = 52 \text{ ft/sec}$ is read as "fifty-two feet per second."

$\dfrac{\$15}{1 \text{ hr}} = \$15/\text{hr}$ is read as "fifteen dollars per hour."

> **Converting a Rate to a Unit Rate**
> To convert a rate to a unit rate, divide the numerator by the denominator and maintain the units of measurement.

| **Example 2** | **Finding Unit Rates** |

Write each rate as a unit rate. Round to three decimal places if necessary.

a. A health club charges $125 for 20 visits. Find the unit rate in dollars per visit.

b. In 1960, Wilma Rudolph won the women's 200-m run in 24 sec. Find her speed in meters per second.

c. During one baseball season, Barry Bonds got 149 hits in 403 at bats. Find his batting average. (*Hint:* Batting average is defined as the number of hits per the number of at bats.)

Solution:

a. The rate of $125 for 20 visits can be expressed as $\dfrac{\$125}{20 \text{ visits}}$.

To convert this to a unit rate, divide $125 by 20 visits.

$\dfrac{\$125}{20 \text{ visits}} = \dfrac{\$6.25}{1 \text{ visit}}$ or $6.25/\text{visit}$

$$
\begin{array}{r}
6.25 \\
20\overline{)125.00} \\
-120 \\
\hline
50 \\
-40 \\
\hline
100 \\
-100 \\
\hline
0
\end{array}
$$

b. The rate of 200 m per 24 sec can be expressed as $\dfrac{200 \text{ m}}{24 \text{ sec}}$.

To convert this to a unit rate, divide 200 m by 24 sec.

$$\frac{200 \text{ m}}{24 \text{ sec}} \approx \frac{8.333 \text{ m}}{1 \text{ sec}} \quad \text{or approximately } 8.333 \text{ m/sec}$$

$$\begin{array}{r} 8.\overline{3} \\ 24\overline{)200.00} \\ -192 \\ \hline 80 \\ -72 \\ \hline 80 \end{array}$$

The quotient repeats.

Wilma Rudolph's speed was approximately 8.333 m/sec.

Avoiding Mistakes

Units of measurement must be included for the answer to be complete.

c. The rate of 149 hits in 403 at bats can be expressed as $\dfrac{149 \text{ hits}}{403 \text{ at bats}}$.

To convert this to a unit rate, divide 149 hits by 403 at bats.

$$\frac{149 \text{ hits}}{403 \text{ at bats}} \approx \frac{0.370 \text{ hit}}{1 \text{ at bat}} \quad \text{or } 0.370 \text{ hit/at bat}$$

3. Unit Cost

A **unit cost** or unit price is the cost per 1 unit of something. At the grocery store, for example, you might purchase meat for $3.79/lb ($3.79 per 1 lb). Unit cost is useful in day-to-day life when we compare prices. Example 3 compares the prices of three different sizes of apple juice.

Example 3   Finding Unit Costs

Apple juice comes in a variety of sizes and packaging options. Find the unit price per ounce and determine which is the best buy.

a. $4.19

b. $4.89

c. $3.55

Apple Juice
46 oz

Apple Juice
64 oz

Apple Juice
8-pack 4 oz each

Skill Practice

10. Gatorade comes in several size packages. Compute the unit price per ounce for each option (round to the nearest thousandth of a dollar). Then determine which is the best buy.

   a. $3.25 for a 64-oz bottle

   b. $9.59 for eight 20-oz bottles

   c. $1.95 for a 32-oz bottle

Classroom Example: p. 306, Exercise 44

Solution:

When we compute a unit cost, the cost is always placed in the numerator of the rate. Furthermore, when we divide the cost by the amount, we need to obtain enough digits in the quotient to see the variation in unit price. In this example, we have rounded to the nearest thousandth of a dollar (nearest tenth of a cent). This means that we use the ten-thousandths-place digit in the quotient on which to base our decision on rounding.

Answers

10. a. $0.051/oz
    b. $0.060/oz
    c. $0.061/oz
    The 64-oz bottle is the best buy.

	Rate	Quotient	Unit Rate (Rounded)
a.	$\dfrac{\$4.19}{46 \text{ oz}}$	$\$4.19 \div 46 \text{ oz} \approx \$0.0911/\text{oz}$	$\$0.091/\text{oz}$ or $9.1\cancel{c}/\text{oz}$
b.	$\dfrac{\$4.89}{64 \text{ oz}}$	$\$4.89 \div 64 \text{ oz} \approx \$0.0764/\text{oz}$	$\$0.076/\text{oz}$ or $7.6\cancel{c}/\text{oz}$
c.	$\dfrac{4 \text{ oz} \times 8 = 32 \text{ oz}}{\dfrac{\$3.55}{32 \text{ oz}}}$	$\$3.55 \div 32 \text{ oz} \approx \$0.1109/\text{oz}$	$\$0.111/\text{oz}$ or $11.1\cancel{c}/\text{oz}$

From the table we see that the most economical buy is the 64-oz size because its unit rate is the least expensive.

4. Applications of Rates

Example 4 uses a unit rate for comparison in an application.

Example 4   Computing Mortality Rates

Mortality rate is defined to be the total number of people who die due to some risk behavior divided by the total number of people who engage in the risk behavior. Based on the following statistics, compare the mortality rate for undergoing heart bypass surgery to the mortality rate of flying on the space shuttle.

   a. Roughly 28 people will die for every 1000 who undergo heart bypass surgery. (*Source:* The Society of Thoracic Surgeons)

   b. As of April 2010, there have been 14 astronauts killed in space shuttle missions out of 912 astronauts who have flown.

Solution:

   a. Mortality rate for heart bypass surgery: $\frac{28}{1000} = 0.028$ death/surgery

   b. Mortality rate for flying on the space shuttle: $\frac{14}{912} \approx 0.015$ deaths/flight

Comparing these rates shows that it is riskier to have heart bypass surgery than to fly on the space shuttle.

Section 5.2   Practice Exercises

Study Skills Exercise

For additional exercises, see Classroom Activities 5.2A–5.2C in the *Student's Resource Manual* at www.mhhe.com/moh.

Budgeting enough time to do homework and to study for a class is one of the most important steps to success in a class. Use a weekly calendar to help you plan your time for your studies this week. Also write in other obligations such as the time required for your job, for your family, for sleeping, and for eating. Be realistic when you estimate the time for each activity.

Vocabulary and Key Concepts

1. a. A ___rate___ is a type of ratio used to compare different types of quantities—for example: 200 mi per 4 hr.

   b. A ___unit___ rate is a rate that has a denominator of 1 unit.

Writing   Translating Expression   Geometry   Scientific Calculator   Video   NS E

Review Exercises

2. Write the ratio $\frac{16}{24}$ in lowest terms. $\frac{2}{3}$

3. Write the ratio 3 to 5 in two other ways.   3:5 and $\frac{3}{5}$

4. Write the ratio 4:1 in two other ways.   4 to 1 and $\frac{4}{1}$

For Exercises 5–8, write the ratio in lowest terms.

5. 36¢ to 27¢ $\frac{4}{3}$

6. $6\frac{3}{4}$ ft to $8\frac{1}{4}$ ft $\frac{9}{11}$

7. 1.08 mi to 2.04 mi $\frac{9}{17}$

8. $28.40 to $20.80 $\frac{71}{52}$

Concept 1: Definition of a Rate

For Exercises 9–20, write each rate in lowest terms. **(See Example 1.)**

9. A type of laminate flooring sells for $32 for 5 square feet (ft²). $\frac{\$32}{5\ \text{ft}^2}$

10. A remote control car can go up to 44 ft in 5 sec. $\frac{44\ \text{ft}}{5\ \text{sec}}$

11. Elaine drives 234 mi in 4 hr. $\frac{117\ \text{mi}}{2\ \text{hr}}$

12. Travis has 14 blooms on 6 of his plants. $\frac{7\ \text{blooms}}{3\ \text{plants}}$

13. Tyler earned $58 in 8 hr. $\frac{\$29}{4\ \text{hr}}$

14. Neil can type only 336 words in 15 min. $\frac{112\ \text{words}}{5\ \text{min}}$

15. A printer can print 13 pages in 26 sec. $\frac{1\ \text{page}}{2\ \text{sec}}$

16. During a bad storm there was 2 in. of rain in 6 hr. $\frac{1\ \text{in.}}{3\ \text{hr}}$

17. There are 130 calories in 8 snack crackers. $\frac{65\ \text{calories}}{4\ \text{crackers}}$

18. The driveway is lined with 14 plants for a length of 22 ft. $\frac{7\ \text{plants}}{11\ \text{ft}}$

19. An advertisement states that TV trays are selling for $30 for a set of 4 trays. $\frac{\$15}{2\ \text{trays}}$

20. There are 50 students assigned to 4 advisers. $\frac{25\ \text{students}}{2\ \text{advisers}}$

Concept 2: Unit Rates

21. Of the following rates, identify those that are unit rates.

a. $\frac{\$0.37}{1\ \text{oz}}$ **b.** $\frac{333.2\ \text{mi}}{14\ \text{gal}}$ **c.** 16 ft/sec   **d.** $\frac{59\ \text{mi}}{1\ \text{hr}}$ a, c, d

22. Of the following rates, identify those that are unit rates.

a. $\frac{3\ \text{lb}}{\$1.00}$ **b.** $\frac{21\ \text{ft}}{1\ \text{sec}}$ **c.** 50 mi/hr   **d.** $\frac{232\ \text{words}}{2\ \text{min}}$ a, b, c

For Exercises 23–28, write each rate as a unit rate. **(See Example 2.)**

23. The Osborne family drove 452 mi in 4 days.   113 mi/day

24. The book of poetry *The Prophet* by Kahlil Gibran has estimated sales of $6,000,000 over an 80-year period. $75,000 per year

25. Philip drove 480 km in 5 hr.   96 km/hr

26. Ian flew 1120 mi in 4 hr.   280 mi/hr

✏ Writing   ←→ Translating Expression   △ Geometry   🖩 Scientific Calculator   ▶ Video   NS & E

27. If Oscar bought an easy chair for $660 and plans to make 12 payments, what is the amount per payment? $55 per payment

28. The jockey David Gall had 7396 wins in 43 years of riding. 172 wins per year

For Exercises 29–32, determine the unit rates and round to the nearest hundredth when necessary.

29. At the market, bananas cost $2.76 for 4 lb. $0.69/lb

30. Ceramic tiles are on sale for $13.08 for a box of 12 tiles. Find the price per tile. $1.09 per tile

31. Lottery prize money of $1,792,000 is for 7 people. $256,000 per person

32. One WeightWatchers group lost 123 lb for its 11 members. 11.18 lb per member

33. A male speed skater skated 500 m in 35 sec. Find the rate in meters per second. (Round to one decimal place.) 14.3 m/sec

34. A female speed skater skated 500 m in 38 sec. Find the rate in meters per second. (Round to one decimal place.) 13.2 m/sec

Concept 3: Unit Cost

For Exercises 35–44, find the unit costs (that is, dollars per unit). Round the answers to three decimal places when necessary. **(See Example 3.)**

35. Gain laundry detergent costs $10.95 for 50 oz.
$0.219 per oz

36. Dove liquid body wash costs $3.49 for 12 oz.
$0.291 per oz

37. Soda costs $1.99 for a 2-L bottle.
$0.995 per liter

38. Four chairs cost $221.00.
$55.25 per chair

39. A set of 4 tires costs $210.
$52.50 per tire

40. A package of 3 shirts costs $64.80.
$21.60 per shirt

41. A package of 6 newborn bodysuits costs $32.50.
$5.417 per bodysuit

42. A package of 8 AAA batteries costs $9.84.
$1.23 per battery

43. **a.** 25 oz of shampoo for $8.35 $0.334/oz

   b. 15 oz of shampoo for $5.01 $0.334/oz

   c. Which is the better buy?
   Both sizes cost the same amount per ounce.

44. **a.** 10 lb of potting soil for $2.99 $0.299/lb

   b. 30 lb of potting soil for $8.97 $0.299/lb

   c. Which is the better buy?
   Both sizes cost the same amount per pound.

45. Creamed corn comes in two size cans, 15 oz and 8.5 oz. The larger can costs $1.85 and the smaller can costs $1.39. Find the unit cost of each can. Which is the better buy? (Round to three decimal places.)
The larger can is $0.123 per ounce. The smaller can is $0.164 per ounce. The larger can is the better buy.

46. Napkins come in a variety of packages. A package of 400 napkins sells for $5.65, and a package of 100 napkins sells for $1.95. Find the unit cost of each package. Which is the better buy? (Round to three decimal places.)
The package of 400 sells for $0.014 per napkin. The package of 100 sells for $0.020 per napkin. The large package is the better buy.

Concept 4: Applications of Rates

47. Carbonated beverages come in different sizes and contain different amounts of sugar. Compute the amount of sugar (in grams) per fluid ounce for each soda. Then determine which has the greatest amount of sugar per fluid ounce. **(See Example 4.)**
Coca-Cola: 3.25 g/fl oz; MelloYello: 3.92 g/fl oz; Ginger Ale: 3 g/fl oz; MelloYello has the greatest amount per fluid oz.

Soda	Amount	Sugar
Coca-Cola	20 fl oz	65 g
Mello Yello	12 fl oz	47 g
Canada Dry Ginger Ale	8 fl oz	24 g

48. Carbonated beverages come in different sizes and have a different number of calories. Compute the number of calories per fluid ounce for each soda. Then determine which has the least number of calories per fluid ounce.
Coca-Cola: 12 cal/fl oz; MelloYello: 14.2 cal/fl oz; Ginger Ale: 11.25 cal/fl oz; Ginger Ale has the least number of calories per fluid oz.

Soda	Amount	Calories
Coca-Cola	20 fl oz	240
Mello Yello	12 fl oz	170
Canada Dry Ginger Ale	8 fl oz	90

 Writing   Translating Expression   Geometry   Scientific Calculator   Video NS𝐸

49. According to the National Institutes of Health, a platelet count below 20,000 per microliter of blood is considered a life-threatening condition. Suppose a patient's test results yield a platelet count of 13,000,000 for 100 microliters. Write this as a unit rate (number of platelets per microliter). Does the patient have a life-threatening condition? 130,000 platelets per microliter; Since the patient's platelet count is above 20,000 per microliter, the patient does *not* have a life-threatening condition.

50. The number of vehicles produced in the United States decreased by 4,260,000 during a recent 12-year period. Compute the rate representing the decrease in the number of vehicles produced per year during this time period. (*Source: Research and Innovative Technology Administration, Bureau of Transportation Statistics*) 355,000 vehicles/year

51. The total number of prisoners in the United States increased steadily by a total of 344,000 in an 8-year period. Compute the rate representing the increase in the number of prisoners per year. 43,000 prisoners per year

52. a. The population of Mexico increased steadily by 22 million people in a 10-year period. Compute the rate representing the increase in the population per year. 2.2 million per year

    b. The population of Brazil increased steadily by 10.2 million in a 5-year period. Compute the rate representing the increase in the population per year. 2.04 million per year

    c. Which country has a greater rate of increase in population per year? Mexico

53. a. The price per share of Microsoft stock rose $18.24 in a 24-month period. Compute the rate representing the increase in the price per month. $0.76 per month

    b. The price per share of IBM stock rose $22.80 in a 12-month period. Compute the rate representing the increase in the price per month. $1.90 per month

    c. Which stock had a greater rate of increase per month? IBM

54. A cheetah can run 120 m in 4.1 sec. An antelope can run 50 m in 2.1 sec. Compare their unit speeds to determine which animal is faster. Round to the nearest whole unit. Cheetah: 29 m/sec; antelope: 24 m/sec. The cheetah is faster.

Calculator Connections

Topic: Applications of Unit Rates

Calculator Exercises

Don Shula coached football for 33 years. He had 328 wins and 156 losses. Tom Landry coached football for 29 years. He had 250 wins and 162 losses. Use this information to answer Exercises 55 and 56.

55. a. Compute a unit rate representing the average number of wins per year for Don Shula. Round to one decimal place. 9.9 wins/year

    b. Compute a unit rate representing the average number of wins per year for Tom Landry. Round to one decimal place. 8.6 wins/year

    c. Which coach had a better rate of wins per year? Shula

56. a. Compute a unit rate representing the number of wins to the number of losses for Don Shula. Round to one decimal place.   2.1 wins/loss

   b. Compute a unit rate representing the number of wins to the number of losses for Tom Landry. Round to one decimal place.   1.5 wins/loss

   c. Which coach had a better win/loss rate?   Shula

57. Compare three brands of soap. Find the price per ounce and determine the best buy. (Round to two decimal places.)

   a. Dove: $9.59 for a 6-bar pack of 4.25-oz bars $0.38 per ounce

   b. Dial: $6.39 for a 8-bar pack of 4.5-oz bars $0.18 per ounce

   c. Irish Spring: $2.55 for a 3-bar pack of 4.5-oz bars $0.19 per ounce; The best buy is Dial.

58. Mayonnaise comes in 48-, 22-, and 18-oz jars. They are priced at $8.69, $6.15, and $4.59, respectively. Find the unit cost of each size jar to find the best buy. (Round to three decimal places.)
   The unit prices are $0.181 per ounce, $0.280 per ounce, and $0.255 per ounce. The best buy is the 48-oz jar.

59. Tuna fish comes in different size cans. Find the unit cost of each package to find the best buy. (Round to three decimal places.)

   a. 2-pack of 4.5-oz cans for $3.61 $0.401/oz

   b. One 12-oz can for $2.00 $0.167/oz

   c. 3-pack of 3.3-oz cans for $3.19 $0.322/oz;
   The best buy is the 12-oz can.

60. A&W root beer is sold in a variety of different packages. Find the unit cost of each package to find the better buy. (Round to three decimal places.)

   a. 6-pack of 8-fl-oz cans for $2.99 $0.062/oz

   b. 12-pack of 12-fl-oz cans for $3.33 $0.023/oz;
   The case of twelve 12-fl-oz cans for $3.33 is the better buy.

Section 5.3 | Proportions

Concepts

1. Definition of a Proportion
2. Determining Whether Two Ratios Form a Proportion
3. Solving Proportions

1. Definition of a Proportion

A statement indicating that two quantities are equal is called an **equation**. In this section, we are interested in a special type of equation called a proportion. A **proportion** states that two ratios or rates are equal. For example:

$$\frac{1}{4} = \frac{10}{40} \text{ is a proportion.}$$

We know that the fractions $\frac{1}{4}$ and $\frac{10}{40}$ are equal because $\frac{10}{40}$ reduces to $\frac{1}{4}$.

We read the proportion $\frac{1}{4} = \frac{10}{40}$ as follows: "1 is to 4 as 10 is to 40."

We also say that the numbers 1 and 4 are *proportional to* the numbers 10 and 40.

Example 1 **Writing Proportions**

Write a proportion for each statement.

a. 5 is to 12 as 30 is to 72.

b. 240 mi is to 4 hr as 300 mi is to 5 hr.

c. The numbers 3 and 7 are proportional to the numbers 12 and 28.

Solution:

a. $\dfrac{5}{12} = \dfrac{30}{72}$ 5 is to 12 as 30 is to 72.

b. $\dfrac{240\text{ mi}}{4\text{ hr}} = \dfrac{300\text{ mi}}{5\text{ hr}}$ 240 mi is to 4 hr as 300 mi is to 5 hr.

c. $\dfrac{3}{7} = \dfrac{12}{28}$ 3 and 7 are proportional to 12 and 28.

2. Determining Whether Two Ratios Form a Proportion

To determine whether two ratios form a proportion, we must determine whether the ratios are equal. Recall from Section 2.3 that two fractions are equal whenever their cross products are equal. That is,

$$\frac{a}{b} \diagdown \frac{c}{d} \quad \text{implies} \quad a \cdot d = b \cdot c \quad \text{(and vice versa).}$$

Example 2 **Determining Whether Two Ratios Form a Proportion**

Determine whether the ratios form a proportion.

a. $\dfrac{3}{5} \overset{?}{=} \dfrac{9}{15}$ **b.** $\dfrac{1\frac{2}{3}}{4} \overset{?}{=} \dfrac{10}{5\frac{1}{2}}$

Solution:

a. $\dfrac{3}{5} \overset{?}{=} \dfrac{9}{15}$

$(3)(15) \overset{?}{=} (5)(9)$ Form an equation from the cross products.

$45 = 45$ ✔

The cross products are equal. Therefore, the ratios form a proportion.

b. $\dfrac{1\frac{2}{3}}{4} \overset{?}{=} \dfrac{10}{5\frac{1}{2}}$

$\left(1\frac{2}{3}\right)\left(5\frac{1}{2}\right) \overset{?}{=} (4)(10)$ Form an equation from the cross products.

$\dfrac{5}{3} \cdot \dfrac{11}{2} \overset{?}{=} 40$ Write the mixed numbers as improper fractions.

$\dfrac{55}{6} \neq 40$ Multiply fractions.

The cross products are not equal. The ratios do not form a proportion.

Classroom Example: p. 314, Exercise 34

Example 3 **Determining Whether Pairs of Numbers Are Proportional**

Determine whether the numbers 2.7 and 5.3 are proportional to the numbers 8.1 and 15.9.

Solution:

Two pairs of numbers are proportional if their ratios are equal.

$$\frac{2.7}{5.3} \overset{?}{=} \frac{8.1}{15.9}$$

$(2.7)(15.9) \overset{?}{=} (5.3)(8.1)$ Form an equation from the cross products.

$42.93 = 42.93 ✔$ Multiply decimals.

The cross products are equal. The ratios form a proportion.

Instructor Note: To prepare students, ask them to estimate how much money a person would earn for 4 hr of work if the person gets paid $7 for 1 hr of work.

3. Solving Proportions

A proportion is made up of four values. If three of the four values are known, we can solve for the fourth.

Consider the proportion $\frac{x}{20} = \frac{3}{4}$. We let the variable x represent the unknown value in the proportion. To solve for x, we can equate the cross products to form an equivalent equation.

$$\frac{x}{20} \overset{\times}{=} \frac{3}{4}$$

$4 \cdot x = 3 \cdot 20$ Form an equation from the cross products.

$4 \cdot x = 60$ To determine the correct value of x, we ask, What number multiplied by 4 equals 60? The answer is 15 and can be found mentally by computing $60 \div 4 = 15$.

$\dfrac{4 \cdot x}{4} = \dfrac{60}{4}$ To show this step in the equation, divide both sides of the equation by 4.

$\dfrac{\overset{1}{4} \cdot x}{\underset{1}{4}} = \dfrac{\overset{15}{60}}{\underset{1}{4}}$ Simplify common factors within each fraction.

$1x = 15$

$x = 15$

We can check the value of x in the original proportion.

Check: $\dfrac{x}{20} = \dfrac{3}{4}$ $\xrightarrow{\text{substitute 15 for } x}$ $\dfrac{15}{20} \overset{?}{=} \dfrac{3}{4}$

$(15)(4) \overset{?}{=} (3)(20)$

$60 = 60 ✔$ The solution 15 checks.

This process uses the following important fact. *We may divide both sides of an equation by the same nonzero number.* By so doing, we produce an equivalent equation that has the same solution. In general, to solve an equation, the goal is to isolate the variable on one side of the equation. In the equation $4 \cdot x = 60$, we must eliminate the factor of 4 next to x. By dividing by 4 we form a ratio $\frac{4 \cdot x}{4}$. This reduces to $\frac{1 \cdot x}{1}$, which in turn simplifies to x.

Answers

6. No   **7.** Divide by 2.
8. Divide by 1.5.   **9.** Divide by 7.

The steps to solve a proportion are summarized next.

Solving a Proportion

Step 1 Set the cross products equal to each other.

Step 2 Divide both sides of the equation by the number being multiplied by the variable.

Step 3 Check the solution in the original proportion.

Example 4   Solving a Proportion

Solve the proportion. $\dfrac{x}{13} = \dfrac{6}{39}$

Solution:

$\dfrac{x}{13} = \dfrac{6}{39}$

$39 \cdot x = (6)(13)$ Set the cross products equal.

$39 \cdot x = 78$ Simplify.

$\dfrac{39 \cdot x}{39} = \dfrac{78}{39}$ Divide both sides by the number being multiplied by x (in this case, 39).

$\dfrac{\overset{1}{39} \cdot x}{39} = \dfrac{\overset{2}{78}}{39}$ Simplify common factors within each fraction.

$x = 2$ Check: $\dfrac{x}{13} = \dfrac{6}{39} \xrightarrow{\overset{\text{substitute}}{x=2}} \dfrac{2}{13} \overset{?}{=} \dfrac{6}{39}$

$(2)(39) \overset{?}{=} (6)(13)$

$78 = 78 \; ✔$

The solution 2 checks in the original proportion.

Skill Practice

Solve the proportion. Be sure to check your answer.

10. $\dfrac{2}{9} = \dfrac{n}{81}$

Classroom Example: p. 315, Exercise 52

Example 5   Solving a Proportion

Solve the proportion. $\dfrac{4}{15} = \dfrac{9}{n}$

Solution:

$\dfrac{4}{15} = \dfrac{9}{n}$ The variable can be represented by any letter.

$4 \cdot n = (9)(15)$ Set the cross products equal.

$4 \cdot n = 135$ Simplify.

$\dfrac{4 \cdot n}{4} = \dfrac{135}{4}$ Divide both sides by the number being multiplied by n (in this case, 4).

$\dfrac{\overset{1}{4} \cdot n}{4} = \dfrac{135}{4}$

$n = \dfrac{135}{4}$ The fraction $\frac{135}{4}$ is in lowest terms.

Avoiding Mistakes

When solving an equation, don't try to "cancel" across the equal sign.

Skill Practice

Solve the proportion. Be sure to check your answer.

11. $\dfrac{3}{w} = \dfrac{21}{77}$

Classroom Example: p. 315, Exercise 62

Answers

10. $n = 18$ **11.** $w = 11$

The solution may be written as $n = \frac{135}{4}$ or $n = 33\frac{3}{4}$ or $n = 33.75$.

To check the solution in the original proportion, we may use any of the three forms of the answer. We will use the decimal form.

Check: $\dfrac{4}{15} = \dfrac{9}{n}$ $\xrightarrow{\text{substitute } n = 33.75}$ $\dfrac{4}{15} \overset{?}{=} \dfrac{9}{33.75}$

$$(4)(33.75) \overset{?}{=} (9)(15)$$

$$135 = 135 \checkmark \quad \text{The solution checks.}$$

Classroom Example: p. 315, Exercise 64

Skill Practice

Solve the proportion. Be sure to check your answer.

12. $\dfrac{0.6}{x} = \dfrac{1.5}{2}$

Example 6 **Solving a Proportion**

Solve the proportion. $\quad \dfrac{0.8}{3.1} = \dfrac{4}{p}$

Solution:

$$\frac{0.8}{3.1} = \frac{4}{p}$$

$(0.8) \cdot p = (4)(3.1)$ Set the cross products equal.

$0.8p = 12.4$ Notice that in this case we dropped the $\cdot$ symbol for multiplication between 0.8 and p. If a variable and a constant are written adjacent to each other without an operator $(+, -, \times, \text{or} \div)$ between them, the operation is understood to be multiplication. That is, $0.8p = 0.8 \times p$.

$$\frac{0.8p}{0.8} = \frac{12.4}{0.8}$$ Divide both sides by the number being multiplied by p (in this case, 0.8).

$$\frac{\overset{1}{\cancel{0.8}}p}{\underset{1}{\cancel{0.8}}} = \frac{12.4}{0.8}$$

$$p = 15.5$$ Divide $12.4 \div 0.8 = 15.5$.

The check is left to the reader.

We chose to give the solution to Example 6 in decimal form because the values in the original proportion are decimal numbers. However, it would be correct to give the solution as a mixed number or fraction. The solution $p = 15.5$ is also equivalent to $p = 15\frac{1}{2}$ or $p = \frac{31}{2}$.

Example 7 **Solving a Proportion**

Solve the proportion. $\quad \dfrac{12}{8} = \dfrac{x}{\frac{2}{3}}$

Solution:

$$\frac{12}{8} = \frac{x}{\frac{2}{3}}$$

$$(12)\left(\frac{2}{3}\right) = 8 \cdot x \qquad \text{Set the cross products equal.}$$

$$\left(\frac{12}{1}\right)\left(\frac{2}{3}\right) = 8x \qquad \text{Write the whole number as an improper fraction.}$$

$$\left(\frac{\overset{4}{12}}{1}\right)\left(\frac{2}{\underset{1}{3}}\right) = 8x \qquad \text{Simplify common factors.}$$

$$\frac{8}{1} = 8x \qquad \text{Multiply fractions.}$$

$$8 = 8x \qquad \text{Simplify.}$$

$$\frac{8}{8} = \frac{8x}{8} \qquad \text{Divide both sides by the number being multiplied by } x \text{ (in this case, 8).}$$

$$\frac{\overset{1}{8}}{\underset{1}{8}} = \frac{\overset{1}{8x}}{\underset{1}{8}} \qquad \text{Simplify common factors.}$$

$$1 = x$$

The solution 1 checks in the original proportion.

Skill Practice

Solve the proportion. Be sure to check your answer.

13. $\dfrac{\frac{1}{2}}{3.5} = \dfrac{x}{14}$

Classroom Example: p. 315, Exercise 68

Answer

13. $x = 2$

Section 5.3 Practice Exercises

For additional exercises, see Classroom Activities 5.3A–5.3B in the *Student's Resource Manual* at www.mhhe.com/moh.

Study Skills Exercise

You should not try to cram for tests. Instead, math is a subject that should be studied every day. This text gives you opportunities to review and practice as you work through the book. Find the page number for the Cumulative Review exercises for this chapter.

Vocabulary and Key Concepts

1. a. A(n) <u>equation</u> is a statement indicating that two quantities are equal.

b. A(n) <u>proportion</u> is an equation indicating that two ratios or rates are equal.

Review Exercises

2. If a box of cereal costs \$4.29 for 16 oz, find the unit price per ounce. Round to three decimal places.
$0.268/oz

For Exercises 3–8, write as a reduced ratio or rate.

3. 3 ft to 45 ft $\dfrac{1}{15}$

4. 3 teachers for 45 students $\dfrac{1 \text{ teacher}}{15 \text{ students}}$

5. 6 apples for 2 pies $\dfrac{3 \text{ apples}}{1 \text{ pie}}$

6. 6 days to 2 days $\dfrac{3}{1}$

7. 264 mi per 36 gal $\dfrac{22 \text{ mi}}{3 \text{ gal}}$

8. \$264 to \$36 $\dfrac{22}{3}$

Concept 1: Definition of a Proportion

For Exercises 9–20, write a proportion for each statement. **(See Example 1.)**

9. 4 is to 16 as 5 is to 20. $\dfrac{4}{16} = \dfrac{5}{20}$

10. 3 is to 18 as 4 is to 24. $\dfrac{3}{18} = \dfrac{4}{24}$

11. 25 is to 15 as 10 is to 6. $\dfrac{25}{15} = \dfrac{10}{6}$

12. 35 is to 14 as 20 is to 8. $\dfrac{35}{14} = \dfrac{20}{8}$

13. The numbers 2 and 3 are proportional to the numbers 4 and 6. $\dfrac{2}{3} = \dfrac{4}{6}$

14. The numbers 2 and 1 are proportional to the numbers 26 and 13. $\dfrac{2}{1} = \dfrac{26}{13}$

15. The numbers 30 and 25 are proportional to the numbers 12 and 10. $\dfrac{30}{25} = \dfrac{12}{10}$

16. The numbers 24 and 18 are proportional to the numbers 8 and 6. $\dfrac{24}{18} = \dfrac{8}{6}$

17. $6.25 per hour is proportional to $187.50 per 30 hr. $\dfrac{\$6.25}{1\ hr} = \dfrac{\$187.50}{30\ hr}$

18. $115 per week is proportional to $460 per 4 weeks. $\dfrac{\$115}{1\ week} = \dfrac{\$460}{4\ weeks}$

19. 1 in. is to 7 mi as 5 in. is to 35 mi. $\dfrac{1\ in.}{7\ mi} = \dfrac{5\ in.}{35\ mi}$

20. 16 flowers is to 5 plants as 32 flowers is to 10 plants. $\dfrac{16\ flowers}{5\ plants} = \dfrac{32\ flowers}{10\ plants}$

Concept 2: Determining Whether Two Ratios Form a Proportion

For Exercises 21–28, determine whether the ratios form a proportion. **(See Example 2.)**

21. $\dfrac{5}{18} \overset{?}{=} \dfrac{4}{16}$
No

22. $\dfrac{9}{10} \overset{?}{=} \dfrac{8}{9}$
No

23. $\dfrac{16}{24} \overset{?}{=} \dfrac{2}{3}$
Yes

24. $\dfrac{4}{5} \overset{?}{=} \dfrac{24}{30}$
Yes

25. $\dfrac{2\frac{1}{2}}{3\frac{2}{3}} \overset{?}{=} \dfrac{15}{22}$
Yes

26. $\dfrac{1\frac{3}{4}}{3} \overset{?}{=} \dfrac{7}{12}$
Yes

27. $\dfrac{2}{3.2} \overset{?}{=} \dfrac{10}{16}$
Yes

28. $\dfrac{4.7}{7} \overset{?}{=} \dfrac{23.5}{35}$
Yes

For Exercises 29–34, determine whether the pairs of numbers are proportional. **(See Example 3.)**

29. Are the numbers 48 and 18 proportional to the numbers 24 and 9? Yes

30. Are the numbers 35 and 14 proportional to the numbers 5 and 2? Yes

31. Are the numbers $2\frac{3}{8}$ and $1\frac{1}{2}$ proportional to the numbers $9\frac{1}{2}$ and 6? Yes

32. Are the numbers $1\frac{2}{3}$ and $\frac{5}{6}$ proportional to the numbers 5 and $2\frac{1}{2}$? Yes

33. Are the numbers 6.3 and 9 proportional to the numbers 12.6 and 16? No

34. Are the numbers 7.1 and 2.4 proportional to the numbers 35.5 and 10? No

Concept 3: Solving Proportions

For Exercises 35–42, what number would you divide by on each side of the equation to solve for the variable?

35. $2x = 8$
Divide by 2

36. $3x = 27$
Divide by 3

37. $5p = 30$
Divide by 5

38. $7w = 49$
Divide by 7

39. $32 = 8m$
Divide by 8

40. $50 = 25y$
Divide by 25

41. $0.15 = 0.6x$
Divide by 0.6

42. $1.4 = 0.4z$
Divide by 0.4

For Exercises 43–46, determine whether the given value is a solution to the proportion.

43. $\dfrac{x}{40} = \dfrac{1}{8}$; $x = 5$ Yes

44. $\dfrac{14}{x} = \dfrac{12}{18}$; $x = 21$ Yes

45. $\dfrac{12.4}{31} = \dfrac{8.2}{y}$; $y = 20$ No

46. $\dfrac{4.2}{9.8} = \dfrac{z}{36.4}$; $z = 15.2$ No

For Exercises 47–70, solve the proportion. Be sure to check your answers. **(See Examples 4–7.)**

47. $\dfrac{12}{16} = \dfrac{3}{x}$
$x = 4$

48. $\dfrac{20}{28} = \dfrac{5}{x}$
$x = 7$

49. $\dfrac{9}{21} = \dfrac{x}{7}$
$x = 3$

50. $\dfrac{15}{10} = \dfrac{3}{x}$
$x = 2$

51. $\dfrac{p}{12} = \dfrac{25}{4}$
$p = 75$

52. $\dfrac{p}{8} = \dfrac{30}{24}$
$p = 10$

53. $\dfrac{6}{n} = \dfrac{4}{8}$
$n = 12$

54. $\dfrac{49}{n} = \dfrac{14}{18}$
$n = 63$

55. $\dfrac{2}{3} = \dfrac{t}{18}$
$t = 12$

56. $\dfrac{34}{51} = \dfrac{2}{t}$
$t = 3$

57. $\dfrac{25}{100} = \dfrac{9}{y}$
$y = 36$

58. $\dfrac{65}{15} = \dfrac{26}{y}$
$y = 6$

59. $\dfrac{17}{12} = \dfrac{4\frac{1}{4}}{x}$
$x = 3$

60. $\dfrac{26}{30} = \dfrac{5\frac{1}{5}}{x}$
$x = 6$

61. $\dfrac{m}{12} = \dfrac{5}{8}$
$m = \dfrac{15}{2}$ or $7\frac{1}{2}$ or 7.5

62. $\dfrac{16}{12} = \dfrac{21}{a}$
$a = \dfrac{63}{4}$ or $15\frac{3}{4}$ or 15.75

63. $\dfrac{3.125}{5} = \dfrac{18.75}{k}$
$k = 30$

64. $\dfrac{4.75}{8} = \dfrac{9.5}{k}$
$k = 16$

65. $\dfrac{0.5}{h} = \dfrac{1.8}{9}$
$h = 2.5$

66. $\dfrac{2.6}{h} = \dfrac{1.3}{0.5}$
$h = 1$

67. $\dfrac{\frac{3}{8}}{6.75} = \dfrac{x}{72}$ $x = 4$

68. $\dfrac{12}{\frac{1}{4}} = \dfrac{120}{y}$ $y = 2.5$

69. $\dfrac{4}{\frac{1}{10}} = \dfrac{\frac{1}{2}}{z}$ $z = \dfrac{1}{80}$

70. $\dfrac{6}{\frac{1}{3}} = \dfrac{\frac{1}{2}}{t}$ $t = \dfrac{1}{36}$

Problem Recognition Exercises

Operations on Fractions versus Solving Proportions

For Exercises 1–6, identify the problem as a proportion or as a product of fractions. Then solve the proportion or multiply the fractions.

1. a. $\dfrac{x}{4} = \dfrac{15}{8}$
Proportion; $\dfrac{15}{2}$

b. $\dfrac{1}{4} \cdot \dfrac{15}{8}$
Product of fractions; $\dfrac{15}{32}$

2. a. $\dfrac{2}{5} \cdot \dfrac{3}{10}$
Product of fractions; $\dfrac{3}{25}$

b. $\dfrac{2}{5} = \dfrac{y}{10}$
Proportion; 4

3. a. $\dfrac{2}{7} \times \dfrac{3}{14}$
Product of fractions; $\dfrac{3}{49}$

b. $\dfrac{2}{7} = \dfrac{n}{14}$
Proportion; 4

4. a. $\dfrac{m}{5} = \dfrac{6}{15}$
Proportion; 2

b. $\dfrac{3}{5} \times \dfrac{6}{15}$
Product of fractions; $\dfrac{6}{25}$

5. a. $\dfrac{48}{p} = \dfrac{16}{3}$
Proportion; 9

b. $\dfrac{48}{8} \cdot \dfrac{16}{3}$
Product of fractions; 32

6. a. $\dfrac{10}{7} \cdot \dfrac{28}{5}$
Product of fractions; 8

b. $\dfrac{10}{7} = \dfrac{28}{t}$
Proportion; $\dfrac{98}{5}$

For Exercises 7–10, solve the proportion or perform the indicated operation on fractions.

7. a. $\dfrac{3}{7} = \dfrac{6}{z}$ 14

b. $\dfrac{3}{7} \div \dfrac{6}{35}$ $\dfrac{5}{2}$

c. $\dfrac{3}{7} + \dfrac{6}{35}$ $\dfrac{3}{5}$

d. $\dfrac{3}{7} \cdot \dfrac{6}{35}$ $\dfrac{18}{245}$

8. a. $\dfrac{4}{5} \div \dfrac{20}{3}$ $\dfrac{3}{25}$

b. $\dfrac{4}{v} = \dfrac{20}{3}$ $\dfrac{3}{5}$

c. $\dfrac{4}{5} \times \dfrac{20}{3}$ $\dfrac{16}{3}$

d. $\dfrac{4}{5} + \dfrac{20}{3}$ $\dfrac{112}{15}$

9. a. $\dfrac{14}{5} \cdot \dfrac{10}{7}$ 4

b. $\dfrac{14}{5} = \dfrac{x}{7}$ $\dfrac{98}{5}$

c. $\dfrac{14}{5} - \dfrac{10}{7}$ $\dfrac{48}{35}$

d. $\dfrac{14}{5} \div \dfrac{10}{7}$ $\dfrac{49}{25}$

10. a. $\dfrac{11}{3} = \dfrac{66}{y}$ 18

b. $\dfrac{11}{3} + \dfrac{66}{11}$ $\dfrac{29}{3}$

c. $\dfrac{11}{3} \div \dfrac{66}{11}$ $\dfrac{11}{18}$

d. $\dfrac{11}{3} \times \dfrac{66}{11}$ 22

Section 5.4 Applications of Proportions and Similar Figures

Concepts

1. Applications of Proportions
2. Similar Figures

1. Applications of Proportions

Proportions are used in a variety of applications. In Examples 1 through 4, we take information from the wording of a problem and form a proportion.

Skill Practice

1. Jacques bought 3 lb of tomatoes for $4.50. At this rate, how much would 7 lb cost?

Classroom Example: p. 322, Exercise 14

TIP: Notice that the two rates have the same units in the numerator (miles) and the same units in the denominator (gallons).

Example 1 Using a Proportion in a Consumer Application

Linda drove her Honda Accord 145 mi on 5 gal of gas. At this rate, how far can she drive on 12 gal?

Solution:

Let x represent the distance Linda can go on 12 gal.

This problem involves two rates. We can translate this to a proportion. Equate the two rates.

$$\text{distance} \longrightarrow \frac{145 \text{ mi}}{5 \text{ gal}} = \frac{x \text{ mi}}{12 \text{ gal}} \longleftarrow \text{distance}$$

$$\text{number of gallons} \longrightarrow \qquad \longleftarrow \text{number of gallons}$$

Solve the proportion.

$(145)(12) = (5) \cdot x$ Form an equation from the cross products.

$1740 = 5x$

$\dfrac{1740}{5} = \dfrac{\overset{1}{5}x}{\underset{1}{5}}$ Divide both sides by 5.

$348 = x$ Divide. $1740 \div 5 = 348$

Linda can drive 348 mi on 12 gal of gas.

Instructor Note: After discussing the different ways to set up the proportion correctly, ask students why $\frac{145 \text{ mi}}{5 \text{ gal}} = \frac{12 \text{ gal}}{x \text{ mi}}$ is *incorrect*.

In Example 1 we could have set up a proportion in many different ways.

$$\frac{145 \text{ mi}}{5 \text{ gal}} = \frac{x \text{ mi}}{12 \text{ gal}} \quad \text{or} \quad \frac{5 \text{ gal}}{145 \text{ mi}} = \frac{12 \text{ gal}}{x \text{ mi}} \quad \text{or}$$

$$\frac{5 \text{ gal}}{12 \text{ gal}} = \frac{145 \text{ mi}}{x \text{ mi}} \quad \text{or} \quad \frac{12 \text{ gal}}{5 \text{ gal}} = \frac{x \text{ mi}}{145 \text{ mi}}$$

Notice that in each case, the cross products produce the same equation. We will generally set up the proportions so that the units in the numerators are the same and the units in the denominators are the same.

Skill Practice

2. It takes 2.5 gal of paint to cover 900 ft² of wall. How much area could be painted with 4 gal of the same paint?

Classroom Example: p. 322, Exercise 16

Example 2 Using a Proportion in a Construction Application

If a cable 25 ft long weighs 1.2 lb, how much will a 120-ft cable weigh?

Solution:

Let w represent the weight of the 120-ft cable.   Label the unknown.

$$\text{length} \longrightarrow \frac{25 \text{ ft}}{1.2 \text{ lb}} = \frac{120 \text{ ft}}{w \text{ lb}} \longleftarrow \text{length}$$

$$\text{weight} \longrightarrow \qquad \longleftarrow \text{weight}$$

Translate to a proportion.

Answers

1. The price for 7 lb would be $10.50.
2. An area of 1440 ft² could be painted.

$$(25) \cdot w = (1.2)(120)$$ Equate the cross products.

$$25w = 144$$

$$\frac{\overset{1}{25}w}{\underset{1}{25}} = \frac{144}{25}$$ Divide both sides by 25.

$$w = 5.76$$ Divide $144 \div 25 = 5.76$.

The 120-ft cable weighs 5.76 lb.

Example 3   **Using Proportions in a Geography Application**

The distance between Phoenix and Los Angeles is 348 mi. On a certain map, this is represented by 6 in. On the same map, the distance between San Antonio and Little Rock is $8\frac{1}{2}$ in. What is the actual distance between San Antonio and Little Rock?

Solution:

Let d represent the distance between San Antonio and Little Rock.

actual distance $\longrightarrow$ $\dfrac{348 \text{ mi}}{6 \text{ in.}} = \dfrac{d \text{ mi}}{8\frac{1}{2} \text{ in.}}$ $\longleftarrow$ actual distance
distance on map $\longrightarrow$ $\longleftarrow$ distance on map

                                Translate to a proportion.

$$(348)(8\tfrac{1}{2}) = (6) \cdot d$$ Equate the cross products.

$$(348)(8.5) = 6d$$ Convert the values to decimal.

$$2958 = 6d$$

$$\frac{2958}{6} = \frac{\overset{1}{6}d}{\underset{1}{6}}$$ Divide both sides by 6.

$$493 = d$$ Divide $2958 \div 6 = 493$.

The distance between San Antonio and Little Rock is 493 mi.

Answer

3. The distance between Seattle and San Francisco is 704 mi.

Skill Practice

4. To estimate the number of fish in a lake, the park service catches 50 fish and tags them. After several months the park service catches a sample of 100 fish and finds that 6 are tagged. Approximately how many fish are in the lake?

Classroom Example: p. 323, Exercise 40

Example 4 Applying a Proportion to Environmental Science

A biologist wants to estimate the number of elk in a wildlife preserve. She sedates 25 elk and clips a small radio transmitter onto the ear of each animal. The elk return to the wild, and after 6 months, the biologist studies a sample of 120 elk in the preserve. Of the 120 elk sampled, 4 have radio transmitters. Approximately how many elk are in the whole preserve?

Solution:

Let n represent the number of elk in the whole preserve.

$$\begin{matrix} & \text{Sample} & & \text{Population} \end{matrix}$$

number of elk in the sample
with radio transmitters ⟶ $\dfrac{4}{120}$ = $\dfrac{25}{n}$ ⟵ number of elk in the population with radio transmitters
total elk in the sample ⟶   ⟵ total elk in the population

$4 \cdot n = (120)(25)$ Equate the cross products.

$4n = 3000$

$\dfrac{4n}{4} = \dfrac{3000}{4}$ Divide both sides by 4.

$n = 750$ Divide $3000 \div 4 = 750$.

There are approximately 750 elk in the wildlife preserve.

Concept Connections

5. Consider the similar triangles shown.

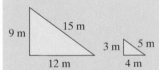

From choices **a, b,** and **c,** select the correct ratio of the lengths of the sides of the larger triangle to the corresponding sides of the smaller triangle.
a. 2 to 1
b. 3 to 1
c. 4 to 1

2. Similar Figures

Two triangles whose corresponding sides are proportional are called **similar triangles**. This means that the corresponding angles have equal measure, but the triangles may be different sizes. The following pairs of triangles are similar (Figure 5-1).

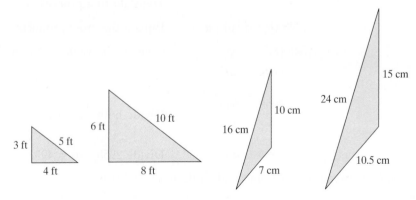

Figure 5-1

Answers

4. There are approximately 833 fish in the lake.
5. b

Consider the triangles:

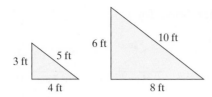

Notice that the ratio formed by the "left" sides of the triangles is $\frac{3}{6}$. This is the same as the ratio formed by the "bottom" sides, $\frac{4}{8}$, and is the same as the ratio formed by the "right" sides, $\frac{5}{10}$. Each ratio simplifies to $\frac{1}{2}$. Because all ratios are equal, the corresponding sides are proportional.

| Example 5 | **Finding the Unknown Sides in Similar Triangles** |

The triangles in Figure 5-2 are similar.

a. Solve for x.   **b.** Solve for y.

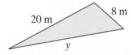

Figure 5-2

Solution:

a. Notice that the lengths of the "left" sides of both triangles are known. This forms a known ratio of $\frac{20}{5}$. Because the triangles are similar, the ratio of the other corresponding sides must be equal to $\frac{20}{5}$. To solve for x, we have

large triangle "left" side ⟶ $\dfrac{20}{5} = \dfrac{8}{x}$ ⟵ large triangle "right" side
small triangle "left" side ⟶   ⟵ small triangle "right" side

$20 \cdot x = (5)(8)$ Equate the cross products.

$20x = 40$

$\dfrac{\overset{1}{\cancel{20}}x}{\underset{1}{\cancel{20}}} = \dfrac{40}{20}$ Divide both sides of the equation by 20.

$x = 2$ Divide. $40 \div 20 = 2$

b. To solve for y, we have

large triangle "left" side ⟶ $\dfrac{20}{5} = \dfrac{y}{4}$ ⟵ large triangle "bottom" side
small triangle "left" side ⟶   ⟵ small triangle "bottom" side

$(20)(4) = 5 \cdot y$ Equate the cross products.

$80 = 5y$

$\dfrac{80}{5} = \dfrac{\overset{1}{\cancel{5}}y}{\underset{1}{\cancel{5}}}$ Divide both sides by 5.

$16 = y$ Divide. $80 \div 5 = 16$

The values for x and y are $x = 2$ m and $y = 16$ m.

Skill Practice

6. The triangles below are similar.

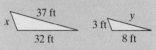

a. Solve for x.
b. Solve for y.

Classroom Example: p. 324, Exercise 46

TIP: In Example 5, notice that the corresponding sides of the triangles form a 4 to 1 ratio.

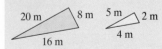

Answer

6. a. $x = 12$ ft   **b.** $y = 9.25$ ft

7. The given triangles are similar. Solve for *y*.

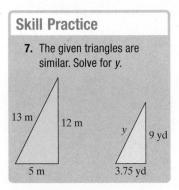

Classroom Example: p. 324, Exercise 48

Example 6 Finding the Unknown Side in a Similar Triangle

The triangles are similar. Solve for *x*.

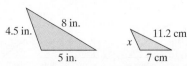

Solution:

The ratio $\dfrac{5}{7}$ can be formed from the lengths of the bottom sides of the triangles.

Then this ratio and the ratio of the left sides of the triangles form a proportion.

$$\text{large triangle bottom} \longrightarrow \frac{5}{7} = \frac{4.5}{x} \longleftarrow \text{large triangle "left" side}$$
small triangle bottom $\qquad\qquad\qquad\qquad$ small triangle "left" side

Notice that the proportion is still true even though the units are not the same.

$$5 \cdot x = (4.5)(7) \qquad \text{Equate the cross products.}$$

$$5x = 31.5$$

$$\frac{\overset{1}{\cancel{5}}x}{\underset{1}{\cancel{5}}} = \frac{31.5}{5} \qquad \text{Divide both sides of the equation by 5.}$$

$$x = 6.3 \qquad \text{Divide. } 31.5 \div 5 = 6.3$$

The value of *x* will be in centimeters. The length of the side is 6.3 cm.

8. Donna has drawn a pattern from which to make a scarf. Assuming that the triangles are similar (the pattern and scarf have the same shape), find the length of side *x* (in yards).

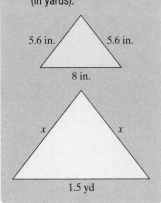

Classroom Example: p. 324, Exercise 50

Instructor Note: Remind students to label the variable by saying let *h* = height of building in feet.

Answers

7. *y* = 9.75 yd
8. Side *x* is 1.05 yd long.

Example 7 Using Similar Triangles in an Application

On a sunny day, a 6-ft man casts a 3.2-ft shadow on the ground. At the same time, a building casts an 80-ft shadow. How tall is the building?

Solution:

We illustrate the scenario in Figure 5-3. Note, however, that the distances are not drawn to scale.

Let *h* represent the height of the building in feet.

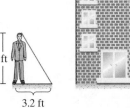
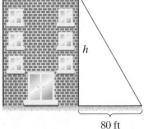

Figure 5-3

We will assume that the triangles are similar because the readings were taken at the same time of day.

$$\frac{6}{h} = \frac{3.2}{80} \qquad \text{Translate to a proportion.}$$

$$(6)(80) = (3.2) \cdot h \qquad \text{Equate the cross products.}$$

$$480 = 3.2h$$

$$\frac{480}{3.2} = \frac{\overset{1}{\cancel{3.2}}h}{\underset{1}{\cancel{3.2}}} \qquad \text{Divide both sides by 3.2.}$$

$$150 = h \qquad \text{Divide. } 480 \div 3.2 = 150$$

The height of the building is 150 ft.

In addition to studying similar triangles, we present similar polygons. Recall from Section 1.2 that a polygon is a flat figure formed by line segments connected at their ends. For a pair of **similar polygons**, the lengths of the corresponding sides of the two polygons are proportional.

Example 8 **Using Similar Polygons in a Photography Application**

A negative for a photograph is 3.5 cm by 2.5 cm. If the width of the resulting picture is 4 in., what is the length of the picture? See Figure 5-4.

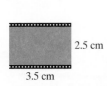

Figure 5-4

Solution:

Let x represent the length of the photo.

The photo and its negative are similar polygons.

$$\frac{3.5 \text{ cm}}{x \text{ in.}} = \frac{2.5 \text{ cm}}{4 \text{ in.}}$$ Translate to a proportion.

$$(3.5)(4) = (2.5) \cdot x$$

$$14 = 2.5x$$

$$\frac{14}{2.5} = \frac{\overset{1}{\cancel{2.5}}x}{\underset{1}{\cancel{2.5}}}$$ Divide both sides by 2.5.

$$5.6 = x$$ Divide. $14 \div 2.5 = 5.6$

The picture is 5.6 in. long.

Skill Practice

9. The first- and second-place plaques for a softball tournament have the same shape but are different sizes. Find the length of side x.

10 in.

x

8 in.   5 in.

Classroom Example: p. 325, Exercise 58

Answer

9. Side x is 6.25 in.

Section 5.4   Practice Exercises

Vocabulary and Key Concepts

For additional exercises, see Classroom Activities 5.4A–5.4B in the *Student's Resource Manual* at www.mhhe.com/moh.

1. **a.** Two triangles are called ___similar___ triangles if the lengths of their corresponding sides are proportional.

   b. If two polygons are similar, the lengths of their corresponding sides are ___proportional___.

Review Exercises

For Exercises 2–5, use = or ≠ to make a true statement.

2. $\dfrac{4}{7} \boxed{=} \dfrac{12}{21}$ 3. $\dfrac{3}{13} \boxed{=} \dfrac{15}{65}$ 4. $\dfrac{2}{5} \boxed{\neq} \dfrac{21}{55}$ 5. $\dfrac{12}{7} \boxed{\neq} \dfrac{35}{19}$

For Exercises 6–12, solve the proportion.

6. $\dfrac{2}{7} = \dfrac{3}{x}$

$x = \dfrac{21}{2}$ or $10\frac{1}{2}$ or 10.5

7. $\dfrac{4}{3} = \dfrac{n}{5}$

$n = \dfrac{20}{3}$ or $6\frac{2}{3}$ or $6.\overline{6}$

8. $\dfrac{p}{9} = \dfrac{1}{6}$

$p = 1\frac{1}{2}$ or $\dfrac{3}{2}$ or 1.5

9. $\dfrac{3\frac{1}{2}}{k} = \dfrac{2\frac{1}{3}}{4}$

$k = 6$

10. $\dfrac{2.4}{3} = \dfrac{m}{5}$

$m = 4$

11. $\dfrac{3}{2.1} = \dfrac{7}{y}$

$y = 4.9$

12. $\dfrac{1.2}{4} = \dfrac{3}{a}$

$a = 10$

Concept 1: Applications of Proportions

13. Pam drives her Toyota Prius 244 mi in city driving on 4 gal of gas. At this rate how many miles can she drive on 10 gal of gas? **(See Example 1.)**   Pam can drive 610 mi on 10 gal of gas.

14. Didi takes her pulse for 10 sec and counts 13 beats. How many beats per minute is this?
This is 78 beats/min.

15. To cement a garden path, it takes crushed rock and cement in a ratio of 3.25 kg of rock to 1 kg of cement. If a 24-kg bag of cement is purchased, how much crushed rock will be needed? **(See Example 2.)**
78 kg of crushed rock will be required.

16. Suppose two adults produce 63.4 lb of garbage in one week. At this rate, how many pounds will 50 adults produce in 1 week?
50 adults will produce 1585 lb of garbage.

17. On a map, the distance from Sacramento, California, to San Francisco, California, is 8 cm. The legend gives the actual distance at 91 mi. On the same map, Faythe measured 7 cm from Sacramento to Modesto, California. What is the actual distance? (Round to the nearest mile.) **(See Example 3.)**
The actual distance is about 80 mi.

18. On a map, the distance from Nashville, Tennessee, to Atlanta, Georgia, is 3.5 in., and the actual distance is 210 mi. If the map distance between Dallas, Texas, and Little Rock, Arkansas, is 4.75 in., what is the actual distance?
The distance is 285 mi.

19. At Central Community College, the ratio of female students to male students is 31 to 19. If there are 6200 female students, how many male students are there?
There are 3800 male students.

20. Evelyn won an election by a ratio of 6 to 5. If she received 7230 votes, how many votes did her opponent receive?
The opponent received 6025 votes.

21. If you flip a coin many times, the coin should come up heads about 1 time out of every 2 times it is flipped. If a coin is flipped 630 times, about how many heads do you expect to come up?
Heads would come up about 315 times.

22. A die is a small cube used in games of chance. It has six sides, and each side has 1, 2, 3, 4, 5, or 6 dots painted on it. If you roll a die, the number 4 should come up about 1 time out of every 6 times the die is rolled. If you roll a die 366 times, about how many times do you expect the number 4 to come up?
The number 4 should come up about 61 times.

23. A pitcher gave up 42 earned runs in 126 innings. Approximately how many earned runs will he give up in one game (9 innings)? This value is called the earned run average.   There would be approximately 3 earned runs for a 9-inning game.

24. In one game Peyton Manning completed 34 passes for 357 yd. At this rate how many yards would be gained for 22 passes?
231 yards would be gained.

25. Pierre bought 38€ (Euros) with $50 American. At this rate, how many Euros can he buy with $900 American?
Pierre can buy 684€.

26. Erik bought $103 Canadian with $100 American. At this rate, how many Canadian dollars can he buy with $235 American?
Erik can buy $242.05 Canadian.

27. According to data collected by the U.S. Consumer Product Safety Commission, for persons age 75 and older, nearly 3 out of 4 hospital emergency room visits are due to falls. If a hospital emergency room sees an average of 60 people 75 years of age or older in a month, how many of these visits are expected to be a result of falls?
45 visits would be a result of falls.

28. Approximately 24 out of 100 Americans over the age of 12 smoke. How many smokers would you expect in a group of 850 Americans over the age of 12?   204 smokers would be expected.

29. A recipe from the National Audubon Society for making hummingbird nectar recommends using a ratio of one part table sugar to four parts water. How much water should be added to one-sixth cup of sugar to create nectar for a hummingbird feeder? $\frac{2}{3}$ cup of water

30. A new car depreciates by a total of $3000 over a 5-year period. How much will the car depreciate in $3\frac{1}{2}$ years? $2100

31. Professor Fusco had 14 students in his online course and received 91 e-mails in one week. He spends about 3 minutes per e-mail with his response.

    a. If he had a class of 30 students, how many e-mails would he expect to receive per week?   195 e-mails

    b. With a class of 30 students, how much time would he expect to spend answering e-mail each week?
    585 min or 9.75 hr

32. A serving of 6 crackers has 225 mg of sodium.

    a. How much sodium would be in 14 crackers?   525 mg

    b. The FDA suggests that the maximum sodium intake per day is 2300 mg. If Donna ate 20 crackers, how much sodium would be allowed that day from other foods?   1550 mg

33. A computer animation that is captured at 30 frames per second takes 2.45 megabytes of memory. If the animation is captured at 12 frames per second, how much memory would be used?   0.98 megabyte or 980 kilobytes

34. Niu earned $312 on an investment of $800. How much would $1100 have earned in the same investment? $429

35. The Willis Tower in Chicago is 1454 ft high. If you made a model in which 50 ft = 1 in., how high would the Willis Tower be in the model?   29.08 in.

36. The Golden Gate Bridge is 8981 ft long. If you made a model in which 100 ft = 1 in., how long would the bridge be in the model?   89.81 in.

37. A chemist mixes 3 parts acid to 5 parts water. How much acid would the chemist need if she used 420 mL of water?   252 mL of acid

38. Mitch mixes 5 parts white paint to 7 parts blue paint. If he has 4 qt of white paint, how much blue paint would he need?   5.6 qt blue paint

39. Park officials stocked a man-made lake with bass last year. To approximate the number of bass this year, a sample of 75 bass is taken out of the lake and tagged. Then later a different sample is taken, and it is found that 21 of 100 bass are tagged. Approximately how many bass are in the lake? Round to the nearest whole unit. **(See Example 4.)**   There are approximately 357 bass in the lake.

40. Laws have been instituted in Florida to help save the manatee. To establish the number in Florida, a sample of 150 manatees was marked and let free. A new sample was taken and found that there were 3 marked manatees out of 40 captured. What is the approximate population of manatees in Florida?
There are approximately 2000 manatees in Florida.

41. Yellowstone National Park in Wyoming has the largest population of free-roaming bison. To approximate the number of bison, 200 are captured and tagged and then let free to roam. Later, a sample of 120 bison is observed and 6 have tags. Approximate the population of bison in the park.
There are approximately 4000 bison in the park.

42. In Cass County, Michigan, there are about 20 white-tailed deer per square mile. If the county covers 492 mi^2, about how many white-tailed deer are in the county?
There are about 9840 white-tailed deer in Cass County.

Concept 2: Similar Figures

For Exercises 43–48, the pairs of triangles are similar. Solve for x and y. **(See Examples 5 and 6.)**

43.

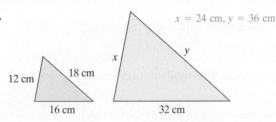

$x = 24$ cm, $y = 36$ cm

44.

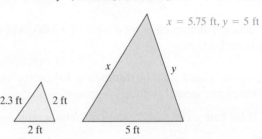

$x = 5.75$ ft, $y = 5$ ft

45.

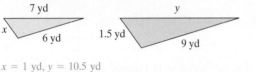

$x = 1$ yd, $y = 10.5$ yd

46.

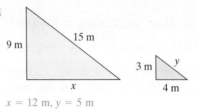

$x = 12$ m, $y = 5$ m

47.

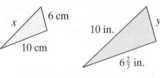

$x = 15$ cm, $y = 4$ in.

48.

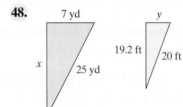

$x = 24$ yd, $y = 5.6$ ft

49. The height of a flagpole can be determined by comparing the shadow of the flagpole and the shadow of a yardstick. From the figure, determine the height of the flagpole. **(See Example 7.)**
The flagpole is 12 ft high.

50. A 15-ft flagpole casts a 4-ft shadow. How long will the shadow be for a 90-ft building?
The shadow will be 24 ft.

51. A person 1.6 m tall stands next to a lifeguard observation platform. If the person casts a shadow of 1 m and the lifeguard platform casts a shadow of 1.5 m, how high is the platform? The platform is 2.4 m high.

52. A 32-ft tree casts a shadow of 18 ft. How long will the shadow be for a 22-ft tree? The shadow will be 12.375 ft.

h

3 ft

$2\frac{1}{2}$ ft   10 ft

Figure for Exercise 49

For Exercises 53–56, the pairs of polygons are similar. Solve for the indicated variables. **(See Example 8.)**

53. 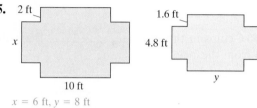 $x = 17.5$ in.

7 in.

x

4 in.   10 in.

54. 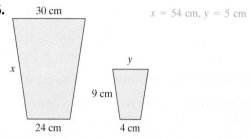 $x = 4$ m, $y = 5\frac{1}{3}$ m

y

4 m

8 m

6 m

3 m x

55.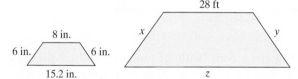

2 ft

x

10 ft

1.6 ft

4.8 ft

y

$x = 6$ ft, $y = 8$ ft

56. 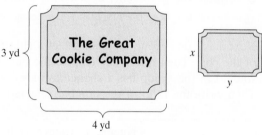 $x = 54$ cm, $y = 5$ cm

30 cm

x

24 cm

y

9 cm

4 cm

57. A carpenter makes a schematic drawing of a porch he plans to build. On the drawing, 2 in. represents 7 ft. Find the dimensions of the porch. $x = 21$ ft; $y = 21$ ft; $z = 53.2$ ft

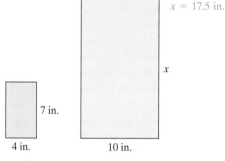

28 ft

8 in.

6 in.   6 in.

x y

15.2 in. z

58. The Great Cookie Company has a sign on the front of its store as shown in the figure. The company would like to put a sign of the same shape in the back, but with dimensions $\frac{1}{3}$ as large. Find the lengths denoted by x and y. $x = 1$ yd, $y = 1\frac{1}{3}$ yd

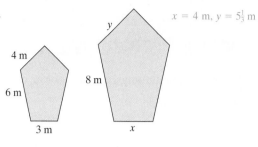

The Great Cookie Company

3 yd

4 yd

x

y

Calculator Connections

Topic: Solving Proportions

Calculator Exercises

59. In a recent year, the annual crime rate for Oklahoma was 4743 crimes per 100,000 people. If Oklahoma had approximately 3,500,000 people at that time, approximately how many crimes were committed? (*Source:* Oklahoma Department of Corrections)   There were approximately 166,005 crimes committed.

60. To measure the height of the Washington Monument, a student 5.5 ft tall measures his shadow to be 3.25 ft. At the same time of day, he measured the shadow of the Washington Monument to be 328 ft long. Estimate the height of the monument to the nearest foot.   The Washington Monument is approximately 555 ft tall.

61. In a recent year, the rate of breast cancer in women was 110 cases per 100,000 women. At that time the state of California had approximately 14,000,000 women. How many women in California would be expected to have breast cancer? (*Source:* Centers for Disease Control)   Approximately 15,400 women would be expected to have breast cancer.

62. In a recent year, the rate of prostate disease in U.S. men was 118 cases per 1000 men. At that time the state of Massachusetts had approximately 2,500,000 men. How many men in Massachusetts would be expected to have prostate disease? (*Source:* National Center for Health Statistics)   Approximately 295,000 men would be expected to have prostate disease.

 Writing   Translating Expression   Geometry   Scientific Calculator   Video   NSE

Group Activity

Investigating Probability

Materials: Paper bags containing 10 white poker chips, 6 red poker chips, and 4 blue poker chips.

Estimated time: 15 minutes

Group Size: 3

1. Each group will receive a bag of poker chips, with 10 white, 6 red, and 4 blue chips.

2. **a.** Write the ratio of red chips in the bag to the total number of chips in the bag. _____ 6:20 or $\frac{6}{20}$ _____ This value represents the *probability* of randomly selecting a red chip from the bag.

   b. Write this fraction in decimal form. _____ 0.30 _____

   c. Write the decimal from step (b) as a percent. _____ 30% _____
   A probability value indicates the likeliness of an event to occur. For example, to interpret this probability, one might say that there is a 30% chance of selecting a red chip at random from the bag.

3. Determine the probability of selecting a white chip from the bag. Interpret your answer.
   $\frac{10}{20}$ or 0.50 or 50%; there is a 50% chance of selecting a white chip.

4. Determine the probability of selecting a blue chip from the bag. Interpret your answer.
   $\frac{4}{20}$ or 0.20 or 20%; there is a 20% chance of selecting a blue chip.

5. Next, have one group member select a chip from the bag at random (without looking), and record the color of the chip. Then return the chip to the bag. Repeat this step for a total of 20 times (be sure that each student in the group has a chance to pick). Record the total number of red, white, and blue chips selected. Then write the ratio of the number selected out of 20.   Answers will vary.

	Number of Times Selected	Ratio of Number Selected Out of 20
Red		
White		
Blue		

6. How well do your experimental results match the theoretical probabilities found in steps 2–4?

7. The instructor will now pool the data from the whole class. Write the total number of times that red, white, and blue chips were selected, respectively, by the whole class. Then write the ratio of these values to the number of total selections made.

8. How well do the experimental results from the whole class match the theoretical probabilities of selecting a red, white, or blue chip?

Chapter 5 Summary

Section 5.1 Ratios

Key Concepts

A **ratio** is a comparison of two quantities.

The ratio of a to b can be written as follows, provided $b \neq 0$.

  1. a to b 2. $a : b$ 3. $\dfrac{a}{b}$

When we write a ratio in fraction form, we generally simplify it to lowest terms.

Ratios that contain mixed numbers, fractions, or decimals can be simplified to lowest terms with whole numbers in the numerator and denominator.

Examples

Example 1

Three forms of a ratio:

4 to 6   4 : 6 $\dfrac{4}{6}$

Example 2

A hockey team won 4 games out of 6. Write a ratio of games won to total games played and simplify to lowest terms.

$$\frac{4 \text{ games won}}{6 \text{ games played}} = \frac{2}{3}$$

Example 3

$$\frac{2\frac{1}{6}}{\frac{2}{3}} = 2\frac{1}{6} \div \frac{2}{3} = \frac{13}{\overset{2}{\cancel{6}}} \cdot \frac{\overset{1}{\cancel{3}}}{2} = \frac{13}{4}$$

Example 4

$$\frac{2.1}{2.8} = \frac{2.1}{2.8} \cdot \frac{10}{10} = \frac{21}{28} = \frac{\overset{3}{\cancel{21}}}{\underset{4}{\cancel{28}}} = \frac{3}{4}$$

Section 5.2 Rates and Unit Cost

Key Concepts

A **rate** compares two different quantities.

Examples

Example 1

New Jersey has 8,470,000 people living in 21 counties. Write a reduced ratio of people per county.

$$\frac{8,470,000 \text{ people}}{21 \text{ counties}} = \frac{1,210,000 \text{ people}}{3 \text{ counties}}$$

A rate having a denominator of 1 unit is called a **unit rate**. To find a unit rate, divide the numerator by the denominator.

A **unit cost** or unit price is the cost per 1 unit, for example, $1.21/lb or 43¢/oz. Comparing unit prices can help determine the best buy.

Example 2

If a race car traveled 1250 mi in 8 hr during a race, what is its speed in miles per hour?

$$\frac{1250 \text{ mi}}{8 \text{ hr}} = 156.25 \text{ mi/hr}$$

Example 3

Tide laundry detergent is offered in two sizes: $20.99 for 150 oz and $15.79 for 100 oz. Find the unit prices to find the best buy.

$$\frac{\$20.99}{150 \text{ oz}} \approx \$0.140/\text{oz}$$

$$\frac{\$15.79}{100 \text{ oz}} = \$0.158/\text{oz}$$

The 150-oz package is the better buy because the unit cost is less.

Section 5.3   Proportions

Key Concepts

A statement indicating that two quantities are equal is called an **equation**.

A **proportion** states that two ratios or rates are equal.

$\frac{14}{21} = \frac{2}{3}$ is a proportion.

To determine if two ratios form a proportion, check to see if the cross products are equal, that is,

$$\frac{a}{b} = \frac{c}{d} \quad \text{implies} \quad a \cdot d = b \cdot c \quad \text{(and vice versa)}$$

Examples

Example 1

Write as a proportion.

56 mi is to 2 gal as 84 mi is to 3 gal.

$$\frac{56 \text{ mi}}{2 \text{ gal}} = \frac{84 \text{ mi}}{3 \text{ gal}}$$

Example 2

$$\frac{3}{8} \overset{?}{=} \frac{2\frac{1}{2}}{6\frac{2}{3}}$$

$$3 \cdot 6\frac{2}{3} \overset{?}{=} 8 \cdot 2\frac{1}{2}$$

$$\frac{3}{1} \cdot \frac{20}{3} \overset{?}{=} \frac{8}{1} \cdot \frac{5}{2}$$

$$20 = 20 \checkmark$$

The ratios form a proportion.

To solve a proportion, solve the equation formed by the cross products.

Example 3

$$\frac{5}{4} = \frac{18}{x} \quad \Rightarrow \quad 5 \cdot x = 4 \cdot 18$$

$$5x = 72$$

$$\frac{\overset{1}{\cancel{5}}x}{\underset{1}{\cancel{5}}} = \frac{72}{5}$$

$$x = \frac{72}{5} \text{ or } 14\frac{2}{5} \text{ or } 14.4$$

Section 5.4 Applications of Proportions and Similar Figures

Key Concepts and Examples

Example 1 demonstrates an application involving a proportion. Example 2 demonstrates the use of proportions involving similar triangles.

Example 1

According to the National Highway Traffic Safety Administration, 2 out of 5 traffic fatalities involve the use of alcohol. If there were 43,200 traffic fatalities in a recent year, how many involved the use of alcohol?

Let n represent the number of traffic fatalities involving alcohol.

Set up a proportion:

$$\frac{2 \text{ traffic fatalities w/alcohol}}{5 \text{ traffic fatalities}} = \frac{n}{43,200}$$

Solve the proportion:

$$2(43,200) = 5n$$

$$86,400 = 5n$$

$$\frac{86,400}{5} = \frac{\overset{1}{\cancel{5}}n}{\underset{1}{\cancel{5}}}$$

$$17,280 = n$$

17,280 traffic fatalities involved alcohol.

Examples

Example 2

On a sunny day, a tree 6-ft tall casts a 4-ft shadow. At the same time a telephone pole casts a 10-ft shadow. What is the height of the telephone pole?

A picture illustrates the situation. Let h represent the height of the telephone pole.

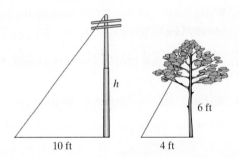

Set up a proportion:

$$\frac{10}{4} = \frac{h}{6}$$

Solve the proportion:

$$(10)(6) = 4h$$

$$60 = 4h$$

$$\frac{60}{4} = \frac{\overset{1}{\cancel{4}}h}{\underset{1}{\cancel{4}}}$$

$$15 = h \qquad \text{The pole is 15 ft high.}$$

Chapter 5   Review Exercises

Section 5.1

For Exercises 1–3, write the ratios in two other ways.

1. 5 : 4

5 to 4 and $\frac{5}{4}$

2. 3 to 1

3 : 1 and $\frac{3}{1}$

3. $\frac{8}{7}$

8 : 7 and 8 to 7

For Exercises 4–6, write the ratios in fraction form.

4. Saul had 3 daughters and 2 sons.

  a. Write a ratio of the number of sons to daughters. $\frac{2}{3}$

  b. Write a ratio of the number of daughters to sons. $\frac{3}{2}$

  c. Write a ratio of the number of daughters to the total number of children. $\frac{3}{5}$

5. In his refrigerator, Jonathan has 4 bottles of soda and 5 bottles of juice.

  a. Write a ratio of bottles of soda to bottles of juice. $\frac{4}{5}$

  b. Write a ratio of bottles of juice to bottles of soda. $\frac{5}{4}$

  c. Write a ratio of bottles of juice to the total number of bottles. $\frac{5}{9}$

6. There are 12 face cards in a regular deck of 52 cards.

  a. Write a ratio of face cards to total cards. $\frac{12}{52}$

  b. Write a ratio of face cards to cards that are not face cards. $\frac{12}{40}$

For Exercises 7–10, write the ratio in lowest terms.

7. 52 cards to 13 cards $\frac{4}{1}$ **8.** \$21 to \$15 $\frac{7}{5}$

9. 80 ft to 200 ft $\frac{2}{5}$ **10.** 7 days to 28 days $\frac{1}{4}$

For Exercises 11–14, write the ratio in lowest terms with whole numbers in the numerator and denominator.

11. $1\frac{1}{2}$ hr to $\frac{1}{3}$ hr $\frac{9}{2}$ **12.** $\frac{2}{3}$ yd to $2\frac{1}{6}$ yd $\frac{4}{13}$

13. \$2.56 to \$1.92 $\frac{4}{3}$ **14.** 42.5 mi to 3.25 mi

$\frac{170}{13}$

15. This year a high school had an increase of 320 students. The enrollment last year was 1200 students.

  a. How many students will be attending this year? This year's enrollment is 1520 students.

  b. Write a ratio of the increase in students to the total enrollment of students this year. Simplify to lowest terms. $\frac{4}{19}$

16. A living room has dimensions of 3.8 m by 2.4 m. Find the ratio of length to width and reduce to lowest terms. $\frac{19}{12}$

For Exercises 17 and 18, refer to the table that shows the number of personnel who smoke in a particular workplace.

	Smokers	Nonsmokers	Totals
Office personnel	12	20	32
Shop personnel	60	55	115

17. Find the ratio of office personnel who smoke to shop personnel who smoke. $\frac{1}{5}$

18. Find the ratio of the total number of personnel who smoke to the total number of personnel. $\frac{24}{49}$

Section 5.2

For Exercises 19–22, write each rate in lowest terms.

19. A concession stand sold 20 hot dogs in 45 min. $\frac{4 \text{ hot dogs}}{9 \text{ min}}$

20. Mike can skate 4 mi in 34 min. $\frac{2 \text{ mi}}{17 \text{ min}}$

21. The CN Tower in Toronto, Canada, weighs 130,000 tons for a height of approximately 1800 ft. $\frac{650 \text{ tons}}{9 \text{ ft}}$

22. Interpol reports that Denmark has a crime rate of 9460 per 100,000 people. $\frac{473 \text{ crimes}}{5000 \text{ people}}$

23. What is the difference between rates in lowest terms and unit rates? All unit rates have a denominator of 1, and reduced rates may not.

For Exercises 24–27, write each rate as a unit rate.

24. A pheasant can fly 44 mi in $1\frac{1}{3}$ hr. 33 mph

25. The temperature dropped 14° in 3.5 hr. 4° per hour

26. A hummingbird can flap its wings 2700 times in 30 sec.   90 times/sec

27. It takes David's lawn company 66 min to cut 6 lawns.   11 min/lawn

For Exercises 28 and 29, find the unit costs. Round the answers to three decimal places when necessary.

28. Body lotion costs $5.99 for 10 oz. $0.599 per ounce

29. Three towels cost $20.00. $6.667 per towel

For Exercises 30 and 31, compute the unit cost (round to three decimal places). Then determine the best buy.

30. a. 32 oz of detergent for $8.39
    $0.262/oz
   b. 45 oz of detergent for $12.59
    $0.280/oz; The 32-oz bottle is the better buy.

31. a. 24 oz of spaghetti sauce for $4.19
    $0.175/oz
   b. 44 oz of spaghetti sauce for $6.99
    $0.159/oz; The 44-oz jar is the better buy.

32. Suntan lotion costs $5.99 for 8 oz. If Jody has a coupon for $2.00 off, what will be the unit cost of the lotion after the coupon has been applied?
    $0.499 per ounce

33. A 6-roll pack of bathroom tissue costs $7.45 without a discount card. The package is advertised at 99¢ per roll if the buyer uses the discount card. What is the difference in price per roll when the buyer uses the discount card? Round to the nearest cent.
    The difference is about 25¢ per roll or $0.25 per roll.

34. In Wilmington, North Carolina, Hurricane Floyd dropped 15.06 in. of rain during a 24-hr period. What was the average rainfall per hour? (*Source:* National Weather Service)   0.6275 in./hr

35. For a recent year, Toyota steadily increased the number of hybrid vehicles for sale in the United States from 130,000 to 250,000.

   a. What was the increase in the number of hybrid vehicles?
    There was in increase of 120,000 hybrid vehicles.
   b. How many additional hybrid vehicles will be available each month?
    There will be 10,000 additional hybrid vehicles per month.

36. In 1990, Americans ate on average 386 lb of vegetables per year. By 2008, this value increased to 449 lb.

   a. What was the increase in the number of pounds of vegetables?
    There was an increase of 63 lb.
   b. How many additional pounds of vegetables did Americans add to their diet per year?
    Americans increased the amount of vegetables in their diet by 3.5 lb per year.

Section 5.3

For Exercises 37–42, write a proportion for each statement.

37. 16 is to 14 as 12 is to $10\frac{1}{2}$. $\dfrac{16}{14} = \dfrac{12}{10\frac{1}{2}}$

38. 8 is to 20 as 6 is to 15. $\dfrac{8}{20} = \dfrac{6}{15}$

39. The numbers 5 and 3 are proportional to the numbers 10 and 6. $\dfrac{5}{3} = \dfrac{10}{6}$

40. The numbers 4 and 3 are proportional to the numbers 20 and 15. $\dfrac{4}{3} = \dfrac{20}{15}$

41. $11 is to 1 hr as $88 is to 8 hr. $\dfrac{\$11}{1\,\text{hr}} = \dfrac{\$88}{8\,\text{hr}}$

42. 2 in. is to 5 mi as 6 in. is to 15 mi. $\dfrac{2\,\text{in.}}{5\,\text{mi}} = \dfrac{6\,\text{in.}}{15\,\text{mi}}$

For Exercises 43–46, determine whether the ratios form a proportion.

43. $\dfrac{64}{81} \overset{?}{=} \dfrac{8}{9}$ No

44. $\dfrac{3\frac{1}{2}}{7} \overset{?}{=} \dfrac{7}{14}$ Yes

45. $\dfrac{5.2}{3} \overset{?}{=} \dfrac{15.6}{9}$ Yes

46. $\dfrac{6}{10} \overset{?}{=} \dfrac{6.3}{10.3}$ No

For Exercises 47–50, determine whether the pairs of numbers are proportional.

47. Are the numbers $2\frac{1}{8}$ and $4\frac{3}{4}$ proportional to the numbers $3\frac{2}{5}$ and $7\frac{3}{5}$?   Yes

48. Are the numbers $5\frac{1}{2}$ and 6 proportional to the numbers $6\frac{1}{2}$ and 7?   No

49. Are the numbers 4.25 and 8 proportional to the numbers 5.25 and 10?   No

50. Are the numbers 12.4 and 9.2 proportional to the numbers 3.1 and 2.3?   Yes

For Exercises 51–56, solve the proportion.

51. $\dfrac{100}{16} = \dfrac{25}{x}$
    $x = 4$

52. $\dfrac{y}{6} = \dfrac{45}{10}$
    $y = 27$

53. $\dfrac{1\frac{6}{7}}{b} = \dfrac{13}{21}$
    $b = 3$

54. $\dfrac{p}{6\frac{1}{3}} = \dfrac{3}{9\frac{1}{2}}$
    $p = 2$

55. $\dfrac{2.5}{6.8} = \dfrac{5}{h}$
    $h = 13.6$

56. $\dfrac{0.3}{1.2} = \dfrac{k}{3.6}$
    $k = 0.9$

Section 5.4

57. One year of a dog's life is about the same as 7 years of a human life. If a dog is 12 years old in dog years, how does that equate to human years?
    The human equivalent is 84 years.

Writing   Translating Expression   Geometry   Scientific Calculator   Video   NSE

58. Lavu bought 9500 Japanese yen with $100 American. At this rate, how many yen can he buy with $450 American?   Lavu can buy 42,750 yen.

59. Recently, the number of births in Alabama was approximately 59,800. If the birthrate was about 13 per 1000, what was the approximate population of Alabama? (Round to the nearest person.)
Alabama had approximately 4,600,000 people.

60. If the tax on a $25.00 item is $1.20, what would be the tax on an item costing $145.00?
The tax would be $6.96.

61. The triangles shown in the figure are similar. Solve for x and y.
$x = 10$ in., $y = 62.1$ in.

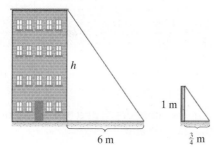

62. The height of a building can be approximated by comparing the shadows of the building and of a meterstick. From the figure, find the height of the building.   The building is 8 m high.

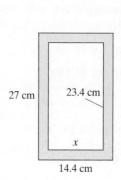

63. The polygons shown in the figure are similar. Solve for x and y. $x = 1.6$ yd, $y = 1.8$ yd

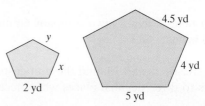

64. The figure shows two picture frames that are similar. Use this information to solve for x and y.
$x = 10.8$ cm, $y = 30$ cm

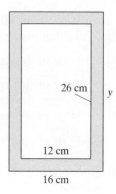

Chapter 5   Test

1. An elementary school has 25 teachers and 521 students. Write a ratio of teachers to students in three different ways.
25 to 521, 25 : 521, $\frac{25}{521}$

2. In a marina, there were 17 sailboats out of a total of 23 boats.

a. Write a ratio of sailboats to total boats. $\frac{17}{23}$

b. Write a ratio of sailboats to boats that are not sailboats.
$\frac{17}{6}$

For Exercises 3 and 4, write as a reduced ratio in fraction form.

3. For the 2010 WNBA season, the Atlanta Dream had a win-loss ratio of 4 to 30. $\frac{2}{15}$

4. For the 2010 WNBA season, the Detroit Shock had a win-loss ratio of 22 to 12. $\frac{11}{6}$

5. Find the ratio of the shortest side to the longest side. Write the ratio in lowest terms. $\frac{5}{8}$

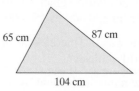

6. a. In a recent year, the number of people in New Mexico whose income was below poverty level was 168 out of every 1000. Write this as a simplified ratio. $\frac{21}{125}$

b. The poverty level in Iowa was 72 people to 1000 people. Write this as a simplified ratio. $\frac{9}{125}$

c. Compare the ratios and comment.
The poverty ratio was greater in New Mexico.

 Writing    Translating Expression   Geometry   Scientific Calculator   Video   NSE

7. Write as a simplified ratio in two ways: 30 sec to $1\frac{1}{2}$ min.

   a. By converting 30 sec to minutes. $\dfrac{\frac{1}{2}}{1\frac{1}{2}} = \dfrac{1}{3}$

   b. By converting $1\frac{1}{2}$ min to seconds. $\dfrac{30}{90} = \dfrac{1}{3}$

For Exercises 8–10, write as a rate, simplified to lowest terms.

8. 255 mi per 6 hr $\dfrac{85\text{ mi}}{2\text{ hr}}$

9. 20 lb in 6 weeks $\dfrac{10\text{ lb}}{3\text{ weeks}}$

10. 4 g of fat in 8 cookies. $\dfrac{1\text{ g}}{2\text{ cookies}}$

For Exercises 11 and 12, write as a unit rate. Round to the nearest hundredth.

11. The element platinum had density of 2145 g per 100 cm^3.   21.45 g/cm^3

12. There are approximately 104.8 oz of iron in 45.8 lb of rocks brought back from the moon. 2.29 oz/lb

13. What is the unit cost for Raid Ant and Roach spray valued at $6.72 for 30 oz? Round to the nearest cent. $0.22 per ounce

14. A package containing 3 toe rings is on sale for 2 packs for $3.00. What is the cost of 1 toe ring? $0.50 per ring

15. A generic pain reliever is on sale for $10.99 for 250 caplets. Advil pain reliever is sold in 24-capsule bottles for $4.29. Find the unit cost of each to determine the better buy. Generic: $0.044 per caplet; Advil: $0.179 per capsule. The generic pain reliever is the better buy.

16. What does it mean for two pairs of numbers to be proportional? They form equal ratios or rates.

For Exercises 17–20, write a proportion for each statement.

17. 42 is to 15 as 28 is to 10. $\dfrac{42}{15} = \dfrac{28}{10}$

18. 20 pages is to 12 min as 30 pages is to 18 min. $\dfrac{20\text{ pages}}{12\text{ min}} = \dfrac{30\text{ pages}}{18\text{ min}}$

19. $15 an hour is proportional to $75 for 5 hr. $\dfrac{\$15}{1\text{ hr}} = \dfrac{\$75}{5\text{ hr}}$

20. Are the numbers 105 and 55 proportional to the numbers 21 and 10?   No

For Exercises 21–24, solve the proportion.

21. $\dfrac{25}{p} = \dfrac{45}{63}$ $p = 35$

22. $\dfrac{32}{20} = \dfrac{20}{x}$ $x = 12.5$

23. $\dfrac{n}{9} = \dfrac{3\frac{1}{3}}{6}$ $n = 5$

24. $\dfrac{y}{14} = \dfrac{7.2}{16.8}$ $y = 6$

25. A computer on dial-up can download 1.6 megabytes (MB) in 2.5 min. How long will it take to download a 4.8-MB file?   It will take 7.5 min.

26. Cherise is an excellent student and studies 7.5 hr outside of class each week for a 3-credit-hour math class. At this rate, how many hours outside of class does she spend on homework if she is taking 12 credit-hours at school?   Cherise spends 30 hr each week on homework outside of class.

27. Ms. Ehrlich wants to approximate the number of goldfish in her backyard pond. She scooped out 8 and marked them. Later she scooped out 10 and found that 3 were marked. Estimate the number of goldfish in her pond. Round to the nearest whole unit.
There are approximately 27 goldfish in her pond.

28. Given that the two triangles are similar, solve for x and y.
$x = 1\frac{1}{2}$ mi,
$y = 8$ mi

29. Maggie takes a brochure that measures 10 cm by 15 cm and enlarges it on a copy machine. If she wants the height to be to be 24 cm, how wide will the new image be?
16 cm

15 cm

10 cm

24 cm

x

Chapters 1–5 Cumulative Review Exercises

1. Write 503,042 in words.
Five hundred three thousand, forty-two

2. Estimate the sum by first rounding the numbers to the nearest hundred.   Approximately 1400

$$251 + 492 + 631$$

3. Multiply. $226 \times 100{,}000$ 22,600,000

4. Divide and write the answer with a remainder.
$355 \div 16$ 22 R 3

5. Divide and write the quotient in decimal form.
$355 \div 16$
22.1875

6. Simplify. $2^2 \times (32 - 11) \div 14$ 6

7. Shade the rectangle to represent the fraction $\frac{3}{7}$.

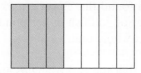

8. Simplify. $\dfrac{245}{175}$ $\dfrac{7}{5}$

9. Multiply. $\dfrac{13}{2} \cdot \dfrac{3}{7}$ $\dfrac{39}{14}$

10. Simplify. $\left(\dfrac{3}{5}\right)^2$ $\dfrac{9}{25}$

11. Bruce decides to share his 6-in. sub sandwich with his friend. If he gives $\frac{1}{4}$ of it to Dennis, how much of the sandwich does he have left?
Bruce has $4\frac{1}{2}$ in. of sandwich left.

12. Simplify. $\dfrac{7}{8} \div \dfrac{3}{4} + \dfrac{5}{6}$ 2

13. Add. $\dfrac{8}{9} + 3$ $\dfrac{35}{9}$ **14.** Subtract. $\dfrac{9}{13} - \dfrac{0}{3}$
$\dfrac{9}{13}$

 15. Emil wants to put a wallpaper border in his bathroom. The border will be on three walls (not in the shower) and not over the door. From the figure, determine how much wallpaper border will be needed.
Emil needs $13\frac{1}{12}$ ft of wallpaper border.

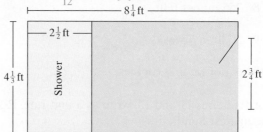

16. Tomoka Consolidated bought $82\frac{1}{4}$ acres of land. If it sold $\frac{3}{4}$ of the land, how much was sold? How much was left?
It sold $61\frac{11}{16}$ acres, and $20\frac{9}{16}$ acres were left.

17. How many ninths are in $6\dfrac{5}{9}$?
There are 59 ninths.

18. Write 1004.701 in words.
One thousand four and seven hundred one thousandths

 19. Add and subtract. $23.88 + 11.3 - 7.123$
28.057

20. Write 4.36 as an improper fraction in lowest terms. $\dfrac{109}{25}$

21. Multiply. 43.923×100
4392.3

22. Divide. $237.9 \div 100$
2.379

 23. Find the perimeter of the figure. Write the answer in decimal form.
130.9 cm

$29\frac{1}{5}$ cm
$26\frac{1}{4}$ cm
10.75 cm
34.2 cm
$30\frac{1}{2}$ cm

24. Americans buy 61 million newspapers each day and throw out 44 million. Write the ratio 61 to 44 in two other ways. $\frac{61}{44}$ or $61:44$

25. For a recent year at Southeastern Community College in North Carolina, there were 1950 students and 150 faculty. Write the student-to-faculty ratio in lowest terms. $\frac{13}{1}$

26. In a recent study of 6000 deaths, 840 were due to cancer. Compute the ratio of deaths due to cancer to the total number of deaths studied. Simplify the ratio and interpret the answer in words. $\frac{7}{50}$; Approximately 7 out of 50 deaths are due to cancer.

27. Oregon has a land area of approximately 9600 mi^2. The population of Oregon is approximately 1,200,000. Compute the population density (recall that population density is a unit rate given by the number of people per square mile). 125 people/mi^2

28. Determine whether the ratios form a proportion.

    a. $\frac{7.5}{10} \stackrel{?}{=} \frac{9}{12}$ Yes   b. $\frac{31}{5} \stackrel{?}{=} \frac{33}{6}$ No

29. Solve the proportion. $\frac{13}{11.7} = \frac{5}{x}$ $x = 4.5$

30. Jim can drive 150 mi on 6 gal of gas. At this rate, how far can he travel on 4 gal?
    Jim can drive 100 mi on 4 gal.

Percents

6

Chapter 6

In this chapter, we present the concept of percent. Percents are used to measure the number of parts per hundred of some whole amount. As a consumer, it is important to have a working knowledge of percents.

Review Your Skills

One way of solving percent equations is by using proportions. To complete the puzzle, first solve the proportions and place the answer in the appropriate circle. A circle from the left column is paired with a circle in the right column. Fill in the circles in each column. For each number in the left column, subtract 4 to determine its partner on the right. For each number in the right column, add 4 to determine its partner on the left. Also note that the sum of the numbers in the first column is 63.

1. $\dfrac{5}{18} = \dfrac{a}{36}$

2. $\dfrac{b}{7} = \dfrac{66}{21}$

3. $\dfrac{48}{20} = \dfrac{12}{c}$

4. $\dfrac{4}{d} = \dfrac{20}{55}$

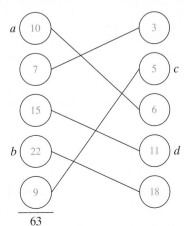

Section 6.1   Percents and Their Fraction and Decimal Forms

Concepts

1. Definition of Percent
2. Converting Percents to Fractions
3. Converting Percents to Decimals
4. Common Percents and Their Fraction and Decimal Forms

1. Definition of Percent

In this chapter, we study the concept of percent. Literally, the word **percent** means *per one hundred*. To indicate percent, we use the percent symbol %. For example, 45% (read as "45 percent") of the land area in South America is rainforest (shaded in green). This means that if South America were divided into 100 squares of equal size, 45 of the 100 squares would cover rainforest. See Figures 6-1 and 6-2.

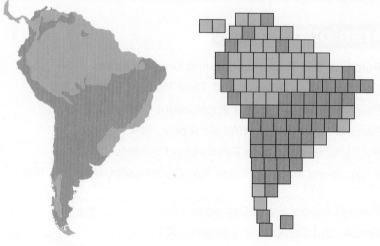

Figure 6-1   Figure 6-2

Instructor Note: To lead into the next concept, ask students what fraction expresses "45 out of 100 squares."

Consider another example. For a recent year, the population of Virginia could be described as follows.

21%	African American	21 out of 100 Virginians are African American
72%	Caucasian (non-Hispanic)	72 out of 100 Virginians are Caucasian (non-Hispanic)
3%	Asian American	3 out of 100 Virginians are Asian American
3%	Hispanic	3 out of 100 Virginians are Hispanic
1%	Other	1 out of 100 Virginians have other backgrounds

Figure 6-3 represents a sample of 100 residents of Virginia.

Concept Connections

1. Shade the portion of the figure represented by 18%.

Answer

1.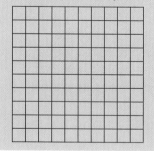

AA	AA	AA	C	C	C	C	C	C	C
AA	AA	C	C	C	C	C	C	C	C
AA	AA	C	C	C	C	C	C	C	C
AA	AA	C	C	C	C	C	C	C	H
AA	AA	C	C	C	C	C	C	C	H
AA	AA	C	C	C	C	C	C	C	H
AA	AA	C	C	C	C	C	C	C	A
AA	AA	C	C	C	C	C	C	C	A
AA	AA	C	C	C	C	C	C	C	A
AA	AA	C	C	C	C	C	C	C	O

AA African American
C   Caucasian (non-Hispanic)
A   Asian American
H   Hispanic
O   Other

Figure 6-3

2. Converting Percents to Fractions

By definition, a percent represents a ratio of parts per 100. Therefore, we can write percents as fractions.

Percent		Fraction	Example/Interpretation
7%	=	$\dfrac{7}{100}$	A sales tax of 7% means that 7 cents in tax is charged for every 100 cents spent.
39%	=	$\dfrac{39}{100}$	To say that 39% of households own a cat means that 39 per every 100 households own a cat.

Notice that $39\% = \dfrac{39}{100} = 39 \times \dfrac{1}{100} = 39 \div 100.$

From this discussion we have the following rule for converting percents to fractions.

Converting Percents to Fractions

Step 1   Replace the symbol % by $\times \frac{1}{100}$ (or by $\div 100$).
Step 2   Simplify the fraction to lowest terms, if possible.

Example 1   Converting Percents to Fractions

Convert each percent to a fraction.

a. 56%   **b.** 60%   **c.** 125%   **d.** 0.4%

Solution:

a. $56\% = 56 \times \dfrac{1}{100}$ Replace the % symbol by $\times \dfrac{1}{100}$.

$= \dfrac{56}{100}$ Multiply.

$= \dfrac{14}{25}$ Simplify to lowest terms.

b. $60\% = 60 \times \dfrac{1}{100}$ Replace the % symbol by $\times \dfrac{1}{100}$.

$= \dfrac{60}{100}$ Multiply.

$= \dfrac{3}{5}$ Simplify to lowest terms.

c. $125\% = 125 \times \dfrac{1}{100}$ Replace the % symbol by $\times \dfrac{1}{100}$.

$= \dfrac{125}{100}$ Multiply.

$= \dfrac{5}{4}$ or $1\dfrac{1}{4}$ Simplify to lowest terms.

Skill Practice

Convert each percent to a fraction.
2. 32%   **3.** 90%
4. 175%   **5.** 0.06%

Classroom Examples: pp. 343–344,
Exercises 18, 24, and 28

Answers

2. $\dfrac{8}{25}$ **3.** $\dfrac{9}{10}$

4. $\dfrac{7}{4}$ **5.** $\dfrac{3}{5000}$

Instructor Note: Convert 0.4%, 4%, 40%, and 400% to fractions to show students that they differ in value.

d. $0.4\% = 0.4 \times \dfrac{1}{100}$ Replace the % symbol by $\times \dfrac{1}{100}$.

$= \dfrac{4}{10} \times \dfrac{1}{100}$ Write 0.4 in fraction form.

$= \dfrac{4}{1000}$ Multiply.

$= \dfrac{1}{250}$ Simplify to lowest terms.

Note that $100\% = 100 \times \frac{1}{100} = 1$. That is, 100% represents 1 whole unit. In Example 1(c), $125\% = \frac{5}{4}$ or $1\frac{1}{4}$. This illustrates that any percent greater than 100% represents a quantity greater than 1 whole. Therefore, its fractional form may be expressed as an improper fraction or as a mixed number.

Note that $1\% = 1 \times \frac{1}{100} = \frac{1}{100}$. In Example 1(d), the value 0.4% represents a quantity less than 1%. Its fractional form is less than one-hundredth.

In Example 2 we convert some common percents to fraction form.

Concept Connections

Determine whether the percent represents a quantity greater than or less than 1 whole.

6. 1.92%

7. 19.2%

8. 192%

Skill Practice

Convert the percents to fractions.

9. 75%

10. 50%

11. $66\dfrac{2}{3}\%$

Classroom Example: p. 344, Exercise 32

Example 2 Converting Percents to Fractions

Convert the percents to fractions.

a. 25%   **b.** 10%   **c.** $33\dfrac{1}{3}\%$

Solution:

a. $25\% = 25 \times \dfrac{1}{100} = \dfrac{25}{100} = \dfrac{1}{4}$ Thus, 25% represents one-quarter of a whole.

b. $10\% = 10 \times \dfrac{1}{100} = \dfrac{10}{100} = \dfrac{1}{10}$ Thus, 10% represents one-tenth of a whole.

c. $33\dfrac{1}{3}\% = 33\dfrac{1}{3} \times \dfrac{1}{100}$ Replace the % symbol by $\times \dfrac{1}{100}$.

$= \dfrac{100}{3} \times \dfrac{1}{100}$ Convert the mixed number to an improper fraction.

$= \dfrac{\overset{1}{\cancel{100}}}{3} \times \dfrac{1}{\underset{1}{\cancel{100}}}$ Simplify common factors.

$= \dfrac{1}{3}$ Thus, $33\dfrac{1}{3}\%$ represents one-third of a whole.

Example 3 Converting Percents to Fractions

Find the fraction form for the percent given in the sentence.

a. Forty-five percent of Americans use the Internet as a resource when planning vacations. (*Source: USA TODAY*)

b. 7.2% of adults suffer from asthma. (*Source:* National Center for Health Statistics)

Answers

6. Less than   **7.** Less than

8. Greater than   **9.** $\dfrac{3}{4}$

10. $\dfrac{1}{2}$ **11.** $\dfrac{2}{3}$

Solution:

a. $45\% = 45 \times \dfrac{1}{100} = \dfrac{45}{100} = \dfrac{9}{20}$

Just under one-half of Americans planning a vacation use the Internet as a resource.

b. $7.2\% = 7.2 \times \dfrac{1}{100} = \dfrac{72}{10} \times \dfrac{1}{100}$

Write 7.2 in fraction form: $\frac{72}{10}$.

$= \dfrac{72}{1000} = \dfrac{9}{125}$

Almost one-tenth of adults have asthma.

Skill Practice

Find the fraction form for the percent given in the sentence.

12. In Pennsylvania, 15% of the residents are 65 or older.

13. One study found that teenage substance abuse rises by 40% during the summer months.

Classroom Example: p. 345, Exercise 76

3. Converting Percents to Decimals

To express part of a whole unit, we can use a percent, a fraction, or a decimal. We would like to be able to convert from one form to another. The procedure for converting a percent to a decimal is the same as that for converting a percent to a fraction. We replace the % symbol by $\times \frac{1}{100}$. However, when converting to a decimal, it is usually more convenient to use the form $\times 0.01$.

Converting Percents to Decimals

Replace the % symbol by $\times 0.01$. (This is equivalent to $\times \frac{1}{100}$ and $\div 100$.)

Note: Multiplying a decimal by 0.01 (or dividing by 100) is the same as moving the decimal point 2 places to the left.

Example 4 Converting Percents to Decimals

Convert each percent to its decimal form.

a. 31%   **b.** 6.5%   **c.** 428%   **d.** $1\dfrac{3}{5}\%$ **e.** 0.05%

Skill Practice

Convert each percent to its decimal form.

14. 67%   **15.** 8.6%

16. 321%   **17.** $6\dfrac{1}{4}\%$

18. 0.7%

Classroom Examples: p. 344, Exercises 40, 46, and 48

Solution:

a. $31\% = 31 \times 0.01$

$= 0.31$

Replace the % symbol by $\times 0.01$.

Move the decimal point 2 places to the left.

b. $6.5\% = 6.5 \times 0.01$

$= 0.065$

Replace the % symbol by $\times 0.01$.

Move the decimal point 2 places to the left.

c. $428\% = 428 \times 0.01$

$= 4.28$

Because 428% is greater than 100% we expect the decimal form to be a number greater than 1.

d. $1\dfrac{3}{5}\% = 1.6 \times 0.01$

$= 0.016$

Convert the mixed number to decimal form.

Because the percent is just over 1%, we expect the decimal form to be just slightly greater than one-hundredth.

e. $0.05\% = 0.05 \times 0.01$

$= 0.0005$

The value 0.05% is less than 1%. We expect the decimal form to be less than one-hundredth.

Answers

12. $\dfrac{3}{20}$ **13.** $\dfrac{2}{5}$ **14.** 0.67

15. 0.086   **16.** 3.21   **17.** 0.0625

18. 0.007

Skill Practice

Find the decimal notation for the percent given in the sentence.
19. The U.S. unemployment rate in 2010 was 9.5%.
20. Satellite Internet subscribers increased by 220% in a 4-year period.

Classroom Example: p. 345, Exercise 72

Example 5 Converting Percents to Decimals

Find the decimal notation for the percent given in the sentence.

a. Recently, forty-eight percent of applicants to U.S. medical schools were female. (*Source:* Association of American Medical Colleges)

b. The price per gallon for regular unleaded gasoline in 2012 was 280% of what it was in 1984.

Solution:

a. $48\% = 48 \times 0.01 = 0.48$ — Just under one-half of the applicants were female.

b. $280\% = 280 \times 0.01 = 2.80$ — The cost of gas in 2012 was almost 3 times as great as in 1984.

Notice from Examples 1–5 that we perform the same procedure to convert a percent to either a decimal or a fraction. In each case we multiply by $\frac{1}{100}$. When converting to a decimal, it is usually easier to use the form $\times\, 0.01$. When converting to a fraction, it is usually easier to use the form $\times\, \frac{1}{100}$. In both cases, this operation is also equivalent to dividing by 100.

4. Common Percents and Their Fraction and Decimal Forms

Table 6-1 shows some common percents and their equivalent fraction and decimal forms.

Table 6-1

Percent	Fraction	Decimal	Example/Interpretation
100%	1	1.00	Of people who give birth, 100% are female.
50%	$\frac{1}{2}$	0.50	Of the population, 50% is male. That is, one-half of the population is male.
25%	$\frac{1}{4}$	0.25	Approximately 25% of the U.S. population smokes. That is, one-quarter of the population smokes.
75%	$\frac{3}{4}$	0.75	Approximately 75% of homes have computers. That is, three-quarters of homes have computers.
10%	$\frac{1}{10}$	0.10	Of the population, 10% is left-handed. That is, one-tenth of the population is left-handed.
1%	$\frac{1}{100}$	0.01	Approximately 1% of babies are born underweight. That is, about 1 in 100 babies is born underweight.
$33\frac{1}{3}\%$	$\frac{1}{3}$	$0.\overline{3}$	A basketball player made $33\frac{1}{3}\%$ of her shots. That is, she made about 1 basket for every 3 shots attempted.
$66\frac{2}{3}\%$	$\frac{2}{3}$	$0.\overline{6}$	Of the population, $66\frac{2}{3}\%$ prefers chocolate ice cream to other flavors. That is, 2 out of 3 people prefer chocolate ice cream.

Answers
19. 0.095   **20.** 2.2

Section 6.1 Practice Exercises

For additional exercises, see Classroom Activities 6.1A–6.1C in the *Student's Resource Manual* at www.mhhe.com/moh.

Study Skills Exercise

> A test is a *grading* tool for your instructor. How can you turn it into a *learning* tool for you?

Vocabulary and Key Concepts

1. The word _____ means per one hundred.   percent

Concept 1: Definition of Percent

2. There are 100 students taking a math course. If 18% earn an A, how many students earned an A?   18

For Exercises 3–8, use a percent to express the shaded portion of each drawing.

3. 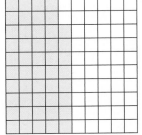   48%

4. 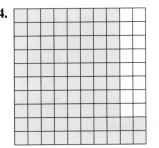   84%

5. 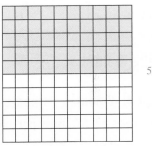   50%

6. 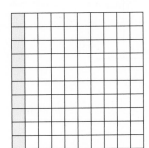   10%

7. 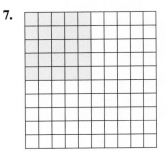   25%

8. 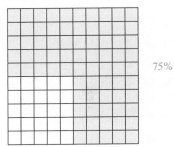   75%

For Exercises 9–12, write a percent for each statement.

9. A bank pays \$2 in interest for every \$100 deposited.   2%

10. In South Dakota, 5 out of every 100 people work in construction.   5%

11. Out of 100 acres, 70 acres were planted with corn.   70%

12. On TV, 26 out of every 100 minutes are filled with commercials.   26%

Concept 2: Converting Percents to Fractions

13. Explain the procedure to change a percent to a fraction.   Replace the symbol % by $\times \frac{1}{100}$ (or $\div$ 100). Then reduce the fraction to lowest terms.

14. What fraction represents 50%.   For example: $\frac{1}{2}$

For Exercises 15–34, change the percent to a simplified fraction or mixed number. **(See Examples 1 and 2.)**

15. 3% $\frac{3}{100}$

16. 7% $\frac{7}{100}$

17. 84% $\frac{21}{25}$

18. 32% $\frac{8}{25}$

19. 25% $\frac{1}{4}$

20. 20% $\frac{1}{5}$

21. 3.4% $\frac{17}{500}$

22. 5.2% $\frac{13}{250}$

23. 115% $\frac{23}{20}$ or $1\frac{3}{20}$ **24.** 150% $\frac{3}{2}$ or $1\frac{1}{2}$ **25.** 175% $\frac{7}{4}$ or $1\frac{3}{4}$ **26.** 120% $\frac{6}{5}$ or $1\frac{1}{5}$

27. 0.5% $\frac{1}{200}$ **28.** 0.2% $\frac{1}{500}$ **29.** 0.25% $\frac{1}{400}$ **30.** 0.75% $\frac{3}{400}$

 **31.** $66\frac{2}{3}$% $\frac{2}{3}$ **32.** $5\frac{1}{6}$% $\frac{31}{600}$ **33.** $24\frac{1}{2}$% $\frac{49}{200}$ **34.** $6\frac{1}{4}$% $\frac{1}{16}$

Concept 3: Converting Percents to Decimals

35. Explain the procedure to change a percent to a decimal.   Replace the % symbol by × 0.01 (or ÷ 100).

For Exercises 36–51, change the percent to a decimal. **(See Example 4.)**

36. 58%   0.58   **37.** 72%   0.72   **38.** 15%   0.15   **39.** 66%   0.66

40. 8.5%   0.085   **41.** 12.9%   0.129   **42.** 72.31%   0.7231   **43.** 41.05%   0.4105

44. 142%   1.42   **45.** 201%   2.01   **46.** 0.55%   0.0055   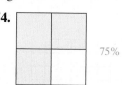 **47.** 0.75%   0.0075

48. $26\frac{2}{5}$%   0.264   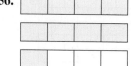 **49.** $16\frac{1}{4}$%   0.1625   **50.** $55\frac{1}{20}$%   0.5505   **51.** $62\frac{1}{5}$%   0.622

Concept 4: Common Percents and Their Fraction and Decimal Forms

For Exercises 52–57, use a percent to express the shaded portion of each drawing.

52.   50%

53.   25%

54.   75%

55.   100%

56.   225%

57.   150%

Mixed Exercises

For Exercises 58–63, match the percent with its fraction form.

58. $66\frac{2}{3}$%   c   **a.** $\frac{3}{2}$

59. 10%   d   **b.** $\frac{3}{4}$

60. 90%   e   **c.** $\frac{2}{3}$

61. 75%   b   **d.** $\frac{1}{10}$

62. 25%   f   **e.** $\frac{9}{10}$

63. 150%   a   **f.** $\frac{1}{4}$

For Exercises 64–69, match the percent with its decimal form.

64. 30%   e   **a.** 0.01

65. $33\frac{1}{3}$%   d   **b.** 0.50

66. 125%   f   **c.** 0.80

67. 50%   b   **d.** $0.\overline{3}$

68. 1%   a   **e.** 0.30

69. 80%   c   **f.** 1.25

70. In which direction do you move the decimal point when you convert a percent to a decimal? By how many places? *Move the decimal point 2 places to the left.*

For Exercises 71–78, find the decimal and fraction equivalent of the percent given in the sentence. **(See Examples 3 and 5.)**

71. Between 2000 and 2010 the population in California grew by 7.6%. $0.076; \frac{19}{250}$

72. Las Vegas is considered the fastest-growing city in the United States. Between 1990 and 2010 its population increased by 75%. $0.75; \frac{3}{4}$

73. For a recent year, the unemployment rate in Kansas was 4.3%. $0.043; \frac{43}{1000}$

74. For a recent year, the unemployment rate in the United States was 5.8%. $0.058; \frac{29}{500}$

75. From 2009 to 2010 there was a drop of 2% in electricity generated by nuclear power in the United States. (*Source:* Energy Information Administration) $0.02; \frac{1}{50}$

76. For a recent year the average U.S. income tax rate was 18.2%. $0.182; \frac{91}{500}$

77. Thirty-five percent of Americans say they entertain at home once or twice a year. (*Source: USA TODAY*) $0.35; \frac{7}{20}$

78. Twenty-nine percent of Americans say they entertain at home once a month. $0.29; \frac{29}{100}$

79. The graph represents the percent of dog owners who participate in certain activities to treat their dogs. Write the decimal and fraction forms of the percents given in the graph. (*Source:* American Animal Hospital Association)

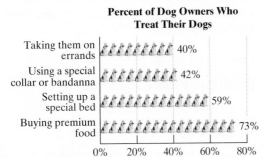

Percent of Dog Owners Who Treat Their Dogs

Taking them on errands — 40%
Using a special collar or bandanna — 42%
Setting up a special bed — 59%
Buying premium food — 73%

$40\% = 0.4$ or $\frac{2}{5}$; $42\% = 0.42$ or $\frac{21}{50}$; $59\% = 0.59$ or $\frac{59}{100}$; $73\% = 0.73$ or $\frac{73}{100}$

80. The graph represents the percent of people with at least a bachelor's degree for selected large cities. Write the decimal and fraction forms of the percents given in the graph. (*Source:* U.S. Census Bureau)

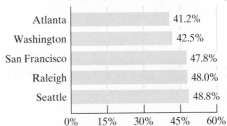

Large Cities with the Highest Percent of Residents with at Least a Bachelor's Degree

Atlanta — 41.2%
Washington — 42.5%
San Francisco — 47.8%
Raleigh — 48.0%
Seattle — 48.8%

$41.2\% = 0.412$ or $\frac{103}{250}$; $42.5\% = 0.425$ or $\frac{17}{40}$; $47.8\% = 0.478$ or $\frac{239}{500}$; $48.0\% = 0.48$ or $\frac{12}{25}$; $48.8\% = 0.488$ or $\frac{61}{125}$

 Writing   Translating Expression   Geometry   Scientific Calculator   Video   NS·E

Section 6.2 Fractions and Decimals and Their Percent Forms

1. Converting Fractions and Decimals to Percents

In Section 6.1, we converted percents to their equivalent fraction and decimal forms. This is done by replacing the % symbol by $\times \frac{1}{100}$. In this section, we reverse the process. We convert fractions and decimals to percents by multiplying by 100 and applying the % symbol.

> **Converting Fractions and Decimals to Percent Form**
> Multiply the fraction or decimal by 100%.
> *Note:* Multiplying a decimal by 100 moves the decimal point 2 places to the right.

Skill Practice

Convert each decimal to its percent form.
1. 0.46
2. 3.25
3. 2
4. 0.0006
5. 2.5

Classroom Examples: p. 351, Exercises 20, 28, and 30

TIP: Multiplying a number by 100% is equivalent to multiplying the number by 1. Thus, the value of the number is not changed.

Example 1   Converting Decimals to Percents

Convert each decimal to its equivalent percent form.

a. 0.62   **b.** 1.75   **c.** 1   **d.** 0.004   **e.** 8.9

Solution:

a. $0.62 = 0.62 \times 100\%$ Multiply by 100%.

$\quad = 62\%$ Multiplying by 100 moves the decimal point 2 places to the right.

b. $1.75 = 1.75 \times 100\%$ Multiply by 100%.

$\quad = 175\%$ The decimal number 1.75 is greater than 1. Therefore, we expect a percent greater than 100%.

c. $1 = 1 \times 100\%$ Multiply by 100%.

$\quad = 100\%$ Recall that 1 whole is equal to 100%.

d. $0.004 = 0.004 \times 100\%$ Multiply by 100%.

$\quad = 00.4\%$ Move the decimal point to the right 2 places.

e. $8.9 = 8.90 \times 100\%$ Multiply by 100%.

$\quad = 890\%$

Answers

1. 46%   **2.** 325%   **3.** 200%
4. 0.06%   **5.** 250%

Example 2 Converting a Fraction to Percent Notation

Convert the fraction to percent notation. $\dfrac{3}{5}$

Solution:

$$\frac{3}{5} = \frac{3}{5} \times 100\% \qquad \text{Multiply by 100\%.}$$

$$= \frac{3}{5} \times \frac{100}{1}\% \qquad \text{Convert the whole number to an improper fraction.}$$

$$= \frac{3}{\overset{}{\underset{1}{5}}} \times \frac{\overset{20}{100}}{1}\% \qquad \text{Multiply fractions and simplify to lowest terms.}$$

$$= 60\%$$

TIP: We could also have converted $\frac{3}{5}$ to decimal form first (by dividing the numerator by the denominator) and then converted the decimal to a percent.

convert to decimal		convert to percent

$$\frac{3}{5} = 0.60 \qquad = \qquad 0.60 \times 100\% = 60\%$$

Skill Practice

Convert the fraction to percent notation.

6. $\dfrac{7}{10}$

Classroom Example: p. 351, Exercise 34

Example 3 Converting a Fraction to Percent Notation

Convert the fraction to percent notation. $\dfrac{2}{3}$

Solution:

$$\frac{2}{3} = \frac{2}{3} \times 100\% \qquad \text{Multiply by 100\%.}$$

$$= \frac{2}{3} \times \frac{100}{1}\% \qquad \text{Convert the whole number to an improper fraction.}$$

$$= \frac{200}{3}\%$$

The number $\frac{200}{3}\%$ can be written as $66\frac{2}{3}\%$ or as $66.\overline{6}\%$.

TIP: First converting $\frac{2}{3}$ to a decimal before converting to percent notation is an alternative approach.

convert to decimal		convert to percent

$$\frac{2}{3} = 0.\overline{6} \qquad = \qquad 0.666 \ldots \times 100\% = 66.\overline{6}\%$$

Skill Practice

Convert the fraction to percent notation.

7. $\dfrac{1}{9}$

Classroom Example: p. 351, Exercise 40

Answers

6. 70%

7. $\dfrac{100}{9}\%$ or $11\dfrac{1}{9}\%$ or $11.\overline{1}\%$

In Example 4, we convert an improper fraction and a mixed number to percent form.

Example 4 Converting Improper Fractions and Mixed Numbers to Percents

Convert to percent notation.

a. $2\frac{1}{4}$ **b.** $\frac{13}{10}$

Solution:

a. $2\frac{1}{4} = 2\frac{1}{4} \times 100\%$ Multiply by 100%.

$= \frac{9}{4} \times \frac{100}{1}\%$ Convert to improper fractions.

$= \frac{9}{\underset{1}{4}} \times \frac{\overset{25}{100}}{1}\%$ Multiply and simplify to lowest terms.

$= 225\%$

b. $\frac{13}{10} = \frac{13}{10} \times 100\%$ Multiply by 100%.

$= \frac{13}{10} \times \frac{100}{1}\%$ Convert the whole number to an improper fraction.

$= \frac{13}{\underset{1}{10}} \times \frac{\overset{10}{100}}{1}\%$ Multiply and simplify to lowest terms.

$= 130\%$

Notice that both answers in Example 4 are greater than 100%. This is reasonable because any number greater than 1 whole unit represents a percent greater than 100%.

2. Approximating Percents

In Example 5 we approximate a percent from its fraction form.

Example 5 Approximating a Percent

Convert the fraction $\frac{5}{13}$ to percent notation rounded to the nearest tenth of a percent.

Solution:

$\frac{5}{13} = \frac{5}{13} \times 100\%$ Multiply by 100%.

$= \frac{5}{13} \times \frac{100}{1}\%$ Write the whole number as an improper fraction.

$= \frac{500}{13}\%$

Avoiding Mistakes

We converted the fraction to percent form *before* dividing and rounding. If you try to convert to decimal form first, you might round too soon.

Answers

8. 170%   **9.** 275%   **10.** 42.9%

To round to the nearest tenth of a percent, we must divide. We will obtain the hundredths-place digit in the quotient on which to base the decision on rounding.

$$38.4\overset{1}{6} \approx 38.5$$

Thus, $\dfrac{5}{13} \approx 38.5\%$.

$$
\begin{array}{r}
38.46 \\
13\overline{)500.00} \\
-39 \\
\hline
110 \\
-104 \\
\hline
60 \\
-52 \\
\hline
80 \\
-78 \\
\hline
2 \\
\end{array}
$$

3. Fractions, Decimals, Percents: A Summary

The diagram in Figure 6-4 illustrates the methods for converting fractions, decimals, and percents.

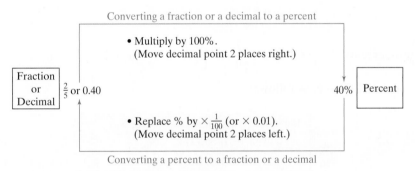

Converting a fraction or a decimal to a percent

• Multiply by 100%.
  (Move decimal point 2 places right.)

Fraction or Decimal $\quad \frac{2}{5}$ or 0.40 $\qquad\qquad$ 40% Percent

• Replace % by $\times \frac{1}{100}$ (or $\times 0.01$).
  (Move decimal point 2 places left.)

Converting a percent to a fraction or a decimal

Figure 6-4

Concept Connections

11. To convert a decimal to a percent, in which direction do you move the decimal point?

12. To convert a percent to a decimal, in which direction do you move the decimal point?

Example 6 Converting Fractions, Decimals, and Percents

Complete the table.

	Fraction	**Decimal**	**Percent**
a.		0.55	
b.	$\frac{1}{200}$		
c.			160%
d.		2.4	
e.			$66\frac{2}{3}\%$
f.	$\frac{2}{9}$		

Solution:

a. 0.55 to fraction: $\quad 0.55 = \dfrac{55}{100} = \dfrac{11}{20}$

0.55 to percent: $\quad 0.55 \times 100\% = 55\%$

Skill Practice

Complete the table.

	Fraction	Decimal	Percent
13.		1.41	
14.	$\frac{1}{50}$		
15.			18%
16.		0.58	
17.			$33\frac{1}{3}\%$
18.	$\frac{7}{9}$		

Classroom Example: p. 353, Exercise 76

b. $\frac{1}{200}$ to decimal: $1 \div 200 = 0.005$

    $\frac{1}{200}$ to percent: $\frac{1}{200} \times 100\% = \frac{100}{200}\% = 0.5\%$

c. 160% to fraction: $160 \times \frac{1}{100} = \frac{160}{100} = \frac{8}{5}$ or $1\frac{3}{5}$

    160% to decimal: $160 \times 0.01 = 1.6$

d. 2.4 to fraction: $\frac{24}{10} = \frac{12}{5}$ or $2\frac{2}{5}$

    2.4 to percent: $2.4 \times 100\% = 240\%$

e. $66\frac{2}{3}\%$ to fraction: $66\frac{2}{3} \times \frac{1}{100} = \frac{\overset{2}{\cancel{200}}}{3} \times \frac{1}{\underset{1}{\cancel{100}}} = \frac{2}{3}$

    $66\frac{2}{3}\%$ to decimal: $66\frac{2}{3} \times 0.01 = 66.\overline{6} \times 0.01 = 0.\overline{6}$

f. $\frac{2}{9}$ to decimal: $2 \div 9 = 0.\overline{2}$

    $\frac{2}{9}$ to percent: $\frac{2}{9} \times 100\% = \frac{2}{9} \times \frac{100}{1}\% = \frac{200}{9}\% = 22\frac{2}{9}\%$ or $22.\overline{2}\%$

The completed table is as follows.

Answers

	Fraction	Decimal	Percent
13.	$\frac{141}{100}$ or $1\frac{41}{100}$	1.41	141%
14.	$\frac{1}{50}$	0.02	2%
15.	$\frac{9}{50}$	0.18	18%
16.	$\frac{29}{50}$	0.58	58%
17.	$\frac{1}{3}$	$0.\overline{3}$	$33\frac{1}{3}\%$
18.	$\frac{7}{9}$	$0.\overline{7}$	$77.\overline{7}\%$

	Fraction	Decimal	Percent
a.	$\frac{11}{20}$	0.55	55%
b.	$\frac{1}{200}$	0.005	0.5%
c.	$\frac{8}{5}$ or $1\frac{3}{5}$	1.6	160%
d.	$\frac{12}{5}$ or $2\frac{2}{5}$	2.4	240%
e.	$\frac{2}{3}$	$0.\overline{6}$	$66\frac{2}{3}\%$
f.	$\frac{2}{9}$	$0.\overline{2}$	$22\frac{2}{9}\%$ or $22.\overline{2}\%$

Section 6.2   Practice Exercises

For additional exercises, see Classroom Activities 6.2A–6.2C in the *Student's Resource Manual* at www.mhhe.com/moh.

Study Skills Exercise

Do you remember your instructor's name, office hours, office location, and office phone? Write them here:

Instructor's name: Instructor's office hours:

Instructor's office location: Instructor's office phone:

Review Exercises

1. Determine whether the given statement is true or false.

    a. $5\% = \frac{1}{2}$ False   **b.** $10\% = \frac{1}{10}$ True   **c.** $200\% = 2$ True

✏ Writing   ← → Translating Expression   ◿ Geometry   🖩 Scientific Calculator   ▶ Video   NS E

For Exercises 2–5, convert the percent to a fraction or mixed number.

2. 60% $\frac{3}{5}$

3. 130% $\frac{13}{10}$ or $1\frac{3}{10}$

4. $16\frac{1}{2}$% $\frac{33}{200}$

5. 0.5% $\frac{1}{200}$

For Exercises 6–9, convert the percent to a decimal.

6. 80% 0.8

7. $6\frac{1}{3}$% $0.06\overline{3}$

8. 143% 1.43

9. 0.3% 0.003

Concept 1: Converting Fractions and Decimals to Percents

For Exercises 10–13, multiply.

10. 0.68×100% 68%

11. 1.62×100% 162%

12. 0.005×100% 0.5%

13. 0.26×100% 26%

14. Write the rule for multiplying a decimal by 100.

When multiplying a decimal by 100, move the decimal point 2 places to the right.

For Exercises 15–18, multiply.

15. $\frac{5}{4} \times 100$% 125%

16. $\frac{2}{5} \times 100$% 40%

17. $\frac{77}{100} \times 100$% 77%

18. $\frac{113}{100} \times 100$% 113%

For Exercises 19–30, convert the decimal to a percent. **(See Example 1.)**

19. 0.27 27%

20. 0.51 51%

21. 0.19 19%

22. 0.33 33%

23. 1.75 175%

24. 2.8 280%

25. 0.124 12.4%

26. 0.277 27.7%

27. 0.006 0.6%

28. 0.0008 0.08%

29. 1.014 101.4%

30. 2.203 220.3%

For Exercises 31–42, convert the fraction to a percent. **(See Examples 2 and 3.)**

31. $\frac{71}{100}$ 71%

32. $\frac{89}{100}$ 89%

33. $\frac{19}{20}$ 95%

34. $\frac{7}{20}$ 35%

35. $\frac{7}{8}$ 87.5% or $87\frac{1}{2}$%

36. $\frac{5}{8}$ 62.5% or $62\frac{1}{2}$%

37. $\frac{13}{16}$ 81.25% or $81\frac{1}{4}$%

38. $\frac{11}{16}$ 68.75% or $68\frac{3}{4}$%

39. $\frac{5}{6}$ $83.\overline{3}$% or $83\frac{1}{3}$%

40. $\frac{5}{12}$ $41.\overline{6}$% or $41\frac{2}{3}$%

41. $\frac{4}{9}$ $44.\overline{4}$% or $44\frac{4}{9}$%

42. $\frac{1}{9}$ $11.\overline{1}$% or $11\frac{1}{9}$%

For Exercises 43–48, write the fraction as a percent.

43. One-quarter of Americans say they entertain at home 2 or more times a month. (*Source: USA TODAY*) 25%

44. According to the Centers for Disease Control (CDC), $\frac{37}{100}$ of U.S. teenage boys say they rarely or never wear their seatbelts. 37%

45. According to the Centers for Disease Control, $\frac{1}{10}$ of teenage girls in the United States say they rarely or never wear their seatbelts. 10%

46. In Italy, $\frac{3}{50}$ of the country's budget comes from tourism. 6%

47. In a recent year, $\frac{2}{3}$ of the beds in U.S. hospitals were occupied. $66.\overline{6}$% or $66\frac{2}{3}$%

Writing   Translating Expression   Geometry   Scientific Calculator   Video   NS E

48. Recently, $\frac{1}{8}$ of U.S. residents between the ages of 55 and 64 were not covered by health insurance. (*Source:* U.S. Bureau of the Census) 12.5%

For Exercises 49–56, convert to percent notation. (See Example 4.)

49. $1\frac{3}{4}$ 175%

50. $\frac{7}{2}$ 350%

51. $\frac{27}{20}$ 135%

52. $2\frac{1}{8}$ 212.5% or $212\frac{1}{2}$%

53. $\frac{11}{9}$ $122.\overline{2}$% or $122\frac{2}{9}$%

54. $1\frac{5}{9}$ $155.\overline{5}$% or $155\frac{5}{9}$%

55. $1\frac{2}{3}$ $166.\overline{6}$% or $166\frac{2}{3}$%

56. $\frac{7}{6}$ $116.\overline{6}$% or $116\frac{2}{3}$%

Concept 2: Approximating Percents

For Exercises 57–64, write the fraction in percent notation to the nearest tenth of a percent. (See Example 5.)

57. $\frac{3}{7}$ 42.9%

58. $\frac{6}{7}$ 85.7%

59. $\frac{1}{13}$ 7.7%

60. $\frac{3}{13}$ 23.1%

61. $\frac{5}{11}$ 45.5%

62. $\frac{8}{11}$ 72.7%

63. $\frac{13}{15}$ 86.7%

64. $\frac{1}{15}$ 6.7%

Concept 3: Fractions, Decimals, Percents: A Summary

65. Explain the difference between $\frac{1}{2}$ and $\frac{1}{2}$%.
The fraction $\frac{1}{2} = 0.5$ and $\frac{1}{2}$% = 0.5% = 0.005.

66. Explain the difference between $\frac{3}{4}$ and $\frac{3}{4}$%.
The fraction $\frac{3}{4} = 0.75$ and $\frac{3}{4}$% = 0.75% = 0.0075.

67. Explain the difference between 25% and 0.25%.
25% = 0.25 and 0.25% = 0.0025

68. Explain the difference between 10% and 0.10%.
10% = 0.1 and 0.10% = 0.1% = 0.001

69. Which of the numbers represent 125%? a, c

   a. 1.25   **b.** 0.125   **c.** $\frac{5}{4}$ **d.** $\frac{5}{4}$%

70. Which of the numbers represent 60%? c, d

   a. 6.0   **b.** 0.60%   **c.** 0.6   **d.** $\frac{3}{5}$

71. Which of the numbers represent 30%? a, c

   a. $\frac{3}{10}$ **b.** $\frac{1}{3}$ **c.** 0.3   **d.** 0.03%

72. Which of the numbers represent 180%? b, c

   a. 18   **b.** 1.8   **c.** $\frac{9}{5}$ **d.** $\frac{9}{5}$%

Mixed Exercises

For Exercises 73–76, complete the table. **(See Example 6.)**

73.

	Fraction	Decimal	Percent
a.	$\frac{1}{4}$	0.25	25%
b.	$\frac{23}{25}$	0.92	92%
c.	$\frac{3}{20}$	0.15	15%
d.	$\frac{8}{5}$ or $1\frac{3}{5}$	1.6	160%
e.	$\frac{1}{100}$	0.01	1%
f.	$\frac{1}{200}$	0.005	0.5%

74.

	Fraction	Decimal	Percent
a.	$\frac{3}{500}$	0.006	0.6%
b.	$\frac{2}{5}$	0.4	40%
c.	2	2	200%
d.	$\frac{1}{2}$	0.5	50%
e.	$\frac{3}{25}$	0.12	12%
f.	$\frac{9}{20}$	0.45	45%

75.

	Fraction	Decimal	Percent
a.	$\frac{7}{50}$	0.14	14%
b.	$\frac{87}{100}$	0.87	87%
c.	1	1	100%
d.	$\frac{1}{3}$	$0.\overline{3}$	$33.\overline{3}\%$ or $33\frac{1}{3}\%$
e.	$\frac{1}{500}$	0.002	0.2%
f.	$\frac{19}{20}$	0.95	95%

76.

	Fraction	Decimal	Percent
a.	$1\frac{3}{10}$ or $\frac{13}{10}$	1.3	130%
b.	$\frac{11}{50}$	0.22	22%
c.	$\frac{3}{4}$	0.75	75%
d.	$\frac{73}{100}$	0.73	73%
e.	$\frac{2}{9}$	$0.\overline{2}$	$22.\overline{2}\%$
f.	$\frac{1}{20}$	0.05	5%

Expanding Your Skills

77. Is the number 1.4 less than or greater than 100%? $1.4 > 100\%$

78. Is the number 0.0087 less than or greater than 1%? $0.0087 < 1\%$

79. Is the number 0.052 less than or greater than 50%? $0.052 < 50\%$

80. Is the number 25 less than or greater than 25%? $25 > 25\%$

Percent Proportions and Applications

Section 6.3

1. Introduction to Percent Proportions

Concepts

1. Introduction to Percent Proportions
2. Identifying the Parts of a Percent Proportion
3. Solving Percent Proportions
4. Applications of Percent Proportions

Recall that a percent is a ratio in parts per 100. For example, $50\% = \frac{50}{100}$. However, a percent can be represented by infinitely many equivalent fractions. Thus,

$$50\% = \frac{50}{100} = \frac{1}{2} = \frac{2}{4} = \frac{3}{6} \quad \text{and infinitely many more.}$$

Equating a percent to an equivalent ratio forms a proportion that we call a **percent proportion**. A percent proportion is a proportion in which one ratio is written with a denominator of 100. For example:

$$\frac{50}{100} = \frac{3}{6} \quad \text{is a percent proportion.}$$

2. Identifying the Parts of a Percent Proportion

We will be using percent proportions to solve a variety of application problems. But first we need to identify and label the parts of a percent proportion.

A percent proportion can be written in the form:

$$\frac{\text{Amount}}{\text{Base}} = p\% \qquad \text{or} \qquad \frac{\text{Amount}}{\text{Base}} = \frac{p}{100}$$

For example:

$$4 \text{ L out of } 8 \text{ L is } 50\% \qquad \frac{4}{8} = 50\% \qquad \text{or} \qquad \frac{4}{8} = \frac{50}{100}$$
amount   base   p

In this example, 8 L is some total (or base) quantity and 4 L is some part (or amount) of that whole. The ratio $\frac{4}{8}$ represents a fraction of the whole equal to 50%. In general, we offer the following guidelines for identifying the parts of a percent proportion.

Identifying the Parts of a Percent Proportion

A percent proportion can be written as

$$\frac{\text{Amount}}{\text{Base}} = p\% \qquad \text{or} \qquad \frac{\text{Amount}}{\text{Base}} = \frac{p}{100}.$$

- The **base** is the total or whole amount being considered. It often appears after the word *of* within a word problem.
- The **amount** is the part being compared to the base. It sometimes appears with the word *is* within a word problem.

Answers

1. Amount = 12; base = 60;
   $p = 20$; $\dfrac{12}{60} = \dfrac{20}{100}$
2. Amount = 84; base = 600;
   $p = 14$; $\dfrac{84}{600} = \dfrac{14}{100}$
3. Amount = 32; base = 2000;
   $p = 1.6$; $\dfrac{32}{2000} = \dfrac{1.6}{100}$

Example 1 Identifying Amount, Base, and *p* for a Percent Proportion

Identify the amount, base, and *p* value, and then set up a percent proportion.

a. 25% of 60 students is 15 students. **b.** $32 is 50% of $64.

c. 5 of 1000 employees is 0.5%.

Solution:

For each problem, we recommend that you identify *p* first. It is the number in front of the symbol %. Then identify the base. In most cases it follows the word *of*. Then, by the process of elimination, find the amount.

a. 25% of 60 students is 15 students.
p   base   amount
(before %   (after the
symbol)   word *of*)

$$\text{amount} \rightarrow \frac{15}{60} = \frac{25}{100} \leftarrow p \atop \leftarrow 100$$
base $\rightarrow$

b. $32 is 50% of $64.
amount   *p*   base

$$\text{amount} \rightarrow \frac{32}{64} = \frac{50}{100} \leftarrow p \atop \leftarrow 100$$
base $\rightarrow$

c. 5 of 1000 employees is 0.5%.
amount   base   *p*

$$\text{amount} \rightarrow \frac{5}{1000} = \frac{0.5}{100} \leftarrow p \atop \leftarrow 100$$
base $\rightarrow$

3. Solving Percent Proportions

In Example 1, we practiced identifying the parts of a percent proportion. Now we consider percent proportions in which one of these numbers is unknown. Furthermore, we will see that the examples come in three types:

- Amount is unknown.
- Base is unknown.
- Value p is unknown.

However, the process for solving in each case is the same.

Example 2 Solving Percent Proportions—Amount Unknown

a. What is 30% of 180?   **b.** 70% of 500 people is how many people?

Solution:

a. What is 30% of 180?
     amount (x) p base

The base and value for p are known.

Let x represent the unknown amount.

$$\frac{x}{180} = \frac{30}{100}$$

Set up a percent proportion.

$$100 \cdot x = (30)(180)$$

Equate the cross products.

$$100x = 5400$$

$$\frac{\overset{1}{\cancel{100}}x}{\underset{1}{\cancel{100}}} = \frac{\overset{}{\cancel{5400}}}{\underset{}{\cancel{100}}}$$

Divide both sides of the equation by 100.

$$x = 54$$

Simplify to lowest terms.

Therefore, 54 is 30% of 180.

> **TIP:** We can check the answer to Example 2(a) as follows. Ten percent of a number is $\frac{1}{10}$ of the number. Furthermore, $\frac{1}{10}$ of 180 is 18. Thirty percent of 180 must be 3 times this amount.
>
> $$3 \times 18 = 54 \quad ✔$$

b. 70% of 500 people is how many people?
   p base   amount (x)

The base and value for p are known.

Let x represent the unknown amount.

$$\frac{x}{500} = \frac{70}{100}$$

Set up a percent proportion.

$$100 \cdot x = (70)(500)$$

Equate the cross products.

$$100x = 35,000$$

$$\frac{\overset{1}{\cancel{100}}x}{\underset{1}{\cancel{100}}} = \frac{35,000}{100}$$

Divide both sides by 100.

$$x = 350$$

Simplify to lowest terms.

Therefore, 70% of 500 people is 350 people.

Classroom Example: p. 361, Exercise 38

Instructor Note: Before proceeding, ask students to estimate the solutions to parts (a) and (b) by using logical thinking.

> **TIP:** To check the solution to Example 2(b), we can compute 10% of 500, which is 50. To find 70%, multiply this by 7. We have $7(50) = 350$. ✔

Answers

4. 205   **5.** $63

Example 3 Solving Percent Proportions—Base Unknown

a. 40% of what number is 25?   **b.** $13.50 is 150% of how many dollars?

Solution:

a. 40% of what number is 25?   The amount and value of p are known.

  p base (x)   amount   Let x represent the unknown base.

$$\frac{25}{x} = \frac{40}{100}$$ Set up a percent proportion.

$(25)(100) = 40 \cdot x$ Equate the cross products.

$2500 = 40x$

$$\frac{2500}{40} = \frac{40x}{40}$$ Divide both sides by 40.

$62.5 = x$ Therefore, 40% of 62.5 is 25.

b. $13.50 is 150% of how many dollars?   The amount and value of p are known.

  amount p base (x)   Let x represent the unknown base.

$$\frac{13.50}{x} = \frac{150}{100}$$ Set up a percent proportion.

$150 \cdot x = (13.50)(100)$ Equate the cross products.

$150x = 1350$

$$\frac{150x}{150} = \frac{1350}{150}$$ Divide both sides by 150.

$x = 9$ Therefore, $13.50 is 150% of $9.

Example 4 Solving Percent Proportions—p Unknown

a. What percent of 80 mi is 12.4 mi?

b. 48 is what percent of 42? Round to the nearest percent.

Solution:

a. What percent of 80 mi is 12.4 mi?   The amount and base are known.

  p base   amount   The value of p is unknown.

$$\frac{12.4}{80} = \frac{p}{100}$$ Set up a percent equation.

$(12.4)(100) = 80 \cdot p$ Equate the cross products.

$1240 = 80p$

$$\frac{1240}{80} = \frac{80p}{80}$$ Divide both sides by 80.

$15.5 = p$ Therefore, 15.5% of 80 mi is 12.4 mi.

b. 48 is what percent of 42? Round to the nearest percent.

amount   p   base

$$\frac{48}{42} = \frac{p}{100}$$ Set up a percent proportion.

$(48)(100) = 42 \cdot p$ Equate the cross products.

$4800 = 42p$

$$\frac{4800}{42} = \frac{42p}{42}$$ Divide both sides by 42.

$114 \approx p$ Note that $4800 \div 42 = 114.\overline{285714}$. Rounded to the nearest whole number, $p \approx 114$.

Therefore, 48 is approximately 114% of 42.

Avoiding Mistakes

Remember that p represents the number of *parts* per 100. However, Example 4 asked us to find the value of p%. Therefore, it was necessary to attach the % symbol to our value of p.
For Example 4(b), $p \approx 114$.
Therefore, $p\% = 114\%$

4. Applications of Percent Proportions

We now use percent proportions to solve application problems involving percents.

Example 5 **Using Percents in Meteorology**

Buffalo, New York, receives an average of 94 in. of snow each year. This year it had 120% of the normal annual snowfall. How much snow did Buffalo get this year?

Solution:

This situation can be translated as:

"The amount of snow Buffalo received is 120% of 94 in."

amount (x)   p   base

$$\frac{x}{94} = \frac{120}{100}$$ Set up a percent proportion.

$100 \cdot x = (120)(94)$ Equate the cross products.

$100x = 11{,}280$

$$\frac{100x}{100} = \frac{11{,}280}{100}$$ Divide both sides by 100.

$x = 112.8$

This year, Buffalo had 112.8 in. of snow.

Skill Practice

10. In a recent year it was estimated that 24.7% of U.S. adults smoked tobacco products regularly. In a group of 2000 adults, how many would be expected to be smokers?

Classroom Example: p. 362, Exercise 70

TIP: In a word problem, it is always helpful to check the reasonableness of your answer. In Example 5, we are looking for 120% of 94 in. But 120% must be *more* than the base amount of 94 in. Therefore, we suspect that our solution is reasonable.

Answer
10. 494 people

| Example 6 | **Using Percents in Statistics** |

Recently, a Harvard University's freshman class had 18% Asian American students. If this represented 380 students, how many students were admitted to the freshman class? Round to the nearest student.

Solution:

This situation can be translated as:

"380 is 18% of what number?"

amount p base (x)

$$\frac{380}{x} = \frac{18}{100}$$ Set up a percent proportion.

$(380)(100) = (18) \cdot x$ Equate the cross products.

$38{,}000 = 18x$

$$\frac{38{,}000}{18} = \frac{18x}{18}$$ Note that $38{,}000 \div 18 \approx 2111.1$. Rounded to the nearest whole unit (whole person), this is 2111.

$2111 \approx x$

The freshman class at Harvard had approximately 2111 students.

TIP: We can check the answer to Example 6 by substituting $x = 2111$ back into the original proportion. The cross products will not be exactly the same because we had to round the value of x. However, the cross products should be *close*.

$$\frac{380}{2111} \stackrel{?}{\approx} \frac{18}{100}$$ Substitute $x = 2111$ into the proportion.

$(380)(100) \stackrel{?}{\approx} (18)(2111)$

$38{,}000 \approx 37{,}998$ ✔   The values are very close.

| Example 7 | **Using Percents in Business** |

Suppose a tennis pro who is ranked 90th in the world on the men's professional tour earns $280,000 per year in tournament winnings and endorsements. He pays his coach $100,000 per year. What percent of his income goes toward his coach? Round to the nearest tenth of a percent.

Solution:

This can be translated as:

"What *percent* of $280,000 is $100,000?"

The value of p is unknown.   base (x)   amount

$$\frac{100,000}{280,000} = \frac{p}{100}$$ Set up a percent proportion.

$$\frac{100,\cancel{000}}{280,\cancel{000}} = \frac{p}{100}$$ The ratio on the left side of the equation can be simplified by a factor of 10,000. "Strike through" four zeros in the numerator and denominator.

$$\frac{10}{28} = \frac{p}{100}$$

$$(10)(100) = (28) \cdot p$$ Equate the cross products.

$$1000 = 28p$$

$$\frac{1000}{28} = \frac{\cancel{28}p}{\cancel{28}}$$ Divide both sides by 28.

$$\frac{1000}{28} = p$$

$$35.7 \approx p$$ Dividing $1000 \div 28$, we get approximately 35.7.

The tennis pro spends about 35.7% of his income on his coach.

```
        35.71
   28)1000.00
      −84
       160
      −140
        200
       −196
         40
        −28
         12
```

Section 6.3   Practice Exercises

Study Skills Exercise

For additional exercises, see Classroom Activities 6.3A–6.3B in the *Student's Resource Manual* at www.mhhe.com/moh.

Do you believe that you have math anxiety? If yes, why do you think so?

Of the list below, circle the activities that you think can help someone with math anxiety.

Deep breathing   Reading a book about math anxiety

Scheduling extra study time   Keeping a positive attitude

Vocabulary and Key Concepts

1. **a.** A ___percent___ proportion is a proportion in which one ratio is written with a denominator of 100.

   b. The first step to solve a percent proportion is to equate the ___cross___ products.

Writing   Translating Expression   Geometry   Scientific Calculator   Video   NS&E

Review Exercises

For Exercises 2–4, convert the decimal to a percent.

2. 0.55   55%   **3.** 1.30   130%   **4.** 0.0006   0.06%

For Exercises 5–7, convert the fraction to a percent.

5. $\dfrac{3}{8}$ 37.5% or $37\frac{1}{2}$%   **6.** $\dfrac{5}{2}$ 250%   **7.** $\dfrac{1}{100}$ 1%

For Exercises 8–10, convert the percent to a fraction.

8. $62\frac{1}{2}$% $\dfrac{5}{8}$ **9.** 2% $\dfrac{1}{50}$ **10.** 77% $\dfrac{77}{100}$

For Exercises 11–13, convert the percent to a decimal.

11. 82%   0.82   **12.** 0.3%   0.003   **13.** 100%   1

Concept 1: Introduction to Percent Proportions

For Example 14–19, determine if the proportion is a percent proportion.

14. $\dfrac{7}{100} = \dfrac{14}{200}$ Yes   **15.** $\dfrac{150}{300} = \dfrac{50}{100}$ Yes   **16.** $\dfrac{5}{7} = \dfrac{25}{35}$ No

17. $\dfrac{2}{3} = \dfrac{6}{9}$ No   **18.** $\dfrac{1\frac{1}{2}}{100} = \dfrac{3}{200}$ Yes   **19.** $\dfrac{\frac{3}{4}}{100} = \dfrac{3}{400}$ Yes

NS E For Exercises 20–24, shade the figure to estimate the amount. The first exercise is given as an example.

Example: Find 60% of 80. (Answer: 48)

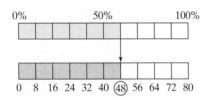

20. Find 40% of 60.   24

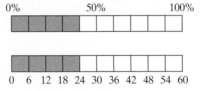

21. Find 75% of 60.   45

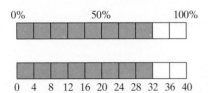

22. Find 15% of 240.   36

23. Find 80% of 40.   32

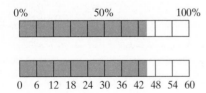

24. Find 10% of 30.   3

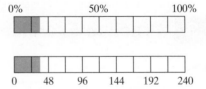

Concept 2: Identifying the Parts of a Percent Proportion

For Exercises 25–30, identify the amount, base, and p value. **(See Example 1.)**

25. 12 balloons is 60% of 20 balloons.
Amount: 12; base: 20; $p = 60$

26. 25% of 400 cars is 100 cars.
Amount: 100; base: 400; $p = 25$

27. $99 of $200 is 49.5%.
Amount: 99; base: 200; $p = 49.5$

28. 45 of 50 children is 90%.
Amount: 45; base: 50; $p = 90$

29. 50 hr is 125% of 40 hr.
Amount: 50; base: 40; $p = 125$

30. 175% of 2 in. of rainfall is 3.5 in.
Amount: 3.5; base: 2; $p = 175$

For Exercises 31–36, write the percent proportion.

31. 10% of 120 trees is 12 trees. $\dfrac{10}{100} = \dfrac{12}{120}$

32. 15% of 20 pictures is 3 pictures. $\dfrac{15}{100} = \dfrac{3}{20}$

33. 72 children is 80% of 90 children. $\dfrac{80}{100} = \dfrac{72}{90}$

34. 21 dogs is 20% of 105 dogs. $\dfrac{20}{100} = \dfrac{21}{105}$

35. 21,684 college students is 104% of 20,850 college students. $\dfrac{104}{100} = \dfrac{21,684}{20,850}$

36. 103% of $40,000 is $41,200. $\dfrac{103}{100} = \dfrac{41,200}{40,000}$

Concept 3: Solving Percent Proportions

For Exercises 37–46, solve the percent problems with an unknown amount. **(See Example 2.)**

37. Compute 54% of 200 employees. 108 employees

38. Find 35% of 412. 144.2

39. What is $\frac{1}{2}$% of 40? 0.2

40. What is 1.8% of 900 grams? 16.2 g

41. Find 112% of 500. 560

42. Compute 106% of 1050. 1113

43. Pedro pays 28% of his salary in income tax. If he makes $72,000 in taxable income, how much income tax does he pay? Pedro pays $20,160 in taxes.

44. A car dealer sets the sticker price of a car by taking 115% of the wholesale price. If a car sells wholesale at $19,000, what is the sticker price?
The sticker price is $21,850.

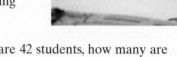

45. A recent study in Missouri showed that over a 2-year period, 72% of the teens (ages 15–19) killed in traffic accidents were not wearing seat belts. If a total of 304 teens were killed, approximately how many were not wearing seat belts? (Round to the nearest whole number.)
Approximately 219 of the 304 teens were not wearing seat belts.

46. In a psychology class, 61.9% of the class consists of freshmen. If there are 42 students, how many are freshmen? Round to the nearest whole unit. There are approximately 26 freshmen.

For Exercises 47–56, solve the percent problems with an unknown base. **(See Example 3.)**

47. 18 is 50% of what number? 36

48. 22% of what length is 44 ft? 200 ft

49. 30% of what weight is 69 lb? 230 lb

50. 70% of what number is 28? 40

51. 9 is $\frac{2}{3}$% of what number? 1350

52. 9.5 is 200% of what number? 4.75

53. Albert saves $120 per month. If this is 7.5% of his monthly income, how much does he make per month?
Albert makes $1600 per month.

54. Janie and Don left their house in South Bend, Indiana, to visit friends in Chicago. They drove 80% of the distance before stopping for lunch. If they had driven 56 mi before lunch, what would be the total distance from their house to their friends' house in Chicago? The total distance is 70 mi.

 Writing  Translating Expression Geometry Scientific Calculator  Video NS E

55. Amiee read 14 e-mails, which was only 40% of her total e-mails. What is her total number of e-mails?
Amiee has a total of 35 e-mails.

56. A recent survey found that 5% of the population of the United States is unemployed. If Charlotte, North Carolina, has 32,000 unemployed, what is the population of Charlotte? The population is approximately 640,000.

For Exercises 57–64, solve the percent problems with *p* unknown. **(See Example 4.)**

57. What percent of $120 is $42? 35%

58. 112 is what percent of 400? 28%

59. 84 is what percent of 70? 120%

60. What percent of 12 letters is 4 letters? $33\frac{1}{3}$%

61. What percent of 320 mi is 280 mi? 87.5%

62. 54¢ is what percent of 48¢? 112.5%

63. A student answered 29 problems correctly on a final exam of 40 problems. What percent of the questions did she answer correctly? She answered 72.5% correctly.

64. During his college basketball season, Jeff made 520 baskets out of 1280 attempts. What was his shooting percentage? Round to the nearest whole percent. Jeff's shooting percentage was about 41%.

For Exercises 65–68, use the table given. The data represent 600 police officers broken down by gender and by the number of officers promoted.

	Promoted	Not Promoted	Total
Male	140	340	480
Female	20	100	120
Total	160	440	600

65. What percent of the officers are female? 20%

66. What percent of the officers are male? 80%

67. What percent of the officers were promoted? Round to the nearest tenth of a percent. 26.7%

68. What percent of the officers were not promoted? Round to the nearest tenth of a percent. 73.3%

Concept 4: Applications of Percent Proportions (Mixed Exercises)

69. The rainfall at Birmingham Airport in the United Kingdom averages 56 mm per month. In August the amount of rain that fell was 125% of the average monthly rainfall. How much rain fell in August? **(See Example 5.)**
70 mm of rain fell in August.

70. In a recent survey 38% of people in the United States say that gas prices have affected the type of vehicle they will buy. In a sample of 500 people who are in the market for a new vehicle, how many would you expect to be influenced by gas prices? 190 people are expected to be influenced.

71. Harvard University reported that 209 African American students were admitted to the freshman class in a recent year. If this represents 11% of the total freshman class, how many freshmen were admitted? **(See Example 6.)**
Approximately 1900 freshmen were admitted.

72. Yellowstone National Park has 3366 mi^2 of undeveloped land. If this represents 99% of the total area, find the total area of the park.
The area of Yellowstone National Park is 3400 mi^2.

73. During the 2009–2010 basketball season, Jason Terry of the Dallas Mavericks made 136 three-point shots out of 373 attempts. To the nearest tenth of a percent, find the percent of three-point shots made.
(See Example 7.) Terry made approximately 36.5% of his three-point shots.

74. During the 2009–2010 football season, Peyton Manning had completed 393 passes out of 571 attempts. Find his completion percentage to the nearest tenth of a percent. Manning's completion percentage was approximately 68.8%.

75. The graph shows the percent of households that own dogs according to the number of people residing in the household. (*Source:* American Veterinary Medical Association)

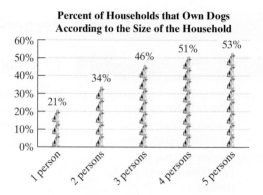

Percent of Households that Own Dogs According to the Size of the Household

   a. If 200 five-person households are surveyed, how many would you expect to own dogs?
   106 five-person households own dogs.

   b. If 50 three-person households are surveyed, how many would you expect to own dogs?
   23 three-person households own dogs.

76. A computer has 74.4 GB (gigabytes) of memory available. If 7.56 GB is used, what percent of the memory is used? Round to the nearest percent.   10% of the memory is used.

A used car dealership sells several makes of vehicles. For Exercises 77–80, refer to the graph. Round the answers to the nearest whole unit.

77. If the dealership sold 215 vehicles in one month, how many were Chevys?   73 were Chevys.

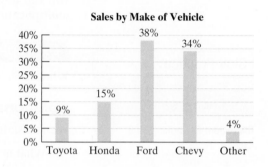

Sales by Make of Vehicle

78. If the dealership sold 182 vehicles in one month, how many were Fords?   69 were Fords.

79. If the dealership sold 27 Hondas in one month, how many total vehicles were sold?   There were 180 total vehicles.

80. If the dealership sold 10 cars in the "Other" category, how many total vehicles were sold?   There were 250 total vehicles.

Expanding Your Skills

81. Carson had $600 and spent 44% of it on clothes. Then he spent 20% of the remaining money on dinner. How much did he spend altogether? $331.20

82. Jasmine took $52 to the mall and spent 24% on makeup. Then she spent one-half of the remaining money on lunch. How much did she spend altogether? $32.24

It is customary to leave a 15–20% tip for the server in a restaurant. However, when you are at a restaurant in a social setting, you probably do not want to take out a pencil and piece of paper to figure out the tip. It is more effective to compute the tip mentally. Try this method.

Step 1: First, if the bill is not a whole dollar amount, simplify the calculations by rounding the bill to the next-higher whole dollar.

Step 2: Take 10% of the bill. This is the same as taking one-tenth of the bill. Move the decimal point to the left 1 place.

Step 3: If you want to leave a 20% tip, double the value found in step 2.

Step 4: If you want to leave a 15% tip, first note that 15% is 5% + 10%. Therefore, add one-half of the value found in step 2 to the number in step 2.

83. Estimate a 20% tip on a bill of $57.65.
(*Hint:* Round up to $58 first.) $11.60

84. Estimate a 20% tip on a bill of $18.79. $3.80

85. Estimate a 15% tip on a dinner bill of $42.00. $6.30

86. Estimate a 15% tip on a luncheon bill of $12.00. $1.80

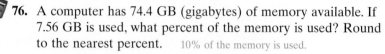
Writing   Translating Expression   Geometry   Scientific Calculator   Video   NS E

Section 6.4   Percent Equations and Applications

1. Solving Percent Equations—Amount Unknown

In this section, we investigate an alternative method to solve applications involving percents. We use percent equations. A **percent equation** represents a percent proportion in an alternative form. For example, recall that we can write a percent proportion as follows:

$$\frac{\text{Amount}}{\text{Base}} = p\% \qquad \text{percent proportion}$$

This is equivalent to writing   Amount $= (p\%) \cdot (\text{base})$ percent equation

To set up a percent equation, it is necessary to translate an English sentence into a mathematical equation. As you read through the examples in this section, you will notice several key words. In the phrase *percent of*, the word *of* implies multiplication. The verb *to be* (am, is, are, was, were, been) often implies =.

Example 1   Solving a Percent Equation—Amount Unknown

What is 30% of 60?

Solution:

We translate the words to mathematical symbols.

What is 30% of 60?

$x\ =\ (30\%) \cdot (60)$

In this context, the word *of* means to multiply.

Let *x* represent the unknown amount.

To find *x*, we must multiply 30% by 60. However, 30% means $\frac{30}{100}$ or 0.30. For the purpose of calculation, we *must* convert 30% to its equivalent decimal or fraction form. The equation becomes

$$x = (0.30)(60)$$
$$= 18$$

> **TIP:** The solution to Example 1 can be checked by noting that 10% of 60 is 6. Therefore, 30% is equal to $(3)(6) = 18$.

The value 18 is 30% of 60.

Skill Practice

Use a percent equation to solve.
1. What is 40% of 90?

Classroom Example: p. 369, Exercise 12

Example 2   Equations—Amount Unknown

142% of 75 amounts to what number?

Solution:

142% of 75 amounts to what number?

$(142\%) \cdot (75)\ =\ x$

$(1.42)(75) = x$

$106.5 = x$

Let *x* represent the unknown amount.
The word *of* implies multiplication.
The phrase *amounts to* implies =.

Convert 142% to its decimal form (1.42).

Multiply.

Therefore, 142% of 75 amounts to 106.5.

Skill Practice

Use a percent equation to solve.
2. 235% of 60 amounts to what number?

Classroom Example: p. 369, Exercise 16

Answers
1. 36   2. 141

Examples 1 and 2 illustrate that the percent equation gives us a quick way to find an unknown amount. For example, because $(p\%) \cdot (\text{base}) = \text{amount}$, we have

$$50\% \text{ of } 80 = 0.50(80) = 40$$

$$25\% \text{ of } 20 = 0.25(20) = 5$$

$$87\% \text{ of } 600 = 0.87(600) = 522$$

$$250\% \text{ of } 90 = 2.50(90) = 225$$

2. Solving Percent Equations—Base Unknown

Examples 3 and 4 illustrate the case in which the base is unknown.

Example 3 Solving a Percent Equation—Base Unknown

225 is 40% of what number?

Solution:

225 is 40% of what number?   Let x represent the base number.

$$225 = (0.40) \cdot x$$

Notice that we immediately converted 40% to its decimal form 0.40 so that we would not forget.

$$225 = 0.40x$$

$$\frac{225}{0.40} = \frac{0.40x}{0.40}$$

Divide both sides of the equation by the number multiplied by the variable x. In this case, divide by 0.40.

$$562.5 = x$$

Divide: $225 \div 0.40 = 562.5$.

The value 225 is 40% of 562.5.

Skill Practice

Use a percent equation to solve.
3. 94 is 80% of what number?

Classroom Example: p. 369, Exercise 28

Example 4 Solving a Percent Equation—Base Unknown

0.19 is 0.2% of what number?

Solution:

0.19 is 0.2% of what number?   Let x represent the base number.

$$0.19 = (0.002) \cdot x$$

Convert 0.2% to its decimal form 0.002.

$$0.19 = 0.002x$$

$$\frac{0.19}{0.002} = \frac{0.002x}{0.002}$$

Divide both sides by 0.002.

$$95 = x$$

Divide: $0.19 \div 0.002 = 95$.

Therefore, 0.19 is 0.2% of 95.

Skill Practice

Use a percent equation to solve.
4. 5.6 is 0.8% of what number?

Classroom Example: p. 369, Exercise 32

Instructor Note: Strongly suggest that students start these problems by labeling the variable.

3. Solving Percent Equations—Percent Unknown

Examples 5 and 6 demonstrate the process to find an unknown percent.

Answers
3. 117.5   **4.** 700

Example 5 **Solving a Percent Equation—Percent Unknown**

75 is what percent of 250?

Solution:

75 is what percent of 250?

$$75 = x \cdot (250) \qquad \text{Let } x \text{ represent the unknown percent.}$$

$$75 = 250x$$

$$\frac{75}{250} = \frac{250x}{250} \qquad \text{Divide both sides by 250.}$$

$$0.3 = x \qquad \text{Divide: } 75 \div 250 = 0.3.$$

At this point, we have $x = 0.3$. To write the value of x in percent form, multiply by 100%.

$$x = 0.3$$

$$= 0.3 \times 100\%$$

$$= 30\%$$

Thus, 75 is 30% of 250.

> **Avoiding Mistakes**
>
> When solving for an unknown percent using a percent equation, it is necessary to convert x to its percent form.

Example 6 **Solving a Percent Equation—Percent Unknown**

What percent of $60 is $92? Round to the nearest tenth of a percent.

Solution:

What percent of $60 is $92?

$$x \cdot (60) = 92 \qquad \text{Let } x \text{ represent the unknown percent.}$$

$$60x = 92$$

$$\frac{60x}{60} = \frac{92}{60} \qquad \text{Divide both sides by 60.}$$

$$x = 1.5\overline{3} \qquad \text{Divide: } 92 \div 60 = 1.5\overline{3}.$$

At this point, we have $x = 1.5\overline{3}$. To convert x to its percent form, multiply by 100%.

$$x = 1.5\overline{3}$$

$$= 1.5\overline{3} \times 100\% \qquad \text{Convert from decimal form to percent form.}$$

$$= (1.53333\ldots) \times 100\%$$

$$= 153.333\ldots\%$$

The hundredths-place digit is less than 5. Discard it and the digits to its right.

Round to the nearest tenth of a percent.

$$\approx 153.3\%$$

> **Avoiding Mistakes**
>
> Notice that in Example 6 we converted the final answer to percent form first *before* rounding. With the number written in percent form, we are sure to round to the nearest tenth of a percent.

Therefore, $92 is approximately 153.3% of $60. (Notice that $92 is just over $1\frac{1}{2}$ times $60, so our answer seems reasonable.)

4. Applications of Percent Equations

In Examples 7, 8, and 9, we use percent equations in application problems. An important part of this process is to extract the base, amount, and percent from the wording of the problem.

Example 7 Using a Percent Equation in Ecology

Forty-six panthers are thought to live in Florida's Big Cypress National Preserve. This represents 53% of the panthers living in Florida. How many panthers are there in Florida? Round to the nearest whole unit. (*Source:* U.S. Fish and Wildlife Services)

Solution:

This problem translates to

"46 is 53% of the number of panthers living in Florida."

$46 = (0.53) \cdot x$ Let x represent the total number of panthers.

$46 = 0.53x$

$\dfrac{46}{0.53} = \dfrac{0.53x}{0.53}$ Divide both sides by 0.53.

$87 \approx x$ Divide: $46 \div 0.53 \approx 87$ (rounded to the nearest whole number).

There are approximately 87 panthers in Florida.

Instructor Note: Point out to students, that in this example, the unknown quantity is the "base."

Example 8 Using a Percent Equation in Sports Statistics

At one time, Steve Young of the San Francisco 49ers was ranked as the NFL's best passer (based on quarterback rating points). For one particular game he completed 23 of 30 passes. What percent of passes did he complete? Round to the nearest tenth of a percent.

Solution:

This problem translates to

"23 is what percent of 30?"

$23 = x \cdot 30$ Let x represent the unknown.

$23 = 30x$

$\dfrac{23}{30} = \dfrac{30x}{30}$ Divide both sides by 30.

$0.767 \approx x$ Divide: $23 \div 30 \approx 0.767$.

Instructor Note: Point out that in this problem, the unknown quantity is the percent.

The decimal value 0.767 has been rounded to 3 decimal places. We did this because the next step is to convert the decimal to a percent. Move the decimal point to the right 2 places and attach the % symbol. We have 76.7% which is rounded to the nearest tenth of a percent.

$$x \approx 0.767$$
$$= 0.767 \times 100\%$$
$$= 76.7\%$$

Steve Young completed approximately 76.7% of his passes.

Example 9 **Using a Percent Equation in Ecology**

Skill Practice

9. In a science class, 85% of the students passed the class. If there were 40 people in the class, how many passed?

Classroom Example: p. 371, Exercise 64

On April 20, 2010, an explosion occurred on an offshore oil rig in the Gulf of Mexico. The explosion left 11 crewmembers dead and an uncontained oil leak that threatened the ecosystem in the Gulf of Mexico and surrounding beaches.

The United States consumes approximately 20 million barrels of oil per day. If the oil obtained from the Gulf of Mexico represents 8% of U.S. daily consumption, how much oil does the United States produce from the Gulf of Mexico?

Instructor Note: Point out that in this problem, the unknown quantity is the "amount."

Solution:

This situation translates to

"What number is 8% of 20?"

$$x = 0.08 \cdot 20 \qquad \text{Write 8\% in decimal form.}$$

$$= (0.08)(20) \qquad \text{Let } x \text{ represent the number of barrels produced by the United States per day in the Gulf of Mexico.}$$

$$= 1.6 \qquad \text{Multiply.}$$

Answer

9. 34 students passed the class.

The United States produces approximately 1.6 million barrels of oil per day from the Gulf of Mexico.

Section 6.4 Practice Exercises

Study Skills Exercise

For additional exercises, see Classroom Activities 6.4A–6.4B in the *Student's Resource Manual* at www.mhhe.com/moh.

There's a saying, "Leave no stone unturned." In math, this means "leave no homework problem undone." Did you do all the assigned homework in Section 6.3? Do you understand the concepts well enough to move on to the homework in this section?

Review Exercises

1. Explain how to solve the equation $26x = 65$.
Divide both sides of the equation by 26 to get $x = 2.5$.

2. Explain how to solve the equation $54 = 6x$.
Divide both sides of the equation by 6 to get $x = 9$.

 Writing   Translating Expression   Geometry   Scientific Calculator   Video   NS E

For Exercises 3–8, solve the equation for the variable.

3. $3x = 27$ $x = 9$

4. $12x = 48$ $x = 4$

5. $0.15x = 45$ $x = 300$

6. $0.32x = 60$ $x = 187.5$

7. $1.02x = 841.5$ $x = 825$

8. $1.06x = 90.1$ $x = 85$

For Exercises 9 and 10, solve the proportion.

9. $\dfrac{165}{100} = \dfrac{693}{x}$ $x = 420$

10. $\dfrac{16}{100} = \dfrac{x}{60}$ $x = 9.6$

Concept 1: Solving Percent Equations—Amount Unknown

↤↦ For Exercises 11–16, write the percent equation. Then solve for the unknown amount. **(See Examples 1 and 2.)**

11. What is 35% of 700?
$x = (0.35)(700); x = 245$

12. Find 12% of 625.
$x = (0.12)(625); x = 75$

13. 0.55% of 900 is what number?
$(0.0055)(900) = x; x = 4.95$

14. What is 0.4% of 75?
$x = (0.004)(75); x = 0.3$

15. Find 133% of 600.
$x = (1.33)(600); x = 798$

16. 120% of 40.4 is what number?
$(1.2)(40.4) = x; x = 48.48$

17. What is a quick way to find 50% of a number?
50% equals one-half of the number. So multiply the number by $\frac{1}{2}$.

18. What is a quick way to find 10% of a number?
10% equals one-tenth of the number. So multiply the number by $\frac{1}{10}$.

19. Compute 200% of 14 mentally. $2 \times 14 = 28$

20. Compute 75% of 80 mentally. $\frac{3}{4} \times 80 = 60$

21. Compute 50% of 40 mentally. $\frac{1}{2} \times 40 = 20$

22. Compute 10% of 32 mentally. $\frac{1}{10} \times 32 = 3.2$

23. Household bleach is 6% sodium hypochlorite (active ingredient). In a 64-oz bottle, how much is active ingredient?
There is 3.84 oz of sodium hypochlorite.

24. One antifreeze solution is 40% alcohol. How much alcohol is in a 12.5-L mixture?   There is 5 L of alcohol in 12.5 L of antifreeze.

25. In football, Dan Marino completed 60% of his passes. If he attempted 8358 passes, how many did he complete? Round to the nearest whole unit.
Marino completed approximately 5015 passes.

26. To pass an exit exam, a student must pass a 60-question test with a score of 80% or better. What is the minimum number of questions she must answer correctly?   She must answer 48 questions correctly to score 80%.

Concept 2: Solving Percent Equations—Base Unknown

↤↦ For Exercises 27–32, write the percent equation. Then solve for the unknown base. **(See Examples 3 and 4.)**

27. 18 is 40% of what number?
$18 = 0.4x; x = 45$

28. 72 is 30% of what number?
$72 = 0.3x; x = 240$

29. 92% of what number is 41.4?
$0.92x = 41.4; x = 45$

30. 84% of what number is 100.8?
$0.84x = 100.8; x = 120$

31. 3.09 is 103% of what number?
$3.09 = 1.03x; x = 3$

32. 189 is 105% of what number?
$189 = 1.05x; x = 180$

33. In tests of a new anti-inflammatory drug, it was found that 47 subjects experienced nausea. If this represents 4% of the sample, how many subjects were tested?   There were 1175 subjects tested.

34. Ted typed 80% of his research paper before taking a break.

a. If he typed 8 pages, how many total pages are in the paper?   There are 10 pages in the paper.

b. How many pages does he have left to type?   Ted has 2 pages left to type.

35. In a recent report, approximately 61.6 million Americans had some form of heart and blood vessel disease. If this represents 22% of the population, approximate the total population of the United States.
At that time, the population was about 280 million.

36. A city has a population of 245,300 which is 110% of the population from the previous year. What was the population the previous year?   The population was 223,000.

 Writing   ↤↦ Translating Expression   Geometry   Scientific Calculator   Video   NS E

Concept 3: Solving Percent Equations—Percent Unknown

For Exercises 37–44, convert the decimal to a percent.

37. 0.13   13%   **38.** 0.4   40%   **39.** 1.08   108%   **40.** 2.2   220%

41. 0.005   0.5%   **42.** 0.007   0.7%   **43.** 0.17   17%   **44.** 0.9   90%

For Exercises 45–50, write the percent equation. Then solve for the unknown percent. Round to the nearest tenth of a percent if necessary. **(See Examples 5 and 6.)**

45. What percent of 480 is 120?
$x \cdot 480 = 120; \ x = 25\%$

46. 180 is what percent of 2000?
$180 = x \cdot 2000; \ x = 9\%$

47. 666 is what percent of 740?
$666 = x \cdot 740; \ x = 90\%$

48. What percent of 60 is 2.88?
$x \cdot 60 = 2.88; \ x = 4.8\%$

49. What percent of 300 is 400?
$x \cdot 300 = 400; \ x = 133.3\%$

50. 28 is what percent of 24?
$28 = x \cdot 24; \ x = 116.7\%$

51. At a softball game, the concession stand had 120 hot dogs and sold 84 of them. What percent was sold?
70% of the hot dogs were sold.

52. The YMCA wants to raise $2500 for its summer program for disadvantaged children. If the YMCA has already raised $900, what percent of its goal has been achieved?   36% of the goal has been achieved.

For Exercises 53 and 54, refer to the table that shows the 1-year absentee record for a business.

53. a. Determine the total number of employees.
There are 80 total employees.
b. What percent missed exactly 3 days of work?
12.5% missed 3 days of work.
c. What percent missed between 1 and 5 days, inclusive?
75% missed 1 to 5 days of work.

54. a. What percent missed at least 4 days?
62.5% missed at least 4 days of work.
b. What percent did not miss any days?
5% did not miss any days.

Number of Days Missed	Number of Employees
0	4
1	2
2	14
3	10
4	16
5	18
6	10
7	6

Concept 4: Applications of Percent Equations (Mixed Exercises)

55. In a recent year, children and adolescents comprised 6.3 million hospital stays. If this represents 18% of all hospital stays, what was the total number of hospital stays? **(See Example 7.)**
There were 35 million total hospital stays that year.

56. One fruit drink advertised that it contained "10% real fruit juice." In one bottle, this was found to be 4.8 oz of real juice.

a. How many ounces of drink does the bottle contain?   48 oz

b. How many ounces is something other than fruit juice?   43.2 oz

57. Of the 87 panthers living in the wild in Florida, 11 are thought to live in Everglades National Park. To the nearest tenth of a percent, what percent is this? (*Source:* U.S. Fish and Wildlife Services) **(See Example 8.)**
Approximately 12.6% of Florida's panthers live in Everglades National Park.

58. Forty-four percent of Americans use online travel sites to book hotel or airline reservations. If 400 people need to make airline or hotel reservations, how many would be expected to use online travel sites?
176 people would be expected to use online travel sites.

59. Fifty-two percent of American parents have started to put money away for their children's college educations. In a survey of 800 parents, how many would be expected to have started saving for their children's education? (*Source: USA TODAY*) **(See Example 9.)**
416 parents would be expected to have started saving for their children's education.

60. The Earth is covered by approximately 360 million km² of water. If the total surface area is 510 million km², what percent is water? (Round to the nearest tenth of a percent.)   70.6% of the Earth is covered by water.

61. Brian has been saving money to buy a 61-in. Samsung Projection HDTV. He has saved $1440 so far, but this is only 60% of the total cost of the television. What is the total cost?   The total cost is $2400.

62. Recently, the number of females that were homeschooled for grades K–12 was 875 thousand. This is 202% of the number of females homeschooled in 1999. How many females were homeschooled in 1999? Round to the nearest thousand. (*Source:* National Center for Educational Statistics)
There were approximately 433 thousand females homeschooled in 1999 for grades K–12.

63. A television station plays commercials for 26% of its air time. In 60 min, how many minutes of commercials would be expected?   15.6 min of commercials would be expected.

64. Sixty-five percent of the human body is water. For a 150-lb person, how much is water?
A 150-lb person has 97.5 lb of water.

For Exercises 65–68, use the graph.

65. If there were 10,000,000 people in the workforce in the 25–34 age group, how many made over $10 per hour?
6,350,000 people ages 25–34 made over $10/hr.

66. If there were 6,600,000 people in the workforce in the 55–64 age group, how many made over $10 per hour?
4,303,200 people ages 55–64 made over $10/hr.

67. If 4,000,000 people in the 16–24 age group made over $10 per hour, how many total workers in this age group are there?
There is a total of 16,000,000 workers in the 16–24 age group.

68. If 9,000,000 people in the 45–54 age group made over $10 per hour, how many total workers in this age group are there?
There are 12,500,000 workers in the 45–54 age group.

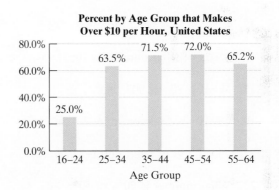

Percent by Age Group that Makes Over $10 per Hour, United States

Expanding Your Skills

The maximum recommended heart rate (in beats per minute) is given by 220 minus a person's age. For aerobic activity, it is recommended that individuals exercise at 60%–85% of their maximum recommended heart rate. This is called the aerobic range. Use this information for Exercises 69 and 70.

69. a. Find the maximum recommended heart rate for a 20-year-old.
200 beats per minute.
   b. Find the aerobic range for a 20-year-old.
Between 120 and 170 beats per minute.

70. a. Find the maximum recommended heart rate for a 42-year-old.
178 beats per minute.
   b. Find the aerobic range for a 42-year-old.
Between 107 and 151 beats per minute.

 Writing   Translating Expression   Geometry    Scientific Calculator   Video   NS E

Problem Recognition Exercises

Percents

For Exercises 1–6, perform the calculations mentally.

1. What is 10% of 82? 8.2

2. What is 5% of 82? 4.1

3. What is 20% of 82? 16.4

4. What is 50% of 82? 41

5. What is 200% of 82? 164

6. What is 15% of 82? 12.3

NS/E **7.** Is 104% of 80 less than or greater than 80?
Greater than

NS/E **8.** Is 8% of 50 less than or greater than 5?
Less than

NS/E **9.** Is 11% of 90 less than or greater than 9?
Greater than

NS/E **10.** Is 52% of 200 less than or greater than 100?
Greater than

For Exercises 11–34, solve the problem by using a percent proportion or a percent equation.

11. 6 is 0.2% of what number? 3000

12. What percent of 500 is 120? 24%

13. 12% of 40 is what number? 4.8

14. 27 is what percent of 180? 15%

15. 150% of what number is 105? 70

16. What number is 30% of 120? 36

17. What is 7% of 90? 6.3

18. 100 is 40% of what number? 250

19. 180 is what percent of 60? 300%

20. 0.5% of 140 is what number? 0.7

21. 75 is 0.1% of what number? 75,000

22. 27 is what percent of 72? 37.5%

23. What number is 50% of 50? 25

24. What number is 15% of 900? 135

25. 50 is 50% of what number? 100

26. 900 is 15% of what number? 6000

27. What percent of 250 is 2? 0.8%

28. 75 is what percent of 60? 125%

29. What number is 10% of 26? 2.6

30. 11 is 55% of what number? 20

31. 186 is what percent of 248? 75%

32. What number is 55% of 11? 6.05

33. 248 is what percent of 186? $133\frac{1}{3}$ %

34. 20 is what percent of 5? 400%

Applications Involving Sales Tax, Commission, Discount, and Markup

Percents are used in an abundance of applications in day-to-day life. In this section, we investigate four common applications of percents:

- Sales tax
- Discount
- Commission
- Markup

Concepts

1. Applications Involving Sales Tax
2. Applications Involving Commission
3. Applications Involving Discount and Markup

1. Applications Involving Sales Tax

The first application involves computing sales tax. **Sales tax** is a tax based on a percent of the cost of merchandise.

Instructor Note: Before starting these application problems, it would be valuable to ask students to review the problem-solving strategies they learned in Section 1.8.

Sales Tax Formula

$$\begin{pmatrix} \text{Amount of} \\ \text{sales tax} \end{pmatrix} = \begin{pmatrix} \text{tax} \\ \text{rate} \end{pmatrix} \cdot \begin{pmatrix} \text{cost of} \\ \text{merchandise} \end{pmatrix}$$

In this formula the tax rate is usually given by a percent. Also note that there are three parts to the formula, just as there are in the general percent equation (see Section 6.4). The sales tax formula is a special case of a percent equation.

Example 1 Computing Sales Tax

Suppose a Toyota Camry sells for $26,000.

a. Compute the sales tax for a tax rate of 5.5%.

b. What is the total price of the car?

Solution:

a. Let x represent the amount of sales tax. Label the unknown.

Tax rate $= 5.5\%$ Identify the parts of the formula.

Cost of merchandise $= \$26,000$

$$\begin{pmatrix} \text{Amount of} \\ \text{sales tax} \end{pmatrix} = \begin{pmatrix} \text{tax} \\ \text{rate} \end{pmatrix} \cdot \begin{pmatrix} \text{cost of} \\ \text{merchandise} \end{pmatrix}$$

$\qquad x \quad = (5.5\%) \cdot \quad (\$26,000)$ Substitute values into the sales tax formula.

$\qquad x = (0.055)(\$26,000)$ Convert the percent to its decimal form.

$\qquad x = \$1430$

The sales tax on the vehicle is $1430.

b. The total price is $26,000 + $1430 = $27,430.

Skill Practice

1. A graphing calculator costs $110. The sales tax rate is 4.5%.
   a. Compute the amount of sales tax.
   b. Compute the total cost.

Classroom Example: p. 380, Exercise 22

Avoiding Mistakes

Notice that we must use the decimal form of the sales tax rate in the calculation.

Answers

1. a. The tax is $4.95.
   b. The total cost is $114.95.

Skill Practice

2. A DVD sells for $15. The sales tax is $0.90. What is the tax rate?

Classroom Example: p. 380, Exercise 24

Example 2 Computing a Sales Tax Rate

Juanita has just moved and must buy a new refrigerator for her home. The refrigerator costs $1200 and the sales tax is $48. Because she is new to the area, she does not know the sales tax rate. Use these figures to compute the tax rate.

Solution:

Let x represent the sales tax rate.   Label the unknown.

Cost of merchandise = $1200   Identify the parts of the formula.

Amount of sales tax = $48

$$\begin{pmatrix} \text{Amount of} \\ \text{sales tax} \end{pmatrix} = \begin{pmatrix} \text{tax} \\ \text{rate} \end{pmatrix} \cdot \begin{pmatrix} \text{cost of} \\ \text{merchandise} \end{pmatrix}$$

$$48 \qquad = \quad x \quad \cdot \quad (1200)$$ Substitute values into the sales tax formula.

$$48 = 1200x$$

$$\frac{48}{1200} = \frac{1200x}{1200}$$ Divide both sides by 1200.

$$0.04 = x$$

The question asks for the tax rate which is given in percent form.

$$x = 0.04$$

$$= 0.04 \times 100\%$$

$$= 4\%$$ The sales tax rate is 4%.

Skill Practice

3. Sales tax on a new lawn mower is $21.50. If the tax rate is 5%, compute the price of the lawn mower before tax.

Classroom Example: p. 380, Exercise 26

Example 3 Finding Cost of Merchandise

The tax on a new CD comes to $1.05. If the tax rate is 6%, find the cost of the CD before tax.

Solution:

Let x represent the cost of the CD.   Label the unknown.

Tax rate = 6%   Identify the parts of the formula.

Amount of tax = $1.05

$$\begin{pmatrix} \text{Amount of} \\ \text{sales tax} \end{pmatrix} = \begin{pmatrix} \text{tax} \\ \text{rate} \end{pmatrix} \cdot \begin{pmatrix} \text{cost of} \\ \text{merchandise} \end{pmatrix}$$

$$1.05 \qquad = (0.06) \cdot \qquad x$$ Notice that we immediately converted 6% to its decimal form.

$$1.05 = 0.06x$$

$$\frac{1.05}{0.06} = \frac{0.06x}{0.06}$$ Divide both sides by 0.06.

$$17.5 = x$$ The CD costs $17.50.

Answers

2. The tax rate is 6%.
3. The mower costs $430 before tax.

2. Applications Involving Commission

Salespeople often receive all or part of their salary in commission. **Commission** is a form of income based on a percent of sales.

> **Commission Formula**
> $$\begin{pmatrix} \text{Amount of} \\ \text{commission} \end{pmatrix} = \begin{pmatrix} \text{commission} \\ \text{rate} \end{pmatrix} \cdot \begin{pmatrix} \text{total} \\ \text{sales} \end{pmatrix}$$

For example, if a realtor gets a 6% commission on the sale of a $200,000 home, then

$$\text{Commission} = (0.06)(\$200,000)$$
$$= \$12,000$$

Example 4 **Finding Commission Rate**

Alexis works in real estate sales.

a. If she sells a $150,000 house and earns a commission of $10,500, what is her commission rate?

b. At this rate, how much will she earn by selling a $200,000 house?

Solution:

a. Let x represent the commission rate.   Label the unknown.

Total sales = $150,000   Identify the parts of the formula.

Amount of commission = $10,500

$$\begin{pmatrix} \text{Amount of} \\ \text{commission} \end{pmatrix} = \begin{pmatrix} \text{commission} \\ \text{rate} \end{pmatrix} \cdot \begin{pmatrix} \text{total} \\ \text{sales} \end{pmatrix}$$

$$10{,}500 = x \cdot (150{,}000)$$ Substitute values into the commission formula.

$$10{,}500 = 150{,}000x$$

$$\frac{10{,}500}{150{,}000} = \frac{150{,}000x}{150{,}000}$$ Divide both sides by 150,000.

$$0.07 = x$$

$$x = 0.07 \times 100\%$$ Convert to percent form.

$$= 7\%$$

The commission rate is 7%.

b. The commission on a $200,000 house is given by

Amount of commission = $(0.07)(\$200,000) = \$14,000$

Alexis will earn $14,000 by selling a $200,000 house.

Skill Practice

4. Trevor sold a home for $160,000 and earned a $6400 commission. What is his commission rate?

Classroom Example: p. 380, Exercise 36

Answer

4. His commission rate is 4%.

Classroom Example: p. 381,
Exercise 38

Skill Practice

5. A sales rep for a pharmaceutical firm makes $50,000 as his base salary. In addition, he makes 6% commission on sales. If his salary for the year amounts to $98,000, what were his total sales?

Example 5 **Finding Sales Base**

Tonya is a real estate agent. She makes $10,000 as her annual base salary for the work she does in the office. In addition, she makes 4% commission on her total sales. If her salary for the year amounts to $106,000, what was her total in sales?

Solution:

First note that her commission is her total salary minus the $10,000 for working in the office. Thus,

Amount of commission = $106,000 − $10,000 = $96,000

Let x represent Tonya's total sales. Label the unknown.

Amount of commission = $96,000 Identify the parts of the formula.

Commission rate = 4%

$$\begin{pmatrix} \text{Amount of} \\ \text{commission} \end{pmatrix} = \begin{pmatrix} \text{commission} \\ \text{rate} \end{pmatrix} \cdot \begin{pmatrix} \text{total} \\ \text{sales} \end{pmatrix}$$

$$96{,}000 = (0.04) \cdot x$$

Substitute values into the commission formula.

$$96{,}000 = 0.04x$$

$$\frac{96{,}000}{0.04} = \frac{\overset{1}{\cancel{0.04}}x}{\underset{1}{\cancel{0.04}}}$$

Divide both sides by 0.04.

$$2{,}400{,}000 = x$$

Tonya's sales totaled $2,400,000 ($2.4 million).

---

3. Applications Involving Discount and Markup

When we go to the store, we often find items discounted or on sale. For example, a printer might be discounted 20%, or a blouse might be on sale for 30% off. We compute the amount of the **discount** (the savings) as follows.

Discount Formulas

$$\begin{pmatrix} \text{Amount of} \\ \text{discount} \end{pmatrix} = \begin{pmatrix} \text{discount} \\ \text{rate} \end{pmatrix} \cdot \begin{pmatrix} \text{original} \\ \text{price} \end{pmatrix}$$

Sale price = original price − amount of discount

Answer

5. He made $800,000 in sales.

Example 6 Computing Discount Rate

A gold chain originally priced $500 is marked down to $375. What is the discount rate?

Solution:

First note that the amount of the discount is given by

Discount = original price − sale price

$$= \$500 - \$375$$

$$= \$125$$

Let x represent the discount rate.   Label the unknown.

Original price = $500   Identify the parts of the formula.

Amount of discount = $125

$$\begin{pmatrix} \text{Amount of} \\ \text{discount} \end{pmatrix} = \begin{pmatrix} \text{discount} \\ \text{rate} \end{pmatrix} \cdot \begin{pmatrix} \text{original} \\ \text{price} \end{pmatrix}$$

$$125 \quad = \quad x \quad \cdot \quad 500 \qquad \text{Substitute values into discount formula.}$$

$$125 = 500x$$

$$\frac{125}{500} = \frac{\overset{1}{\cancel{500}}x}{\underset{1}{\cancel{500}}} \qquad \text{Divide both sides by 500.}$$

$$0.25 = x$$

Converting $x = 0.25$ to percent form, we have $x = 25\%$. The chain has been discounted 25%.

Retailers often buy goods from manufacturers or wholesalers. To make a profit, the retailer must increase the cost of the merchandise before reselling it. This is called **markup**.

> **Markup Formulas**
>
> $$\begin{pmatrix} \text{Amount of} \\ \text{markup} \end{pmatrix} = \begin{pmatrix} \text{markup} \\ \text{rate} \end{pmatrix} \cdot \begin{pmatrix} \text{original} \\ \text{price} \end{pmatrix}$$
>
> Retail price = original price + amount of markup

Example 7 Computing Markup

A college bookstore marks up the price of books 40%.

a. What is the markup for a grammar handbook that has a manufacturer price of $66?

b. What is the retail price of the book?

c. If there is a 6% sales tax, how much will the book cost to take home?

Solution:

a. Let x represent the amount of markup.   Label the unknown.

Markup rate $= 40\%$ Identify parts of the formula.

Original price $= \$66$

$$\begin{pmatrix} \text{Amount of} \\ \text{markup} \end{pmatrix} = \begin{pmatrix} \text{markup} \\ \text{rate} \end{pmatrix} \cdot \begin{pmatrix} \text{original} \\ \text{price} \end{pmatrix}$$

$$\begin{matrix} \downarrow & & \downarrow & & \downarrow \\ x & = & (0.40) & \cdot & (\$66) \end{matrix}$$ Use the decimal form of 40%.

$$x = (0.40)(\$66)$$

$$= \$26.40$$

The amount of markup is $26.40.

b. Retail price $=$ original price $+$ markup

$$= \$66 + \$26.40$$

$$= \$92.40$$

The retail price is $92.40.

c. Next we must find the amount of the sales tax. This value is added to the cost of the book. The sales tax rate is 6%.

$$\begin{pmatrix} \text{Amount of} \\ \text{sales tax} \end{pmatrix} = \begin{pmatrix} \text{tax} \\ \text{rate} \end{pmatrix} \cdot \begin{pmatrix} \text{cost of} \\ \text{merchandise} \end{pmatrix}$$

$$\text{Tax} = (0.06) \cdot (\$92.40)$$

$$\approx \$5.54$$ Round the tax to the nearest cent.

The total cost of the book is $92.40 + $5.54 = $97.94.

It is important to note the similarities in the formulas presented in this section. To find the amount of sales tax, commission, discount, or markup, we multiply a rate (percent) by some base value.

Summary Formulas for Sales Tax, Commission, Discount, and Markup

$$\text{Amount} = \underline{\text{rate}} \times \underline{\text{base value}}$$

Sales tax: $\begin{pmatrix} \text{Amount of} \\ \text{sales tax} \end{pmatrix} = \begin{pmatrix} \text{tax} \\ \text{rate} \end{pmatrix} \cdot \begin{pmatrix} \text{cost of} \\ \text{merchandise} \end{pmatrix}$

Commission: $\begin{pmatrix} \text{Amount of} \\ \text{commission} \end{pmatrix} = \begin{pmatrix} \text{commission} \\ \text{rate} \end{pmatrix} \cdot \begin{pmatrix} \text{total} \\ \text{sales} \end{pmatrix}$

Discount: $\begin{pmatrix} \text{Amount of} \\ \text{discount} \end{pmatrix} = \begin{pmatrix} \text{discount} \\ \text{rate} \end{pmatrix} \cdot \begin{pmatrix} \text{original} \\ \text{price} \end{pmatrix}$

Markup: $\begin{pmatrix} \text{Amount of} \\ \text{markup} \end{pmatrix} = \begin{pmatrix} \text{markup} \\ \text{rate} \end{pmatrix} \cdot \begin{pmatrix} \text{original} \\ \text{price} \end{pmatrix}$

Section 6.5 Practice Exercises

Study Skills Exercise

For additional exercises, see Classroom Activities 6.5A–6.5C in the *Student's Resource Manual* at www.mhhe.com/moh.

Which of the following strategies can help you study for a test? Check all that apply.

☐ Read the Chapter Summary. The Chapter Summary for this chapter is on page ___.

☐ Do the Review Exercises at the end of the chapter. The Review Exercises for this chapter are on page ___.

☐ Do the Chapter Test at the end of the chapter. The Chapter Test for this chapter is on page ___.

Vocabulary and Key Concepts

1. **a.** Write a formula to compute the amount of sales tax.

    b. Write a formula to compute the amount of commission.

    c. Write a formula to compute the amount of discount.

    d. Write a formula to compute the amount of markup.

a. $\begin{pmatrix} \text{Sales} \\ \text{tax} \end{pmatrix} = \begin{pmatrix} \text{sales tax} \\ \text{rate} \end{pmatrix} \cdot \begin{pmatrix} \text{cost of} \\ \text{merchandise} \end{pmatrix}$

b. $(\text{Commission}) = \begin{pmatrix} \text{commission} \\ \text{rate} \end{pmatrix} \cdot \begin{pmatrix} \text{total} \\ \text{sales} \end{pmatrix}$

c. $(\text{Discount}) = \begin{pmatrix} \text{discount} \\ \text{rate} \end{pmatrix} \cdot \begin{pmatrix} \text{original} \\ \text{price} \end{pmatrix}$

d. $(\text{Markup}) = \begin{pmatrix} \text{markup} \\ \text{rate} \end{pmatrix} \cdot \begin{pmatrix} \text{original} \\ \text{price} \end{pmatrix}$

Review Exercises

2. **a.** Write 82% in decimal form.   0.82   **b.** Write 0.003 as a percent.   0.3%

For Exercises 3–6, find the answer mentally.

3. What is 15% of 80?   12

4. 20 is what percent of 60? $33\frac{1}{3}$%

5. 14 is 50% of what number?   28

6. What percent of 6 is 12?   200%

For Exercises 7–12, solve the percent problem by using either method from Sections 6.3 and 6.4.

7. 52 is 0.2% of what number?   26,000

8. What is 225% of 36?   81

9. 6 is what percent of 25?   24%

10. 18 is 75% of what number?   24

11. What is 1.6% of 550?   8.8

12. 32.2 is what percent of 28?   115%

Concept 1: Applications Involving Sales Tax

For Exercises 13–20, complete the table.

	Cost of Merchandise	Sales Tax Rate	Amount of Tax	Total Cost
13.	$ 20.00	5%	$ 1.00	$ 21.00
14.	$ 56.00	6%	$ 3.36	$ 59.36
15.	$ 12.50	4%	$ 0.50	$ 13.00
16.	$212.00	7%	$14.84	$ 226.84
17.	$ 110.00	2.5%	$ 2.75	$ 112.75
18.	$ 600	3%	$18.00	$ 618.00
19.	$ 55.00	6%	$ 3.30	$ 58.30
20.	$214.00	3%	$ 6.42	$220.42

21. A new coat costs $68.25. If the sales tax rate is 5%, what is the total bill? **(See Example 1.)**
The total bill is $71.66.

22. Sales tax for a county in Oklahoma is 4.5%. Compute the amount of tax on a new personal MP3 player that sells for $64.
The tax is $2.88.

23. The sales tax on a set of luggage is $16.80. If the luggage cost before tax is $240.00, what is the sales tax rate? **(See Example 2.)**   The tax rate is 7%.

24. A new shirt is labeled at $42.00. Jon purchased the shirt and paid $44.10.

   a. How much was the sales tax?   The tax is $2.10.

   b. What is the sales tax rate?
   The tax rate is 5%.

25. The 6% sales tax on a fruit basket came to $2.67. What is the price of the fruit basket? **(See Example 3.)**
The price is $44.50.

26. The sales tax on a bag of groceries came to $1.50. If the sales tax rate is 6%, what was the price of the groceries before tax?   The price is $25.00.

Concept 2: Applications Involving Commission

For Exercises 27–32, complete the table.

	Total Sales	Commission Rate	Amount of Commission
27.	$ 20,000.00	5%	$ 1000.00
28.	$ 540.00	16%	$ 86.40
29.	$125,000.00	8%	$10,000.00
30.	$ 800.00	3%	$ 24.00
31.	$ 5400.00	10%	$ 540.00
32.	$ 1060.00	15%	$ 159.00

33. Zach works in an insurance office. He receives a commission of 7% on new policies. How much did he make last month in commission if he sold $48,000 in new policies?   Zach made $3360 in commission.

34. Marisa makes a commission of 15% on sales over $400. One day she sells $750 worth of merchandise.

   a. How much over $400 did Marisa sell?   Marisa sold $350 over $400.

   b. How much did she make in commission that day?   She earned $52.50 in commission.

35. In one week, Rodney sold $2000.00 worth of sports equipment. He received $300.00 in commission. What is his commission rate? **(See Example 4.)**
Rodney's commission rate is 15%.

36. A realtor sold a townhouse for $95,000. If he received a commission of $7600, what is his commission rate?
His commission rate is 8%.

37. A realtor makes an annual salary of $25,000 plus a 3% commission on sales. If a realtor's salary is $67,000, what was the amount of her sales? **(See Example 5.)**   Her sales were $1,400,000.

38. A salesperson receives a weekly salary of $100, plus a 5.5% commission on sales. Her salary last week was $1090. What were her sales that week?   She had $18,000 in sales.

39. Kabir works as a pharmaceutical representative. He receives a 6% monthly commission on sales up to $60,000. He receives an 8.5% commission on all sales above $60,000. If he sold $86,000 worth of his product one month, how much did he receive in commission?   Kabir's commission totaled $5810.00.

Concept 3: Applications Involving Discount and Markup

For Exercises 40–47, complete the table.

	Original Price	Discount Rate	Amount of Discount	Sale Price
40.	$ 56.00	20%	$ 11.20	$ 44.80
41.	$175.00	15%	$ 26.25	$ 148.75
42.	$900.00	$33\frac{1}{3}$%	$300.00	$600.00
43.	$900.00	30%	$270.00	$630.00
44.	$ 85.00	10%	$ 8.50	$ 76.50
45.	$ 110.00	30%	$ 33.00	$ 77.00
46.	$ 76.00	50%	$ 38.00	$ 38.00
47.	$ 58.40	40%	$ 23.36	$ 35.04

48. Hospital employees get a 15% discount at the hospital cafeteria. If the lunch bill originally comes to $5.60, what is the price after the discount?   The discounted lunch bill is $4.76.

49. A health club membership costs $550 for 1 year. If a member pays up front in a lump sum, the member will receive a 10% discount.

  a. How much money is discounted?
  The discount is $55.
  b. How much will the yearly membership cost with the discount?
  The discounted yearly membership will cost $495.

50. A bathing suit is on sale for $45. If the regular price is $60, what is the discount rate?   The discount rate is 25%.

51. A printer that sells for $229 is on sale for $183.20. What is the discount rate?
(See Example 6.)   The discount rate is 20%.

52. Find the discount and the sale price of the tent in the given advertisement.
The discount is $80.70 and the sale price is $188.30.

53. A set of dishes had an original price of $112. Then it was discounted 50%. A week later, the new sale price was discounted another 50%. At that time, was the set of dishes free? Explain why or why not.
The set of dishes is not free. After the first discount, the price was 50% or one-half of $112, which is $56. Then the second discount is 50% or one-half of $56, which is $28.

Explorer 4-person tent
On Sale 30% OFF

Was $269

54. Find the discount and the sale price of the bike in the given advertisement.
The discount is $11.00, and the sale price is $98.99.

**Huffy Chopper Bike $109.99
Now 10% off.**

55. Find the discount and the discount rate of the chair from the given advertisement.
The discount is $47.00, and the discount rate is 20%.

**Accent Chair was $235.00
and is now $188.00**

56. Find the discount and the discount rate of the watch from the given advertisement. Round the discount rate to the nearest tenth of a percent.
The discount is $20.00, and the rate is approximately 28.6%.

For Exercises 57–64, complete the table.

	Original Price	Markup Rate	Amount of Markup	Retail Price
57.	$ 92.00	5%	$ 4.60	$ 96.60
58.	$ 25.00	10%	$ 2.50	$ 27.50
59.	$110.00	8%	$ 8.80	$118.80
60.	$ 50.00	15%	$ 7.50	$ 57.50
61.	$ 325.00	30%	$ 97.50	$422.50
62.	$ 700.00	25%	$175.00	$875.00
63.	$ 45.00	20%	$ 9.00	$ 54.00
64.	$ 175.00	18%	$ 31.50	$ 206.50

65. A business suit has a wholesale price of $150.00. A department store's markup rate is 18%. **(See Example 7.)**

a. What is the markup for this suit?   The markup is $27.00.

b. What is the retail price?   The retail price is $177.00.

c. If Antonio buys this suit including a 7% sales tax, how much will he pay?   The total price is $189.39.

66. An import/export business marks up imported merchandise by 110%. If a wicker chair imported from Singapore originally costs $84 from the manufacturer, what is the retail price?
The retail price is $176.40.

67. A table is purchased from the manufacturer for $300 and is sold retail at $375. What is the markup rate?
The markup rate is 25%.

68. A $60 hairdryer is sold for $69. What is the markup rate?
The markup rate is 15%.

69. A campus bookstore adds $43.20 to the cost of a science text. If the final cost is $123.20, what is the markup rate?
The markup rate is 54%.

70. The retail price of a golf club is $420.00. If the golf store has marked up the price by $70, what is the markup rate?
The markup rate is 20%.

Percent Increase and Decrease

1. Definition of Percent Increase and Decrease

Two important applications of percents are finding percent increase and percent decrease. For example:

- The price of gas increased 40% in 4 years.
- After taking a new drug for 3 months, a patient's cholesterol decreased by 35%.

When we compute **percent increase** or **percent decrease**, we are comparing the *change* between two given amounts to the *original value*. The change (amount of increase or decrease) is found by subtraction. To compute the percent increase or decrease, we use the following formulas.

Percent Increase or Percent Decrease

$$\left(\begin{array}{c}\text{Percent}\\\text{increase}\end{array}\right) = \left(\frac{\text{amount of increase}}{\text{original value}}\right) \times 100\%$$

$$\left(\begin{array}{c}\text{Percent}\\\text{decrease}\end{array}\right) = \left(\frac{\text{amount of decrease}}{\text{original value}}\right) \times 100\%$$

The formulas for percent increase and percent decrease are derived from the standard percent equation.

$$(\text{Amount of increase}) = (\text{rate of increase})(\text{original value})$$

We can divide both sides of the equation by original value:

$$\frac{\text{Amount of increase}}{\text{Original value}} = \text{rate of increase}$$

The rate of increase is a rate given in decimal form. To convert it to a percent, we multiply both sides by 100%.

$$\left(\frac{\text{Amount of increase}}{\text{Original value}}\right) \times 100\% = \text{percent increase}$$

2. Computing Percent Increase

In Example 1, we apply the percent increase formula.

Example 1   **Computing Percent Increase**

The price of heating oil rose from $2.20 per gallon to $2.75 per gallon in 1 year. Compute the percent increase.

Solution:

The original price was $2.20 — Identify the parts of the formula.

The final price after the increase is $2.75.

The amount of increase is determined by subtraction.

Skill Practice

1. After a raise, Denisha's salary increased from $42,500 to $44,200. What is the percent increase?

Classroom Example: p. 386, Exercise 24

Answer

1. Denisha's salary increased by 4%.

Amount of increase = $2.75 − $2.20

$$= \$0.55$$

There was a $0.55 increase in price.

$$\left(\begin{array}{c}\text{Percent}\\\text{increase}\end{array}\right) = \left(\dfrac{\text{amount of increase}}{\text{original value}}\right) \times 100\%$$

$$= \dfrac{0.55}{2.20} \times 100\%$$

Apply the percent increase formula.

$$= 0.25 \times 100\%$$

$$= 25\%$$

There was a 25% increase in the price of heating oil.

Example 1 could also have been solved by using a percent proportion.

$$\frac{p}{100} = \frac{\text{amount of increase}}{\text{original value}}$$

$$\frac{p}{100} = \frac{0.55}{2.20}$$

Substitute values into the proportion.

$$2.20p = (0.55)(100)$$

Equate the cross products.

$$2.20p = 55$$

$$\frac{\overset{1}{\cancel{2.20}}p}{\underset{1}{\cancel{2.20}}} = \frac{55}{2.20}$$

Divide both sides by 2.20.

$$p = 25$$

There was a 25% increase in heating oil.

3. Computing Percent Decrease

Example 2 Finding Percent Decrease in an Application

The graph in Figure 6-5 represents the closing price of Time Warner stock for a 5-day period. Compute the percent decrease between the first day and the fifth day.

Skill Practice

2. Refer to the graph in Example 2. Compute the percent decrease between day 2 and day 3.

Classroom Example: p. 387, Exercise 34

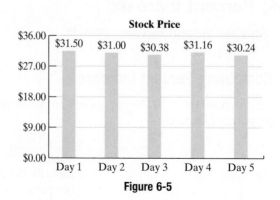

Figure 6-5

Solution:

The amount of decrease is given by $31.50 - $30.24 = $1.26

The original value is the closing price on day 1: $31.50

$$\binom{\text{Percent}}{\text{decrease}} = \left(\frac{\text{amount of decrease}}{\text{original value}}\right) \times 100\%$$

$$= \frac{\$1.26}{\$31.50} \times 100\%$$

$$= 0.04 \times 100\%$$

$$= 4\%$$

The stock fell by 4%.

It is very important to note that percent increase and percent decrease are based on the original value or starting point. Suppose that a 125-lb college student gains 25 lb in her first year of college. We would say that she had a percent increase of $\frac{25}{125} = 20\%$. If she then loses 25 lb during her second year, her percent decrease would not be 20% but rather $\frac{25}{150} = 16.\overline{6}\%$. The percent decrease must be calculated from her new starting weight of 150 lb.

Section 6.6 Practice Exercises

Study Skills Exercise

For additional exercises, see Classroom Activities 6.6A–6.6B in the *Student's Resource Manual* at www.mhhe.com/moh.

When you are taking a test, which of the following should you do? Check all that apply. Can you add any other suggestions to this list?

☐ Answer each question.

☐ Write neatly.

☐ Check your work if time allows.

☐ Show all work.

☐ Include your scratch paper with your test.

☐ Correct your test after you get it back so that you do not make the same errors again.

Vocabulary and Key Concepts

1. **a.** Write a formula to compute percent increase. $\binom{\text{Percent}}{\text{increase}} = \left(\frac{\text{amount of increase}}{\text{original value}}\right) \times 100\%$

   b. Write a formula to compute percent decrease. $\binom{\text{Percent}}{\text{decrease}} = \left(\frac{\text{amount of decrease}}{\text{original value}}\right) \times 100\%$

Review Exercises

2. **a.** What percent is 12 of 60?   20%

   b. What percent is 80 of 50?   160%

3. The price of a sports coat is $65.

   a. If Damien buys the sports coat today, what will the final price be after he adds a 5% sales tax?
   The total price will be $68.25.

   b. Tomorrow the coat goes on sale for 20% off. How much would he spend tomorrow (including the 5% tax)?
   The total price with the 20% discount would be $54.60.

   c. How much will Damien save by waiting until tomorrow to buy the coat?   Damien will save $13.65.

Writing   Translating Expression   Geometry   Scientific Calculator   Video   NS&E

4. The price of a jogging suit is $32.50.

   a. If Kira buys the jogging suit today, what will the final price be after a 6% sales tax is added?
   The total price will be $34.45.

   b. Tomorrow the suit goes on sale for 25% off. How much would she spend tomorrow (including the 6% tax)?
   The total price with the 25% discount would be $25.84.

   c. How much will Kira save by waiting until tomorrow to buy the suit?   Kira will save $8.61.

5. Pablo earns a commission of 14% on all merchandise that he sells over $200. What is his commission if he sells a total of $425 of merchandise?   Pablo's commission is $31.50.

6. Sean earns a salary of $12 per hour and a commission of 3% on all sales. If Sean works a 30-hr week and sells $1290.00 of merchandise, what is his salary for that week?   Sean's salary is $398.70.

7. Explain how to change a decimal into a percent.
   Multiply the decimal by 100% by moving the decimal point 2 places to the right and attaching the % sign.

For Exercises 8–11, change the decimal to a percent.

8. 0.23   23%   **9.** 0.05   5%   **10.** 0.88   88%   **11.** 0.12   12%

Concept 1: Definition of Percent Increase and Decrease

12. **a.** To find the amount of increase in the price of a product, should you subtract the original price from the increased price or the increased price from the original price?   Subtract the original price from the increased price.

    b. To find the amount of decrease in the price of a product, should you subtract the original price from the decreased price or the decreased price from the original price?   Subtract the decreased price from the original price.

For Exercises 13–20, (a) identify if there is an increase or decrease and (b) find the amount of increase or decrease.

13. 48 to 59   a. Increase   **14.** 78 to 123   a. Increase   **15.** 145 to 135   a. Decrease   **16.** 190 to 109   a. Decrease
              b. 11   b. 45   b. 10   b. 81

17. 654 to 645   a. Decrease   **18.** 24 to 42   a. Increase   **19.** 67 to 79   a. Increase   **20.** 205 to 105   a. Decrease
                 b. 9   b. 18   b. 12   b. 100

Concept 2: Computing Percent Increase

21. Select the correct percent increase for a price that is double the original amount. For example, a book that originally cost $30 now costs $60.   c

   a. 200%   **b.** 2%

   c. 100%   **d.** 150%

22. Select the correct percent increase for a price that is greater by $\frac{1}{2}$ of the original amount. For example, an employee made $20 per hour and now makes $30 per hour.   b

   a. 150%   **b.** 50%

   c. $\frac{1}{2}$%   **d.** 200%

23. The U.S. government classified 8 million documents as secret in 2001. By 2003 (2 years after the attacks on 9-11), this number had increased to 14 million. What is the percent increase? (*Source: Time,* April 12, 2004)
   (See Example 1.)   75%

24. One of the top-selling cars is the Honda Civic. In 1 year, the sales increased from 300,000 to 309,000. Compute the percent increase.   3%

 Writing   Translating Expression   Geometry   Scientific Calculator   Video   NS E

25. The number of accidents from all-terrain vehicles that required emergency room visits for children under 16 increased from 21,000 to 42,000 in a 10-year period. What was the percent increase?   100%

26. The number of deaths from alcohol-induced causes rose from approximately 20,200 to approximately 20,700 in a 10-year period. (*Source:* Centers for Disease Control) What is the percent increase? Round to the nearest tenth of a percent.   2.5%

27. Robin's health care premium increased from $5000 per year to $5500 per year. What is the percent increase?   10%

28. The yearly deductible for Diane's health care plan rose from $800 to $1000. What is the percent increase?   25%

29. Lynn is an accountant and charges $165 per hour. If she raises her hourly rate to $170 per hour, what is the percent increase? Round to the nearest percent.   3%

30. Joel's yearly salary went from $42,000 to $45,000. What is the percent increase? Round to the nearest percent.   7%

Concept 3: Computing Percent Decrease

31. Select the correct percent decrease for a price that is one-half of the original amount. For example, a bathing suit that originally cost $62 now costs $31.   a

   a. 50%　　　　**b.** $\frac{1}{2}$%　　　　　　　**c.** 100%　　　　　**d.** 5%

32. Select the correct percent decrease for a price that is one-quarter of the original amount. For example, a T-shirt that originally cost $40 now costs $10.   d

   a. $\frac{1}{4}$%　　　　**b.** 25%　　　　　　　**c.** $\frac{3}{4}$%　　　　　**d.** 75%

33. A stock closed at $12.60 per share on Monday. By Friday, the closing price was $11.97 per share. What was the percent decrease? **(See Example 2.)**   5%

34. During a 5-year period, the number of participants collecting food stamps went from 27 million to 17 million. What is the percent decrease? Round to the nearest whole percent. (*Source:* U.S. Department of Agriculture)   37%

35. Shanti bought a new water efficient toilet for her house. Her old toilet used 5 gal of water per flush. The new toilet uses only 1.6 gal of water per flush. What is the percent decrease in water per flush?   68%

36. Rafu put new insulation in his attic and discovered that his heating bill for December decreased from $160 to $140. What is the percent decrease?   12.5%

37. A paper shredder was marked down from $79 to $59. What is the percent decrease in price? Round to the nearest tenth of a percent.   25.3%

38. A 19-in. computer monitor is marked down from $279 to $249. What is the percent decrease in price? Round to the nearest tenth of a percent.   10.8%

39. Gus, the cat, originally weighed 12 lb. He was diagnosed with a thyroid disorder, and Dr. Smith the veterinarian found that his weight had decreased to 10.2 lb. What percent of his body weight did Gus lose?   15%

40. To lose weight, Tamir reduced his caloric intake from 3000 Calories per day to 1800 Calories per day. What is the percent decrease in Calories?   40%

Calculator Connections

Topic: Using a Calculator to Compute Percent Increase and Percent Decrease

Calculator Exercises

For Exercises 41–44, round the percent increase or decrease to the nearest tenth of a percent.

	Country	Population in 2005 (Millions)	Population in 2010 (Millions)	Change (Millions)	Percent Increase or Decrease
41.	Mexico	110.8	112.3	1.5	1.4% increase
42.	France	61.4	62.6	1.2	2.0% increase
43.	Bulgaria	8.11	7.78	0.33	4.1% decrease
44.	Trinidad	1.075	1.047	0.028	2.6% decrease

45. Unemployment in the United States is shown in the table.

Year	Number of Unemployed (in millions)
2008	8.92
2009	15.27
2010	14.84

Source: U.S. Bureau of Labor Statistics

a. What is the change between the years 2008 and 2009? Does this present an increase or decrease?
An increase of 6.35 million people
b. Determine the percent increase or decrease from 2008 to 2009.
Between 2008 and 2009, there was a 71.2% increase in unemployment.
c. What is the change between the years 2009 and 2010? Does this present an increase or decrease?
A decrease of 0.43 million people
d. Determine the percent increase or decrease from 2009 to 2010.
Between 2009 and 2010, there was a 2.8% decrease in unemployment.

46. Carbon dioxide emissions are anticipated to reach 6320 million metric tons by the year 2035. If emissions totaled 5814 million metric tons in 2008, determine the percent increase. 8.7%

Section 6.7 Simple and Compound Interest

Concepts

1. Simple Interest
2. Compound Interest
3. Using the Compound Interest Formula

1. Simple Interest

In this section, we use percents to compute simple and compound interest on an investment or a loan.

Banks hold large quantities of money for their customers. They keep some cash for day-to-day transactions, but invest the remaining portion of the money. As a result, banks often pay customers interest.

When making an investment, **simple interest** is the money that is earned on principal. **Principal** is the original amount of money. When people take out a loan, the amount borrowed is the principal. The interest is a percent of the amount borrowed that you must pay back in addition to the principal.

The following formulas can be used to compute simple interest for an investment or a loan and to compute the total amount in the account.

 Writing Translating Expression Geometry  Scientific Calculator Video NS

Simple Interest Formulas

Simple interest = principal × rate × time $I = Prt$

Total amount = principal + interest $A = P + I$

where I = amount of interest
P = amount of principal
r = annual interest rate (in decimal form)
t = time (in years)
A = total amount in an account

The time, t, is expressed in years because the rate, r, is an *annual* interest rate. If we were given a monthly interest rate, then the time, t, should be expressed in months.

Example 1   Computing Simple Interest

Suppose $2000 is invested in an account that earns 7% simple interest.

a. How much interest is earned after 3 years?

b. What is the total value of the account after 3 years?

Solution:

a. Principal: $P = \$2000$ Identify the parts of the formula.

Annual interest rate: $r = 7\%$

Time (in years): $t = 3$

Let I represent the amount of interest.   Label the unknown.

$I = Prt$

$= (\$2000)(7\%)(3)$ Substitute values into the formula.

$= (2000)(0.07)(3)$ Convert 7% to decimal form.

$= 140(3)$ Apply the order of operations.

$= 420$ Multiply from left to right.

The amount of interest earned is $420.

b. The total amount in the account is given by

$A = P + I$

$= \$2000 + \$420$

$= \$2420$

The total amount in the account is $2420.

Avoiding Mistakes

It is important to use the decimal form of the interest rate when calculating interest.

When applying the simple interest formula, it is important that time be expressed in years. This is demonstrated in Example 2.

4. Morris takes out a loan for $10,000. He pays simple interest at 7% for 66 months.
   a. Write 66 months in terms of years.
   b. How much money does he pay in interest?
   c. How much total money must he repay to pay off the loan?

Classroom Example: p. 395, Exercise 18

Example 2 Computing Simple Interest

Clyde takes out a loan for $3500. He pays simple interest at a rate of 6% for 4 years 3 months.

   a. How much money does he pay in interest?

   b. How much total money must he pay to pay off the loan?

Solution:

a. $P = \$3500$ Identify parts of the formula.

     $r = 6\%$

     $t = 4\frac{1}{4}$ years or 4.25 years   3 months $= \frac{3}{12}$ year $= \frac{1}{4}$ year

     $I = Prt$

       $= (\$3500)(0.06)(4.25)$ Substitute values into the interest formula. Convert 6% to decimal form 0.06.

       $= \$892.50$ Multiply.

     The interest paid is $892.50.

b. To find the total amount that must be paid, we have

     $A = P + I$

       $= \$3500 + \$892.50$

       $= \$4392.50$ The total amount that must be paid is $4392.50.

2. Compound Interest

Simple interest is based only on a percent of the original principal. However, many day-to-day applications involve compound interest. **Compound interest** is based on both the original principal and the interest earned.

     To compare the difference between simple and compound interest, consider this scenario in Example 3.

5. Suppose $2000 is invested at 5% interest for 3 years.
   a. Compute the total amount after 3 years, using simple interest.
   b. Compute the total amount after 3 years of compounding interest annually.

Classroom Example: p. 396, Exercise 28

Answers

4. **a.** 66 months = 5.5 years
   b. Morris pays $3850 in interest.
   c. Morris must pay a total of $13,850 to pay off the loan.
5. **a.** $2300   **b.** $2315.25

Example 3 Comparing Simple Interest and Compound Interest

Suppose $1000 is invested at 8% interest for 3 years.

   a. Compute the total amount in the account after 3 years, using simple interest.

   b. Compute the total amount in the account after 3 years of compounding interest annually.

Solution:

a. $P = \$1000$

     $r = 8\%$

     $t = 3$ years

     $I = Prt$

       $= (\$1000)(0.08)(3)$

       $= \$240$

The amount of simple interest earned is $240. The total amount in the account is $1000 + $240 = $1240.

b. To compute interest compounded annually over a period of 3 years, compute the interest earned in the first year. Then add the principal plus the interest earned in the first year. This value then becomes the principal on which to base the interest earned in the second year. We repeat this process, finding the interest for the second and third years based on the principal and interest earned in the preceding years. This process is outlined in a table.

Year	Interest Earned $I = Prt$	Total Amount in Account
First year	$I = (\$1000)(0.08)(1) = \$80$	$\$1000 + \$80 = \$1080$
Second year	$I = (\$1080)(0.08)(1) = \$86.40$	$\$1080 + \$86.40 = \$1166.40$
Third year	$I = (\$1166.40)(0.08)(1) = \$93.31$	$\$1166.40 + 93.31 = \textbf{\$1259.71}$

The total amount in the account by compounding interest annually is $1259.71.

Instructor Note: After working both parts of the example, ask students to observe that in part b, the interest for the previous year became part of the principal for the next year.

Notice that in Example 3 the final amount in the account is greater for the situation where interest is compounded. The difference is $1259.71 − $1240 = $19.71. By compounding interest we earn more money.

Interest may be compounded more than once per year.

Annually	1 time per year
Semiannually	2 times per year
Quarterly	4 times per year
Monthly	12 times per year
Daily	365 times per year

To compute compound interest, the calculations become very tedious. Banks use computers to perform the calculations quickly. You may want to use a calculator if calculators are allowed in your class.

3. Using the Compound Interest Formula

As you can see from Example 3, computing compound interest by hand is a cumbersome process. Can you imagine computing daily compound interest (365 times a year) by hand!

We now use a formula to compute compound interest. This formula requires the use of a scientific or graphing calculator. In particular, the calculator must have an exponent key y^x , x^y , or $\wedge$.

Let A = total amount in an account

P = principal

r = annual interest rate

t = time in years

n = number of compounding periods per year

Then $A = P \cdot \left(1 + \dfrac{r}{n}\right)^{n \cdot t}$ computes the total amount in an account.

To use this formula, note the following guidelines:

- Rate r must be expressed in decimal form.
- Time t must be the total time of the investment in *years*.
- Number n is the number of compounding periods per year.
  Annual $n = 1$
  Semiannual $n = 2$
  Quarterly $n = 4$
  Monthly $n = 12$
  Daily $n = 365$

Skill Practice

6. Suppose $2000 is invested at 5% interest compounded annually for 3 years. Use the formula.

$$A = P \cdot \left(1 + \frac{r}{n}\right)^{n \cdot t}$$

to find the total amount after 3 years. Compare the answer to Skill Practice exercise 5(b).

Classroom Example: p. 397, Exercise 34

Example 4 **Computing Compound Interest by Using the Compound Interest Formula**

Suppose $1000 is invested at 8% interest compounded annually for 3 years. Use the compound interest formula to find the total amount in the account after 3 years. Compare the result to the answer from Example 3(b).

Solution:

$P = \$1000$ Identify the parts of the formula.

$r = 8\%$ (0.08 in decimal form)

$t = 3$ years

$n = 1$ (annual compound interest is compounded 1 time per year)

$$A = P \cdot \left(1 + \frac{r}{n}\right)^{n \cdot t}$$

$= 1000 \cdot \left(1 + \dfrac{0.08}{1}\right)^{1 \cdot 3}$ Substitute values into the formula.

$= 1000(1 + 0.08)^3$ Apply the order of operations. Divide within parentheses. Simplify the exponent.

$= 1000(1.08)^3$ Add within parentheses.

$= 1000(1.259712)$ Evaluate $(1.08)^3 = (1.08)(1.08)(1.08) = 1.259712$.

$= 1259.712$ Multiply.

≈ 1259.71 Round to the nearest cent.

The total amount in the account after 3 years is $1259.71. This is the same value obtained in Example 3(b).

Answer

6. $2315.25; this is the same as the result in margin exercise 5(b).

Calculator Connections

Topic: Using a Calculator to Compute Compound Interest

To enter the expression from Example 4 into a calculator, follow these keystrokes.

Expression	Keystrokes	Result
$1000 \cdot (1 + 0.08)^3$	1000 ☒ (1 ➕ 0.08) y^x 3 ═	1259.712
or	1000 ☒ (1 ➕ 0.08) ^ 3 ENTER	1259.712

| Example 5 | Computing Compound Interest by Using the Compound Interest Formula |

Suppose $8000 is invested in an account that earns 5% interest compounded quarterly for $1\frac{1}{2}$ years. Use the compound interest formula to compute the total amount in the account after $1\frac{1}{2}$ years.

Solution:

$P = \$8000$ Identify the parts of the formula.

$r = 5\%$ (0.05 in decimal form)

$t = 1.5$ years

$n = 4$ (quarterly interest is compounded 4 times per year)

$A = P \cdot \left(1 + \dfrac{r}{n}\right)^{n \cdot t}$

$\quad = 8000 \cdot \left(1 + \dfrac{0.05}{4}\right)^{(4)(1.5)}$ Substitute values into the formula.

$\quad = 8000(1 + 0.0125)^6$ Apply the order of operations. Divide within parentheses. Simplify the exponent.

$\quad = 8000(1.0125)^6$ Add within parentheses.

$\quad \approx 8000(1.077383181)$ Evaluate $(1.0125)^6$. If your teacher allows the use of a calculator, consider using the exponent key.

$\quad \approx 8619.07$ Multiply and round to the nearest cent.

The total amount in the account after $1\frac{1}{2}$ years is $8619.07.

Skill Practice

7. Suppose $5000 is invested at 9% interest compounded monthly for 30 years. Use the formula for compound interest to find the total amount in the account after 30 years.

Classroom Example: p. 397, Exercise 38

Animation

Calculator Connections

Topic: Using a Calculator to Compute Compound Interest

To enter the expression from Example 5 into a calculator, follow these keystrokes.

Expression   Keystrokes   Result

$8000 \cdot \left(1 + \dfrac{0.05}{4}\right)^{(4)(1.5)}$ 8000 $\boxed{\times}$ $\boxed{(}$ 1 $\boxed{+}$ 0.05 $\boxed{\div}$ 4 $\boxed{)}$ $\boxed{y^x}$ $\boxed{(}$ 4 $\boxed{\times}$ 1.5 $\boxed{)}$ $\boxed{=}$ $\boxed{8619.065444}$

or   8000 $\boxed{\times}$ $\boxed{(}$ 1 $\boxed{+}$ 0.05 $\boxed{\div}$ 4 $\boxed{)}$ $\boxed{\wedge}$ $\boxed{(}$ 4 $\boxed{\times}$ 1.5 $\boxed{)}$ $\boxed{\text{ENTER}}$ $\boxed{8619.065444}$

Note: It is mandatory to insert parentheses ( ) around the product in the exponent.

Answer

7. The account is worth $73,652.88 after 30 years.

Section 6.7 Practice Exercises

Study Skills Exercise

For additional exercises, see Classroom Activities 6.7A–6.7C in the *Student's Resource Manual* at www.mhhe.com/moh.

Instructors vary in what they emphasize on tests. For example, test material may come from the textbook, notes, handouts, homework, etc. What do you find that your instructor emphasizes?

Vocabulary and Key Concepts

1. a. ___Simple___ interest is the money that is earned (or owed) on an original amount of money invested (or borrowed). The original amount of money invested (or borrowed) is called the ___principal___.

b. Write a formula that represents the amount of simple interest I earned on an investment of P dollars at an annual interest rate r for t years. $I = Prt$

c. ___Compound___ interest is interest earned on both the original principal and interest already earned.

d. Write a formula that represents the amount of money A in an account based on P dollars of principal compounded n times per year at an annual interest rate r for t years. $A = P \cdot \left(1 + \dfrac{r}{n}\right)^{n \cdot t}$

Review Exercises

For Exercises 2–6, find the percent increase or the percent decrease. Round to the nearest whole percent.

	U.S. National Parks	Visitors in 2000 (Thousands)	Visitors in 2004 (Thousands)	Change	Percent Increase or Decrease
2.	Everglades, FL	995	1181	186	19% increase
3.	Bryce Canyon, UT	1099	987	112	10% decrease
4.	Petrified Forest, AZ	605	580	25	4% decrease
5.	Denali, AK	364	404	40	11% increase
6.	Glacier, MT	1729	2034	305	18% increase

Concept 1: Simple Interest

For Exercises 7–14, find the simple interest and the total amount including interest. **(See Examples 1 and 2.)**

	Principal	Annual Interest Rate	Time, Years	Interest	Total Amount
7.	$6,000	5%	3	$900	$6900
8.	$4,000	3%	2	$240	$4240
9.	$5,050	6%	4	$1212	$6262
10.	$4,800	4%	3	$576	$5376

 Writing Translating Expression Geometry Scientific Calculator Video NS E

	Principal	Annual Interest Rate	Time, Years	Interest	Total Amount
11.	$12,000	4%	$4\frac{1}{2}$	$2160	$14,160
12.	$6,230	7%	$6\frac{1}{3}$	$2761.97	$8991.97
13.	$10,500	4.5%	4	$1890.00	$12,390.00
14.	$9,220	8%	4	$2950.40	$12,170.40

15. Leon deposited $2500 in an account that pays $3\frac{1}{2}$% simple interest for 4 years.

    a. How much interest will he earn in 4 years? $350

    b. What will be the total value of the account after 4 years? $2850

16. Selena invested $3400 at 4% simple interest for 5 years.

    a. How much interest will she earn in 5 years? $680

    b. What will be the total value of the account after 5 years? $4080

17. Gloria borrowed $400 for 18 months at 8% simple interest.

    a. How much interest will Gloria have to pay? $48

    b. What will be the total amount that she has to pay back? $448

18. Floyd borrowed $1000 for 2 years 3 months at 8% simple interest.

    a. How much interest will Floyd have to pay? $180

    b. What will be the total amount that he has to pay back? $1180

19. Jozef deposited $10,300 into an account paying 4% simple interest 5 years ago. If he withdraws the entire amount of money, how much will he have? $12,360

20. Heather invested $20,000 in an account that pays 6% simple interest. If she invests the money for 10 years, how much will she have? $32,000

21. Mercedes borrowed $4500 from a bank that charges 10% simple interest. If she repays the loan in $2\frac{1}{2}$ years, how much will she have to pay back? $5625

22. Dan borrowed $750 from his brother who is charging 8% simple interest. If Dan pays his brother back in 6 months, how much does he have to pay back?
$780

Concept 2: Compound Interest

23. If a bank compounds interest semiannually for 3 years, how many total compounding periods are there?
There are 6 total compounding periods.

24. If a bank compounds interest quarterly for 2 years, how many total compounding periods are there?
There are 8 total compounding periods.

25. If a bank compounds interest monthly for 2 years, how many total compounding periods are there?
There are 24 total compounding periods.

26. If a bank compounds interest monthly for $1\frac{1}{2}$ years, how many total compounding periods are there?
There are 18 total compounding periods.

27. Mary Ellen deposited $500 in a bank. **(See Example 3.)**

    a. If the bank offers 4% simple interest, compute the amount in the account after 3 years. $560

    b. Now suppose the bank offers 4% interest compounded annually. Complete the table to determine the amount in the account after 3 years.

Year	Interest Earned	Total Amount in Account
1	$20.00	$520.00
2	20.80	540.80
3	21.63	**562.43**

28. Fatima deposited $12,000 in an account.

a. If the bank offers 5% simple interest, compute the amount in the account after 3 years. $13,800

b. Now suppose the bank offers 5% interest compounded annually, Complete the table to determine the amount in the account after 3 years.

Year	Interest Earned	Total Amount in Account
1	$600.00	$12,600.00
2	630.00	13,230.00
3	661.50	**13,891.50**

29. The amount of $8000 is invested at 4% for 3 years.

a. Compute the ending balance if the bank calculates simple interest. $8960

b. Compute the ending balance if the bank calculates interest compounded annually. $8998.91

Year	Interest Earned	Total Amount in Account
1	$320.00	$8320.00
2	332.80	8652.80
3	346.11	**8998.91**

c. How much more interest is earned in the account with compound interest? $38.91

30. The amount of $12,000 is invested at 8% for 1 year.

a. Compute the ending balance if the bank calculates simple interest. $12,960

b. Compute the ending balance if the bank calculates interest compounded quarterly. $12,989.19

Year	Interest Earned	Total Amount in Account
1st	$240.00	$12,240.00
2nd	244.80	12,484.80
3rd	249.70	12,734.50
4th	254.69	**12,989.19**

c. How much more interest is earned in the account with compound interest? $29.19

Concept 3: Using the Compound Interest Formula

31. For the formula $A = P \cdot \left(1 + \dfrac{r}{n}\right)^{n \cdot t}$, identify what each variable means.

A = total amount in the account; P = principal; r = annual interest rate; n = number of compounding periods per year; t = time in years

32. If $1000 is deposited in an account paying 8% interest compounded monthly for 3 years, label the following variables: P, r, n, and t. $P = \$1000$; $r = 0.08$; $n = 12$; $t = 3$

 Writing   Translating Expression   Geometry   Scientific Calculator   Video   NS&E

Calculator Connections

Topic: Computing Compound Interest

Calculator Exercises

For Exercises 33–40, find the total amount for the investment, using compound interest. **(See Examples 4 and 5.)**

	Principal	Annual Interest Rate	Time, Years	Compounded	Total Amount
33.	$5,000	4.5%	5	Annually	$6230.91
34.	$12,000	5.25%	4	Annually	$14,725.49
35.	$6,000	5%	2	Semiannually	$6622.88
36.	$4,000	3%	3	Semiannually	$4373.77
37.	$10,000	6%	$1\frac{1}{2}$	Quarterly	$10,934.43
38.	$9,000	4%	$2\frac{1}{2}$	Quarterly	$9941.60
39.	$14,000	4.5%	3	Monthly	$16,019.47
40.	$9,000	8%	2	Monthly	$10,555.99

Group Activity

Credit Card Interest

Materials: Calculator.

Estimated time: 15 minutes

Group Size: 4, two pairs

Suppose that an individual plans to open a new credit card and wants to compare the costs associated with two different cards. Have one pair of students fill out the information for Credit Card A and the other pair fill out the information for Credit Card B. Then compare the results. Assume that in each case the opening balance is $0 for a new card.

Credit Card A

1. Find the balance at the end of the month after charging $1538.50 in purchases along with a $200 cash advance. $1738.50

2. This credit card requires a minimum monthly payment of 2% of the monthly balance. What is the minimum payment for the balance calculated in Question 1? $34.77

Writing   Translating Expression   Geometry   Scientific Calculator   Video   NS E

3. The interest rate on the monthly balance is different for purchases than for cash advances. Suppose Credit Card A has the following interest rates.

APR (annual percentage rate) for purchases: 16.24%

APR for cash advances: 19.24%

Determine the amount of interest that will be charged.
(*Hint:* Use the formula $I = Prt$ with $t = \frac{1}{12}$.)

	Amount Charged on Card	**Interest**
Purchases	$1538.50	$20.82
Cash Advance	$200.00	$3.21
Total:	$1738.50	$24.03

4. Suppose that the individual pays only the minimum payment. The credit card company will use this money to pay the interest first, and then use the remaining amount to reduce the balance on the amount owed.

   a. By how much will the balance be reduced? $10.74

   b. What is the new balance that will be carried forward to the next month? $1727.76

Credit Card B

1. Find the balance at the end of the month after charging $1538.50 in purchases along with a $200 cash advance. $1738.50

2. This credit card requires a minimum monthly payment of 2.2% of the monthly balance. What is the minimum payment for the balance calculated in Question 1? $38.25

3. The interest rate on the monthly balance is different for purchases than for cash advances. Suppose Credit Card B has the following interest rates.

APR for purchases: 11.15%

APR for cash advances: 16.15%

Determine the amount of interest that will be charged.
(*Hint:* Use the formula $I = Prt$ with $t = \frac{1}{12}$.)

	Amount Charged on Card	**Interest**
Purchases	$1538.50	$14.30
Cash Advance	$200.00	$2.69
Total:	$1738.50	$16.99

4. Suppose that the individual pays only the minimum payment. The credit card company will use this money to pay the interest first, and then use the remaining amount to reduce the balance on the amount owed.

   a. By how much will the balance be reduced? $21.26

   b. What is the new balance that will be carried forward to the next month? $1717.24

Chapter 6   Summary

Section 6.1   Percents and Their Fraction and Decimal Forms

Key Concepts

The word **percent** means *per one hundred.*

Converting Percents to Fractions

1. Replace the % symbol by $\times \frac{1}{100}$ (or by $\div 100$).
2. Simplify the fraction to lowest terms, if possible.

Converting Percents to Decimals

Replace the % symbol by $\times 0.01$.
(This is equivalent to $\times \frac{1}{100}$ and $\div 100$.)

Note: Multiplying a decimal by 0.01 is the same as moving the decimal point 2 places to the left.

Examples

Example 1

40% means 40 per 100 or $\frac{40}{100}$.

Example 2

$84\% = 84 \times \frac{1}{100} = \frac{84}{100} = \frac{21}{25}$

Example 3

$24.5\% = 24.5 \times 0.01 = 0.245$

Example 4

$0.07\% = 0.07 \times 0.01 = 0.0007$

(Move the decimal point 2 places to the left.)

Section 6.2   Fractions and Decimals and Their Percent Forms

Key Concepts

Converting Fractions and Decimals to Percent Form

Multiply the fraction or decimal by 100%.
($100\% = 1$.)

Examples

Example 1

$\frac{1}{5} = \frac{1}{5} \times 100\% = \frac{100}{5}\% = 20\%$

Example 2

$1.14 = 1.14 \times 100\% = 114\%$

Example 3

$\frac{2}{3} = 0.\overline{6} \times 100\% = 66.\overline{6}\% \text{ or } 66\frac{2}{3}\%$

Section 6.3 Percent Proportions and Applications

Key Concepts and Examples

A **percent proportion** is a proportion that equates a percent to an equivalent ratio.

A percent proportion can be written in the form

$$\frac{\text{Amount}}{\text{Base}} = p\% \quad \text{or} \quad \frac{\text{Amount}}{\text{Base}} = \frac{p}{100}$$

The **base** is the total or whole amount being considered. The **amount** is the part being compared to the base.

Example 1

$\dfrac{36}{100} = \dfrac{9}{25}$ is a percent proportion.

Example 2

For the percent proportion $\dfrac{12}{200} = \dfrac{6}{100}$,
12 is the amount, 200 is the base, and $p = 6$.

To solve a percent proportion, equate the cross products and divide by the factor with the variable. The variable can represent the amount, base, or p. Examples 3–5 demonstrate each type of percent problem.

Examples

Example 3

44% of what number is 275?

Solve the proportion: $\dfrac{275}{x} = \dfrac{44}{100}$

$$44x = 275 \cdot 100$$

$$\frac{\overset{1}{\cancel{44}}x}{\underset{1}{\cancel{44}}} = \frac{27{,}500}{44}$$

$$x = 625$$

Example 4

Of a sample of 400 people, 85% found relief using a particular pain reliever. How many people found relief?

Solve the proportion: $\dfrac{x}{400} = \dfrac{85}{100}$

$$85 \cdot 400 = 100x$$

$$\frac{34{,}000}{100} = \frac{\cancel{100}x}{\cancel{100}}$$

$$340 = x$$

340 people found relief.

Example 5

There are approximately 750,000 career employees in the U.S. Postal Service. If 60,000 are mail handlers, what percent does this represent?

Solve the proportion: $\dfrac{60{,}000}{750{,}000} = \dfrac{p}{100}$

$$750{,}000p = 60{,}000 \cdot 100$$

$$\frac{\overset{1}{\cancel{750{,}000}}p}{\underset{1}{\cancel{750{,}000}}} = \frac{6{,}000{,}000}{750{,}000}$$

$$p = 8$$

Of postal employees, 8% are mail handlers.

Section 6.4 Percent Equations and Applications

Key Concepts and Examples

A **percent equation** represents a percent proportion in an alternative form:

Amount = $(p\%) \cdot$ (base)

Examples 1–3 demonstrate three types of percent problems.

Example 1

Of the car repairs performed on a certain day, 21 were repairs on transmissions. If 60 cars were repaired, what percent involved transmissions?

Solve the equation: $21 = 60x$

$$\frac{21}{60} = \frac{\overset{1}{\cancel{60}}x}{\underset{1}{\cancel{60}}}$$

$$0.35 = x$$

Because the problem asks for a percent, we have $x = 0.35$

$$= 0.35 \times 100\%$$

$$= 35\%$$

Therefore, 35% of cars repaired involved transmissions.

Examples

Example 2

Of all breast cancer cases, 99% occur in women. Out of 2700 cases of breast cancer reported, how many are expected to occur in women?

Solve the equation: $x = (0.99)(2700)$

$$x = 2673$$

About 2673 cases are expected to occur in women.

Example 3

There are 599 endangered plants in the United States. This represents 60.7% of the total number of endangered species. Find the total number of endangered species. Round to the nearest whole number.

Solve the equation: $599 = 0.607x$

$$\frac{599}{0.607} = \frac{\overset{1}{\cancel{0.607}}x}{\underset{1}{\cancel{0.607}}}$$

$$987 \approx x$$

There is a total of approximately 987 endangered species in the United States.

Section 6.5   Applications Involving Sales Tax, Commission, Discount, and Markup

Key Concepts

To find **sales tax**, use the formula

$$\begin{pmatrix} \text{Amount of} \\ \text{sales tax} \end{pmatrix} = \begin{pmatrix} \text{tax} \\ \text{rate} \end{pmatrix} \cdot \begin{pmatrix} \text{cost of} \\ \text{merchandise} \end{pmatrix}$$

To find a **commission**, use the formula

$$\begin{pmatrix} \text{Amount of} \\ \text{commission} \end{pmatrix} = \begin{pmatrix} \text{commission} \\ \text{rate} \end{pmatrix} \cdot \begin{pmatrix} \text{total} \\ \text{sales} \end{pmatrix}$$

To find **discount** and sale price, use the formulas

$$\begin{pmatrix} \text{Amount of} \\ \text{discount} \end{pmatrix} = \begin{pmatrix} \text{discount} \\ \text{rate} \end{pmatrix} \cdot \begin{pmatrix} \text{original} \\ \text{price} \end{pmatrix}$$

    Sale price = original price − amount of discount

To find **markup** and retail price, use the formulas

$$\begin{pmatrix} \text{Amount of} \\ \text{markup} \end{pmatrix} = \begin{pmatrix} \text{markup} \\ \text{rate} \end{pmatrix} \cdot \begin{pmatrix} \text{original} \\ \text{price} \end{pmatrix}$$

    Retail price = original price + amount of markup

Examples

Example 1

A Blu-ray DVD is priced at $23.00, and the total amount paid is $24.84. To find the sales tax rate, first find the amount of tax.

$24.84 − $23.00 = $1.84

To compute the sale tax rate, solve:

$$1.84 = x \cdot 23.00$$

$$\frac{1.84}{23.00} = \frac{x \cdot \cancel{23.00}}{\cancel{23.00}}$$

$$0.08 = x$$

The sales tax rate is 8%.

Example 2

Fletcher makes 13% commission on the sale of all merchandise. If he sells $11,290 worth of merchandise, find how much Fletcher will earn.

$$x = (0.13)(11,290)$$

$$= 1467.7$$

Fletcher will earn $1467.70 in commission.

Example 3

Margaret found a ring that was originally $425 but is on sale for 30% off. To find the sale price, first find the amount of discount.

$$a = (0.30) \cdot (425)$$

$$= 127.5$$

The sale price is $425 − $127.50 = $297.50.

Example 4

A wholesale coat company marks up its coats 20% before selling them retail. To find the retail price of a $340 coat, first find the amount of markup.

$$a = (0.20) \cdot (340)$$

$$= 68$$

The retail price is $340 + $68 = $408.

Section 6.6 Percent Increase and Decrease

Key Concepts

Percent increase or **percent decrease** compares the *change* between two given amounts to the *original value.*

Computing Percent Increase or Decrease

$$\left(\begin{array}{c}\text{Percent}\\\text{increase}\end{array}\right) = \left(\frac{\text{amount of increase}}{\text{original value}}\right) \times 100\%$$

$$\left(\begin{array}{c}\text{Percent}\\\text{decrease}\end{array}\right) = \left(\frac{\text{amount of decrease}}{\text{original value}}\right) \times 100\%$$

Examples

Example 1

In one year a child grows from 35 in. to 42 in. The increase is $42 - 35 = 7$ in. The percent increase is

$$\frac{7}{35} \times 100\% = 0.20 \times 100\%$$

$$= 20\%$$

Section 6.7 Simple and Compound Interest

Key Concepts

To find the **simple interest** made on a certain **principal**, use the formula $I = Prt$

where
    I = amount of interest

    P = amount of principal

    r = annual interest rate

    t = time, years

The formula for the total amount in an account is $A = P + I$, where A = total amount in an account.

Many day-to-day applications involve compound interest. **Compound interest** is based on both the original principal and the interest earned.

The formula $A = P \cdot \left(1 + \dfrac{r}{n}\right)^{n \cdot t}$ computes the total amount in an account that uses compound interest

where
    A = total amount in an account

    P = principal

    r = annual interest rate

    t = time, years

    n = number of compounding periods per year

Examples

Example 1

Betsey deposited $2200 in her account which pays 1.2% simple interest. To find the simple interest she will earn after 4 years, use the formula $I = Prt$ and solve for I.

$$I = (2200)(0.012)(4)$$

$$= 105.6 \qquad \text{She will earn \$105.60 interest.}$$

To find the balance or total amount of her account, apply the formula $A = P + I$.

$$A = 2200 + 105.6$$

$$= 2305.6 \qquad \text{Betsey's balance will be \$2305.60.}$$

Example 2

Gene borrows $1000 at 6% interest compounded semiannually. If he pays off the loan in 3 years, how much will he have to pay?

We are given $P = 1000$, $r = 0.06$, $n = 2$ (semiannually means twice a year), and $t = 3$.

$$A = 1000\left(1 + \frac{0.06}{2}\right)^{2 \cdot 3}$$

$$= 1000(1.03)^6$$

$$\approx 1194.05$$

Gene will have to pay $1194.05 to pay off the loan with interest.

Chapter 6   Review Exercises

Section 6.1

For Exercises 1–4, use a percent to express the shaded portion of each drawing.

1.

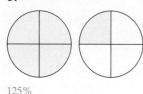

75%

2.

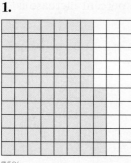

33%

3.

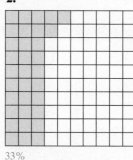

125%

4.

50%

 5. 68% can be expressed as which of the following forms? Identify all that apply.   b, c

    a. $\dfrac{68}{1000}$ **b.** $\dfrac{68}{100}$

    c. 0.68   **d.** 0.068

6. 0.4% can be expressed as which of the following forms? Identify all that apply.   c, d

    a. $\dfrac{4}{100}$ **b.** 0.04

    c. $\dfrac{0.4}{100}$ **d.** 0.004

For Exercises 7–12, match the percent with its fraction form.

7. 30%   f   **a.** $\dfrac{33}{100}$

8. $33\frac{1}{3}\%$ d   **b.** $\dfrac{1}{2}$

9. 33%   a   **c.** $\dfrac{2}{3}$

10. 50%   b   **d.** $\dfrac{1}{3}$

11. $66\frac{2}{3}\%$ c   **e.** $\dfrac{3}{5}$

12. 60%   e   **f.** $\dfrac{3}{10}$

For Exercises 13–18, match the percent with its decimal form.

13. 7.5%   e   **a.** 1

14. 75%   c   **b.** 0.25

15. 50%   f   **c.** 0.75

16. 100%   a   **d.** 0.0025

17. 0.25%   d   **e.** 0.075

18. 25%   b   **f.** 0.5

For Exercises 19 and 20, write the percent as a fraction and as a decimal.

19. Out of all the phone calls received per week, 42% were from solicitors. $\dfrac{21}{50}$; 0.42

20. In an hour-long TV show, 20% of the time is devoted to commercials. $\dfrac{1}{5}$; 0.20

The graph represents the average annual population growth rate for four of the fastest-growing cities. Use this graph for Exercises 21–24.

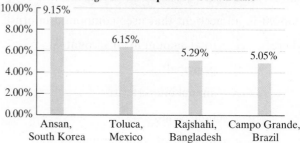

For Exercises 1–4, use a percent to express the shaded portion of each drawing.

21. Write the rate for Toluca, Mexico, in decimal form.
0.0615

22. Write the rate for Rajshahi, Bangladesh, in decimal form.   0.0529

23. Write the rate for Ansan, South Korea, in fraction form. $\frac{183}{2000}$

24. Write the rate for Campo Grande, Brazil, in fraction form. $\frac{101}{2000}$

Section 6.2

For Exercises 25–28, convert the fraction to a percent.

25. $\frac{17}{100}$ 17%

26. $\frac{22}{50}$ 44%

27. $\frac{4}{5}$ 80%

28. $\frac{7}{4}$ 175%

For Exercises 29–32, convert the decimal to a percent.

29. 0.12   12%

30. 1.1   110%

31. 0.005   0.5%

32. 0.4   40%

For Exercises 33–36, write the fraction as a percent.

33. In a classroom the ratio of girls to boys is $\frac{14}{16}$.
87.5%

34. Recently $\frac{19}{25}$ of the U.S. population used the Internet daily.   76%

35. In 2010, $\frac{3}{5}$ of the population were between the ages of 18 and 64 years.   60%

36. Video game sales are expected to decline $\frac{1}{10}$ after the holidays.   10%

For Exercises 37–42, complete the table.

	Fraction	Decimal	Percent
37.	$\frac{9}{20}$	0.45	45%
38.	1	1	100%
39.	$\frac{3}{50}$	0.06	6%
40.	$\frac{6}{5}$	1.2	120%
41.	$\frac{9}{1000}$	0.009	0.9%
42.	$\frac{3}{4}$	0.75	75%

Section 6.3

For Exercises 43–46, identify the amount, base, and p values for the percent proportion $\dfrac{\text{Amount}}{\text{Base}} = \dfrac{p}{100}$.

43. 45% of $150 is $67.50.
Amount: 67.50; base: 150; $p = 45$

44. 360 births is 12% of 3000 births.
Amount: 360; base: 3000; $p = 12$

45. 30.24 m² of 144 m² is 21%.
Amount: 30.24; base: 144; $p = 21$

46. 106% of 30 gal is 31.8 gal.
Amount: 31.8; base: 30; $p = 106$

For Exercises 47–50, write the percent proportion.

47. 6 books of 8 books is 75%. $\frac{6}{8} = \frac{75}{100}$

48. 15% of 180 lb is 27 lb. $\frac{27}{180} = \frac{15}{100}$

49. 200% of $420 is $840. $\frac{840}{420} = \frac{200}{100}$

50. 6 pine trees out of 2000 pine trees is 0.3%.
$\frac{6}{2000} = \frac{0.3}{100}$

For Exercises 51–56, solve the percent problems, using proportions.

51. What is 12% of 50?   6

52. $5\frac{3}{4}$% of 64 is what number?   3.68

53. 11 is what percent of 88?   12.5%

54. 8 is what percent of 2500?   0.32%

55. 13 is $33\frac{1}{3}$% of what number?   39

56. 24 is 120% of what number?   20

57. Based on recent statistics, one airline expects that 4.2% of its customers will be "no-shows." If the airline sold 260 seats, how many people would the airline expect as no-shows? Round to the nearest whole unit.
Approximately 11 people would be no-shows.

58. In a survey of college students, 58% said that they wore their seat belts regularly. If this represents 493 people, how many people were surveyed?
850 people were surveyed.

59. Victoria spends $720 per month on rent. If her monthly take-home pay is $1800, what percent does she pay in rent? Round to the nearest whole percent.
Victoria spends 40% on rent.

60. Of the rental cars at the U-Rent-It company, 40% are compact cars. If this represents 26 cars, how many cars are on the lot?   There are 65 cars.

Section 6.4

→ For Exercises 61–66, write as a percent equation and solve.

61. 18% of 900 is what number?
$0.18 \cdot 900 = x; \; x = 162$

62. What number is 29% of 404?
$x = 0.29 \cdot 404; \; x = 117.16$

63. 18.90 is what percent of 63?
$18.90 = x \cdot 63; \; x = 30\%$

64. What percent of 250 is 86?
$x \cdot 250 = 86; \; x = 34.4\%$

65. 30 is 25% of what number?
$30 = 0.25 \cdot x; \; x = 120$

66. 26 is 130% of what number?
$26 = 1.30 \cdot x; \; x = 20$

67. A student buys a used book for $54.40. This is 80% of the original price. What was the original price? Round to the nearest cent.
The original price is $68.00.

68. Veronica has read 330 pages of a 600-page novel. What percent of the novel has she read?
Veronica read 55% of the novel.

69. Elaine tries to keep her fat intake to no more than 30% of her total calories. If she has a 2400-calorie diet, how many fat calories can she consume to stay within her goal?
Elaine can consume 720 fat calories.

70. In 2010, 13% of Americans were over the age of 65. By 2050 that number could rise to 20%. Suppose that the U.S. population is 300,000,000 in 2010 and 404,000,000 in 2050.

   a. Find the number of Americans over 65 in the year 2010.   39,000,000

   b. Find the number of Americans over 65 in the year 2050.   80,800,000

Section 6.5

For Exercises 71–74, solve the problem involving sales tax.

71. A Plasma TV costs $1279. Find the sales tax if the rate is 6%.   The sales tax is $76.74.

72. The sales tax on sofa is $47.95. If the sofa costs $685.00 before tax, what is the sales tax rate?
The sales tax rate is 7%.

73. The price of a digital camera four-band memory card is $90.25. The total bill is $97.47.

   a. How much is the tax?   The tax is $7.22.

   b. What is the sales tax rate?   The tax rate is 8%.

74. A resort hotel charges an 11% resort tax along with the 6% sales tax. If the hotel's one-night accommodation is $225.00, what will a tourist pay for 4 nights, including tax?
The total amount for 4 nights will be $1053.00.

For Exercises 75–78, solve the problems involving commission.

75. At a recent auction, *Boy with a Pipe*, an early work by Pablo Picasso, sold for $104 million. The commission for the sale of the work was $11 million. What was the rate of commission? Round to the nearest tenth of a percent. (*Source: The New York Times*)
The commission rate was approximately 10.6%.

76. Andre earns a commission of 4% on sales of restaurant supplies. If he sells $4075 in one week, how much commission will he earn?
Andre earned $163 in commission.

77. Sela sells sportswear at a department store. She earns an hourly wage of $15, and she gets a 5% commission on all merchandise that she sells over $200. If Sela works an 8-hr day and sells $420 of merchandise, how much will she earn that day?
Sela will earn $131 that day.

78. A house is sold for $160,000, and the real estate agent earned $5600 in commission. What is the commission rate?
The commission rate is 3.5%.

For Exercises 79–82, solve the problems involving discount and markup.

79. Find the discount and the sale price of the movie if the regular price is $28.95.
The discount is $8.69. The sale price is $20.26.

80. This notebook computer was originally priced at $1747. How much is the discount? After the $50 rebate, how much will a person pay for this computer?
The discount is $174.70. The final price is $1522.30.

81. A rug manufacturer sells a rug to a retail store for $160. The store then marks up the rug to $208. What is the markup rate?
The markup rate is 30%.

82. Peg sold some homemade baskets to a store for $50 each. The store marks up all merchandise by 18%. What will be the retail price of the baskets after the markup?
The baskets will sell for $59 each.

Section 6.6

For Exercises 83 and 84, (**a**) identify if there is an increase or decrease and (**b**) find the percent of increase or decrease.

83. 86 to 107.5
a. Increase
b. 25%

84. 410 to 82
a. Decrease
b. 80%

85. The number of species of animals on the endangered species list went from 263 in 1990 to 574 in 2010. Find the percent increase. Round to the nearest tenth of a percent. (*Source:* U.S. Fish and Wildlife Services) 118.3%

86. In 2 years, the number of Madden NFL video games (for PS2) sold went from 3.2 million to 1.8 million. Find the percent decrease. 43.75%

The number of subscribers to cellular telephones has increased since 1985. Use the information in the table to answer Exercises 87 and 88. In each case, round to the nearest whole percent.

Year	Number of Subscribers (1000s)
1985	300
1993	16,000
2001	128,000
2009	285,000

87. What was the percent increase in cellular phone subscribers between 2001 and 2009? 123%

88. What was the percent increase in cellular phone subscribers between 1985 and 1993? 5233%

Section 6.7

For Exercises 89 and 90, find the simple interest and the total amount including interest.

	Principal	Annual Interest Rate	Time, Years	Interest	Total Amount
89.	$10,200	3%	4	$1224	$11,424
90.	$7000	4%	5	$1400	$8400

91. Jean-Luc borrowed $2500 at 5% simple interest. What is the total amount that he will pay back at the end of 18 months (1.5 years)?
Jean-Luc will have to pay $2687.50.

92. Kyle loaned his brother $800 for 2 years. He is charging 2.5% simple interest. How much will his brother owe Kyle in 2 years?
Kyle's brother will owe him $840.

93. Sydney deposited $6000 in a certificate of deposit that pays 4% interest compounded annually. Complete the table to determine her balance after 3 years.

Year	Interest	Total
1	$240.00	$6240.00
2	249.60	6489.60
3	259.58	**6749.18**

94. Nell deposited $10,000 in a money market account that pays 3% interest compounded semiannually. Complete the table to find her balance after 2 years.

Compound Periods	Interest	Total
Period 1 (end of first 6 months)	$150.00	$10,150.00
Period 2 (end of year 1)	152.25	10,302.25
Period 3 (end of 18 months)	154.53	10,456.78
Period 4 (end of year 2)	156.85	10,613.63

For Exercises 95–98, find the total amount for the investment, using compound interest. Use the formula
$$A = P \cdot \left(1 + \frac{r}{n}\right)^{n \cdot t}.$$

	Principal	Annual Interest Rate	Time in Years	Compounded	Total Amount
95.	$850	8%	2	Quarterly	$995.91
96.	$2050	5%	5	Semiannually	$2624.17
97.	$11,000	7.5%	6	Annually	$16,976.32
98.	$8200	4.5%	4	Monthly	$9813.88

Chapter 6 Test

1. Write a percent to express the shaded portion of the figure.   22%

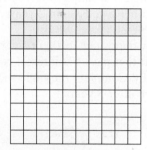

2. Shade the figure so that it represents 85%.

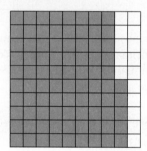

3. Write the percent in decimal form and in fraction form.

   a. For a recent year, the unemployment rate of Illinois was 5.4%. $0.054; \frac{27}{500}$

   b. The incidence of breast cancer increased by 0.15% between 2003 and 2005. $0.0015; \frac{3}{2000}$

   c. For a certain city, gas prices increased by 170% in 10 years. $1.70; \frac{17}{10}$

4. Write the following percents in fraction form.

   a. 1% $\frac{1}{100}$ **b.** 25% $\frac{1}{4}$

   c. $33\frac{1}{3}$% $\frac{1}{3}$ **d.** 50% $\frac{1}{2}$

   e. $66\frac{2}{3}$% $\frac{2}{3}$ **f.** 75% $\frac{3}{4}$

   g. 100%   1   **h.** 150% $\frac{3}{2}$

For Exercises 5 and 6, write the percent as a decimal and as a fraction.

5. The incidence of colon/rectal cancer in men decreased by 2.8% over the past year.

   $0.028; \frac{7}{250}$

6. In 1950, 9.9% of the U.S. population was made up of African Americans. $0.099; \frac{99}{1000}$

7. Explain the process to write a fraction as a percent.   Multiply the fraction by 100%.

For Exercises 8–11, write the fraction as a percent. Round to the nearest tenth of a percent if necessary.

8. $\frac{3}{5}$ 60%   **9.** $\frac{1}{250}$ 0.4%

10. $\frac{7}{4}$ 175%   **11.** $\frac{5}{7}$ 71.4%

12. Explain the process to write a decimal as a percent.   Multiply the decimal by 100%.

For Exercises 13–16, write the decimal as a percent.

13. 0.32   32%   **14.** 0.052   5.2%

15. 1.3   130%   **16.** 0.006   0.6%

For Exercises 17–22, solve the percent problems.

17. What is 24% of 150?   36

18. What is 120% of 16?   19.2

19. 21 is 6% of what number?   350

20. 40% of what number is 80?   200

21. What percent of 220 is 198?   90%

22. 75 is what percent of 150?   50%

23. At McDonald's, a side salad without dressing has 10 mg of sodium. With a serving of Newman's Own Low-Fat Balsamic Dressing, the sodium content of the salad is 740 mg. (*Source:* www.mcdonalds.com)

   a. How much sodium is in the dressing itself?   730 mg

   b. What percent of the sodium content in a side salad with dressing is from the dressing? Round to the nearest tenth of a percent.   98.6%

The composition of the lower level of the Earth's atmosphere is given in the figure (other gases are present in minute quantities). For Exercises 24 and 25, use the information in the graph.

Composition of the Earth's Atmosphere

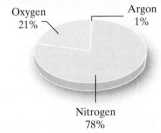

Oxygen 21%

Argon 1%

Nitrogen 78%

24. How much nitrogen would be expected in 500 m³ of atmosphere?
390 m³

25. How much oxygen would be expected in 2000 m³ of atmosphere?
420 m³

26. Darell bought a pair of blue jeans that cost $30.00. He wrote his check for $32.10.

   a. What is the amount of sales tax that he paid?
   The amount of sales tax is $2.10.
   b. What is the sales tax rate?
   The sales tax rate is 7%.

27. Charles earns a salary of $400 per week and gets a bonus of 6% commission on all merchandise that he sells. If Charles sells $3500 worth of merchandise, how much will he earn in that week?
Charles will earn $610.

28. Find the discount rate of the product in the advertisement.
The discount rate of this product is 60%.

29. Over a 2-year period, the price of a Canon PowerShot digital camera went from $240.00 to $169.00. What is the percent decrease? (Round to the nearest tenth of a percent.)   29.6%

30. Maury borrowed $5000 at 8% simple interest. He plans to pay back the loan in 3 years.

   a. How much interest will he have to pay? $1200
   b. What is the total amount that he has to pay back? $6200

31. Use the formula $A = P \cdot \left(1 + \dfrac{r}{n}\right)^{n \cdot t}$ to calculate the total amount in an account that began with $25,000 invested at 4.5% compounded quarterly for 5 years. $31,268.76

Chapters 1–6   Cumulative Review Exercises

1. What is the name of the place value for the digit 6 in the number 26,009,235?
Millions place

2. Fill in the table with either the word name for the number or the number in standard form.

	Country	Area (mi²)	
		Standard Form	**Words**
a.	United States	3,539,245	Three million, five hundred thirty-nine thousand, two hundred forty-five
b.	Saudi Arabia	830,000	Eight hundred thirty thousand
c.	Falkland Islands	4,700	Four thousand, seven hundred
d.	Colombia	401,044	Four hundred one thousand, forty-four

3. Multiply: 34,882
3,488,200 $\times 100$

4. Divide: $9\overline{)783}$ 87

5. Add: 234 + 44 + 6 + 2901   3185

6. Simplify: $\sqrt{16} - 6 \div 3 + 3^2$ 11

7. Identify whether the fraction is proper or improper.

   a. $\dfrac{6}{6}$ Improper
   b. $\dfrac{10}{7}$ Improper
   c. $\dfrac{7}{10}$ Proper
   d. $\dfrac{1}{100}$ Proper

For Exercises 8–11, multiply or divide as indicated. Simplify the fraction to lowest terms.

8. $\dfrac{3}{8} \times \dfrac{32}{9}$ $\dfrac{4}{3}$ or $1\dfrac{1}{3}$

9. $\dfrac{42}{25} \div \dfrac{7}{100}$ 24

10. $\dfrac{21}{2} \div 7$ $\dfrac{3}{2}$ or $1\dfrac{1}{2}$

11. $16 \times \dfrac{1}{24}$ $\dfrac{2}{3}$

12. Find the area of the figure. $\dfrac{15}{32}$ yd^2

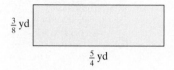

$\frac{3}{8}$ yd

$\frac{5}{4}$ yd

13. Find the perimeter of the figure.   9 km

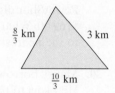

$\frac{8}{3}$ km   3 km

$\frac{10}{3}$ km

14. Add: $\dfrac{3}{10} + \dfrac{17}{100} + \dfrac{3}{1000}$ $\dfrac{473}{1000}$

15. The value of the Daxor stock rose $1\frac{2}{5}$ from $13\frac{7}{10}$. What is the current price? $15\dfrac{1}{10}$

16. A sheet of paper has the dimensions $13\frac{1}{2}$ in. by 17 in. What is the area of the paper? $\dfrac{459}{2}$ or $229\dfrac{1}{2}$ in.^{2}

17. a. List four multiples of 18.   18, 36, 54, 72

   b. List all factors of 18.   1, 2, 3, 6, 9, 18

   c. Write the prime factorization of 18. $2 \cdot 3^2$

18. Write a fraction that represents the shaded portion of each figure.

   a.   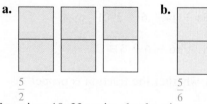   b.

   $\dfrac{5}{2}$ $\dfrac{5}{6}$

For Exercises 19–22, write the fraction as a decimal.

19. $\dfrac{3}{8}$ 0.375

20. $\dfrac{4}{3}$ $1.\overline{3}$

21. $\dfrac{7}{9}$ $0.\overline{7}$

22. $\dfrac{3}{4}$ 0.75

For Exercises 23 and 24, refer to the chart.

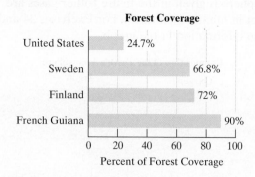

Forest Coverage

United States   24.7%
Sweden   66.8%
Finland   72%
French Guiana   90%

Percent of Forest Coverage

23. What is the difference in the percent of forest coverage in French Guiana and the United States?   65.3%

24. What is the difference in the percent of forest coverage in Sweden and the United States?   42.1%

For Exercises 25–28, multiply and divide by powers of 10.

25. 85×0.001 0.085

26. 85×100 8500

27. $85 \div 10$ 8.5

28. $85 \div 0.0001$ 850,000

For Exercises 29–32, solve the proportions.

29. $\dfrac{3}{4} = \dfrac{15}{p}$ $p = 20$

30. $\dfrac{2.5}{6} = \dfrac{p}{9}$ $p = 3.75$

31. $\dfrac{4\frac{1}{3}}{p} = \dfrac{12}{18}$ $p = 6\frac{1}{2}$

32. $\dfrac{p}{100} = \dfrac{21.87}{81}$ $p = 27$

33. If it takes $\frac{1}{2}$ hr to read 1 chapter of a book, how long will it take to read 5 chapters?
   It will take $2\dfrac{1}{2}$ hr.

34. A 9.2-oz bag of candy costs $2.30. Find the unit cost, that is, the price per ounce.
   The unit price is $0.25 per ounce.

35. A computer can download 1.6 megabytes (MB) in 2.5 min. How long will it take to download a 4.6-MB file? Round to the nearest tenth of a minute.   It will take about 7.2 min.

36. A DC-10 aircraft flew 1799 mi in 3.5 hr. Find the unit rate in miles per hour.
   The DC-10 flew 514 mph.

37. The world's consumption of energy was 398,134 quadrillion Btu in the year 2000. By 2009 it had increased to 449,265 quadrillion Btu. Determine the percent increase. Round to the nearest whole percent. (*Source:* U.S. Department of Energy)   13%

38. New York City has the greatest population of any U.S. city. In 1900 it had 3.4 million people. By 2010, it had 8.4 million.

  a. Compute the difference in population between the year 2010 and the year 1900.
  5 million

  b. Assuming a steady increase, compute the rate representing the increase in population per year. (*Source:* U.S. Bureau of the Census)
  Approximately 0.045 million people per year or 45,000 people per year

39. Kevin deposited $13,000 into a certificate of deposit that pays 3.2% simple interest. How much will Kevin have if he keeps the certificate for 5 years?   Kevin will have $15,080.

40. A mortgage charges 8% interest that is compounded monthly. Use the formula

$$A = P \cdot \left(1 + \frac{r}{n}\right)^{n \cdot t}$$ to find the total amount of

interest paid for a 10-year mortgage on $75,000.
There is $91,473.02 paid in interest.

Measurement

Chapter 7

In this chapter, we present units of measurement in both the U.S. Customary System of measure and the metric system. It is useful to have a mental concept of these measurements for estimating in many real-world situations. We will also learn how to convert between different units of measurement.

Review Your Skills

Familiarize yourself with several common units of measurement by locating these terms in the word search puzzle.

ton   pound   ounce   quart   gallon   cup
pint   mile   foot   yard   inch

Section 7.1 Converting U.S. Customary Units of Length

1. U.S. Customary Units of Length

In many applications in day-to-day life, we need to measure things. To measure an object means to assign it a number and a **unit of measure**. We first present common units of measure for length which include inches (in.), feet (ft), yards (yd), and miles (mi). These units are part of the **U.S. Customary System of measurement** (sometimes called the English system of measurement).

The U.S. Customary units of length and some common equivalents are given in Table 7-1.

Table 7-1   U.S. Customary Units of Length and Their Equivalents

1 ft = 12 in.	1 in. = $\frac{1}{12}$ ft
1 yd = 3 ft	1 ft = $\frac{1}{3}$ yd
1 mi = 5280 ft	1 ft = $\frac{1}{5280}$ mi
1 mi = 1760 yd	1 yd = $\frac{1}{1760}$ mi

Instructor Note: To prepare students, ask them to write down their heights. They will probably record the number in units of feet and inches. Then ask them to state how tall they are in inches.

Note that sometimes units of feet are denoted with the ′ symbol. That is, 3 ft = 3′. Similarly, sometimes units of inches are denoted with the ″ symbol. That is, 4 in. = 4″.

2. Converting U.S. Customary Units of Length by Using Substitution

Example 1 demonstrates how we can use substitution to convert between two units of length.

Skill Practice

Use substitution to convert units.
1. 6 ft = _____ in.
2. 12.6 ft = _____ yd

Classroom Examples: p. 420, Exercises 18 and 24

Example 1   Converting Units of Length

a. 4 ft = _____ in.   **b.** 66 in. = _____ ft

Solution:

a. First note that 4 ft = 4 × 1 ft. Then we can substitute 1 ft = 12 in.

$4 \text{ ft} = 4 \times 1 \text{ ft}$

$\qquad = 4 \times 12 \text{ in.}$ Substitute 1 ft = 12 in.

$\qquad = 48 \text{ in.}$

b. 66 in. = 66 × 1 in.

$\qquad = 66 \times \frac{1}{12} \text{ ft}$ Substitute 1 in. = $\frac{1}{12}$ ft.

$\qquad = \frac{66}{12} \text{ ft}$ Multiply.

$\qquad = \frac{11}{2} \text{ ft or } 5\frac{1}{2} \text{ ft or } 5.5 \text{ ft}$

TIP: In Example 1(a) we converted feet to inches by multiplying by 12. In Example 1(b) we reversed this process to convert inches to feet. In this case we multiplied by $\frac{1}{12}$ which is the same as *dividing* by 12.

Answers
1. 72 in.   **2.** 4.2 yd

3. Converting U.S. Customary Units of Length by Using Conversion Factors

Instructor Note: Remind students to check whether their answer is reasonable. For instance, a certain number of feet will convert to a smaller number of yards but a larger number of inches.

Example 1 demonstrates how substitution can be used to convert between two units of measure. Another method is to multiply by a **conversion factor**. A conversion factor is a ratio of equivalent measures.

For example, note that 1 yd = 3 ft. Therefore, $\dfrac{1 \text{ yd}}{3 \text{ ft}} = 1$ and $\dfrac{3 \text{ ft}}{1 \text{ yd}} = 1$.

These conversion factors are unit ratios or unit fractions because the quotient is 1. To convert from one unit of measure to another, we can multiply by an appropriate conversion factor. We offer these guidelines to determine the proper conversion factor to use.

Choosing a Conversion Factor

In a conversion factor,
- The unit of measure in the numerator should be the new unit you want to convert *to*.
- The unit of measure in the denominator should be the original unit you want to convert *from*.

Concept Connections

Complete the conversion factor.

3. $\dfrac{1 \text{ ft}}{\text{in.}}$ **4.** $\dfrac{\text{yd}}{3 \text{ ft}}$

Example 2 Converting Units of Length by Using Conversion Factors

a. 1500 ft = ____ yd   **b.** 9240 yd = ____ mi   **c.** 8.2 mi = ____ ft

Solution:

a. From Table 7-1, we have 1 yd = 3 ft.

$1500 \text{ ft} = 1500 \text{ ft} \cdot \dfrac{1 \text{ yd}}{3 \text{ ft}}$ ← new unit to convert to
    ← unit to convert from

$= \dfrac{1500 \text{ ft}}{1} \cdot \dfrac{1 \text{ yd}}{3 \text{ ft}}$ Notice that the original units of ft divide out in the same way as simplifying common factors. The unit yd remains in the final answer.

$= \dfrac{1500}{3} \text{ yd}$

$= 500 \text{ yd}$

b. From Table 7-1, we have 1 mi = 1760 yd.

$9240 \text{ yd} = 9240 \text{ yd} \cdot \dfrac{1 \text{ mi}}{1760 \text{ yd}}$ ← new unit to convert to
    ← unit to convert from

$= \dfrac{9240 \text{ yd}}{1} \cdot \dfrac{1 \text{ mi}}{1760 \text{ yd}}$ The units of yd divide out, leaving the answer in miles.

$= \dfrac{9240}{1760} \text{ mi}$ Multiply fractions.

$= 5.25 \text{ mi}$ Simplify.

Skill Practice

Convert using conversion factors.

5. 720 in. = ____ ft
6. 4224 ft = ____ mi
7. 8 mi = ____ yd

Classroom Examples: p. 420, Exercises 36 and 40

TIP: It is important to write the units associated with the numbers. The units can help you select the correct conversion factor.

Answers

3. $\dfrac{1 \text{ ft}}{12 \text{ in.}}$ **4.** $\dfrac{1 \text{ yd}}{3 \text{ ft}}$ **5.** 60 ft
6. 0.8 mi **7.** 14,080 yd

c. From Table 7-1 we have 1 mi = 5280 ft.

$$8.2 \text{ mi} = 8.2 \text{ mi} \cdot \frac{5280 \text{ ft}}{1 \text{ mi}} \quad \longleftarrow \text{ new unit to convert to}$$
$$\longleftarrow \text{ unit to convert from}$$

$$= \frac{8.2 \text{ mi}}{1} \cdot \frac{5280 \text{ ft}}{1 \text{ mi}} \qquad \text{The units of mi divide out, leaving the answer in feet.}$$

$$= 43,296 \text{ ft}$$

Skill Practice

Convert using conversion factors.

8. 6.2 yd = _____ in.

9. 6336 in. = _____ mi

Classroom Example: p. 420, Exercise 46

Example 3 | **Making Multiple Conversions of Length**

a. 0.25 mi = _____ in. **b.** 22 in. = _____ yd

Solution:

a. To convert miles to inches, we use two conversion factors. The first converts miles to feet. The second converts feet to inches.

converts mi to ft | converts ft to in.

$$0.25 \text{ mi} = 0.25 \text{ mi} \cdot \frac{5280 \text{ ft}}{1 \text{ mi}} \cdot \frac{12 \text{ in.}}{1 \text{ ft}}$$

$$= \frac{0.25 \text{ mi}}{1} \cdot \frac{5280 \text{ ft}}{1 \text{ mi}} \cdot \frac{12 \text{ in.}}{1 \text{ ft}} \qquad \text{The units mi and ft divide out, leaving the answer in inches.}$$

$$= 15,840 \text{ in.}$$

converts in. to ft | converts ft to yd

b. $22 \text{ in.} = 22 \text{ in.} \cdot \dfrac{1 \text{ ft}}{12 \text{ in.}} \cdot \dfrac{1 \text{ yd}}{3 \text{ ft}}$ | Multiply by two conversion factors. The first converts inches to feet. The second converts feet to yards.

$$= \frac{22 \text{ in.}}{1} \cdot \frac{1 \text{ ft}}{12 \text{ in.}} \cdot \frac{1 \text{ yd}}{3 \text{ ft}} \qquad \text{The units in. and ft divide out, leaving the answer in yards.}$$

$$= \frac{22}{36} \text{ yd} \qquad \text{Multiply fractions.}$$

$$= \frac{11}{18} \text{ yd} \quad \text{or} \quad 0.6\overline{1} \text{ yd} \qquad \text{Simplify.}$$

4. Adding and Subtracting Mixed Units

To add and subtract measurements, we must have like units. For example:

$$3 \text{ ft} + 8 \text{ ft} = 11 \text{ ft}$$

Sometimes, however, measurements have mixed units. For example, a drainpipe might be 4 ft 6 in. long or symbolically 4′6″. Measurements and calculations with mixed units can be handled in much the same way as mixed numbers.

Answers

8. 223.2 in. **9.** 0.1 mi

| **Example 4** | **Adding and Subtracting Mixed Units of Measurement** |

a. Add 4′6″ + 2′9″.

b. Subtract 8′2″ − 3′6″.

Solution:

a.
$$
\begin{array}{r}
4'6'' + 2'9'' = \quad 4\ \text{ft} + \ 6\ \text{in.} \\
+2\ \text{ft} + \ 9\ \text{in.} \\
\hline
6\ \text{ft} + 15\ \text{in.} \\
\end{array}
$$
   Add like units.

= 6 ft + 1 ft + 3 in.   Because 15 in. is more than 1 ft, we can write 15 in. = 1 ft + 3 in.

= 7 ft 3 in.   or   7′3″

b.
$$
8'2'' - 3'6'' = \quad \begin{array}{r} 8\ \text{ft} + 2\ \text{in.} \\ -(3\ \text{ft} + 6\ \text{in.}) \end{array} = \quad \begin{array}{r} \overset{7}{8}\ \text{ft} + \overset{12\,\text{in.}}{2}\ \text{in.} \\ -(3\ \text{ft} + 6\ \text{in.}) \end{array}
$$
   Borrow 1 ft = 12 in.

$$
= \quad \begin{array}{r} 7\ \text{ft} + 14\ \text{in.} \\ -(3\ \text{ft} + \ 6\ \text{in.}) \\ \hline 4\ \text{ft} + \ 8\ \text{in.} \end{array} \quad \text{or}\quad 4'8''
$$

5. Applications

In Example 5 we convert to a common unit of measurement to find the perimeter of an object.

| **Example 5** | **Finding Perimeter** |

Find the perimeter in feet.

2.2 mi

1056 ft

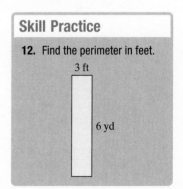
Solution:

The perimeter is found by adding the lengths of the sides. Before we begin, the length and width must have the same units. We will convert 2.2 mi to feet.

First note that $2.2\ \text{mi} = 2.2\ \cancel{\text{mi}} \cdot \dfrac{5280\ \text{ft}}{1\ \cancel{\text{mi}}}$

2.2 mi = 11,616 ft

1056 ft

= 11,616 ft

Perimeter = 11,616 ft + 1056 ft + 11,616 ft + 1056 ft

= 25,344 ft

Skill Practice

13. Shemika cut 10 pieces of rope that are each 2 ft 3 in. How much total rope was cut?

Classroom Example: p. 421, Exercise 78

Example 6 Applying U.S. Customary Units of Length

A plumber needs 5 pipes that are 8 ft 4 in. each. What is the total length of piping?

Solution:

First, note that 8 ft 4 in. = 8 ft + 4 in. Second, recall that multiplication represents repeated addition. Therefore, the total length is 5(8 ft + 4 in.). See Figure 7-1. We can apply the distributive property introduced in Section 1.5. That is, $a(b + c) = ab + ac$.

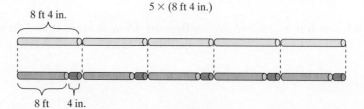

Figure 7-1

$$5(8 \text{ ft} + 4 \text{ in.}) = 5(8 \text{ ft}) + 5(4 \text{ in.})$$
$$= 40 \text{ ft} + 20 \text{ in.}$$
$$= 40 \text{ ft} + \overbrace{1 \text{ ft} + 8 \text{ in.}} \qquad \text{Note that } 20 \text{ in.} = 1 \text{ ft} + 8 \text{ in.}$$
$$= 41 \text{ ft} + 8 \text{ in.}$$

The total length of piping is 41 ft 8 in.

TIP: Example 6 could have been solved by first converting 8 ft 4 in. to either inches or feet and then multiplying by 5. Converting to inches, we have

$$8 \text{ ft } 4 \text{ in.} = 96 \text{ in.} + 4 \text{ in.} = 100 \text{ in.}$$

Now multiply by 5. $5(100 \text{ in.}) = 500 \text{ in.}$

$$= 500 \text{ in.} \cdot \left(\frac{1 \text{ ft}}{12 \text{ in.}} \right) \longleftarrow \text{Convert back to feet.}$$

$$= \frac{500}{12} \text{ ft}$$

$$= 41 \text{ ft } 8 \text{ in.} \qquad \begin{array}{l} \text{Convert to mixed} \\ \text{units by dividing.} \end{array}$$

$$\begin{array}{r} 41 \\ 12\overline{)500} \\ \underline{48} \\ 20 \\ \underline{12} \\ 8 \end{array}$$

Example 7 Applying U.S. Customary Units of Length in Construction

A carpenter has a board 6′10″ in length. He must cut the board into two pieces of equal length. How long is each piece?

Solution:

The distance 6′10″ is equal to 6 ft 10 in. or 6 ft + 10 in. We can divide this total distance in half by dividing each individual unit by 2. That is,

$$(6 \text{ ft} + 10 \text{ in.}) \div 2 = \frac{6 \text{ ft}}{2} + \frac{10 \text{ in.}}{2} = 3 \text{ ft} + 5 \text{ in.} \quad \text{or} \quad 3'5''$$

Skill Practice

14. A 12′6″ length of ribbon is cut into 3 pieces of equal length. Find the length of each piece.

Classroom Example: p. 422, Exercise 84

Answer

14. 4 ft 2 in.

Section 7.1 Practice Exercises

Study Skills Exercise

For additional exercises, see Classroom Activities 7.1A–7.1B in the *Student's Resource Manual* at www.mhhe.com/moh.

Careless mistakes are usually caused by losing focus on what you are doing. What are some of the distractions that you encounter when doing homework?

List some ways you can avoid distractions while doing your homework.

Vocabulary and Key Concepts

1. a. To measure an object means to assign it a number and a unit of __measure__.

   b. A fraction containing two equivalent units of measure is called a __conversion__ factor, which is used to convert units of measurement.

Concept 1: U.S. Customary Units of Length

2. What is wrong with the following statement? The statement is ambiguous because no units of measurement are given. For example, it is not clear whether the table is 4.2 ft long, 4.2 in. long, 4.2 yd long, and so on.

   "The length of a table is 4.2."

For Exercises 3–8, fill in the blanks with the correct units. Refer to Table 7-1.

3. 5280 ft = 1 ____ 1 mi

4. $\frac{1}{12}$ ft = 1 ____ 1 in.

5. 1 yd = 3 ____ 3 ft

6. 1 ft = 12 ____ 12 in.

7. 1 ft = $\frac{1}{3}$ ____ $\frac{1}{3}$ yd

8. 1 ft = $\frac{1}{5280}$ ____ $\frac{1}{5280}$ mi

For Exercises 9–14, select the most reasonable measurement.

9. A shoe box is approximately ____ b ____ long.

   a. 6 in.   **b.** 14 in.

   c. 2 ft   **d.** 1 yd

10. A dining room table is ____ d ____ long.

   a. 2 ft   **b.** 20 in.

   c. 10 yd   **d.** $1\frac{1}{2}$ yd

11. The height of a grown man is ____ a ____.

   a. 72 in.   **b.** 18 in.

   c. 18 ft   **d.** 3 yd

12. A typical ceiling height is ____ c ____.

   a. 5 ft   **b.** 7 yd

   c. 8 ft   **d.** 65 in.

 Writing   Translating Expression   Geometry   Scientific Calculator   Video   NS E

13. The height of a one-story house is approximately ____c____.

 a. 30 ft   **b.** 60 in.

 c. 15 ft   **d.** 20 yd

14. The length of a typical computer mouse is ____a____.

 a. 5 in.   **b.** 10 in.

 c. 2 in.   **d.** $\frac{2}{3}$ yd

Concept 2: Converting U.S. Customary Units of Length by Using Substitution

For Exercises 15–26, convert the units of length by using substitution. **(See Example 1.)**

15. 2 yd = ____ ft   6 ft

16. 2 mi = ____ yd   3520 yd

17. 6 ft = ____ in.   72 in.

18. 1.25 mi = ____ ft   6600 ft

19. 2 mi = ____ ft   10,560 ft

20. 5 ft = ____ in.   60 in.

21. 24 ft = ____ yd   8 yd

22. 36 in. = ____ ft   3 ft

23. 9 in. = ____ ft $\frac{3}{4}$ ft

24. 10 ft = ____ yd $3\frac{1}{3}$ yd

25. 1760 ft = ____ mi $\frac{1}{3}$ mi

26. 880 yd = ____ mi $\frac{1}{2}$ mi

Concept 3: Converting U.S. Customary Units of Length by Using Conversion Factors

27. Identify an appropriate ratio to convert feet to inches by using multiplication.   b

 a. $\dfrac{12\ \text{ft}}{1\ \text{in.}}$ **b.** $\dfrac{12\ \text{in.}}{1\ \text{ft}}$ **c.** $\dfrac{1\ \text{ft}}{12\ \text{in.}}$ **d.** $\dfrac{1\ \text{in.}}{12\ \text{ft}}$

28. Identify an appropriate ratio to convert yards to miles by using multiplication.   c

 a. $\dfrac{1\ \text{mi}}{5280\ \text{ft}}$ **b.** $\dfrac{1760\ \text{mi}}{1\ \text{yd}}$ **c.** $\dfrac{1\ \text{mi}}{1760\ \text{yd}}$ **d.** $\dfrac{1760\ \text{yd}}{1\ \text{mi}}$

29. Identify an appropriate ratio to convert yards to feet by using multiplication.   a

 a. $\dfrac{3\ \text{ft}}{1\ \text{yd}}$ **b.** $\dfrac{1\ \text{yd}}{3\ \text{ft}}$ **c.** $\dfrac{3\ \text{yd}}{1\ \text{ft}}$ **d.** $\dfrac{1\ \text{ft}}{3\ \text{yd}}$

30. Identify an appropriate ratio to convert feet to miles by using multiplication.   d

 a. $\dfrac{5280\ \text{ft}}{1\ \text{mi}}$ **b.** $\dfrac{1\ \text{ft}}{5280\ \text{mi}}$ **c.** $\dfrac{5280\ \text{mi}}{1\ \text{ft}}$ **d.** $\dfrac{1\ \text{mi}}{5280\ \text{ft}}$

For Exercises 31–42, convert the units of length by using conversion factors. **(See Example 2.)** Animation

31. 9 ft = ____ yd   3 yd

32. $2\frac{1}{3}$ yd = ____ ft   7 ft

33. 3.5 ft = ____ in.   42 in.

34. $4\frac{1}{2}$ in. = ____ ft $\frac{3}{8}$ ft

35. 11,880 ft = ____ mi $2\frac{1}{4}$ mi

36. 0.75 mi = ____ ft   3960 ft

37. 6 yd = ____ ft   18 ft

38. 5280 yd = ____ mi   3 mi

39. 14 ft = ____ yd $4\frac{2}{3}$ yd

40. 75 in. = ____ ft $6\frac{1}{4}$ ft

41. 320 mi = ____ yd   563,200 yd

42. $3\frac{1}{4}$ ft = ____ in.   39 in.

For Exercises 43–51, convert the units of length, involving multiple conversions. **(See Example 3.)**

43. 171 in. = ____ yd $4\frac{3}{4}$ yd

44. 0.3 mi = ____ in.   19,008 in.

45. 2 yd = ____ in.   72 in.

46. 12,672 in. = ____ mi $\frac{1}{5}$ mi or 0.2 mi

47. 0.8 mi = ____ in.   50,688 in.

48. 900 in. = ____ yd   25 yd

49. 12,672 in. = ____ mi   0.2 mi

50. 6 yd = ____ in.   216 in.

51. 1.6 mi = ____ in.   101,376 in.

Concept 4: Adding and Subtracting Mixed Units

52. a. Convert 6′4″ to inches.   76 in.

   b. Convert 6′4″ to feet. $6\frac{1}{3}$ ft

 53. a. Convert 10 ft 8 in. to inches.   128 in.

   b. Convert 10 ft 8 in. to feet. $10\frac{2}{3}$ ft

54. a. Convert 2 yd 2 ft to feet.   8 ft

   b. Convert 2 yd 2 ft to yards. $2\frac{2}{3}$ yd

55. a. Convert 3′6″ to feet.   3.5 ft

   b. Convert 3′6″ to inches.   42 in.

For Exercises 56–64, add or subtract as indicated. **(See Example 4.)**

56. 1′3″ + 6′4″   7′7″

57. 6′2″ + 4′6″   10′8″

58. 2 ft 8 in. + 3 ft 4 in.   6 ft

59. 5 ft 2 in. + 6 ft 10 in.   12 ft

60. 4′10″ + 6′4″   11′2″

61. 4′9″ + 3′9″   8′6″

 62. 8 ft 8 in. − 5 ft 4 in.   3 ft 4 in.

63. 3 ft 2 in. − 1 ft 5 in.   1 ft 9 in.

64. 9′2″ − 4′10″   4′4″

For Exercises 65–72, multiply or divide as indicated.

65. 2(4 ft 5 in.)   8 ft 10 in.

66. 4(5 ft 1 in.)   20 ft 4 in.

67. 6(4 ft 8 in.)   28 ft

68. 8(2 ft 5 in.)   19 ft 4 in.

69. (6′4″) ÷ 2   3′2″

70. (16′8″) ÷ 8   2′1″

71. $\dfrac{18 \text{ ft } 3 \text{ in.}}{3}$ 6 ft 1 in.

72. $\dfrac{10 \text{ ft } 10 \text{ in.}}{5}$ 2 ft 2 in.

Concept 5: Applications (Mixed Exercises)

73. Find the perimeter in feet. **(See Example 5.)**

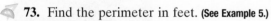
$5\frac{1}{2}$ ft

74. Find the perimeter in yards.   6 yd   16 yd

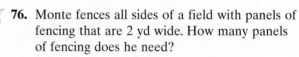

75. The garden pictured needs a decorative border. The border comes in pieces that are 1.5 ft long. How many pieces of border are needed?

18 pieces of border are needed.

76. Monte fences all sides of a field with panels of fencing that are 2 yd wide. How many panels of fencing does he need?

59 panels of fencing are needed.

77. A carpenter needs eight pieces of molding that are 6 ft 4 in. each. What is the total length? **(See Example 6.)**   50 ft 8 in. is needed.

78. To gift wrap a package, 3′2″ of ribbon is needed. How much ribbon is needed to wrap 10 packages?   31 ft 8 in. is needed.

79. A plumber used two pieces of pipe for a job. One piece was 4′6″ and the other was 2′8″. How much pipe was used?

The plumber used 7′2″ of pipe.

80. A carpenter needs to put wood molding around three sides of a room. Two sides are 6′8″ long, and the third side is 10′ long. How much molding should the carpenter purchase?   The carpenter needs 23′4″ of molding.

✏ Writing   ←→ Translating Expression   ◤ Geometry   🖩 Scientific Calculator   ▶ Video   NS&E

81. If you have 4 yd of rope and you use 5 ft, how much is left over? Express the answer in feet.   7 ft is left over.

82. In 2002, the Blaisdell Arena football field, home of the Hawaiian Islanders football team, was discovered to be smaller than the official dimensions. The width was measured to be 82 ft 10 in. Regulation width is 85 ft. What is the difference between the widths of a regulation field and the field at Blaisdell Arena?   The arena is 2 ft 2 in. shorter than regulation.

83. A cable 6 ft 9 in. is cut into three pieces of equal length. How long is each piece? **(See Example 7.)**   Each piece is 2 ft 3 in. long.

84. A piece of rope 8′6″ in length is cut in half. How long is each piece?   Each piece is 4′3″ long.

85. A picnic table requires 5 boards that are 6′ long, 4 boards that are 3′3″ long, and 2 boards that are 18″ long. Find the total length of lumber required.   The total length is 46′.

86. A roll of ribbon is 60 yd. If you wrap 12 packages that each use 2.5 ft of ribbon, how much ribbon is left over?   There is 50 yd left over.

87. Jessica wants to put adhesive outdoor tread on the 14 steps leading to her front door. Her plan is to center two strips of tread, 4 in. from the edges of each step, as shown in the figure. If the tread is packaged in 5-yd rolls, how many rolls should she buy?   4 rolls

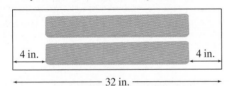

4 in. ──── 4 in.
──────── 32 in. ────────

88. Each year the Moon's orbit moves 1.5 in. farther away from Earth. At this rate, how far will the Moon's orbit move in the next 100 years? Express your answer in feet.   12.5 ft

Expanding Your Skills

In Section 1.5 we learned that area is measured in square units such as in.^{2}, ft^2, yd^2, and mi^2. Converting square units involves a different set of conversion factors. For example, 1 yd = 3 ft, but 1 yd^2 = 9 ft^2. To understand why, recall that the formula for the area of a rectangle is $A = l \times w$. In a square, the length and the width are the same distance s. Therefore, the area of a square is given by the formula $A = s \times s$, where s is the length of a side. Thus,

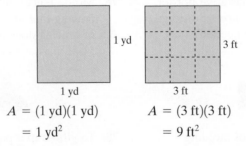

1 yd

3 ft

1 yd

3 ft

$$A = (1 \text{ yd})(1 \text{ yd})$$
$$= 1 \text{ yd}^2$$

$$A = (3 \text{ ft})(3 \text{ ft})$$
$$= 9 \text{ ft}^2$$

Instead of learning a new set of conversion factors, we can use multiples of the conversion factors that we mastered in Section 7.1.

Example: Converting Area

Convert 4 yd^2 to square feet

Solution: We will use the conversion factor of $\dfrac{3 \text{ ft}}{1 \text{ yd}}$ twice.

$$\frac{4 \text{ yd}^2}{1} \cdot \underbrace{\frac{3 \text{ ft}}{1 \text{ yd}} \cdot \frac{3 \text{ ft}}{1 \text{ yd}}}_{\text{Multiply first.}} = \frac{4 \text{ yd}^2}{1} \cdot \frac{9 \text{ ft}^2}{1 \text{ yd}^2} = 36 \text{ ft}^2$$

> **TIP:** We use the conversion factor $\frac{3 \text{ ft}}{1 \text{ yd}}$ twice because there are two dimensions. Length and width must both be converted to feet.

For Exercises 89–96, convert the units of area by using multiple factors of the given conversion factor.

89. $54 \text{ ft}^2 = \underline{\hspace{0.5in}} \text{ yd}^2$ (Use two factors of the ratio $\dfrac{1 \text{ yd}}{3 \text{ ft}}$.) $\quad 6 \text{ yd}^2$

90. $108 \text{ ft}^2 = \underline{\hspace{0.5in}} \text{ yd}^2 \quad 12 \text{ yd}^2$

91. $432 \text{ in.}^2 = \underline{\hspace{0.5in}} \text{ ft}^2$ (Use two factors of the ratio $\dfrac{1 \text{ ft}}{12 \text{ in.}}$.) $\quad 3 \text{ ft}^2$

92. $720 \text{ in}^2 = \underline{\hspace{0.5in}} \text{ ft}^2 \quad 5 \text{ ft}^2$

93. $5 \text{ ft}^2 = \underline{\hspace{0.5in}} \text{ in.}^2$ (Use two factors of $\dfrac{12 \text{ in.}}{1 \text{ ft}}$.) $\quad 720 \text{ in.}^2$

94. $7 \text{ ft}^2 = \underline{\hspace{0.5in}} \text{ in.}^2 \quad 1008 \text{ in.}^2$

95. $3 \text{ yd}^2 = \underline{\hspace{0.5in}} \text{ ft}^2$ (Use two factors of $\dfrac{3 \text{ ft}}{1 \text{ yd}}$.) $\quad 27 \text{ ft}^2$

96. $10 \text{ yd}^2 = \underline{\hspace{0.5in}} \text{ ft}^2 \quad 90 \text{ ft}^2$

Converting U.S. Customary Units of Time, Weight, and Capacity

1. U.S. Customary Units of Time, Weight, and Capacity

In this section we convert U.S. Customary units of time, weight, and capacity. Table 7-2 summarizes several common units and their equivalent measures. These relationships should be memorized as common knowledge.

Concepts

1. U.S. Customary Units of Time, Weight, and Capacity
2. Converting Units of Time
3. Converting U.S. Customary Units of Weight
4. Converting U.S. Customary Units of Capacity

Table 7-2 **Summary of U.S. Customary Units of Length, Time, Weight, and Capacity**

Length	Time
1 foot (ft) = 12 inches (in.)	1 year (yr) = 365 days
1 yard (yd) = 3 feet (ft)	1 week (wk) = 7 days
1 mile (mi) = 5280 feet (ft)	1 day = 24 hours (hr)
1 mile (mi) = 1760 yards (yd)	1 hour (hr) = 60 minutes (min)
	1 minute (min) = 60 seconds (sec)
Capacity	**Weight**
3 teaspoons (tsp) = 1 tablespoon (T)	1 pound (lb) = 16 ounces (oz)
1 cup (c) = 8 fluid ounces (fl oz)	1 ton = 2000 pounds (lb)
1 pint (pt) = 2 cups (c)	
1 quart (qt) = 2 pints (pt)	
1 quart (qt) = 4 cups (c)	
1 gallon (gal) = 4 quarts (qt)	

2. Converting Units of Time

In Example 1, we convert from one unit of time to another.

> **Example 1** Converting Units of Time
>
> **a.** 32 hr = ___ days   **b.** 36 hr = ___ sec
>
> **Solution:**
>
> **a.** $32 \text{ hr} = \dfrac{32 \text{ hr}}{1} \cdot \dfrac{1 \text{ day}}{24 \text{ hr}}$ ← new unit to convert to / ← unit to convert from   Recall that 1 day = 24 hr.
>
> $= \dfrac{32}{24} \text{ days}$ Multiply fractions.
>
> $= \dfrac{4}{3} \text{ days or } 1\dfrac{1}{3} \text{ days}$ Simplify.
>
> converts hr to min   converts min to sec
>
> **b.** $36 \text{ hr} = \dfrac{36 \text{ hr}}{1} \cdot \dfrac{60 \text{ min}}{1 \text{ hr}} \cdot \dfrac{60 \text{ sec}}{1 \text{ min}}$ Multiply by two conversion factors.
>
> $= 129{,}600 \text{ sec}$ Simplify.

> **Example 2** Converting Units of Time
>
> After running a marathon, Dave crossed the finish line and noticed that the race clock read 2:20:30. Convert this time to minutes.
>
> **Solution:**
>
> The notation 2:20:30 means 2 hr 20 min 30 sec. We must convert 2 hr to minutes and 30 sec to minutes. Then we add the total number of minutes.
>
> $$2 \text{ hr} = \dfrac{2 \text{ hr}}{1} \cdot \dfrac{60 \text{ min}}{1 \text{ hr}} = 120 \text{ min}$$
>
> $$30 \text{ sec} = \dfrac{30 \text{ sec}}{1} \cdot \dfrac{1 \text{ min}}{60 \text{ sec}} = \dfrac{30}{60} \text{ min} = \dfrac{1}{2} \text{ min} \quad \text{or} \quad 0.5 \text{ min}$$
>
> The total number of minutes is 120 min + 20 min + 0.5 min = 140.5 min. Dave finished the race in 140.5 min.

3. Converting U.S. Customary Units of Weight

Measurements of weight record the force of an object subject to gravity. In Example 3 we convert from one unit of weight to another.

Example 3 **Converting Units of Weight**

a. The average weight of an adult male African elephant is 12,400 lb. Convert this value to tons.

b. Convert the weight of a 7-lb 3-oz baby to ounces.

Solution:

a. Recall that 1 ton = 2000 lb.

$$12{,}400 \text{ lb} = \frac{12{,}400 \text{ lb}}{1} \cdot \frac{1 \text{ ton}}{2000 \text{ lb}}$$

$$= \frac{12{,}400}{2000} \text{ tons} \quad \begin{array}{l}\text{Multiply}\\\text{fractions.}\end{array}$$

$$= \frac{31}{5} \text{ tons} \quad \text{or} \quad 6.2 \text{ tons}$$

An adult male African elephant weighs 6.2 tons.

b. To convert 7 lb 3 oz to ounces, we must convert 7 lb to ounces.

$$7 \text{ lb} = \frac{7 \text{ lb}}{1} \cdot \frac{16 \text{ oz}}{1 \text{ lb}} \qquad \text{Recall that 1 lb} = 16 \text{ oz.}$$

$$= 112 \text{ oz}$$

The baby's total weight is 112 oz + 3 oz = 115 oz.

Skill Practice

4. The blue whale is the largest animal on Earth. It is so heavy that it would crush under its own weight if it were taken from the water. An average adult blue whale weighs 120 tons. Convert this to pounds.
5. A box of apples weighs 5 lb 12 oz. Convert this to ounces.

Classroom Examples: p. 428, Exercises 44 and 46

Example 4 **Applying U.S. Customary Units of Weight**

Rashona lifts four boxes of books. The boxes have the following weights: 16 lb 4 oz, 18 lb 8 oz, 12 lb 5 oz, and 22 lb 9 oz. How much weight did she lift altogether?

Solution:

```
   16 lb   4 oz                    Add like units in columns.
   18 lb   8 oz
   12 lb   5 oz
 + 22 lb   9 oz
   68 lb  26 oz  = 68 lb + 26 oz
```

$$= 68 \text{ lb} + \overbrace{1 \text{ lb} + 10 \text{ oz}} \qquad \text{Recall that 1 lb} = 16 \text{ oz.}$$

$$= 69 \text{ lb} \ 10 \text{ oz}$$

Rashona lifted 69 lb 10 oz of books.

Skill Practice

6. A set of triplets weighed 4 lb 3 oz, 3 lb 9 oz, and 4 lb 5 oz. What is the total weight of all three babies?

Classroom Example: p. 428, Exercise 50

4. Converting U.S. Customary Units of Capacity

A typical can of soda contains 12 fl oz. This is a measure of capacity. Capacity is the volume or amount that a container can hold. The U.S. Customary units of capacity are fluid ounces (fl oz), cup (c), pint (pt), quart (qt), and gallon (gal).

Answers
4. 240,000 lb   **5.** 92 oz
6. 12 lb 1 oz

One fluid ounce is approximately the amount of liquid that two large spoonfuls will hold. One cup is the amount in an average-size cup of tea. While Table 7-2 summarizes the relationships among units of capacity, we also offer an illustration (Figure 7-2).

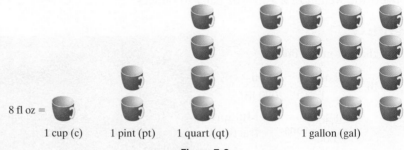

8 fl oz =

1 cup (c)   1 pint (pt)   1 quart (qt)   1 gallon (gal)

Figure 7-2

Example 5 Converting Units of Capacity

a. $1.25 \text{ pt} = \underline{\hspace{1cm}} \text{ qt}$ **b.** $2 \text{ gal} = \underline{\hspace{1cm}} \text{ c}$ **c.** $48 \text{ fl oz} = \underline{\hspace{1cm}} \text{ gal}$

Solution:

a. $1.25 \text{ pt} = \dfrac{1.25 \text{ pt}}{1} \cdot \dfrac{1 \text{ qt}}{2 \text{ pt}}$ Recall that 1 qt = 2 pt.

$= \dfrac{1.25}{2} \text{ qt}$ Multiply fractions.

$= 0.625 \text{ qt}$ Simplify.

b. $2 \text{ gal} = 2 \text{ gal} \cdot \dfrac{4 \text{ qt}}{1 \text{ gal}} \cdot \dfrac{4 \text{ c}}{1 \text{ qt}}$ Use two conversion factors. The first converts gallons to quarts. The second converts quarts to cups.

$= \dfrac{2 \text{ gal}}{1} \cdot \dfrac{4 \text{ qt}}{1 \text{ gal}} \cdot \dfrac{4 \text{ c}}{1 \text{ qt}}$

$= 32 \text{ c}$ Multiply.

c. $48 \text{ fl oz} = \dfrac{48 \text{ fl oz}}{1} \cdot \dfrac{1 \text{ c}}{8 \text{ fl oz}} \cdot \dfrac{1 \text{ qt}}{4 \text{ c}} \cdot \dfrac{1 \text{ gal}}{4 \text{ qt}}$ Convert from fluid ounces to cups, from cups to quarts, and from quarts to gallons.

$= \dfrac{48}{128} \text{ gal}$

$= \dfrac{3}{8} \text{ gal} \quad \text{or} \quad 0.375 \text{ gal}$

Example 6 Applying Units of Capacity

A recipe calls for $1\frac{3}{4}$ c of chicken broth. A can of chicken broth holds 14.5 fl oz. Is there enough chicken broth in the can for the recipe?

Solution:

We need to convert each measurement to the same unit of measure for comparison. Converting $1\frac{3}{4}$ c to fluid ounces, we have

$$1\frac{3}{4} \text{ c} = \frac{7}{4} \cancel{c} \cdot \frac{8 \text{ fl oz}}{1 \cancel{c}} \quad \text{Recall that 1 c = 8 fl oz.}$$

$$= \frac{56}{4} \text{ fl oz} \quad \text{Multiply fractions.}$$

$$= 14 \text{ fl oz} \quad \text{Simplify.}$$

The recipe calls for $1\frac{3}{4}$ c or 14 fl oz of chicken broth. The can of chicken broth holds 14.5 fl oz which is enough.

Classroom Example: p. 429, Exercise 72

Skill Practice

12. A recipe calls for $3\frac{1}{2}$ c of tomato sauce. A jar of sauce holds 24 fl oz. Is there enough sauce in the jar for the recipe?

Answer

12. No, $3\frac{1}{2}$ c is equal to 28 fl oz. The jar only holds 24 fl oz.

Section 7.2 Practice Exercises

Review Exercises

For additional exercises, see Classroom Activities 7.2A–7.2C in the *Student's Resource Manual* at www.mhhe.com/moh.

1. Which ratio is a conversion factor? c

   a. $\frac{3 \text{ ft}}{12 \text{ in.}}$ **b.** $\frac{3 \text{ yd}}{1 \text{ ft}}$ **c.** $\frac{3 \text{ ft}}{1 \text{ yd}}$ **d.** $\frac{1 \text{ yd}}{12 \text{ in.}}$

For Exercises 2–8, complete the table.

	Object	in.	ft	yd	mi
2.	Distance to work	190,080 in.	15,840 ft	5,280 yd	3 mi
3.	Length of a hallway	144 in.	12 ft	4 yd	
4.	Length of a car	168 in.	14 ft	$4\frac{2}{3}$ yd	
5.	Height of a tree	216 in.	18 ft	6 yd	
6.	A basketball player's height	76 in.	$6\frac{1}{3}$ ft	$2\frac{1}{9}$ yd	
7.	Perimeter of a backyard	1,800 in.	150 ft	50 yd	
8.	Distance to the gas station	95,040 in.	7,920 ft	2,640 yd	$1\frac{1}{2}$ mi

Concept 1: U.S. Customary Units of Time, Weight, and Capacity

For Exercises 9–16, fill in the blanks with the correct units.

9. 1 qt = 2 ___
   2 pt

10. 1 c = 8 ___
   8 fl oz

11. 1 lb = 16 ___
   16 oz

12. 1 pt = 2 ___
   2 c

13. 1 yr = 365 ___
   365 days

14. 1 ___ = 2000 lb
   1 ton

15. 1 gal = 4 ___
   4 qt

16. 1 ___ = 24 hr
   1 day

NS&E For Exercises 17–20, estimate the correct measurement by selecting the best answer.

17. A healthy newborn baby weighs _____d_____.

    a. 25 lb   **b.** 50 oz

    c. 25 oz   **d.** 112 oz

18. A car weighs _____c_____.

    a. 100 lb   **b.** 100 tons

    c. 2000 lb   **d.** 500 lb

19. A bowl of soup contains _____b_____.

    a. 1 qt   **b.** 12 fl oz

    c. $\frac{1}{2}$ gal   **d.** 2 fl oz

20. A glass of water contains _____d_____.

    a. 30 fl oz   **b.** 42 fl oz

    c. 3 fl oz   **d.** 10 fl oz

Concept 2: Converting Units of Time

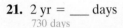

For Exercises 21–32, convert the units of time. **(See Example 1.)**

21. 2 yr = ___ days
730 days

22. $1\frac{1}{2}$ days = ___ hr
36 hr

23. 90 min = ___ hr
$1\frac{1}{2}$ hr

24. 3 wk = ___ days
21 days

25. 180 sec = ___ min
3 min

26. $3\frac{1}{2}$ hr = ___ min
210 min

27. 72 hr = ___ days
3 days

28. 28 days = ___ wk
4 wk

29. 3600 sec = ___ hr
1 hr

30. 168 hr = ___ wk
1 wk

31. 9 wk = ___ hr
1512 hr

32. 1680 hr = ___ wk
10 wk

For Exercises 33–36, convert the time given as hr:min:sec to minutes. **(See Example 2.)**

33. 1:20:30
80.5 min

34. 3:10:45
190.75 min

35. 2:55:15
175.25 min

36. 1:40:30
100.5 min

For Exercises 37 and 38, refer to the table.

37. Gil is a distance runner. The durations of his training runs for one week are given in the table. Find the total time that Gil ran that week, and express the answer by using mixed units. Gil ran for 5 hr 35 min.

38. Find the difference between the amount of time Gil trained on Monday and the amount of time he trained on Friday. The difference is 20 min.

Day	Time
Mon.	1 hr 10 min
Tues.	45 min
Wed.	1 hr 20 min
Thur.	30 min
Fri.	50 min
Sat.	Rest
Sun.	1 hr

39. In a team triathlon, Torie swam $\frac{1}{2}$ mi in 15 min 30 sec. David rode his bicycle 20 mi in 50 min 20 sec. Emilie ran 4 mi in 28 min 10 sec. Find the total time for the team. The total time was 1 hr 34 min.

40. Keiji competes in a biathlon. He runs 5 mi in 32 min 8 sec. He rides his bike 25 mi in 1 hr 2 min 40 sec. Find the total time for his race. Keiji's total time is 1 hr 34 min 48 sec.

Concept 3: Converting U.S. Customary Units of Weight

For Exercises 41–48, convert the units of weight. **(See Example 3.)**

41. 32 oz = ___ lb
2 lb

42. 2500 lb = ___ tons
$1\frac{1}{4}$ tons or 1.25 tons

43. 2 tons = ___ lb
4000 lb

44. 8 oz = ___ lb
$\frac{1}{2}$ lb or 0.5 lb

45. 4 lb = ___ oz
64 oz

46. $3\frac{1}{4}$ tons = ___ lb
6500 lb

47. 3000 lb = ___ tons
$1\frac{1}{2}$ tons or 1.5 tons

48. 6 lb = ___ oz
96 oz

For Exercises 49–54, add or subtract as indicated. **(See Example 4.)**

49. 6 lb 10 oz + 3 lb 14 oz
10 lb 8 oz

50. 12 lb 11 oz + 13 lb 7 oz
26 lb 2 oz

51. 30 lb 10 oz − 22 lb 8 oz
8 lb 2 oz

52. 5 lb − 2 lb 5 oz
2 lb 11 oz

53. 10 lb − 3 lb 8 oz
6 lb 8 oz

54. 20 lb 3 oz + 15 oz
21 lb 2 oz

55. Byron lays sod in his backyard. Each piece of sod weighs 6 lb 4 oz. If he puts down 50 pieces, find the total weight.
The total weight is 312 lb 8 oz.

56. A can of paint weighs 2 lb 4 oz. How much would 6 cans weigh?
The total weight of 6 cans is 13 lb 8 oz.

57. Two and one-half tons of trash must be moved. If a truck can handle 2500 lb in one trip, how many trips will the truck need to make to remove all the trash?
The truck will have to make 2 trips.

58. A box that contains 6 textbooks weighs 18 lb 12 oz. What is the average weight of each textbook?
Each textbook weighs 3 lb 2 oz.

Concept 4: Converting U.S. Customary Units of Capacity

For Exercises 59–70, convert the units of capacity. **(See Example 5.)**

59. 16 fl oz = _____ c
2 c

60. 5 pt = _____ c
10 c

61. 6 gal = _____ qt
24 qt

62. 8 pt = _____ qt
4 qt

63. 1 gal = _____ c
16 c

64. 1 T = _____ tsp
3 tsp

65. 2 qt = _____ gal
$\frac{1}{2}$ gal

66. 2 qt = _____ c
8 c

67. 1 pt = _____ fl oz
16 fl oz

68. 32 fl oz = _____ qt
1 qt

69. 2 T = _____ tsp
6 tsp

70. 2 gal = _____ pt
16 pt

71. A recipe for minestrone soup calls for 3 c of spaghetti sauce. If a jar of sauce has 48 fl oz, is there enough for the recipe?
(See Example 6.)
Yes, 3 c is 24 fl oz, so the 48-fl-oz jar will suffice.

72. A recipe for punch calls for 6 c of apple juice. A bottle of juice has 2 qt. Is there enough juice in the bottle for the recipe?
Yes, 6 c is 1.5 qt, so the 2-qt bottle will suffice.

73. A 24-fl-oz jar of spaghetti sauce sells for $2.69. Another jar that holds 1 qt of sauce sells for $3.29. Which is a better buy? Explain.
The unit price for the 24-fl-oz jar is about $0.112 per ounce, and the unit price for the 1-qt jar is about $0.103 per ounce; therefore the 1-qt jar is the better buy.

74. Tatiana went to purchase bottled water. She found the following options: a 12-pack of 1-pt bottles; a 1-gal jug; and a 6-pack of 24-fl-oz bottles. Which option should Tatiana choose to get the most water? Explain.
The 12-pack of 1-pt bottles is 1.5 gal, and the 6-pack is 1.125 gal. The 12-pack contains the most water.

Calculator Connections

Topic: Converting Units of Capacity

For Exercises 75–83, complete the table.

	Object	fl oz	c	pt	qt	gal
75.	Bottle of canola oil	32 fl oz	4 c	2 pt	1 qt	0.25 gal
76.	Can of soda	12 fl oz	1.5 c	0.75 pt	0.375 qt	0.09375 gal
77.	Laundry detergent	128 fl oz	16 c	8 pt	4 qt	1 gal
78.	Container of gasoline	640 fl oz	80 c	40 pt	20 qt	5 gal
79.	Bottle of Gatorade	16 fl oz	2 c	1 pt	0.5 qt	0.125 gal
80.	Container of orange juice	24 fl oz	3 c	1.5 pt	0.75 qt	0.1875 gal
81.	Bottle of spring water	8 fl oz	1 c	0.5 pt	0.25 qt	0.0625 gal
82.	Milkshake	16 fl oz	2 c	1 pt	0.5 qt	0.125 gal
83.	Jug of maple syrup	64 fl oz	8 c	4 pt	2 qt	0.5 gal

 Writing Translating Expression  Geometry Scientific Calculator Video NSE

Section 7.3 | Metric Units of Length

1. Introduction to the Metric System

Throughout history the lack of standard units of measure led to much confusion in trade between countries. In 1790 the French Academy of Sciences adopted a simple, decimal-based system of units. This system is known today as the **metric system**. The metric system is the predominant system of measurement used in science.

The simplicity of the metric system is a result of having one basic unit of measure for each type of quantity (length, mass, and capacity). The base units are the *meter* for length, the *gram* for mass, and the *liter* for capacity. Other units of length, mass, and capacity in the metric system are products of the base unit and a power of 10.

2. Metric Units of Length

The **meter** (m) is the basic unit of length in the metric system. A meter is slightly longer than a yard.

| 1 meter | 1 m ≈ 39 in. |

| 1 yard | 1 yd = 36 in. |

The meter was defined in the late 1700s as one ten-millionth of the distance along the Earth's surface from the North Pole to the Equator through Paris, France. Today the meter is defined as the distance traveled by light in a vacuum during $\frac{1}{299,792,458}$ sec.

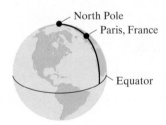

Six other common metric units of length are given in Table 7-3. Notice that each unit is related to the meter by a power of 10. This makes it particularly easy to convert from one unit to another.

TIP: The units of hectometer, dekameter, and decimeter are not frequently used.

Table 7-3 **Metric Units of Length and Their Equivalents**

1 kilometer (km) = 1000 m	
1 hectometer (hm) = 100 m	
1 dekameter (dam) = 10 m	
1 meter (m) = 1 m	
1 decimeter (dm) = 0.1 m	$(\frac{1}{10}$ m$)$
1 centimeter (cm) = 0.01 m	$(\frac{1}{100}$ m$)$
1 millimeter (mm) = 0.001 m	$(\frac{1}{1000}$ m$)$

TIP: In addition to the key facts presented in Table 7-3, the following equivalences are useful.

100 cm = 1 m
1000 mm = 1 m

Notice that each unit of length has a prefix followed by the word *meter* (kilometer, for example). You should memorize these prefixes along with their multiples of the basic unit, the meter. Furthermore, it is generally easiest to memorize the prefixes in order.

kilo-	hecto-	deka-	meter	deci-	centi-	milli-
× 1000	× 100	× 10	× 1	× 0.1	× 0.01	× 0.001

As you familiarize yourself with the metric units of length, it is helpful to have a sense of the distance represented by each unit (Figure 7-3).

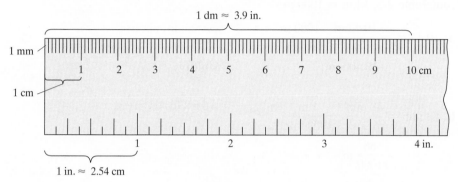

Figure 7-3

- 1 *millimeter* is approximately the thickness of five sheets of paper.
- 1 *centimeter* is approximately the width of a key on a calculator.
- 1 *decimeter* is approximately 4 in.
- 1 *meter* is just over 1 yd.
- A *kilometer* is used to express longer distances in much the same way we use miles. The distance between Los Angeles and San Diego is about 193 km.

Example 1 **Measuring Distances in Metric Units**

Approximate the distance in centimeters and in millimeters.

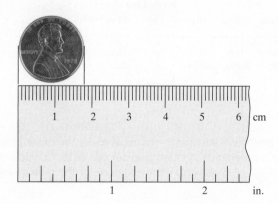

Solution:

The numbered lines on the ruler are units of centimeters. Each centimeter is divided into 10 mm. We see that the width of the penny is not quite 2 cm. We can approximate this distance as 1.8 cm or equivalently 18 mm.

Concept Connections

Fill in the blank with < or >.

1. 1 km ☐ 1 m

2. 1 cm ☐ 1 m

3. 1 dam ☐ 1 dm

Select the most reasonable value.

4. The length of a fork is
  a. 20 m   **b.** 20 km
  c. 20 cm   **d.** 20 mm

5. The length of a city block is
  a. $\frac{1}{2}$ km   **b.** $\frac{1}{2}$ cm
  c. $\frac{1}{2}$ m   **d.** $\frac{1}{2}$ mm

Skill Practice

6. Approximate the length of the pin in centimeters and in millimeters.

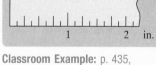

Classroom Example: p. 435, Exercise 14

Answers

1. >   **2.** <   **3.** >   **4.** c
5. a   **6.** 3.2 cm or 32 mm

3. Converting Metric Units of Length

In Example 2, we convert metric units of length by using conversion factors.

Example 2 **Converting Metric Units of Length**

a. 10.4 km = _____ m

b. 88 mm = _____ m

Solution:

From Table 7-3, 1 km = 1000 m.

a. $10.4 \text{ km} = \dfrac{10.4 \text{ km}}{1} \cdot \dfrac{1000 \text{ m}}{1 \text{ km}}$ ← new unit to convert to
← unit to convert from

$= 10{,}400 \text{ m}$ Multiply.

b. $88 \text{ mm} = \dfrac{88 \text{ mm}}{1} \cdot \dfrac{1 \text{ m}}{1000 \text{ mm}}$ ← new unit to convert to
← unit to convert from

$= \dfrac{88}{1000} \text{ m}$

$= 0.088 \text{ m}$

Recall that the place positions in our numbering system are based on powers of 10. For this reason, when we multiply a number by 10, 100, or 1000, we move the decimal point 1, 2, or 3 places, respectively, to the right. Similarly when we multiply by 0.1, 0.01, or 0.001, we move the decimal point to the left 1, 2, or 3 places, respectively.

Since the metric system is also based on powers of 10, we can convert between two metric units of length by moving the decimal point. The direction and number of place positions to move are based on the metric **prefix line**, shown in Figure 7-4.

Prefix Line

1000 m	100 m	10 m	1 m	0.1 m	0.01 m	0.001 m
km	hm	dam	m	dm	cm	mm
kilo-	hecto-	deka-		deci-	centi-	milli-

Figure 7-4

TIP: To use the prefix line effectively, you must know the order of the metric prefixes. Sometimes a mnemonic (memory device) can help. Consider the following sentence. The first letter of each word represents one of the metric prefixes.

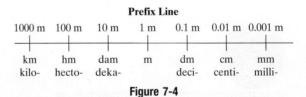

kids	have	doughnuts	until	dad	calls	mom.
kilo-	hecto-	deka-	unit	deci-	centi-	milli-

↑
represents the main
unit of measurement
(meter, liter, or gram)

Using the Prefix Line to Convert Metric Units

Step 1 To use the prefix line, begin at the point on the line corresponding to the original unit you are given.

Step 2 Then count the number of positions you need to move to reach the new unit of measurement.

Step 3 Move the decimal point in the original measured value the same direction and same number of places as on the prefix line.

Step 4 Replace the original unit with the new unit of measure.

Example 3 **Using the Prefix Line to Convert Metric Units of Length**

Use the prefix line for each conversion.

a. 0.0413 m = _____ cm   **b.** 4700 cm = _____ km

Solution:

a.

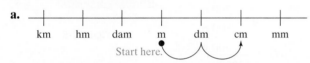

0.0413 m = 4.13 cm   From the prefix line, move the decimal point two places to the right.

b.

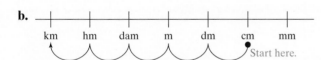

4700 cm = 04700. cm = 0.047 km   From the prefix line, move the decimal point five places to the left.

Skill Practice

Convert.

9. 864 cm = _____ m
10. 8.2 km = _____ m

Classroom Examples: p. 437, Exercises 40 and 44

Example 4 **Converting Metric Units of Length**

a. Mt. Everest, the highest mountain in the world, is 8850 m. How many kilometers is this?

b. Shaquille O'Neal is 2.159 m tall. How many millimeters is this?

Solution:

a. $8850 \text{ m} = \dfrac{8850 \text{ m}}{1} \cdot \dfrac{1 \text{ km}}{1000 \text{ m}}$ ← new unit to convert to
    ← unit to convert from

$= 8.85 \text{ km}$

8850 m = 8.85 km

Skill Practice

11. The Marianas Trench in the Pacific Ocean is 11,033 m deep. How many kilometers is this?

12. The length of an adult blue whale, the largest animal on Earth, is 24 m. How many centimeters is this?

Classroom Example: p. 437, Exercise 52

Answers

9. 8.64 m   **10.** 8200 m
11. 11.033 km   **12.** 2400 cm

b. $2.159 \text{ m} = \dfrac{2.159 \text{ m}}{1} \cdot \dfrac{1000 \text{ mm}}{1 \text{ m}}$ ⟵ new unit to convert to
⟵ unit to convert from

$= 2159 \text{ mm}$

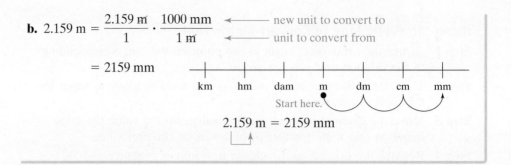

$2.159 \text{ m} = 2159 \text{ mm}$

4. Applications

Example 5 **Applying Metric Units of Length**

A rectangular field is 800 m by 300 m. If a farmer has 2 km of fencing, will this be enough to surround the field (see Figure 7-5)?

Solution:

The perimeter of the field is given by

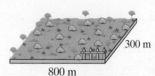

Figure 7-5

$$
\begin{array}{r}
800 \text{ m} \\
300 \text{ m} \\
800 \text{ m} \\
+ \; 300 \text{ m} \\
\hline
2200 \text{ m}
\end{array}
$$

Converting this to kilometers, we have: 2200 m = 2.2 km.

Therefore, 2 km of fencing will not be enough to enclose the field.

Skill Practice

13. Mikael is on the swim team at the college. For his last three workouts, he swam distances of 4000 m, 3500 m, and 2500 m. How many kilometers did he swim?

Classroom Example: p. 437, Exercise 58

Answer

13. 10 km

Section 7.3 Practice Exercises

Study Skills Exercise

For additional exercises, see Classroom Activity 7.3A in the *Student's Resource Manual* at www.mhhe.com/moh.

Label the statements as True or False.

To do math, you must be born with a special skill. _____

There is only one way to solve a math problem. _____

Two answers can look different but are both correct. _____

Vocabulary and Key Concepts

1. a. A decimal-based system of units called the ____metric____ system was first proposed by the French Academy of Sciences in 1790 and is now the predominant system of measurement used in the sciences.

b. A metric ____prefix____ line is a figure used to order metric units for the purpose of counting the number of place positions needed to convert between two metric units of measurement.

c. The ____meter____ is the fundamental unit of length in the metric system and is denoted by ____m____.

 Writing   Translating Expression   Geometry   Scientific Calculator   Video   NS E

Review Exercises

For Exercises 2–9, convert the units of measurements.

2. 4.2 tons = _____ lb
8400 lb

3. 2200 yd = _____ mi
1.25 mi

4. 8 c = _____ pt
4 pt

5. 48 oz = _____ lb
3 lb

6. 2 yd = _____ in.
72 in.

7. 1 day = _____ min
1440 min

8. 160 fl oz = _____ gal
1.25 gal

9. 3.5 lb = _____ oz
56 oz

Concept 1: Introduction to the Metric System

10. Identify the units that apply to length. Circle all that apply.   a, d, i, j

   a. Yard   **b.** Ounce   **c.** Fluid ounce   **d.** Meter   **e.** Quart

   f. Gram   **g.** Pound   **h.** Liter   **i.** Mile   **j.** Inch

11. Identify the units that apply to weight or mass. Circle all that apply.   b, f, g

   a. Yard   **b.** Ounce   **c.** Fluid ounce   **d.** Meter   **e.** Quart

   f. Gram   **g.** Pound   **h.** Liter   **i.** Mile   **j.** Inch

12. Identify the units that apply to capacity. Circle all that apply.   c, e, h

   a. Yard   **b.** Ounce   **c.** Fluid ounce   **d.** Meter   **e.** Quart

   f. Gram   **g.** Pound   **h.** Liter   **i.** Mile   **j.** Gram

Concept 2: Metric Units of Length

For Exercises 13–16, approximate each distance in centimeters and millimeters. **(See Example 1.)**

13.
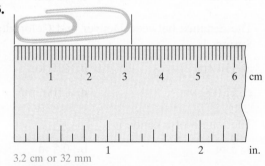
3.2 cm or 32 mm

14.
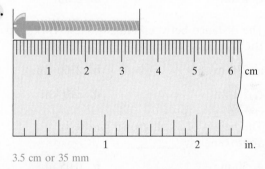
3.5 cm or 35 mm

15.
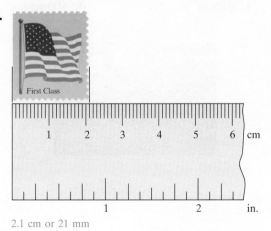
2.1 cm or 21 mm

16.
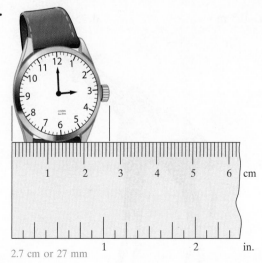
2.7 cm or 27 mm

 Writing   Translating Expression   Geometry   Scientific Calculator   Video   NS E

For Exercises 17 and 18, use a metric ruler to measure the dimensions of each figure to the closest centimeter. Then find the perimeter and area.

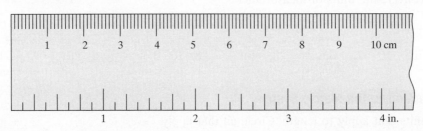

17. a. Length = <u>5 cm</u>   **b.** Width = <u>2 cm</u>   **18. a.** Length = <u>3 cm</u>   **b.** Width = <u>3 cm</u>

    c. Perimeter = <u>14 cm</u>   **d.** Area = <u>10 cm²</u>   **c.** Perimeter = <u>12 cm</u>   **d.** Area = <u>9 cm²</u>

NS E For Exercises 19–24, select the most reasonable measurement.

19. A table is _____ long.   a

  a. 2 m   **b.** 2 cm

  c. 2 km   **d.** 2 hm

20. A picture frame is _____ wide.   a

  a. 22 cm   **b.** 22 mm

  c. 22 m   **d.** 22 km

21. The distance between Albany, New York, and Buffalo, New York, is _____.   d

  a. 210 m   **b.** 2100 cm

  c. 2.1 km   **d.** 210 km

22. The distance between Denver and Colorado Springs is _____.   b

  a. 110 cm   **b.** 110 km

  c. 11,000 km   **d.** 1100 mm

23. The height of a full-grown giraffe is approximately _____.   d

  a. 50 m   **b.** 0.05 m

  c. 0.5 m   **d.** 5 m

24. The length of a canoe is _____.   a

  a. 5 m   **b.** 0.5 m

  c. 0.05 m   **d.** 500 m

Concept 3: Converting Metric Units of Length

For Exercises 25–30, complete the conversion factors.

25. $\dfrac{1 \text{ km}}{\_\_ \text{ m}}$ $\dfrac{1 \text{ km}}{1000 \text{ m}}$

26. $\dfrac{1 \text{ hm}}{\_\_ \text{ m}}$ $\dfrac{1 \text{ hm}}{100 \text{ m}}$

27. $\dfrac{1 \text{ m}}{\_\_ \text{ cm}}$ $\dfrac{1 \text{ m}}{100 \text{ cm}}$

28. $\dfrac{1 \text{ m}}{\_\_ \text{ mm}}$ $\dfrac{1 \text{ m}}{1000 \text{ mm}}$

29. $\dfrac{1 \text{ m}}{\_\_ \text{ dm}}$ $\dfrac{1 \text{ m}}{10 \text{ dm}}$

30. $\dfrac{1 \text{ dam}}{\_\_ \text{ m}}$ $\dfrac{1 \text{ dam}}{10 \text{ m}}$

For Exercises 31–50, convert metric units of length by using conversion factors or the prefix line. **(See Examples 2 and 3.)**

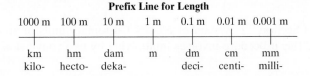

Prefix Line for Length

1000 m	100 m	10 m	1 m	0.1 m	0.01 m	0.001 m
km	hm	dam	m	dm	cm	mm
kilo-	hecto-	deka-		deci-	centi-	milli-

31. 2430 m = ____ km   2.43 km

32. 52 hm = ____ m   5200 m

33. 103 dm = ____ m   10.3 m

34. 1251 mm = ____ m   1.251 m

35. 50 m = ____ mm   50,000 mm

36. 1.3 m = ____ mm   1300 mm

37. 4 km = ____ m   4000 m

38. 5 m = ____ cm   500 cm

39. 4.31 cm = ____ mm   43.1 mm

40. 18 cm = ____ mm   180 mm

41. 3328 dm = ____ km   0.3328 km

42. 128 hm = ____ km   12.8 km

43. 345 mm = ____ m   0.345 m

44. 450 mm = ____ dm   4.5 dm

45. 0.25 km = ____ m   250 m

46. 3 hm = ____ m   300 m

47. 4003 cm = ____ dm   400.3 dm

48. 6.8 m = ____ cm   680 cm

49. 0.07 mm = ____ cm   0.007 cm

50. 8 m = ____ cm   800 cm

51. One of the tallest sand sculptures was 20.91 m tall. How many centimeters is this? **(See Example 4.)**   2091 cm

52. The smallest steam engine is 16.24 mm long. How many centimeters is this? (*Source: Guinness Book of World Records*)   1.624 cm

53. The lowest point in Antarctica is 2538 m below sea level. How many kilometers is this?   2.538 km

54. The lowest point in South America is 40 m below sea level. How many kilometers is this?   0.04 km

55. The Renaissance Tower in Dallas is 270 m tall. How many kilometers is this?   0.27 km

56. The Trump Building in New York is 283 m tall. How many kilometers is this?   0.283 km

Concept 4: Applications

57. Veronique has a piece of framing 1 m long. Does she have enough to cut four pieces to frame the picture shown in the figure? **(See Example 5.)**   No, she needs 1.04 m of framing.

58. Rosanna has material 1.5 m long for a window curtain. If the window is 90 cm and she needs 10 cm for a hem at the bottom and 12 cm for finishing the top, does Rosanna have enough material?   Yes, she only needs 1.12 m of material.

59. A square tile is 110 mm in length. If they are placed side by side, how many tiles will it take to cover a length of wall 1.43 m long?   It will take 13 tiles.

60. Two Olympic speed skating races for women are 500 m and 5 km. What is the difference (in meters) between the lengths of these races?   The difference is 4500 m.

12 cm

40 cm

Writing   Translating Expression   Geometry   Scientific Calculator   Video   NS E

61. A parking area in an apartment complex is 0.108 km long. If parking spaces are 4.5 m wide, how many can fit along the length of the lot?

There can be 24 parking spaces.

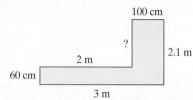

4.5 m

0.108 km

62. Find the missing length.

150 cm or 1.5 m

100 cm

?

2.1 m

2 m

60 cm

3 m

Expanding Your Skills

In the Expanding Your Skills of Section 7.1, we converted U.S. Customary units of area. We use the same procedure to convert metric units of area. This procedure involves multiplying by two conversion factors of length.

Example: Converting area

Convert 1000 mm² to square centimeters.

Solution: $\dfrac{1000 \text{ mm}^2}{1} \cdot \dfrac{1 \text{ cm}}{10 \text{ mm}} \cdot \dfrac{1 \text{ cm}}{10 \text{ mm}} = \dfrac{1000 \text{ mm}^2}{1} \cdot \dfrac{1 \text{ cm}^2}{100 \text{ mm}^2} = \dfrac{1000 \text{ cm}^2}{100} = 10 \text{ cm}^2$

Multiply first.

For Exercises 63–66, convert the units of area, using two of the given conversion factors.

63. 30,000 mm² = _____ cm² $\left(\text{Use } \dfrac{1 \text{ cm}}{10 \text{ mm}}.\right)$

300 cm²

64. 65,000,000 m² = _____ km² $\left(\text{Use } \dfrac{1 \text{ km}}{1000 \text{ m}}.\right)$

65 km²

65. 4.1 m² = _____ cm² $\left(\text{Use } \dfrac{100 \text{ cm}}{1 \text{ m}}.\right)$

41,000 cm²

66. 5600 cm² = _____ m² $\left(\text{Use } \dfrac{1 \text{ m}}{100 \text{ cm}}.\right)$

0.56 m²

Section 7.4   Metric Units of Mass, Capacity, and Medical Applications

Concepts

1. Converting Metric Units of Mass
2. Converting Metric Units of Capacity
3. Summary of Metric Conversions
4. Medical Applications

1. Converting Metric Units of Mass

In Section 7.2 we learned that the pound and ton are two measures of weight in the U.S. Customary System. Measurements of weight give the force of an object under the influence of gravity. The mass of an object is related to its weight, however, mass is not affected by gravity. Thus, the weight of an object will be different on Earth than on the Moon because the effect of gravity is different. However, the mass of the object will stay the same.

The fundamental unit of mass in the metric system is the **gram** (g). A penny is approximately 2.5 g (Figure 7-6). A paper clip is approximately 1 g (Figure 7-7).

Concept Connections

1. Which object could have a mass of 2 g?
   a. Rubber band
   b. Can of tuna fish
   c. Cell phone

≈ 2.5 g

Figure 7-6

≈ 1 g

Figure 7-7

Answer

1. a

Other common metric units of mass are given in Table 7-4. Once again, notice that the metric units of mass are related to the gram by powers of 10.

Table 7-4 **Metric Units of Mass and Their Equivalents**

1 kilogram (kg) = 1000 g	
1 hectogram (hg) = 100 g	
1 dekagram (dag) = 10 g	
1 gram (g) = 1 g	
1 decigram (dg) = 0.1 g	$\left(\frac{1}{10}\text{ g}\right)$
1 centigram (cg) = 0.01 g	$\left(\frac{1}{100}\text{ g}\right)$
1 milligram (mg) = 0.001 g	$\left(\frac{1}{1000}\text{ g}\right)$

TIP: In addition to the key facts presented in Table 7-4, the following equivalences are useful.

$$100 \text{ cg} = 1 \text{ g}$$
$$1000 \text{ mg} = 1 \text{ g}$$

Concept Connections

Fill in the blank with $<$ or $>$.

2. 1 g ☐ 1 kg

3. 1 g ☐ 1 cg

On the surface of Earth, 1 kg of mass is equivalent to approximately 2.2 lb of weight. Therefore, a 180-lb man has approximately 81.8 kg of mass.

$$180 \text{ lb} \approx 81.8 \text{ kg}$$

The metric prefix line for mass is shown in Figure 7-8. This can be used to convert from one unit of mass to another.

Prefix Line

1000 g	100 g	10 g	1 g	0.1 g	0.01 g	0.001 g
kg	hg	dag	g	dg	cg	mg
kilo-	hecto-	deka-		deci-	centi-	milli-

Figure 7-8

Example 1 **Converting Metric Units of Mass**

a. 1.6 kg = _____ g   **b.** 1400 mg = _____ g

Solution:

a. $1.6 \text{ kg} = \dfrac{1.6 \text{ kg}}{1} \cdot \dfrac{1000 \text{ g}}{1 \text{ kg}}$ ← new unit to convert to
← unit to convert from

$= 1600 \text{ g}$

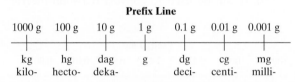

Start here.

$$1.6 \text{ kg} = 1.600 \text{ kg} = 1600 \text{ g}$$

b. $1400 \text{ mg} = \dfrac{1400 \text{ mg}}{1} \cdot \dfrac{1 \text{ g}}{1000 \text{ mg}}$ ← new unit to convert to
← unit to convert from

$= \dfrac{1400}{1000} \text{ g}$

$= 1.4 \text{ g}$

Start here.

$$1400 \text{ mg} = 1400 \text{ mg} = 1.4 \text{ g}$$

Skill Practice

Convert.

4. 80 kg = _____ g

5. 49 cg = _____ g

Classroom Examples: p. 443, Exercises 10 and 16

Answers

2. $<$ **3.** $>$
4. 80,000 g   **5.** 0.49 g

2. Converting Metric Units of Capacity

The basic unit of capacity in the metric system is the **liter** (L). One liter is slightly more than 1 qt. Other common units of capacity are given in Table 7-5.

Table 7-5 **Metric Units of Capacity and Their Equivalents**

1 kiloliter (kL) = 1000 L	
1 hectoliter (hL) = 100 L	
1 dekaliter (daL) = 10 L	
1 liter (L) = 1 L	
1 deciliter (dL) = 0.1 L	$(\frac{1}{10}$ L$)$
1 centiliter (cL) = 0.01 L	$(\frac{1}{100}$ L$)$
1 milliliter (mL) = 0.001 L	$(\frac{1}{1000}$ L$)$

TIP: In addition to the key facts presented in Table 7-5, the following equivalences are useful.

$$100 \text{ cL} = 1 \text{ L}$$
$$1000 \text{ mL} = 1 \text{ L}$$

Concept Connections

Fill in the blank with $<$, $>$, or $=$.

6. 1 kL ☐ 1 cL

7. 1 cL ☐ 1 L

8. 1 mL ☐ 1 cc

1 mL is also equivalent to a **cubic centimeter** (**cc** or **cm³**). The unit cc is often used to measure dosages of medicine. For example, after having an allergic reaction to a bee sting, a patient might be given 1 cc of adrenalin.

The metric prefix line for capacity is similar to that of length and mass (Figure 7-9). It can be used to convert between metric units of capacity.

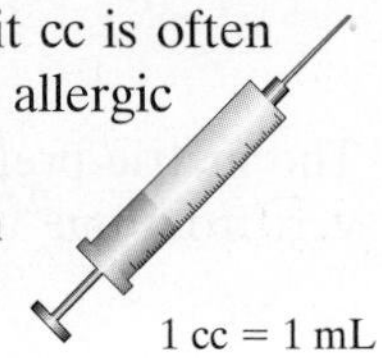

1 cc = 1 mL

Prefix Line

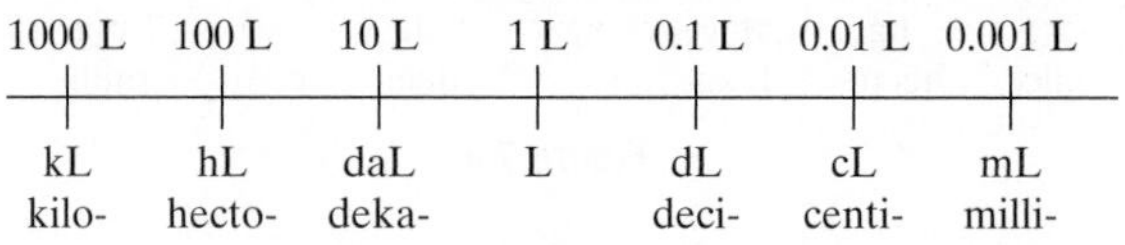

1000 L	100 L	10 L	1 L	0.1 L	0.01 L	0.001 L
kL	hL	daL	L	dL	cL	mL
kilo-	hecto-	deka-		deci-	centi-	milli-

Figure 7-9

Skill Practice

Convert.

9. 10.2 L = _____ cL

10. 150,000 mL = _____ kL

Classroom Example: p. 444, Exercise 38

Example 2 **Converting Metric Units of Capacity**

a. 5.5 L = _____ mL   **b.** 150 cL = _____ L

Solution:

a. $5.5 \text{ L} = \dfrac{5.5 \text{ L}}{1} \cdot \dfrac{1000 \text{ mL}}{1 \text{ L}}$ ← new unit to convert to
                                          ← unit to convert from

$= 5500 \text{ mL}$

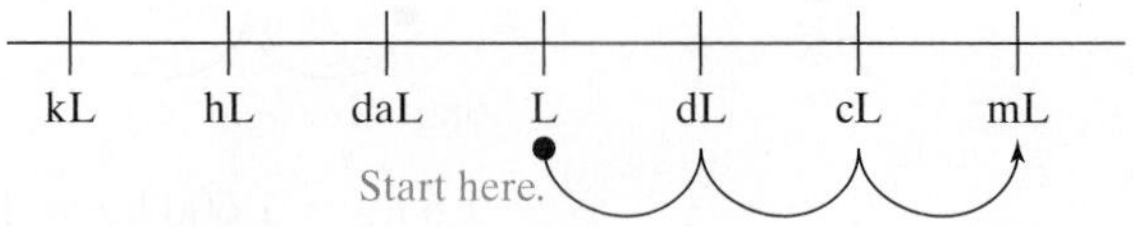

kL	hL	daL	L	dL	cL	mL

Start here.

$$5.5 \text{ L} = 5.500 \text{ L} = 5500 \text{ mL}$$

b. $150 \text{ cL} = \dfrac{150 \text{ cL}}{1} \cdot \dfrac{1 \text{ L}}{100 \text{ cL}}$ ← new unit to convert to
                                          ← unit to convert from

$= \dfrac{150}{100} \text{ L}$

kL	hL	daL	L	dL	cL	mL

$= 1.5 \text{ L}$

Start here.

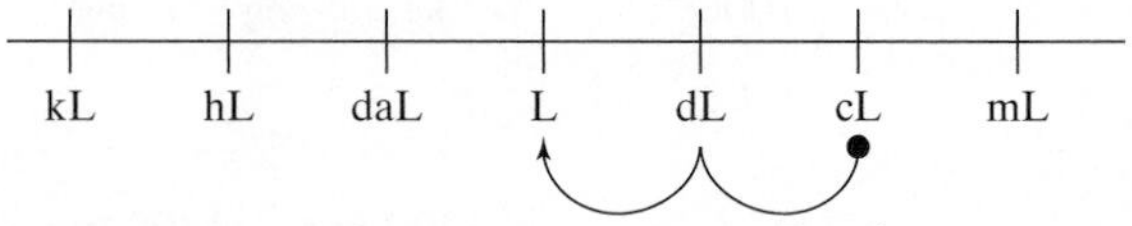

$$150 \text{ cL} = 150. \text{ cL} = 1.5 \text{ L}$$

Answers

6. $>$ **7.** $<$ **8.** $=$

9. 1020 cL   **10.** 0.15 kL

Example 3 **Converting Metric Units of Capacity**

a. $15 \text{ cc} = \underline{\hspace{2cm}} \text{ mL}$ b. $0.8 \text{ cL} = \underline{\hspace{2cm}} \text{ cc}$

Solution:

a. Recall that 1 cc = 1 mL. Therefore 15 cc = 15 mL.

b. We must convert from centiliters to milliliters, and then from milliliters to cubic centimeters.

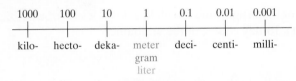

$$0.8 \text{ cL} = \frac{0.8 \text{ cL}}{1} \cdot \frac{10 \text{ mL}}{1 \text{ cL}}$$

$$= 8 \text{ mL}$$

$$= 8 \text{ cc} \qquad \text{Recall that 1 mL = 1 cc.}$$

kL   hL   daL   L   dL   cL   mL

Start here.

$0.8 \text{ cL} = 8 \text{ mL} = 8 \text{ cc}$

Skill Practice

Convert.
11. $0.5 \text{ cc} = \underline{\hspace{1cm}} \text{ mL}$
12. $0.04 \text{ L} = \underline{\hspace{1cm}} \text{ cc}$

Classroom Example: p. 444,
Exercise 44

3. Summary of Metric Conversions

The prefix line in Figure 7-10 summarizes the relationships learned thus far.

1000	100	10	1	0.1	0.01	0.001
kilo-	hecto-	deka-	meter gram liter	deci-	centi-	milli-

Figure 7-10

Example 4 **Converting Metric Units**

a. The distance between San Jose and Santa Clara is 26 km. Convert this to meters.

b. A bottle of canola oil holds 946 mL. Convert this to liters.

c. The mass of a bag of rice is 90,700 cg. Convert this to grams.

d. A dose of an antiviral medicine is 0.5 cc. Convert this to milliliters.

Solution:

a. $26 \text{ km} = 26.000 \text{ km} = 26{,}000 \text{ m}$

kilo-   hecto-   deka-   meter
gram
liter   deci-   centi-   milli-

Start here.

b. $946 \text{ mL} = 946. \text{ mL} = 0.946 \text{ L}$

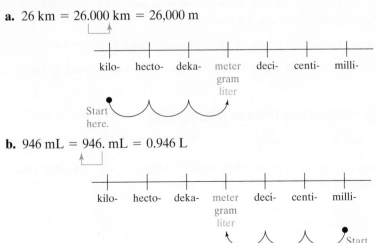

kilo-   hecto-   deka-   meter
gram
liter   deci-   centi-   milli-

Start here.

Skill Practice

13. The distance between Savannah and Hinesville is 64 km. How many meters is this?
14. A bottle of water holds 1420 mL. How many liters is this?
15. The mass of a box of cereal is 680 g. Convert this to kilograms.
16. A cat receives 1 mL of an antibiotic solution. Convert this to cc.

Classroom Example: p. 445,
Exercise 60

Answers

11. 0.5 mL   **12.** 40 cc
13. 64,000 m   **14.** 1.42 L
15. 0.68 kg   **16.** 1 cc

c. 90,700 cg = 90700. cg = 907 g

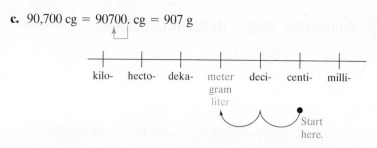

d. Recall that 1 cc = 1 mL. Therefore, 0.5 cc = 0.5 mL.

4. Medical Applications

Example 5 Applying Metric Units of Measure to Medicine

A doctor orders the antibiotic oxacillin for a child. The dosage is 12.5 mg of the drug per kilogram of the child's body mass. This dosage is given 4 times a day.

a. How much of the drug should a 24-kg child get in one dose?

b. How much of the drug would the child get if she were on a 10-day course of the antibiotic?

Solution:

a. We need to multiply the unit rate of 12.5 mg per kilogram times the child's body mass.

$$\text{Single dose} = (12.5 \text{ mg/kg})(24 \text{ kg})$$
$$= 300 \text{ mg}$$

b. For a 10-day course, we need to multiply 300 g by the number of doses per day (4), and the total number of days (10).

$$\text{Total amount of drug} = (300 \text{ mg})(4)(10)$$
$$= 12{,}000 \text{ mg} \quad \text{or equivalently 12 g.}$$

Sometimes doctors prescribe medicines in very small amounts. In these cases, it is sometimes more convenient to use units of **micrograms**. The abbreviation for microgram is mcg or sometimes μg. Furthermore,

| 1000 mcg = 1 mg | It takes 1 thousand micrograms to equal 1 milligram. |
| 1,000,000 μg = 1 g | It takes 1 million micrograms to equal 1 gram. |

Example 6 Converting Units of Micrograms

a. Convert 0.85 mg = _____ mcg

b. A doctor gives a heart patient an initial dose of 200 μg of nitroglycerin. How many milligrams is this?

Solution:

a. $0.85 \text{ mg} = \dfrac{0.85 \text{ mg}}{1} \cdot \dfrac{1000 \text{ mcg}}{1 \text{ mg}}$ ← new unit to convert to
                                            ← unit to convert from

            $= 850 \text{ mcg}$

b. $200 \text{ μg} = \dfrac{200 \text{ μg}}{1} \cdot \dfrac{1 \text{ mg}}{1000 \text{ μg}}$ ← new unit to convert to
                                            ← unit to convert from

            $= 0.2 \text{ mg}$

Section 7.4   Practice Exercises

Study Skills Exercise

For additional exercises, see Classroom Activities 7.4A–7.4C in the *Student's Resource Manual* at www.mhhe.com/moh.

To help you stay motivated for this class, list three reasons why you are taking the class.

Vocabulary and Key Concepts

1. a. The ___gram___ is the fundamental unit of mass in the metric system and is denoted by ___g___.

   b. The ___liter___ is the fundamental unit of capacity in the metric system and is denoted by ___L___.

   c. The ___cubic___ centimeter or cc is a unit of capacity equivalent to 1 mL.

   d. A unit of capacity equivalent to $\frac{1}{1,000,000}$ of a gram is called the ___microgram___ and is denoted by μg.

Review Exercises

2. Determine which distance is greatest.

   a. 40,000 cm   **b.** 40 km   **c.** 4000 m   b

For Exercises 3–8, complete the table.

	Object	mm	cm	m	km
3.	Distance between Orlando and Miami	670,000,000	67,000,000	670,000	670
4.	Length of the Mississippi River	3,766,000,000	376,600,000	3,766,000	3766
5.	Length of a screw	25	2.5	0.025	0.000025
6.	Thickness of a pizza	32	3.2	0.032	0.000032
7.	Thickness of a dime	1.35	0.135	0.00135	0.00000135
8.	Diameter of a quarter	24.3	2.43	0.0243	0.0000243

Concept 1: Converting Metric Units of Mass

For Exercises 9–18, convert the units of mass. **(See Example 1.)**

9. 539 g = _____ kg   0.539 kg

10. 328 mg = _____ g   0.328 g

11. 2.5 kg = _____ g   2500 g

12. 2011 g = _____ kg   2.011 kg

13. 0.0334 g = _____ mg   33.4 mg

14. 0.38 dag = _____ dg   38 dg

15. 90 hg = _____ kg   9 kg

16. 0.003 kg = _____ g   3 g

17. 45 dg = _____ g   4.5 g

18. 409 cg = _____ g   4.09 g

  ✎ Writing   ↔ Translating Expression   ◁ Geometry   🖩 Scientific Calculator   ▶ Video   NS&E

For Exercises 19–26, complete the table.

	Object	mg	cg	g	kg
19.	Bag of cat food	1,580,000	158,000	1580	1.58
20.	Bag of flour	2,260,000	226,000	2260	2.26
21.	Can of tuna	170,000	17,000	170	0.17
22.	Bag of rice	907,000	90,700	907	0.907
23.	Box of raisins	425,000	42,500	425	0.425
24.	Hockey puck	170,000	17,000	170	0.17
25.	Dose of acetaminophen	325	32.5	0.325	0.000325
26.	Olive	12	1.2	0.012	0.000012

NS&E Concept 2: Converting Metric Units of Capacity

For Exercises 27–32, fill in the blank with >, <, or =.

27. 1 cL ____ 1 L   <

28. 1 L ____ 1 mL   >

29. 1 mL ____ 1 cc   =

30. 1 L ____ 1 cc   >

31. 1 cL ____ 1 kL   <

32. 1 mL ____ 1 cL   <

33. What does the abbreviation *cc* represent?   Cubic centimeter

34. Which of the following are measures of capacity? Circle all that apply.   b, c

    a. cm   **b.** cc   **c.** cL   **d.** cg

For Exercises 35–44, convert the units of capacity. **(See Examples 2 and 3.)**

35. 3200 mL = _____ L   3.2 L

36. 280 L = _____ kL   0.28 kL

37. 7 L = _____ cL   700 cL

38. 0.52 L = _____ mL   520 mL

39. 42 mL = _____ dL   0.42 dL

40. 0.88 L = _____ hL   0.0088 hL

41. 64 cc = _____ mL   64 mL

42. 125 mL = _____ cc   125 cc

43. 0.04 L = _____ cc   40 cc

44. 38 cc = _____ L   0.038 L

For Exercises 45–52, complete the table.

	Object	mL	cL	L	kL
45.	1 Tablespoon	15	1.5	0.015	0.000015
46.	Bottle of vanilla extract	59	5.9	0.059	0.000059
47.	Bottle of vinegar	355	35.5	0.355	0.000355
48.	Bottle of soy sauce	296	29.6	0.296	0.000296
49.	Bottle of soda pop	2,000	200	2	0.002
50.	Bottle of water	1,000	100	1	0.001
51.	Capacity of a cooler	37,700	3,770	37.7	0.0377
52.	Capacity of a gasoline tank	75,700	7,570	75.7	0.0757

/ Writing   ↔ Translating Expression   ◣ Geometry   ▦ Scientific Calculator   ▶▥ Video   NS&E

Concept 3: Summary of Metric Conversions (Mixed Exercises)

53. Identify the unit that applies to length.   c

   a. L   **b.** g   **c.** m

54. Identify the unit that applies to capacity.   a

   a. L   **b.** g   **c.** m

55. Identify the unit that applies to mass.   b

   a. L   **b.** g   **c.** m

56. Identify the units that apply to length.   b, f

   a. mL   **b.** mm   **c.** hg   **d.** cc   **e.** kg   **f.** hm   **g.** cL

57. Identify the units that apply to capacity.   c, d

   a. kg   **b.** km   **c.** cL   **d.** cc   **e.** hm   **f.** dag   **g.** mm

58. Identify the units that apply to mass.   a, f

   a. dg   **b.** hm   **c.** kL   **d.** cc   **e.** dm   **f.** kg   **g.** cL

For Exercises 59–64, convert the metric units as indicated. **(See Example 4.)**

59. The height of the tallest living tree is 112.014 m. Convert this to dekameters.
   11.2014 dm

60. The Congo River is 4669 km long. Convert this to meters.   4,669,000 m

61. There is 600 mg of calcium in a multivitamin. Convert this to grams.   0.6 g

62. A can of soup contains 305 g. Convert this to milligrams.   305,000 mg

63. A gasoline can has a capacity of 19 L. Convert this to kiloliters.   0.019 kL

64. The capacity of a coffee cup is 0.25 L. Convert this to milliliters.   250 mL

65. In one day, Stacy gets 600 mg of calcium in her daily vitamin, 500 mg in her calcium supplement, and 250 mg in the dairy products she ingests. How many grams of calcium will she ingest in one week?   Stacy gets 9.45 g per week.

66. Cliff drives his children to their sports activities outside of school. When he drives his son to baseball practice, it is a 6-km round trip. When he drives his daughter to basketball practice, it is a 1800-m round trip. If basketball practice is 3 times a week and baseball practice is twice a week, how many kilometers does Cliff drive?
   Cliff drives 17.4 km per week.

67. A gas tank holds 45 L. If it costs $74.25 to fill up the tank, what is the price per liter?   The price is $1.65 per liter.

68. A can of paint holds 120 L. How many kiloliters are contained in 8 cans?
   8 cans hold 0.96 kL.

69. A bottle of water holds 710 mL. How many liters are in a 6-pack?   A 6-pack contains 4.26 L.

70. A bottle of olive oil has 33 servings of 15 mL each. How many centiliters of oil does the bottle contain?   The bottle contains 49.5 cL.

71. A quart of milk has 130 mg of sodium per cup. How much sodium is in the whole bottle?
520 mg of sodium per 1-qt bottle

72. A ½-c serving of cereal has 180 mg of potassium. This is 5% of the recommended daily allowance of potassium. How many cups of cereal are needed to get 100% of the recommended daily allowance of potassium?
10 cups of cereal give 100% of the recommended daily allowance of potassium.

Concept 4: Medical Applications

73. The drug amoxicillin is an antibiotic used to treat bacterial infections. A doctor orders 250 mg every 8 hr. How many grams of the drug would be given in 1 wk? 5.25 g of the drug would be given in 1 wk.

74. The drug acetaminophen is used as a pain reliever. A patient takes 325 mg every 6 hr. How many grams of the drug would be taken in a 3-day period?
3.9 g of the drug would be taken in a 3-day period.

75. Dr. Boyd gives a patient 2 cc of Zantac. How many milliliters is this? 2 mL

76. If a nurse mixed 11.5 mL of sterile water with 1.5 mL of oxacillin, how many cubic centimeters will this produce? 13 cc

77. A tetanus vaccine was purchased by a group of family practice doctors. They purchased 1 L of the vaccine. How many patients can be vaccinated if the normal dose is 2 cc? 500 people

78. A pharmacist has a 1-L bottle of cough syrup. How many 20 cL bottles can she make? 5 bottles

79. A doctor orders 0.2 mg of a drug per kilogram of a patient's body mass. How much of the drug should be given to a patient who is 48 kg? 9.6 mg

80. The dosage for a painkiller is 0.05 mg per kilogram of a patient's body mass. How much of the drug should be administered to a patient who is 90 kg? 4.5 mg

81. The drug Zovirax is sometimes used to treat chicken pox in children. One doctor recommended 20 mg of the drug per kilogram of the child's body mass, 4 times daily. **(See Example 5.)**

a. How much of the drug should a 20-kg child receive for one dose? 400 mg

b. How much of the drug would be given over a 5-day period? 8000 mg or 8 g

82. The drug Amoxil is sometimes used to treat children with bacterial infections. One doctor prescribed 40 mg of the drug per kilogram of the child's body mass, 3 times daily.

a. How much of the drug should a 15-kg child receive for one dose? 600 mg

b. How much of the drug would be given to the child over a 10-day period? 18,000 mg or 18 g

83. Convert 0.01 mg to micrograms. **(See Example 6.)** 10 mcg

84. Convert 0.0004 cg to micrograms. 4 mcg

85. A doctor orders 0.2 mg of the drug atropine given by injection. How many micrograms is this? 200 mcg

86. The drug Synthroid is used to treat thyroid disease. A patient is sometimes started on a dose of 0.05 mg/day. How many micrograms is this? 50 mcg/day

87. The drug cyanocobalamin is prescribed by one doctor in the amount of 1000 μg. How many milligrams is this?   1 mg

88. An injection of naloxone is given in the amount of 800 μg. How many milligrams is this?   0.8 mg

89. A nurse must administer 45 mg of a drug. The drug is available in a liquid form with a concentration of 15 mg per milliliter of the solution. How many milliliters of the solution should the nurse give?
3 mL

90. A patient must receive 500 mg of medication in a solution that has a strength of 250 mg per 5 milliliter of solution. How many milliliters of solution should be given?
10 mL

Expanding Your Skills

91. A normal value of hemoglobin in the blood for an adult male is 18 g/dL (that is, 18 grams per deciliter). How much hemoglobin would be expected in 20 mL of a males's blood?   3.6 g

92. A normal value of hemoglobin in the blood for an adult female is 15 g/dL (that is, 15 grams per deciliter). How much hemoglobin would be expected in 40 mL of a female's blood?   6 g

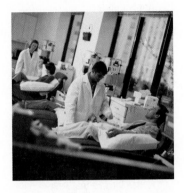

In the U.S. Customary System of measurement, 1 ton = 2000 lb. In the metric system, 1 metric ton = 1000 kg. Use this information to answer Exercises 93–96.

93. Convert 3300 kg to metric tons.   3.3 metric tons

94. Convert 5780 kg to metric tons.   5.78 metric tons

95. Convert 10.9 metric tons to kilograms.
10,900 kg

96. Convert 8.5 metric tons to kilograms.
8500 kg

Problem Recognition Exercises

U.S. Customary and Metric Conversions

For Exercises 1–28, convert the units as indicated.

1. 36 c = ____ qt
9 qt

2. 220 cm = ____ m
2.2 m

3. $\frac{3}{4}$ lb = ____ oz
12 oz

4. 0.3 L = ____ mL
300 mL

5. 12 ft = ____ yd
4 yd

6. 6.03 kg = ____ g
6030 g

7. 9 in. = ____ ft
$\frac{3}{4}$ ft

8. $\frac{1}{2}$ mi = ____ ft
2640 ft

9. 6000 lb = ____ tons
3 tons

10. 8 pt = ____ qt
4 qt

11. 1.5 tsp = ____ T
$\frac{1}{2}$ T

12. 21 m = ____ km
0.021 km

13. 36 mL = ____ cc
36 cc

14. 64 oz = ____ lb
4 lb

15. 4322 g = ____ kg
4.322 kg

16. 5 m = ____ mm
5000 mm

17. 20 fl oz = ____ c
2.5 c

18. 510 sec = ____ min
8.5 min

19. 4 pt = ____ gal
0.5 gal

20. 26 fl oz = ____ c
3.25 c

21. 5.46 kg = ____ g
5460 g

22. 9.02 L = ____ cL
902 cL

23. 9.1 mi = ____ yd
16,016 yd

24. 48 oz = ____ lb
3 lb

25. 1.62 tons = ____ lb
3240 lb

26. 4.6 km = ____ m
4600 m

27. 60 hr = ____ days
2.5 days

28. 8 cc = ____ mL
8 mL

 Writing   Translating Expression   Geometry   Scientific Calculator   Video   NS E

Section 7.5 Converting Between U.S. Customary and Metric Units

Concepts

1. Summary of U.S. Customary and Metric Unit Equivalents
2. Converting U.S. Customary and Metric Units
3. Applications
4. Units of Temperature

1. Summary of U.S. Customary and Metric Unit Equivalents

In this section, we learn how to convert between U.S. Customary and metric units of measure. Suppose, for example, that you take a trip to Europe. A street sign indicates that the distance to Paris is 45 km (Figure 7-11). This distance may be unfamiliar to you until you convert to miles.

Instructor Note: Since students may be familiar with 10-k races and often know how to convert 10 km to miles, this might be a good start-up activity.

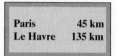

Paris	**45 km**
Le Havre	**135 km**

Figure 7-11

Skill Practice

1. Use the fact that 1 mi ≈ 1.61 km to convert 184 km to miles. Round to the nearest mile.

Classroom Example: p. 454, Exercise 16

Example 1 Converting Metric Units to U.S. Customary Units

Use the fact that 1 mi ≈ 1.61 km to convert 45 km to miles. Round to the nearest mile.

Solution:

$$45 \text{ km} \approx \frac{45 \text{ km}}{1} \cdot \frac{1 \text{ mi}}{1.61 \text{ km}}$$ Set up a conversion factor to convert kilometers to miles.

$$= \frac{45}{1.61} \text{ mi}$$ Multiply fractions.

$$\approx 28 \text{ mi}$$ Divide and round to the nearest mile.

The distance of 45 km to Paris is approximately 28 mi.

Table 7-6 summarizes some common metric and U.S. Customary equivalents.

Instructor Note: Ask students to convert their heights in inches to centimeters.

Table 7-6

Length	Weight/Mass (on Earth)	Capacity
1 in. = 2.54 cm	1 lb ≈ 0.45 kg	1 qt ≈ 0.95 L
1 ft ≈ 0.305 m	1 oz ≈ 28 g	1 fl oz ≈ 30 mL = 30 cc
1 yd ≈ 0.914 m		
1 mi ≈ 1.61 km		

2. Converting U.S. Customary and Metric Units

Using the U.S. Customary and metric equivalents given in Table 7-6, we can create conversion factors to convert between units.

Answer

1. 114 mi

Example 2 **Converting Units of Length**

Fill in the blanks. Round to two decimal places, if necessary.

  a. 18 cm = _____ in.   **b.** 15 yd ≈ _____ m   **c.** 8.2 m ≈ _____ ft

Solution:

a. $18 \text{ cm} = \dfrac{18 \text{ cm}}{1} \cdot \dfrac{1 \text{ in.}}{2.54 \text{ cm}}$ From Table 7-6, we know 1 in. = 2.54 cm.

$= \dfrac{18}{2.54} \text{ in.}$ Multiply fractions.

$\approx 7.09 \text{ in.}$ Divide and round to two decimal places.

b. $15 \text{ yd} \approx \dfrac{15 \text{ yd}}{1} \cdot \dfrac{0.914 \text{ m}}{1 \text{ yd}}$ From Table 7-6, we know 1 yd ≈ 0.914 m.

$= 13.71 \text{ m}$ Multiply.

c. $8.2 \text{ m} \approx \dfrac{8.2 \text{ m}}{1} \cdot \dfrac{1 \text{ ft}}{0.305 \text{ m}}$ From Table 7-6, we know 1 ft ≈ 0.305 m.

$= \dfrac{8.2}{0.305} \text{ ft}$ Multiply.

$\approx 26.89 \text{ ft}$ Divide and round to two decimal places.

Example 3 **Converting Units of Weight and Mass**

Fill in the blank. Round to one decimal place, if necessary.

  a. 180 g ≈ _____ oz   **b.** 5.25 tons ≈ _____ kg

Solution:

a. $180 \text{ g} \approx \dfrac{180 \text{ g}}{1} \cdot \dfrac{1 \text{ oz}}{28 \text{ g}}$ From Table 7-6, we know 1 oz ≈ 28 g.

$= \dfrac{180}{28} \text{ oz}$

$\approx 6.4 \text{ oz}$ Divide and round to one decimal place.

b. We can first convert 5.25 tons to pounds. Then we can use the fact that 1 lb ≈ 0.45 kg.

$5.25 \text{ tons} = \dfrac{5.25 \text{ tons}}{1} \cdot \dfrac{2000 \text{ lb}}{1 \text{ ton}}$ Convert tons to pounds.

$= 10{,}500 \text{ lb}$

$\approx 10{,}500 \text{ lb} \cdot \dfrac{0.45 \text{ kg}}{1 \text{ lb}}$ Convert pounds to kilograms.

$= 4725 \text{ kg}$

Answers

2. 13.1 ft   **3.** 7.6 cm   **4.** 5941 m
5. 17.8 lb   **6.** 3600 kg

Skill Practice

Convert. Round to one decimal place, if necessary.

7. 120 mL ≈ _____ fl oz

8. 4 qt ≈ _____ L

Classroom Example: p. 455, Exercise 34

Example 4 Converting Units of Capacity

For each of the following conversions, round to two decimal places, if necessary.

a. 75 mL ≈ _____ fl oz   **b.** 3 qt ≈ _____ L

Solution:

a. $75 \text{ mL} \approx \dfrac{75 \text{ mL}}{1} \cdot \dfrac{1 \text{ fl oz}}{30 \text{ mL}}$ From Table 7-6, we know 1 fl oz ≈ 30 mL.

     $= \dfrac{75}{30} \text{ fl oz}$ Multiply fractions.

     $= 2.5 \text{ fl oz}$ Divide.

b. $3 \text{ qt} \approx \dfrac{3 \text{ qt}}{1} \cdot \dfrac{0.95 \text{ L}}{1 \text{ qt}}$ From Table 7-6, we know 1 qt ≈ 0.95 L.

     $= 2.85 \text{ L}$ Multiply.

3. Applications

Example 5 Converting Units in an Application

A 2-L bottle of soda sells for $2.19. A 32-fl-oz bottle of soda sells for $1.59. Compare the price per quart of each bottle to determine the better buy.

Solution:

Note that 1 qt = 2 pt = 4 c = 32 fl oz. So a 32-fl-oz bottle of soda costs $1.59 per quart. Next, if we can convert 2 L to quarts, we can compute the unit cost per quart and compare the results.

     $2 \text{ L} \approx \dfrac{2 \text{ L}}{1} \cdot \dfrac{1 \text{ qt}}{0.95 \text{ L}}$ Recall that 1 qt = 0.95 L.

     $\approx \dfrac{2}{0.95} \text{ qt}$ Multiply fractions.

     $\approx 2.11 \text{ qt}$ Divide and round to 2 decimal places.

Now find the cost per quart. $\dfrac{\$2.19}{2.11 \text{ qt}} \approx \$1.04 \text{ per quart}$

The cost for the 2-L bottle is $1.04 per quart, whereas the cost for 32 fl oz is $1.59 per quart. Therefore, the 2-L bottle is the better buy.

Skill Practice

9. A 720-mL bottle of water sells for $0.79. A 32-fl-oz bottle of water sells for $1.29. Compare the price per ounce to determine the better buy.

Classroom Example: p. 455, Exercise 40

Answers

7. 4 fl oz   **8.** 3.8 L
9. 720 mL is 24 oz. The cost per ounce is $0.033. The unit price for the 32-fl-oz bottle is $0.040 per ounce. The 720-mL bottle is the better buy.

Example 6 **Converting Units in an Application**

In track and field, the 1500-m race is slightly less than 1 mi. How many yards less is it? Round to the nearest yard.

1 mi = 1760 yd

1500 m = ? yd

Solution:

We know that 1 mi = 1760 yd. If we can convert 1500 m to yards, then we can subtract the results.

$$1500 \text{ m} \approx \frac{1500 \text{ m}}{1} \cdot \frac{1 \text{ yd}}{0.914 \text{ m}} \qquad \text{Recall that 1 yd = 0.914 m.}$$

$$\approx \frac{1500}{0.914} \text{ yd} \qquad \text{Multiply fractions.}$$

$$\approx 1641 \text{ yd} \qquad \text{Divide and round to the nearest yard.}$$

Therefore, the difference between 1 mi and 1500 m is:

(1 mi) − (1500 m)

1760 yd − 1641 yd = 119 yd

Skill Practice

10. In track and field, the 800-m race is slightly shorter than a half-mile race. How many yards less is it? Round to the nearest yard.

Classroom Example: p. 455, Exercise 44

4. Units of Temperature

In the United States, the **Fahrenheit** scale is used most often to measure temperature. On this scale, water freezes at 32°F and boils at 212°F. The symbol ° stands for "degrees," and °F means "degrees Fahrenheit."

Another scale used to measure temperature is the **Celsius** temperature scale. On this scale, water freezes at 0°C and boils at 100°C. The symbol °C stands for "degrees Celsius."

Figure 7-12 shows the relationship between the Celsius scale and the Fahrenheit scale.

Concept Connections

Use Figure 7-12 to select the best choice.

11. The high temperature in Dallas for a day in July is
   a. 34°C   **b.** 34°F
12. The high temperature in Minneapolis for a day in March is
   a. 34°C   **b.** 34°F

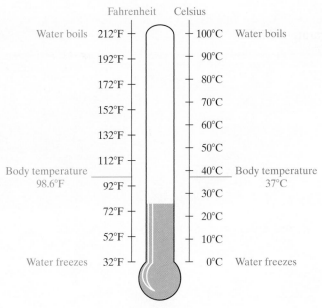

Figure 7-12

Answers

10. 5 yd   **11.** a   **12.** b

To convert back and forth between the Fahrenheit and Celsius scales, we use the following formulas.

Instructor Note: Ask students to use the two conversion formulas to show that 0°C = 32°F and that 212°F = 100°C.

Conversions for Temperature Scale

To convert from °C to °F: To convert from °F to °C:

$$F = \frac{9}{5}C + 32 \qquad\qquad C = \frac{5}{9}(F - 32)$$

Note: Using decimal notation we can write the formulas as

$$F = 1.8C + 32 \qquad\qquad C = \frac{F - 32}{1.8}$$

Example 7 **Converting Units of Temperature**

Convert a body temperature of 98.6°F to degrees Celsius.

Solution:

Because we want to convert degrees Fahrenheit to degrees Celsius, we use the formula $C = \frac{5}{9}(F - 32)$.

$$C = \frac{5}{9}(F - 32)$$

$$= \frac{5}{9}(98.6 - 32) \qquad \text{Substitute } F = 98.6.$$

$$= \frac{5}{9}(66.6) \qquad\qquad \text{Perform the operation inside parentheses first.}$$

$$= \frac{(5)(66.6)}{9}$$

$$= 37$$

Body temperature is 37°C.

Skill Practice

13. The ocean temperature in the Caribbean in August averages 84°F. Convert this to degrees Celsius and round to one decimal place.

Classroom Example: p. 456, Exercise 60

Answer

13. 28.9°C

Example 8 **Converting Units of Temperature**

Convert the temperature inside a refrigerator, 5°C, to degrees Fahrenheit.

Solution:

Because we want to convert degrees Celsius to degrees Fahrenheit, we use the formula $F = \dfrac{9}{5}C + 32$

$$F = \frac{9}{5}C + 32$$

$$= \frac{9}{5} \cdot 5 + 32 \qquad \text{Substitute } C = 5.$$

$$= \frac{9}{\cancel{5}} \cdot \frac{\cancel{5}}{1} + 32$$

$$= 9 + 32$$

$$= 41$$

The temperature inside the refrigerator is 41°F.

Skill Practice

14. The high temperature on a day in March for Raleigh, North Carolina, was 10°C. Convert this to degrees Fahrenheit.

Classroom Example: p. 456, Exercise 62

Answer
14. 50°F

Section 7.5 Practice Exercises

For additional exercises, see Classroom Activities 7.5A–7.5B in the *Student's Resource Manual* at www.mhhe.com/moh.

Study Skills Exercise

Make a list of all the section titles in the chapter that you are studying. Write each section title on a separate sheet of paper or index card. Go back and fill in the list of objectives under each section title. When you are studying for the test, try to make up an exercise that corresponds to each objective and then work the exercise. To get started, write a problem for the objective of converting metric units of capacity from Section 7.4.

Vocabulary and Key Concepts

1. a. In the United States, the ___Fahrenheit___ scale is used most often to measure temperature. Using this scale, water freezes at __32__ °F and boils at __212__ °F.

   b. In the sciences, the ___Celsius___ scale is used most often to measure temperature. Using this scale, water freezes at __0__ °C and boils at __100__ °C.

Review Exercises

2. Identify the unit of measure as a unit of length, capacity, or mass.

   a. mm **b.** cL **c.** kg **d.** cc   a. Length b. Capacity c. Mass d. Capacity

For Exercises 3–6, select the equivalent amounts of mass. (*Hint:* There may be more than one answer for each exercise.)

3. 500 g   d, f

4. 500 mg   g, h

5. 500 cg   b, e

6. 500 kg   a, c

   a. 500,000 g

   b. 5 g

   c. 500,000,000 mg

   d. 0.5 kg

   e. 5000 mg

   f. 50,000 cg

   g. 0.5 g

   h. 50 cg

 Writing   Translating Expression   Geometry   Scientific Calculator   Video   NSE

For Exercises 7–10, select the equivalent amounts of capacity.

7. 200 L   c, f

    a. 2000 mL

    b. 200 cc

8. 200 kL   d, e

    c. 0.2 kL

    d. 20,000,000 cL

9. 200 mL   b, g

    e. 200,000 L

    f. 200,000 mL

10. 200 cL   a, h

    g. 0.2 L

    h. 2 L

Concept 1: Summary of U.S. Customary and Metric Unit Equivalents

11. Identify an appropriate ratio to convert 5 yards to meters by using multiplication.   b

    a. $\dfrac{1 \text{ yd}}{0.914 \text{ m}}$ **b.** $\dfrac{0.914 \text{ m}}{1 \text{ yd}}$ **c.** $\dfrac{0.914 \text{ yd}}{1 \text{ m}}$ **d.** $\dfrac{1 \text{ m}}{0.914 \text{ yd}}$

12. Identify an appropriate ratio to convert 3 pounds to kilograms by using multiplication.   d

    a. $\dfrac{0.45 \text{ lb}}{1 \text{ kg}}$ **b.** $\dfrac{1 \text{ kg}}{0.45 \text{ lb}}$ **c.** $\dfrac{1 \text{ lb}}{0.45 \text{ kg}}$ **d.** $\dfrac{0.45 \text{ kg}}{1 \text{ lb}}$

13. Identify an appropriate ratio to convert 2 quarts to liters by using multiplication.   a

    a. $\dfrac{0.95 \text{ L}}{1 \text{ qt}}$ **b.** $\dfrac{1 \text{ qt}}{0.95 \text{ L}}$ **c.** $\dfrac{0.95 \text{ qt}}{1 \text{ L}}$ **d.** $\dfrac{1 \text{ L}}{0.95 \text{ qt}}$

14. Identify an appropriate ratio to convert 10 miles to kilometers by using multiplication.   c

    a. $\dfrac{1 \text{ mi}}{1.61 \text{ km}}$ **b.** $\dfrac{1 \text{ km}}{1.61 \text{ mi}}$ **c.** $\dfrac{1.61 \text{ km}}{1 \text{ mi}}$ **d.** $\dfrac{1.61 \text{ mi}}{1 \text{ km}}$

Concept 2: Converting U.S. Customary and Metric Units

For Exercises 15–23, convert the units of length. Round the answer to one decimal place, if necessary.
(See Examples 1 and 2.)

15. 2 in. ≈ _____ cm   5.1 cm

16. 120 km ≈ _____ mi   74.5 mi

17. 8 m ≈ _____ yd   8.8 yd

18. 4 ft ≈ _____ m   1.2 m

 19. 400 ft ≈ _____ m   122 m

20. 0.75 m ≈ _____ yd   0.8 yd

21. 45 in ≈ _____ m   1.1 m

22. 150 cm ≈ _____ ft   4.9 ft

23. 0.5 ft ≈ _____ cm   15.2 cm

For Exercises 24–32, convert the units of weight and mass. Round the answer to one decimal place, if necessary. **(See Example 3.)**

24. 6 oz ≈ _____ g   168 g

25. 6 lb ≈ _____ kg   2.7 kg

26. 4 kg ≈ _____ lb   8.9 lb

27. 10 g ≈ _____ oz   0.4 oz

28. 14 g ≈ _____ oz   0.5 oz

29. 0.54 kg ≈ _____ lb   1.2 lb

 30. 0.3 lb ≈ _____ kg   0.1 kg

31. 2.2 tons ≈ _____ kg   1980 kg

32. 4500 kg ≈ _____ tons   5 tons

For Exercises 33–38, convert the units of capacity. Round the answer to one decimal place, if necessary. **(See Example 4.)**

33. 6 qt ≈ _____ L   5.7 L

34. 5 fl oz ≈ _____ mL   150 mL

35. 120 mL ≈ _____ fl oz   4 fl oz

36. 19 L ≈ _____ qt   20 qt

37. 960 cc ≈ _____ fl oz   32 fl oz

38. 0.5 fl oz ≈ _____ cc   15 cc

Concept 3: Applications

For Exercises 39–54, refer to Table 7-6 on page 448.

39. A 2-lb box of sugar costs $3.19. A box that contains single-serving packets contains 354 g and costs $1.49. Find the unit costs in dollars per ounce to determine the better buy. **(See Example 5.)**
The box of sugar costs $0.100 per ounce, and the packets cost $0.118 per ounce. The 2-lb box is the better buy.

40. At the grocery store, Debbie compares the prices of a 2-L bottle of water and a 6-pack of bottled water. The 2-L bottle is priced at $1.59. The 6-pack costs $3.60 and each bottle in the package contains 24 fl oz. Compare the cost of water per quart to determine which is a better buy.
The 2-L bottle costs $0.76/qt. The 6-pack costs $0.80/qt. The 2-L bottle is a better buy.

41. A cross-country skiing race is 30 km long. Is this length more or less than 18 mi? **(See Example 6.)**
18 mi is about 28.98 km. Therefore the 30-km race is longer than 18 mi.

42. A can of cat food is 85 g. How many ounces is this? Round to the nearest ounce.   The can contains approximately 3 oz of cat food.

43. Carly Patterson of the U.S. Olympic gymnastic team weighed 97 lb when she was the Olympic All-Around champion. How many kilograms is this?
97 lb is approximately 43.65 kg.

44. Warren's dad ran the 100-yd dash when he was in high school. Suppose Warren runs the 100-*meter* dash on his track team.

a. Who runs the longer race?   Warren runs the longer race.

b. Find the difference between the lengths in meters.   8.6 m longer

45. In a recent year, the price of gas in Germany was $1.90 per liter. What is the price per gallon?   The price is approximately $7.22 per gallon.

46. A jar of spaghetti sauce is 2 lb 8 oz. How many kilograms is this? Round to two decimal places.   This jar contains about 1.13 kg of spaghetti sauce.

47. The thickness of a hockey puck is 2.54 cm. How many inches is this?
A hockey puck is 1 in. thick.

48. A bottle of grape juice contains 1.9 L of juice. Is there enough juice to fill 10 glasses that hold 6 oz each? If yes, how many ounces will be left over?
Yes. There will be approximately 4 oz left over.

49. Football player Tony Romo weighs 99,790 g. How many pounds is this? Round to the nearest pound.   Tony weighs about 222 lb.

50. The distance between two lightposts is 6 m. How many feet is this? Round to the nearest foot.   The distance is approximately 20 ft.

51. A nurse gives a patient 45 cc of saline solution. How many fluid ounces does this represent?
45 cc is 1.5 fl oz.

52. Cough syrup comes in a bottle that contains 4 fl oz. How many milliliters is this?   4 fl oz is 120 mL.

53. Suppose this figure is a drawing of a room where 1 cm represents 2 ft. If you were to install molding along the edge of the floor, how many feet would you need?   40.8 ft

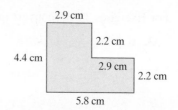

2.9 cm
2.2 cm
4.4 cm
2.9 cm
2.2 cm
5.8 cm

54. Suppose this figure is a drawing of a room where 1 cm represents 3 ft. If you were to install molding along the edge of the floor, how many feet would you need?   45 ft

4.3 cm
3.2 cm

Concept 4: Units of Temperature

For Exercises 55–60, convert the temperatures by using the appropriate formula: $F = \frac{9}{5}C + 32$ or $C = \frac{5}{9}(F - 32)$.
(See Examples 7 and 8.)

55. $25°C =$ _____ $°F$ 77°F

56. $113°F =$ _____ $°C$ 45°C

57. $68°F =$ _____ $°C$ 20°C

58. $15°C =$ _____ $°F$ 59°F

59. $30°C =$ _____ $°F$ 86°F

60. $104°F =$ _____ $°C$ 40°C

61. The boiling point of the element boron is 4000°C. Find the boiling point in degrees Fahrenheit.   7232°F

62. The melting point of the element copper is 1085°C. Find the melting point in degrees Fahrenheit.   1985°F

63. If the outdoor temperature is 35°C, is it a hot day or a cold day?   It is a hot day. The temperature is 95°F.

64. If the temperature is 25° in Miami Beach, Florida, is it a warm or a cold day? Explain.
In Miami, the Fahrenheit scale is used so 25°F would be a cold day.

65. If the temperature is 25° in Florence, Italy, is it a warm or a cold day? Explain.
In Italy, the Celsius scale is used. Converting 25°C to Fahrenheit gives 77°F, which would be a warm day.

66. The high temperature in London, England, on a typical September day was 18°C, and the low was 13°C. Convert these temperatures to degrees Fahrenheit.
The high temperature was 64.4°F, and the low temperature was 55.4°F.

67. Use the fact that water boils at 100°C to show that the boiling point is 212°F.
$F = \frac{9}{5}C + 32 = \frac{9}{5} \cdot 100 + 32 = 9(20) + 32 = 180 + 32 = 212$

68. Use the fact that water freezes at 0°C to show that the temperature at which water freezes is 32°F.
$F = \frac{9}{5}C + 32 = \frac{9}{5} \cdot 0 + 32 = 0 + 32 = 32$

Expanding Your Skills

In the U.S. Customary System of measurement, 1 ton = 2000 lb. In the metric system, 1 metric ton = 1000 kg. Use this information to answer Exercises 69–72.

69. A Lincoln Navigator weighs 5700 lb. How many metric tons is this?   The Navigator weighs approximately 2.565 metric tons.

70. An elevator has a maximum capacity of 1200 lb. How many metric tons is this?   The maximum capacity is about 0.54 metric ton.

71. The average mass of a blue whale (the largest mammal in the world) is approximately 108 metric tons. How many pounds is this?
The average weight of the blue whale is approximately 240,000 lb.

72. The mass of a Mini-Cooper is 1.25 metric tons. How many pounds is this? Round to the nearest pound.   The weight of the Mini-Cooper is approximately 2778 lb.

Group Activity

Remodeling the Classroom

Materials: A tape measure for measuring the size of the classroom.
Advertisements from the newspaper for carpet and paint.

Estimated time: 30 minutes

Group Size: 3–4

In this activity, your group will determine the cost for updating your classroom with new paint and new carpet.

Answers will vary.

1. Measure and record the dimensions of the room and also the height of the walls. You may want to sketch the floor and walls and then label their dimensions on the figure.

2. Calculate and record the area of the floor.

3. Calculate and record the total area of the walls. Subtract any area taken up by doors, windows, and chalkboards.

4. Look through advertisements for carpet that would be suitable for your classroom. You may have to look online for a better choice. Choose a carpet.

5. Calculate how much carpet is needed based on your measurements. To do this, take the area of the floor found in step 2 and add 10% of that figure to allow for waste.

6. Determine the cost to carpet the classroom. Do not forget to include carpet padding and labor to install the carpet. You may have to look online to find prices for padding and installation if they are not included in the price. Also include sales tax for your area.

7. Look through advertisements for paint that would be suitable for your classroom. You may have to look online for a better choice. Choose a paint.

8. Calculate the number of gallons of paint needed for your classroom. You may assume that 1 gal of paint will cover 400 ft^2. Calculate the cost of paint for the classroom. Include sales tax, but do not include labor costs for painting. You will do the painting yourself!

9. Calculate the total cost for carpeting and painting the classroom.

Chapter 7   Summary

Section 7.1   Converting U.S. Customary Units of Length

Key Concepts

The **U.S. Customary System** for measuring length is commonly used in the United States. Conversion of length can be done by substitution using these equivalents.

1 ft = 12 in.	1 in. = $\frac{1}{12}$ ft
1 yd = 3 ft	1 ft = $\frac{1}{3}$ yd
1 mi = 5280 ft	1 ft = $\frac{1}{5280}$ mi
1 mi = 1760 yd	1 yd = $\frac{1}{1760}$ mi

Conversion of length can also be done by multiplying by an appropriate conversion factor, such as $\frac{12\,\text{in.}}{1\,\text{ft}}$ or $\frac{1\,\text{mi}}{5280\,\text{ft}}$.

To choose a **conversion factor**, follow these guidelines:

- The unit of measure in the numerator is the new unit you want to convert *to*.
- The unit of measure in the denominator is the original unit you want to convert *from*.

When adding and subtracting measurements, add or subtract like units.

Examples

Example 1

To convert 18 in. to feet, write

$$18 \text{ in.} = 18 \times 1 \text{ in.}$$

$$= \frac{18}{1} \times \frac{1}{12} \text{ ft}$$

$$= \frac{18}{12} \text{ ft} = \frac{3}{2} \text{ ft or } 1\frac{1}{2} \text{ ft}$$

Example 2

To convert 8 yd to feet, multiply.

$$8 \text{ yd} \cdot \frac{3 \text{ ft}}{1 \text{ yd}} = \frac{8 \text{ yd}}{1} \cdot \frac{3 \text{ ft}}{1 \text{ yd}} = 24 \text{ ft}$$

Example 3

To add 3 ft 9 in. + 2 ft 10 in., add like terms.

$$\begin{array}{r} 3 \text{ ft} + 9 \text{ in.} \\ + 2 \text{ ft} + 10 \text{ in.} \\ \hline 5 \text{ ft} + 19 \text{ in.} = 5 \text{ ft} + 1 \text{ ft} + 7 \text{ in.} \end{array}$$

$$= 6 \text{ ft } 7 \text{ in.}$$

Section 7.2   Converting U.S. Customary Units of Time, Weight, and Capacity

Key Concepts

Several common U.S. Customary units—time, weight, and capacity—are given.

Time

1 year = 365 days

1 week = 7 days

1 day = 24 hours (hr)

1 hour (hr) = 60 minutes (min)

1 minute (min) = 60 seconds (sec)

Examples

Example 1

To convert 200 min to hours, multiply.

$$200 \text{ min} \cdot \frac{1 \text{ hr}}{60 \text{ min}}$$

$$= \frac{200 \text{ min}}{1} \cdot \frac{1 \text{ hr}}{60 \text{ min}}$$

$$= \frac{200}{60} \text{ hr}$$

$$= \frac{10}{3} \text{ hr or } 3\frac{1}{3} \text{ hr}$$

Weight

1 pound (lb) = 16 ounces (oz)

1 ton = 2000 pounds (lb)

Example 2

To convert 6 lb to ounces, multiply.

$$6 \text{ lb} \cdot \frac{16 \text{ oz}}{1 \text{ lb}}$$

$$= \frac{6 \text{ lb}}{1} \cdot \frac{16 \text{ oz}}{1 \text{ lb}}$$

$$= \frac{96 \text{ oz}}{1} \text{ or } 96 \text{ oz}$$

Capacity

1 cup (c) = 8 fluid ounces (fl oz)

1 pint (pt) = 2 cups (c)

1 quart (qt) = 2 pints (pt)

1 gallon (gal) = 4 quarts (qt)

Example 3

To convert 40 cups to gallons, multiply.

$$40 \text{ c} \cdot \frac{1 \text{ pt}}{2 \text{ c}} \cdot \frac{1 \text{ qt}}{2 \text{ pt}} \cdot \frac{1 \text{ gal}}{4 \text{ qt}}$$

$$= \frac{40 \text{ c}}{1} \cdot \frac{1 \text{ pt}}{2 \text{ c}} \cdot \frac{1 \text{ qt}}{2 \text{ pt}} \cdot \frac{1 \text{ gal}}{4 \text{ qt}}$$

$$= \frac{40}{16} \text{ gal}$$

$$= \frac{5}{2} \text{ gal or } 2\frac{1}{2} \text{ gal}$$

Section 7.3   Metric Units of Length

Key Concepts

The **metric system** offers other units for measuring length, mass, and capacity. The base units are the **meter** for length, the **gram** for mass, and the **liter** for capacity. Other units of length, mass, and capacity in the metric system are powers of 10 of the base unit.

Metric units of length and their equivalents are given.

1 kilometer (km) = 1000 m

1 hectometer (hm) = 100 m

1 dekameter (dam) = 10 m

1 meter (m) = 1 m

1 decimeter (dm) = 0.1 m $\quad (\frac{1}{10} \text{ m})$

1 centimeter (cm) = 0.01 m $\quad (\frac{1}{100} \text{ m})$

1 millimeter (mm) = 0.001 m $\quad (\frac{1}{1000} \text{ m})$

Examples

Example 1

To convert 2 km to meters, we can use a conversion factor:

$$2 \text{ km} = \frac{2 \text{ km}}{1} \cdot \frac{1000 \text{ m}}{1 \text{ km}} \quad \begin{array}{l} \leftarrow \text{ new unit} \\ \leftarrow \text{ original unit} \end{array}$$

$$= 2000 \text{ m}$$

Or we can use the prefix line.

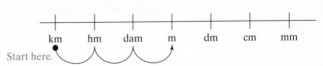

2 km = 2.000 km = 2000 m

Section 7.4 Metric Units of Mass, Capacity, and Medical Applications

Key Concepts

The prefix line can be used to convert metric units for mass, capacity, and length.

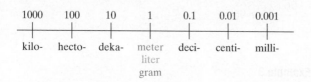

The metric unit conversions for mass are given in **grams**.

1 kilogram (kg) = 1000 g

1 hectogram (hg) = 100 g

1 dekagram (dag) = 10 g

1 gram (g) = 1 g

1 decigram (dg) = 0.1 g $\left(\frac{1}{10} \text{ g}\right)$

1 centigram (cg) = 0.01 g $\left(\frac{1}{100} \text{ g}\right)$

1 milligram (mg) = 0.001 g $\left(\frac{1}{1000} \text{ g}\right)$

The metric unit conversions for capacity are given in **liters**.

1 kiloliter (kL) = 1000 L

1 hectoliter (hL) = 100 L

1 dekaliter (daL) = 10 L

1 liter (L) = 1 L

1 deciliter (dL) = 0.1 L $\left(\frac{1}{10} \text{ L}\right)$

1 centiliter (cL) = 0.01 L $\left(\frac{1}{100} \text{ L}\right)$

1 milliliter (mL) = 0.001 L $\left(\frac{1}{1000} \text{ L}\right)$

Note that 1 mL = 1 cc.

In medical applications the **microgram** is often used.

1000 µg = 1 mg

Examples

Example 1

To convert 0.962 kg to grams, we can use a conversion factor:

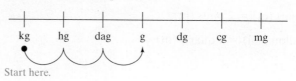

$$0.962 \text{ kg} = \frac{0.962 \text{ kg}}{1} \cdot \frac{1000 \text{ g}}{1 \text{ kg}} \quad \longleftarrow \text{ new unit} \\ \longleftarrow \text{ original unit}$$

$$= 962 \text{ g}$$

Or we can use the prefix line.

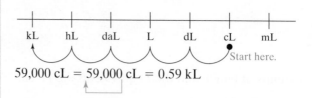

$$0.962 \text{ kg} = 0.962 \text{ kg} = 962 \text{ g}$$

Example 2

To convert 59,000 cL to kL, we can use conversion factors:

$$59,000 \text{ cL} = \frac{59,000 \text{ c}\cancel{L}}{1} \cdot \frac{1 \cancel{L}}{100 \text{ c}\cancel{L}} \cdot \frac{1 \text{ kL}}{1000 \cancel{L}}$$

$$= 0.59 \text{ kL}$$

Or we can use the prefix line.

$$59,000 \text{ cL} = 59,000 \text{ cL} = 0.59 \text{ kL}$$

Section 7.5 Converting Between U.S. Customary and Metric Units

Key Concepts

The common conversions between the U.S. Customary and metric systems are given.

Length

1 in. = 2.54 cm

1 ft ≈ 0.305 m

1 yd ≈ 0.914 m

1 mi ≈ 1.61 km

Weight/Mass (on Earth)

1 lb ≈ 0.45 kg

1 oz ≈ 28 g

Capacity

1 qt ≈ 0.95 L

1 fl oz ≈ 30 mL = 30 cc

To convert U.S. Customary units to metric or metric units to U.S. Customary units, use conversion factors.

The U.S. Customary System uses the **Fahrenheit** scale (°F) to measure temperature. The metric system uses the **Celsius** scale (°C). The conversions are given.

To convert from °C to °F: $F = \dfrac{9}{5}C + 32$

To convert from °F to °C: $C = \dfrac{5}{9}(F - 32)$

Examples

Example 1

Convert 1200 yd to meters by using a conversion factor.

$$1200 \text{ yd} \approx \frac{1200 \text{ yd}}{1} \cdot \frac{0.914 \text{ m}}{1 \text{ yd}} = 1096.8 \text{ m}$$

Example 2

To convert 900 cc to fluid ounces, recall that 1 cc = 1 mL. Therefore, 900 cc = 900 mL. Then use a conversion factor to convert to fluid ounces.

$$900 \text{ mL} \approx \frac{900 \text{ mL}}{1} \cdot \frac{1 \text{ fl oz}}{30 \text{ mL}} = 30 \text{ fl oz}$$

Example 3

The average January temperature in Havana, Cuba, is 21°C. The average January temperature in Johannesburg, South Africa, is 69°F. Which temperature is warmer?

Convert 21°C to degrees Fahrenheit:

$$F = \frac{9}{5}C + 32$$

$$= \frac{9}{5}(21) + 32$$

$$= 37.8 + 32 = 69.8$$

The value 21°C = 69.8°F, which is 0.8 degree warmer than the temperature in Johannesburg.

Chapter 7   Review Exercises

Section 7.1

For Exercises 1–8, convert the units of length.

1. 48 in. = ___ ft
4 ft

2. $3\frac{1}{4}$ ft = ___ in.
39 in.

3. 2 mi = ___ yd
3520 yd

4. 2200 yd = ___ mi
$1\frac{1}{4}$ mi

5. 7040 ft = ___ mi
$1\frac{1}{3}$ mi

6. $\frac{1}{2}$ mi = ___ ft
2640 ft

7. 2 yd = ___ in.
72 in.

8. 6336 in. = ___ mi
0.1 mi

For Exercises 9–16, perform the indicated operations.

9. 3 ft 9 in. + 5 ft 6 in.
9 ft 3 in.

10. 4′11″ + 1′5″
6′4″

11. 5′3″ − 2′5″
2′10″

12. 12 ft 7 in. − 8 ft 10 in.
3 ft 9 in.

13. 4 × (5′3″)
21′

14. 2 × (4 ft 8 in.)
9 ft 4 in.

15. 6 ft 3 in. ÷ 3
2 ft 1 in.

16. 6 yd 2 ft ÷ 2
3 yd 1 ft

17. Find the perimeter in feet.
$7\frac{1}{2}$ ft

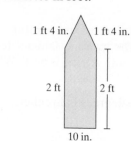
1 ft 4 in.   1 ft 4 in.
2 ft   2 ft
10 in.

18. A roll of wire contains 50 yd of wire. If Ivan uses 48 ft, how much wire is left?
There is 102 ft or 34 yd of wire left.

Section 7.2

For Exercises 19–30, convert the units of time, weight, and capacity.

19. 72 hr = ___ days
3 days

20. 6 min = ___ sec
360 sec

21. 5 lb = ___ oz
80 oz

22. 1 wk = ___ hr
168 hr

23. 12 fl oz = ___ c
$1\frac{1}{2}$ c

24. 0.25 ton = ___ lb
500 lb

25. 3500 lb = ___ tons
$1\frac{3}{4}$ tons

26. 150 min = ___ hr
2.5 hr

27. 1800 sec = ___ hr
0.5 hr

28. 2 gal = ___ pt
16 pt

29. 12 oz = ___ lb
$\frac{3}{4}$ lb

30. 16 qt = ___ gal
4 gal

31. A runner finished a race with a time of 2:24:30. Convert the time to minutes.
144.5 min

32. Margaret Johansson gave birth to triplets who weighed 3 lb 10 oz, 4 lb 2 oz, and 4 lb 1 oz. What was the total weight of the triplets?
The total weight was 11 lb 13 oz.

33. One and one-half tons of dirt are to be equally distributed to 8 locations. How many pounds will go to each location?
375 lb will go to each location.

34. A case of 12-fl-oz cans of soda contains 24 cans. How many gallons of soda are in the case?
There are $2\frac{1}{4}$ gal of soda.

Section 7.3

NS&E For Exercises 35–38, select the most reasonable measurement.

35. A pencil is _____ long.   b
a. 16 mm   **b.** 16 cm
c. 16 m   **d.** 16 km

36. A mosquito is _____ long.   a
a. 12 mm   **b.** 12 cm
c. 12 m   **d.** 12 km

37. A two-story house is _____ tall.   c
a. 9 mm   **b.** 9 cm
c. 9 m   **d.** 9 km

38. The distance between Houston and Dallas is _____ .   d
a. 362 mm   **b.** 362 cm
c. 362 m   **d.** 362 km

For Exercises 39–48, convert the metric units of length.

39. 52 cm = ___ mm
520 mm

40. 91 mm = ___ cm
9.1 cm

41. 2.338 km = ___ m
2338 m

42. 93 m = ___ km
0.093 km

43. 34 dm = ___ m
3.4 m

44. 2.1 m = ___ dam
0.21 dam

45. 4 cm = ___ m
0.04 m

46. 3 m = ___ cm
300 cm

47. 1.2 m = ___ mm
1200 mm

48. 4023 hm = ___ km
402.3 km

49. The highest point in Arizona is 3.851 km above sea level. The highest point in Louisiana is 163 m. What is the difference between these elevations in meters?
The difference is 3688 m.

 50. Determine the perimeter of the triangle in centimeters.
22.6 cm

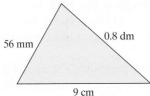

56 mm 0.8 dm 9 cm

Section 7.4

For Exercises 51–56, convert the metric units of mass.

51. 6.1 g = ___ cg
610 cg

52. 420 g = ___ kg
0.42 kg

53. 3212 mg = ___ g
3.212 g

54. 0.7 hg = ___ g
70 g

55. 5 cg = ___ mg
50 mg

56. 0.1 dag = ___ cg
100 cg

For Exercises 57–62, convert the metric units of capacity.

57. 300 mL = ___ L
0.3 L

58. 2.4 hL = ___ L
240 L

59. 830 cL = ___ L
8.3 L

60. 124 mL = ___ cc
124 cc

61. 225 cc = ___ cL
22.5 cL

62. 0.49 kL = ___ L
490 L

63. The dimensions of a dining room table are 2 m by 125 cm. Convert the units to meters and find the perimeter and area of the tabletop.
Perimeter: 6.5 m; area: 2.5 m²

64. A bottle of apple juice contains 1.2 L of juice. If a glass holds 24 cL, how many glasses can be filled from this bottle?
5 glasses can be filled.

65. An adult has a mass of 68 kg. A baby has a mass of 3200 g. What is the difference in their masses, in kilograms?
The difference is 64.8 kg.

66. From a wooden board 2 m long, Jesse needs to cut 3 pieces that are each 75 cm long. Is the 2-m length of board long enough for the 3 pieces?
No, the board is 25 cm too short.

67. A physician prescribes a drug based on a patient's mass. The dosage is given as 0.04 mg of the drug per kilogram of the patient's mass.

   a. How much would the physician prescribe for an 80-kg patient? 3.2 mg

   b. If the dosage was to be given twice a day, how much of the drug would the patient take in a week? 44.8 mg

68. Convert 0.45 mg to micrograms (mcg). 450 mcg

69. A standard hypodermic syringe holds 3 cc of fluid. If a nurse uses 1.8 mL of the fluid, how much is left in the syringe?
There is 1.2 cc or 1.2 mL of fluid left.

70. A medication comes in 250-mg capsules. If Clayton took 3 capsules a day for 10 days, how many grams of the medication did he take?
Clayton took 7.5 g.

Section 7.5

For Exercises 71–80, refer to page 448. Convert the units of length, capacity, mass, and weight. Round to the nearest hundredth, if necessary.

71. 6.2 in. ≈ _____ cm
15.75 cm

72. 75 mL ≈ _____ fl oz
2.5 fl oz

73. 140 g ≈ _____ oz
5 oz

74. 5 L ≈ _____ qt
5.26 qt

75. 3.4 ft ≈ _____ m
1.04 m

76. 100 lb ≈ _____ kg
45 kg

77. 120 km ≈ _____ mi
74.53 mi

78. 6 qt ≈ _____ L
5.7 L

79. 1.5 fl oz ≈ _____ cc
45 cc

80. 12.5 tons ≈ _____ kg
11,250 kg

81. The height of a computer desk is 30 in. The height of the chair is 38 cm. What is the difference in height between the desk and chair, in centimeters?
The difference in height is 38.2 cm.

82. A bag of snack crackers contains 7.2 oz. If one serving is 30 g, approximately how many servings are in one bag?
There are approximately 6.72 servings.

83. A prescription for cough syrup indicates that 30 mL should be taken twice a day for 7 days. What is the total amount, in liters, of cough syrup to be taken?
The total amount of cough syrup is approximately 0.42 L.

84. The Boston Marathon is 42.195 km long. Convert this distance to miles. Round to the tenths place.
The marathon is approximately 26.2 mi.

85. Write the formula to convert degrees Fahrenheit to degrees Celsius. $C = \frac{5}{9}(F - 32)$

86. When roasting a turkey, the meat thermometer should register between 180°F and 185°F to indicate that the turkey is done. Convert these temperatures to degrees Celsius. Round to the nearest tenth, if necessary.
82.2°C to 85°C

87. Write the formula to convert degrees Celsius to degrees Fahrenheit.
$F = \frac{9}{5}C + 32$

88. The average October temperature for Toronto, Ontario, Canada, is 8°C. Convert this temperature to degrees Fahrenheit. 46.4°F

Chapter 7 Test

1. Identify the units that apply to measuring length. Circle all that apply.   c, d, g, j

    a. Pound   **b.** Ounce   **c.** Meter

    d. Mile   **e.** Gram   **f.** Pint

    g. Feet   **h.** Liter   **i.** Fluid ounce

    j. Kilometer

2. Identify the units that apply to measuring capacity. Circle all that apply.   f, h, i

    a. Pound   **b.** Ounce   **c.** Meter

    d. Mile   **e.** Gram   **f.** Pint

    g. Foot   **h.** Liter   **i.** Fluid ounce

    j. Kilometer

3. Identify the units that apply to measuring mass or weight. Circle all that apply.   a, b, e

    a. Pound   **b.** Ounce   **c.** Meter

    d. Mile   **e.** Gram   **f.** Pint

    g. Foot   **h.** Liter   **i.** Fluid ounce

    j. Kilometer

4. A backyard needs 25 ft of fencing. How many yards is this? $8\frac{1}{3}$ yd

5. It's estimated that an adult *Tyrannosaurus Rex* weighed approximately 11,000 lb. How many tons is this?
5.5 tons

6. Two exits on the highway are 52,800 ft apart. How many miles is this?
10 mi

7. A recipe for brownies calls for $\frac{3}{4}$ c of milk and 4 fl oz of water. What is the total amount of liquid in fluid ounces?
10 fl oz of liquid

8. A television show has 1200 sec of commercials. How many minutes is this?
20 min

9. Find the perimeter of the rectangle in feet.
9′

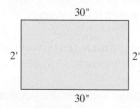

30″

2′   2′

30″

10. A decorator wraps a gift, using two pieces of ribbon. One is 1′10″ and the other is 2′4″. Find the total length of ribbon used.
4′2″

11. When Stephen was born, he weighed 8 lb 1 oz. When he left the hospital, he weighed 7 lb 10 oz. How much weight did he lose after he was born?
He lost 7 oz.

12. A decorative pillow requires 3 ft 11 in. of fringe around the perimeter of the pillow. If 5 pillows are produced, how much fringe is required?
19 ft 7 in.

13. Josh ran a race and finished with the time of 1:15:15. Convert this time to minutes.
75.25 min

14. Approximate the width of the nut in centimeters and millimeters.
2.4 cm or 24 mm

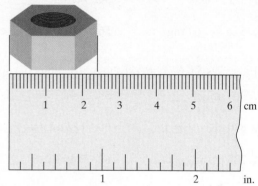

15. Select the most reasonable measurement for the length of a living room.   c

    a. 5 mm   **b.** 5 cm   **c.** 5 m   **d.** 5 km

16. The span of the Mackinac Bridge in Michigan is 1158 m. What is this length in kilometers?
1.158 km

17. A tablespoon (1 T) contains 0.015 L. How many milliliters is this?
15 mL

18. **a.** What does the abbreviation *cc* stand for?
Cubic centimeters
    b. Convert 235 mL to cubic centimeters.
235 cc
    c. Convert 1 L to cubic centimeters.
1000 cc

19. A can of diced tomatoes is 411 g. Convert 411 g to centigrams.
41,100 cg

20. A box of crackers is 210 g. If a serving of crackers is 30,000 mg, how many servings are in the box?
7 servings

For Exercises 21–26, refer to Table 7-6 on page 448.

21. A bottle of Sprite contains 2 L. What is the capacity in quarts? Round to the nearest tenth.
2.1 qt

22. Maurice Greene was one of the premier sprinters in U.S. track and field. His best race is the 100-m dash. How many yards is this? Round to the nearest yard.
109 yd

23. The distance between two exits on a highway is 4.5 km. How far is this in miles? Round to the nearest tenth.
2.8 mi

24. Breckenridge, Colorado, is 9603 ft above sea level. What is this height in meters? Round to the nearest meter.
2929 m

25. A snowy egret stands about 20 in. tall and has a 38-in. wingspan. Convert both values to centimeters.
50.8 cm tall and 96.52-cm wingspan

26. The mass of a laptop computer is 5000 g. What is the weight in pounds? Round to the nearest pound.
11 lb

27. The oven temperature needed to bake cookies is 375°F. What is this temperature in degrees Celsius? Round to the nearest tenth.
190.6°C

28. The average January temperature in Albuquerque, New Mexico, is 2°C. Convert this temperature to degrees Fahrenheit.
35.6°F

29. A patient is supposed to get 0.1 mg of a drug for every kilogram of body mass, four times a day. How much of the drug would a 70-kg woman get each day?   28 mg

30. Gus, the cat, had an overactive thyroid gland. The vet prescribed 0.125 mg of Methimazole every 12 hr. How many micrograms is this per week?
1750 mcg per week

Chapters 1-7   Cumulative Review Exercises

1. Round each number to the indicated place.

 a. 2499; thousands place   **b.** 42,099; tens place
 2000   42,100

For Exercises 2 and 3, refer to the figure.

2. Find the perimeter.
56 cm

3. Find the area.
180 cm^2

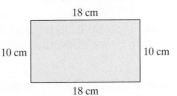

18 cm
10 cm   10 cm
18 cm

4. Simplify. $144 \div 9 \div (17 - 3 \cdot 5)^2$
4

5. The table gives the amount of money spent on research and development for five major U.S. companies. (*Source:* The Financial Times Ltd.)

Amount Spent for Research and Development

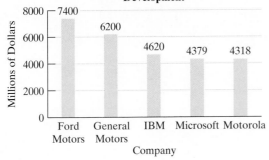

a. Which company spends the most on research and development? What is that amount?
Ford Motor Company spends the most. That amount is $7400 million or $7,400,000,000.

b. What is the difference between the amount spent at IBM and the amount spent at Motorola?   The difference between IBM and Motorola is $302 million or $302,000,000.

c. What is the total amount spent for research and development for these five companies?
The total amount spent is $26,917 million or $26,917,000,000.

6. Write an equivalent fraction with the indicated denominator. $\frac{2}{13} = \frac{}{39}$ $\frac{6}{39}$

7. Is 32,542 divisible by 3? Explain why or why not.
The number 32,542 is not divisible by 3 because the sum of the digits (16) is not divisible by 3.

8. Write the prime factorization of 108.
$2 \cdot 2 \cdot 3 \cdot 3 \cdot 3$

9. Find the area of the triangle.
540 in.^{2}

20 in.
54 in.

10. Keesha wants to bake oatmeal cookies, but she has only 1 c of oatmeal. The recipe calls for 3 c. Does Keesha have enough oatmeal to make $\frac{1}{4}$ of the recipe? $\frac{1}{4}$ of the recipe would call for $\frac{3}{4}$ c of oatmeal. This is less than 1 c so Keesha does have enough.

Writing   Translating Expression   Geometry   Scientific Calculator   Video   NSE

11. Find the LCD of $\frac{1}{2}, \frac{6}{5}$, and $\frac{3}{10}$. 10

12. Simplify. $\frac{1}{2} + \frac{6}{5} - \frac{3}{10}$ $\frac{7}{5}$

For Exercises 13–16, perform the indicated operations with mixed numbers.

13. $6\frac{2}{3} + 2\frac{5}{6}$ $9\frac{1}{2}$

14. $6\frac{2}{3} \cdot 2\frac{5}{6}$ $18\frac{8}{9}$

15. $6\frac{2}{3} \div 2\frac{5}{6}$ $2\frac{6}{17}$

16. $6\frac{2}{3} - 2\frac{5}{6}$ $3\frac{5}{6}$

For Exercises 17–22, complete the table.

	Fraction	Decimal
17.	$\frac{1}{3}$	$0.\overline{3}$
18.	$\frac{9}{20}$	0.45
19.	$\frac{5}{4}$	1.25
20.	$\frac{7}{2}$	3.5
21.	$\frac{3}{8}$	0.375
22.	$\frac{1}{25}$	0.04

23. A football team won 6 games and lost 5.

   a. Write a ratio of wins to losses. $\frac{6}{5}$

   b. Write a ratio of wins to total games played. $\frac{6}{11}$

24. A pharmacist has a 1-L bottle of an antacid solution. How many smaller bottles can he fill if they contain 25 mL each?   40 bottles

25. If a car dealership sells 18 cars in a 5-day period, how many cars can the dealership expect to sell in 25 days?
90 cars

26. A hospital ward has 40 beds and employs 6 nurses. What is the unit rate of beds per nurse? Round to the nearest tenth.
6.7 beds per nurse

27. Are 6 and 8 proportional to 2 and 3?   No

28. Four-fifths of the drinks sold at a movie theater were soda. What percent is this?   80%

29. Assume that these figures are similar. Solve for x.

$x = \frac{16}{3}$ yd

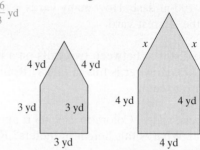

30. Of the trees in a forest, 62% were saved from a fire. If this represents 1420 trees, how many trees were originally in the forest? Round to the nearest whole tree.
2290 trees

31. Of 60 people, 45% had eaten at least one meal at McDonald's this week. How many people ate at McDonald's?
27 people

32. The sales tax on a $21 meal is $1.26. What is the sales tax rate?
6%

33. The commission rate that Yvonne earns is 14% of sales. If she earned $2100 in commission, how much did she sell?
$15,000 in sales

34. If $5000 is invested at 3.4% simple interest for 6 years, how much interest is earned?
$1020 in interest

For Exercises 35–40, convert the units of capacity, length, mass, and weight. Round to one decimal place, if necessary.

35. 5800 g = ___ kg
5.8 kg

36. 5.8 kg ≈ ___ lb
12.9 lb

37. 72 in. ≈ ___ cm
182.9 cm

38. 72 in. = ___ ft
6 ft

39. $3\frac{1}{2}$ qt = ___ pt
7 pt

40. $3\frac{1}{2}$ qt ≈ ___ L
3.3 L

Geometry

Chapter 8

In this chapter, we study geometry. We begin by categorizing familiar figures such as squares, rectangles, triangles, and circles. Then we study the concepts of perimeter and area. We identify familiar three-dimensional figures and learn how to find their volumes.

Review Your Skills

Earlier in this text we introduced formulas for perimeter and area. Review these formulas by matching each formula with the appropriate figure.

1. $A = \frac{1}{2}bh$ _____e_____

2. $P = a + b + c$ _____d_____

3. $P = 2l + 2w$ _____c_____

4. $A = s^2$ _____a_____

5. $P = 4s$ _____f_____

6. $A = lw$ _____b_____

a. Area

b. Area

c. Perimeter

d. Perimeter

e. Area

f. Perimeter

Section 8.1 Lines and Angles

1. Basic Definitions

In this chapter, we will introduce some basic concepts of geometry.

A **point** is a specific location in space. We often symbolize a point by a dot and label it with a capital letter such as *P*.

$\bullet$
P

A **line** consists of infinitely many points that follow a straight path. A line extends forever in both directions. This is illustrated by arrowheads at both ends. Figure 8-1 shows a line through points *A* and *B*. The line can be represented as $\overleftrightarrow{AB}$ or as $\overleftrightarrow{BA}$.

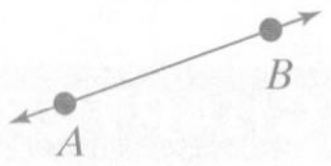

Figure 8-1

Instructor Note: Point out to students that a line does not have finite length, but a line segment does have finite length.

A **line segment** is a part of a line between and including two distinct endpoints. A line segment with endpoints *P* and *Q* can be denoted $\overline{PQ}$ or $\overline{QP}$. See Figure 8-2.

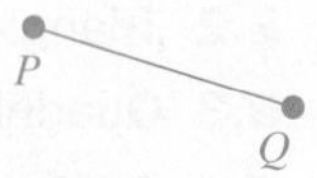

Figure 8-2

Concept Connections

1. Are the rays $\overrightarrow{AB}$ and $\overrightarrow{BA}$ the same?
2. Are the lines $\overleftrightarrow{PQ}$ and $\overleftrightarrow{QP}$ the same?

A **ray** is the part of a line that includes an endpoint and all points on one side of the endpoint. In Figure 8-3, ray $\overrightarrow{PQ}$ is named by using the endpoint *P* and another point *Q* on the ray. Notice that the rays $\overrightarrow{PQ}$ and $\overrightarrow{QP}$ are different because they extend in different directions.

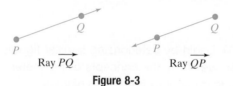

Ray $\overrightarrow{PQ}$ Ray $\overrightarrow{QP}$

Figure 8-3

TIP: A ray has only one endpoint, which is always written first.

Skill Practice

Identify each as a point, line, line segment, or ray.

3. $\overline{RS}$ 4. $\bullet Q$
5. $\overrightarrow{XY}$ 6. $\overleftrightarrow{TV}$

Classroom Examples: p. 474, Exercises 6 and 8

Example 1 Identifying Points, Lines, Line Segments, and Rays

Identify each as a point, line, line segment, or ray.

a. $\overleftrightarrow{MN}$ b. $\overrightarrow{NM}$ c. $\overline{MN}$ d. $\bullet S$

Solution:

a. The double arrowheads indicate that $\overleftrightarrow{MN}$ is a line.

b. The single arrowhead indicates that $\overrightarrow{NM}$ is a ray with endpoint *N*.

c. The bar drawn above the letters $\overline{MN}$ indicates a line segment with endpoints *M* and *N*.

d. The dot represents a point.

2. Naming and Measuring Angles

An **angle** is a geometric figure formed by two rays that share a common endpoint. The common endpoint is called the **vertex** of the angle. In Figure 8-4, the rays $\overrightarrow{PR}$ and $\overrightarrow{PQ}$ share the endpoint *P*. These rays form the sides of the angle and the angle is denoted $\angle QPR$ or $\angle RPQ$. Notice that when we name an angle, the vertex must

Answers

1. No   2. Yes
3. Line segment
4. Point   5. Ray   6. Line

be the middle letter. Sometimes a small arc ⟩ is drawn to illustrate the location of an angle. In such a case, the angle may be named by using the symbol ∠ along with the letter of the vertex.

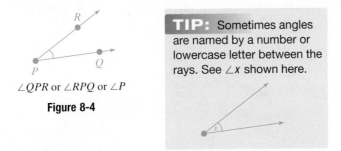

$\angle QPR$ or $\angle RPQ$ or $\angle P$

Figure 8-4

> **TIP:** Sometimes angles are named by a number or lowercase letter between the rays. See $\angle x$ shown here.

The most common unit to measure an angle is the degree, denoted by °. To become familiar with the measure of angles, consider the following benchmarks. Two rays that form a quarter turn of a circle make a 90° angle. A 90° angle is called a **right angle** and is often depicted with a □ symbol. Two rays that form a half turn of a circle make a 180° angle. A 180° angle is called a **straight angle** because it appears as a straight line. A full circle has 360°. For example, the second hand of a clock sweeps out an angle of 360° in 1 minute. See Figure 8-5.

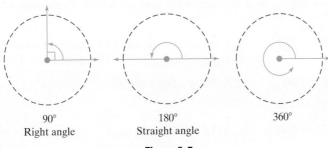

| 90° | 180° | 360° |
| Right angle | Straight angle | |

Figure 8-5

We can approximate the measure of an angle by using a tool called a *protractor*, shown in Figure 8-6. A protractor uses equally spaced tick marks around a semicircle to measure angles from 0° to 180°.

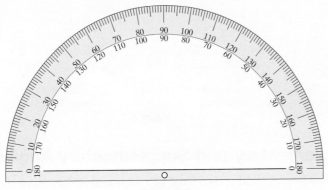

Figure 8-6

Example 2 shows how we can use a protractor to measure several angles. To denote the measure of an angle, we use the symbol m, written in front of the name of the angle. For example, if the measure of angle A is 30°, we write $m(\angle A) = 30°$.

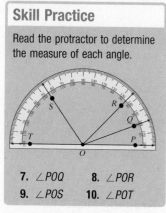
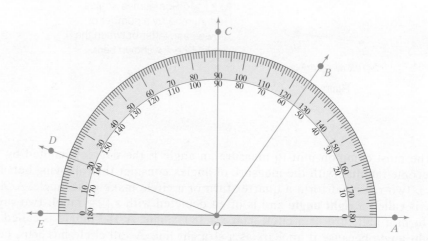

Example 2   Measuring Angles

Read the protractor to determine the measure of each angle.

a. ∠AOB   **b.** ∠AOC   **c.** ∠AOD   **d.** ∠AOE

Solution:

We will use the inner scale on the protractor. This is done because we are measuring the angles in a counterclockwise direction, beginning at 0° along ray $\overrightarrow{OA}$.

a. $m(\angle AOB) = 55°$ On the inner scale, ray $\overrightarrow{OA}$ is aligned with 0° and ray $\overrightarrow{OB}$ passes through 55°. Therefore, $m(\angle AOB) = 55°$.

b. $m(\angle AOC) = 90°$ ∠AOC is a right angle.

c. $m(\angle AOD) = 160°$

d. $m(\angle AOE) = 180°$ ∠AOE is a straight angle.

Concept Connections

Answer true or false.
11. An angle whose measure is 102° is obtuse.
12. An angle whose measure is 98° is acute.

An angle is said to be an **acute angle** if its measure is between 0° and 90°. An angle is said to be an **obtuse angle** if its measure is between 90° and 180°. See Figure 8-7.

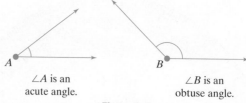

∠A is an   ∠B is an
acute angle.   obtuse angle.

Figure 8-7

3. Complementary and Supplementary Angles

- Two angles are said to be equal or **congruent** if they have the same measure.
- Two angles are said to be **complementary** if the sum of their measures is 90°. In Figure 8-8, the complement of a 60° angle is a 30° angle, and vice versa.

- Two angles are said to be **supplementary** if the sum of their measures is 180°. In Figure 8-9, the supplement of a 60° angle is a 120° angle, and vice versa.

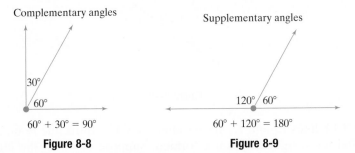

Complementary angles

30°
60°
60° + 30° = 90°

Figure 8-8

Supplementary angles

120° 60°
60° + 120° = 180°

Figure 8-9

Example 3 Identifying Supplementary and Complementary Angles

a. What is the supplement of a 105° angle?

b. What is the complement of a 12° angle?

Solution:

a. The sum of a 105° angle and its supplement must equal 180°. Writing the related subtraction problem, we have

$$180° - 105° = 75°$$ The supplement is a 75° angle.

b. The sum of a 12° angle and its complement must equal 90°. Writing the related subtraction problem, we have

$$90° - 12° = 78°$$ The complement is a 78° angle.

Skill Practice

13. What is the supplement of a 35° angle?

14. What is the complement of a 52° angle?

Classroom Examples: pp. 474–475, Exercises 36 and 42

4. Parallel and Perpendicular Lines

Two lines may intersect (cross) or may be parallel. **Parallel lines** lie on the same flat surface, but never intersect. See Figure 8-10.

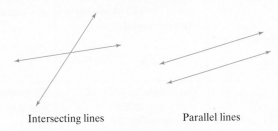

Intersecting lines

Parallel lines

Figure 8-10

Notice that two intersecting lines form four angles. In Figure 8-11, $\angle a$ and $\angle c$ are **vertical angles**. They appear on opposite sides of the vertex. Likewise, $\angle b$ and $\angle d$ are vertical angles. Vertical angles are equal in measure. That is $m(\angle a) = m(\angle c)$ and $m(\angle b) = m(\angle d)$.

Angles that share a side are called *adjacent* angles. One pair of adjacent angles in Figure 8-11 is $\angle a$ and $\angle b$.

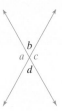

b
a c
d

Figure 8-11

Answers

13. 145°   14. 38°

If two lines intersect at a right angle, they are **perpendicular lines**. See Figure 8-12.

TIP: Sometimes we use the symbol ⊥ to denote perpendicular lines.

Figure 8-12

In Figure 8-13, lines L_1 and L_2 are parallel lines. If a third line m intersects the two parallel lines, eight angles are formed. Suppose we label the eight angles formed by lines L_1, L_2, and m with the numbers 1–8.

Instructor Note: You may want to tell your students that the line passing through the parallel lines is called a transversal.

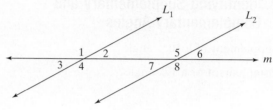

Figure 8-13

These angles have the special properties found in Table 8-1.

Table 8-1

Lines L_1 and L_2 are Parallel; Line m is an Intersecting Line	Name of Angles	Property
	The following pairs of angles are called **alternate interior angles**: ∠2 and ∠7 ∠4 and ∠5	**Alternate interior angles are equal in measure.** $m(\angle 2) = m(\angle 7)$ $m(\angle 4) = m(\angle 5)$
	The following pairs of angles are called **alternate exterior angles**: ∠1 and ∠8 ∠3 and ∠6	**Alternate exterior angles are equal in measure.** $m(\angle 1) = m(\angle 8)$ $m(\angle 3) = m(\angle 6)$
	The following pairs of angles are called **corresponding angles**: ∠1 and ∠5 ∠2 and ∠6 ∠3 and ∠7 ∠4 and ∠8	**Corresponding angles are equal in measure.** $m(\angle 1) = m(\angle 5)$ $m(\angle 2) = m(\angle 6)$ $m(\angle 3) = m(\angle 7)$ $m(\angle 4) = m(\angle 8)$

Example 4 **Finding the Measure of Angles in a Diagram**

Assume that lines L_1 and L_2 are parallel. Find the measure of each angle, and explain how the angle is related to the given angle of 65°.

a. $\angle a$

b. $\angle b$

c. $\angle c$

d. $\angle d$

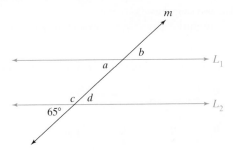

Skill Practice

Assume that lines L_1 and L_2 are parallel. Find the measure of each angle. Explain how the angle is related to the given angle.

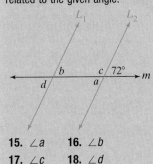

15. $\angle a$ 16. $\angle b$

17. $\angle c$ 18. $\angle d$

Solution:

a. $m(\angle a) = 65°$ $\angle a$ is a corresponding angle to the given angle.

b. $m(\angle b) = 65°$ $\angle b$ is an alternate exterior angle to the given angle.

c. $m(\angle c) = 115°$ $\angle c$ is the supplement to the given angle.

d. $m(\angle d) = 65°$ $\angle d$ and the given angle are vertical angles.

Classroom Example: p. 476, Exercise 72

Answers

15. 72°; vertical angles
16. 72°; corresponding angles
17. 108°; supplementary angles
18. 72°; alternate exterior angles

Section 8.1 Practice Exercises

Study Skills Exercise

For additional exercises, see Classroom Activities 8.1A–8.1B in the *Student's Resource Manual* at www.mhhe.com/moh.

For your next test, make a memory sheet: On a 3 × 5 card (or several 3 × 5 cards), write all the formulas and rules that you need to know. Memorize all this information. Then when your instructor hands you the test, write down all the information that you can remember before you begin the test. Then you can take the test without worrying that you will forget something important. This process is referred to as a "memory dump." What important definitions and concepts have you learned in this section of the text?

Vocabulary and Key Concepts

1. a. A(n) __point__ is a specific location in space.

   b. A(n) __line__ consists of infinitely many points that follow a straight path.

   c. A line __segment__ is part of a line between and including two distinct endpoints.

   d. The ray $\overrightarrow{PQ}$ is the set of points beginning at point __P__ and extending along the line in the direction of point __Q__.

   e. A(n) __angle__ is a geometric figure formed by two rays that share a common endpoint. The common endpoint is called the __vertex__.

   f. A(n) __right__ angle has a measure of 90°. A straight angle has a measure of __180__°.

   g. A(n) __protractor__ is a tool used to measure an angle.

   h. A(n) __acute__ angle measures less than 90° and a(n) __obtuse__ angle measures between 90° and 180°.

   i. Two angles are called __complementary__ angles if the sum of their measures is 90°. Two angles are called __supplementary__ angles if the sum of their measures is 180°.

   j. Two lines that lie on the same flat surface but never intersect are called __parallel__ lines.

   k. Two lines that intersect at a 90° angle are called __perpendicular__ lines.

Concept 1: Basic Definitions

2. What point represents the vertex of angle ∠*ABC*?   Point *B*

3. Explain the difference between a line and a line segment.   A line extends forever in both directions. A line segment is a portion of a line between and including two endpoints.

4. Explain the difference between a line and a ray.   A line extends forever in both directions. A ray begins at an endpoint and extends forever in one direction.

For Exercises 5–10, identify each figure as a line, line segment, ray, or point. **(See Example 1.)**

5. *K   H*   Ray   **6.** *K   H*   Line segment   **7.** *H*   Point

8. *H   K*   Ray   **9.** *H   K*   Line   **10.** *K*   Point

For Exercises 11–14, draw a figure that represents the expression.

11. $\overline{XY}$ For example: *X   Y* or *Y   X*

12. A point named *X*   •*X*

13. $\overleftrightarrow{YX}$ For example: *X   Y* or *Y   X*

14. $\overrightarrow{XY}$ For example: *X   Y* or *Y   X*

Concept 2: Naming and Measuring Angles

15. Sketch a right angle.

16. Sketch a straight angle.

17. Sketch an acute angle.

18. Sketch an obtuse angle.

For Exercises 19–24, use the protractor to determine the measure of each angle. **(See Example 2.)**

19. $m(\angle AOB)$ 20°

20. $m(\angle AOC)$ 52°

21. $m(\angle AOD)$ 90°

22. $m(\angle AOE)$ 115°

23. $m(\angle AOF)$ 148°

24. $m(\angle AOG)$ 170°

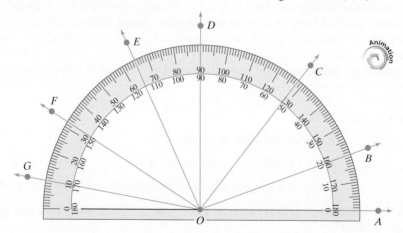

For Exercises 25–32, label each as an obtuse angle, acute angle, right angle, or straight angle.

25. $m(\angle A) = 90°$ Right   **26.** $m(\angle E) = 91°$ Obtuse   **27.** $m(\angle B) = 98°$ Obtuse   **28.** $m(\angle F) = 30°$ Acute

29. $m(\angle C) = 2°$ Acute   **30.** $m(\angle G) = 130°$ Obtuse   **31.** $m(\angle D) = 180°$ Straight   **32.** $m(\angle H) = 45°$ Acute

Concept 3: Complementary and Supplementary Angles

For Exercises 33–40, the measure of an angle is given. Find the measure of the complement. **(See Example 3.)**

33. 80°   10°   **34.** 5°   85°   **35.** 27°   63°   **36.** 64°   26°

37. 29.5°   60.5°   **38.** 13.2°   76.8°   **39.** 89°   1°   **40.** 1°   89°

For Exercises 41–48, the measure of an angle is given. Find the measure of the supplement. **(See Example 3.)**

41. 80° 100°   **42.** 5° 175°   **43.** 127° 53°   **44.** 124° 56°

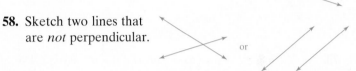 **45.** 37.4° 142.6°   **46.** 173.9° 6.1°   **47.** 179° 1°   **48.** 1° 179°

49. Can two supplementary angles both be obtuse? Why or why not?
No, because the sum of two angles that are both greater than 90° will be more than 180°.

50. Can two supplementary angles both be acute? Why or why not? No, because the sum of two angles that are both less than 90° will be less than 180°.

51. Can two complementary angles both be acute? Why or why not?
Yes. For two angles to add to 90°, the angles themselves must both be less than 90°.

52. Can two complementary angles both be obtuse? Why or why not? No, because the sum of complementary angles is 90°, neither angle can be more than 90°.

53. What angle is its own supplement?
A 90° angle

54. What angle is its own complement?
A 45° angle

Concept 4: Parallel and Perpendicular Lines

55. Sketch two lines that are parallel.

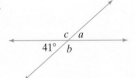

56. Sketch two lines that are *not* parallel.

57. Sketch two lines that are perpendicular.

58. Sketch two lines that are *not* perpendicular.

or

For Exercises 59–62, find the measure of angles *a*, *b*, *c*, and *d*.

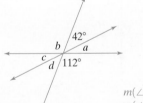 **59.**

$m(\angle a) = 41°; m(\angle b) = 139°;$
$m(\angle c) = 139°$

60.

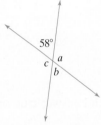

$m(\angle a) = 122°; m(\angle b) = 58°;$
$m(\angle c) = 122°$

61.

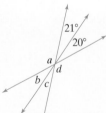

$m(\angle a) = 26°; m(\angle b) = 112°;$
$m(\angle c) = 26°; m(\angle d) = 42°$

62.

$m(\angle a) = 139°; m(\angle b) = 20°;$
$m(\angle c) = 21°; m(\angle d) = 139°$

63. If two intersecting lines form vertical angles and each angle measures 90°, what can you say about the lines?
The two lines are perpendicular.

64. Can two adjacent angles formed by two intersecting lines be complementary, supplementary, or neither? Supplementary

For Exercises 65 and 66, refer to the figure.

65. Describe the pair of angles, $\angle a$ and $\angle c$, as complementary, vertical, or supplementary angles.
Vertical angles

66. Describe the pair of angles, $\angle b$ and $\angle c$, as complementary, vertical, or supplementary angles.
Supplementary angles

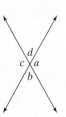

For Exercises 67–70, refer to the figure.

67. Identify a pair of vertical angles.
    a, c or *b, h* or *e, g* or *f, d*

68. Identify a pair of alternate interior angles.
    d, h or *g, c*

69. Identify a pair of alternate exterior angles.
    a, e or *f, b*

70. Identify a pair of corresponding angles.
    f, h or *g, a* or *b, d* or *c, e*

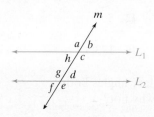

For Exercises 71–74, find the measure of angles *a*–*g* in the figure. Assume that L_1 and L_2 are parallel and that *m* is an intersecting line. **(See Example 4.)**

71.   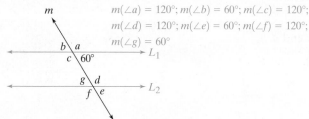 $m(\angle a) = 55°; m(\angle b) = 125°;$
    $m(\angle c) = 55°; m(\angle d) = 55°;$
    $m(\angle e) = 125°; m(\angle f) = 55°;$
    $m(\angle g) = 125°$

72.   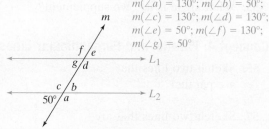 $m(\angle a) = 130°; m(\angle b) = 50°;$
    $m(\angle c) = 130°; m(\angle d) = 130°;$
    $m(\angle e) = 50°; m(\angle f) = 130°;$
    $m(\angle g) = 50°$

73.   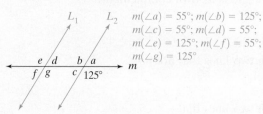 $m(\angle a) = 120°; m(\angle b) = 60°; m(\angle c) = 120°;$
    $m(\angle d) = 120°; m(\angle e) = 60°; m(\angle f) = 120°;$
    $m(\angle g) = 60°$

74.   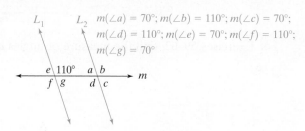 $m(\angle a) = 70°; m(\angle b) = 110°; m(\angle c) = 70°;$
    $m(\angle d) = 110°; m(\angle e) = 70°; m(\angle f) = 110°;$
    $m(\angle g) = 70°$

For Exercises 75–84, refer to the figure and answer true or false.

75. $\overleftrightarrow{AC}$ and $\overleftrightarrow{BC}$ are perpendicular lines.   True

76. $\overleftrightarrow{AB}$ and $\overleftrightarrow{AC}$ are perpendicular lines.   False

77. $\angle GBF$ is an acute angle.   True

78. $\angle EAD$ is an acute angle.   True

79. $\angle EAD$ and $\angle DAC$ are complementary angles.   False

80. $\angle GBF$ and $\angle FBA$ are complementary angles.
    False

81. $\angle EAD$ and $\angle CAB$ are vertical angles.   True

82. $\angle ABC$ and $\angle FBG$ are vertical angles.   True

83. The point *B* is on $\overline{GA}$.   True

84. The point *C* is on $\overline{BH}$.   True

For Exercises 85–88, find the measure of $\angle XYZ$.

85.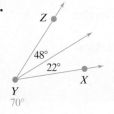
    70°

86.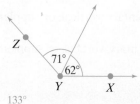
    133°

87.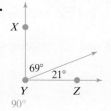
    90°

88.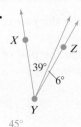
    45°

Writing   Translating Expression   Geometry   Scientific Calculator   Video   NS E

89. Use the figure to find the measure of each angle.

  a. ∠AOB   48°

  b. ∠EOD   48°

  c. ∠AOE   132°

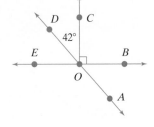

90. Use the figure to find the measure of each angle.

  a. ∠AOB   90°

  b. ∠EOD   23°

  c. ∠AOE   90°

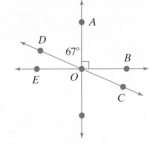

Expanding Your Skills

The second hand on a clock sweeps out a complete circle in 1 min. A circle forms a 360° arc. Use this information for Exercises 91–94.

  91. How many degrees does a second hand on a clock move in 30 sec?   180°

  92. How many degrees does a second hand on a clock move in 15 sec?   90°

  93. How many degrees does a second hand on a clock move in 20 sec?   120°

  94. How many degrees does a second hand on a clock move in 45 sec?   270°

Triangles and the Pythagorean Theorem

Section 8.2

1. Categorizing Triangles

Recall that a **polygon** is a flat figure formed by line segments connected at their ends. A triangle is a three-sided polygon. Furthermore, the sum of the measures of the angles within a triangle is 180°. Teachers often demonstrate this fact by tearing a triangular sheet of paper as shown in Figure 8-14. Then they align the **vertices** (points) of the triangle to form a straight angle.

Concepts

1. Categorizing Triangles
2. Square Roots
3. Pythagorean Theorem and Applications

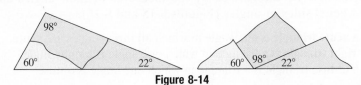

Figure 8-14

Angles of a Triangle

The sum of the measures of the angles of a triangle equals 180°.

1.

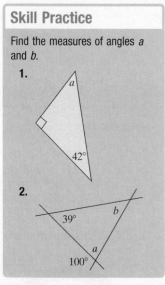

2.

Classroom Example: p. 483,
Exercise 14

Concept Connections

Match the triangle with the appropriate category, **a**, **b**, or **c**.

3.   **a.** Acute triangle

4.   **b.** Right triangle

5. **c.** Obtuse triangle

| **Example 1** | **Finding the Measure of Angles within a Triangle** |

Find the measure of angles a and b.

a.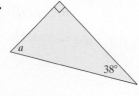

b.

Solution:

a. Recall that the □ symbol represents a 90° angle.

$38° + 90° + m(\angle a) = 180°$ The sum of the angles within a triangle is 180°.

$128° + m(\angle a) = 180°$ Add the measures of the two known angles.

$m(\angle a) = 180° - 128°$ To solve this equation isolate the variable $m(\angle a)$. From the addition equation, we can write a related subtraction equation.

$m(\angle a) = 52°$

b. $\angle a$ is the supplement of the 130° angle. Thus $m(\angle a) = 50°$.

$43° + 50° + m(\angle b) = 180°$ The sum of the angles within a triangle is 180°.

$93° + m(\angle b) = 180°$ Add the measures of the two known angles.

$m(\angle b) = 180° - 93°$ Write the related subtraction problem to find $m(\angle b)$.

$m(\angle b) = 87°$

Triangles may be categorized by the measures of their angles and by the number of equal sides or angles (Figures 8-15 and 8-16).

- An **acute triangle** is a triangle in which all three angles are acute.
- A **right triangle** is a triangle in which one angle is a right angle.
- An **obtuse triangle** is a triangle in which one angle is obtuse.

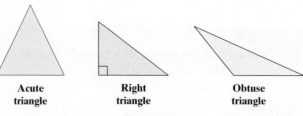

Acute triangle   Right triangle   Obtuse triangle

Figure 8-15

Answers

1. $m(\angle a) = 48°$
2. $m(\angle a) = 80°$
    $m(\angle b) = 61°$
3. b   **4.** a   **5.** c

- An **equilateral triangle** is a triangle in which all three sides (and all three angles) are equal in measure.
- An **isosceles triangle** is a triangle in which two sides are equal in length (the angles opposite the equal sides are also equal in measure).
- A **scalene triangle** is a triangle in which no sides (or angles) are equal in measure.

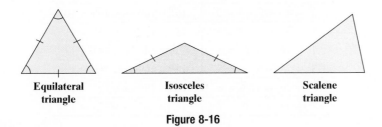

Equilateral Isosceles Scalene
triangle   triangle   triangle

Figure 8-16

TIP: Sometimes we use tick marks / to denote segments of equal length. Similarly, we sometimes use a small arc ⟩ to denote angles of equal measure.

2. Square Roots

In this section we present an important theorem called the Pythagorean theorem. To understand this theorem, we first need some background definitions.

 Recall from Section 1.7 that to square a number means to find the product of the number and itself. Thus, $b^2 = b \cdot b$. For example:

$$6^2 = 6 \cdot 6 = 36$$

We now want to reverse this process by finding a square root of a number. Recall that this is denoted by the radical sign $\sqrt{}$. For example, $\sqrt{36}$ reads as "the positive square root of 36." Thus,

$$\sqrt{36} = 6 \qquad \text{because } 6^2 = 6 \cdot 6 = 36.$$

Example 2   **Evaluating Squares and Square Roots**

Simplify.

 a. $\sqrt{64}$ **b.** $\sqrt{100}$ **c.** 100^2 **d.** $\sqrt{1}$

Solution:

 a. $\sqrt{64} = 8$ because $8 \cdot 8 = 64$

 b. $\sqrt{100} = 10$ because $10 \cdot 10 = 100$

 c. $100^2 = 100 \cdot 100$

     $= 10{,}000$

 d. $\sqrt{1} = 1$ because $1 \cdot 1 = 1$

Table 8-2 gives a list of several whole numbers, their squares, and the corresponding square roots.

Table 8-2

$0^2 = 0 \rightarrow \sqrt{0} = 0$	$8^2 = 64 \rightarrow \sqrt{64} = 8$
$1^2 = 1 \rightarrow \sqrt{1} = 1$	$9^2 = 81 \rightarrow \sqrt{81} = 9$
$2^2 = 4 \rightarrow \sqrt{4} = 2$	$10^2 = 100 \rightarrow \sqrt{100} = 10$
$3^2 = 9 \rightarrow \sqrt{9} = 3$	$11^2 = 121 \rightarrow \sqrt{121} = 11$
$4^2 = 16 \rightarrow \sqrt{16} = 4$	$12^2 = 144 \rightarrow \sqrt{144} = 12$
$5^2 = 25 \rightarrow \sqrt{25} = 5$	$13^2 = 169 \rightarrow \sqrt{169} = 13$
$6^2 = 36 \rightarrow \sqrt{36} = 6$	$14^2 = 196 \rightarrow \sqrt{196} = 14$
$7^2 = 49 \rightarrow \sqrt{49} = 7$	$15^2 = 225 \rightarrow \sqrt{225} = 15$

Concept Connections

13. Label the triangle with the words *leg, leg,* and *hypotenuse.*

14. Label the triangle with the letters *a, b,* and *c,* where *c* represents the hypotenuse.

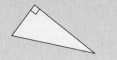

3. Pythagorean Theorem and Applications

Recall that a right triangle is a triangle with a 90° angle. The two sides forming the right angle are called the **legs**. The side opposite the right angle is called the **hypotenuse**. Note that the hypotenuse is always the longest side. See Figure 8-17. We often use the letters *a* and *b* to represent the legs of a right triangle. The letter *c* is used to label the hypotenuse.

Figure 8-17

Instructor Note: Mention to students that this relationship of $a^2 + b^2 = c^2$ does *not* imply that $a + b = c$.

Instructor Note: Explain the difference in the notation between angle *a* and side *a*.

For any right triangle, the **Pythagorean theorem** gives us the following important relationship among the lengths of the sides.

Answers

13.

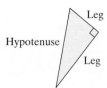

14.

Pythagorean Theorem

For any right triangle,

$$(\text{Leg})^2 + (\text{Leg})^2 = (\text{Hypotenuse})^2$$

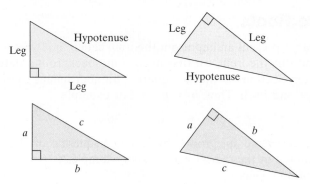

Using the letters *a*, *b*, and *c* to represent the legs and hypotenuse, respectively, we have

$$a^2 + b^2 = c^2$$

Example 3 Finding the Length of the Hypotenuse of a Right Triangle

Find the length of the hypotenuse of the right triangle.

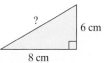

Solution:

The lengths of the legs are given.

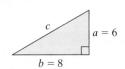

Label the triangle, using a, b, and c. It does not matter which leg is labeled a and which is labeled b.

$a^2 + b^2 = c^2$	Apply the Pythagorean theorem.
$(6)^2 + (8)^2 = c^2$	Substitute $a = 6$ and $b = 8$.
$36 + 64 = c^2$	Simplify.
$100 = c^2$	The solution to this equation is the positive number, c, that when squared equals 100.
$\sqrt{100} = c$	
$10 = c$	Simplify the square root of 100.

The solution may be checked using the Pythagorean theorem.

$$a^2 + b^2 = c^2$$
$$(6)^2 + (8)^2 \stackrel{?}{=} (10)^2$$
$$36 + 64 = 100 ✔$$

The hypotenuse is 10 cm long.

Skill Practice

15. Find the length of the hypotenuse.

Classroom Example: p. 484, Exercise 40

In Example 4, we solve for one of the legs of a right triangle when the other leg and the hypotenuse are known.

Example 4 Finding the Length of a Leg in a Right Triangle

Find the length of the unknown side of the right triangle.

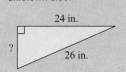

Solution:

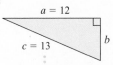

Label the triangle, using a, b, and c. One of the legs is unknown. It doesn't matter whether we call the unknown leg a or b.

$a^2 + b^2 = c^2$	Apply the Pythagorean theorem.
$(12)^2 + b^2 = (13)^2$	Substitute $a = 12$ and $c = 13$.
$144 + b^2 = 169$	Simplify.
$b^2 = 169 - 144$	Write a related subtraction equation to find b^2.
$b^2 = 25$	The solution to this equation is the positive number b that when squared equals 25.
$b = \sqrt{25}$	
$b = 5$	Simplify the square root of 25.

Skill Practice

16. Find the length of the unknown side.

Classroom Example: p. 484, Exercise 44

Avoiding Mistakes

Always remember that the hypotenuse (longest side) is given the letter "c" when applying the Pythagorean theorem.

Answers

15. 5 ft   **16.** 10 in.

The solution may be checked by using the Pythagorean theorem.

$$a^2 + b^2 = c^2$$

$$(12)^2 + (5)^2 \stackrel{?}{=} (13)^2$$

$$144 + 25 = 169 \checkmark$$

The length of the side is 5 m.

In Example 5 we use the Pythagorean theorem in an application.

Example 5 **Using the Pythagorean Theorem in an Application**

When Barb swam across a river, the current carried her 300 yd downstream from her starting point. If the river is 400 yd wide, how far did Barb swim?

Solution:

We first familiarize ourselves with the problem and draw a diagram (Figure 8-18). The distance Barb actually swims is the hypotenuse of the right triangle. Therefore, we label this distance c.

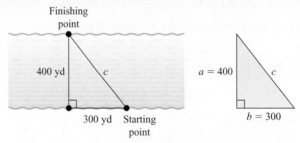

Figure 8-18

$$a^2 + b^2 = c^2 \qquad \text{Apply the Pythagorean theorem.}$$

$$(400)^2 + (300)^2 = c^2 \qquad \text{Substitute } a = 400 \text{ and } b = 300.$$

$$160{,}000 + 90{,}000 = c^2 \qquad \text{Simplify.}$$

$$250{,}000 = c^2 \qquad \text{Add. The solution to this equation is the positive number } c \text{ that when squared equals 250,000.}$$

$$\sqrt{250{,}000} = c$$

$$500 = c \qquad \text{Simplify.}$$

The distance that Barb swims is 500 yd.

Skill Practice

17. The bottom of a 17-ft ladder is placed 8 ft from the bottom of a building. How far up the building is the top of the ladder?

17 ft

8 ft

Classroom Example: p. 485, Exercise 50

Answer

17. 15 ft

Section 8.2 Practice Exercises

Study Skills Exercise

For additional exercises, see Classroom Activities 8.2A–8.2C in the *Student's Resource Manual* at www.mhhe.com/moh.

When solving an application involving geometry, draw a picture of the situation and label the known quantities with numbers and the unknown quantities with variables. This will help to solve the problem. After reading this section, what geometric figure do you think you will be drawing most often in this section?

 Writing Translating Expression Geometry Scientific Calculator Video NSE

Vocabulary and Key Concepts

1. **a.** The sum of the measures of the angles of a triangle is _____180_____°.

   b. A(n) _____acute_____ triangle is a triangle in which all three angles measure less than 90°. A(n) _____right_____ triangle is a triangle in which one angle measures 90°. A(n) _____obtuse_____ triangle is a triangle in which one angle measures more than 90°.

   c. A(n) _____equilateral_____ triangle is a triangle in which all three sides have equal measure.

   d. A(n) _____isosceles_____ triangle is a triangle in which two sides have equal measure.

   e. A(n) _____scalene_____ triangle is a triangle in which no sides have equal measure.

   f. In a right triangle, the _____hypotenuse_____ is the name given to the longest side. The two shorter sides are called the _____legs_____ of the right triangle.

   g. Given a triangle with legs of lengths a and b and hypotenuse of length c, the _____Pythagorean_____ theorem states that $a^2 + b^2 =$ _____c^2_____.

Review Exercises

2. **a.** What is the supplement of a 75° angle?   105°   **b.** What is the complement of a 75° angle?   15°

3. Do $\angle ACB$ and $\angle BCA$ represent the same angle?   Yes

4. Is line segment $\overline{MN}$ the same as the line segment $\overline{NM}$?   Yes

5. Is ray $\overrightarrow{AB}$ the same as ray $\overrightarrow{BA}$?   No

6. Is the line $\overleftrightarrow{PQ}$ the same as the line $\overleftrightarrow{QP}$?   Yes

7. Is a right angle an obtuse angle?   No

8. Can two acute angles be supplementary?   No

Concept 1: Categorizing Triangles

For Exercises 9–16, find the measures of angles a and b. **(See Example 1.)**

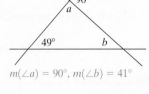

9.
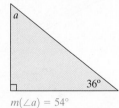
$m(\angle a) = 54°$

10.
$m(\angle b) = 105°$

11.
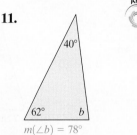
$m(\angle b) = 78°$

12.
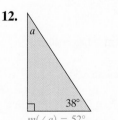
$m(\angle a) = 52°$

13.
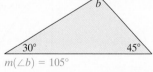
$m(\angle a) = 60°, m(\angle b) = 80°$

14.
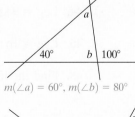
$m(\angle a) = 90°, m(\angle b) = 41°$

15.
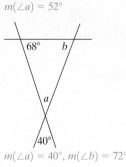
$m(\angle a) = 40°, m(\angle b) = 72°$

16.
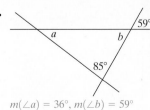
$m(\angle a) = 36°, m(\angle b) = 59°$

For Exercises 17–22, choose all figures that apply. The tick marks / denote segments of equal length, and small arcs ⌒ denote angles of equal measure.

17. Acute triangle   c, f

18. Obtuse triangle   a, e

19. Right triangle   b, d

20. Scalene triangle   a, d, f

21. Isosceles triangle   b, c, e

22. Equilateral triangle   c

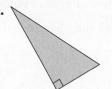

a.

b.

c.

d.

e.

f.

Concept 2: Square Roots

For Exercises 23–38, simplify the squares and square roots. **(See Example 2.)**

23. $\sqrt{49}$ 7

24. $\sqrt{64}$ 8

25. 7^2 49

26. 8^2 64

27. 4^2 16

28. 5^2 25

29. $\sqrt{16}$ 4

30. $\sqrt{25}$ 5

31. $\sqrt{36}$ 6

32. $\sqrt{100}$ 10

33. 6^2 36

34. 10^2 100

35. 9^2 81

36. 3^2 9

37. $\sqrt{81}$ 9

38. $\sqrt{9}$ 3

Concept 3: Pythagorean Theorem and Applications

For Exercises 39–42, find the length of the unknown side. **(See Examples 3 and 4.)**

39.

3 m

c

4 m

$c = 5$ m

40.

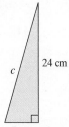

c

24 cm

7 cm $c = 25$ cm

41.

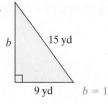

b

15 yd

9 yd $b = 12$ yd

42.

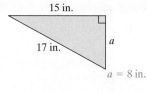

15 in.

17 in.

a

$a = 8$ in.

For Exercises 43–46, find the length of the unknown leg or hypotenuse.

43. Leg = 24 ft, hypotenuse = 26 ft   Leg = 10 ft

44. Leg = 9 km, hypotenuse = 41 km   Leg = 40 km

45. Leg = 32 in., leg = 24 in.   Hypotenuse = 40 in.

46. Leg = 16 m, leg = 30 m   Hypotenuse = 34 m

47. Find the length of the supporting brace.
(See Example 5.)
The brace is 20 in. long.

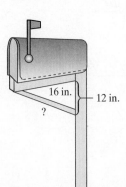

16 in.

12 in.

?

48. Find the length of the ramp.

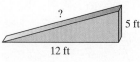

?

5 ft

12 ft

The ramp is 13 ft long.

49. Find the height of the airplane above the ground.

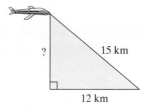

The height is 9 km.

50. A 25-in. television measures 25 in. across the diagonal. If the width is 20 in., find the height.

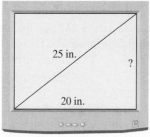

The height is 15 in.

51. A car travels east 24 mi and then south 7 mi. How far is the car from its starting point?

The car is 25 mi from the starting point.

52. A 26-ft-long wire is to be tied from a stake in the ground to the top of a 24-ft pole. How far from the bottom of the pole should the stake be placed? The stake should be 10 ft from the pole.

For Exercises 53–56, find the perimeter.

53.

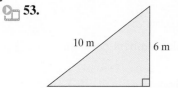

54.

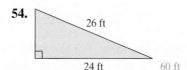

55.

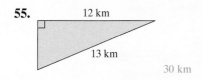

56.

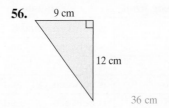

Expanding Your Skills

57. Find the length of side *c* by dividing this figure into a rectangle and a right triangle. Then find the perimeter of the figure.

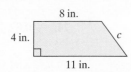

c = 5 in.; perimeter = 28 in.

58. Find the perimeter of the figure.

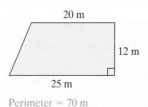

Perimeter = 70 m

59. Find the perimeter of the figure.

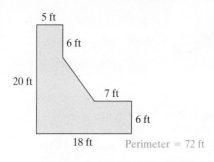

Perimeter = 72 ft

60. Tyler drives 20 mi north, 8 mi east, 4 mi south, and 4 mi east. How far is Tyler from his starting point?

Tyler is 20 mi from the starting point.

Calculator Connections

Topic: Entering Square Roots on a Calculator

Many square roots cannot be written as a whole number. For example, there is no whole number that when squared equals 26. However, we might speculate that $\sqrt{26}$ is a number slightly greater than 5 because $\sqrt{25} = 5$. A decimal approximation can be made by using a calculator.

$$\sqrt{26} \approx 5.099 \qquad \text{because } 5.099^2 = 25.999801 \approx 26$$

To enter a square root on a calculator, use the $\boxed{\sqrt{x}}$ key. On some calculators, the $\boxed{\sqrt{x}}$ function is associated with the x^2 key. In such a case, it is necessary to press $\boxed{2^{nd}}$ or $\boxed{\text{SHIFT}}$ first, followed by the $\boxed{x^2}$ key. Some calculators require the square root key to be entered first, before the number, while with others we enter the number first followed by $\boxed{\sqrt{x}}$.

Expression	Keystrokes	Result
$\sqrt{26}$	26 $\boxed{\sqrt{x}}$　or　$\boxed{\sqrt{x}}$ 26 $\boxed{=}$	5.099019514
$\sqrt{9325}$	9325 $\boxed{\sqrt{x}}$　or　$\boxed{\sqrt{x}}$ 9325 $\boxed{=}$	96.56603958
$\sqrt{100}$	100 $\boxed{\sqrt{x}}$　or　$\boxed{\sqrt{x}}$ 100 $\boxed{=}$	10

For Exercises 61–66, complete the table. For the estimate, find two consecutive whole numbers between which the square root lies. The first row is done for you.

	Square Root	Estimate	Calculator Approximation (Round to 3 Decimal Places)
	$\sqrt{50}$	is between 7 and 8	7.071
61.	$\sqrt{10}$	is between _3_ and _4_	3.162
62.	$\sqrt{90}$	is between _9_ and _10_	9.487
63.	$\sqrt{116}$	is between _10_ and _11_	10.770
64.	$\sqrt{65}$	is between _8_ and _9_	8.062
65.	$\sqrt{5}$	is between _2_ and _3_	2.236
66.	$\sqrt{48}$	is between _6_ and _7_	6.928

For Exercises 67–74, use a calculator to approximate the positive square root to three decimal places.

67. $\sqrt{427.75}$　20.682　　**68.** $\sqrt{3184.75}$　56.434　　**69.** $\sqrt{1{,}246{,}000}$　1116.244　　**70.** $\sqrt{50{,}416{,}000}$
　　7100.423

71. $\sqrt{0.49}$　0.7　　**72.** $\sqrt{0.25}$　0.5　　**73.** $\sqrt{0.56}$　0.748　　**74.** $\sqrt{0.82}$　0.906

Topic: Pythagorean Theorem

For Exercises 75–80, find the length of the unknown side. Round to three decimal places if necessary.

75.

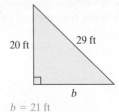

$b = 21$ ft

76.

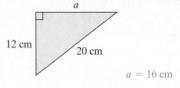

$a = 16$ cm

77. Leg = 5 mi, leg = 10 mi
Hypotenuse = 11.180 mi

78. Leg = 2 m, leg = 8 m
Hypotenuse = 8.246 m

79. Leg = 12 in., hypotenuse = 22 in.
Leg = 18.439 in.

80. Leg = 15 ft, hypotenuse = 18 ft
Leg = 9.950 ft

Writing　　Translating Expression　　Geometry　　Scientific Calculator　　Video　　NS&E

81. A square tile is 1 ft on each side. What is the length of the diagonal? Round to the nearest hundredth of a foot. The diagonal length is 1.41 ft.

82. A tennis court is 120 ft long and 60 ft wide. What is the length of the diagonal? Round to the nearest hundredth of a foot. The length of the diagonal is 134.16 ft.

83. A contractor plans to construct a cement patio for one of the houses that he is building. The patio will be a square, 25 ft by 25 ft. After the contractor builds the frame for the cement, he checks to make sure that it is square by measuring the diagonals. Use the Pythagorean theorem to determine what the length of the diagonals should be if the contractor has constructed the frame correctly. Round to the nearest hundredth of a foot. The length of the diagonal is 35.36 ft.

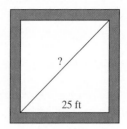

Quadrilaterals, Perimeter, and Area

Section 8.3

1. Quadrilaterals

Concepts

1. Quadrilaterals
2. Perimeter
3. Area
4. Applications of Area

A four-sided polygon is called a **quadrilateral**. Some quadrilaterals fall in the following categories.

A **parallelogram** is a quadrilateral with opposite sides parallel. It follows that opposite sides must be equal in length.

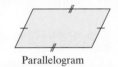

Parallelogram

A **rectangle** is a parallelogram with four right angles.

Rectangle

A **square** is a rectangle with sides of equal length.

Square

A **rhombus** is a parallelogram with sides of equal length. The angles are not necessarily equal.

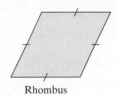

Rhombus

A **trapezoid** is a quadrilateral with one pair of parallel sides.

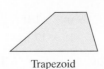

Trapezoid

Notice that some figures belong to more than one category. For example, a square is also a rectangle and a parallelogram.

Writing   Translating Expression   Geometry   Scientific Calculator   Video   NSE

2. Perimeter

Recall that the **perimeter** of a polygon is the distance around the figure. For example, we use perimeter to find the amount of fencing needed to enclose a yard. The perimeter of a polygon is found by adding the lengths of the sides. However, with some geometric figures we can shorten the process by using a formula.

Perimeter of a Square

$$P = s + s + s + s$$
$$= 4s$$

Perimeter of a Rectangle

$$P = l + l + w + w$$
$$= 2l + 2w$$

We usually do not give perimeter formulas for other polygons. It is generally easier simply to add the lengths of the sides than to memorize numerous formulas.

Example 1 **Finding Perimeter**

Use an appropriate formula to find the perimeter.

8.2 yd

Solution:

$P = 4s$	The figure is a square. Use $P = 4s$.
$= 4(8.2 \text{ yd})$	Substitute $s = 8.2$ yd.
$= 32.8 \text{ yd}$	Simplify.

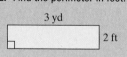
Example 2 **Finding Perimeter**

Use an appropriate formula to find the perimeter.

9 in.

6 ft

Solution:

First note that to add the lengths of the sides, we must have like units.

$$9 \text{ in.} = \frac{9 \text{ in.}}{1} \cdot \frac{1 \text{ ft}}{12 \text{ in.}} = \frac{9}{12} \text{ ft} = \frac{3}{4} \text{ ft or } 0.75 \text{ ft}$$

$P = 2l + 2w$	The figure is a rectangle. Use $P = 2l + 2w$.
$= 2(6 \text{ ft}) + 2(0.75 \text{ ft})$	Substitute $l = 6$ ft and $w = 0.75$ ft.
$= 12 \text{ ft} + 1.5 \text{ ft}$	Simplify.
$= 13.5 \text{ ft}$	

3. Area

The area of a region is the number of square units that can be enclosed within the region. The rectangle shown in Figure 8-19 encloses 6 square inches (in.^{2}). We would compute area, for example, if we wanted to determine how much sod was needed to cover a yard.

We offer the following convenient formulas to find the area enclosed within various figures. We begin with the familiar formulas for the area of a rectangle and square. These formulas were first introduced in Section 1.5.

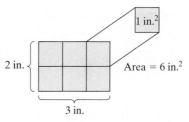

Figure 8-19

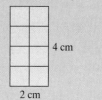
Area of a Rectangle

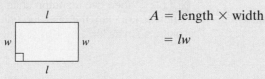

$A = \text{length} \times \text{width}$

$\quad = lw$

Area of a Square

A square is also a rectangle. Therefore, the area of a square is

$A = \text{length} \times \text{width}$

$\quad = s \cdot s$

$\quad = s^2$

To find the area of a parallelogram, consider cutting it into two pieces, as shown in Figure 8-20. Then realign the pieces to form a rectangle.

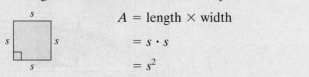

$A = b \cdot h \qquad\qquad A = b \cdot h$

Figure 8-20

The area of a parallelogram is the corresponding area of the rectangle: $A = \text{base} \times \text{height}$, or simply $A = bh$. The height h is the distance between the base and its opposite side.

TIP: The height h of a parallelogram can also be drawn *outside* the parallelogram.

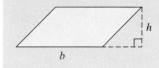

Area of a Parallelogram

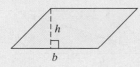

$A = \text{base} \times \text{height}$

$\quad = bh$

Answer

3. 8 cm^2

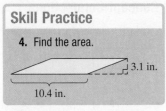
Example 3 **Finding Area**

Find the area of the field.

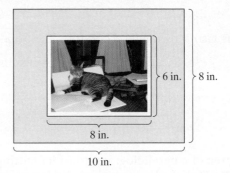

Solution:

The field is in the shape of a parallelogram. The base is 0.6 km and the height is 1.8 km.

$A = bh$ Area formula for a parallelogram.

$= (0.6 \text{ km})(1.8 \text{ km})$ Substitute $b = 0.6$ km and $h = 1.8$ km.

$= 1.08 \text{ km}^2$

The field is 1.08 km^2.

TIP: When two common units are multiplied, such as km · km, the resulting units are square units, such as km^2.

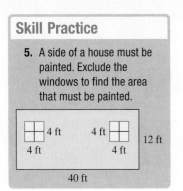
Example 4 **Finding Area**

Find the area of the matting.

Solution:

To find the area of the matting only, we can subtract the inner 6-in. by 8-in. area from the outer 8-in. by 10-in. area.

$$\text{Area of matting} = \overset{\text{outer area}}{(10 \text{ in.})(8 \text{ in.})} - \overset{\text{inner area}}{(8 \text{ in.})(6 \text{ in.})}$$

$$= 80 \text{ in.}^2 - 48 \text{ in.}^2$$

$$= 32 \text{ in.}^2$$

The matting is 32 in.^2

The formula for the area of a triangle was first introduced in Section 2.4. The formula is $A = \frac{1}{2}bh$, where b is the base of the triangle and h is the height. Notice that this formula gives one-half the value of the area of a parallelogram (Figure 8-21).

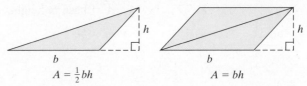

Figure 8-21

Area of a Triangle

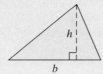

$$A = \tfrac{1}{2} \times \text{base} \times \text{height}$$
$$= \tfrac{1}{2}bh$$

Animation

Example 5 **Finding Area**

Find the area of the region.

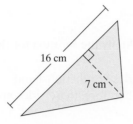

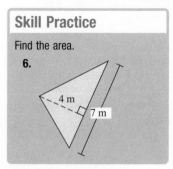
Solution:

$A = \dfrac{1}{2}bh$ Apply the formula for the area of a triangle.

$= \dfrac{1}{2}(16 \text{ cm})(7 \text{ cm})$ Substitute $b = 16$ cm and $h = 7$ cm.

$= \dfrac{1}{2}\left(\dfrac{\overset{8}{16}}{1} \text{ cm}\right)\left(\dfrac{7}{1} \text{ cm}\right)$ Multiply fractions.

$= 56 \text{ cm}^2$

    The last formula we present here is the formula to find the area of a trapezoid. To develop the formula, we place two identical trapezoids next to each other where one is inverted (Figure 8-22). Together they form a parallelogram with base $a + b$. Notice that the height of a trapezoid is the distance between the two parallel sides.

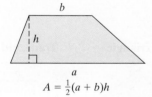

$A = \tfrac{1}{2}(a + b)h$

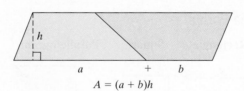

$A = (a + b)h$

Figure 8-22

The area of the trapezoid is one-half the area of the parallelogram.

Area of a Trapezoid

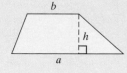

$$A = \tfrac{1}{2} \times (\text{sum of the parallel sides}) \times \text{height}$$
$$= \tfrac{1}{2}(a + b)h$$

Animation

Answer

6. 14 m²

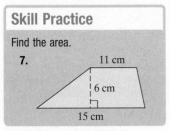
Classroom Example: p. 496, Exercise 36

Instructor Note: Before starting Example 6 ask students which line segment represents height. Be prepared to explain that the height is 5 ft, not 11 ft.

Example 6 **Finding Area**

Find the area of the region.

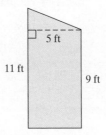

Solution:

$$A = \frac{1}{2}(a + b)h \qquad \text{Apply the formula for the area of a trapezoid.}$$

In this case, the two parallel sides are the left-hand side and the right-hand side. Therefore, these sides are the two bases, a and b.

The "height" is the distance between the two parallel sides.

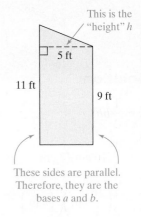

$$A = \frac{1}{2}(11\text{ ft} + 9\text{ ft})(5\text{ ft}) \qquad \begin{array}{l}\text{Substitute } a = 11\text{ ft,}\\ b = 9\text{ ft, and } h = 5\text{ ft.}\end{array}$$

$$= \frac{1}{2}(20\text{ ft})(5\text{ ft}) \qquad \begin{array}{l}\text{Simplify within}\\ \text{parentheses first.}\end{array}$$

$$= \frac{1}{2}\left(\frac{\overset{10}{20}}{1}\text{ ft}\right)\left(\frac{5}{1}\text{ft}\right) \qquad \text{Multiply fractions.}$$

$$= 50\text{ ft}^2$$

Here is a summary of the area formulas presented thus far. These should be memorized as common knowledge.

Area Formulas

Rectangle	Square	Parallelogram	Triangle	Trapezoid
$A = lw$	$A = s^2$	$A = bh$	$A = \frac{1}{2}bh$	$A = \frac{1}{2}(a + b)h$

Answer

7. 78 cm^2

4. Applications of Area

Example 7 Finding the Area of a Composite Figure

Find the area of the shaded region.

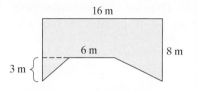

Solution:

The shaded region can be thought of as a large "outer" rectangle with a trapezoidal piece removed (Figure 8-23).

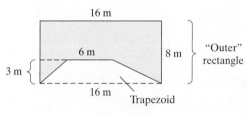

Figure 8-23

Therefore, we can subtract the area of the trapezoid from the area of the outer rectangle.

Area of rectangle: $A = lw$

$$= (16 \text{ m})(8 \text{ m})$$

$$= 128 \text{ m}^2$$

Area of trapezoid: $A = \frac{1}{2}(a + b)h$

$$= \frac{1}{2}(16 \text{ m} + 6 \text{ m})(3 \text{ m})$$

$$= \frac{1}{2}(22 \text{ m})(3 \text{ m})$$

$$= 33 \text{ m}^2$$

The area of the shaded region is the area of the rectangle minus the area of the trapezoid.

Area of shaded region $= 128 \text{ m}^2 - 33 \text{ m}^2$

$$= 95 \text{ m}^2$$

Example 8 Finding Area for a Landscaping Application

Sod can be purchased in palettes for $225. If a palette contains 240 ft^2 of sod, how much will it cost to cover the area in Figure 8-24?

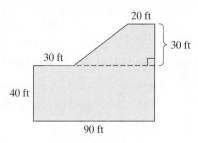

Figure 8-24

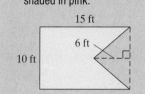

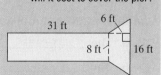

Solution:

To find the total cost, we need to know the total number of square feet. Then we can determine how many 240-ft² palettes are required.

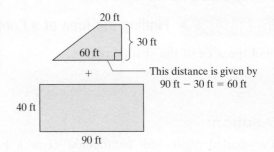

The total area is given by

$$\underset{\substack{\text{area of} \\ \text{trapezoid}}}{A} = \underset{\substack{\downarrow}}{\tfrac{1}{2}(a + b)h} + \underset{\substack{\text{area of} \\ \text{rectangle} \\ \downarrow}}{lw}$$

$$= \tfrac{1}{2}(60 \text{ ft} + 20 \text{ ft})(30 \text{ ft}) + (90 \text{ ft})(40 \text{ ft})$$

$$= \tfrac{1}{2}(80 \text{ ft})(30 \text{ ft}) + 3600 \text{ ft}^2$$

$$= 1200 \text{ ft}^2 + 3600 \text{ ft}^2$$

$$= 4800 \text{ ft}^2 \qquad \text{The total area is } 4800 \text{ ft}^2.$$

To determine how many 240-ft² palettes of sod are required, divide the total area by 240 ft².

Number of palettes: $4800 \text{ ft}^2 \div 240 \text{ ft}^2 = 20$

The total cost for 20 palettes is   ($225 per palette) × 20 palettes = $4500

The cost for the sod is $4500.

Section 8.3   Practice Exercises

Study Skills Exercise

For additional exercises, see Classroom Activities 8.3A–8.3C in the *Student's Resource Manual* at www.mhhe.com/moh.

It may help to remember formulas if you understand how they were derived. For example, the perimeter of a square has the formula $P = 4s$. It was derived from the fact that perimeter measures the distance around a figure. Observe:

Explain how the formula for the perimeter of a rectangle ($P = 2l + 2w$) was derived.

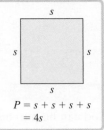

$$P = s + s + s + s$$
$$= 4s$$

Vocabulary and Key Concepts

1. a. The _____perimeter_____ of a polygon is the distance around the polygon.

    b. The _____area_____ of a region is the number of square units that can be enclosed within the region.

Review Exercises

2. Which of the units are units of area?

    a. cm²   **b.** cm   **c.** ft   **d.** yd²   **e.** mi²   a, d, e

For Exercises 3–8, select two terms that apply to the triangle. Choose from acute triangle, obtuse triangle, right triangle, scalene triangle, isosceles triangle, and equilateral triangle.

3.

a. _____acute triangle_____

b. _____scalene triangle_____

4.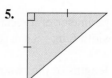

a. _____obtuse triangle_____

b. _____scalene triangle_____

5.

a. _____right triangle_____

b. _____isosceles triangle_____

6.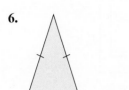

a. _____acute triangle_____

b. _____isosceles triangle_____

7.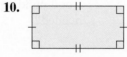

a. _____obtuse triangle_____

b. _____isosceles triangle_____

8.

a. _____acute triangle_____

b. _____equilateral triangle_____

Concept 1: Quadrilaterals

For Exercises 9–14, select all terms that apply to each figure. There may be more than one selection.

a. quadrilateral b. polygon c. square d. rectangle

e. parallelogram f. triangle g. trapezoid h. rhombus

9. a, b, c, d, e, h

10. a, b, d, e

11. a, b, e

12. 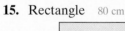 a, b, g

13.  a, b, e, h

14. b, f

Concept 2: Perimeter

For Exercises 15–20, find the perimeter. **(See Examples 1 and 2.)**

15. Rectangle 80 cm

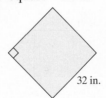

15 cm

25 cm

16. Square 128 in.

32 in.

17. Square 260 mm

65 mm

18. Rectangle 18.4 yd

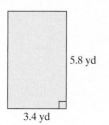

5.8 yd

3.4 yd

19. Trapezoid 10.7 m

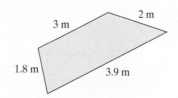

3 m

2 m

1.8 m

3.9 m

20. Trapezoid 237 cm

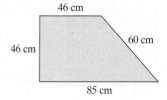

46 cm

60 cm

46 cm

85 cm

21. Find the perimeter of a triangle with sides 3 ft 8 in., 2 ft 10 in., and 4 ft.   10 ft 6 in.

22. Find the perimeter of a triangle with sides 4 ft 2 in., 3 ft, and 2 ft 9 in.   9 ft 11 in.

23. Find the perimeter of a rectangle with length 2 ft and width 6 in.   5 ft or 60 in.

24. Find the perimeter of a rectangle with length 4 m and width 85 cm.   9.7 m or 970 cm

25. Find the lengths of the two missing sides labeled x and y. Then find the perimeter of the figure.

$x = 550$ mm; $y = 3$ dm; perimeter $= 26$ dm or 2600 mm

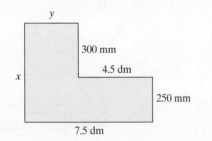

26. Find the lengths of the two missing sides labeled a and b. Then find the perimeter of the figure.

$a = 1\frac{1}{2}$ ft; $b = \frac{3}{4}$ ft; perimeter $= 6\frac{1}{2}$ ft

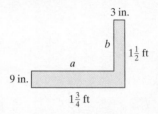

27. Rain gutters are going to be installed around the perimeter of the house. How many feet of rain gutters are needed?   280 ft of rain gutters is needed.

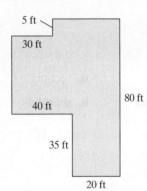

28. Wood molding needs to be installed around the perimeter of a living room floor. With no wood molding needed in the doorway, how much wood molding is needed?   53.5 yd of molding is needed.

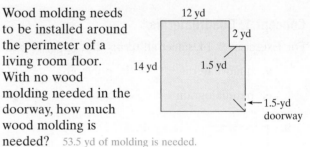

Concept 3: Area

For Exercises 29–38, find the area. (See Examples 3, 5, and 6.)

29.   576 yd²

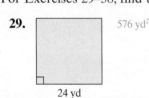

30.   14 ft²

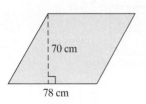

31.   54 m²

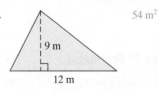

32.   5460 cm²

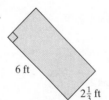

33.   656 in.²

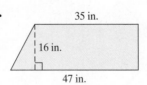

34.   1.5 km²

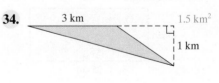

35.   18.4 ft²

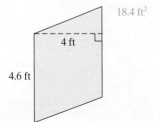

36.   654 cm²

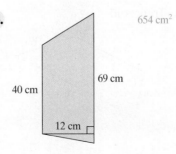

37. Write the answer in square feet.
12.375 ft²

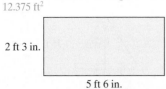

2 ft 3 in.

5 ft 6 in.

38. Write the answer in square feet. 12.25 ft²

3'6"

Concept 4: Applications of Area

For Exercises 39–44, find the area of the shaded region. **(See Examples 4 and 7.)**

39.

7.5 yd   148.5 yd²

4 yd

18.2 yd   3 yd

40.   217.54 ft²

5.8 ft

12.8 ft   3.1 ft

18.4 ft

41.   280 mm²

9 mm

16 mm

22 mm

42.   24 in.   348 in.²

16 in.

3 in.

⊢ 12 in. ⊣⊦ 12 in. ⊣

43.   60 in.²

5 in.   2 in.

10 in.

5 in.   2 in.

8 in.

44.   12 in.   139.16 in.²

2.2 in.

2.2 in.

12 in.

2.2 in.

2.2 in.

45. A rectangular living room is all to be carpeted except for the tiled portion in front of the fireplace. What is the area to be carpeted? What is the area to be tiled? The area to be carpeted is 382.5 ft². The area to be tiled is 13.5 ft².

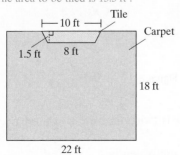

⊢ 10 ft ⊣   Tile

Carpet

1.5 ft   8 ft

18 ft

22 ft

46. A patio area is to be covered with outdoor tile. What is the area of the patio? The area is 276 m².

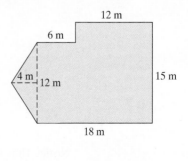

12 m

6 m

4 m 12 m   15 m

18 m

47. Find the area of the sign. **(See Example 8.)**
The area of the sign is 16.5 yd².

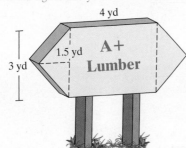

4 yd

1.5 yd

3 yd

A+
Lumber

48. Find the area of the kite. The area is 1.625 ft².

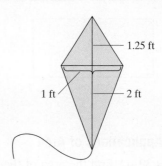

1.25 ft

1 ft

2 ft

49. The King family plans to paint their garage floor with paint that resists gas, oil, and dirt from tires. The garage is 21 ft wide and 23 ft long. The paint kit they plan to use will cover approximately 250 square feet.

a. What is the area of the garage floor? The area is 483 ft².

b. How many kits will be needed to paint the entire garage floor? They will need 2 paint kits.

Expanding Your Skills

50. If the lengths of the sides of a square are doubled, by how many times is the area increased? The area is increased by 4 times.

51. If the lengths of the sides of a square are tripled, by how many times is the area increased? The area is increased by 9 times.

For Exercises 52–55, answer true or false.

52. All rectangles are parallelograms. True

53. All trapezoids are parallelograms. False

54. All rhombi (plural of rhombus) are squares. False

55. All squares are rhombi. True

Section 8.4 | Circles, Circumference, and Area

Concepts

1. Basic Definitions
2. Circumference
3. Area
4. Applications

1. Basic Definitions

A **circle** is a figure consisting of all points located the same distance r from a fixed point C. The point C is called the **center** of the circle. The distance r is the length of a radius of the circle. A **radius** of a circle is a line segment drawn from the center of a circle to a point on the circle.

A line segment connecting two points on a circle and passing through the center is called a **diameter** of the circle. In Figure 8-25, $\overline{AC}$ is a radius of the circle, and $\overline{AB}$ is a diameter.

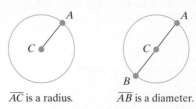

$\overline{AC}$ is a radius. $\overline{AB}$ is a diameter.

Figure 8-25

Notice that the length of a diameter is twice the radius. Therefore, we have

$$d = 2r \quad \text{or equivalently} \quad r = \frac{1}{2}d$$

Example 1 **Finding Diameter and Radius**

a. Find the length of a radius.

4.6 cm

b. Find the length of a diameter.

$\frac{3}{4}$ ft

Solution:

a. $r = \frac{1}{2}d = \frac{1}{2}(4.6 \text{ cm}) = 2.3 \text{ cm}$

b. $d = 2r = 2\left(\frac{3}{4}\text{ ft}\right) = \frac{3}{2}$ ft or 1.5 ft

2. Circumference

The distance around a circle is called the **circumference**. Early mathematicians discovered that if you measure the circumference (C) and diameter (d) of any circle, the ratio of C to d always has the same value. We say that $\frac{C}{d}$ is constant. This is true for any size circle. For example, the circumference of a can of beans is approximately 9.25 in., and its diameter is 3 in. The ratio $\frac{C}{d} = \frac{9.25 \text{ in.}}{3 \text{ in.}} \approx 3.1$. The same ratio is found for a can of paint. See Figure 8-26.

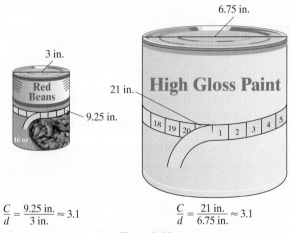

$\frac{C}{d} = \frac{9.25 \text{ in.}}{3 \text{ in.}} \approx 3.1$
$\frac{C}{d} = \frac{21 \text{ in.}}{6.75 \text{ in.}} \approx 3.1$

Figure 8-26

The constant value given by $\frac{C}{d}$ is called the number π (read "pi"). The number π in decimal form is 3.1415926535..., which goes on forever without a repeating pattern. We approximate π by 3.14 or $\frac{22}{7}$ to make it easier to use in calculations. The relationship between the circumference and diameter of a circle gives us the following formulas for the circumference of a circle.

Concept Connections

3. What is the meaning of the circumference of a circle?

Instructor Note: Mention that the number represented by π is irrational, a term to be defined later. Therefore its value can only be approximated.

Answers

1. 5.2 m **2.** $\frac{5}{4}$ yd
3. The circumference of a circle is the distance around the circle.

Circumference of a Circle

The circumference C of a circle is given by

$$C = \pi d \quad \text{or} \quad C = 2\pi r$$

where π is approximately 3.14 or $\frac{22}{7}$.

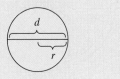

Example 2 **Finding Circumference**

Find the circumference.

a. Give the exact answer in terms of π.

b. Approximate the answer using 3.14 for π.

8 m

Solution:

a. The diameter is given, $d = 8$ m.

$$C = \pi d$$
$$= \pi(8 \text{ m}) \qquad \text{Substitute } d = 8 \text{ m.}$$
$$= 8\pi \text{ m} \qquad \text{The circumference is exactly } 8\pi \text{ meters.}$$

b. $C = 8\pi$ m

$$\approx 8(3.14) \text{ m} \qquad \text{Approximate } \pi \text{ with 3.14.}$$
$$= 25.12 \text{ m} \qquad \text{The circumference is approximately 25.12 meters.}$$

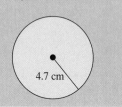
Example 3 **Finding Circumference**

Find the circumference. Give the exact answer in terms of π, and then approximate the answer using 3.14 for π.

5.1 ft

Solution:

The radius is given, $r = 5.1$ ft.

$$C = 2\pi r$$
$$= 2\pi(5.1 \text{ ft}) \qquad \text{Substitute } r = 5.1 \text{ ft.}$$
$$= 10.2\,\pi \text{ ft} \qquad \text{This is the exact circumference.}$$
$$\approx 10.2(3.14 \text{ ft}) \qquad \text{Approximate by using 3.14 for } \pi.$$
$$= 32.028 \text{ ft} \qquad \text{The circumference is approximately 32.028 ft.}$$

3. Area

The circumference of a circle is given by $C = 2\pi r$. The length of a **semicircle** (one-half of a circle) is one-half of this amount: $\frac{1}{2}2\pi r = \pi r$. To visualize the formula for the area of a circle, consider the bottom half and top half of a circle cut into pie-shaped wedges. Unfold the figure as shown (Figure 8-27).

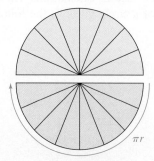

Figure 8-27

The resulting figure is nearly a parallelogram, with base $\approx \pi r$ and height approximately equal to the radius of the circle. The area is (base) $\times$ (height) $\approx$ $(\pi r) \cdot r = \pi r^2$. This is the area formula for a circle.

Area of a Circle

The area A of a circle is given by

$$A = \pi r^2$$

Example 4 Finding the Area of a Circle

Find the area.

a. Give the exact answer in terms of π.

b. Approximate the answer by using $\frac{22}{7}$ for π.

14 m

Solution:

a. $A = \pi r^2$

$\quad = \pi (14 \text{ m})^2$ Substitute $r = 14$ m.

$\quad = \pi (196 \text{ m}^2)$ Square the radius.

$\quad = 196\pi \text{ m}^2$ The exact area is 196π m².

b. $A = 196\pi \text{ m}^2$

$\quad \approx 196\left(\dfrac{22}{7}\right) \text{m}^2$ Approximate π with $\dfrac{22}{7}$.

$\quad = \dfrac{\overset{28}{196}}{1} \cdot \dfrac{22}{\underset{1}{7}} \text{ m}^2$ Multiply the fractions.

$\quad = 616 \text{ m}^2$ The area is approximately 616 m².

Skill Practice

6. Find the area of the circle.

    a. Give the exact answer in terms of π.

    b. Approximate the answer by using $\frac{22}{7}$ for π.

21 in.

Example 5 Finding the Area of a Circle

Find the area of a cover for a wading pool that has diameter 4 ft. Give the exact answer and then approximate using 3.14 for π.

4 ft

Solution:

To compute the area, we first find the radius of the circle.

$$r = \frac{1}{2}d = \frac{1}{2}(4 \text{ ft}) = 2 \text{ ft}$$

$A = \pi r^2$

$\quad = \pi (2 \text{ ft})^2$ Substitute $r = 2$ ft.

$\quad = \pi (4 \text{ ft}^2)$ Square the radius.

$\quad = 4\pi \text{ ft}^2$ The exact area is 4π ft².

$\quad \approx 4(3.14) \text{ ft}^2$ Approximate π with 3.14.

$\quad = 12.56 \text{ ft}^2$ The area is approximately 12.56 ft².

Skill Practice

7. Find the area of the clock face. Give the exact answer, and then approximate by using 3.14 for π. Round to the nearest whole unit.

9.4 in.

Answers

6. a. 441π in.²   **b.** ≈ 1386 in.²

7. 22.09π in.² ≈ 69 in.²

4. Applications

Example 6 Finding Perimeter and Area
of a Composite Figure

The region shown is formed by a rectangle and a semicircle. Find the perimeter and area. Use 3.14 for π.

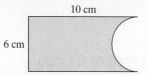

Solution:

The figure consists of a rectangle and a semicircle. We can label the figure further to identify the length of each side and the radius of the semicircle.

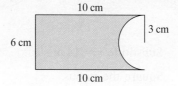

The perimeter is the sum of the three sides of the rectangle and the distance around the semicircle.

$$\underbrace{}_{\text{sum of the 3 sides}} \qquad \underbrace{}_{\text{distance around semicircle } (\frac{1}{2} \cdot 2\pi r)}$$

$$P \approx 6\text{ cm} + 10\text{ cm} + 10\text{ cm} + \frac{1}{2} \cdot 2(3.14)(3\text{ cm})$$

$$= 26\text{ cm} + \frac{1}{2} \cdot 2(3.14)(3\text{ cm}) \qquad \text{Multiply fractions.}$$

$$= 26\text{ cm} + 9.42\text{ cm} \qquad \text{Add.}$$

$$= 35.42\text{ cm} \qquad \text{The perimeter is approximately } 35.42\text{ cm.}$$

The area is the difference of the area of the rectangle and the area of the semicircle.

$$\overbrace{}^{\text{area of rectangle}} \quad \overbrace{}^{\text{area of semicircle } (\frac{1}{2}\pi r^2)}$$

$$A \approx (10\text{ cm})(6\text{ cm}) - \frac{1}{2}(3.14)(3\text{ cm})^2$$

$$= 60\text{ cm}^2 - \frac{1}{2}(3.14)(9\text{ cm}^2) \qquad \text{Multiply.}$$

$$= 60\text{ cm}^2 - (1.57)(9\text{ cm}^2)$$

$$= 60\text{ cm}^2 - 14.13\text{ cm}^2 \qquad \text{Subtract like units.}$$

$$= 45.87\text{ cm}^2 \qquad \text{The area is approximately } 45.87\text{ cm}^2.$$

Answer

8. Perimeter $\approx$ 42.84 cm;
area $\approx$ 43.74 cm^2

Section 8.4 Practice Exercises

Study Skills Exercise

For additional exercises, see Classroom Activities 8.4A–8.4B in the *Student's Resource Manual* at www.mhhe.com/moh.

To help remember the formulas for circles, list them together and note the similarities and differences in the formulas.

Circumference = _____

Area = _____

Similarities: Differences:

Vocabulary and Key Concepts

1. a. A(n) _____radius_____ of a circle is the distance from center of the circle to any point on the circle.

b. A(n) _____diameter_____ of a circle is the length of a line segment connecting two points on the circle and passing through the center of the circle.

c. The ___circumference___ of a circle is the distance around the circle.

d. The number π is the exact value of the ratio of the circumference of a circle to the _____diameter_____ of the circle.

e. The number π is often approximated by the decimal number _____3.14_____ or by the fraction _____$\frac{22}{7}$_____.

f. Which formula can be used to find the circumference C of a circle with radius r and diameter d?
$C = 2\pi r$ or $C = \pi d$ Either formula can be used.

2. The circumference of a circle is measured in linear units such as yd, ft, m, and so on. The area of a circle is measured in square units such as yd^2, ft^2, m^2, and so on. How can the units of measurement help you remember that circumference is given by $2\pi r$, whereas area is given by πr^2.
The formula for circumference $C = 2\pi r$ shows the radius raised to the first power, which is consistent with linear units of measurement such as yd, ft, and m. The formula for area $A = \pi r^2$ shows the radius squared, which is consistent with square units of measurement such as yd^2, ft^2, and m^2.

Review Exercises

3. Find the area of a rectangle with length 42 cm and width 30 cm.   1260 cm^2

4. Find the area of a parallelogram with base 42 cm and height 30 cm.   1260 cm^2

5. Find the area of a triangle with base 42 cm and height 30 cm.   630 cm^2

6. How do the areas found in Exercises 3–5 compare to one another?
The areas of the rectangle and the parallelogram are the same, and the area of the triangle is one-half that area.

7. Could the formula $A = bh$ apply to finding the area of a rectangle? Explain.   Yes. Since a rectangle is a special type of parallelogram (one that contains four right angles), the area formula for a parallelogram applies to a rectangle.

Concept 1: Basic Definitions

8. How does the length of a radius of a circle compare to the length of a diameter?
The length of a radius is one-half the length of a diameter.

For Exercises 9–12, find the length of a diameter. **(See Example 1.)**

9.   12 in.   **10.**   88 mm   **11.**   3 m   **12.** $\frac{1}{2}$ yd

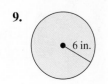   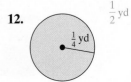

For Exercises 13–16, find the length of a radius. **(See Example 1.)**

13. 4 in.

8 in.

14. 10 cm

20 cm

15. 8.3 m

16.6 m

16. 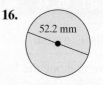 26.1 mm

52.2 mm

Concept 2: Circumference

17. Circumference is similar to which type of measure? (Circle the correct answer.) c

  a. Area   **b.** Capacity   **c.** Perimeter   **d.** Weight

18. Indicate the type of units that could be associated with measuring circumference. Circle all that apply. a, d, g, h

  a. ft   **b.** m^2 **c.** Liters   **d.** Meters   **e.** Grams

  f. $in.^2$ **g.** Miles   **h.** Kilometers   **i.** Cubic centimeters

19. Define π in terms of the circumference and diameter of a circle.

  π is the circumference divided by the diameter. That is, $\pi = \frac{C}{d}$.

20. Which of the following are *not* good approximations for π? Circle all that apply. a, d

  a. 31.4   **b.** 3.14   **c.** $\frac{22}{7}$ **d.** $22\frac{1}{7}$

For Exercises 21–28, find the circumference of the circle. (a) Give the exact answer in terms of π and (b) approximate the answer by using 3.14 for π. **(See Examples 2 and 3.)**

21. a. 4π m
b. 12.56 m

2 m

22. a. 10π ft
b. 31.4 ft

5 ft

23. a. 20π cm
b. 62.8 cm

20 cm

24. a. 12π yd
b. 37.68 yd

12 yd

25. a. 4.2π cm
b. 13.188 cm

2.1 cm

26. a. 12.6π in.
b. 39.564 in.

6.3 in.

27. a. 5π km
b. 15.7 km

$2\frac{1}{2}$ km

28. a. 2.5π m
b. 7.85 m

$1\frac{1}{4}$ m

For Exercises 29–34, use 3.14 for π.

29. Find the circumference of the can of soda.

18.84 cm

├— 6 cm —┤

30. Find the circumference of a can of tuna.

26.69 cm

TUNA

├— 8.5 cm —┤

31. Find the circumference of a compact disk.

14.13 in.

├———— 4.5 in. ————┤

32. Find the outer circumference of a pipe with 3.5-in. diameter.

10.99 in.

3.5 in.

33. Find the outer circumference of a washer 2.2 cm in diameter. 6.908 cm

34. Find the circumference of a pencil 5 mm in diameter. 15.7 mm

Concept 3: Area

For Exercises 35–38, find the area of the circle, (a) give the exact answer in terms of π and (b) approximate the answer by using $\frac{22}{7}$ for π. **(See Examples 4 and 5.)**

 35.
a. 49π m^2
b. 154 m^2

36.
a. $\frac{49}{16}\pi$ km^2
b. $\frac{77}{8}$ km^2 or $9\frac{5}{8}$ km^2

37.
a. 441π in.^{2}
b. 1386 in.^{2}

38.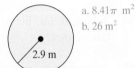
a. $\frac{441}{4}\pi$ cm^2
b. $\frac{693}{2}$ cm^2 or $346\frac{1}{2}$ cm^2

For Exercises 39–42, find the area, (a) give the exact answer in terms of π and (b) approximate the answer by using 3.14 for π. Round to the nearest whole unit.

 39.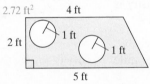
a. 156.25π mm^2
b. 491 mm^2

40.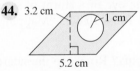
a. 25π ft^2
b. 79 ft^2

41.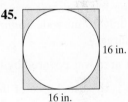
a. 38.44π ft^2
b. 121 ft^2

42.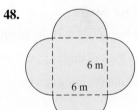
a. 8.41π m^2
b. 26 m^2

Concept 4: Applications (Mixed Exercises)

 For Exercises 43–51, find the area of the shaded region. Use 3.14 for π. **(See Example 6.)**

43. 2.72 ft^2
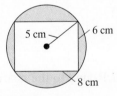

44. 3.2 cm   13.5 cm^2
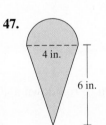

45. 55.04 in.^{2}

46. 30.5 cm^2
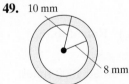

47. 18.28 in.^{2}
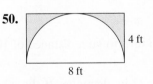

48. 92.52 m^2

49. 10 mm   113.04 mm^2

50. 6.88 ft^2

51. 222.39 in.^{2}

For Exercises 52–57, use 3.14 for π.

52. A roller hockey rink is a rectangle with a semicircle at each end.

a. How much will it cost to put up a rail around the rink if railing costs $2.59 per *foot*? $1051.75

b. How much will it cost to put down the floor if flooring costs $8.00 per square yard? $9377.28

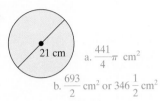

 Writing   Translating Expression   Geometry   Scientific Calculator   Video   NS&E

53. The Large Hadron Collider (LHC) is a particle accelerator and collider built to detect subatomic particles. The accelerator is in a huge circular tunnel that straddles the border between France and Switzerland. The diameter of the tunnel is 5.3 mi. Find the circumference of the tunnel. (*Source:* European Organization for Nuclear Research) 16.642 mi

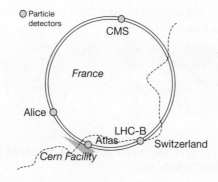

54. A ceiling fan blade rotates in a full circle. If the fan blades are 2 ft long, what is the area covered by the fan blades? 12.56 ft²

55. An outdoor torch lamp shines light a distance of 30 ft in all directions. What is the total ground area lighted? 2826 ft²

56. How many times larger is the area of Circle 2 than Circle 1? **(See figure.)**
The area of the larger circle is 4 times larger.

57. Hurricane Katrina's eye was 32 mi wide. The eye of a storm of similar intensity is usually only 10 mi wide. (*Source:* Associated Press 10/8/05 "Mapping Katrina's Storm Surge")

    a. What area was covered by the eye of Katrina? Round to the nearest square mile. ≈ 804 mi²

    b. What is the usual area of the eye of a similar storm? ≈ 79 mi²

Expanding Your Skills

58. A hula hoop has a 20-in. diameter.

    a. Find the circumference. Use 3.14 for π. 62.8 in.

    b. How many times will the hula hoop have to turn to roll down a 40-yd driveway? Round to the nearest whole unit. 23 times

59. A bicycle wheel has a 26-in. diameter.

    a. Find the circumference. Use 3.14 for π. 81.64 in.

    b. How many times will the wheel have to turn to go a distance of 1000 ft (12,000 in.)? Round to the nearest whole unit. 147 times

60. Latasha has a bicycle, and the wheel has a 22-in. diameter. If the wheels of the bike turned 1000 times, how far did she travel? Use 3.14 for π. Give the answer to the nearest inch and to the nearest foot.
69,080 in. or 5757 ft

61. The exercise wheel for Al's dwarf hamster has a diameter of 6.75 in.

    a. Find the circumference. Use 3.14 for π and round to the nearest inch. 21 in.

    b. How far does Al's hamster travel if he completes 25 revolutions? Write the answer in feet and round to one decimal place. 43.8 ft

Calculator Connections

Topic: Using the π Key

When finding the circumference or the area of a circle, we can use the $\boxed{\pi}$ key on the calculator to lend more accuracy to our calculations. If you press the $\boxed{\pi}$ key on the calculator, the display will show 3.141592654. This number is not the exact value of π (remember that in decimal form π is a nonterminating and nonrepeating decimal). However, using the $\boxed{\pi}$ key provides more accuracy than by using 3.14. For example, suppose we want to find the area of a circle of radius 3 ft. Compare the values by using 3.14 for π versus using the $\boxed{\pi}$ key.

Expression	Keystrokes	Result
$3.14 \cdot 3^2$	$3.14 \;\boxed{\times}\; 3 \;\boxed{x^2}\; \boxed{=}$	28.26
$\pi \cdot 3^2$	$\boxed{\pi} \;\boxed{\times}\; 3 \;\boxed{x^2}\; \boxed{=}$	28.27433388

Again, it is important to note that neither of these answers is the exact area. The only way to write the exact value is to express the answer in terms of π. The exact area is 9π ft^2.

For Exercises 62–65, find the area and circumference rounded to four decimals places. Use the $\boxed{\pi}$ key on your calculator.

62.

12.83 cm

Area ≈ 517.1341 cm^2;
circumference ≈ 80.6133 cm

63.
5.1 ft

Area ≈ 81.7128 ft^2;
circumference ≈ 32.0442 ft

64.
9.5 in

Area ≈ 70.8822 in.^{2};
circumference ≈ 29.8451 in.

65.
103.24 mm

Area ≈ 8371.1644 mm^2;
circumference ≈ 324.3380 mm

66. Find the area for each pizza using the π key on your calculator. Then compute the unit cost per square inch. If the pizzas have the same thickness, which pizza is a better buy?

Diameter	Cost	Area	Cost per in.^{2}
8 in.	$ 6.50	50.27 in.^{2}	$ 0.129
12 in.	12.40	113.10 in.^{2}	0.110

The 12-in. is the better buy.

 Writing   Translating Expression   Geometry   Scientific Calculator   Video   NS E

Problem Recognition Exercises

Area, Perimeter, and Circumference

For Exercises 1–15, find the area and the perimeter or circumference for each figure.

1.
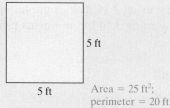
5 ft

5 ft

Area = 25 ft²;
perimeter = 20 ft

2.
12 m

12 m

Area = 144 m²;
perimeter = 48 m

3.
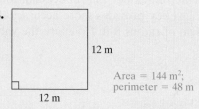
300 cm

4 m

Area = 12 m² or 120,000 cm²;
perimeter = 14 m or 1400 cm

4.
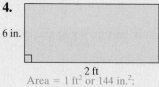
6 in.

2 ft

Area = 1 ft² or 144 in.²;
perimeter = 5 ft or 60 in.

5.
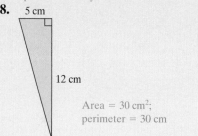
1 ft   1.5 ft

1 yd

Area = $\frac{1}{3}$ yd² or 3 ft²;
perimeter = 3 yd or 9 ft

6.
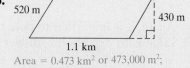
520 m   430 m

1.1 km

Area = 0.473 km² or 473,000 m²;
perimeter = 3.24 km or 3240 m

7.
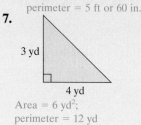
3 yd

4 yd

Area = 6 yd²;
perimeter = 12 yd

8.
5 cm

12 cm

Area = 30 cm²;
perimeter = 30 cm

9.
14 m

5 m   4 m   5 m

8 m

Area = 44 m²;
perimeter = 32 m

10.
8 in.

10 in.   8 in.

14 in.

Area = 88 in.²;
perimeter = 40 in.

11. Use 3.14 for π.

6 yd

Area ≈ 28.26 yd²;
circumference ≈ 18.84 yd

12. Use 3.14 for π.

40 cm

Area ≈ 1256 cm²;
circumference ≈ 125.6 cm

13. Use $\frac{22}{7}$ for π.

7 cm

Area ≈ 154 cm²;
circumference ≈ 44 cm

14. Use $\frac{22}{7}$ for π.
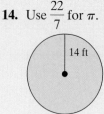
14 ft

Area ≈ 616 ft²;
circumference ≈ 88 ft

15. Use 3.14 for π.
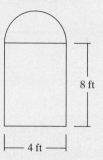
8 ft

4 ft

Area ≈ 38.28 ft²;
perimeter ≈ 26.28 ft

Volume

1. Introduction to Volume

In this section, we learn how to compute volume. Volume is another word for capacity. We use volume, for example, to determine how much can be held in a moving van.

In addition to the units of capacity learned in Sections 7.2 and 7.4, volume can be measured in cubic units. For example, a cube that is 1 cm on a side has a volume of 1 cubic centimeter (1 cm³ or cc). A cube that is 1 in. on a side has a volume of 1 cubic inch (1 in.³). See Figure 8-28. Additional units of volume include cubic feet (ft³), cubic yards (yd³), cubic meters (m³), and so on.

Concepts

1. Introduction to Volume
2. Volume Formulas for Selected Solids
3. Volumes of Composite Figures

Instructor Note: Remind students that, as is mentioned in the introduction to Chapter 8, these figures are three-dimensional solids.

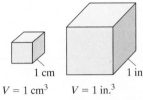

1 cm   1 in.
$V = 1 \text{ cm}^3$ $V = 1 \text{ in.}^3$

Figure 8-28

TIP: Recall that 1 cubic centimeter can also be denoted as 1 cc. Furthermore, 1 cc = 1 mL.

2. Volume Formulas for Selected Solids

The formulas used to compute the volume of several common solids are given.

Rectangular Solid	Cube	Right Circular Cylinder
$V = lwh$	$V = s^3$	$V = \pi r^2 h$

Notice that the volume formulas for these three figures are given by the product of the area of the base and the height of the figure:

$$V = lw\mathbf{h} \qquad V = s{\cdot}s{\cdot}\mathbf{s} \qquad V = \pi r^2\mathbf{h}$$

| area of rectangular base | area of square base | area of circular base |

Example 1   Finding Volume

Find the volume. Round to the nearest whole unit.

Solution:

$V = lwh$ Use the volume formula for a rectangular solid. Identify the length, width, and height.

$= (4 \text{ in.})(3 \text{ in.})(5 \text{ in.})$ $l = 4$ in., $w = 3$ in., and

$= 60 \text{ in.}^3$ $h = 5$ in.

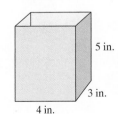

5 in.

3 in.

4 in.

We can visualize the volume by "layering" cubes that are each 1 in. high (Figure 8-29). The number of cubes in each layer is equal to $4 \times 3 = 12$. Each layer has 12 cubes, and there are 5 layers. Thus, the total number of cubes is $12 \times 5 = 60$ for a volume of 60 in.^{3}

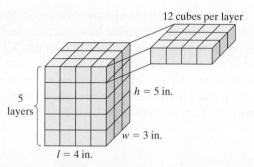

12 cubes per layer

5 layers

$h = 5$ in.

$w = 3$ in.

$l = 4$ in.

Figure 8-29

Example 2 **Finding Volume of a Cylinder**

Find the volume. Use 3.14 for π. Round to the nearest whole unit.

3.7 cm

11.2 cm

Solution:

$V = \pi r^2 h$ — Use the formula for the volume of a right circular cylinder.

$\approx (3.14)(3.7 \text{ cm})^2(11.2 \text{ cm})$ — Substitute 3.14 for π, $r = 3.7$ cm, and $h = 11.2$ cm.

$= (3.14)(13.69 \text{ cm}^2)(11.2 \text{ cm})$ — Simplify exponents first.

$= 481.44992 \text{ cm}^3$ — Multiply from left to right.

$\approx 481 \text{ cm}^3$ — Round to the nearest whole unit.

A right circular cone has the shape of a party hat. A sphere has the shape of a ball. To compute the volume of a cone and a sphere, we use the following formulas.

Right Circular Cone

h

r

$V = \frac{1}{3}\pi r^2 h$

Sphere

r

$V = \frac{4}{3}\pi r^3$

Answers

1. 176 in.^{3} **2.** ≈ 56.52 in.^{3}

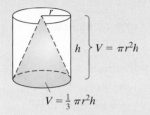

TIP: Notice that the formula for the volume of a right circular cone is $\frac{1}{3}$ that of a right circular cylinder.

$h \left.\right\} V = \pi r^2 h$

$V = \frac{1}{3}\pi r^2 h$

TIP: A **hemisphere** is one-half of a sphere. Therefore, the volume of a hemisphere is one-half that of a full sphere.

Hemisphere
$V = \frac{1}{2} \cdot \left(\frac{4}{3}\pi r^3\right)$

Example 3 Finding the Volume of a Sphere

Find the volume. Use 3.14 for π. Round to one decimal place.

$r = 6$ in.

Solution:

$V = \dfrac{4}{3}\pi r^3$ Use the formula for the volume of a sphere.

$\approx \dfrac{4}{3}(3.14)(6 \text{ in.})^3$ Substitute 3.14 for π and $r = 6$ in.

$= \dfrac{4}{3}(3.14)(216 \text{ in.}^3)$ Simplify exponents first.
                     $(6 \text{ in.})^3 = (6 \text{ in.})(6 \text{ in.})(6 \text{ in.}) = 216 \text{ in.}^3$

$= \dfrac{4}{3}\left(\dfrac{3.14}{1}\right)\left(\dfrac{216 \text{ in.}^3}{1}\right)$ Multiply fractions.

$= \dfrac{4}{\underset{1}{\cancel{3}}}\left(\dfrac{3.14}{1}\right)\left(\dfrac{\overset{72}{\cancel{216}} \text{ in.}^3}{1}\right)$ Simplify to lowest terms.

$= 904.32 \text{ in.}^3$ Multiply from left to right.

$\approx 904.3 \text{ in.}^3$ Round to one decimal place.

Skill Practice

Find the volume. Use 3.14 for π.

3. $r = 3$ cm

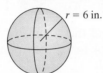

Classroom Example: p. 514, Exercise 20

Example 4 Finding the Volume of a Cone

Find the volume. Use 3.14 for π. Round to one decimal place.

Solution:

$V = \dfrac{1}{3}\pi r^2 h$ Use the formula for the volume of a right circular cone.

8 in.

5 in.

Skill Practice

Find the volume. Use 3.14 for π.

4.

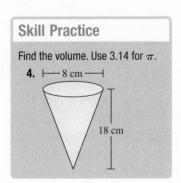

8 cm

18 cm

Classroom Example: p. 514, Exercise 22

Answers

3. 113.04 cm³   **4.** 301.44 cm³

To find the radius we have $r = \frac{1}{2}d = \frac{1}{2}(5 \text{ in.}) = 2.5 \text{ in.}$

$V \approx \frac{1}{3}(3.14)(2.5 \text{ in.})^2(8 \text{ in.})$ Substitute 3.14 for π, $r = 2.5$ in., and $h = 8$ in.

$= \frac{1}{3}\left(\frac{3.14}{1}\right)\left(\frac{6.25 \text{ in.}^2}{1}\right)\left(\frac{8 \text{ in.}}{1}\right)$ Simplify exponents first.

$= \frac{157}{3} \text{ in.}^3$ Multiply fractions.

$\approx 52.3 \text{ in.}^3$ Round to one decimal place.

3. Volumes of Composite Figures

Example 5 **Finding the Volume of a Composite Figure**

Find the volume of the HEPA filter (Figure 8-30). Use 3.14 for π and round the answer to the nearest whole unit.

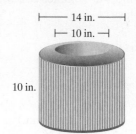

Figure 8-30

Solution:

The solid consists of an outer cylinder with a cylindrical core cut out of the center.

To find the volume, we can find the volume of the outer cylinder and subtract the volume of the inner cylinder.

The radius of the outer cylinder is

$r = \frac{1}{2}d = \frac{1}{2}(14 \text{ in.}) = 7 \text{ in.}$

The radius of the inner cylinder is

$r = \frac{1}{2}d = \frac{1}{2}(10 \text{ in.}) = 5 \text{ in.}$

The volume of the outer cylinder is

$V = \pi r^2 h$

$\approx (3.14)(7 \text{ in.})^2(10 \text{ in.})$

$= (3.14)(49 \text{ in.}^2)(10 \text{ in.})$

$= 1538.6 \text{ in.}^3$

The volume of the inner cylinder is

$V = \pi r^2 h$

$\approx (3.14)(5 \text{ in.})^2(10 \text{ in.})$

$= (3.14)(25 \text{ in.}^2)(10 \text{ in.})$

$= 785 \text{ in.}^3$

The volume of the HEPA filter is the difference of the outer cylinder and the inner cylinder.

volume of outer cylinder   volume of inner cylinder

Volume of filter $= 1538.6 \text{ in.}^3 - 785 \text{ in.}^3$

$= 753.6 \text{ in.}^3$

$\approx 754 \text{ in.}^3$ Round to the nearest whole unit.

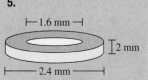

Answers

5 a. 1.2 mm
b. 0.8 mm
c. $\approx 5 \text{ mm}^3$

Section 8.5 Practice Exercises

Study Skills Exercise

For additional exercises, see Classroom Activities 8.5A–8.5B in the *Student's Resource Manual* at www.mhhe.com/moh.

Apply what you have learned to real-life situations. This can help you remember formulas and methods as well as give some meaning to math. Write down one real-life application of geometry.

Vocabulary and Key Concepts

1. **a.** The volume of a cube with sides of length s is given by $V = \underline{\quad s^3 \quad}$.

    b. The volume of a rectangular solid with sides of lengths l, w, and h is given by $V = \underline{\quad lwh \quad}$.

    c. The volume of a right circular cylinder with radius r and height h is given by $V = \underline{\quad \pi r^2 h \quad}$.

    d. The formula $V = \frac{1}{3}\pi r^2 h$ gives the volume of a (sphere/cone) whereas the formula $V = \frac{4}{3}\pi r^3$ gives the volume of a (sphere/cone).   cone; sphere

Review Exercises

2. If one angle in a right triangle is 12°, determine the measure of the other two angles.   90° and 78°

For Exercises 3 and 4, find the circumference, C, and area, A, of the circles. Use 3.14 for π.

3.
    4 in.
    $C \approx 25.12$ in;
    $A \approx 50.24$ in.^{2}

4.
    10 cm
    $C \approx 62.8$ cm;
    $A \approx 314$ cm^2

5. Find the area of the shaded region.

    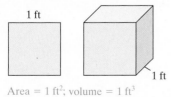
    24 cm
    8 cm   8 cm
    12 cm
    187.52 cm^2

6. Find the area.

    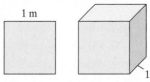
    24 in.
    12 in.
    14 in.
    228 in.^{2}

Concept 1: Introduction to Volume

7. Which of the units denote volume? Circle all that apply.   b, d

    a. ft^2 **b.** m^3 **c.** in. **d.** cc **e.** mi

8. Which of the units denote volume? Circle all that apply.   c, e

    a. yd **b.** yd^2 **c.** yd^3 **d.** km **e.** km^3

For Exercises 9–12, determine the area of the square and the volume of the cube.

9.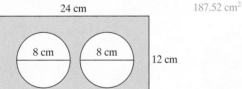
    1 ft
    1 ft
    Area = 1 ft^2; volume = 1 ft^3

10.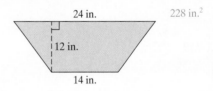
    1 m
    1 m
    Area = 1 m^2; volume = 1 m^3

 Writing   Translating Expression   Geometry   Scientific Calculator   Video   NS E

11.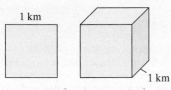

Area = 1 km²; volume = 1 km³

12.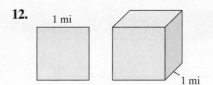

Area = 1 mi²; volume = 1 mi³

Concept 2: Volume Formulas for Selected Solids

For Exercises 13–24, find the volume. Use 3.14 for π where necessary. **(See Examples 1–4.)**

13. 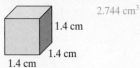   2.744 cm³

1.4 cm
1.4 cm
1.4 cm

14. 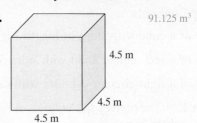   91.125 m³

4.5 m
4.5 m
4.5 m

15. 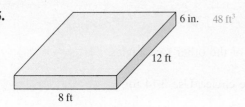   6 in.   48 ft³

12 ft
8 ft

16. 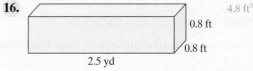   4.8 ft³

0.8 ft
0.8 ft
2.5 yd

17. 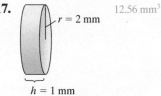   12.56 mm³

r = 2 mm
h = 1 mm

18. 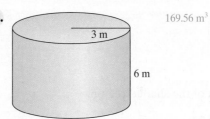   169.56 m³

3 m
6 m

19. 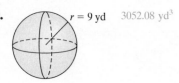   r = 9 yd   3052.08 yd³

20. 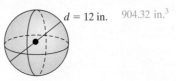   d = 12 in.   904.32 in.³

21. 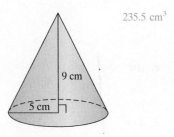   235.5 cm³

9 cm
5 cm

22. 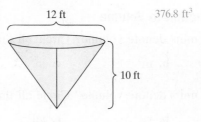   12 ft   376.8 ft³

10 ft

23. 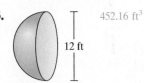   452.16 ft³

12 ft

24. 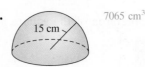   7065 cm³

15 cm

For Exercises 25–32, use 3.14 for π. Round each value to the nearest whole unit.

25. The diameter of a volleyball is 8.2 in. Find the volume.   289 in.³

26. The diameter of a basketball is 9 in. Find the volume.   382 in.³

27. Find the volume of the sand pile.   314 ft³

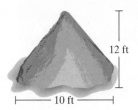

12 ft

10 ft

28. In decorating cakes, many people use an icing bag that has the shape of a cone. Find the volume of the icing bag.   754 cm³

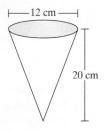

12 cm

20 cm

29. Find the volume of water (in cubic feet) that the pipe can hold.   10 ft³

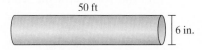

50 ft

6 in.

30. Find the volume of the wastebasket that has the shape of a cylinder with the height of 2 ft and diameter of 3 ft.   9 ft³

3 ft

2 ft

31. Sam bought an above ground circular swimming pool with diameter 27 ft and height 54 in.

a. Approximate the volume of the pool in cubic feet using 3.14 for π.   2575 ft³

b. How many gallons of water will it take to fill the pool? (*Hint:* 1 gal ≈ 0.1337 ft³.)
19,260 gal

32. Richard needs 3 in. of topsoil for his vegetable garden that is in the shape of a rectangle, 15 ft by 20 ft.

a. Find the amount of topsoil needed in cubic feet.   75 ft³

b. If top soil can be purchased in bags containing 2 ft³, how many bags must Richard purchase?   38 bags

Concept 3: Volumes of Composite Figures

For Exercises 33–38, use 3.14 for π. Round each value to the nearest whole unit.

33. A machine part is in the shape of a cylinder with a hole drilled through the center. Find the volume of the machine part.
(See Example 5.)   502 mm³

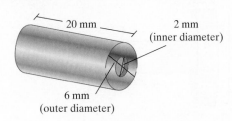

20 mm

2 mm (inner diameter)

6 mm (outer diameter)

34. To insulate pipes, a cylinder of Styrofoam has a hole drilled through it to fit around a pipe. What is the volume of this piece of insulation?   471 in.³

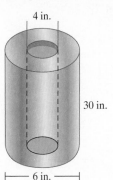

4 in.

30 in.

6 in.

35. A gasoline storage tank is in the shape of a cylinder with hemispheres on each end. Find the volume. 32 ft³

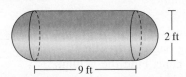

2 ft

9 ft

36. A silo is in the shape of a cylinder with a hemisphere on the top. Find the volume. 622 ft³

20 ft

6 ft

37. An ice cream cone is in the shape of a cone with a sphere on top. Assuming that ice cream is packed inside the cone, find the volume of the ice cream. 56 in.³

2 in.

5.5 in.

38. A birdbath is made from a hemisphere on a pedestal. Find the volume of water that the birdbath will hold. 22,290 cm³

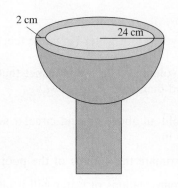

2 cm

24 cm

For Exercises 39–44, find the volume of the shaded region. Use 3.14 for π if necessary.

39.

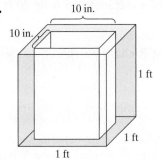

10 in.

10 in.

1 ft $\frac{11}{36}$ ft³ or 0.306 ft³ or 528 in.³

1 ft

1 ft

40.

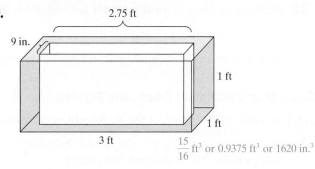

2.75 ft

9 in.

1 ft

1 ft

3 ft $\frac{15}{16}$ ft³ or 0.9375 ft³ or 1620 in.³

41.

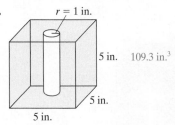

r = 1 in.

5 in.   109.3 in.³

5 in.

5 in.

42.

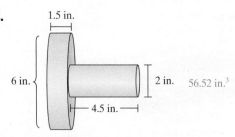

1.5 in.

6 in.

2 in.   56.52 in.³

4.5 in.

43.

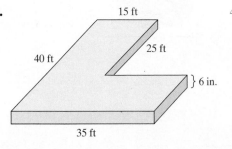

15 ft   450 ft³

25 ft

40 ft

6 in.

35 ft

44.

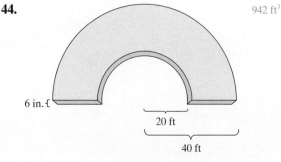

942 ft³

6 in.

20 ft

40 ft

Expanding Your Skills

The volume formulas for right circular cylinders and right circular cones are the same for slanted cylinders and cones. For Exercises 45–48, find the volume. Use 3.14 for π.

45.

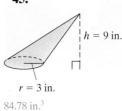

$h = 9$ in.

$r = 3$ in.

84.78 in.^{3}

46.

$h = 12$ mm

$r = 5$ mm

942 mm^3

47.

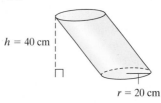

$h = 40$ cm

$r = 20$ cm

50,240 cm^3

48.

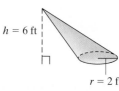

$h = 6$ ft

$r = 2$ ft

25.12 ft^3

Group Activity

Constructing a Golden Rectangle

Materials: Ruler, calculator, and a compass

Estimated time: 20 minutes

Group Size: 2

Consider a line segment divided into two pieces and labeled as follows.

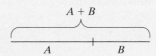

$A + B$

A B

The Golden Ratio is given by: $\dfrac{A}{B} = \dfrac{A + B}{A}$

Although it is beyond the scope of this text, we can solve for A to get, $A = \left(\dfrac{\sqrt{5} + 1}{2} \right) \cdot B$ or $A \approx 1.618 \cdot B$.

1. This ratio occurs in nature and is used in art and architecture. Measure the length of the hand. Then multiply by 1.618 and compare to the length of the arm.

2. Notre Dame in Paris, France, is constructed with many golden ratios. Check by measuring the blue lines and multiplying each length by 1.618 to get the measure of the corresponding white lines.

A Golden Rectangle is one in which the ratio of length to width is approximately 1.618. This rectangle is thought to be most pleasing to the eye.

 Writing   Translating Expression   Geometry   Scientific Calculator  Video NS&E

3. In Leonardo da Vinci's *Mona Lisa*, the face forms a golden rectangle. Measure the sides of the rectangle and find the ratio of length to width.

Also notice that the width times 1.618 should approximate the length.

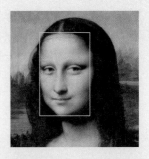

4. Construct a golden rectangle with a width of 1 inch. (The length will be approximately 1.618 in.)

5. Construct a golden rectangle with a width of 2.5 in. (Compute the length by multiplying the width by 1.618.)

6. A golden rectangle can be constructed from a width of any measure by following these steps.

Step 1: Construct a square.

Step 2: Divide the square in half.

Step 3: Draw a diagonal from the center line to a corner.

Step 4: Drop this diagonal line to the horizontal.
(Use a compass to maintain its length.)

Step 5: Use the horizontal line as the length of the rectangle.

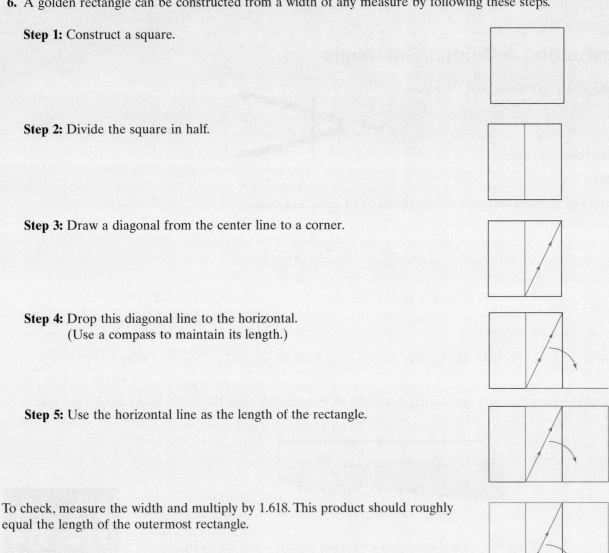

To check, measure the width and multiply by 1.618. This product should roughly equal the length of the outermost rectangle.

Chapter 8   Summary

Section 8.1   Lines and Angles

Key Concepts

An **angle** is a geometric figure formed by two rays that share a common endpoint. The common endpoint is called the **vertex** of the angle.

An angle is **acute** if its measure is between 0° and 90°. An angle is **obtuse** if its measure is between 90° and 180°.

Two angles are said to be **complementary** if the sum of their measures is 90°. Two angles are said to be **supplementary** if the sum of their measures is 180°.

Given two intersecting lines, **vertical angles** are angles that appear on opposite sides of the vertex.

When two parallel lines are crossed by another line eight angles are formed.

Examples

Example 1

$\angle DEF$

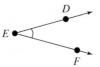

Example 2

 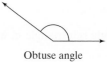

Acute angle   Obtuse angle

Example 3

The complement of a 32° angle is a 58° angle. The supplement of a 32° angle is a 148° angle.

Example 4

Intersecting lines:

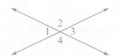

$\angle 1$ and $\angle 3$ are vertical angles and are congruent. Also $\angle 2$ and $\angle 4$ are vertical angles and are congruent.

Example 5

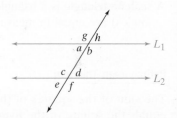

$m(\angle a) = m(\angle d)$ because they are **alternate interior angles**.
$m(\angle e) = m(\angle h)$ because they are **alternate exterior angles**.
$m(\angle c) = m(\angle g)$ because they are **corresponding angles**.

Section 8.2 Triangles and the Pythagorean Theorem

Key Concepts

The sum of the measures of the angles of any triangle is 180°.

An **acute triangle** is a triangle in which all three angles are acute.

A **right triangle** is a triangle in which one angle is a right angle.

An **obtuse triangle** is a triangle in which one angle is obtuse.

An **equilateral triangle** is a triangle in which all three sides (and all three angles) are equal in measure.

An **isosceles triangle** is a triangle in which two sides are equal in length (the angles opposite the equal sides are also equal in measure).

A **scalene triangle** is a triangle in which no sides (or angles) are equal in measure.

Pythagorean Theorem

The sum of the squares of the **legs of a right triangle** equals the square of the **hypotenuse**.

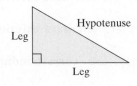

$$a^2 + b^2 = c^2$$

Examples

Example 1

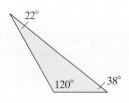

$$22° + 120° + 38° = 180°$$

Example 2

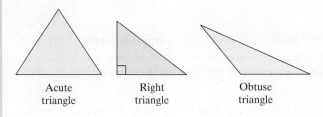

Acute triangle   Right triangle   Obtuse triangle

Example 3

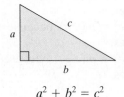

Equilateral triangle   Isosceles triangle   Scalene triangle

Example 4

To find the length of the hypotenuse, solve for c.

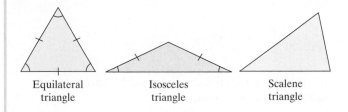

$$6^2 + 8^2 = c^2$$

$$36 + 64 = c^2$$

$$100 = c^2$$

$$\sqrt{100} = c$$ c is the positive number that when squared equals 100.

$$10 = c$$

The length of the hypotenuse is 10 cm.

Section 8.3   Quadrilaterals, Perimeter, and Area

Key Concepts

A four-sided polygon is called a **quadrilateral**.

A **parallelogram** is a quadrilateral with opposite sides parallel.

A **rectangle** is a parallelogram with four right angles.

A **square** is a rectangle with sides equal in length.

A **rhombus** is a parallelogram with sides equal in length.

A **trapezoid** is a quadrilateral with one pair of parallel sides.

Examples

Example 1

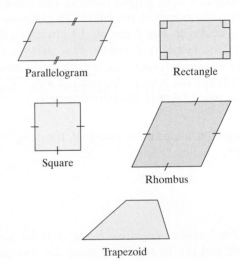

Parallelogram   Rectangle

Square

Rhombus

Trapezoid

Perimeter is the distance around a figure.

Perimeter of a square: $P = 4s$

Perimeter of a rectangle: $P = 2l + 2w$

Example 2

Given:

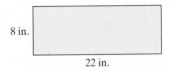

8 in.

22 in.

The perimeter of the rectangle is

$P = 2(22 \text{ in.}) + 2(8 \text{ in.})$

$= 44 \text{ in.} + 16 \text{ in.}$

$= 60 \text{ in.}$

The perimeter is 60 in.

Area is the number of square units that can be enclosed by a figure.

Area of a rectangle: $A = lw$

Area of a square: $A = s^2$

Area of a parallelogram: $A = bh$

Area of a triangle: $A = \frac{1}{2}bh$

Area of a trapezoid: $A = \frac{1}{2}(a + b)h$

Example 3

Given:

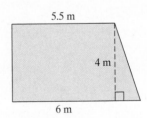

5.5 m

4 m

6 m

The area of the trapezoid is

$A = \frac{1}{2}(6 \text{ m} + 5.5 \text{ m})(4 \text{ m})$

$= \frac{1}{2}(11.5 \text{ m})(4 \text{ m})$

$= 23 \text{ m}^2$

The area is 23 m².

Section 8.4   Circles, Circumference, and Area

Key Concepts

A **circle** is a figure consisting of all points located the same distance r from a fixed point C. The fixed point C is called the **center** of the circle. The line segment from the center to any point on the circle is called a **radius** of the circle. A **diameter** of a circle is a line segment whose endpoints are on the circle and that passes through the center. The number $\pi = \frac{\text{circumference}}{\text{diameter}}$. We often use 3.14 or $\frac{22}{7}$ to approximate π.

The length of a radius is one-half the length of a diameter.

$$r = \frac{1}{2}d \quad \text{or} \quad d = 2r$$

The **circumference** of a circle is the distance around the circle and can be found by using the formula

$$C = 2\pi r \quad \text{or} \quad C = \pi d$$

The area of a circle is found by using the formula

$$A = \pi r^2$$

Examples

Example 1

20 in.

a. Find the radius.

$$r = \frac{1}{2}d = \frac{1}{2}(20 \text{ in.}) = 10 \text{ in.}$$

b. Find the circumference.

$$
\begin{aligned}
C &= 2\pi r \\
&= 2\pi \, (10 \text{ in.}) \\
&= 20\pi \text{ in.} \quad \text{(exact value)} \\
&\approx 20(3.14) \text{ in.} \\
&= 62.8 \text{ in.} \quad \text{(approximate value)}
\end{aligned}
$$

c. Find the area.

$$
\begin{aligned}
A &= \pi r^2 \\
&= \pi(10 \text{ in.})^2 \\
&= \pi(100 \text{ in.}^2) \\
&= 100\pi \text{ in.}^2 \quad \text{(exact value)} \\
&\approx 100(3.14) \text{ in.}^2 \\
&= 314 \text{ in.}^2 \quad \text{(approximate value)}
\end{aligned}
$$

Section 8.5   Volume

Key Concepts

Volume is another word for capacity.

Formulas for selected solids are given.

Rectangular solid

$V = lwh$

Cube

$V = s^3$

Right circular cylinder

$V = \pi r^2 h$

Right circular cone

$V = \dfrac{1}{3}\pi r^2 h$

Sphere

$V = \dfrac{4}{3}\pi r^3$

Examples

Example 1

Find the volume of a tissue box with dimensions 23.5 cm by 12 cm by 12 cm.

Volume of a rectangular solid: $V = lwh$

$V = (23.5 \text{ cm})(12 \text{ cm})(12 \text{ cm})$

$\quad = 3384 \text{ cm}^3$

The volume is 3384 cm³.

Example 2

Find the volume of the cone.

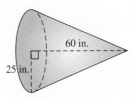

$V = \dfrac{1}{3}\pi r^2 h$

$\quad \approx \dfrac{1}{3}(3.14)(25 \text{ in.})^2(60 \text{ in.})$

$\quad = 39{,}250 \text{ in.}^3$

The volume is approximately 39,250 in.³

Chapter 8   Review Exercises

Section 8.1

For Exercises 1–4, match the symbol with a description.

1. $\overleftrightarrow{AB}$ d   a. Ray AB

2. $\overrightarrow{AB}$ a   b. Line segment AB

3. $\overrightarrow{BA}$ c   c. Ray BA

4. $\overline{AB}$ b   d. Line AB

5. Describe the measure of an acute angle.
   The measure of an acute angle is between 0° and 90°.

6. Describe the measure of an obtuse angle.
   The measure of an obtuse angle is between 90° and 180°.

7. Describe the measure of a straight angle.
   The measure of a straight angle is 180°.

8. Describe the measure of a right angle.
   The measure of a right angle is 90°.

9. Let $m(\angle X) = 33°$.

   a. Find the complement of $\angle X$.   57°

   b. Find the supplement of $\angle X$.   147°

10. Let $m(\angle T) = 20°$.

    a. Find the complement of $\angle T$.   70°

    b. Find the supplement of $\angle T$.   160°

For Exercises 11–14, refer to the figure to determine the measure of the indicated angle.

11. $m(\angle ABE)$ 60°

12. $m(\angle DBC)$ 90°

13. $m(\angle ABG)$ 175°

14. $m(\angle ABC)$ 180°

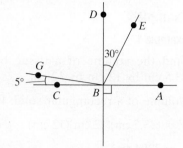

For Exercises 15–17, select the figure or figures that apply.

15. Two lines that are parallel.   b

16. Two lines that are *not* perpendicular.   b, c

17. Two lines that are intersecting.   a, c

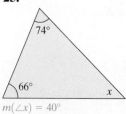

a.   **b.**   **c.**

For Exercises 18–24, refer to the figure. Find the measures of the angles.

18. $m(\angle a)$ 62°

19. $m(\angle b)$ 118°

20. $m(\angle c)$ 118°

21. $m(\angle d)$ 62°

22. $m(\angle e)$ 62°

23. $m(\angle f)$ 118°   **24.** $m(\angle g)$ 118°

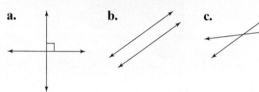

L_1 is parallel to L_2

Section 8.2

For Exercises 25 and 26, find the measures of the angles x and y.

25.

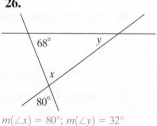

74°

66° x

$m(\angle x) = 40°$

26.

68° y

x

80°

$m(\angle x) = 80°$; $m(\angle y) = 32°$

For Exercises 27–32, describe the characteristics of each type of triangle.

27. Obtuse triangle   An obtuse triangle has one obtuse angle.

28. Equilateral triangle   An equilateral triangle has three sides of equal length and three angles of equal measure.

29. Right triangle   A right triangle has a right (90°) angle.

30. Acute triangle   An acute triangle has three acute angles.

31. Isosceles triangle   An isosceles triangle has two sides of equal length and two angles of equal measure.

32. Scalene triangle   A scalene triangle has no sides or angles of equal measure.

For Exercises 33–36, simplify the square roots.

33. $\sqrt{25}$ 5   **34.** $\sqrt{49}$ 7

35. $\sqrt{100}$ 10   **36.** $\sqrt{64}$ 8

37. State the Pythagorean theorem in words.
The sum of the squares of the legs of a right triangle equals the square of the hypotenuse.

For Exercises 38 and 39, find the length of the unknown side.

38.   **39.**

b

25 cm   24 cm   12 ft c

$b = 7$ cm   16 ft

$c = 20$ ft

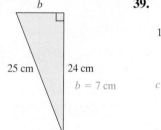

40. Kayla is flying a kite. At one point the kite is 5 m from Kayla horizontally and 12 m above her (see figure). How much string will be extended at this time? (Assume there is no slack.)
13 m of string is extended.

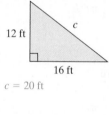

12 m

5 m

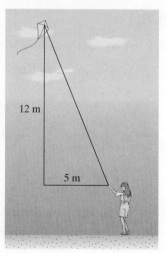

Section 8.3

For Exercises 41–44, indicate the similarities and differences of the quadrilaterals.

41. A rhombus and a square   They both have sides of equal length, but a square also has four right angles.

42. A trapezoid and a parallelogram   A parallelogram must have both pairs of opposite sides parallel.

43. A rectangle and a square
A square is a rectangle with four sides of equal length.

44. A rectangle and a parallelogram
A rectangle is a parallelogram with four right angles.

45. Find the perimeter of the figure.   90 cm

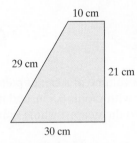

10 cm

29 cm

21 cm

30 cm

46. Find the perimeter of a triangle with sides of 4.2 m, 6.1 m, and 7.0 m.   17.3 m

47. Find the perimeter of a rectangle with length 16 mi and width 12 mi.   56 mi

48. How much fencing is required to put up a chain link fence around a 120-yd by 80-yd playground?
400 yd

49. The perimeter of a rectangle is 120 ft. The two shorter sides add up to 36 ft. What is the length of each of the longer sides?   42 ft

50. The perimeter of a square is 62 ft. What is the length of each side?   15.5 ft

51. Find the area of the triangle.   20 in.²

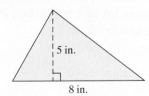

5 in.

8 in.

52. Fatima has a Persian rug 8.5 ft by 6 ft. What is the area?   51 ft²

53. A lot is 150 ft by 80 ft. Within the lot, there is a 12-ft easement along all edges. An easement is the portion of the lot on which nothing may be built. What is the area of the portion that may be used for building?   7056 ft²

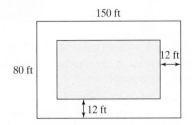

150 ft

80 ft

12 ft

12 ft

54. Find the area and the perimeter of the shaded triangle.   The area is 36 m². The perimeter is 36 m.

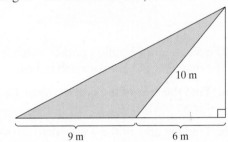

10 m

9 m

6 m

Section 8.4

55. Find the diameter of a circle whose radius is 45 mm.   90 mm

56. Find the diameter of a circle whose radius is 3.2 ft.   6.4 ft

57. Find the radius of a circle whose diameter is 45 mm.   22.5 mm

58. Find the radius of a circle whose diameter is 3.2 ft.   1.6 ft

For Exercises 59–62, find the circumference C and the area A of the circle.

59. Use 3.14 for π.   **60.** Use $\dfrac{22}{7}$ for π.

8 m   2.1 yd

$C = 50.24$ m; $A = 200.96$ m² $C = 13.2$ yd; $A = 13.86$ yd²

61. Use $\frac{22}{7}$ for π.

$C = 440$ in.; $A = 15{,}400$ in.^{2}

62. Use 3.14 for π.

$C = 125.6$ ft; $A = 1256$ ft^2

63. Find the area of the shaded region. Use 3.14 for π. 134.88 in.^{2}

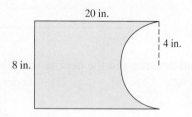

64. The diameter of Emilio's pocket watch is 6 cm. The diameter of his wristwatch is 3 cm.

   a. Find the area of the pocket watch. Use 3.14 for π. 28.26 cm^2

   b. Find the area of the wristwatch. 7.065 cm^2

   c. Is the area of the pocket watch twice the area of the wristwatch? No

65. A sign is constructed from a square with a side length of 2 yd. The square has a semicircle on top with diameter the same length as the side of the square. What is the area of the sign? 5.57 yd^2

Section 8.5

For Exercises 66–69, find the volume. Use 3.14 for π.

66.

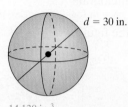

25,000 cm^3

67.

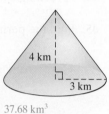

226.08 ft^3

68.

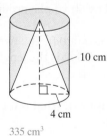

14,130 in.^{3}

69.

37.68 km^3

70. Find the volume of a can of paint if the can is a cylinder with radius 6.5 in. and height 7.5 in. Round to the nearest whole unit. 995 in.^{3}

71. Find the volume of a ball if the diameter of the ball is approximately 6 in. Round to the nearest whole unit. 113 in.^{3}

For Exercises 72 and 73, find the volume of the shaded region. Use 3.14 for π if necessary. Round to the nearest whole unit.

72.

10 cm

4 cm

335 cm^3

73.

15 in.

5 ft

50 in.

54 in.

10 in.

28,500 in.^{3}

74. A microwave oven is a rectangular solid with dimensions 1 ft by 1 ft 9 in. by 1 ft 4 in. Find the volume in cubic feet. $2\frac{1}{3}$ ft^3

Chapter 8 Test

1. Which is a correct representation of the line shown?   d

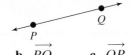

   a. $\overline{PQ}$ **b.** $\overrightarrow{PQ}$ **c.** $\overrightarrow{QP}$ **d.** $\overleftrightarrow{PQ}$

2. Which is a correct representation of the ray pictured?   c

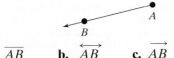

   a. $\overline{AB}$ **b.** $\overleftrightarrow{AB}$ **c.** $\overrightarrow{AB}$ **d.** $\overrightarrow{BA}$

3. What is the complement of a 16° angle?   74°

4. What is the supplement of a 147° angle?   33°

5. Find the missing angle.   103°

For Exercises 6 and 7, refer to the figure.

6. Find the diameter. $\frac{5}{2}$ ft   or $2\frac{1}{2}$ ft

7. Find the circumference. Use $\frac{22}{7}$ for π. $\frac{55}{7}$ ft or $7\frac{6}{7}$ ft

8. A farmer uses a rotating sprinkler to water his crops. If the spray of water extends 150 ft, find the area of one such region. Use 3.14 for π.
   70,650 ft²

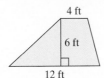

9. Find the area of the shaded region.   48 ft²

10. Simplify.

   a. $\sqrt{4}$ 2   **b.** 4^2 16

For Exercises 11–14, identify the angles as acute, obtuse, right, or straight.

11. $m(\angle A) = 100°$
    Obtuse

12. $m(\angle C) = 73°$
    Acute

13. $m(\angle E) = 90°$
    Right

14. $m(\angle G) = 180°$
    Straight

15. Determine the measure of angles x and y. Assume that line 1 is parallel to line 2.
    $m(\angle x) = 125°, m(\angle y) = 55°$

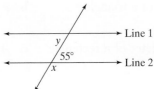

16. Given that the lengths of $\overline{AB}$ and $\overline{BC}$ are equal, what are the measures of $\angle A$ and $\angle C$?   They are each 45°.

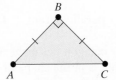

17. From the figure, determine $m(\angle S)$.   49°

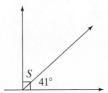

18. What is the sum of all the angles of a triangle?
    180°

19. What is the measure of $\angle A$? $m(\angle A) = 80°$

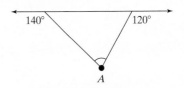

20. A firefighter places a 13-ft ladder against a wall of a burning building. If the bottom of the ladder is 5 ft from the base of the building, how far up the building will the ladder reach?   12 ft

21. José is a landscaping artist and wants to make a walkway through a rectangular garden, as shown. What is the length of the walkway?   100 m

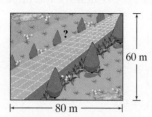

60 m

80 m

For Exercises 22–27, match the formula with the description.

22. Area of a trapezoid   d   **a.** $A = lw$

23. Area of a triangle   c   **b.** $P = 4s$

24. Perimeter of a rectangle   f   **c.** $A = \frac{1}{2}bh$

25. Perimeter of a square   b   **d.** $A = \frac{1}{2}(a + b) \cdot h$

26. Area of a rectangle   a   **e.** $A = bh$

27. Area of a parallelogram   e   **f.** $P = 2l + 2w$

28. A *regular* octagon has eight sides of equal length. A stop sign is in the shape of a regular octagon. Find the perimeter of the stop sign.   96 in.

12 in.

29. Jayne wants to put up a wallpaper border for the perimeter of a 12-ft by 15-ft room. The border comes in 6-yd rolls. How many rolls would be needed?   3 rolls are needed.

30. Find the area of the ceiling fan blade shown in the figure.   The area is 72 in.²

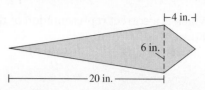

4 in.

6 in.

20 in.

31. Which gives more pizza, a 12-in. by 8-in. rectangular pizza or a 12-in.-diameter round pizza? By how much? Assume that the pizzas have equal thicknesses. Use 3.14 for π.   The area of the rectangular pizza is 96 in.² The area of the round pizza is approximately 113.04 in.² The round pizza is larger by about 17 in.²

32. Find the volume of the child's wading pool shown in the figure. Use 3.14 for π and round the answer to the nearest whole unit.   The volume is about 151 ft³.

3 ft

8 ft

33. Find the volume of the briefcase excluding the handle.   The volume is 1260 in.³

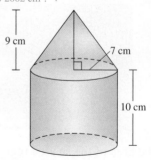

14 in.

18 in.

5 in.

34. Find the volume of the solid. Use $\frac{22}{7}$ for π.

The volume is 2002 cm³.

9 cm

7 cm

10 cm

Chapters 1–8   Cumulative Review Exercises

For Exercises 1–3, divide.

1. $80,535 \div 21$
3835

2. $0 \div 21$
0

3. $21 \div 0$
Undefined

For Exercises 4 and 5, refer to the table.

State	Population
Maine	1,275,000
New Hampshire	1,236,000
Vermont	609,000

4. Find the difference in the populations of Maine and Vermont.   666,000

5. Find the sum of the populations of Maine and New Hampshire.   2,511,000

6. Rank the fractions from least to greatest. $\frac{2}{3}, \frac{5}{6},$ and $\frac{3}{5}$ $\frac{3}{5}, \frac{2}{3}, \frac{5}{6}$

7. There is 14 fl oz of ketchup in a bottle. If $\frac{1}{4}$ is used, how many ounces are left?   There is $10\frac{1}{2}$ fl oz left.

For Exercises 8–10, simplify the expressions.

8. $6 \div \frac{1}{3}$ 18

9. $\frac{1}{3} \div 6$ $\frac{1}{18}$

10. $\frac{2}{7} \div \frac{3}{7} \cdot \frac{9}{5}$ $\frac{6}{5}$

11. Find the LCM of 6, 4, and 10.   60

12. Add: $\frac{1}{6} + \frac{1}{4} + \frac{7}{10}$ $\frac{67}{60}$

13. Subtract: $\frac{13}{6} - \frac{3}{4} - \frac{3}{10}$ $\frac{67}{60}$

14. Write $\frac{132}{8}$ as a mixed number. $16\frac{1}{2}$

15. Write $5\frac{1}{9}$ as an improper fraction. $\frac{46}{9}$

16. The price of a collectible Three Stooges glass is $11.99. How much will a set of four cost?   Four glasses cost $47.96.

17. A sale advertises "Buy 2 get 1 free." If Geraldo buys three shirts and spends $26.98, how much money is he saving?   Geraldo will save the cost of one shirt which is $13.49.

For Exercises 18–20, complete the table.

	Fraction	Decimal
18.	$\frac{3}{8}$	0.375
19.	$\frac{2}{9}$	$0.\overline{2}$
20.	$\frac{1}{50}$	0.02

21. Simplify the ratio: $\frac{2\frac{1}{2}}{3\frac{3}{4}}$ $\frac{2}{3}$

22. Solve the proportion: $\frac{2}{9} = \frac{8.3}{n}$ $n = 37.35$

23. A party consisting of 25 people requires about 7 pizzas. How many pizzas should be purchased if 60 people are expected? (Round to the nearest whole pizza.)   17 pizzas

24. Diane drives her Honda Civic 408 mi on one tank of gas. If a tank contains 12 gal, write the unit rate that gives the miles per gallon.
34 mpg

25. The operating cost for a Boeing 727 aircraft is approximately $8590 for 2.5 hr of operation. Find the unit rate in dollars per hour.
$3436 per hour

26. What is 22% of 240?   52.8

27. 65% of what number is 46.8?   72

28. 65 is what percent of 50?   130%

29. A park bench costs $150 and is marked up to sell for $180. What is the percent markup?
20% markup

30. A jacket normally sells for $85. If it is on sale for $71.40, what is the percent discount?
16% discount

31. Find the perimeter in feet.   10 ft

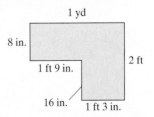

32. A piece of material is 60 in. wide. A sewing pattern requires a width of $4\frac{1}{2}$ ft. Is the material wide enough?   Yes, $4\frac{1}{2}$ ft is 54 in.

33. A recipe requires $\frac{1}{2}$ cup (c) of milk and 6 fl oz of pineapple juice. How many total cups of liquid are required for this recipe?   There is a total of $1\frac{1}{4}$ c or 10 fl oz of liquid.

34. In Canada, just outside of London, Ontario, the speed limit is posted as 100 kilometers per hour (kph). Convert 100 km to miles to find the equivalent speed in miles per hour (1 mi = 1.61 km). Round to the nearest whole unit.   100 kph ≈ 62 mph

35. If the temperature in Paris on a winter day is 5°C, what is the temperature in Fahrenheit? $\left(F = \frac{9}{5}C + 32\right)$ 41°F

 For Exercises 36 and 37, find the perimeter.

36.   13.3 m

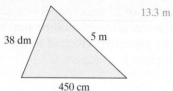

38 dm

5 m

450 cm

37.

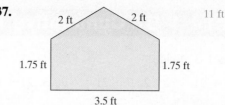

2 ft   2 ft   11 ft

1.75 ft   1.75 ft

3.5 ft

 For Exercises 38 and 39, find the area. Use 3.14 for π if necessary.

38.   1256 cm²

40 cm

39.   3 yd² or 27 ft²

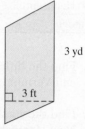

3 yd

3 ft

 40. Find the volume of the hemisphere. Use 3.14 for π and round to the nearest whole unit.   452 in.³

$r = 6$ in.

Introduction to Statistics

Chapter 9

This chapter introduces the study of statistics. This includes interpreting and constructing a variety of statistical graphs such as bar graphs, line graphs, circle graphs, pictographs, and histograms. We also learn how to compute the mean, median, and mode to measure the "center" of a set of values. We conclude with an introduction to probability, which measures the likelihood of an event to occur.

Review Your Skills

In Chapters 1 and 6 we introduced data in the form of tables and graphs. Take a minute to review interpreting data from tables and graphs.

The information in this graph was taken from the financial page of a newspaper. Fill in the blanks.

Average CD Yields	
a. 6-month	0.19%
1-year	0.31%
2.5-year	b. 0.46%
5-year	1.02%

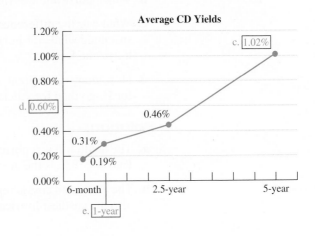

Average CD Yields

Section 9.1 Tables, Bar Graphs, Pictographs, and Line Graphs

1. Introduction to Data and Tables

Statistics is the branch of mathematics that involves collecting, organizing, and analyzing **data** (information). One method to organize data is by using tables. A **table** uses rows and columns to reference information. The individual entries within a table are called **cells**.

Skill Practice

For Exercises 1–4, use the information in Table 9-1.

1. Which hurricane had the highest sustained wind speed at landfall?
2. Which hurricane was the most recent?
3. What was the difference in the death toll for Fran and Alicia?
4. How many times greater was the cost for Katrina than for Fran?

Classroom Example: p. 538, Exercise 6

Example 1 Interpreting Data in a Table

Table 9-1 summarizes the maximum wind speed, number of reported deaths, and estimated cost for recent hurricanes that made landfall in the United States. (*Source:* National Oceanic and Atmospheric Administration)

Table 9-1

Hurricane	Date	Landfall	Maximum Sustained Winds at Landfall (mph)	Number of Reported Deaths	Estimated Cost ($ Billions)
Katrina	2005	Louisiana, Mississippi	125	1836	81.2
Fran	1996	North Carolina, Virginia	115	37	5.8
Andrew	1992	Florida, Louisiana	145	61	35.6
Hugo	1989	South Carolina	130	57	10.8
Alicia	1983	Texas	115	19	5.9

a. Which hurricane caused the greatest number of deaths?

b. Which hurricane was the most costly?

c. What was the difference in the maximum sustained winds for hurricane Andrew and hurricane Katrina?

d. How many times greater was the death toll for Hugo than for Alicia?

Solution:

a. The death toll is reported in the 5th column. The death toll for hurricane Katrina, 1836, is the greatest value.

b. The estimated cost is reported in the 6th column. Hurricane Katrina was also the costliest hurricane at $81.2 billion.

Answers

1. Andrew 2. Katrina
3. 18 deaths 4. 14 times greater

c. The wind speeds are given in the 4th column. The difference in the wind speed for hurricane Andrew and hurricane Katrina is 145 mph − 125 mph = 20 mph.

d. There were 57 deaths from Hugo and 19 from Alicia. The ratio of deaths from Hugo to deaths from Alicia is given by

$$\frac{57}{19} = 3$$

There were 3 times as many deaths from Hugo as from Alicia.

Example 2 **Constructing a Table from Observed Data**

The following data were taken by a student conducting a study for a statistics class. The student observed the type of vehicle and gender of the driver for 18 vehicles in the school parking lot. Complete the table.

Male–car	Female–truck	Male–truck
Male–truck	Female–car	Male–truck
Female–car	Male–truck	Male–motorcycle
Female–car	Male–car	Female–car
Male–motorcycle	Female–car	Female–car
Female–motorcycle	Male–car	Male–car

Driver \ Vehicle	Car	Truck	Motorcycle
Male			
Female			

Solution:

We need to count the number of data values that fall in each of the six cells. One method is to go through the list of data one by one. For each value place a tally mark | in the appropriate cell. For example, the first data value male–car would go in the cell in the first row, first column.

Driver \ Vehicle	Car	Truck	Motorcycle
Male	IIII	IIII	II
Female	IIII I	I	I

To form the completed table, count the number of tally marks in each cell. See Table 9-2.

Table 9-2

Driver \ Vehicle	Car	Truck	Motorcycle
Male	4	4	2
Female	6	1	1

2. Bar Graphs

In Table 9-1, we see that hurricane Andrew had the greatest wind speed of those listed. This can be visualized in a graph. Figure 9-1 shows a bar graph of the wind speed for the hurricanes listed in Table 9-1. Notice that the bar showing wind speed for Andrew is the highest.

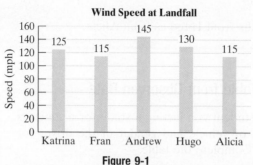

Figure 9-1

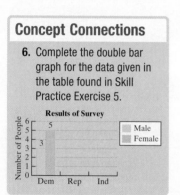
Notice that the **bar graph** compares the data values through the height of each bar. The bars in a bar graph may also be presented horizontally. For example, the double bar graph in Figure 9-2 illustrates the data from Example 2.

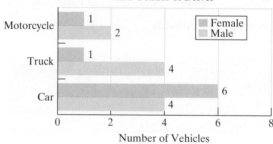

Figure 9-2

When constructing a graph, be sure to include

- A title.
- Labels on the vertical and horizontal axes.
- An appropriate range and scale.

Skill Practice

7. The amount of sodium in milligrams (mg) per $\frac{1}{2}$-c serving for three different brands of cereal is given in the table. Construct a bar graph with horizontal bars.

Brand/Flavor	Amount of Sodium (mg)
Post Grape Nuts	310
Post Cranberry Almond	190
Post Great Grains	200

Classroom Example: p. 539, Exercise 16

Answers

6.
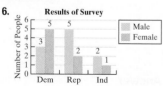

7.
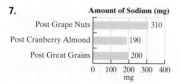

Example 3   Constructing a Bar Graph

The number of fat grams for five different ice cream brands and flavors is given in Table 9-3. Each value is based on a $\frac{1}{2}$-c serving. Construct a bar graph with vertical bars to depict this information.

Table 9-3

Brand/Flavor	Number of Fat Grams (g) per $\frac{1}{2}$-c Serving
Breyers Strawberry	6
Edy's Grand Light Mint Chocolate Chip	4.5
Healthy Choice Chocolate Fudge Brownie	2
Ben and Jerry's Chocolate Chip Cookie Dough	15
Häagen-Dazs Vanilla Swiss Almond	20

Solution:

First draw a horizontal line and label the different food categories. Then draw a vertical line on the left-hand side of the graph as in Figure 9-3. The vertical line represents the number of grams of fat. The vertical scale must extend to at least 20 to accommodate the largest value in the table. In Figure 9-3, the vertical scale ranges from 0 to 24 in multiples (or steps) of 4.

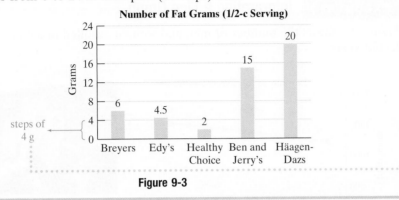

Figure 9-3

> **Avoiding Mistakes**
>
> The scale on the *y*-axis should begin at 0 and increase by equal intervals.

3. Pictographs

Sometimes a bar graph might use an icon or small image to convey a unit of measurement. This type of bar graph is called a **pictograph**.

Example 4   Interpreting a Pictograph

For a recent year, California led the United States in hybrid vehicle sales (83,000 sold). The graph displays the hybrid vehicle sales for five other states for the same year (Figure 9-4). (*Source:* HybridCars.com)

a. What is the value of each car icon in the graph?

b. From the graph, estimate the number of hybrids sold in the state of Washington.

c. For which state were approximately 10,000 hybrid vehicles sold?

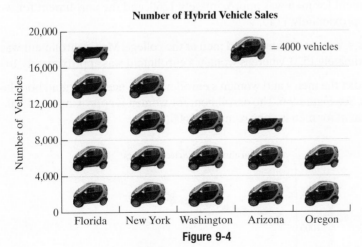

Figure 9-4

Solution:

a. The legend indicates that 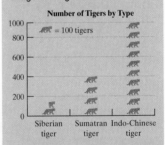 = 4000 vehicles sold.

b. The height of the "bar" for Washington is given by 3 car icons. This represents 3×4000 vehicles. Therefore, 12,000 hybrid vehicles were sold in Washington.

c. The bar containing $2\frac{1}{2}$ icons represents 10,000 hybrids sold. This corresponds to the state of Arizona.

> **Skill Practice**
>
> For Exercises 8–10, refer to the pictograph showing the number of tigers living in the wild.
>
> **Number of Tigers by Type**
>
> [AME] = 100 tigers
>
> **8.** What is the value of each tiger icon?
>
> **9.** From the graph, estimate the number of Sumatran tigers.
>
> **10.** Estimate the number of Siberian tigers.

Classroom Example: p. 541, Exercise 22

Answers

8. 100 tigers   **9.** 400 Sumatran tigers
10. 150 Siberian tigers

4. Line Graphs

Line graphs are often used to track how one variable changes with respect to a second variable. For example, a line graph may illustrate a pattern or trend of a variable over time and allow us to make predictions.

| Example 5 | **Interpreting a Line Graph** |

Figure 9-5 shows the number of men and women enrolled in a college for selected years.

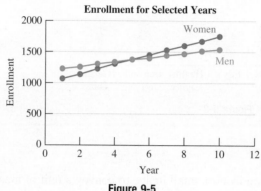

Figure 9-5

a. In year 2, were there more men or more women enrolled in the college?

b. In year 9, were there more men or more women enrolled in the college?

c. Use the trends in the graph to predict the number of men and women enrolled in the college for year 12.

Solution:

a. The blue graph represents men's enrollment and the red graph represents women's enrollment. In year 2, men's enrollment was greater. The enrollment for men was approximately 1260, and the enrollment for women was approximately 1140.

b. In year 9, women outnumbered men at the college. Men's enrollment was approximately 1520, whereas women's enrollment was approximately 1670.

c. To predict the men's and women's enrollment, we need to extend both line graphs. See Figure 9-6. The enrollment for women is approximately 1900. The enrollment for men is approximately 1600.

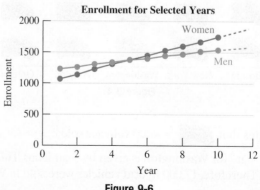

Figure 9-6

Example 6 Constructing a Line Graph

Table 9-4 gives the number of major oil spills in the United States for several years. The amount of oil or fuel is given in millions of gallons.

a. Use the data given in the table to create a line graph.

Table 9-4

Year	Number of Gallons of Oil or Fuel (in millions)
2004	0.34
2005	7.00
2006	2.98
2007	0.50
2008	0.42
2009	0.05

b. In 2010, the British Petroleum (BP) company had a leak from a deep water drilling rig that caused an estimated 206 million gallons of oil to spill into the Gulf of Mexico. If this information were included in the graph, discuss how it would affect the graph. (*Source:* www.washingtonpost.com)

Solution:

a. First draw a horizontal line and label the year. Then draw a vertical line on the left-hand side of the graph, as in Figure 9-7. The vertical line represents the number of gallons of oil or fuel (in millions). In Figure 9-7, the vertical scale ranges from 0 to 8 in steps of 1.0 million. For each year, plot a point corresponding to the number of millions of gallons for that year. Then connect the points.

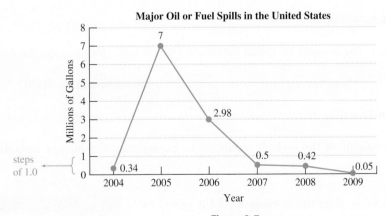

Figure 9-7

In Figure 9-7, we labeled the value at each data point because the exact values are difficult to read from the graph.

b. The vertical scale would have to accommodate the value 206. This would make it difficult to see the vertical values of the other data points because they are much smaller by contrast.

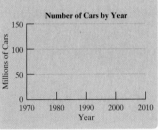
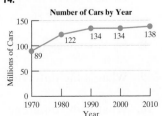

Section 9.1   Practice Exercises

Study Skills Exercise

For additional exercises, see Classroom Activities 9.1A–9.1B in the *Student's Resource Manual* at www.mhhe.com/moh.

> List three benefits of successfully completing this course.

Vocabulary and Key Concepts

1. a. _____Statistics_____ is the branch of mathematics that involves collecting, organizing, and analyzing information.

     b. A _____table_____ uses rows and columns to reference information. The individual entries within a table are called _____cells_____.

     c. A _____pictograph_____ is a type of bar graph that uses an icon or small image to convey a unit of measurement.

Concept 1: Introduction to Data and Tables

For Exercises 2–6, refer to the table. The table represents the Seven Summits (the highest peaks from each continent). **(See Example 1.)**

Mountain	Continent	Height (ft)
Mt. Kilimanjaro	Africa	19,340
Elbrus	Europe	18,510
Aconcagua	South America	22,834
Denali	North America	20,320
Vinson Massif	Antarctica	16,864
Mt. Kosciusko	Australia	7,310
Mt. Everest	Asia	29,035

2. In which continent does the highest mountain lie?   Asia

3. Which mountain among those listed is the lowest? In which continent does it lie?   Mt. Kosciusko; Australia

4. What is the difference between the heights of the highest mountain and the lowest mountain listed? 21,725 ft

5. How much higher is Aconcagua than Denali?   2514 ft

6. What is the difference between the heights of the highest mountain in Europe and the highest mountain in Australia?   11,200 ft

For Exercises 7–12 refer to the table. The table gives the average ages (in years) for U.S. women and men married for the first time for selected years. (*Source:* U.S. Census Bureau)

	Men	Women
1940	24.3	21.5
1960	22.8	20.3
1980	24.7	22.0
2000	26.8	25.1

7. By how much has the average age for women increased between 1940 and 2000?   3.6 yr

8. By how much has the average age for men increased between 1940 and 2000? 2.5 yr

9. What is the difference between the men's and women's average age at first marriage in 1940?   2.8 yr

10. What is the difference between the men's and women's average age at first marriage in 2000?   1.7 yr

11. Which group, men or women, had the consistently higher age at first marriage? Men

12. Which group, men or women, had a greater increase in age between 1940 and 2000? Women

13. The following data were taken from a survey of a third-grade class. The survey denotes the gender of a student and whether the student owned a dog, a cat, or neither. Complete the table. Be sure to label the rows and columns. **(See Example 2.)**

	Dog	**Cat**	**Neither**
Boy	4	1	3
Girl	3	4	5

Boy–dog	Boy–dog	Boy–cat	Boy–neither
Girl–dog	Girl–neither	Boy–dog	Girl–cat
Girl–neither	Girl–neither	Girl–dog	Girl–cat
Boy–dog	Girl–cat	Boy–neither	Girl–dog
Boy–neither	Girl–neither	Girl–cat	Girl–neither

14. In a group of 20 women, 10 were given an experimental drug to lower cholesterol. The other 10 were given a placebo. The letter "D" indicates that the person got the drug, and the letter "P" indicates that the person received the placebo. The values "yes" or "no" indicate whether the person's cholesterol was lowered. Complete the table.

	Yes	**No**
Drug (D)	7	3
Placebo (P)	4	6

D–yes	D–yes	P–no	D–no	P–yes	D–yes	P–no	P–no	D–yes	D–no
P–yes	P–no	D–yes	P–yes	P–yes	D–no	D–yes	P–no	D–yes	P–no

Concept 2: Bar Graphs

15. A study done in a suburban area reported the percentage of Internet users in various age groups. The data are given in the table. **(See Example 3.)**

 a. For which age group is the percentage of Internet users the greatest?
 The 18- to 29-year age group has the greatest percentage of Internet users.
 b. Draw a bar graph with vertical bars to illustrate these data.

Age Group (in years)	**Percentage of Internet Users**
18–29	93%
30–49	51%
50–64	70%
65+	38%

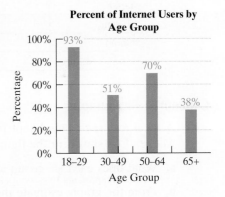

16. The number of new jobs for selected industries are given in the table. (*Source:* Bureau of Labor Statistics)

 a. Which category has the greatest number of new jobs? How many new jobs is this?
 The health care industry has 219,400 new jobs, which is the greatest number.
 b. Draw a bar graph with vertical bars to illustrate these data.

Industry	**Number of New Jobs**
Health care	219,400
Temporary help	212,000
Construction	173,000
Food service	167,600
Retail	78,600

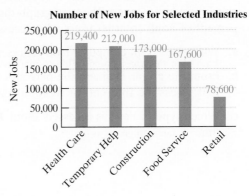

17. The table shows the percentage of broadband subscribers in various countries. Construct a bar graph with horizontal bars. The length of each bar should represent the percentage of broadband subscribers for the corresponding country. (*Source*: International Telecommunications Union, www.itu.int)

Country	Percent of Subscribers
Sweden	37.3%
Canada	29.0%
United Kingdom	28.3%
United States	25.6%
Japan	23.5%

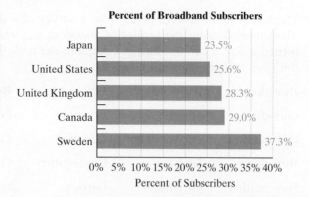

18. The table represents the world's major consumers of primary energy for a recent year. All measurements are in quadrillions of Btu. *Note:* 1 quadrillion = 1,000,000,000,000,000. (*Source:* Energy Information Administration, U.S. Department of Energy) Construct a bar graph using horizontal bars. The length of each bar gives the amount of energy consumed for that country.

Country	Amount of Energy Consumed (Quadrillions of Btu)
Germany	14
Japan	22
Russia	28
China	37
United States	99

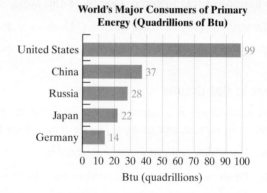

Concept 3: Pictographs

19. A local ice cream stand kept track of its ice cream sales for one weekend, as shown in the figure. (See Example 4.)

   a. What does each ice cream icon represent?
   One icon represents 100 servings sold.
   b. From the graph, estimate the number of servings of ice cream sold on Saturday.   About 450 servings

   c. Which day had approximately 275 servings of ice cream sold?
   Sunday

20. Adults access the Internet to see weather updates and check on current news. The pictograph displays the percent of adult Internet users who access these topics.

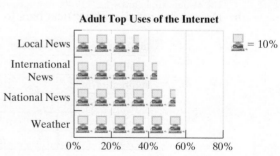

   a. What does each computer icon represent?
   Each computer icon represents 10% of adults who access the Internet.
   b. From the graph, estimate the percent of adult users that access the Internet for weather.   About 60%

   c. Which type of news is accessed about 45% of the time?   International news

 Writing    Translating Expression   Geometry   Scientific Calculator   Video   NS E

21. The figure displays the annual sales of books for three major companies prior to the closing of Borders.

   a. Estimate the sales for the company with the greatest annual sales.
   Barnes & Noble/B. Dalton has approximately $4.5 billion in book sales.

   b. Estimate the total sales for all three companies. There is approximately $11.5 billion in book sales.

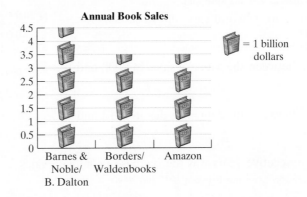

22. Recently the largest populations of senior citizens were in California, Florida, New York, Texas, and Pennsylvania, as shown in the figure. (*Source:* U.S. Bureau of the Census)

   a. Estimate the number of senior citizens living in Texas. About 2.25 million

   b. How many more senior citizens are living in California than in Pennsylvania? About 2 million

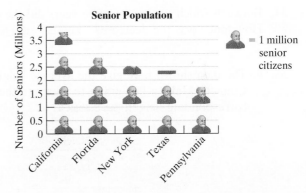

Concept 4: Line Graphs

For Exercises 23–28, use the line graph provided. The graph shows the trend depicted by the percent of men and women over the age of 65 in the labor force in a selected metropolitan area. (*Source:* Bureau of the Census)
(See Example 5.)

23. What was the difference in the percent of men and the percent of women over 65 in the labor force in the year 1920? 48.4%

24. What was the difference in the percent of men and the percent of women over 65 in the labor force in the year 2000? 8.6%

25. What was the overall trend in the percent of women over 65 in the labor force for the years shown in the graph?
The trend for women over 65 in the labor force shows a slight increase.

26. What was the overall trend in the percent of men over 65 in the labor force for the years shown in the graph?
The trend for men over 65 in the labor force shows a significant decrease until 1980 and then levels off.

27. Use the graph to predict the number of men over 65 in the labor force in the year 2020. Answers will vary.
For example: 18%

28. Use the graph to predict the number of women over 65 in the labor force in the year 2020. Answers will vary. For example: 10.5%

For Exercises 29–34, refer to the graph representing the number of hybrid cars sold in January in the United States for the given years. (*Source*: U.S. Energy Information Administration, www.eia.gov)

29. In which year were the most hybrid cars sold in January? How many were sold?
The most cars were sold in 2008. 22,400 cars were sold.

30. In which year were the fewest hybrid cars sold in January? How many were sold?
The fewest cars were sold in 2009. 15,400 cars were sold.

31. What is the difference between the January sales in 2008 and 2007?
4800 cars

32. What is the difference between the January sales in 2012 and 2011?
2300 cars

33. Between which two consecutive years was the increase in January sales the greatest?   The greatest increase was between 2007 and 2008.

34. Between which two consecutive years did the January sales decrease? How much was the decrease?   Sales decreased between 2008 and 2009 by 7000 cars.

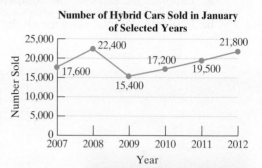

Number of Hybrid Cars Sold in January of Selected Years

35. a. The data shown here give the average height for girls based on age. (*Source:* National Parenting Council) Make a line graph to illustrate these data. For each age value, plot a point for the corresponding height. **(See Example 6.)**

Age	Height (in.)
2	35
3	38.5
4	41.5
5	44
6	46
7	48
8	50.5
9	53

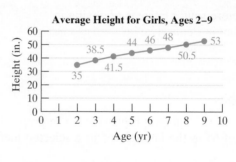

Average Height for Girls, Ages 2–9

b. Predict the height of a 10-year-old girl based on the graph in part (a).   Approximately 56 in.

36. a. The data shown here give the average height for boys based on age. (*Source:* National Parenting Council) Make a line graph to illustrate these data. For each age value, plot a point for the corresponding height.

Age	Height (in.)
2	36
3	39
4	42
5	44
6	46.75
7	49
8	51
9	53.5

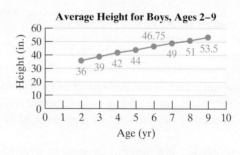

Average Height for Boys, Ages 2–9

b. Predict the height of a 10-year-old boy based on the graph in part (a).   Approximately 56 in.

 Writing   Translating Expression   Geometry   Scientific Calculator   Video   NS E

Expanding Your Skills

All packaged food items have to display nutritional facts so that the consumer can make informed choices. For Exercises 37–40, refer to the nutritional chart for Breyers French Vanilla ice cream.

Nutrition Facts
Serving Size $\frac{1}{2}$ cup (68 g)
Servings per Container 14

Amount per Serving	
Calories 150	Calories from Fat 80

		% Daily Value
Total Fat	8 g	13%
Saturated fat	5 g	25%
Cholesterol	50 mg	17%
Sodium	45 mg	2%
Total Carbohydrate		
Dietary fiber	0 g	
Sugars	15 g	
Protein	3 g	

37. How many servings are there per container? How much total fat is in one container of this ice cream? There are 14 servings per container, which means that there is 8 g × 14 = 112 g of fat in one container.

38. How much total sodium is in one container of this ice cream? There is 630 mg of sodium in one container.

39. If 8 g of fat is 13% of the daily value, what is the daily value of fat? Round to 1 decimal place. The daily value of fat is approximately 61.5 g.

40. If 50 mg of cholesterol is 17% of the daily value, what is the daily value of cholesterol? Round to the nearest whole unit. The daily value of cholesterol is about 294 mg.

Frequency Distributions and Histograms

Section 9.2

1. Frequency Distributions

Concepts

1. Frequency Distributions
2. Histograms

The medical community agrees that sodium intake is one of the variables contributing to high blood pressure. The amount of sodium (in mg) for a 1-cup serving of processed breakfast cereals is given for 40 cereals.

146	143	316	269	243	249	213	160	269	288
299	234	267	275	263	171	202	120	146	139
241	210	150	197	359	238	249	200	178	219
204	203	253	165	207	362	250	223	255	224

Suppose we wanted to organize this information. One way is to create a frequency distribution. A **frequency distribution** is a table displaying the number of data values that fall within specified categories. When the categories represent a range of numerical values, we call the categories **class intervals**. This is demonstrated in Example 1.

Example 1   Creating a Frequency Distribution

Complete the table to form a frequency distribution for the number of milligrams of sodium for forty 1-cup servings of breakfast cereals.

Class Intervals Amount of Sodium (mg)	Tally	Frequency Number of Cereals
100–149		
150–199		
200–249		
250–299		
300–349		
350–399		

Answers

Solution:

The classes in the first column represent different amounts of sodium. Go through the list of data and use tally marks to track the number of cereals that fall into each class. Tally marks are shown in red for the first six values in the first row: 146, 143, 316, 269, 243, and 249.

Table 9-5

Class Intervals Amount of Sodium (mg)	Tally	Frequency Number of Cereals
100–149	IIII	5
150–199	IIII I	6
200–249	IIII IIII IIII I	16
250–299	IIII IIII	10
300–349	I	1
350–399	II	2

The frequency is a count of the tally marks within each class. See Table 9-5.

Example 2 Interpreting a Frequency Distribution

Consider the frequency distribution in Table 9-5.

a. Which class has the most values?

b. How many values are represented in the table?

c. What percent of cereals have 300 mg or more of sodium per 1-cup serving?

Solution:

a. The 200- to 249-mg class has 16 data values. This is the greatest frequency.

b. The number of data values is given by the sum of the frequencies.

$$\text{Total number of values} = 5 + 6 + 16 + 10 + 1 + 2$$
$$= 40$$

c. One cereal belongs to the 300- to 349-mg class, and two cereals belong to the 350- to 399-mg class. So there are 3 cereals among the group of 40 studied that have 300 mg or more of sodium. The percentage is given by

$$\frac{3}{40} \approx 0.075 \quad \text{or} \quad 7.5\%$$

Therefore, approximately 7.5% of processed breakfast cereals have 300 mg or more of sodium per cup.

When creating a frequency distribution, keep these important guidelines in mind.

- The classes should represent intervals of equal length. For instance in Example 1, we would not want one class to represent a 50-mg interval, and another to represent a 100-mg interval.
- The classes should not overlap. That is, a data value should belong to one and only one class.
- In general, we usually create a frequency distribution with between 5 and 15 classes, inclusive.

2. Histograms

A **histogram** is a special bar graph that illustrates data given in a frequency distribution. The class intervals are given on the horizontal scale. The height of each bar in a histogram represents the frequency for each class. Furthermore, the bars of a histogram touch with no space between the bars.

Example 3 **Constructing a Histogram**

Construct a histogram for the frequency distribution given in Example 1.

Solution:

To create a histogram, list the classes (amount of sodium) on the horizontal scale. Then on the vertical scale, we represent the frequency (Figure 9-8).

Class Intervals Amount of Sodium (mg)	Frequency Number of Cereals
100–149	5
150–199	6
200–249	16
250–299	10
300–349	1
350–399	2

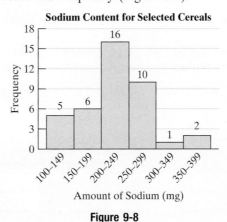

Sodium Content for Selected Cereals

Figure 9-8

Answer

5.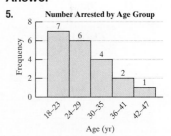

Number Arrested by Age Group

Section 9.2 Practice Exercises

For additional exercises, see Classroom Activity 9.2A in the *Student's Resource Manual* at www.mhhe.com/moh.

Study Skills Exercise

It is always helpful to read the material in a section and make notes before it is presented in class. Writing notes ahead of time will free you to listen more in class and to pay special attention to the concepts that need clarification. Refer to your class syllabus and list the next two sections that will be covered in class and a time that you can read them beforehand.

Vocabulary and Key Concepts

1. a. A _____frequency_____ distribution is a table displaying the number of values that fall within categories called class intervals.

   b. A _____histogram_____ is a bar graph that illustrates data given in a frequency distribution.

Concept 1: Frequency Distributions

2. Determine if the statement is true or false.

   a. A frequency distribution usually contains between 5 and 15 classes, inclusive.   True

   b. The class intervals in a frequency distribution should overlap.   False

   c. The class intervals in a frequency distribution should be equal in length.   True

3. From the frequency distribution, determine the total number of data. There are 72 data.

Class Intervals	Frequency
1–4	14
5–8	18
9–12	24
13–16	10
17–20	6

4. From the frequency distribution, determine the total number of data. There are 185 data.

Class Intervals	Frequency
1–50	29
51–100	12
101–150	6
151–200	22
201–250	56
251–300	60

5. For the table in Exercise 3, which category contains the most data? 9–12

6. For the table in Exercise 4, which category contains the most data? 251–300

7. The retirement age (in years) for 20 college professors is given. Complete the frequency distribution. **(See Examples 1 and 2.)**

67	56	68	70	60	65	73	72	56	65
71	66	72	69	65	65	63	65	68	70

Class Intervals (Age in Years)	Tally	Frequency (Number of Professors)
56–58	‖	2
59–61	∣	1
62–64	∣	1
65–67	⣼ ‖	7
68–70	⣼	5
71–73	‖‖	4

a. Which class has the most values? The class of 65–67 has the most values.

b. How many data values are represented in the table? There are 20 values represented in the table.

c. What percent of the professors retire when they are 68 to 70 years old? Of the professors, 25% retire when they are 68 to 70 years old.

8. The number of miles run in one day by 16 selected runners is given. Complete the frequency distribution.

2	4	7	3	8	4	5	7
4	6	4	3	4	2	4	10

Class Intervals (Number of Miles)	Tally	Frequency (Number of Runners)
1–2	‖	2
3–4	⣼ ‖‖	8
5–6	‖	2
7–8	‖‖	3
9–10	∣	1

a. Which class has the most values? The 3–4 class has the highest frequency.

b. How many data values are represented in the table? There are 16 data values represented in the table.

c. What percent of the runners run 3 to 4 mi/day? Of the runners, 50% ran 3 to 4 mi that day.

9. The number of gallons of gas purchased by 16 customers at a certain gas station is given. Complete the frequency distribution.

12.7	13.1	9.8	12.0	10.4	9.8	14.2	8.6
19.2	8.1	14.0	15.4	12.8	18.2	15.1	13.0

Class Intervals (Amount in Gal)	Tally	Frequency (Number of Customers)
8.0–9.9	IIII	4
10.0–11.9	I	1
12.0–13.9	IIHI	5
14.0–15.9	IIII	4
16.0–17.9		0
18.0–19.9	II	2

a. Which class has the most values?
   The 12.0–13.9 class has the highest frequency.
b. How many data values are represented in the table?
   There are 16 data values represented in the table.
c. What percent of the customers purchased 18 to 19.9 gal of gas?
   Of the customers, 12.5% purchased 18 to 19.9 gal of gas.

10. The hourly salaries (in dollars) for 15 student employees at Miami-Dade College are given. Complete the frequency distribution.

7.95	8.00	9.20	8.15	7.85
7.95	8.50	9.00	8.25	8.95
9.25	9.50	10.05	10.00	8.30

Class Intervals (Hourly Wage, $)	Tally	Frequency (Number of Employees)
7.50–7.99	III	3
8.00–8.49	IIII	4
8.50–8.99	II	2
9.00–9.49	III	3
9.50–9.99	I	1
10.00–10.49	II	2

a. Which class has the most values?
   The 8.00–8.49 class has the highest frequency.
b. How many data values are represented in the table?
   There are 15 data values represented in the table.
c. What percent of the employees earn $9.00 or more?
   Of the employees, 40% earn $9.00 or more.

11. Explain what is wrong with the following class intervals.   The class widths are not the same.

Class	Tally	Frequency
0–4		
5–10		
11–17		
18–25		
26–34		

12. Explain what is wrong with the following class intervals.   The class widths are not the same.

Class	Tally	Frequency
1–6		
7–11		
12–17		
18–23		
24–28		

13. Explain what is wrong with the following class intervals.   There are too few classes.

Class	Tally	Frequency
1–20		
21–40		

14. Explain what is wrong with the following class intervals.   There are too few classes.

Class	Tally	Frequency
1–33		
34–66		
67–99		

Writing   Translating Expression   Geometry   Scientific Calculator   Video   NS E

15. Explain what is wrong with the following class intervals.

Class	Tally	Frequency
10–12		
12–14		
14–16		
16–18		
18–20		

The class intervals overlap. For example, it is unclear whether the data value 12 should be placed in the first class or the second class.

16. Explain what is wrong with the following class intervals.

Class	Tally	Frequency
1–5		
5–10		
10–15		
15–20		
20–25		
25–30		

The class intervals overlap. For example, it is unclear whether the data value 5 should be placed in the first class or the second class.

17. The heights of 20 students at Valencia College are given. Complete the frequency distribution.

70	71	73	62	65	70	69	70
64	66	73	63	68	67	69	72
64	66	67	69				

Class Interval (Height, in.)	Frequency (Number of Students)
62–63	2
64–65	3
66–67	4
68–69	4
70–71	4
72–73	3

18. The amount withdrawn in dollars from a certain ATM is given for 20 customers. Construct a frequency distribution.

40	50	200	200	100	120	200
50	100	60	100	100	30	40
100	100	50	200	150	200	

Class Interval (Amount, $)	Frequency (Number of Customers)
0–49	3
50–99	4
100–149	7
150–199	1
200–249	5

Concept 2: Histograms

19. Construct a histogram for the frequency table in Exercise 17. **(See Example 3.)**

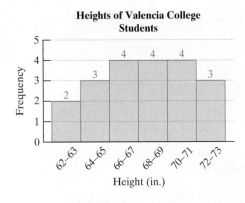

Heights of Valencia College Students

20. Construct a histogram for the frequency table in Exercise 18.

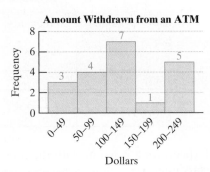

Amount Withdrawn from an ATM

21. Construct a histogram, using the given data. Each number represents the number of Calories in a 100-g serving for selected fruits.

59	65	48	49	105	47	92	43
52	59	56	49	35	55	72	30
67	44	32	32	61	29	30	

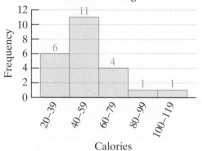

Number of Calories in 100 g of Selected Fruits

22. The list of data gives the number of children of the Presidents of the United States (in no particular order). Construct a histogram.

0	4	3	3	2	2	2	10	0	5	2
4	5	7	4	3	6	4	0	7	5	6
1	4	3	0	4	3	2	6	4	6	8
3	2	1	0	2	6	0	2	2	8	2

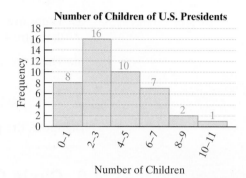

Number of Children of U.S. Presidents

Circle Graphs

Section 9.3

1. Interpreting Circle Graphs

Thus far we have used bar graphs, line graphs, and histograms to visualize data. A **circle graph** (or pie graph) is another type of graph used to show how a whole amount is divided into parts. Each part of the circle, called a **sector**, is like a slice of pie. The size of each piece relates to the fraction of the whole it represents.

Concepts

1. Interpreting Circle Graphs
2. Circle Graphs and Percents
3. Constructing Circle Graphs

Example 1 Interpreting a Circle Graph

The grade distribution for a math test is shown in the circle graph (Figure 9-9).

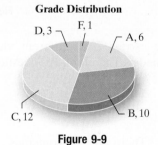

Grade Distribution

D, 3 — F, 1 — A, 6
C, 12 — B, 10

Figure 9-9

a. How many total grades are represented?

b. How many grades are B's?

c. How many times more C's are there than D's?

d. What percent of the grades were A's?

Skill Practice

A used car dealership sells cars and trucks. Use the circle graph to answer Exercises 1–4.

Vehicle Distribution by Make

Toyota, 13 — Ford, 10 — Chevy, 6 — Dodge, 3 — Honda, 18

1. How many total vehicles are represented?

2. How many are Toyotas?

3. How many times more Hondas are there than Dodges?

4. What percent are Fords?

Classroom Examples: p. 553, Exercises 4 and 8

Answers

1. 50   **2.** 13
3. There are 6 times more Hondas than Dodges.
4. 20% are Fords.

Solution:

a. The total number of grades is equal to the sum of the number of grades from each category.

$$\text{Total number of grades} = 6 + 10 + 12 + 3 + 1$$
$$= 32$$

b. The number of B's is represented by the red portion of the graph. There are 10 B's.

c. There are 12 C's and 3 D's. The ratio of C's to D's is $\frac{12}{3} = 4$. Therefore, there are 4 times as many C's as D's.

d. There are 6 A's. The percent of A's is given by

$$\frac{6}{32} = 0.1875$$

Therefore, the percent of A's is 18.75%.

2. Circle Graphs and Percents

Sometimes circle graphs show data in percent form. This is illustrated in Example 2.

Classroom Example: p. 554, Exercise 18

Skill Practice

For Exercises 5 and 6, refer to Figure 9-10.

5. How many videos are action?

6. How many videos are general interest or children's?

Instructor Note: Remind students that in order for the graph to make sense, each video can be placed into exactly one category.

| Example 2 | Calculating Amounts by Using a Circle Graph |

A certain video rental store carries 2000 different videos. It groups its video collection by the categories shown in the graph (Figure 9-10).

a. How many videos are comedy?

b. How many videos are action or horror?

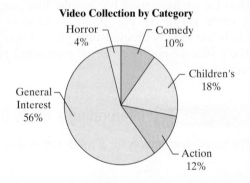

Video Collection by Category

Horror 4% — Comedy 10% — Children's 18% — Action 12% — General Interest 56%

Figure 9-10

Solution:

a. First note that the store carries 2000 different videos. From the graph we know that 10% are comedies. Therefore, this question can be interpreted as

What is 10% of 2000?
$$x = (0.10) \cdot (2000)$$
$$= 200 \qquad \text{There are 200 comedies.}$$

b. From the graph we know that 12% of the videos are action and 4% are horror. This accounts for 16% of the total video collection. Therefore, this question asks

What is 16% of 2000?
$$x = (0.16) \cdot (2000)$$
$$= 320 \qquad \text{There are 320 videos that are action or horror.}$$

Answers

5. There are 240 action videos.
6. There are 1480 general interest or children's videos.

3. Constructing Circle Graphs

Recall that a full circle is a 360° arc. To draw a circle graph, we must compute the number of degrees of arc for each sector. In Example 2, 10% of the videos are comedies. To draw the sector for this category, we must determine 10% of 360°.

$$10\% \text{ of } 360° = 0.10(360°) = 36°$$

The sector representing comedies should be drawn with a 36° angle. To do this, we can use a protractor (Figure 9-11).

To draw a sector with a 36° arc, first draw a circle. Place the hole in the protractor over the center of the circle. Using the inner scale on the protractor, place a tick mark at 0° and at 36°. Use a straightedge to draw two line segments from the center of the circle to each tick mark. See Figure 9-11.

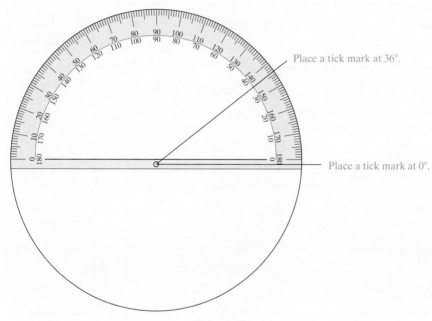

Place a tick mark at 36°.

Place a tick mark at 0°.

Figure 9-11

In Example 3, we use this technique to construct a circle graph.

| Example 3 | **Constructing a Circle Graph** |

A teacher earns a monthly salary of $3600 after taxes. Her monthly budget is broken down in Table 9-6.

Table 9-6

Budget Item	Monthly Value ($)
Rent	1260
Utilities	315
Car expenses	765
Groceries	540
Savings	450
Other	270

Construct a circle graph illustrating the information in this table. Label each sector of the graph with the percent that it represents.

Answers
7. c
8. a

Skill Practice

9. Voters in Oregon were asked to identify the political party to which they belonged. Construct a circle graph. Label each sector of the graph with the percent that it represents.

Political Affiliation	Number
Democrat	900
Republican	720
Libertarian	36
Green Party	144

Classroom Example: p. 555, Exercise 38

Solution:

This problem calls for two types of calculations: (1) For each budget item, we must compute the percent of the whole that it represents. (2) We must determine the number of degrees for each category. We can use a table to help organize our calculations.

Budget Item	Monthly Value ($)	Percent	Number of Degrees
Rent	1260	$=\dfrac{1260}{3600} = 0.35$ or 35%	35% of 360° $= 0.35(360°)$ $= 126°$
Utilities	315	$=\dfrac{315}{3600} = 0.0875$ or 8.75%	8.75% of 360° $= 0.0875(360°)$ $= 31.5°$
Car	765	$=\dfrac{765}{3600} = 0.2125$ or 21.25%	21.25% of 360° $= 0.2125(360°)$ $= 76.5°$
Groceries	540	$=\dfrac{540}{3600} = 0.15$ or 15%	15% of 360° $= 0.15(360°)$ $= 54°$
Savings	450	$=\dfrac{450}{3600} = 0.125$ or 12.5%	12.5% of 360° $= 0.125(360°)$ $= 45°$
Other	270	$=\dfrac{270}{3600} = 0.075$ or 7.5%	7.5% of 360° $= 0.075(360°)$ $= 27°$

Now construct the circle graph. Use the degree measures found in the table for each sector. Label the graph with the percent for each sector (Figure 9-12).

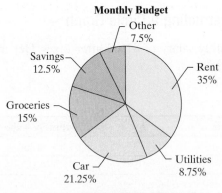

Monthly Budget

Figure 9-12

Answer

9. **Distribution by Political Party**

Republican 40%
Democrat 50%
Libertarian 2%
Green Party 8%

Section 9.3   Practice Exercises

Study Skills Exercise

For additional exercises, see Classroom Activities 9.3A–9.3B in the *Student's Resource Manual* at www.mhhe.com/moh.

Some instructors are available to answer questions during evening hours via e-mail. Find out if you can contact your instructor by e-mail during evening hours or weekends, and write down the e-mail address.

Vocabulary and Key Concepts

1. A _____circle_____ graph or pie graph illustrates how a whole amount is divided into parts. The individual parts (or wedges) of the graph are called _____sectors_____.

Concept 1: Interpreting Circle Graphs

2. **a.** If a sector in a circle graph is 90°, what percent of the graph is represented by this sector?   25%

 b. If a sector in a circle graph is 180°, what percent of the graph is represented by this sector?   50%

For Exercises 3–10, refer to the graph. The graph represents the number of traffic fatalities by age group in the United States. (*Source:* U.S. Bureau of the Census) **(See Example 1.)**

Number of U.S. Traffic Fatalities

65 yr and older (9600)
55–64 yr (6400)
45–54 yr (9600)
35–44 yr (10,880)
25–34 yr (11,520)
15–24 yr (16,000)

3. What is the total number of traffic fatalities?   64,000

4. Which of the age groups has the most fatalities?   15–24 years

5. How many more people died in the 25–34 age group than in the 35–44 age group?   640

6. How many more people died in the 45–54 age group than in the 55–64 age group?   3200

7. What percent of the deaths were from the 15–24 age group?   25%

8. What percent of the deaths were from the 65 and older age group?   15%

9. How many times more deaths were from the 15–24 age group than the 55–64 age group?   2.5 times

10. How many times more deaths were from the 25–34 age group than from the 65 and older age group?   1.2 times

For Exercises 11–16, refer to the figure representing the average number of viewers for five daytime dramas for one week.

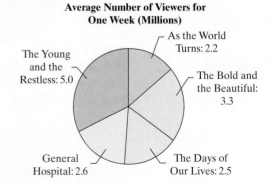

Average Number of Viewers for One Week (Millions)

The Young and the Restless: 5.0
As the World Turns: 2.2
The Bold and the Beautiful: 3.3
General Hospital: 2.6
The Days of Our Lives: 2.5

11. How many viewers are represented?
 There were 15.6 million viewers represented.

12. How many viewers does the most popular daytime drama have?
 The most popular drama has 5 million (5,000,000) viewers.

13. How many times more viewers does *The Young and the Restless* have than *The Days of Our Lives*?
 The Young and the Restless has 2 times as many viewers.

14. How many times more viewers does *The Bold and the Beautiful* have than *As the World Turns*?
 The Bold and the Beautiful has 1.5 times as many viewers.

15. What percent of the viewers watch *General Hospital*? (Round to the nearest percent.)
 Of the viewers, approximately 17% watch *General Hospital*.

16. What percent of the viewers watch *The Young and the Restless*? (Round to the nearest percent.)   Of the viewers, approximately 32% watch *The Young and the Restless*.

  Writing   Translating Expression   Geometry   Scientific Calculator   Video   NS&E

Concept 2: Circle Graphs and Percents

For Exercises 17–20, use the graph representing the type of music CDs found in a store containing approximately 8000 CDs. **(See Example 2.)**

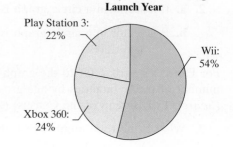

CD Collection by Category

17. How many CDs are musica Latina?   There are 960 Latina CDs.

18. How many CDs are rap?   There are 1600 rap CDs.

19. How many CDs are jazz or classical?
There are 640 CDs that are classical or jazz.

20. How many CDs are *not* Pop/R&B?
There are 4000 CDs that are not Pop/R&B.

For Exercises 21–24, use the graph representing the number of the three most popular game systems sold during their launch year. Twenty-five million game systems were sold.

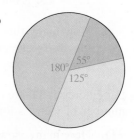

Number of Systems Sold During Launch Year

21. How many Wii systems were sold during the launch year?
13.5 million Wii systems were sold.

22. How many Xbox 360 systems were sold during the launch year?
6 million Xbox 360 systems were sold.

23. How many Play Station 3 systems were sold during the launch year?
5.5 million Play Station 3 systems were sold.

24. How many systems were sold that were not Wii systems?
There were 11.5 million systems sold that were not Wii.

Concept 3: Constructing Circle Graphs

For Exercises 25–32, use a protractor to construct an angle of the given measure.

25. 20°

26. 70°

27. 125°

28. 270°

29. 195°

30. 5°

31. 300°

32. 90°

33. Draw a circle and divide it into sectors of 30°, 60°, 100°, and 170°.

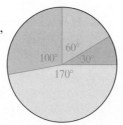

34. Draw a circle and divide it into sectors of 125°, 180°, and 55°.

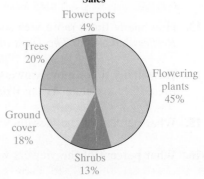

35. The Sunshine Nursery sells flowering plants, shrubs, ground cover, trees, and assorted flower pots. Construct a pie graph to show the distribution of the types of purchases.

Sunshine Nursery Distribution of Sales

Types of Purchases	Percent of Distribution
Flowering plants	45%
Shrubs	13%
Ground cover	18%
Trees	20%
Flower pots	4%

 Writing   Translating Expression   Geometry   Scientific Calculator   Video   NS&E

36. The party affiliation of registered Latino voters for a recent year is as follows:

45% Democrat   20% Republican

13% Other   22% Independent

Construct a circle graph from this information.

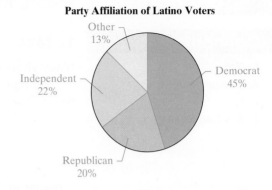

Party Affiliation of Latino Voters

37. The table provided gives the expenses for one semester at college. **(See Example 3.)**

a. Complete the table.

b. Construct a circle graph to display the college expenses. Label the graph with percents.

	Expenses	Percent	Number of Degrees
Tuition	$9000	75%	270°
Books	600	5%	18°
Housing	2400	20%	72°

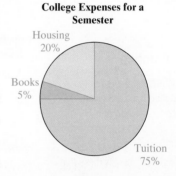

College Expenses for a Semester

38. The table provided gives the number of establishments of the three largest pizza chains.

a. Complete the table.

b. Construct a circle graph. Label the graph with percents.

	Number of Stores	Percent	Number of Degrees
Pizza Hut	8100	45%	162°
Domino's	7200	40%	144°
Papa Johns	2700	15%	54°

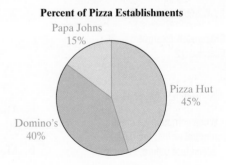

Percent of Pizza Establishments

Section 9.4   Mean, Median, and Mode

Concepts

1. Mean
2. Median
3. Mode
4. Weighted Mean

Instructor Note: Ask students to compute the mean of their quiz scores or test scores for this class, as a warm-up exercise.

1. Mean

When given a list of numerical data, it is often desirable to obtain a single number that represents the central value of the data. In this section, we introduce three such values called the mean, median, and mode. The first calculation we present is the mean (or average) of a list of data values.

Mean

The **mean** (or average) of a set of numbers is the sum of the values divided by the number of values. We can write this as a formula.

$$\text{Mean} = \frac{\text{sum of the values}}{\text{number of values}}$$

Skill Practice

Housing prices for five homes in one neighborhood are given.

$108,000 $149,000
$164,000 $118,000
$144,000

1. Find the mean of these five houses.
2. Suppose a new home is built in the neighborhood for $1.3 million ($1,300,000). Find the mean price of all six homes.

Classroom Example: p. 561, Exercise 12

Avoiding Mistakes

When computing a mean remember that the data are added first before dividing.

Example 1 Finding the Mean of a Data Set

A small business employs five workers. Their yearly salaries are

$42,000 $36,000 $45,000 $35,000 $38,000

a. Find the mean yearly salary for the five employees.

b. Suppose the owner of the business makes $218,000 per year. Find the mean salary for all six individuals (that is, include the owner's salary).

Solution:

a. Mean salary of five employees

$$= \frac{42{,}000 + 36{,}000 + 45{,}000 + 35{,}000 + 38{,}000}{5}$$

$$= \frac{196{,}000}{5} \quad \text{Add the data values.}$$

$$= 39{,}200 \quad \text{Divide.}$$

The mean salary for employees is $39,200.

b. Mean of all six individuals

$$= \frac{42{,}000 + 36{,}000 + 45{,}000 + 35{,}000 + 38{,}000 + 218{,}000}{6}$$

$$= \frac{414{,}000}{6}$$

$$= 69{,}000$$

The mean salary with the owner's salary included is $69,000.

Answers

1. $136,600   **2.** $330,500

2. Median

In Example 1, you may have noticed that the mean salary was greatly affected by the unusually high value of $218,000. For this reason, you may want to use a different measure of "center" called the median. The **median** is the "middle" number in an ordered list of numbers.

Finding the Median

To compute the median of a list of numbers, first arrange the numbers in order from least to greatest.

- If the number of data values in the list is *odd*, then the median is the middle number in the list.
- If the number of data values is *even*, there is no single middle number. Therefore, the median is the mean (average) of the two middle numbers in the list.

Example 2 **Finding the Median of a Data Set**

Consider the salaries of the five workers from Example 1.

$42,000 $36,000 $45,000 $35,000 $38,000

a. Find the median salary for the five workers.

b. Find the median salary including the owner's salary of $218,000.

Solution:

a. 35,000   36,000   38,000   42,000   45,000   Arrange the data in order.

Because there are five data values (an *odd* number), the median is the middle number.

The median is $38,000.

b. Now consider the scores of all six individuals (including the owner). Arrange the data in order.

35,000   36,000   38,000   42,000   45,000   218,000

There are six data values (an *even* number). The median is the average of the two middle numbers.

$$\frac{38,000 + 42,000}{2}$$

$$= \frac{80,000}{2}$$ Add the two middle numbers.

$$= 40,000$$ Divide.

The median of all six salaries is $40,000.

Skill Practice

3. Find the median of the five housing prices given in Skill Practice Exercise 1.

$108,000 $149,000
$164,000 $118,000
$144,000

4. Find the median of the six housing prices given in Skill Practice Exercise 2.

$108,000 $149,000
$164,000 $118,000
$144,000 $1,300,000

Classroom Examples: p. 562, Exercises 18 and 20

Answers

3. $144,000   **4.** $146,500

In Examples 1 and 2, the mean of all six salaries is $69,000, whereas the median is $40,000. These examples show that the median is a better representation for a central value when the data list has an unusually high (or low) value.

5. The monthly rainfall for Houston, Texas, is given in the table. Find the median rainfall amount.

Month	Rainfall (in.)
Jan.	4.5
Feb.	3.0
March	3.2
April	3.5
May	5.1
June	6.8
July	4.3
Aug.	4.5
Sept.	5.6
Oct.	5.3
Nov.	4.5
Dec.	3.8

Classroom Example: p. 562, Exercise 26

Example 3	**Finding the Median of a Data Set**

For a recent year, the student-to-teacher ratio for elementary schools is shown in Table 9-7. Find the median student-to-teacher ratio.

Table 9-7

State	Student-to-Teacher Ratio
California	20.6
Illinois	16.1
Indiana	16.1
Maine	12.5
Mississippi	16.1
New Hampshire	14.5
North Dakota	13.4
Rhode Island	14.8
Utah	21.9
Wisconsin	14.1

Source: National Center for Education Statistics

Solution:

First arrange the numbers in order from least to greatest:

ME	ND	WI	NH	RI	IL	IN	MS	CA	UT
12.5	13.4	14.1	14.5	14.8	16.1	16.1	16.1	20.6	21.9

$$\text{Median} = \frac{14.8 + 16.1}{2} = 15.45$$

There are 10 data values (an *even* number). Therefore, the median is the average of the middle two numbers. The median student-to-teacher ratio is 15.45. This indicates that there are approximately 15 or 16 students per teacher.

TIP: The median may not be one of the original data values. This was true in Example 3.

3. Mode

A third representative value for a list of data is called the mode.

Mode

The **mode** of a set of data is the value or values that occur most often.

- If two values occur most often we say the data are **bimodal**.
- If more than two values occur most often, we say there is no mode.

Answer

5. 4.5 in.

Example 4 Finding the Mode of a Data Set

Find the mode of the student-to-teacher ratios from Example 3.

Solution:

12.5   13.4   14.1   14.5   14.8   16.1   16.1   16.1   20.6   21.9

The data value 16.1 appears the most often. Therefore, the mode is 16.1.

Example 5 Finding the Mode of a Data Set

Find the mode of the list of average monthly temperatures for Albany, New York. Values are in °F.

Jan.	Feb.	March	April	May	June	July	Aug.	Sept.	Oct.	Nov.	Dec.
22	25	35	47	58	66	71	69	61	49	39	26

Solution:

No data value occurs most often. There is no mode for this set of data.

Example 6 Finding the Mode of a Data Set

The grades for a quiz in college algebra are as follows. The scores are out of a possible 10 points. Find the mode.

9	4	6	9	9	8	2	1	4	9
5	10	10	5	7	7	9	8	7	3
9	7	10	7	10	1	7	4	5	6

Solution:

Sometimes arranging the data in order makes it easier to find the repeated values.

1	1	2	3	4	4	4	5	5	5
6	6	7	7	7	7	7	7	8	8
9	9	9	9	9	9	10	10	10	10

The score of 9 occurs 6 times. The score of 7 occurs 6 times. There are two modes, 9 and 7, because these scores both occur more than any other score. We say that these data are *bimodal*.

TIP: To remember the difference between median and mode, think of the *median* of a highway that goes down the *middle*. Think of the word *mode* as sounding similar to the word *most*.

4. Weighted Mean

Sometimes data values in a list appear multiple times. In such a case, we can compute a **weighted mean**. In Example 7, each data value is "weighted" by the number of times it appears in the list.

Skill Practice

6. Find the mode of the rainfall amounts from Skill Practice Exercise 5.

4.5	3.0	3.2
3.5	5.1	6.8
4.3	4.5	5.6
5.3	4.5	3.8

Classroom Example: p. 563, Exercise 30

Skill Practice

7. Find the mode of the weights in pounds of babies born one day at Brackenridge Hospital in Austin, Texas.

7.2	8.1	6.9
9.3	8.3	7.7
7.9	6.4	7.5

Classroom Example: p. 563, Exercise 32

Skill Practice

8. The ages of children participating in an after-school sports program are given. Find the mode(s).

13	15	17	15
14	15	16	16
15	16	12	13
15	14	16	15
15	16	16	13
16	13	14	18

Classroom Example: p. 563, Exercise 34

Answers

6. 4.5 in.   7. No mode
8. There are two modes, 15 and 16.

Example 7 — Computing a Weighted Mean

Donations are made to a certain charitable organization in increments of $25, $50, $75, and $100, as shown in Figure 9-13. Find the mean amount donated.

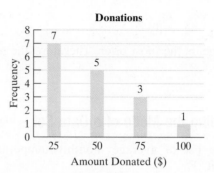

Figure 9-13

Solution:

Notice that there are 16 data values represented in the graph:

7 of these $\begin{cases} \$25 \\ \$25 \\ \$25 \\ \$25 \\ \$25 \\ \$25 \\ \$25 \end{cases}$ 5 of these $\begin{cases} \$50 \\ \$50 \\ \$50 \\ \$50 \\ \$50 \end{cases}$ 3 of these $\begin{cases} \$75 \\ \$75 \\ \$75 \end{cases}$ 1 of these $\{\$100$

The data value $25 occurs seven times. Rather than adding $25 seven times, we can find the sum by multiplying $25(7) = $175. Similarly, the value $50 occurs 5 times for a sum of $50(5) = $250, and so on. To find the sum of all the values, multiply each data value by the number of times it occurs (its frequency). Then add the results. This process can be organized easily in a table.

Amount Donated ($)	Frequency	Product ($)
25	7	(25)(7) = 175
50	5	(50)(5) = 250
75	3	(75)(3) = 225
100	1	(100)(1) = 100
Total:	**16**	**750**

← sum of all data values

↑ total number of donations made

The mean is the sum of all donations divided by the total number of donations made (total frequency).

$$\text{Mean} = \frac{750}{16} = 46.875$$

The mean amount donated is $46.88.

Section 9.4   Practice Exercises

Study Skills Exercise

For additional exercises, see Classroom Activities 9.4A–9.4B in the *Student's Resource Manual* at www.mhhe.com/moh.

Most people cannot concentrate on studying for more than 1 hr without taking a break. To make the most of your time, write down a schedule for your next study session. Include breaks where you can eat a meal, walk the dog, or perform other simple tasks that need to be completed.

Vocabulary and Key Concepts

1. **a.** The _____mean_____ or average of a set of numbers is the sum of the values divided by the number of values.

   b. The _____median_____ of an *odd* number of data values ranked in order from least to greatest is the "middle" number in the list.

   c. To find the median of an *even* number of data values ranked in order from least to greatest, find the _____mean_____ (or average) of the two middle numbers in the list.

   d. The _____mode_____ of a list of data values is the value that occurs most often.

   e. A _____weighted_____ mean is a mean where each data value is weighted according to the number of times it appears in the list.

Concept 1: Mean

2. Find the mean test score for the tests you have taken thus far in this class.   Answers will vary.

For Exercises 3–8, find the mean of each set of numbers. **(See Example 1.)**

3. $4, 6, 5, 10, 4, 5, 8$ 6

4. $3, 8, 5, 7, 4, 2, 7, 4$ 5

5. $0, 5, 7, 4, 7, 2, 4, 3$ 4

6. $7, 6, 5, 10, 8, 4, 8, 6, 0$ 6

7. $10, 13, 18, 20, 15$ 15.2

8. $22, 14, 12, 16, 15$ 15.8

9. The wingspan of five butterflies is given in the table. Find the mean wingspan.   8.76 in.

Butterfly	Wingspan (in.)
Queen Alexandra's birdwing	11.0
African giant swallowtail	9.1
Goliath birdwing	8.3
Buru opalescent birdwing	7.9
Chimaera birdwing	7.5

10. The number of wins in the American Baseball League, Central Division, for a recent year is given in the table. Find the mean number of wins.   84.2 wins

Team	Number of Wins
Minnesota Twins	96
Chicago White Sox	90
Kansas City Royals	62
Cleveland Indians	78
Detroit Tigers	95

11. The flight times in hours for six flights between New York and Los Angeles are given. Find the mean flight time. Round to the nearest tenth of an hour.

    $5.5, \ 6.0, \ 5.8, \ 5.8, \ 6.0, \ 5.6$ 5.8 hr

12. A nurse takes the temperature of a patient every 10 min and records the temperatures as follows: 98°F, 98.4°F, 98.9°F, 100.1°F, and 99.2°F. Find the patient's mean temperature.   98.92°F

13. The number of Calories for six different chicken sandwiches and chicken salads is given in the table.

    a. What is the mean number of Calories for a chicken sandwich? Round to the nearest whole unit.   397 Cal

    b. What is the mean number of Calories for a salad with chicken? Round to the nearest whole unit.   386 Cal

    c. What is the difference in the means?   There is only an 11-Cal difference in the means.

Chicken Sandwiches	Salads with Chicken
360	310
370	325
380	350
400	390
400	440
470	500

14. The heights of the players from two NBA teams are given in the table. All heights are in inches.

Philadelphia 76ers' Height (in.)	Milwaukee Bucks' Height (in.)
83	70
83	83
72	82
79	72
77	82
84	85
75	75
76	75
82	78
79	77

   a. Find the mean height for the players on the Philadelphia 76ers.   79 in.

   b. Find the mean height for the players on the Milwaukee Bucks.   77.9 in.

   c. What is the difference in the mean heights?
   The mean height for the 76ers is slightly higher by 1.1 in.

15. Zach received the following scores for his first four tests: 98%, 80%, 78%, 90%.

   a. Find Zach's mean test score.   86.5%

   b. Zach got a 59% on his fifth test. Find the mean of all five tests.   81%

   c. How did the low score of 59% affect the overall mean of five tests?
   The low score of 59% decreased Zach's average by 5.5%.

16. The prices of four steam irons are $50, $30, $25, and $45.

   a. Find the mean of these prices. $37.50

   b. An iron that costs $140 is added to the list. What is the mean of all five irons? $58.00

   c. How does the expensive iron affect the mean?   By including the iron for $140, the mean increased by $20.50.

Concept 2: Median

For Exercises 17–22, find the median for each set of numbers. **(See Examples 2 and 3.)**

17. 16, 14, 22, 13, 20, 19, 17
   17

18. 32, 35, 22, 36, 30, 31, 38
   32

19. 109, 118, 111, 110, 123, 100
   110.5

20. 134, 132, 120, 135, 140, 118
   133

21. 58, 55, 50, 40, 40, 55
   52.5

22. 82, 90, 99, 82, 88, 87
   87.5

23. The infant mortality rates for five countries are given in the table. Find the median.
3.93 deaths per 1000

Country	Infant Mortality Rate (Deaths per 1000)
Sweden	3.93
Japan	4.10
Finland	3.82
Andorra	4.09
Singapore	3.87

24. The inflation rates for five countries are given in the table. Find the median.   133%

Country	Inflation Rate (%)
Angola	1700
Sudan	133
Turkey	80
Venezuela	103
Bulgaria	311

25. The ages (in years) of the last 10 presidents at the time of their inauguration are given. Find the median age.

46, 64, 69, 52, 61, 56, 55, 43, 62, 60
58 years old

26. A list of the number of commuter rail stations from eight systems is given. Find the median number of stations.

124, 227, 108, 167, 121, 177, 49, 18
122.5 stations

27. The number of passengers (in millions) on nine leading airlines for a recent year is listed. Find the median number of passengers. (*Source: International Airline Transport Association*)

48.3, 42.4, 91.6, 86.8, 46.5, 71.2, 45.4, 56.4, 51.7
51.7 million passengers

28. For a recent year the number of albums sold (in millions) is listed for the 10 best sellers. Find the median number of albums sold.

2.7, 3.0, 4.8, 7.4, 3.4, 2.6, 3.0, 3.0, 3.9, 3.2
3.1 million albums

Concept 3: Mode

For Exercises 29–34, find the mode(s) for each set of numbers. **(See Examples 4–6.)**

29. 4, 5, 3, 8, 4, 9, 4, 2, 1, 4   4

30. 12, 14, 13, 17, 19, 18, 19, 17, 17   17

31. 90%, 89%, 91%, 77%, 88%   No mode

32. 132, 253, 553, 255, 552, 234   No mode

33. 28, 21, 24, 23, 24, 30, 21   21, 24

34. 45, 42, 40, 41, 49, 49, 42   42, 49

35. The table gives the price of seven "smart" cell phones. Find the mode. $300

Brand and Model	Price ($)
Samsung	300
Kyocera	400
Sony Ericsson	200
PalmOne	300
Motorola	300
Siemens	600

36. The table gives the number of hazardous waste sites for selected states. Find the mode.   39

State	Number of Sites
Florida	51
New Jersey	112
Michigan	67
Wisconsin	39
California	96
Pennsylvania	94
Illinois	39
New York	90

37. The unemployment rates in percent for nine countries are given. Find the mode.

6.3%, 7.0%, 5.8%, 9.1%, 5.2%, 8.8%, 8.4%, 5.4%, 5.2%   5.2%

38. The list gives the number of children who were absent from class for a 10-day period. Find the mode.

1, 6, 2, 2, 4, 4, 2, 2, 3, 2   2 children

39. The prices for five different brands of paper towel are given in the list. Find the mode.

$2.49, $2.39, $2.51, $2.49, $2.51

These data are bimodal: $2.49 and $2.51.

40. The length of time (in minutes) of eight TV commercials is given. Find the mode.

1.00, 0.50, 1.00, 1.25, 2.00, 0.50, 1.00, 0.50

These data are bimodal: 1.00 and 0.50 min.

Mixed Exercises

41. Six test scores for Jonathan's history class are listed. Find the mean and median. Round to the nearest tenth if necessary. Did the mean or median give a better overall score for Jonathan's performance?

92%, 98%, 43%, 98%, 97%, 85%

Mean: 85.5%; median: 94.5%; The median gave Jonathan a better overall score.

42. Nora's math test results are listed. Find the mean and median. Round to the nearest tenth if necessary. Did the mean or median give a better overall score for Nora's performance?

52%, 85%, 89%, 90%, 83%, 89%

Mean: 81.3%; median: 87%; The median gave Nora a better overall score.

43. Listed below are monthly costs for seven health insurance companies for a self-employed person, 55 years of age, and in good health. Find the mean, median, and mode (if one exists). Round to the nearest dollar. (*Source:* eHealth Insurance Company)

$312, $225, $221, $256, $308, $280, $147

Mean: $250; median: $256; mode: There is no mode.

44. The salaries for seven Associate Professors at the University of Michigan are listed. These are salaries for 9-month contracts in a recent year. Find the mean, median, and mode (if one exists). Round to the nearest dollar. (*Source:* University of Michigan, University Library, Issue 1)

$104,000, $107,000, $67,750, $82,500, $73,500, $88,300, $104,000

Mean: $89,579; median: $88,300; mode: $104,000

Writing   Translating Expression   Geometry   Scientific Calculator   Video   NS&E

45. The prices of 10 single-family, 3-bedroom homes for sale in Santa Rosa, California, are listed. Find the mean, median, and mode (if one exists).

$850,000, $835,000, $839,000, $829,000,

$850,000, $850,000, $850,000, $847,000,

$1,850,000, $825,000

Mean: $942,500; median: $848,500; mode: $850,000

46. The prices of 10 single-family, 3-bedroom homes for sale in Boston, Massachusetts, are listed. Find the mean, median, and mode (if one exists).

$300,000, $2,495,000, $2,120,000, $220,000,

$194,000, $391,000, $315,000, $330,000,

$435,000, $250,000

Mean: $705,000; median: $322,500; mode: There is no mode.

Concept 4: Weighted Mean

47. There are 20 students enrolled in a high school Algebra 2 class. The graph displays the number of students by age. First complete the table, and then find the mean. **(See Example 7.)**

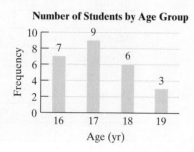

Number of Students by Age Group

Age (yr)	Number of Students	Product
16	7	112
17	9	153
18	6	108
19	3	57
Total:	25	430

The mean age is 17.2 years.

48. A survey was made in a neighborhood of 37 houses. The graph represents the number of residents who live in each house. Complete the table and determine the mean number of residents per house. Round to one decimal place.

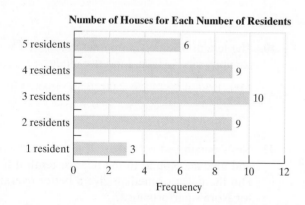

Number of Houses for Each Number of Residents

Number of Residents in Each House	Number of Houses	Product
1	3	3
2	9	18
3	10	30
4	9	36
5	6	30
Total:	37	117

The mean number of residents is approximately 3.2.

49. The data in the graph represent the number of classes at a college based on the size of the initial enrollment in the class. Find the weighted mean representing the mean number of students initially enrolled per class. Round to the nearest whole unit.   The mean number of students per class is approximately 28.

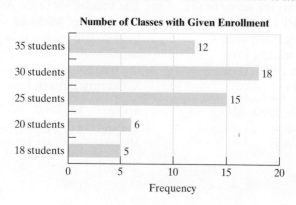

Number of Classes with Given Enrollment

At most colleges and universities, weighted means are used to compute students' grade point averages (GPAs). At one college, the grades A–F are assigned numerical values as follows:

A	= 4.0	C	= 2.0
B+	= 3.5	D+	= 1.5
B	= 3.0	D	= 1.0
C+	= 2.5	F	= 0.0

Grade point average is a weighted mean where the "weights" for each grade are the number of credit-hours for that class. Use this information to answer Exercises 50–53.

50. Compute the GPA for the following grades. Round to the nearest hundredth. 2.38

Course	Grade	Number of Credit-Hours (Weights)
Intermediate Algebra	B	4
Theater	C	1
Music Appreciation	A	3
World History	D	5

51. Compute the GPA for the following grades. Round to the nearest hundredth. 3.59

Course	Grade	Number of Credit-Hours (Weights)
General Psychology	B+	3
Beginning Algebra	A	4
Student Success	A	1
Freshman English	B	3

52. Compute the GPA for the following grades. Round to the nearest hundredth. 2.77

Course	Grade	Number of Credit-Hours (Weights)
Business Calculus	B+	3
Biology	C	4
Library Research	F	1
American Literature	A	3

53. Compute the GPA for the following grades. Round to the nearest hundredth. 2.73

Course	Grade	Number of Credit-Hours (Weights)
University Physics	C+	5
Calculus I	A	4
Computer Programming	D	3
Swimming	A	1

Introduction to Probability

Section 9.5

1. Basic Definitions

Concepts

1. Basic Definitions
2. Probability of an Event
3. Estimating Probabilities from Observed Data
4. Complementary Events

The probability of an event measures the likelihood of the event to occur. It is of particular interest because of its application to everyday life.

- The probability of picking the winning six-number combination for the New York lotto grand prize is $\frac{1}{45,057,474}$.
- Genetic DNA analysis can be used to determine the risk that a child will be born with cystic fibrosis. If both parents test positive, the probability is 25% that a child will be born with cystic fibrosis.

To begin our discussion, we must first understand some basic definitions.

An activity with observable outcomes such as flipping a coin or rolling a die is called an **experiment**. The collection (or set) of all possible outcomes of an experiment is called the **sample space** of the experiment.

Example 1   Determining the Sample Space of an Experiment

a. Suppose a single die is rolled. Determine the sample space of the experiment.

b. Suppose a coin is flipped. Determine the sample space of the experiment.

Solution:

a. A die is a single six-sided cube on which each side has between 1 and 6 dots painted on it. When the die is rolled, any of the six sides may come up.

The sample space is $\{1, 2, 3, 4, 5, 6\}$.   Notice that the symbols { } (called *set braces*) are used to enclose the elements.

b. The coin may land as a head H or as a tail T. The sample space is $\{H, T\}$.

2. Probability of an Event

Any part of a sample space is called an **event**. For example, if we roll a die, the event of rolling number 5 or a greater number consists of the outcomes 5 and 6. In mathematics, we measure the likelihood of an event to occur by its probability.

Probability of an Event

$$\text{Probability of an event} = \frac{\text{number of elements in event}}{\text{number of elements in sample space}}$$

Example 2   Finding the Probabilities of Events

a. Find the probability of rolling a 5 or greater on a die.

b. Find the probability of flipping a coin and having it land as heads.

Solution:

a. The event can occur in 2 ways: The die lands as a 5 or 6. The sample space has 6 elements: 1, 2, 3, 4, 5, and 6.

The probability of rolling a 5 or greater: $\dfrac{2}{6}$ ← number of ways to roll a 5 or greater
← number of elements in the sample space

$$= \frac{1}{3} \qquad \text{Simplify to lowest terms.}$$

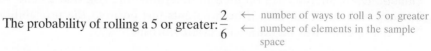

b. The event can occur in 1 way (the coin lands head side up).
The sample space has 2 outcomes: heads or tails.

The probability of flipping a head on a coin: $\dfrac{1}{2}$ ← number of ways to get heads

← number of elements in the
sample space

The value of a probability can be written as a fraction, as a decimal, or as a percent.
For example, the probability of a coin landing as heads is $\frac{1}{2}$ or 0.5 or 50%. In words,
this means that if we flip a coin many times, theoretically we expect one-half (50%)
of the outcomes to land as heads.

Example 3 **Finding Probabilities**

A class has 4 freshmen, 12 sophomores, and 6 juniors. If one individual is
selected at random from the class, find the probability of selecting

a. A sophomore

b. A junior

c. A senior

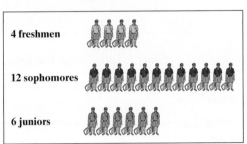

4 freshmen

12 sophomores

6 juniors

Solution:

In this case, there are 22 members of the class (4 freshmen + 12 sophomores + 6
juniors). This means that the sample space has 22 elements.

a. There are 12 sophomores in the class. The probability of selecting a
sophomore is

$\dfrac{12}{22}$ There are 12 sophomores out of 22 people in the sample space.

$= \dfrac{6}{11}$ Simplify to lowest terms.

b. There are 6 juniors out of 22 people in the sample space. The probability of
selecting a junior is

$\dfrac{6}{22}$ or $\dfrac{3}{11}$

c. There are no seniors in the class. The probability of selecting a senior is

$\dfrac{0}{22}$ or   0

A probability of 0 indicates that the event is impossible. It is impossible to select
a senior from a class that has no seniors.

From the definition of the probability of an event, it follows that the value of a probability must be between 0 and 1, inclusive. An event with a probability of 0 is called
an *impossible event*. An event with a probability of 1 is called a *certain event*.

Skill Practice

A group of registered voters has
9 Republicans, 8 Democrats, and
3 Independents. Suppose one
person from the group is selected
at random.

5. What is the probability that
the person is a Democrat?

6. What is the probability
that the person is an
Independent?

7. What is the probability that
the person is registered with
the Libertarian Party?

Classroom Examples: p. 570,
Exercises 22 and 24

Answers

5. $\dfrac{2}{5}$ or 0.4   **6.** $\dfrac{3}{20}$ or 0.15   **7.** 0

3. Estimating Probabilities from Observed Data

We were able to compute the probabilities in Examples 2 and 3 because the sample space was known. Sometimes we need to collect information to help us estimate probabilities.

Example 4 Estimating Probabilities from Observed Data

A dental hygienist records the number of times a day her patients say that they brush their teeth. Table 9-8 displays the results.

Table 9-8

Number of Times of Brushing Teeth per Day	Frequency
1	6
2	10
3	4
More than 3	1

If one of her patients is selected at random,

a. What is the probability of selecting a patient who brushes only one time a day?

b. What is the probability of selecting a patient who brushes more than once a day?

Solution:

a. The table shows that there are 6 patients who brush once a day. To get the total number of patients we add all of the frequencies ($6 + 10 + 4 + 1 = 21$). The probability of selecting a patient who brushes only once a day is

$$\frac{6}{21} \quad \text{or} \quad \frac{2}{7}$$

b. To find the number of patients who brush more than once a day, we add the frequencies for the patients who brush 2 times, 3 times, and more than 3 times ($10 + 4 + 1 = 15$). The probability of selecting a patient who brushes more that once a day is

$$\frac{15}{21} \quad \text{or} \quad \frac{5}{7}$$

4. Complementary Events

The events in Example 4(a) and 4(b) are called complementary events. The **complement of an event** is the set of all elements in the sample space that are not in the event. In this case, the number of patients who brush once a day and the number of patients who brush more than once a day make up the entire sample space, yet do not overlap. For this reason, the probability of an event plus the probability of its complement is 1. For Example 4, we have $\frac{2}{7} + \frac{5}{7} = \frac{7}{7} = 1$.

Skill Practice

Refer to Table 9-8 in Example 4.

8. What is the probability of selecting a patient who brushes twice a day?

9. What is the probability of selecting a patient who brushes more than twice a day?

Classroom Example: p. 572, Exercise 36

Answers

8. $\frac{10}{21}$ **9.** $\frac{5}{21}$

Example 5 Finding the Probability of Complementary Events

Find the indicated probability.

a. The probability of getting a winter cold is $\frac{3}{10}$. What is the probability of *not* getting a winter cold?

b. If the probability that a washing machine will break before the end of the warranty period is 0.0042, what is the probability that a washing machine will *not* break before the end of the warranty period?

Solution:

a. The probability of an event plus the probability of its complement must add up to 1. Therefore, we have an addition problem with a missing addend. This may also be expressed as subtraction.

$$\frac{3}{10} + ? = 1 \qquad \text{or equivalently} \qquad 1 - \frac{3}{10} = ?$$

$$\frac{10}{10} - \frac{3}{10} = \frac{7}{10} \quad \text{Find a common denominator and subtract.}$$

There is a $\frac{7}{10}$ chance (70% chance) of *not* getting a winter cold.

b. The probability that a washing machine will break before the end of the warranty period is 0.0042. Then the probability that a machine will *not* break before the end of the warranty period is given by

$$1 - 0.0042 = 0.9958 \text{ or equivalently } 99.58\%$$

Section 9.5   Practice Exercises

For additional exercises, see Classroom Activities 9.5A–9.5B in the *Student's Resource Manual* at www.mhhe.com/moh.

Study Skills Exercise

A good way to determine what will be on a test is to look at both your notes and the exercises assigned by your instructor. List five kinds of problems that you think will be on the test for this chapter.

Vocabulary and Key Concepts

1. a. An activity with observable outcomes is called a(n) ____experiment____.

b. The set of all possible outcomes of an experiment is called the ____sample____ space.

c. The ____probability____ of an event is the ratio of the number of elements in the event to the number of elements in the sample space.

d. The ____complement____ of an event is the set of all elements in the sample space that are not in the event.

e. The sum of the probability of an event and the probability of its complement is ____1____.

Review Exercises

2. Which value, the mean or the median, will be more affected by the unusually large number in the list of values?   The mean.

5, 6, 3, 8, 4, 5, 6, 4, 165

 Writing   Translating Expression   Geometry   Scientific Calculator   Video   NS E

For Exercises 3–8, find the mean, median, and mode (if one exists).

3. 13, 16, 22, 25, 10
Mean: 17.2; median: 16; no mode

4. 62, 64, 62, 67, 40
Mean: 59; median: 62; mode: 62

5. 8, 9, 10, 7, 8, 8, 11, 10
Mean: 8.875; median: 8.5; mode: 8

6. 96%, 88%, 89%, 90%, 88%, 50%
Mean: 83.5%; median: 88.5%; mode: 88%

7. 20, 20, 18, 17, 19, 5
Mean: 16.5; median: 18.5; mode: 20

8. 100, 90, 95, 98, 90, 10
Mean: 80.5; median: 92.5; mode: 90

Concept 1: Basic Definitions

9. A card is chosen from a deck consisting of 10 cards numbered 1–10. Determine the sample space of this experiment. **(See Example 1.)**
{1, 2, 3, 4, 5, 6, 7, 8, 9, 10}

10. A marble is chosen from a jar containing a yellow marble, a red marble, a blue marble, a green marble, and a white marble. Determine the sample space of this experiment. {yellow, red, blue, green, white}

11. Two dice are thrown, and the sum of the top sides is observed. Determine the sample space of this experiment. {2, 3, 4, 5, 6, 7, 8, 9, 10, 11, 12}

12. A coin is tossed twice. Determine the sample space of this experiment. {TT, TH, HT, HH}

13. If a die is rolled, in how many ways can an odd number come up? 3 ways

14. If a die is rolled, in how many ways can a number less than 6 come up? 5 ways

Concept 2: Probability of an Event

15. Which of the values can represent the probability of an event? c, d, g, h

   a. 1.62   **b.** $-\dfrac{7}{5}$ **c.** 0   **d.** 1

   e. 200%   **f.** 4.5   **g.** 4.5%   **h.** 0.87

16. Which of the values can represent the probability of an event? b, c, d, g, h

   a. 1.5   **b.** 0   **c.** $\dfrac{2}{3}$ **d.** 1

   e. 150%   **f.** 3.7   **g.** 3.7%   **h.** 0.92

17. If a single die is rolled, what is the probability that it will come up as a number less than 3? **(See Example 2.)** $\dfrac{2}{6} = \dfrac{1}{3}$

18. If a single die is rolled, what is the probability that it will come up as a number greater than 5? $\dfrac{1}{6}$

19. If a single die is rolled, what is the probability that it will come up with an even number? $\dfrac{3}{6} = \dfrac{1}{2}$

20. If a single die is rolled, what is the probability that it will come up as an odd number? $\dfrac{3}{6} = \dfrac{1}{2}$

For Exercises 21–24, refer to the figure. A sock drawer contains 2 white socks, 5 black socks, and 1 blue sock. **(See Example 3.)**

21. What is the probability of choosing a black sock from the drawer? $\dfrac{5}{8}$

22. What is the probability of choosing a white sock from the drawer? $\dfrac{2}{8} = \dfrac{1}{4}$

23. What is the probability of choosing a blue sock from the drawer? $\dfrac{1}{8}$

24. What is the probability of choosing a purple sock from the drawer? 0

25. If a die is tossed, what is the probability that a number from 1 to 6 will come up? 1

26. If a die is tossed, what is the probability of getting a 7? 0

Writing   Translating Expression   Geometry   Scientific Calculator   Video   NS E

27. What is an impossible event? An impossible event is one in which the probability is 0.

28. What is the sum of the probabilities of an event and its complement? 1

29. In a deck of cards there are 12 face cards and 40 cards with numbers. What is the probability of selecting a face card from the deck? $\frac{12}{52} = \frac{3}{13}$

30. In a deck of cards, 13 are diamonds, 13 are spades, 13 are clubs, and 13 are hearts. Find the probability of selecting a diamond from the deck. $\frac{13}{52} = \frac{1}{4}$

31. A jar contains 7 yellow marbles, 5 red marbles, and 4 green marbles. What is the probability of selecting a red marble or a yellow marble? $\frac{12}{16} = \frac{3}{4}$

32. A jar contains 10 black marbles, 12 white marbles, and 4 blue marbles. What is the probability of selecting a blue marble or a black marble? $\frac{14}{26} = \frac{7}{13}$

Concept 3: Estimating Probabilities from Observed Data

33. The table displays the length of stay for vacationers at a small motel. **(See Example 4.)**

Length of Stay in Days	Frequency
2	14
3	13
4	18
5	28
6	11
7	30
8	6

a. What is the probability that a vacationer will stay for 4 days? $\frac{18}{120} = \frac{3}{20}$

b. What is the probability that a vacationer will stay for less than 4 days? $\frac{27}{120} = \frac{9}{40}$

c. Based on the information from the table, what percent of vacationers stay for more than 6 days? 30%

34. A number of students at a large university were asked if they owned a car. The table shows the results.

	Number of Car Owners	Number Who Do Not Own a Car
Dorm resident	32	88
Lives off campus	59	26

a. What is the probability that a student selected at random lives in a dorm? $\frac{120}{205} = \frac{24}{41}$

b. What is the probability that a student selected at random does not own a car? $\frac{114}{205}$

35. A survey was made of 60 participants, asking if they drive an American-made car, a Japanese car, or a car manufactured in another foreign country. The table displays the results.

	Frequency
American	21
Japanese	30
Other	9

a. What is the probability that a randomly selected car is manufactured in America? $\frac{21}{60} = \frac{7}{20}$

b. What percent of cars is manufactured in some country other than Japan? 50%

36. The number of customer complaints for service representatives is given in the table. If one representative is picked at random, find the probability that the representative received

a. Exactly 3 complaints $\frac{10}{80} = \frac{1}{8}$

b. Between 1 and 5 complaints, inclusive $\frac{60}{80} = \frac{3}{4}$

c. At least 4 complaints $\frac{50}{80} = \frac{5}{8}$

d. More than 5 complaints or fewer than 2 complaints $\frac{22}{80} = \frac{11}{40}$

Number of Complaints	Number of Representatives
0	4
1	2
2	14
3	10
4	16
5	18
6	10
7	6

37. Mr. Gutierrez noted the times in which his students entered his classroom and constructed a chart.

a. What is the probability that a student will be early to class? $\frac{7}{29}$

b. What is the probability that a student will be late to class? $\frac{11}{29}$

c. What percent of students arrive on time or early? Round to the nearest whole percent. 62%

Time	Number of Students
About 10 min early	1
About 5 min early	6
On time	11
About 5 min late	7
About 10 min late	3
About 15 min late	1

38. Each person at an office party purchased a raffle ticket. The graph shows the results of the raffle.

a. How many people bought raffle tickets? 60 people

b. What is the probability of winning the TV set? $\frac{1}{60}$

c. What percent of people won some type of prize? 15%

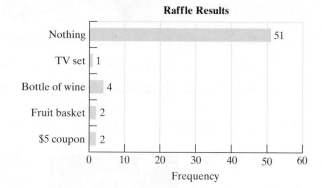

Raffle Results

Nothing 51
TV set 1
Bottle of wine 4
Fruit basket 2
$5 coupon 2

Frequency

Concept 4: Complementary Events

39. If the probability of the horse Sugar 'N Spice to win is $\frac{2}{11}$, what is the probability that he will not win? **(See Example 5.)** $1 - \frac{2}{11} = \frac{9}{11}$

40. If the probability of being hit by lightning is $\frac{1}{1,000,000}$, what is the probability of not getting hit by lightning? $1 - \frac{1}{1,000,000} = \frac{999,999}{1,000,000}$

41. If the probability of having twins is 1.2%, what is the probability of not having twins? $100\% - 1.2\% = 98.8\%$

42. The probability of a woman's surviving breast cancer is 88%. What is the probability that a woman would not survive breast cancer? $100\% - 88\% = 12\%$

Group Activity

Creating a Statistical Report

Materials: A computer with Internet access or the local newspaper

Estimated Time: 20–30 minutes

Group Size: 4

The group members will collect numerical data from the Internet or the newspaper. The data will be analyzed using the statistical techniques learned in this chapter. Here is one suggested project.

1. Record the age and gender of the individuals who were arrested in your town during the past week. This can often be found in the local section of the newspaper. For example, you can visit the website for the Daytona Beach *News-Journal* and select "local news" and then "news of record." Record 20 or 30 data values.

2. Compute the mean, median, and mode for the ages of men arrested. Compute the mean, median, and mode for the ages of women arrested. Do the statistics suggest a difference in the average age of arrest for men versus women?

3. Determine the percentage of men and the percentage of women in the sample. Does there appear to be a significant difference?

4. Organize the data by age group and construct a frequency distribution and histogram.

Note: The steps given in this project offer suggestions for organizing and analyzing the data you collect. These steps outline standard statistical techniques that apply to a variety of data sets. You might consider doing a different project that investigates a topic of interest to you. Here are some other ideas.

- Collect the weight and gender of babies born in the local hospital.

- Collect the age and gender of students who take classes at night versus those who take classes during the day.

- Collect stock prices for a 2- or 3-week period.

Can you think of other topics for a project?

Chapter 9   Summary

Section 9.1	Tables, Bar Graphs, Pictographs, and Line Graphs

Key Concepts and Examples

Statistics is the branch of mathematics that involves collecting, organizing, and analyzing **data** (information). Information can often be organized in tables and graphs. The individual entries within a table are called **cells**.

Example 1

The data in the table give the number of Calories for a 1-c serving of selected vegetables.

Vegetable (1 c)	Number of Calories
Corn	85
Green beans	35
Eggplant	25
Peas	125
Spinach	40

A **pictograph** uses an icon or small image to convey a unit of measurement.

Example 3

What is the value of each icon in the graph?

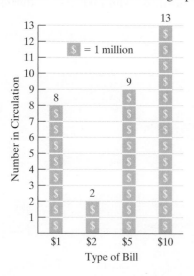

Each icon is worth 1,000,000 bills in circulation.

Examples

Example 2

Construct a bar graph for the data in Example 1.

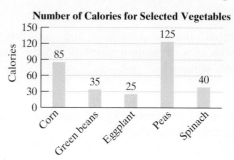

Line graphs are often used to track how one variable changes with respect to the change in a second variable.

Example 4

In what year were there 52.3 million married-couple households?

From the graph, the year 1990 corresponds to 52.3 million married-couple households.

Section 9.2 — Frequency Distributions and Histograms

Key Concepts

A **frequency distribution** is a table displaying the number of data values that fall within specified categories. When the categories represent a range of numerical values, we call the categories **class intervals**.

When constructing a frequency distribution, keep these important guidelines in mind.

- The classes should represent intervals of equal length.
- The classes should not overlap.
- In general, use between 5 and 15 classes, inclusive.

A **histogram** is a special bar graph that illustrates data given in a frequency distribution. The class intervals are given on the horizontal scale. The height of each bar in a histogram measures the frequency for each class.

Examples

Example 1

Create a frequency distribution for the following data.

50	53
54	51
50	40
50	47
53	36
44	34
52	32
42	30

Class Intervals	Tally	Frequency
30–34	III	3
35–39	I	1
40–44	III	3
45–49	I	1
50–54	IIII III	8

Example 2

Create a histogram for the data in Example 1.

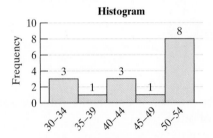

Section 9.3 — Circle Graphs

Key Concepts

A **circle graph** (or pie graph) is a type of graph used to show how a whole amount is divided into parts. Each part of the circle, called a **sector**, is like a slice of pie.

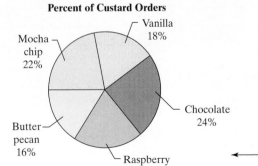

Percent of Custard Orders

Examples

Example 1

At Ritter's Frozen Custard, the flavors for the day are given with the number of orders for each flavor.

Flavor	Number of Orders
Vanilla	180
Chocolate	240
Raspberry	200
Butter pecan	160
Mocha chip	220

Construct a circle graph for the data given. Label the sectors with percents.

Section 9.4   Mean, Median, and Mode

Key Concepts

The **mean** (or average) of a set of numbers is the sum of the values divided by the number of values.

$$\text{Mean} = \frac{\text{sum of the values}}{\text{number of values}}$$

The **median** is the "middle" number in an ordered list of numbers. For an ordered list of numbers:

- If the number of data values is *odd*, then the median is the middle number in the list.
- If the number of data values is *even*, the median is the mean of the two middle numbers in the list.

The **mode** of a set of data is the value or values that occur most often.

When data values in a list appear multiple times, we can compute a **weighted mean**.

Examples

Example 1

Find the mean test score: 92, 100, 86, 60, 90

$$\text{Mean} = \frac{92 + 100 + 86 + 60 + 90}{5}$$

$$= \frac{428}{5} = 85.6$$

Example 2

Find the median: 12   18   6   10   5

First order the list: 5   6   10   12   18
The median is the middle number, 10.

Example 3

Find the median: 15   20   20   32   40   45

The median is the average of 20 and 32:

$$\frac{20 + 32}{2} = \frac{52}{2} = 26 \qquad \text{The median is 26.}$$

Example 4

Find the mode: 7   2   5   7   7   4   6   10

The value 7 is the mode because it occurs most often.

Example 5

The ages of children in a day-care center are given in the table. Find the mean age.

Age (yr)	Frequency	Product
3	5	$3 \cdot 5 = 15$
4	10	$4 \cdot 10 = 40$
5	3	$5 \cdot 3 = 15$
6	2	$6 \cdot 2 = 12$
Total	**20**	**82**

$$\text{Mean} = \frac{82}{20} = 4.1$$

The mean age is 4.1 yr.

Section 9.5   Introduction to Probability

Key Concepts

The collection (or set) of all possible outcomes of an experiment is called the **sample space**.

The **probability of event** is given by:

$$\frac{\text{number of elements in event}}{\text{number of elements in sample space}}$$

The probability of an event cannot be greater than 1 nor less than 0.

The **complement of an event** is the set of all elements in the sample space that are not in the event.

Examples

Example 1

Define the sample space for selecting a colored ball.

Sample space = {red, blue, yellow, green}

Example 2

What is the probability of selecting a yellow ball from Example 1?

Let A represent the event of picking a yellow ball. Then A = {yellow}.

$$P(A) = \frac{1}{4} \quad\begin{array}{l}\leftarrow \text{number of yellow balls}\\ \leftarrow \text{number of balls in box}\end{array}$$

Example 3

49 CDs are in a shopping cart.

10 Rap
24 Rock
12 Latina
3 Classical

If one CD is selected at random, find the probability that

a. A rock CD is selected. $\dfrac{24}{49}$

b. A rock CD is *not* selected. This is the complementary event to part (a).

$$1 - \frac{24}{49} = \frac{25}{49}$$

Chapter 9   Review Exercises

Section 9.1

For Exercises 1–4, refer to the table. The table gives the number of Calories and the amount of fat, cholesterol, sodium, and total carbohydrates for a single $\frac{1}{2}$-c serving of chocolate ice cream.

Ice Cream	Calories	Fat(g)	Cholesterol (mg)	Sodium (mg)	Carbohydrate (g)
Breyers	150	8	20	35	17
Häagen-Dazs	270	18	115	60	22
Edy's Grand	150	8	25	35	17
Blue Bell	160	8	35	70	18
Godiva	290	18	65	50	28

1. Which ice cream has the most calories?
   Godiva

2. Which ice cream has the least amount of cholesterol?   Breyers

3. How many more times the sodium does Blue Bell have per serving than Edy's Grand?
   Blue Bell has 2 times more sodium than Edy's Grand.

4. What is the difference in the amount of carbohydrate for Godiva and Blue Bell?
   There is a 10-g difference.

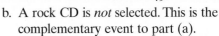

Writing   Translating Expression   Geometry   Scientific Calculator   Video   NS&E

Since 1940 the number of U.S. farms has decreased. However, the average size of the farms has increased. The graph shows the average size of U.S. farms for selected years. Refer to the graph for Exercises 5–8. (*Source:* U.S. Department of Agriculture)

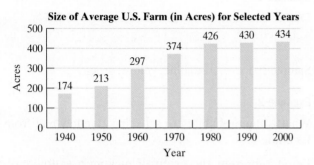

Size of Average U.S. Farm (in Acres) for Selected Years

5. What was the average size of the farms in 1970?
374 acres

6. What is the difference between the average size farm in the year 2000 compared to 1940?
The difference is 260 acres.

7. What is the difference between the average size farm in the year 1990 compared to 1980?
The difference is 4 acres.

8. In which 10-year interval was the increase the greatest?
The greatest increase was between 1950 and 1960.

For Exercises 9–12, refer to the pictograph. The graph represents the number of tornadoes during four months with active weather.

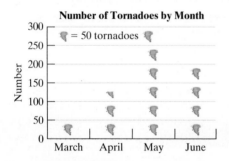

Number of Tornadoes by Month

9. What does each icon represent?
1 icon represents 50 tornadoes.

10. From the graph, estimate the number of tornadoes in May.
300

11. Which month had approximately 200 tornadoes?
June

12. Estimate the difference in the number of tornadoes in April and the number in March.
75

For Exercises 13–16, refer to the graph. The graph represents the number of liver transplants in the United States for selected years. (*Source:* U.S. Department of Health and Human Services)

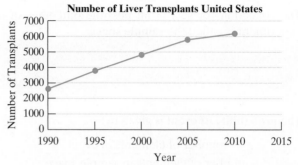

Number of Liver Transplants United States

13. In which year did the greatest number of liver transplants occur? 2010

14. Approximate the number of liver transplants for the year 2000. 4900

15. Does the trend appear to be increasing or decreasing? Increasing

16. Extend the graph to predict the number of liver transplants for the year 2015. ≈ 7000

17. The table shows five of the highest grossing movies in the United States. Construct a bar graph using horizontal bars. The length of each bar should represent the amount of money in millions that each movie grossed. (*Source: Washington Post*)

Movie Title	Gross ($ in millions)	Year
Avatar	749	2009
Titanic	601	1997
The Dark Knight	533	2008
Star Wars	461	1977
Shrek 2	437	2004

High Grossing Movies in the United States

Shrek 2 — 437
Star Wars — 461
The Dark Night — 533
Titanic — 601
Avatar — 749

0 100 200 300 400 500 600 700 800
Dollars (in millions)

Section 9.2

The ages of students in a Spanish class are given.

18 22 19 26 31 20 40 24 43 22
29 28 35 42 29 30 24 31 23 21

Use these data for Exercises 18 and 19.

18. Complete the frequency table.

Class Intervals (Age)	Frequency
18–21	4
22–25	5
26–29	4
30–33	3
34–37	1
38–41	1
42–45	2

19. Construct a histogram of the data in Exercise 18.

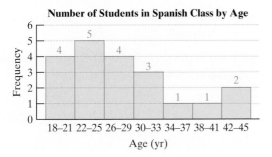

Number of Students in Spanish Class by Age

Section 9.3

The pie graph describes the types of subs offered at Larry's Sub Shop. Use the information in the graph for Exercises 20–22.

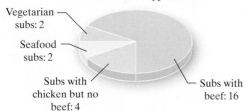

Number of Certain Types of Subs

Vegetarian subs: 2
Seafood subs: 2
Subs with chicken but no beef: 4
Subs with beef: 16

20. How many types of subs are offered at Larry's?
There are 24 types of subs.

21. What fraction of the subs at Larry's is made with beef? $\frac{2}{3}$ of the subs contain beef.

22. What fraction of the subs at Larry's is not made with beef? $\frac{1}{3}$ of the subs do not contain beef.

23. A survey was conducted with 200 people, and they were asked their highest level of education. The results of the survey are given in the table.

a. Complete the table.

Education Level	Number of People	Percent	Number of Degrees
Grade school	10	5%	18°
High school	50	25%	90°
Some college	60	30%	108°
Four-year degree	40	20%	72°
Postgraduate	40	20%	72°

b. Construct a circle graph using percents from the information in the table.

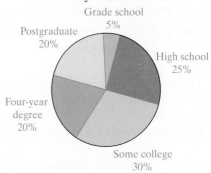

Percent by Education Level

Grade school 5%
Postgraduate 20%
High school 25%
Four-year degree 20%
Some college 30%

Section 9.4

24. For the list of quiz scores, find the mean, median, and mode(s).

20, 20, 18, 16, 18, 17, 16, 10, 20, 20, 15, 20
Mean: 17.5; median: 18; mode: 20

25. Juanita kept track of how many milligrams of calcium she took each day through vitamins and dairy products. Determine the mean number of milligrams of calcium per day. Round to the nearest 10 (mg).
The mean daily calcium intake is approximately 1060 mg.

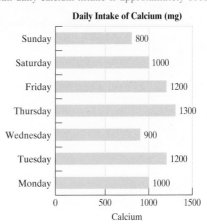

Daily Intake of Calcium (mg)

Sunday 800
Saturday 1000
Friday 1200
Thursday 1300
Wednesday 900
Tuesday 1200
Monday 1000

Calcium

 Writing   Translating Expression   Geometry   Scientific Calculator   Video   NS E

26. The seating capacity for five arenas used by the NBA is given in the table. Find the median number of seats.

The median is 20,562 seats.

Arena	Number of Seats
Phelps Arena, Atlanta	20,000
Fleet Center, Boston	18,624
Chevrolette Coliseum, Charlotte	23,799
United Center, Chicago	21,500
Gund Arena, Cleveland	20,562

27. The manager of a restaurant had his customers fill out evaluations on the service that they received. A scale of 1 to 5 was used, where 1 represents very poor service and 5 represents excellent service. Given the list of responses, determine the mode(s).

4 5 3 4 4 3 2 5 5 1 4 3 4 4 5
2 5 4 4 3 2 5 5 1 4
4

28. There are 20 children participating in an afternoon fitness program. The graph displays the number of children by age. Complete the table and then find the mean age of the children.

The mean age is 10.95 years.

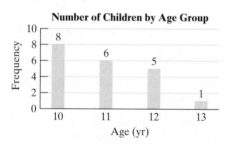

Number of Children by Age Group

Age (yr)	Number of Children	Product
10	8	80
11	6	66
12	5	60
13	1	13

Section 9.5

29. Roberto has six pairs of socks, each a different color: blue, green, brown, black, gray, and white. If Roberto randomly chooses a pair of socks, write the sample space for this event.

{blue, green, brown, black, gray, white}

30. Refer to Exercise 29. What is the probability that Roberto will select a pair of gray socks? $\frac{1}{6}$

31. Which of the following numbers could represent a probability? a, c, d, e, g

a. $\frac{1}{2}$ **b.** $\frac{5}{4}$ **c.** 0   **d.** 1

e. 25%   **f.** 2.5   **g.** 6%   **h.** 6

32. A bicycle shop sells a child's tricycle in three colors: red, blue, and pink. In the warehouse there are 8 red tricycles, 6 blue tricycles, and 2 pink tricycles.

a. If Kevin selects a tricycle at random, what is the probability that he will pick a red tricycle? $\frac{1}{2}$

b. What is the probability that he will not pick a red tricycle? $\frac{1}{2}$

c. What is the probability of Kevin's selecting a green tricycle? 0

Chapter 9   Test

1. The table represents the world's major producers of primary energy for a recent year. All measurements are in quadrillions of Btu.

Note: 1 quadrillion = 1,000,000,000,000,000. (*Source:* Energy Information Administration, U.S. Dept. of Energy)

Country	Amount of Energy Produced (quadrillions of BTUs)
United States	72
Russia	43
China	35
Saudi Arabia	43
Canada	18

Construct a bar graph using horizontal bars. The length of each bar corresponds to the amount of energy produced for each country.

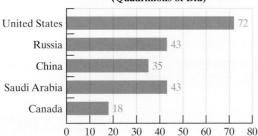

World's Major Producers of Primary Energy (Quadrillions of Btu)

For Exercises 2–4, refer to the following information.

Of the approximately 2.9 million workers in 1820 in the United States, 71.8% were employed in farm occupations. Since then, the percent of U.S. workers in farm occupations has declined. The table shows the percent of total U.S. workers who worked in farm-related occupations for selected years. (*Source:* U.S. Department of Agriculture)

Year	Percent of U.S. Workers in Farm Occupations
1820	72%
1860	59%
1900	38%
1940	17%
1980	3%

2. Which year had the greatest percent of U.S. workers employed in farm occupations? What is the value of the greatest percent?
The year 1820 had the greatest percent of workers employed in farm occupations. This was 72%.

3. Make a line graph with the year on the horizontal scale and the percent on the vertical scale.

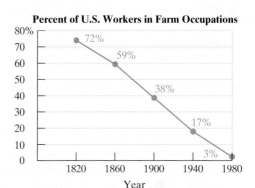

Percent of U.S. Workers in Farm Occupations

4. Based on the graph, estimate the percent of U.S. workers employed in farm occupations for the year 1960.
Approximately 10% of U.S. workers were employed in farm occupations in the year 1960.

For Exercises 5–7, refer to the pictograph. The pictograph shows the flower sales for the first 5 months of the year for a flower shop.

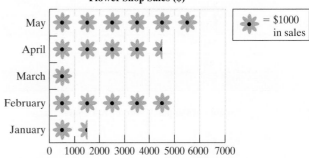

Flower Shop Sales ($)

= $1000 in sales

5. What is the value of each flower icon? $1000

6. From the graph, estimate the sales for the month of April. $4500

7. Which month brought in sales of $5000?
February

For Exercises 8–10, refer to the table. The rainfall amounts for Salt Lake City, Utah, and Seattle, Washington, are given in the table for selected months. All values are in inches. (*Source:* National Oceanic and Atmospheric Administration)

	April	**May**	**June**	**July**
Salt Lake City	2.02	2.09	0.77	0.72
Seattle	2.75	2.03	2.5	0.92

8. Which city is generally wetter?   Seattle

9. What is the difference in the amount of rainfall in Seattle and Salt Lake City during June?   1.73 in.

10. In what month does Salt Lake City have a greater rainfall amount than Seattle?
May

11. A cellular phone company questioned 20 people at a mall, to determine approximately how many minutes each individual spent on the cell phone each month. Using the list of results, complete the frequency distribution and construct a histogram.

100	120	250	180	300	200	250	175
110	280	330	280	300	325	60	75
100	350	60	90

Number of Minutes Used Monthly	**Tally**	**Frequency**
51–100	||||	6
101–150	||	2
151–200	|||	3
201–250	||	2
251–300	||||	4
301–350	|||	3

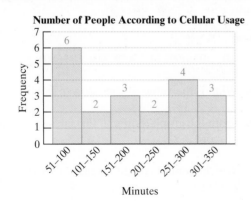

Number of People According to Cellular Usage

For Exercises 12–14, refer to the circle graph. The circle graph shows the percent of homes having different types of flooring in the living room area.

Percent of Types of Floor Covering

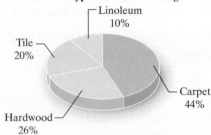

12. If 150 people were questioned, how many would be expected to have carpet on their living room floor?   66 people would have carpet.

13. If 200 people were questioned, how many would be expected to have tile on their living room floor?   40 people would have tile.

14. If 300 people were questioned, how many would not be expected to have linoleum on their living room floor?
270 people would have something other than linoleum.

For Exercises 15–17, refer to the table. The table represents the heights of the Seven Summits (the highest peaks from each continent).

Mountain	**Continent**	**Height (ft)**
Mt. Kilimanjaro	Africa	19,340
Elbrus	Europe	18,510
Aconcagua	South America	22,834
Denali	North America	20,320
Vinson Massif	Antarctica	16,864
Mt. Kosciusko	Australia	7,310
Mt. Everest	Asia	29,035

15. What is the mean height of the Seven Summits? Round to the nearest whole unit.
19,173 ft

16. What is the median height?
19,340 ft

17. Is there a mode?
There is no mode.

18. Mike and Darcy listed the amount of money paid for going to the movies for the past 3 months. This list contains the amount for 2 tickets. Find the mean, median, and mode.

$11 $14 $11 $16 $15 $16 $12 $16 $15 $20
Mean: $14.60; median: $15; mode $16

19. A board game has a die with eight sides with the numbers 1–8 printed on each side.

a. What is the sample space for rolling the die one time? {1, 2, 3, 4, 5, 6, 7, 8}

b. What is the probability of rolling a 6? $\frac{1}{8}$

c. What is the probability of rolling an even number? $\frac{1}{2}$

d. What is the probability of rolling a number less than 3? $\frac{1}{4}$

20. At a party there is a cooler filled with ice and soft drinks: 6 cans of diet cola, 4 cans of ginger ale, 2 cans of root beer, and 2 cans of cream soda. A person takes a can of soda at random.

a. What is the probability that the person selects a can of ginger ale? $\frac{2}{7}$

b. What is the probability of the person not selecting a can of ginger ale? $\frac{5}{7}$

21. Compute the GPA for the following grades. Round to the nearest hundredth. Use this scale: 3.09

A	= 4.0	C	= 2.0
B+	= 3.5	D+	= 1.5
B	= 3.0	D	= 1.0
C+	= 2.5	F	= 0.0

Course	Grade	Number of Credit-Hours (Weights)
Art Appreciation	B	4
College Algebra	A	3
English II	C	3
Physical Fitness	A	1

NS&E **22.** Which of the following is not a reasonable value for a probability? c

a. 0.36   **b.** $\frac{3}{4}$ **c.** 1.5

Chapters 1–9   Cumulative Review Exercises

1. Identify the place value of the underlined digit.

a. 23,9<u>9</u>0,192   **b.** 5,981,9<u>0</u>2   **c.** 3,019,2<u>2</u>6
Millions   Ten-thousands   Hundreds

2. Add.   2087 + 53 + 10,499 + 6
12,645

3. Estimate the product by first rounding each number to the nearest hundred.
700 × 1200 = 840,000
687 × 1243

4. Divide 651 by 23. Identify the divisor, dividend, whole part of the quotient, and remainder.
Divisor: 23; dividend: 651; quotient: 28; remainder: 7

5. What fraction of this circle is shaded?
$\frac{3}{8}$

For Exercises 6–8, multiply or divide as indicated. Reduce the answers to lowest terms.

6. $\frac{12}{7} \cdot \frac{14}{36}$ **7.** $\frac{105}{96} \div \frac{7}{16}$ **8.** $\frac{5}{8} \div \frac{6}{15} \cdot \frac{24}{25}$
$\frac{2}{3}$ $\frac{5}{2}$ $\frac{3}{2}$

For Exercises 9–11, add or subtract as indicated. Reduce to lowest terms.

9. $\frac{97}{102} - \frac{63}{102}$ **10.** $\frac{3}{10} + \frac{7}{100}$ **11.** $\frac{1}{2} + \frac{5}{3} - \frac{1}{6}$
$\frac{1}{3}$ $\frac{37}{100}$ 2

12. Simplify, using the order of operations.
$2\frac{1}{8} \div 17 + \frac{7}{12} \cdot \frac{9}{14} - \frac{1}{3}$ $\frac{1}{6}$

13. The table gives the prices of certain stocks and their increase or decrease from one day to the next. Complete the table.

Stock	Yesterday's Closing Price ($)	Increase/ Decrease	Today's Closing Price ($)
RylGold	13.28	0.27	13.55
NetSolve	9.51	−0.17	9.34
Metals USA	14.35	0.10	14.45
PAM Transpt	18.09	0.09	18.18
Steel Tch	21.63	−0.37	21.26

For Exercises 14–16, multiply or divide by powers of 10.

14. 68.412×100
6841.2

15. 68.412×0.1
6.8412

16. $68.412 \div 0.001$
68,412

17. The estimated forest cover in the Brazilian Amazon in 1970 was approximately 3.7 million square kilometers. By 2010, the amount dropped to 2.95 million km². (*Source:* National Geographic Society)

 a. By how many square kilometers had the Brazilian Amazon forest cover decreased?
 0.75 million km² or equivalently, 750,000 km²

 b. Compute the percent decrease in forest cover. Round to the nearest tenth of a percent. 20.3%

18. Quick Cut Lawn Company can service 5 customers in $2\frac{3}{4}$ hr. Speedy Lawn Company can service 6 customers in 3 hr. Find the unit rate in time per customer for both lawn companies and decide which company is faster. Quick Cut Lawn Company's rate is 0.55 hr per customer. Speedy Lawn Company's rate is 0.5 hr per customer. Speedy Lawn Company is faster.

19. If Rosa can type a 4-page English paper in 50 min, how long will it take her to type a 10-page term paper?
125 min or 2 hr 5 min

20. Find the values of x and y, assuming that the two triangles are similar. $x = 5$ m, $y = 22.4$ m

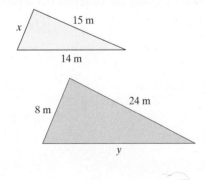

21. Out of a group of people, 95 said that they brushed their teeth twice a day. If this number represents 78% of the people surveyed, how many were surveyed? Round to the nearest whole unit.
122 people

22. The number of children accompanying their parents on business trips has jumped 230% in the last 10 years. If the number of children 10 years ago was 7.4 million, determine the present number.
17.02 million

23. Out of 120 people, 78 wear glasses. What percent does this represent?
65%

24. A savings account pays 3.4% simple interest. If $1200 is invested for 5 years, what will be the balance?
$1404

25. Convert 2 ft 5 in. to inches.
29 in.

26. Convert $4\frac{1}{2}$ gal to quarts.
18 qt

27. Add. 3 yd 2 ft + 5 yd 2 ft
9 yd 1 ft

28. Subtract. 12 km − 2360 m
9.64 km or 9640 m

29. Divide 16 lb 12 oz by 4.
4 lb 3 oz

For Exercises 30–32, identify the type of angle. Choose from acute, obtuse, right, or straight.

30.
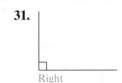
Obtuse

31.

Right

32.

Acute

33. Find the area.   Area: 8 ft²

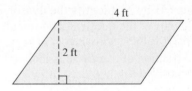

34. Find the volume. Use $\pi \approx \frac{22}{7}$.   66 m³

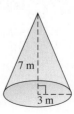

35. The data shown in the table give the average weight for boys based on age. (*Source:* National Parenting Council) Make a line graph to illustrate these data.

Age	Weight (lb)
5	44.5
6	48.5
7	54.5
8	61.25
9	69
10	74.5
11	85
12	89

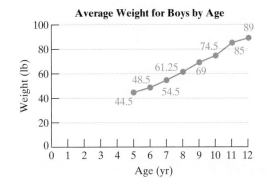

Average Weight for Boys by Age

36. The monthly number of deaths resulting from tornadoes for a recent year are given. Find the mean and median. Round to the nearest whole unit.   Mean: 105; median: 123

33,   10, 62, 132, 123, 316,

123, 133, 18, 150,   26, 138

37. Simplify the expression.

$$30 - 3(5 - 2)^2 \quad 3$$

A game has a spinner with four sections of equal size. Refer to the spinner for Exercises 38–40.

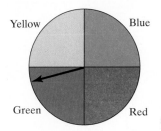

38. If a person spins the pointer once, determine the sample space.   {yellow, blue, red, green}

39. What is the probability of the pointer landing on green? $\frac{1}{4}$

40. What is the probability of the pointer *not* landing on green? $\frac{3}{4}$

Real Numbers

CHAPTER OUTLINE

Chapter 10

In this chapter, we begin our study of algebra by learning how to add, subtract, multiply, and divide positive and negative numbers.

Review Your Skills

Before working with positive and negative numbers, review operations on whole numbers by completing the puzzle.

Across

1. $13 + 4 \cdot 7$

3. $7^2 + 1$

5. $186 - 72$

7. $(16 - 3)(27 - 2)$

8. $12^2 - 2(8 - 6)$

Down

2. $22{,}000 \div 5 \cdot 2 + 80^2 + 3$

4. $8[142 + 3(18 - 2)] + 2999$

5. $\dfrac{520(8^2 + 236)}{5^2 - 15}$

6. $6^3 - 2^5$

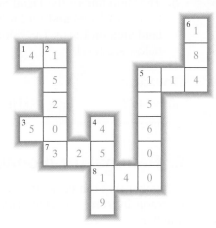

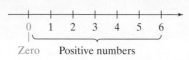

Real Numbers and the Real Number Line

Concepts

1. Integers
2. Rational, Irrational, and Real Numbers
3. Absolute Value
4. Opposite

1. Integers

Thus far in the text we have worked with the number zero and numbers greater than zero. Numbers greater than zero are called **positive numbers**. Positive numbers lie to the right of zero on a number line (Figure 10-1).

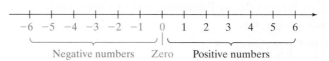

Figure 10-1

In some applications of mathematics, we need to use *negative* numbers. For example:

• On a winter day in Detroit, the low temperature was 5 degrees below zero: $-5°$
• A golfer's score in the U.S. Open was 3 below par: -3
• Maria is $128 overdrawn on her checking account. Her balance is: $-\$128$

The values $-5°$, -3, and $-\$128$ are negative numbers. **Negative numbers** lie to the left of zero on a number line (Figure 10-2).

Figure 10-2

The numbers . . . $-3, -2, -1, 0, 1, 2, 3, . . .$ and so on are called **integers**.

Skill Practice

Write an integer that denotes each number.

1. The average temperature at the South Pole in July is 65°C below zero.
2. Sylvia's checking account is overdrawn by $156.

Classroom Example: p. 594, Exercise 4

Example 1   Writing Integers

Write an integer that denotes each numerical value.

a. Liquid nitrogen freezes at 346°F below zero.

b. The shoreline of the Dead Sea on the border of Israel and Jordan is the lowest land area on Earth. The "altitude" is 1300 ft below sea level.

Solution:

a. $-346°F$ b. -1300 ft

2. Rational, Irrational, and Real Numbers

A number that can be written as a ratio of two integers is called a **rational number** (division by zero is excluded). For example, the following numbers are rational numbers.

$\dfrac{2}{3}$ because it is a ratio of 2 and 3.

$\dfrac{-5}{7}$ because it is a ratio of -5 and 7.

8   because it is a ratio of 8 and 1. That is, $8 = \frac{8}{1}$.
             This shows that an integer is also a rational number.

0.25   because it is a ratio of 25 and 100. That is, $0.25 = \frac{25}{100}$.
             This shows that a terminating decimal is a rational number.

$0.\overline{3}$ because it is a ratio of 1 and 3. That is, $0.\overline{3} = \frac{1}{3}$.
             This shows that a repeating decimal is a rational number.

Rational numbers consist of all numbers that can be expressed as a terminating decimal or as a repeating decimal.

- All numbers that can be expressed as *repeating decimals* are rational numbers.

     For example, $0.\overline{3} = 0.3333\ldots$ is a rational number.

- All numbers that can be expressed as *terminating decimals* are rational numbers.

     For example, 0.25 is a rational number.

- A number that *cannot* be expressed as a repeating or terminating decimal is not a rational number. These are called **irrational numbers**. An example of an irrational number is $\sqrt{2}$.

     For example, $\sqrt{2} \approx \underline{1.41421356237\ldots}$
                        The digits never repeat
                        and never stop.

The rational numbers and the irrational numbers together make up the **set of real numbers**. The relationship among these sets of numbers, along with the set of integers and whole numbers, is illustrated in Figure 10-3.

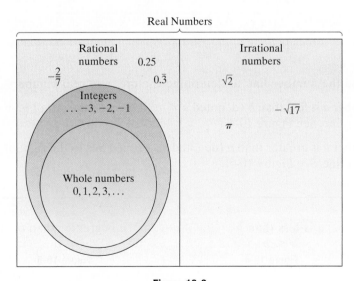

Figure 10-3

Furthermore, every real number can be matched up with a point on the **real number line**.

TIP: In Chapter 8 we used another irrational number, π. Recall that we approximated π with 3.14 or $\frac{22}{7}$ when calculating area and circumference of a circle.

Answers

3. $\frac{4}{5}$ is the ratio of the integers 4 and 5.
4. $6 = \frac{6}{1}$ is the ratio of the integers 6 and 1.
5. $0.3 = \frac{3}{10}$ is the ratio of the integers 3 and 10.
6. $0.\overline{6} = \frac{2}{3}$ is the ratio of the integers 2 and 3.

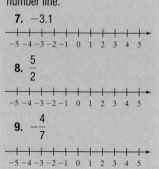
| **Example 2** | **Locating Numbers on a Number Line** |

Locate each number on a number line.

a. -2 **b.** -4.2 **c.** $\dfrac{15}{4}$ **d.** $-\dfrac{1}{3}$

Solution:

a. Because -2 is negative, it is located to the left of zero on the number line.

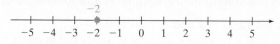

b. Draw a dot two-tenths of a unit to the left of -4.

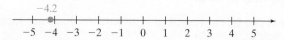

c. Write $\frac{15}{4}$ as a mixed number or decimal.

$$\frac{15}{4} = 3\frac{3}{4} \text{ or } 3.75$$

Draw a dot three-fourths of a unit to the right of 3 on the number line.

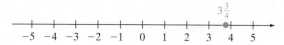

d. $-\dfrac{1}{3}$

Draw a dot one-third of a unit to the left of zero.

We can use the number line to determine the order of real numbers.

- A number a is less than b (denoted $a < b$) if a lies to the left of b on the number line. See Figure 10-4.

- A number a is greater than b (denoted $a > b$) if a lies to the right of b on the number line. See Figure 10-5.

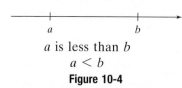

a is less than b
$a < b$
Figure 10-4

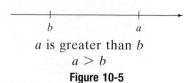

a is greater than b
$a > b$
Figure 10-5

Answers

7.

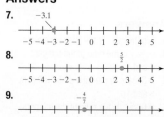

Example 3 Determining Order

Fill in the blank with $<$ or $>$.

a. $-3 \square -5$ **b.** $-2.7 \square 4.1$ **c.** $-\frac{4}{7} \square -\frac{3}{5}$

Solution:

a. $-3 \boxed{>} -5$ because -3 lies to the right of -5

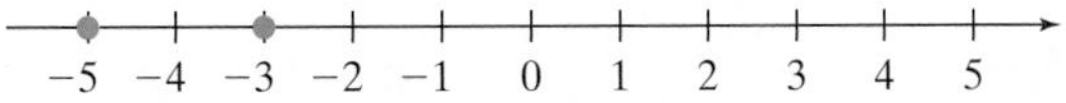

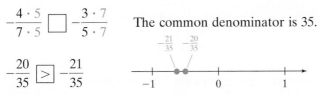

b. $-2.7 \boxed{<} 4.1$ because -2.7 lies to the left of 4.1

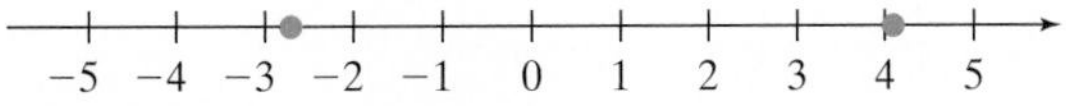

c. To determine order between two fractions, write the fractions with a common denominator.

$-\frac{4 \cdot 5}{7 \cdot 5} \square -\frac{3 \cdot 7}{5 \cdot 7}$ The common denominator is 35.

$-\frac{20}{35} \boxed{>} -\frac{21}{35}$

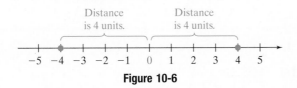

Notice that on the negative side of zero, -20 lies to the right of -21. Therefore, $-\frac{20}{35}$ is greater than $-\frac{21}{35}$.

Skill Practice

Fill in the blank with $<$ or $>$.

10. $-4 \square -6$

11. $5.1 \square -6.1$

12. $-\frac{2}{3} \square -\frac{3}{5}$

Classroom Examples: p. 595, Exercises 40 and 46

3. Absolute Value

Notice on the number line that pairs of numbers such as 4 and -4 are the same distance from 0 (Figure 10-6). The distance between a number and zero on the number line is called its **absolute value**.

Instructor Note: Remind students that "distance" is positive or zero.

Figure 10-6

Absolute Value

The absolute value of a number a is denoted $|a|$. The value of $|a|$ is the distance between a and 0 on the number line.

From the number line we see that $|-4| = 4$ and $|4| = 4$.

Answers

10. $>$ **11.** $>$ **12.** $<$

Example 4 **Finding Absolute Value**

Determine the absolute value.

a. $|-5|$ **b.** $|2.1|$ **c.** $\left|-\dfrac{3}{2}\right|$ **d.** $|0|$

Solution:

a. $|-5| = 5$ The number -5 is 5 units from 0 on the number line.

b. $|2.1| = 2.1$ The number 2.1 is 2.1 units from 0 on the number line.

c. $\left|-\dfrac{3}{2}\right| = \dfrac{3}{2}$ The number $-\dfrac{3}{2}$ is $\dfrac{3}{2}$ units (or 1.5 units) from 0.

d. $|0| = 0$ The number 0 is 0 units from 0 on the number line.

4. Opposite

Two numbers that are the same distance from zero on the number line, but on opposite sides of zero are called **opposites**. For example, the numbers -2 and 2 are opposites (see Figure 10-7).

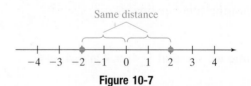

Figure 10-7

The opposite of a number a is denoted $-(a)$.

Original Number a	Opposite $-(a)$	Simplified Form	
5	$-(5)$	-5	The opposite of a positive number is a negative number.
2.4	$-(2.4)$	-2.4	
-7	$-(-7)$	7	The opposite of a negative number is a positive number.
$-\frac{1}{2}$	$-(-\frac{1}{2})$	$\frac{1}{2}$	

Concept Connections

17. True or false. For any real number a, the number $-a$ is negative.

The opposite of a negative number is a positive number. That is, for a positive number, a, the value of $-a$ is negative and $-(-a) = a$. This is sometimes called the *double-negative property*.

Example 5 Finding the Opposite of a Real Number

Find the opposite of each number.

a. 4   **b.** 6.3   **c.** -11 **d.** $-\dfrac{5}{2}$

Solution:

a. The opposite of 4 is -4.

b. The opposite of 6.3 is -6.3.

c. The opposite of -11 is 11.

d. The opposite of $-\frac{5}{2}$ is $\frac{5}{2}$.

> **TIP:** To find the opposite of a number, we can simply change the sign.

Skill Practice

Find the opposite of each number.

18. 8   **19.** -4.8

20. -2 **21.** $\dfrac{5}{3}$

Classroom Examples: p. 596, Exercises 74 and 78

Example 6 Simplifying Expressions

Simplify.

a. $-|-12|$ **b.** $-(-12)$

Solution:

a. $-|-12|$ This expression represents the opposite of the absolute value of -12.

     $= -12$

b. $-(-12)$ This expression represents the opposite of -12.

     $= 12$

Skill Practice

Simplify.

22. $-|-23|$ **23.** $-(-23)$

Classroom Examples: p. 596, Exercises 100 and 102

Answers

17. False. For example, if $a = 0$, then $-a = 0$, and if $a = -6$, then $-a = -(-6) = 6$.

18. -8 **19.** 4.8   **20.** 2

21. $-\dfrac{5}{3}$ **22.** -23 **23.** 23

Section 10.1   Practice Exercises

Study Skills Exercise

For additional exercises, see Classroom Activities 10.1A–10.1B in the *Student's Resource Manual* at www.mhhe.com/moh.

When working with signed numbers, keep a simple example in your mind, such as temperature. We understand that 10 degrees below zero is colder than 2 degrees below zero so the inequality $-10 < -2$ makes sense. Write down another example involving signed numbers that you can easily remember.

Vocabulary and Key Concepts

1. **a.** On a number line, _____positive_____ numbers lie to the right of zero and _____negative_____ numbers lie to the left of zero.

   b. The numbers … $-3, -2, -1, 0, 1, 2, 3, …$ are called _____integers_____ .

   c. A number that can be written as a ratio of two integers is called a(n) _____rational_____ number. A number that cannot be written as a ratio of two integers is called a(n) _____irrational_____ number.

   d. The distance between a number and zero on the number line is called its _____absolute_____ value.

   e. Two numbers that are the same distance from zero on the number line but on opposite sides of zero are called _____opposites_____ .

Concept 1: Integers

For Exercises 2–12, write an integer that represents each numerical value. **(See Example 1.)**

2. A submarine dove to a depth of 340 ft below sea level. $-340\,\text{ft}$

3. Death Valley, California, has elevation of 86 m below sea level. $-86\,\text{m}$

4. In a card game, Jack lost $45. $-\$45$

5. Playing *Wheel of Fortune,* Sally won $3800. $\$3800$

6. Jim's golf score is 5 over par.   5

7. Rena lost $500 in the stock market in one month. $-\$500$

8. LaTonya earned $23 in interest on her savings account. $\$23$

9. Patrick lost 14 lb on a diet. $-14\,\text{lb}$

10. Emilie's score on a video game was 5000 points higher than that of the past record holder.   5000

11. The number of Internet users rose by about 140,000.   140,000

12. A small business experienced a loss of $20,000 last year. $-\$20,000$

Concept 2: Rational, Irrational, and Real Numbers

For Exercises 13–24, locate the numbers on the number line. **(See Example 2.)**

13. $-2, 4$

14. $3, -1$

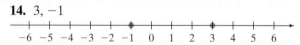

15. $2, -3$

16. $-1, 5$

17. $-\frac{7}{2}, \frac{17}{4}$

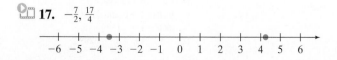

18. $\frac{4}{3}, -\frac{2}{3}$

19. $\frac{11}{4}$, $-\frac{7}{8}$

20. $-\frac{1}{4}$, $\frac{27}{5}$

21. 4.1, -0.8

22. 0, -3.1

23. -2.5, 1.6

24. -1.9, 4.2

For Exercises 25–36, identify the number as rational or irrational.

25. $-\frac{2}{5}$ Rational   **26.** $-\frac{1}{9}$ Rational   **27.** 5   Rational   **28.** 3   Rational

29. 3.5   Rational   **30.** 1.1   Rational   **31.** $\sqrt{7}$ Irrational   **32.** $\sqrt{11}$ Irrational

33. π Irrational   **34.** 2π Irrational   **35.** $\sqrt{4}$ Rational   **36.** $\sqrt{9}$ Rational

For Exercises 37–52, place the correct symbol, $>$ or $<$, between the two numbers. **(See Example 3.)**

37. 0 ☐ -3 $>$ **38.** -1 ☐ 0 $<$ **39.** -8 ☐ -9 $>$ **40.** -5 ☐ -2 $<$

41. -9.1 ☐ 2.2 $<$ **42.** -1.5 ☐ 1.5 $<$ **43.** -3.35 ☐ -3.3 $<$ **44.** 0.9 ☐ -0.5 $>$

45. $-\frac{2}{3}$ ☐ $-\frac{5}{6}$ $>$ **46.** $-\frac{1}{5}$ ☐ $-\frac{1}{4}$ $>$ **47.** $\frac{7}{8}$ ☐ $-\frac{1}{9}$ $>$ **48.** $\frac{1}{3}$ ☐ $-\frac{3}{2}$ $>$

49. 0 ☐ $\frac{1}{10}$ $<$ **50.** $-\frac{8}{7}$ ☐ 1 $<$ **51.** $-\frac{6}{5}$ ☐ -1 $<$ **52.** -1 ☐ $-\frac{10}{11}$ $<$

Concept 3: Absolute Value

For Exercises 53–64, determine the absolute value. **(See Example 4.)**

53. $|-2|$ 2   **54.** $|-6|$ 6   **55.** $|4.5|$ 4.5   **56.** $|2.9|$ 2.9

57. $\left|-\frac{5}{2}\right|$ $\frac{5}{2}$ **58.** $\left|-\frac{4}{9}\right|$ $\frac{4}{9}$ **59.** $|0|$ 0   **60.** $|6|$ 6

61. $|-3.2|$ 3.2   **62.** $|-0.4|$ 0.4   **63.** $|21|$ 21   **64.** $|8|$ 8

65. a. Which is greater, -12 or -8? -8
     b. Which is greater, $|-12|$ or $|-8|$? $|-12|$

66. a. Which is greater, -14 or -20? -14
     b. Which is greater, $|-14|$ or $|-20|$? $|-20|$

67. a. Which is greater, 5.2 or 7.8?   7.8
     b. Which is greater, $|5.2|$ or $|7.8|$? $|7.8|$

68. a. Which is greater, 3.89 or 4.29?   4.29
     b. Which is greater, $|3.89|$ or $|4.29|$? $|4.29|$

69. Which is greater, $-\frac{4}{5}$ or $\left|-\frac{4}{5}\right|$? $\left|-\frac{4}{5}\right|$

70. Which is greater, $-\frac{3}{8}$ or $\left|-\frac{3}{8}\right|$? $\left|-\frac{3}{8}\right|$

71. Which is greater, 10 or $|10|$?
     Neither, they are equal.

72. Which is greater, 256 or $|256|$?
     Neither, they are equal.

Writing   Translating Expression   Geometry   Scientific Calculator   Video   NS E

Concept 4: Opposite

For Exercises 73–84, find the opposite. **(See Example 5.)**

73. 5   _{−5}

74. 31   _{−31}

 75. −12   ₁₂

76. −25   ₂₅

77. $-\dfrac{1}{6}$ $\frac{1}{6}$

78. $-\dfrac{4}{7}$ $\frac{4}{7}$

79. $\dfrac{2}{11}$ $-\frac{2}{11}$

80. $\dfrac{14}{15}$ $-\frac{14}{15}$

81. 8.1   _{−8.1}

82. 9.5   _{−9.5}

83. −1.14   _{1.14}

84. −2.25   _{2.25}

For Exercises 85–96, write in symbols, do not simplify.

85. The opposite of 6 -6

86. The opposite of 23 -23

87. The opposite of negative 2 $-(-2)$

88. The opposite of negative 9 $-(-9)$

89. The absolute value of 7 $|7|$

90. The absolute value of 11 $|11|$

91. The absolute value of negative 3 $|-3|$

92. The absolute value of negative 10 $|-10|$

93. The opposite of the absolute value of 14 $-|14|$

94. The opposite of the absolute value of 42 $-|42|$

95. The opposite of the absolute value of negative 30 $-|-30|$

96. The opposite of the absolute value of negative 5 $-|-5|$

For Exercises 97–108, simplify the expression. **(See Example 6.)**

97. $-|2|$ -2

98. $-|9|$ -9

99. $-|-5.3|$ -5.3

100. $-|-6.9|$ -6.9

101. $-(-15)$ 15

102. $-(-4)$ 4

103. $|-4.7|$ 4.7

104. $|-9.5|$ 9.5

105. $-\left|-\dfrac{12}{17}\right|$ $-\frac{12}{17}$

106. $-\left|-\dfrac{1}{7}\right|$ $-\frac{1}{7}$

107. $-\left(-\dfrac{3}{8}\right)$ $\frac{3}{8}$

108. $-\left(-\dfrac{4}{9}\right)$ $\frac{4}{9}$

Section 10.2   Addition of Real Numbers

Concepts

1. Addition of Integers by Using a Number Line
2. Addition of Real Numbers
3. Translating English Phrases to Mathematical Expressions

1. Addition of Integers by Using a Number Line

Addition of real numbers can be visualized on a number line. To do so, we locate the first addend on the number line. Then to add a positive number, we move to the right on the number line. To add a negative number, we move to the left on the number line. This is demonstrated in Example 1.

Example 1   Using a Number Line to Add Integers

Use a number line to add.

   a. $5 + 3$ **b.** $-5 + 3$

Solution:

   a. $5 + 3 = 8$

Move 3 units *right.*

Start here.

Begin at 5. Then because we are adding *positive* 3, move to the *right* 3 units. The sum is 8.

---

b. $-5 + 3 = -2$

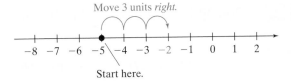

Move 3 units *right.*

Start here.

Begin at -5. Then because we are adding *positive* 3, move to the *right* 3 units. The sum is -2.

Example 2   **Using a Number Line to Add Integers**

Use a number line to add.

a. $5 + (-3)$ **b.** $-5 + (-3)$

Solution:

a. $5 + (-3) = 2$

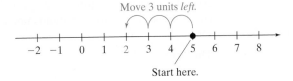

Move 3 units *left.*

Start here.

Begin at 5. Then because we are adding *negative* 3, move to the *left* 3 units. The sum is 2.

b. $-5 + (-3) = -8$

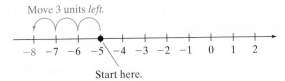

Move 3 units *left.*

Start here.

Begin at -5. Then because we are adding *negative* 3, move to the *left* 3 units. The sum is -8.

TIP: In Example 2(a) and 2(b), parentheses are inserted for clarity. The parentheses separate the number -3 from the $+$ symbol for addition.

$$5 + (-3) \quad \text{and} \quad -5 + (-3)$$

2. Addition of Real Numbers

It is inconvenient to draw a number line each time we want to add signed numbers. Therefore, we offer two rules for adding real numbers. The first rule is used when the addends have the *same* sign (that is, if the addends are both positive or both negative).

Adding Numbers with the Same Sign

To add two numbers with the *same* sign, add their absolute values and apply the common sign.

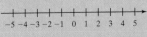

Example 3 Adding Real Numbers with the Same Sign

Add.

a. $-2 + (-4)$ b. $-12 + (-37)$ c. $10 + 66$

Solution:

a. $-2 + (-4)$ First find the absolute value of the addends.
                         $|-2| = 2$ and $|-4| = 4$.

   $= -(2 + 4)$ Add their absolute values and apply the common sign
                         (in this case, the common sign is negative).
   Common sign is negative.

   $= -6$ The sum is -6.

b. $-12 + (-37)$ First find the absolute value of the addends.
                         $|-12| = 12$ and $|-37| = 37$.

   $= -(12 + 37)$ Add their absolute values and apply the common sign
                         (in this case, the common sign is negative).
   Common sign is negative.

   $= -49$ The sum is -49.

c. $10 + 66$ First find the absolute value of the addends.
                         $|10| = 10$ and $|66| = 66$.

   $= +(10 + 66)$ Add their absolute values and apply the common sign
                         (in this case, the common sign is positive).
   Common sign is positive.

   $= 76$ The sum is 76.

The next rule helps us add two numbers with different signs.

Adding Numbers with Different Signs
To add two numbers with *different* signs, subtract the smaller absolute value from the larger absolute value. Then apply the sign of the number having the larger absolute value.

Example 4 Adding Real Numbers with Different Signs

Add.

a. $2 + (-7)$ b. $-6 + 24$ c. $-8 + 8$

Solution:

a. $2 + (-7)$ First find the absolute value of the addends.
                     $|2| = 2$ and $|-7| = 7$.

                     Note: The absolute value of -7 is greater than the absolute
                     value of 2. Therefore, the sum is negative.

   $= -(7 - 2)$ Next, subtract the smaller absolute value from the larger
                     absolute value.
   Apply the sign of the number with the larger absolute value.

   $= -5$

b. $-6 + 24$ First find the absolute value of the addends.
                    $|-6| = 6$ and $|24| = 24$.

                    Note: The absolute value of 24 is greater than the absolute value of -6. Therefore, the sum is positive.

$= +(24 - 6)$ Next, subtract the smaller absolute value from the larger absolute value.

    ⌐ Apply the sign of the number with the larger absolute value.

$= 18$

c. $-8 + 8$ First find the absolute value of the addends.
                   $|-8| = 8$ and $|8| = 8$.

$= (8 - 8)$ The absolute values are equal. Therefore, their difference is 0. The number zero is neither positive nor negative.

$= 0$

Example 4(c) illustrates that the sum of a number and its opposite is zero. For example,

$$-8 + 8 = 0 \qquad -12 + 12 = 0 \qquad 6 + (-6) = 0$$

Example 5   **Adding Signed Numbers**

Simplify.

a. $-10 + 4 + (-16)$ **b.** $-30 + (-12) + 4 + (-10) + 6$

Solution:

a. $-10 + 4 + (-16)$

$= -6 + (-16)$ Apply the order of operations by adding from left to right.

$= -22$

b. $-30 + (-12) + 4 + (-10) + 6$ Add from left to right.

$= -42 + 4 + (-10) + 6$

$= -38 + (-10) + 6$

$= -48 + 6$

$= -42$

TIP: When several numbers are added, we can reorder and regroup the addends by using the commutative property and associative property of addition. In particular, we can group all the positive addends, and we can group all the negative addends. This makes the arithmetic easier. For example:

        positive   negative
        addends   addends

$$-30 + (-12) + 4 + (-10) + 6 = 4 + 6 + (-30) + (-12) + (-10)$$

$$= 10 + (-52)$$

$$= -42$$

Answers

13. -13 **14.** -52

Skill Practice

Add.

15. $-16.8 + 14.3$

Classroom Example: p. 603,
Exercise 78

Example 6 **Adding Real Numbers**

Add. $\quad -2.73 + 4.81$

Solution:

$-2.73 + 4.81$ Find the absolute value of the addends.
$|-2.73| = 2.73$ and $|4.81| = 4.81$.

The sum is positive because 4.81 has a greater absolute value than -2.73.

$= +(4.81 - 2.73)$ Subtract the smaller absolute value from the larger absolute value.

$\quad$ Apply the sign of the number with the larger absolute value.

$= 2.08$

Skill Practice

Add.

16. $-\dfrac{6}{8} + \dfrac{5}{8}$

17. $2\dfrac{7}{15} + \left(-1\dfrac{4}{5}\right)$

Classroom Example: p. 603,
Exercise 82

Example 7 **Adding Real Numbers**

Add. $\quad$ **a.** $-\dfrac{6}{11} + \left(-\dfrac{3}{11}\right)$ $\quad$ **b.** $1\dfrac{2}{15} + \left(-2\dfrac{4}{5}\right)$

Solution:

a. $-\dfrac{6}{11} + \left(-\dfrac{3}{11}\right)$ Find the absolute value of the addends.

$$\left|-\frac{6}{11}\right| = \frac{6}{11} \text{ and } \left|-\frac{3}{11}\right| = \frac{3}{11}.$$

$= -\left(\dfrac{6}{11} + \dfrac{3}{11}\right)$ Add their absolute values and apply the common sign (in this case, the common sign is negative).

$\quad$ Common sign is negative.

$= -\dfrac{9}{11}$

b. $1\dfrac{2}{15} + \left(-2\dfrac{4}{5}\right)$

$= \dfrac{17}{15} + \left(-\dfrac{14}{5}\right)$ Convert the mixed numbers to improper fractions.

Avoiding Mistakes

Do not forget to get a common denominator before adding or subtracting fractions.

$= \dfrac{17}{15} + \left(-\dfrac{14 \cdot 3}{5 \cdot 3}\right)$ The least common denominator is 15.

$= \dfrac{17}{15} + \left(-\dfrac{42}{15}\right)$ Find the absolute value of each addend.

$$\left|\frac{17}{15}\right| = \frac{17}{15} \text{ and } \left|-\frac{42}{15}\right| = \frac{42}{15}$$

$= -\left(\dfrac{42}{15} - \dfrac{17}{15}\right)$ Subtract the smaller absolute value from the larger.

$\quad$ Apply the sign of the number with the larger absolute value.

$= -\dfrac{25}{15}$ Subtract.

$= -\dfrac{5}{3}$ Simplify to lowest terms. $-\dfrac{\overset{5}{\cancel{25}}}{\underset{3}{\cancel{15}}} = -\dfrac{5}{3}$

$= -1\dfrac{2}{3}$ Write as a mixed number.

Answers

15. -2.5 **16.** $-\dfrac{1}{8}$ **17.** $\dfrac{2}{3}$

3. Translating English Phrases to Mathematical Expressions

Recall from Section 1.2 that several key words imply addition: *sum; added to; increased by; more than; plus;* and *total of.*

Example 8 **Translating to a Mathematical Expression**

Translate to a mathematical expression. Then simplify the expression.

The sum of -14.1, 8.7, and 12.9

Solution:

$-14.1 + 8.7 + 12.9$	Translate the English phrase.
$= -5.4 + 12.9$	Simplify. Add from left to right.
$= \quad 7.5$	

Skill Practice

Translate to a mathematical expression. Then simplify.

18. The sum of -6.4, 7.2, and -3.1

Classroom Example: p. 603, Exercise 88

Example 9 **Translating to a Mathematical Expression**

Translate to a mathematical expression. Then simplify the expression.

The number $-\dfrac{1}{6}$ added to $\dfrac{2}{3}$

Solution:

$\dfrac{2}{3} + \left(-\dfrac{1}{6}\right)$	Translate the English phrase.
$= \dfrac{2 \cdot 2}{3 \cdot 2} + \left(-\dfrac{1}{6}\right)$	Write the first fraction with a least common denominator of 6.
$= \dfrac{4}{6} + \left(-\dfrac{1}{6}\right)$	Find the absolute value of the addends.
	$\left|\dfrac{4}{6}\right| = \dfrac{4}{6}$ and $\left|-\dfrac{1}{6}\right| = \dfrac{1}{6}$.
	The absolute value of $\dfrac{4}{6}$ is greater than the absolute value of $-\dfrac{1}{6}$. Therefore, the sum is positive.
$= +\left(\dfrac{4}{6} - \dfrac{1}{6}\right)$	Subtract the smaller absolute value from the larger absolute value.
⌐ Apply the sign of the number with the larger absolute value.	
$= \dfrac{3}{6}$	Subtract.
$= \dfrac{1}{2}$	Simplify to lowest terms.

Skill Practice

Translate to a mathematical expression. Then simplify.

19. $\dfrac{1}{5}$ added to $-\dfrac{3}{10}$

Classroom Example: p. 603, Exercise 90

Answers

18. $-6.4 + 7.2 + (-3.1)$; -2.3

19. $-\dfrac{3}{10} + \dfrac{1}{5}$; $-\dfrac{1}{10}$

Section 10.2   Practice Exercises

For additional exercises, see Classroom Activities 10.2A–10.2B in the *Student's Resource Manual* at www.mhhe.com/moh.

Study Skills Exercise

When you are trying to memorize rules for signed numbers, 3×5 cards come in handy. Write the rule you want to memorize on one side and an example of the rule on the other side. Keep these cards handy so that when you have a few minutes (such as waiting for the doctor), you can pull them out and quiz yourself. Write the rules from this section that would be good to put on cards.

Vocabulary and Key Concepts

1. a. The sum of a number and its opposite is ___0___ .

b. The sum of two negative numbers is (positive/negative).   negative
The sum of two positive numbers is (positive/negative).   positive

c. Explain how to find the sum of two numbers with different signs.   Subtract the smaller absolute value from the larger absolute value. The sum takes on the sign of the addend with the greater absolute value.

Review Exercises

For Exercises 2–8, place the correct symbol ($>$, $<$, or $=$) between the two numbers.

2. $-6 \;\square\; -5$ $<$

3. $-\dfrac{2}{3} \;\square\; -\dfrac{11}{12}$ $>$

4. $|-2.4| \;\square\; -|2.4|$ $>$

5. $|6| \;\square\; |-6|$ $=$

6. $0 \;\square\; -0.6$ $>$

7. $-|-10| \;\square\; 10$ $<$

8. $-(-2) \;\square\; 2$ $=$

Concept 1: Addition of Integers by Using a Number Line

For Exercises 9–20, refer to the number line to add the integers. **(See Examples 1 and 2.)**

9. $2 + (-4)$ -2

10. $5 + (-1)$ 4

11. $-3 + 5$ 2

12. $-6 + 3$ -3

13. $-4 + (-4)$ -8

14. $-2 + (-5)$ -7

15. $-3 + 9$ 6

16. $-1 + 5$ 4

17. $0 + (-7)$ -7

18. $(-5) + 0$ -5

19. $-1 + (-3)$ -4

20. $-4 + 3$ -1

Concept 2: Addition of Real Numbers

21. Explain the process to add two numbers with the same sign.
To add two numbers with the same sign, add their absolute values and apply the common sign.

For Exercises 22–29, add the numbers with the same sign. **(See Example 3.)**

22. $23 + 12$ 35

23. $12 + 3$ 15

24. $-70 + (-15)$ -85

25. $-40 + (-33)$ -73

26. $-6 + (-10)$ -16

27. $-100 + (-24)$ -124

28. $23 + 50$ 73

29. $44 + 45$ 89

30. Explain the process to add two numbers with different signs.   To add two numbers with different signs, subtract the smaller absolute value from the larger absolute value. Then apply the sign of the number having the larger absolute value.

For Exercises 31–42, add the numbers with different signs. **(See Example 4.)**

31. $75 + (-23)$ 52

32. $12 + (-7)$ 5

33. $-34 + 12$ -22

34. $-88 + 35$ -53

35. $-90 + 66$ -24

36. $-23 + 49$ 26

37. $78 + (-33)$ 45

38. $10 + (-23)$ -13

39. $2 + (-2)$ 0

40. $-6 + 6$ 0

41. $-1.3 + 1.3$ 0

42. $4.5 + (-4.5)$ 0

 Writing   Translating Expression   Geometry   Scientific Calculator   Video   NS&E

Mixed Exercises

For Exercises 43–66, simplify. **(See Example 5.)**

43. $12 + (-3)$ 9

44. $-33 + (-1)$ -34

45. $-23 + (-3)$ -26

46. $-5 + 15$ 10

47. $4 + (-45)$ -41

48. $-13 + (-12)$ -25

49. $(-103) + (-47)$ -150

50. $119 + (-59)$ 60

51. $0 + (-17)$ -17

52. $-29 + 0$ -29

53. $-19 + (-22)$ -41

54. $-300 + (-24)$ -324

55. $6 + (-12) + 8$ 2

56. $20 + (-12) + (-5)$ 3

57. $-33 + (-15) + 18$ -30

58. $3 + 5 + (-1)$ 7

59. $7 + (-3) + 6$ 10

60. $12 + (-6) + (-9)$ -3

61. $-10 + (-3) + 5$ -8

62. $-23 + (-4) + (-12) + (-5)$ -44

63. $-18 + (-5) + 23$ 0

64. $14 + (-15) + 20 + (-42)$ -23

65. $4 + (-12) + (-30) + 16 + 10$ -12

66. $24 + (-5) + (-19)$ 0

For Exercises 67–84, add the real numbers. **(See Examples 6 and 7.)**

67. $23.9 + 2.1$ 26

68. $10.9 + 6.3$ 17.2

69. $-34.2 + (-4.1)$ -38.3

70. $-8.6 + (-12)$ -20.6

71. $-\dfrac{3}{4} + \left(-\dfrac{5}{4}\right)$ -2

72. $-\dfrac{2}{5} + \left(-\dfrac{1}{10}\right)$ $-\dfrac{1}{2}$

73. $-1\dfrac{1}{6} + \left(-2\dfrac{1}{3}\right)$ $-3\dfrac{1}{2}$

74. $-3\dfrac{2}{15} + \left(-4\dfrac{1}{5}\right)$ $-7\dfrac{1}{3}$

75. $34.8 + (-45)$ -10.2

76. $90 + (-12.3)$ 77.7

77. $-23.1 + 24.5$ 1.4

78. $-12.2 + 10.9$ -1.3

79. $\dfrac{3}{8} + \left(-\dfrac{3}{16}\right)$ $\dfrac{3}{16}$

80. $\dfrac{1}{3} + \left(-\dfrac{7}{9}\right)$ $-\dfrac{4}{9}$

81. $\left(-\dfrac{5}{6}\right) + \dfrac{1}{4}$ $-\dfrac{7}{12}$

82. $\left(-\dfrac{4}{5}\right) + \dfrac{7}{20}$ $-\dfrac{9}{20}$

83. $1\dfrac{3}{10} + \left(-2\dfrac{4}{5}\right)$ $-1\dfrac{1}{2}$

84. $-7\dfrac{3}{4} + 5\dfrac{1}{12}$ $-2\dfrac{2}{3}$

Concept 3: Translating English Phrases to Mathematical Expressions

85. Give at least two words or phrases that would indicate addition.   Sum, added to, increased by, more than, plus, total

For Exercises 86–95, translate to a mathematical expression. Then simplify the expression. **(See Examples 8 and 9.)**

86. The sum of -23 and 49 $-23 + 49$; 26

87. The sum of 89 and -11 $89 + (-11)$; 78

88. The total of 3, -10, and 5 $3 + (-10) + 5$; -2

89. The total of -2, -4, 14, and 20 $-2 + (-4) + 14 + 20$; 28

90. The number -4.2 is added to -2.2 $-2.2 + (-4.2)$; -6.4

91. The number -4.5 is added to -12 $-12 + (-4.5)$; -16.5

92. 8 more than $-\dfrac{1}{4}$ $-\dfrac{1}{4} + 8$; $\dfrac{31}{4}$

93. 2 more than $-\dfrac{1}{3}$ $-\dfrac{1}{3} + 2$; $\dfrac{5}{3}$

94. $-\dfrac{3}{4}$ increased by 6 $-\dfrac{3}{4} + 6$; $\dfrac{21}{4}$

95. $-\dfrac{1}{5}$ increased by 1 $-\dfrac{1}{5} + 1$; $\dfrac{4}{5}$

Writing   Translating Expression   Geometry   Scientific Calculator   Video   NS&E

96. At 6:00 A.M. the temperature was −4°F. By noon, the temperature had risen by 12°F. What was the temperature at noon? 8°F

97. At noon the temperature was 14°F. By midnight, the temperature had fallen 20°F. What was the temperature at midnight? −6°F

98. Jorge's checking account is overdrawn. His beginning balance was −$56.52. If he deposits his paycheck for $389.81, what is his new balance? $333.29

99. Ellen's checking account balance is $23.89. If she writes a check for $40.00, what is her balance? −$16.11

100. Ron bought a new pair of glasses. A bill for $320.50 was sent to his insurance company. If his insurance paid only $150, what is Ron's balance? $170.50

101. Savannah was in a minor accident with her car. To repair the damaged car, she paid $570.32. Her insurance company paid for the repair except for a $250 deductible. How much did the insurance company pay? $320.32

Expanding Your Skills

102. Find two integers whose sum is −10. Answers may vary. For example: −12 + 2

103. Find two integers whose sum is −14. Answers may vary. For example: −6 + (−8)

104. Find two integers whose sum is −2. Answers may vary. For example: −1 + (−1)

105. Find two integers whose sum is 0. Answers may vary. For example: 5 + (−5)

Calculator Connections

Topic: Adding Real Numbers on a Calculator

To enter negative numbers on a calculator, use the $(-)$ key or the $+\circ-$ key. To use the $(-)$ key, enter the number the same way that it is written. That is, enter the negative sign first and then the number, such as $(-)$ 5. If your calculator has the $+\circ-$ key, type the number first, followed by the $+\circ-$ key. Thus, −5 is entered as 5 $+\circ-$.

Try entering the expressions below to determine which method your calculator uses.

Expression	Keystrokes	Result
−10 + (−3)	$(-)$ 10 $+$ $(-)$ 3 **ENTER** or 10 $+\circ-$ $+$ 3 $+\circ-$ $=$	−13
−4.2 + 6.7	$(-)$ 4.2 $+$ 6.7 **ENTER** or 4.2 $+\circ-$ $+$ 6.7 $=$	2.5

Calculator Exercises

For Exercises 106–111, add by using a calculator.

106. 302 + (−422)
−120

107. −900 + 334
−566

108. −23.991 + (−44.23)
−68.221

109. −103.4 + (−229.1)
−332.5

110. 23 + (−125) + 912 + (−99)
711

111. 891 + 12 + (−223) + (−341)
339

Subtraction of Real Numbers

1. Definition of Subtraction

In Section 10.2, we learned the rules for adding real numbers. Subtraction of real numbers is defined in terms of the addition process. For example, consider the following subtraction problem. The corresponding addition problem produces the same result.

$$6 - 4 = 2 \quad \Leftrightarrow \quad 6 + (-4) = 2$$

In each case, we start at 6 on the number line and move to the *left* 4 units. Adding the *opposite* of 4 produces the same result as subtracting 4. This is true in general. To subtract two real numbers, add the opposite of the second number to the first number.

Concepts

1. Definition of Subtraction
2. Subtraction of Real Numbers
3. Applying the Order of Operations
4. Applications of Subtraction

Subtracting Real Numbers

If a and b are real numbers, then $a - b = a + (-b)$.
Therefore, to perform subtraction, follow these steps:

Step 1 Leave the first number (the minuend) unchanged.
Step 2 Change the subtraction sign to an addition sign.
Step 3 Add the opposite of the second number (the subtrahend).

For example,

$$\left. \begin{array}{l} 10 - 4 = 10 + (-4) = 6 \\ -10 - 4 = -10 + (-4) = -14 \end{array} \right\}$$ Subtracting 4 is the same as adding -4.

$$\left. \begin{array}{l} 10 - (-4) = 10 + (4) = 14 \\ -10 - (-4) = -10 + (4) = -6 \end{array} \right\}$$ Subtracting -4 is the same as adding 4.

Concept Connections

Fill in the blank to change subtraction to addition of the opposite.

1. $9 - 3 = 9 + \boxed{}$
2. $-9 - 3 = -9 + \boxed{}$
3. $9 - (-3) = 9 + \boxed{}$
4. $-9 - (-3) = -9 + \boxed{}$

Example 1   Subtracting Real Numbers

Subtract.

a. $15 - 20$ **b.** $-7 - 12$ **c.** $40 - (-8)$

Solution:

Add the opposite of 20.

a. $15 - \underset{\uparrow}{\boxed{20}} = 15 + (-20) = -5$ Rewrite the subtraction in terms of addition of the opposite.

Change subtraction to addition.

b. $-7 - 12 = -7 + (-12)$ Rewrite as addition of the opposite.

$\qquad = -19$

c. $40 - (-8) = 40 + (8)$ Rewrite as addition of the opposite.

$\qquad = 48$

Skill Practice

Subtract.

5. $12 - 19$
6. $-8 - 14$
7. $30 - (-3)$

Classroom Examples: p. 609, Exercises 20 and 22

Answers

1. -3 **2.** -3 **3.** 3   **4.** 3
5. -7 **6.** -22 **7.** 33

Recall from Section 1.3 that several key words imply subtraction.

Word or Phrase	Example	In Symbols
a minus b	-15 minus 10	$-15 - 10$
the difference of a and b	the difference of 10 and -2	$10 - (-2)$
a decreased by b	9 decreased by 1	$9 - 1$
a less than b	-12 less than 5	$5 - (-12)$
subtract a from b	subtract -3 from 8	$8 - (-3)$

Skill Practice

Translate to a mathematical expression. Then simplify.

8. The difference of -16 and 4
9. -8 decreased by -9

Classroom Examples: p. 610,
Exercises 50 and 52

Example 2 **Translating to a Mathematical Expression**

Translate each English phrase to a mathematical expression. Then simplify.

a. The difference of -52 and 10

b. -35 decreased by -6

Solution:

a. the difference of
$$-52 \;\overset{\downarrow}{-}\; 10 \qquad \text{Translate: The difference of } -52 \text{ and } 10$$
$$= -52 + (-10) \qquad \text{Rewrite as addition of the opposite.}$$
$$= -62$$

b. decreased by
$$-35 \;\overset{\downarrow}{-}\; (-6) \qquad \text{Translate: } -35 \text{ decreased by } -6.$$
$$= -35 + (6) \qquad \text{Rewrite subtraction in terms of addition.}$$
$$= -29$$

Skill Practice

Translate to a mathematical expression. Then simplify.

10. 6 less than 2
11. Subtract -4 from -1.

Classroom Examples: p. 610,
Exercises 54 and 56

Example 3 **Translating to a Mathematical Expression**

Translate each English phrase to a mathematical expression. Then simplify.

a. 12 less than -8

b. Subtract 27 from 5.

Solution:

a. To translate "12 less than -8," we must *start* with -8 and then subtract 12.
$$-8 - 12 \qquad \text{Translate: 12 less than } -8.$$
$$= -8 + (-12) \qquad \text{Rewrite as addition of the opposite.}$$
$$= -20$$

b. To translate "Subtract 27 from 5," we must *start* with 5, then subtract 27.
$$5 - 27 \qquad \text{Translate: Subtract 27 from 5.}$$
$$= 5 + (-27) \qquad \text{Rewrite as addition of the opposite.}$$
$$= -22$$

Answers

8. $-16 - 4$; -20
9. $-8 - (-9)$; 1
10. $2 - 6$; -4
11. $-1 - (-4)$; 3

2. Subtraction of Real Numbers

Example 4 Subtracting Real Numbers

Subtract.

a. $-6.7 - 4.2$ **b.** $\dfrac{9}{4} - \dfrac{7}{4}$ **c.** $2\dfrac{3}{10} - \left(-1\dfrac{7}{30}\right)$

Solution:

a. $-6.7 - 4.2$

$= -6.7 + (-4.2)$ Rewrite as addition of the opposite.

$= -10.9$

b. $\dfrac{9}{4} - \dfrac{7}{4}$

$= \dfrac{9}{4} + \left(-\dfrac{7}{4}\right)$ Rewrite as addition of the opposite.

$= \dfrac{2}{4}$ Add.

$= \dfrac{1}{2}$ Simplify.

c. $2\dfrac{3}{10} - \left(-1\dfrac{7}{30}\right)$

$= \dfrac{23}{10} - \left(-\dfrac{37}{30}\right)$ Convert the mixed numbers to improper fractions.

$= \dfrac{23}{10} + \dfrac{37}{30}$ Rewrite as addition of the opposite.

$= \dfrac{23 \cdot 3}{10 \cdot 3} + \dfrac{37}{30}$ The least common denominator is 30.

$= \dfrac{69}{30} + \dfrac{37}{30}$

$= \dfrac{106}{30}$ Add.

$= \dfrac{53}{15}$ Simplify to lowest terms. $\dfrac{\overset{53}{\cancel{106}}}{\underset{15}{\cancel{30}}} = \dfrac{53}{15}$

$= 3\dfrac{8}{15}$ Write as a mixed number.

Skill Practice

Subtract.

12. $-4.6 - 7.1$

13. $-\dfrac{13}{15} + \dfrac{7}{15}$

14. $2\dfrac{5}{6} - \left(-9\dfrac{1}{2}\right)$

Classroom Examples: p. 610, Exercises 58 and 68

Answers

12. -11.7 **13.** $-\dfrac{2}{5}$ **14.** $12\dfrac{1}{3}$

TIP: Refer to page 65 to review the order of operations.

Skill Practice

Simplify.

15. $-8 - 10 + (-6) - (-1)$

16. $(-4 - 2)^2$

Classroom Examples: p. 610, Exercises 76 and 78

3. Applying the Order of Operations

In Example 5, we revisit the order of operations.

Example 5 Applying the Order of Operations

Simplify.   **a.** $-4 - 6 + (-3) - 5 + 8$ **b.** $[3 - (-4)]^2$

Solution:

a. $-4 - 6 + (-3) - 5 + 8$

$= \underbrace{-4 + (-6)} + (-3) + (-5) + 8$ Rewrite all subtractions in terms of addition.

$= \underbrace{-10 + (-3)} + (-5) + 8$ Add from left to right.

$= \underbrace{-13 + (-5)} + 8$

$= \underbrace{-18 + 8}$

$= -10$

b. $[3 - (-4)]^2$ Perform operations inside grouping symbols.

$= [3 + (4)]^2$ Rewrite as addition of the opposite.

$= [7]^2$ Add.

$= 49$ Simplify.

4. Applications of Subtraction

Example 6 Applying Subtraction of Real Numbers

A helicopter is hovering at a height of 200 ft above the ocean. A submarine is directly below the helicopter 125 ft below sea level. Find the difference in elevation between the helicopter and the submarine.

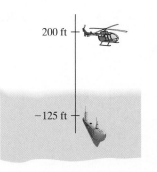

Skill Practice

17. The highest point in California is Mt. Whitney at 14,494 ft above sea level. The lowest point in California is Death Valley at an "altitude" of -282 ft (282 ft below sea level). Find the difference in the elevation between the highest point and lowest point in California.

Classroom Example: p. 611, Exercise 94

Solution:

$$\begin{pmatrix} \text{Difference between} \\ \text{elevation of helicopter} \\ \text{and submarine} \end{pmatrix} = \begin{pmatrix} \text{elevation of} \\ \text{helicopter} \end{pmatrix} - \begin{pmatrix} \text{"elevation" of} \\ \text{submarine} \end{pmatrix}$$

$= 200 \text{ ft} - (-125 \text{ ft})$

$= 200 \text{ ft} + (125 \text{ ft})$ Rewrite as addition.

$= 325 \text{ ft}$

The helicopter and submarine are 325 ft apart.

Answers

15. -23 **16.** 36   **17.** 14,776 ft

Section 10.3 Practice Exercises

Study Skills Exercise

For additional exercises, see Classroom Activities 10.3A–10.3B in the *Student's Resource Manual* at www.mhhe.com/moh.

In this section we find that the symbol $-$ has more than one meaning. It can mean minus, opposite, or negative. Write yourself an explanation and example for each of these meanings of $-$.

Minus: Opposite: Negative:

Vocabulary and Key Concepts

1. a. If a and b are real numbers, then $a - b = a +$ _____$(-b)$_____.

b. To write $-5 - (-4)$ as an equivalent statement using addition, we write _____$-5 + 4$_____.

Review Exercises

For Exercises 2–9, add the numbers.

2. $34 + (-13)$ 21

3. $-34 + (-13)$ -47

4. $-34 + 13$ -21

5. $-\dfrac{5}{9} + \dfrac{7}{12}$ $\dfrac{1}{36}$

6. $\dfrac{5}{9} + \left(-\dfrac{7}{12}\right)$ $-\dfrac{1}{36}$

7. $-\dfrac{5}{9} + \left(-\dfrac{7}{12}\right)$ $-\dfrac{41}{36}$

8. $3 + (-4.2) + \left(-\dfrac{2}{5}\right) + 5$ 3.4

9. $\left(-\dfrac{1}{2}\right) + 6.5 + (-8) + 2 + (-4)$ -4

Concept 1: Definition of Subtraction

10. Explain the process to subtract signed numbers.

To subtract two numbers, add the opposite of the second number to the first.

For Exercises 11–18, rewrite the subtraction problem as an equivalent addition problem. Then simplify. **(See Example 1.)**

11. $2 - 9 =$ ____ $+$ ____ $=$ ____
$2 + (-9); -7$

12. $5 - 11 =$ ____ $+$ ____ $=$ ____
$5 + (-11); -6$

13. $4 - (-3) =$ ____ $+$ ____ $=$ ____
$4 + 3; 7$

14. $12 - (-8) =$ ____ $+$ ____ $=$ ____
$12 + 8; 20$

15. $-3 - 15 =$ ____ $+$ ____ $=$ ____
$-3 + (-15); -18$

16. $-7 - 21 =$ ____ $+$ ____ $=$ ____
$-7 + (-21); -28$

17. $-11 - (-13) =$ ____ $+$ ____ $=$ ____
$-11 + 13; 2$

18. $-23 - (-9) =$ ____ $+$ ____ $=$ ____
$-23 + 9; -14$

For Exercises 19–42, subtract the integers.

19. $35 - (-17)$ 52

20. $23 - (-12)$ 35

21. $-24 - 9$ -33

22. $-5 - 15$ -20

23. $50 - 62$ -12

24. $38 - 46$ -8

25. $-17 - (-25)$ 8

26. $-2 - (-66)$ 64

27. $-8 - (-8)$ 0

28. $-14 - (-14)$ 0

29. $120 - (-41)$ 161

30. $91 - (-62)$ 153

31. $-15 - 19$ -34

32. $-82 - 44$ -126

33. $3 - 25$ -22

34. $6 - 33$ -27

35. $-13 - 13$ -26

36. $-43 - 43$ -86

37. $24 - 25$ -1

38. $43 - 98$ -55

39. $-6 - (-38)$ 32

40. $-75 - (-21)$ -54

41. $-48 - (-33)$ -15

42. $-29 - (-32)$ 3

43. State at least two words or phrases that would indicate subtraction. Minus, difference, decreased, less than, subtract from

44. Is subtraction commutative? For example, does $3 - 7 = 7 - 3$? Subtraction is not commutative. $3 - 7 \neq 7 - 3$.

Writing   Translating Expression   Geometry   Scientific Calculator   Video   NS E

↔ For Exercises 45–56, translate each English phrase to a mathematical expression. Then simplify. **(See Examples 2 and 3.)**

45. 14 minus 23 $14 - 23; -9$

46. 27 minus 40 $27 - 40; -13$

47. Subtract 12 from 5. $5 - 12; -7$

48. Subtract 10 from 16.
$16 - 10; 6$

49. The difference of 105 and 110
$105 - 110; -5$

50. The difference of 70 and 98
$70 - 98; -28$

51. 320 decreased by -20
$320 - (-20); 340$

52. 150 decreased by 75
$150 - 75; 75$

53. Subtract 24 from -35.
$-35 - 24; -59$

54. Subtract 189 from 175.
$175 - 189; -14$

55. 21 less than -34
$-34 - 21; -55$

56. 22 less than -90
$-90 - 22; -112$

Concept 2: Subtraction of Real Numbers

For Exercises 57–72, subtract the real numbers. **(See Example 4.)**

57. $5.2 - 13.5$
-8.3

58. $4.4 - 10.2$
-5.8

59. $-2.3 - 1.9$
-4.2

60. $-4.1 - 2.1$
-6.2

61. $-3.6 - (-9.1)$
5.5

62. $-8.9 - (-10.5)$
1.6

63. $5.5 - (-2.8)$
8.3

64. $11.9 - (-4.3)$
16.2

65. $\frac{2}{3} - \left(-\frac{1}{6}\right)$ $\frac{5}{6}$

66. $\frac{5}{9} - \left(-\frac{2}{9}\right)$ $\frac{7}{9}$

67. $-\frac{3}{10} - \left(-\frac{7}{10}\right)$ $\frac{2}{5}$

68. $-\frac{5}{8} - \left(-\frac{3}{4}\right)$ $\frac{1}{8}$

69. $1\frac{3}{14} - 2\frac{5}{7}$ $-1\frac{1}{2}$

70. $2\frac{3}{20} - 5\frac{9}{10}$ $-3\frac{3}{4}$

71. $-\frac{1}{2} - \frac{5}{4}$ $-\frac{7}{4}$

72. $-\frac{11}{15} - \frac{3}{5}$ $-\frac{4}{3}$

Concept 3: Applying the Order of Operations

For Exercises 73–84, simplify by using the order of operations. **(See Example 5.)**

73. $2 + 5 - (-3) - 10$ 0

74. $4 - 8 + 12 - (-1)$ 9

75. $-5 + 6 + (-7) - 4 - (-9)$ -1

76. $-2 - 1 + (-11) + 6 - (-8)$
0

77. $[-2 - (-6)]^2$ 16

78. $[-1 - (-4)]^2$ 9

79. $[-5 - (-6)]^3$ 1

80. $[-3 - (-5)]^3$ 8

81. $25 - 13 - (-40)$ 52

82. $-35 + 15 - (-28)$ 8

83. $5.5 - \left(\frac{1}{2} - \frac{1}{5}\right)$ 5.2

84. $-6.8 - \left(\frac{2}{5} + \frac{3}{10}\right)$ -7.5

Concept 4: Applications of Subtraction

85. The liquid hydrogen in the Space Shuttle main engine is $-423°$F. The temperature in the engine's combustion chamber reaches 6000°F. Find the difference between the temperature in the combustion chamber and the temperature of the liquid hydrogen. (*Source:* NASA, www.nasa.gov)   **(See Example 6.)** $6423°$F

86. The Moon does not have an atmosphere, so temperatures range from $-184°$C during its night to 214°C during its day. Find the difference between the highest temperature on the Moon and the lowest temperature. $398°$C

87. The Campus Food Court reports its total profit or loss balances each day. The table shows reported profits/losses for a 1-week period.

If the Campus Food Court's profit/loss balance was $17,476.55 at the beginning of the week, what is the balance at the end of the reported week?
The balance was $18,085.51.

Monday	+$1786.84
Tuesday	−$2342.47
Wednesday	−$754.32
Thursday	+$321.63
Friday	+$1597.28

88. For nine holes of golf, Ernie Els made the following scores: $-1, 0, 0, -1, 0, 1, 0, -2, 0$. A negative number represents a score *below* par and a positive number represents a score *above* par. What is his total score after nine holes?
Ernie's score is -3, which means 3 below par.

For Exercises 89 and 90, refer to the graph indicating the change in value of a particular stock for one week.

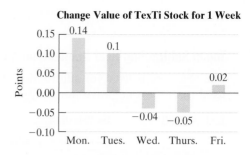

Change Value of TexTi Stock for 1 Week

89. What is the difference between the change in value of TexTi stock on Monday compared to its change in value on Wednesday? The difference is 0.18 point.

90. What is the difference between the change in value of TexTi stock on Tuesday and its change in value on Thursday?
The difference is 0.15 point.

91. Ivan owes $320 on his credit card; that is, his balance is −$320. If he charges $55 for a night out, what is his new balance?
His new balance is −$375.

92. If Justin's balance on his credit card was −$210 and he made the minimum payment of $25, what is his new balance? His balance is −$185.

93. The height of Mount Everest is 29,029 ft. The lowest point on the surface of the Earth is −35,798 ft (that is, 35,798 ft below sea level), occurring at the Mariana Trench on the Pacific Ocean floor. What is the difference in altitude between the height of Mt. Everest and the Mariana Trench? 64,827 ft

94. Mt. Rainier is 4392 m at its highest point. Death Valley, California, is 86 m below sea level (−86) at the basin, Badwater. What is the difference between the altitude of Mt. Rainier and the altitude at Badwater? 4478 m

In a statistics class, a student learns that the **range** of a set of data is the difference between the highest and lowest values. That is, range = highest − lowest.

For Exercises 95 and 96, find the range.

95. Low temperatures (°C) for one week in Anchorage, Alaska: −4°, −8°, 0°, 3°, −8°, −1°, 2° The range is 3° − (−8°) = 11°.

96. Low temperatures (°C) for one week in Fargo, North Dakota: −6°, −2°, −10°, −4°, −12°, −1°, −3° The range is −1° − (−12°) = 11°.

97. Find two integers whose difference is −6. Answers may vary. For example, 4 − 10

98. Find two integers whose difference is −20. Answers may vary. For example, 10 − 30

Expanding Your Skills

For Exercises 99–102, write the next three numbers in the sequence.

99. 5, 1, −3, −7, ___, ___, ___ −11, −15, −19

100. −13, −18, −23, −28, ___, ___, ___ −33, −38, −43

101. $\frac{1}{3}$, 0, $-\frac{1}{3}$, $-\frac{2}{3}$, ___, ___, ___ $-1, -\frac{4}{3}, -\frac{5}{3}$

102. $\frac{1}{4}$, 0, $-\frac{1}{4}$, $-\frac{1}{2}$, ___, ___, ___ $-\frac{3}{4}, -1, -\frac{5}{4}$

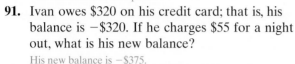 For Exercises 103–110, assume $a > 0$ (this means that a is positive) and $b < 0$ (this means that b is negative). Find the sign of each expression.

103. $a - b$ Positive

104. $b - a$ Negative

105. $|a| + |b|$ Positive

106. $|a + b|$ Positive or zero

107. $-|a|$ Negative

108. $-|b|$ Negative

109. $-(a)$ Negative

110. $-(b)$ Positive

Calculator Connections

Topic: Subtracting Real Numbers on a Calculator

The $\boxed{-}$ key is used for subtraction. This should not be confused with the $\boxed{(-)}$ key or $\boxed{+\circ\text{-}}$ key, which is used to enter a negative number.

Expression	Keystrokes		Result
$-7 - 4$	$\boxed{(-)}$ 7 $\boxed{-}$ 4 $\boxed{\text{ENTER}}$	or 7 $\boxed{+\circ\text{-}}$ $\boxed{-}$ 4 $\boxed{=}$	-11
$-4.2 - (-6.8)$	$\boxed{(-)}$ 4.2 $\boxed{-}$ $\boxed{(-)}$ 6.8 $\boxed{\text{ENTER}}$	or 4.2 $\boxed{+\circ\text{-}}$ $\boxed{-}$ 6.8 $\boxed{+\circ\text{-}}$ $\boxed{=}$	2.6

Calculator Exercises

For Exercises 111–116, subtract the real numbers by using a calculator.

111. $-190 - 223$ -413

112. $-288 - 145$ -433

113. $-23.24 - (-90.01)$ 66.77

114. $-14.93 - (-34.99)$ 20.06

115. $89.2 - (-23.6)$ 112.8

116. $104.9 - (-24.8)$ 129.7

Problem Recognition Exercises

Addition and Subtraction of Real Numbers

For Exercises 1–40, perform the indicated operations.

1. $-7 - 5$ -12 **2.** $-7 - (-5)$ -2 **3.** $-7 + (-5)$ -12 **4.** $-7 + 5$ -2

5. $10 - (-45)$ 55 **6.** $10 - 45$ -35 **7.** $10 + (-45)$ -35 **8.** $10 + 45$ 55

9. $-31.2 - (-52.6)$ 21.4 **10.** $-31.2 + (-52.6)$ -83.8 **11.** $-31.2 - 52.6$ -83.8 **12.** $-31.2 + (52.6)$ 21.4

13. $-19.5 + (21.5)$ 2 **14.** $-19.5 - 21.5$ -41 **15.** $-19.5 + (-21.5)$ -41 **16.** $-19.5 - (-21.5)$ 2

17. $|-12 + 8|$ 4 **18.** $|12 - 8|$ 4 **19.** $|-12 - 8|$ 20 **20.** $|12 - (-8)|$ 20

21. $\dfrac{1}{8} - \dfrac{5}{4}$ $-\dfrac{9}{8}$ or $-1\dfrac{1}{8}$ **22.** $-\dfrac{1}{8} - \dfrac{5}{4}$ $-\dfrac{11}{8}$ or $-1\dfrac{3}{8}$ **23.** $\dfrac{1}{8} + \left(-\dfrac{5}{4}\right)$ $-\dfrac{9}{8}$ or $-1\dfrac{1}{8}$ **24.** $-\dfrac{1}{8} - \left(-\dfrac{5}{4}\right)$ $\dfrac{9}{8}$ or $1\dfrac{1}{8}$

25. $-\dfrac{7}{9} - \dfrac{1}{6}$ $-\dfrac{17}{18}$ **26.** $-\dfrac{7}{9} + \dfrac{1}{6}$ $-\dfrac{11}{18}$ **27.** $-\dfrac{7}{9} - \left(-\dfrac{1}{6}\right)$ $-\dfrac{11}{18}$ **28.** $\dfrac{7}{9} - \dfrac{1}{6}$ $\dfrac{11}{18}$

29. $2\dfrac{1}{4} - 5\dfrac{1}{2}$ $-\dfrac{13}{4}$ or $-3\dfrac{1}{4}$ **30.** $4\dfrac{1}{3} + \left(-\dfrac{5}{6}\right)$ $\dfrac{7}{2}$ or $3\dfrac{1}{2}$ **31.** $-1\dfrac{2}{5} - 3\dfrac{1}{10}$ $-\dfrac{9}{2}$ or $-4\dfrac{1}{2}$ **32.** $2\dfrac{5}{6} - \left(-1\dfrac{1}{6}\right)$ 4

33. $-\dfrac{3}{4} + 3$ $\dfrac{9}{4}$ or $2\dfrac{1}{4}$ **34.** $-\dfrac{4}{5} - 1$ $-\dfrac{9}{5}$ or $-1\dfrac{4}{5}$ **35.** $-2 + 0.001$ -1.999 **36.** $4 - 5.987$ -1.987

37. $-56 + 56$ 0 **38.** $14 + (-14)$ 0 **39.** $-56 - 56$ -112 **40.** $14 - (-14)$ 28

✎ Writing   ↔ Translating Expression   ◿ Geometry   📱 Scientific Calculator   Video   NS␣E

Multiplication and Division of Real Numbers

1. Multiplication of Real Numbers

We know from our knowledge of arithmetic that the product of two positive numbers is a positive number. This can be shown by using repeated addition.

For example: $3(4) = 4 + 4 + 4 = 12$

Now consider a product of numbers with different signs.

For example: $3(-4) = -4 + (-4) + (-4) = -12$ (3 times -4)

These examples suggest that the product of a positive number and a negative number is *negative*.

Now what if we have a product of two negative numbers? To determine the sign, consider the following pattern of products.

$$3 \times -4 = -12$$
$$2 \times -4 = -8$$
$$1 \times -4 = -4$$
$$0 \times -4 = 0$$
$$-1 \times -4 = 4$$
$$-2 \times -4 = 8$$
$$-3 \times -4 = 12$$

The pattern increases by 4 with each row.

The product of two negative numbers is *positive.*

From the first four rows, we see that the product increases by 4 for each row. For the pattern to continue, it follows that the product of two negative numbers must be *positive*.

Multiplying Real Numbers

- The product of two real numbers with the *same* sign is positive.

    Examples: $(5)(6) = 30$

    $(-5)(-6) = 30$

- The product of two real numbers with *different* signs is negative.

    Examples: $4(-10) = -40$

    $-4(10) = -40$

- The product of any real number and zero is zero.

    Examples: $3(0) = 0$

    $0(5) = 0$

Concepts

1. Multiplication of Real Numbers
2. Multiplying Many Factors
3. Exponential Expressions
4. Division of Real Numbers

Concept Connections

1. Write $4(-5)$ as repeated addition.

Avoiding Mistakes

Do not confuse the rule for multiplying two negative numbers with the rule for adding two negative numbers.

- The product of two negative numbers is positive.
- The sum of two negative numbers is negative.

Answer

1. $-5 + (-5) + (-5) + (-5)$

Example 1 **Multiplying Real Numbers**

Multiply.   **a.** $-8(-7)$ **b.** $-5 \cdot 10$ **c.** $(18)(-2)$

Solution:

a. $-8(-7) = 56$ Same signs. Product is positive.

b. $-5 \cdot 10 = -50$ Different signs. Product is negative.

c. $(18)(-2) = -36$ Different signs. Product is negative.

Example 2 **Multiplying Real Numbers**

Multiply.   **a.** $(-3.1)(-4.6)$ **b.** $-\dfrac{4}{7} \cdot \left(5\dfrac{5}{6}\right)$ **c.** $-12 \cdot \left(-\dfrac{7}{3}\right)$ **d.** $-\dfrac{3}{4} \cdot 0$

Solution:

a. $(-3.1)(-4.6) = 14.26$ Same signs. Product is positive.

b. $-\dfrac{4}{7} \cdot \left(5\dfrac{5}{6}\right) = -\dfrac{4}{7} \cdot \dfrac{35}{6}$ Convert the mixed number to an improper fraction.

$= -\dfrac{\overset{2}{4}}{7} \cdot \dfrac{\overset{5}{35}}{\underset{3}{6}}$ Multiply fractions.

$= -\dfrac{10}{3}$ or $-3\dfrac{1}{3}$ Different signs. Product is negative.

c. $-12 \cdot \left(-\dfrac{7}{3}\right) = -\dfrac{12}{1} \cdot \left(-\dfrac{7}{3}\right)$ Write the whole number as a fraction.

$= -\dfrac{\overset{4}{12}}{1} \cdot \left(-\dfrac{7}{\underset{1}{3}}\right)$

$= \dfrac{28}{1}$ Same signs. Product is positive.

$= 28$

d. $-\dfrac{3}{4} \cdot 0 = 0$ The product of any number and zero is zero.

Recall that the terms *product*, *multiply*, and *times* imply multiplication.

Example 3 **Translating to an Algebraic Expression**

Translate each English phrase to an algebraic expression. Then simplify.

a. The product of 7 and -5 **b.** -3 times -11

Solution:

a. $7(-5)$ Translate: The product of 7 and -5.
$= -35$ Different signs. Product is negative.

b. $(-3)(-11)$ Translate: -3 times -11.
$= 33$ Same signs. Product is positive.

2. Multiplying Many Factors

In each of the following products, we can apply the order of operations and multiply from left to right.

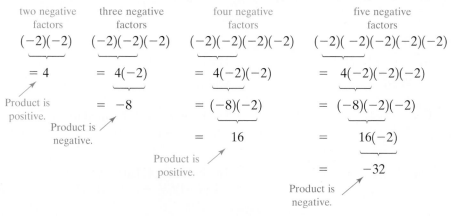

Concept Connections

Multiply.

11. $(-1)(-1)$
12. $(-1)(-1)(-1)$
13. $(-1)(-1)(-1)(-1)$
14. $(-1)(-1)(-1)(-1)(-1)$

These products indicate the following rules.

- The product of an *even* number of negative factors is *positive*.

- The product of an *odd* number of negative factors is *negative*.

Example 4   **Multiplying Several Factors**

Multiply.

a. $(-2)(-5)(-7)$ **b.** $(-4)(2)(-1)(5)$

Solution:

a. $(-2)(-5)(-7)$ This product has an odd number of negative factors.

$= -70$ The product is negative.

b. $(-4)(2)(-1)(5)$ This product has an even number of negative factors.

$= 40$ The product is positive.

Skill Practice

Multiply.

15. $(-3)(-4)(-8)(-1)$
16. $(-1)(-4)(-6)(5)$

Classroom Example: p. 620,
Exercise 48

3. Exponential Expressions

Be particularly careful when evaluating exponential expressions involving negative numbers. An exponential expression with a negative base is written with parentheses around the base, such as $(-3)^4$.

To evaluate $(-3)^4$, the base -3 is multiplied 4 times:

$$(-3)^4 = (-3)(-3)(-3)(-3) = 81$$

If parentheses are *not* used, the expression -3^4 has a different meaning:

- The expression -3^4 has a base of 3 (not -3) and can be interpreted as $-1 \cdot 3^4$. Hence,

$$-3^4 = -1 \cdot (3)(3)(3)(3) = -81$$

- The expression -3^4 can also be interpreted as "the opposite of 3^4." Hence,

$$-3^4 = -(3 \cdot 3 \cdot 3 \cdot 3) = -81$$

Answers

11. 1   **12.** -1 **13.** 1   **14.** -1
15. 96   **16.** -120

Example 5 **Simplifying Exponential Expressions**

Simplify.

a. $(-4)^2$ **b.** -4^2 **c.** $-(-4)^2$ **d.** $\left(-\dfrac{1}{2}\right)^4$

Solution:

a. $(-4)^2 = (-4)(-4)$ The base is -4.

         $= 16$ Multiply.

b. $-4^2 = -(4)(4)$ The base is 4. This is equal to $-1 \cdot 4^2 = -1 \cdot (4)(4)$.

        $= -16$

c. $-(-4)^2 = -(-4)(-4)$ The base is -4. This is equal to $-1 \cdot (-4)^2 = -1 \cdot (-4)(-4)$.

          $= -16$

d. $\left(-\dfrac{1}{2}\right)^4 = \left(-\dfrac{1}{2}\right)\left(-\dfrac{1}{2}\right)\left(-\dfrac{1}{2}\right)\left(-\dfrac{1}{2}\right)$ The base is $\left(-\dfrac{1}{2}\right)$.

          $= \dfrac{1}{16}$ Multiply.

Example 6 **Simplifying Exponential Expressions**

Simplify.

a. $(-5)^3$ **b.** -5^3 **c.** $-(-5)^3$

Solution:

a. $(-5)^3 = (-5)(-5)(-5)$ The base is -5.

         $= -125$ Multiply.

b. $-5^3 = -(5)(5)(5)$ The base is 5. This is equal to $-1 \cdot 5^3 = -1 \cdot (5)(5)(5)$.

         $= -125$ Multiply.

c. $-(-5)^3 = -(-5)(-5)(-5)$ The base is -5. This is equal to $-1 \cdot (-5)^3 = -1 \cdot (-5)(-5)(-5)$.

          $= -(-125)$

          $= 125$ Multiply.

4. Division of Real Numbers

Recall from Section 2.5 that two numbers are **reciprocals** if their product is 1. In particular, the reciprocal of a negative number must also be a negative number to form a product of 1.

Number	Reciprocal	Product
$\dfrac{2}{3}$	$\dfrac{3}{2}$	$\dfrac{2}{3} \cdot \dfrac{3}{2} = 1$
$-\dfrac{4}{7}$	$-\dfrac{7}{4}$	$-\dfrac{4}{7} \cdot \left(-\dfrac{7}{4}\right) = 1$
-5	$-\dfrac{1}{5}$	$-5 \cdot \left(-\dfrac{1}{5}\right) = 1$

Also recall from Section 2.5 that division can be expressed in terms of multiplication. To divide two real numbers, we multiply the first number (the dividend) by the reciprocal of the second number (the divisor).

$$\overset{\text{multiply}}{12 \div (-2)} = 12 \cdot \left(-\frac{1}{2}\right) = \frac{\overset{6}{\cancel{12}}}{1} \cdot \left(-\frac{1}{\underset{1}{\cancel{2}}}\right) = -6$$

by the reciprocal
of the divisor

Because division of real numbers can be expressed in terms of multiplication, the sign rules that apply to multiplication also apply to division.

Dividing Real Numbers

- The quotient of two real numbers with the *same* sign is positive.

  Examples: $\dfrac{20}{5} = 4$ and $\dfrac{-20}{-5} = 4$

- The quotient of two real numbers with different signs is negative.

  Examples: $\dfrac{16}{-8} = -2$ and $\dfrac{-16}{8} = -2$

- Zero divided by any nonzero number is zero.

  Examples: $\dfrac{0}{-3} = 0$ and $\dfrac{0}{12} = 0$

- Any nonzero number divided by zero is undefined.

  Example: $\dfrac{5}{0}$ is undefined.

Example 7 **Dividing Real Numbers**

Divide.

a. $50 \div (-5)$ **b.** $\dfrac{-42}{-7}$ **c.** $\dfrac{-39}{3}$

Solution:

a. $50 \div (-5) = -10$ Different signs. The quotient is negative.

b. $\dfrac{-42}{-7} = 6$ Same signs. The quotient is positive.

c. $\dfrac{-39}{3} = -13$ Different signs. The quotient is negative.

Concept Connections

24. If $a \cdot b = 1$, what do you know about a and b?

For each number, give the reciprocal, if possible.

25. $\dfrac{1}{7}$ **26.** $-\dfrac{3}{5}$

27. -6 **28.** 0

Skill Practice

Divide.

29. $-40 \div 10$ **30.** $\dfrac{-36}{-12}$

31. $\dfrac{18}{-2}$

Classroom Example: p. 621, Exercise 74

Answers

24. They are reciprocals. **25.** 7

26. $-\dfrac{5}{3}$ **27.** $-\dfrac{1}{6}$ **28.** Not possible

29. -4 **30.** 3 **31.** -9

Example 8 **Dividing Real Numbers**

Divide.

a. $-9.45 \div (-2.7)$ **b.** $\dfrac{4}{15} \div \left(-\dfrac{8}{25}\right)$

Solution:

a. $-9.45 \div (-2.7) = 3.5$ Same signs. The quotient is positive.

b. $\dfrac{4}{15} \div \left(-\dfrac{8}{25}\right) = \dfrac{4}{15} \cdot \left(-\dfrac{25}{8}\right)$ Multiply by the reciprocal of the divisor.

$= \dfrac{\overset{1}{4}}{\underset{3}{15}} \cdot \left(-\dfrac{\overset{5}{25}}{\underset{2}{8}}\right)$ Multiply fractions.

$= -\dfrac{5}{6}$ Different signs. The quotient is negative.

When we use fraction notation to divide a positive and negative number, there are several forms in which we can write the quotient. For example, the following are all equal.

$$\dfrac{-2}{3} = \dfrac{2}{-3} = -\dfrac{2}{3}$$ By convention, we usually write the quotient with the negative sign in front of the fraction, $-\frac{2}{3}$.

If we use fraction notation to divide two negative numbers, we know that the quotient must be positive. For example:

$$\dfrac{-5}{-9} = \dfrac{5}{9}$$ The quotient is positive.

Example 9 **Dividing Real Numbers**

Divide, if possible.

a. $\dfrac{-7}{-4}$ **b.** $15 \div (-25)$ **c.** $\dfrac{0}{-7}$ **d.** $-6.1 \div 0$

Solution:

a. $\dfrac{-7}{-4} = \dfrac{7}{4}$ Same signs. The quotient is positive.

In this example, 7 and 4 share no common factors. Therefore, the fraction cannot be simplified further. However, the quotient may be written in several forms: $\frac{7}{4}$ or $1\frac{3}{4}$ or 1.75.

b. $15 \div (-25)$ The number 25 does not divide evenly into 15. However, the expression can be written as a fraction and then simplified to lowest terms.

$$= \frac{15}{-25}$$

$$= \frac{\overset{3}{15}}{\underset{5}{-25}}$$ Simplify to lowest terms.

$$= -\frac{3}{5}$$ Different signs. The quotient is negative.

c. $\dfrac{0}{-7} = 0$ Zero divided by any nonzero number is 0.

d. $-6.1 \div 0$ Any number divided by zero is undefined.

Section 10.4   Practice Exercises

Study Skills Exercise

For additional exercises, see Classroom Activities 10.4A–10.4C in the *Student's Resource Manual* at www.mhhe.com/moh.

Often students learn a rule about signs that states, "Two negatives are a positive." This rule is incomplete and therefore not always true. Note the following combinations of two negatives:

$-2 + (-4)$	the sum of two negatives
$-(-5)$	the opposite of a negative
$-|-10|$	the opposite of an absolute value
$(-3)(-6)$	the product of two negatives

Determine which of the following are negative and which are positive. Then write the rule for multiplying two numbers with the same sign.

$-2 + (-4)$ _____

$-(-5)$ _____

$-|-10|$ _____

$(-3)(-6)$ _____

When multiplying two numbers with the same sign, the product is _____.

Vocabulary and Key Concepts

1. a. The product of two numbers with the same sign is (positive/negative).   positive
The product of two numbers with different signs is (positive/negative).   negative

b. The quotient of two numbers with the same sign is (positive/negative).   positive
The quotient of two numbers with different signs is (positive/negative).   negative

c. Two numbers are called _____ if their product is 1.   reciprocals

Review Exercises

2. Perform the indicated operations.

a. $\dfrac{2}{3} \cdot \dfrac{3}{2}$ 1   **b.** $7 \cdot \dfrac{1}{14}$ $\dfrac{1}{2}$ **c.** $\dfrac{3}{5} \div \dfrac{6}{25}$ $\dfrac{5}{2}$

For Exercises 3–8, add or subtract as indicated.

3. $14 - (-5)$ 19

4. $-24 - 50$ −74

5. $-33 + (-11)$ −44

6. $-7 - (-23)$ 16

7. $23 - 12 + (-4) - (-10)$ 17

8. $9 + (-12) - 17 - 4 - (-15)$ −9

Concept 1: Multiplication of Real Numbers

For Exercises 9–36, multiply the real numbers. **(See Examples 1 and 2.)**

9. $-3(5)$ −15

10. $-2(13)$ −26

11. $-12 \cdot 4$ −48

12. $-6 \cdot 11$ −66

13. $-15(-3)$ 45

14. $-3(-25)$ 75

 **15.** $9(-8)$ −72

16. $8(-3)$ −24

17. $(-1.2)(-3.2)$ 3.84

18. $(-3.3)(-2.5)$ 8.25

19. $-6(0.4)$ −2.4

20. $-8(1.3)$ −10.4

21. $7(-1.1)$ −7.7

22. $5(-3.4)$ −17

23. $-14 \cdot 0$ 0

24. $-8 \cdot 0$ 0

 **25.** $\left(-\frac{2}{3}\right)\left(-\frac{6}{7}\right)$ $\frac{4}{7}$

26. $\left(-\frac{8}{9}\right)\left(-\frac{3}{4}\right)$ $\frac{2}{3}$

27. $\frac{3}{5}\left(-\frac{5}{21}\right)$ $-\frac{1}{7}$

28. $\frac{5}{12}\left(-\frac{4}{7}\right)$ $-\frac{5}{21}$

29. $6 \cdot \left(-\frac{5}{12}\right)$ $-\frac{5}{2}$ or $-2\frac{1}{2}$

30. $4 \cdot \left(-\frac{1}{16}\right)$ $-\frac{1}{4}$

31. $\left(-2\frac{3}{5}\right)\left(-1\frac{2}{3}\right)$ $\frac{13}{3}$ or $4\frac{1}{3}$

32. $\left(-3\frac{1}{3}\right)\left(-2\frac{1}{5}\right)$ $\frac{22}{3}$ or $7\frac{1}{3}$

33. $\left(-\frac{8}{9}\right) \cdot 0$ 0

34. $\left(-\frac{2}{11}\right) \cdot 0$ 0

35. $(-3.5)(-1.4)$ 4.9

36. $(-1.6)(-6.5)$ 10.4

For Exercises 37–42, translate the English phrase to an algebraic expression. Then simplify. **(See Example 3.)**

37. Multiply -3 and -1.
    $-3(-1)$; 3

38. Multiply -12 and -4.
    $-12(-4)$; 48

39. The product of -5 and 3
    $-5 \cdot 3$; −15

40. The product of 9 and -2
    $9(-2)$; −18

41. 1.3 times -3
    $1.3(-3)$; −3.9

42. -2.3 times 6
    $(-2.3)(6)$; −13.8

43. During the downturn of the economy, a restaurant owner noticed that for 3 weeks, there were 12 fewer patrons per week. Write an expression that describes the loss of customers for that 3-week period. Simplify to determine the loss in customers. $3(-12)$; −36 customers

44. In the stock market, Quinn lost $15 a day for a 5-day business week. Write an expression that describes Quinn's loss for the week. Simplify the expression to determine the total loss for the week. $5(-15)$; −$75

45. The Detroit Lions football team lost 8 yd for each of four consecutive plays. Write an expression that describes the loss of yards after these 4 plays. Simplify to determine the total lost yardage from these plays. $4(-8)$; −32 yd

46. The temperature outside an airplane decreases by 20°F for each kilometer the plane ascends. Write an expression that describes the change in temperature if the plane ascends 11 km. Simplify the expression for the total change in temperature.
    $11(-20)$; −220°F

Concept 2: Multiplying Many Factors

For Exercises 47–56, multiply. **(See Example 4.)**

47. $(5)(-2)(4)(-10)$ 400

48. $(-3)(-5)(-2)(4)$ −120

 **49.** $(-11)(-4)(-2)$ −88

50. $(20)(-3)(-1)$ 60

51. $(24)(-2)(0)(-3)$ 0

52. $(3)(0)(-13)(22)$ 0

53. $(-1)(-1)(-1)(-1)(-1)(-1)$ 1

54. $(-1)(-1)(-1)(-1)(-1)(-1)(-1)$ −1

55. $(-1)(-1)(-1)(-1)(-1)$ −1

56. $(-1)(-1)(-1)(-1)$ 1

Concept 3: Exponential Expressions

For Exercises 57–72, simplify. **(See Examples 5 and 6.)**

57. -10^2 $\quad$ –100

58. -8^2 $\quad$ –64

59. $(-10)^2$ $\quad$ 100

60. $(-8)^2$ $\quad$ 64

61. -3^3 $\quad$ –27

62. -4^3 $\quad$ –64

63. $(-3)^3$ $\quad$ –27

64. $(-4)^3$ $\quad$ –64

65. -0.2^3 $\quad$ –0.008

66. -0.4^3 $\quad$ –0.064

67. $\left(-\dfrac{2}{3}\right)^3$ $\quad$ $-\dfrac{8}{27}$

68. $\left(-\dfrac{3}{5}\right)^3$ $\quad$ $-\dfrac{27}{125}$

69. $(-6)^2$ $\quad$ 36

70. $(-6)^3$ $\quad$ –216

71. $-(-6)^2$ $\quad$ –36

72. $-(-6)^3$ $\quad$ 216

Concept 4: Division of Real Numbers

For Exercises 73–96, divide the real numbers, if possible. **(See Examples 7–9.)**

73. $\dfrac{-15}{5}$ $\quad$ –3

74. $\dfrac{30}{-6}$ $\quad$ –5

75. $\dfrac{56}{-8}$ $\quad$ –7

76. $\dfrac{-48}{3}$ $\quad$ –16

77. $\dfrac{-25}{-15}$ $\quad$ $\dfrac{5}{3}$

78. $\dfrac{-6}{-18}$ $\quad$ $\dfrac{1}{3}$

79. $\dfrac{-2}{-3}$ $\quad$ $\dfrac{2}{3}$

80. $\dfrac{-9}{-8}$ $\quad$ $\dfrac{9}{8}$

81. $\dfrac{13}{0}$ $\quad$ Undefined

82. $\dfrac{-41}{0}$ $\quad$ Undefined

83. $\dfrac{0}{-2}$ $\quad$ 0

84. $\dfrac{0}{5}$ $\quad$ 0

85. $(-20) \div (-5)$ $\quad$ 4

86. $(-10) \div (-2)$ $\quad$ 5

87. $-0.91 \div -0.7$ $\quad$ 1.3

88. $-1.3 \div -0.5$ $\quad$ 2.6

89. $\left(\dfrac{8}{7}\right) \div \left(-\dfrac{4}{5}\right)$ $\quad$ $-\dfrac{10}{7}$

90. $\left(-\dfrac{2}{5}\right) \div \left(\dfrac{8}{15}\right)$ $\quad$ $-\dfrac{3}{4}$

91. $\left(-\dfrac{1}{6}\right) \div 0$ $\quad$ Undefined

92. $\left(\dfrac{2}{11}\right) \div 0$ $\quad$ Undefined

93. $\dfrac{-5}{-8}$ $\quad$ $\dfrac{5}{8}$

94. $\dfrac{-1}{-3}$ $\quad$ $\dfrac{1}{3}$

95. $-18 \div 24$ $\quad$ $-\dfrac{3}{4}$

96. $12 \div (-30)$ $\quad$ $-\dfrac{2}{5}$

For Exercises 97–102, translate the English phrase into a mathematical expression. Then simplify.

97. The quotient of -100 and 20 $\quad$ $-100 \div 20;\ -5$

98. The quotient of 46 and -23 $\quad$ $46 \div (-23);\ -2$

99. -32 divided by -64 $\quad$ $-32 \div (-64);\ \dfrac{1}{2}$

100. 108 divided by -24 $\quad$ $108 \div (-24);\ -\dfrac{9}{2}$

101. 13 divided into -52 $\quad$ $-52 \div 13;\ -4$

102. -15 divided into -45 $\quad$ $-45 \div (-15);\ 3$

Mixed Exercises

For Exercises 103–118, perform the indicated operation.

103. $8 + (-6)$ $\quad$ 2

104. $8 - (-6)$ $\quad$ 14

105. $8(-6)$ $\quad$ –48

106. $8 \div (-6)$ $\quad$ $-\dfrac{4}{3}$

107. $-9 - (-12)$ $\quad$ 3

108. $(-9)(-12)$ $\quad$ 108

109. $-36 \div (-12)$ $\quad$ 3

110. $-36 + (-12)$ $\quad$ –48

111. $(-5)(-4)$ $\quad$ 20

112. $-90 \div (-6)$ $\quad$ 15

113. $0 + (-15)$ $\quad$ –15

114. $0 - (-15)$ $\quad$ 15

115. $\dfrac{1}{3} \div \left(-\dfrac{5}{6}\right)$ $\quad$ $-\dfrac{2}{5}$

116. $\dfrac{1}{3} + \left(-\dfrac{5}{6}\right)$ $\quad$ $-\dfrac{1}{2}$

117. $\dfrac{1}{3} - \left(-\dfrac{5}{6}\right)$ $\quad$ $\dfrac{7}{6}$

118. $\dfrac{1}{3}\left(-\dfrac{5}{6}\right)$ $\quad$ $-\dfrac{5}{18}$

Expanding Your Skills

119. Which is greater, $(-2)^{50}$ or $(-2)^{51}$? $\quad$ $(-2)^{50}$

120. Which is greater, $(-3)^{20}$ or $(-3)^{21}$? $\quad$ $(-3)^{20}$

121. Which is greater, $(5)^{40}$ or $(5)^{41}$? $\quad$ $(5)^{41}$

122. Which is greater, $(6)^{10}$ or $(6)^{11}$? $\quad$ $(6)^{11}$

NS $\frac{Q}{E}$ For Exercises 123–126, assume $a > 0$ (this means that a is positive) and $b < 0$ (this means that b is negative). Find the sign of each expression.

123. $a \cdot b$ Negative   **124.** $b \div a$ Negative   **125.** $-a \div (b)$ Positive   **126.** $a(-b)$ Positive

Calculator Connections

Topic: Multiplying and Dividing Real Numbers on a Calculator

Knowing the sign of the product or quotient can make using the calculator easier. For example, look at $\frac{-78}{-26}$. Note the keystrokes if we enter this into the calculator as written:

78 [+○−] [÷] 26 [+○−] [=] | 3 |

But because we know that the quotient of two negative numbers is positive, we can simply enter:

78 [÷] 26 [=] | 3 |

Calculator Exercises:

For Exercises 127–132, use a calculator to perform the indicated operations.

127. $(-413)(871)$ −359,723   **128.** $-6125 \cdot (-97)$ 594,125   **129.** $(-52.12)(-101.5)$ 5290.18

130. $\dfrac{-576,828}{-10,682}$ 54   **131.** $5,945,308 \div (-9452)$ −629   **132.** $\dfrac{-301,224}{9128}$ −33

Problem Recognition Exercises

Operations on Real Numbers

For Exercises 1–40, perform the indicated operations.

1. $15 - (-5)$ 20

2. $15(-5)$ −75

3. $15 + (-5)$ 10

4. $15 \div (-5)$ −3

5. $-36(-2)$ 72

6. $-36 - (-2)$ −34

7. $\dfrac{-36}{-2}$ 18

8. $-36 + (-2)$ −38

9. $20(-4)$ −80

10. $-20(-4)$ 80

11. $-20(4)$ −80

12. $20(4)$ 80

13. $-5 - 9 - 2$ −16

14. $-4(-9)(-2)$ −72

15. $10 + (-3) + (-12)$ −5

16. $10 - (-3) - (-12)$ 25

17. $(-1)(-2)(-3)(-4)$ 24

18. $(-1)(-2)(3)(4)$ 24

19. $(-1)(-2)(-3)(4)$ −24

20. $(-1)(2)(3)(4)$ −24

21. $\dfrac{3}{5} \div \left(-\dfrac{10}{9}\right)$ $-\dfrac{27}{50}$

22. $\dfrac{3}{5}\left(-\dfrac{10}{9}\right)$ $-\dfrac{2}{3}$

23. $\dfrac{3}{5} + \left(-\dfrac{10}{9}\right)$ $-\dfrac{23}{45}$

24. $\dfrac{3}{5} - \left(-\dfrac{10}{9}\right)$ $\dfrac{77}{45}$

25. $-\dfrac{2}{3} + \left(-\dfrac{7}{9}\right)$ $-\dfrac{13}{9}$

26. $-\dfrac{2}{3} \div \left(-\dfrac{7}{9}\right)$ $\dfrac{6}{7}$

27. $\left(-2\dfrac{1}{4}\right)\left(1\dfrac{4}{9}\right)$ $-\dfrac{13}{4}$ or $-3\dfrac{1}{4}$

28. $-2\dfrac{1}{4} + 1\dfrac{4}{9}$ $-\dfrac{29}{36}$

29. $41.5 - (-13.6)$ 55.1

30. $-13.56 + 4.12$ −9.44

31. $-60.41 - 33.50$ −93.91

32. $-0.06 - (-0.04)$ −0.02

33. $\dfrac{-12}{-11}$ $\dfrac{12}{11}$

34. $\dfrac{5}{-30}$ $-\dfrac{1}{6}$

35. $\dfrac{0}{-8}$ 0

36. $-4.5 \div 0$ Undefined

37. $42 \div (-0.002)$ −21,000

38. $-360 \div (-0.009)$ 40,000

39. $-44 - (-44)$ 0

40. $-60 \cdot \left(-\dfrac{1}{60}\right)$ 1

 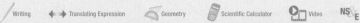

Order of Operations

1. Order of Operations

The order of operations was first introduced in Section 1.7 and then used throughout the text. The order of operations still applies when simplifying expressions with real numbers.

> **Order of Operations**
>
> **Step 1** Perform all operations inside parentheses and other grouping symbols first.
>
> **Step 2** Simplify any expressions containing exponents or square roots.
>
> **Step 3** Perform multiplication or division in the order that they appear from left to right.
>
> **Step 4** Perform addition or subtraction in the order that they appear from left to right.

Example 1 Applying the Order of Operations

Simplify.

   a. $-12 - 6(7 - 5)$ **b.** $3^2 - 10^2 \div (-1 - 4)$

Solution:

a. $-12 - 6(7 - 5)$

   $= -12 - 6(2)$ Simplify within parentheses first.

   $= -12 - 12$ Multiply before subtracting.

   $= -24$ Subtract. *Note:* $-12 - 12 = -12 + (-12) = -24$.

b. $3^2 - 10^2 \div (-1 - 4)$ Simplify within parentheses.
                                       Note: $-1 - 4 = -1 + (-4) = -5$.

   $= 3^2 - 10^2 \div (-5)$ Simplify exponents.
                                         Note: $3^2 = 3 \cdot 3 = 9$ and $10^2 = 10 \cdot 10 = 100$.

   $= 9 - 100 \div (-5)$ Divide before subtracting.
                                         Note: $100 \div (-5) = -20$.

   $= 9 - (-20)$ Subtract. *Note:* $9 - (-20) = 9 + (20) = 29$.

   $= 29$

Example 2 Applying the Order of Operations

Simplify. $\dfrac{1}{30} - \left(-\dfrac{1}{3}\right)^2 \cdot \dfrac{3}{5}$

Solution:

$\dfrac{1}{30} - \left(-\dfrac{1}{3}\right)^2 \cdot \dfrac{3}{5}$

$= \dfrac{1}{30} - \dfrac{1}{9} \cdot \dfrac{3}{5}$ Simplify exponents. *Note:* $\left(-\dfrac{1}{3}\right) \cdot \left(-\dfrac{1}{3}\right) = \dfrac{1}{9}.$

$= \dfrac{1}{30} - \dfrac{1}{\underset{3}{9}} \cdot \dfrac{\overset{1}{3}}{5}$ Multiply fractions.

$= \dfrac{1}{30} - \dfrac{1}{15}$ The least common denominator is 30.

$= \dfrac{1}{30} - \dfrac{1 \cdot 2}{15 \cdot 2}$ Write each fraction with the LCD.

$= \dfrac{1}{30} - \dfrac{2}{30}$

$= \dfrac{1}{30} + \left(-\dfrac{2}{30}\right)$ Write the subtraction in terms of addition.

$= -\dfrac{1}{30}$

Example 3 Applying the Order of Operations

Simplify. $3 - [-6 - (5 - 7)]$

Solution:

$3 - [-6 - (5 - 7)]$

$= 3 - [-6 - (5 + (-7))]$ Write the subtraction in terms of addition.

$= 3 - [-6 - (-2)]$ Simplify within the innermost parentheses.

$= 3 - [-6 + (2)]$ Write the subtraction in terms of addition.

$= 3 - [-4]$ Simplify within the parentheses by adding.

$= 3 + (4)$ Write the subtraction in terms of addition.

$= 7$ Add.

Answers

3. $\dfrac{1}{18}$ **4.** -6

Example 4 Applying the Order of Operations

Simplify. $\dfrac{|-7+3|-9}{3-2\cdot4}$

Solution:

$\dfrac{|-7+3|-9}{3-2\cdot4}$ Simplify the numerator and denominator separately.

$=\dfrac{|-4|-9}{3-8}$ Numerator: Add within the absolute values.
Denominator: Multiply before subtracting.

$=\dfrac{4+(-9)}{3+(-8)}$ Numerator: Evaluate the absolute value and write the subtraction as addition.
Denominator: Write the subtraction as addition.

$=\dfrac{-5}{-5}$ Add.

$=1$ Divide.

Section 10.5   Practice Exercises

For additional exercises, see Classroom Activity 10.5A in the *Student's Resource Manual* at www.mhhe.com/moh.

Study Skills Exercise

Make a list of resources that are available to you at times when you are not on campus and you need help with your algebra studies (for example, websites, online tutoring, and classmates). Write down Web addresses and phone numbers and keep them handy.

Review Exercises

For Exercises 1–8, multiply or divide as indicated.

1. $-28+72$ 44

2. $-100-(-4)$ -96

3. $\left(-\dfrac{2}{9}\right)\div\left(\dfrac{8}{27}\right)$ $-\dfrac{3}{4}$

4. $10\cdot\left(-\dfrac{3}{5}\right)$ -6

5. $-2.8(-1.1)$ 3.08

6. $5.5\div(-0.5)$ -11

7. $(-1)(-5)(-8)(3)$ -120

8. $-1+(-5)+(-8)+3$ -11

Concept 1: Order of Operations

For Exercises 9–62, simplify by using the order of operations. **(See Examples 1–4.)**

9. $5+2(3-5)$ 1

10. $6-4(8-10)$ 14

11. $-8-6^2$ -44

12. $-10-5^2$ -35

13. $4+(3-8)^2$ 29

14. $5+(2-9)^2$ 54

15. $120\div(-4)(5)$ -150

16. $36\div(-2)(3)$ -54

17. $-2.1-6\div5$ -3.3

18. $-8.3-10\div8$ -9.55

19. $[5.3-(-2.7)]^2$ 64

20. $(-7.1-1.9)^2$ 81

21. $-2(3-6)+10$ 16

22. $-4\div(1-3)-8$ -6

23. $-16\div(-4)(-5)$ -20

24. $-12(-1)\div6$ 2

25. $8-(-3)(-2)^3$ -16

26. $1-(-5)(-3)^2$ 46

27. $12+(14-16)^2\div(-4)$ 11

28. $-7+(1-5)^2\div4$ -3

29. $-48\div12\div(-2)$ 2

30. $-100 \div (-5) \div (-5)$ -4 **31.** $90 \div (-3)(-1) \div (-6)$ -5 **32.** $64 \div (-4)2 \div (-16)$ 2

33. $[9^2 - (-7)^2] \div (-4)$ -8 **34.** $|(-8)^2 - 5^2| \div (-3)$ -13 **35.** $2 + 2^3 - |10 - 12|$ 8

36. $14 - 4^2 + 2 - 10$ -10 **37.** $-6(48 \div 12)^2$ -96 **38.** $-5(35 \div 5)^2$ -245

39. $\left(-\dfrac{1}{2}\right) \cdot \dfrac{1}{3} \div \dfrac{1}{12}$ -2 **40.** $\dfrac{2}{9} \div \left(-\dfrac{1}{3}\right) \cdot \dfrac{6}{5}$ $-\dfrac{4}{5}$ **41.** $\dfrac{1}{6} + \left(-\dfrac{5}{4}\right) \cdot \dfrac{4}{3}$ $-\dfrac{3}{2}$

42. $-\dfrac{3}{8} + \dfrac{5}{24} \div \left(-\dfrac{5}{6}\right)$ $-\dfrac{5}{8}$ **43.** $\left(-\dfrac{2}{3}\right)^2 - \left(\dfrac{5}{21}\right) \div \dfrac{15}{7}$ $\dfrac{1}{3}$ **44.** $\left(-\dfrac{1}{3}\right)^3 + \left(\dfrac{2}{9}\right) \cdot \dfrac{5}{6}$ $\dfrac{4}{27}$

45. $2\dfrac{1}{2} \cdot \left(1\dfrac{1}{2}\right)^2 + \left(-\dfrac{1}{2}\right)^3$ $5\dfrac{1}{2}$ **46.** $\left(-2\dfrac{1}{3}\right) \div \left(1\dfrac{1}{3}\right)^2 - \left(\dfrac{1}{2}\right)^4$ $-1\dfrac{3}{8}$ **47.** $21 - [4 - (5 - 8)]$ 14

48. $15 - [10 - (20 - 25)]$ 0 **49.** $-17 - 2[18 \div (-3)]$ -5 **50.** $-8 - 5(-45 \div 15)$ 7

51. $4 + 2[9 + (-4 + 12)]$ 38 **52.** $-13 + 3[11 + (-15 + 10)]$ 5 **53.** $2^2 - |-3 + 9|$ -2

54. $5^2 - |10 + (-8)|$ 23 **55.** $\dfrac{|3 + (-5)|}{4 - (3)(-2)}$ $\dfrac{1}{5}$ **56.** $\dfrac{|-11 + 7|}{8 - 4(-1)}$ $\dfrac{1}{3}$

57. $\dfrac{13 - (2)(4)}{-1 - 2^2}$ -1 **58.** $\dfrac{10 - (-3)(5)}{-9 - 4^2}$ -1 **59.** $\dfrac{1 - 4(3 - 5)}{5^2 - 2^2}$ $\dfrac{3}{7}$

60. $\dfrac{-3 - (2 + 4)}{6^2 - 3^2}$ $-\dfrac{1}{3}$ **61.** $\dfrac{6 - 3^2}{(5 - 2)^2}$ $-\dfrac{1}{3}$ **62.** $\dfrac{-2 - 5^2}{(6 - 3)^2}$ -3

63. Find the average temperature: $-8°, -11°, -4°, 1°, 9°, 4°, -5°$ $-2°$

64. Find the average temperature: $15°, 12°, 10°, 3°, 0°, -2°, -3°$ $5°$

65. Find the average golf score: $-8, -8, -6, -5, -2, 3, 3, 0, -4$ -3

66. Find the average golf score: $-6, -2, 5, 1, 0, -3, 7, 2, -4$ 0

67. According to Ask A Scientist©, the coldest temperature ever recorded was in Vostok, Antarctica, on July 31, 1973. The temperature was recorded at $-89.6°C$. Use the following formula to convert this temperature to Fahrenheit. $-129.28°F$

$$F = \dfrac{9}{5}C + 32$$

68. The BBC reports that the coldest temperature in the United Kingdom was $-27.2°C$ in Altnaharra, Highland, in 1995. Using the formula from Exercise 67, convert this temperature to Fahrenheit. $-16.96°F$

Expanding Your Skills

For Exercises 69–72, simplify the expressions containing both fractions and decimals.

69. $\left(\dfrac{1}{2}\right)^2 \div 0.05 + \left(-\dfrac{3}{4} \cdot \dfrac{8}{3}\right)$ 3 **70.** $-0.8 - \dfrac{19}{20} + \left(\dfrac{4}{5} \div \dfrac{1}{2}\right) - (-0.15)$ 0

71. $2\left(\dfrac{7}{8} - \dfrac{1}{4}\right) - (-1.5)^2$ -1 **72.** $-2.1 + 4\left(\dfrac{7}{16} - 0.0375\right)$ -0.5 or $-\dfrac{1}{2}$

Group Activity

Checking Weather Predictions

Materials: A computer with online access

Estimated Time: 2–3 minutes each day for 10 days

Group Size: 3

Instructor Note: This activity may be performed over multiple days. It will require about 5–10 min on the first day for students to look up and record the predicted temperatures for a 10-day period. Then, each day for 10 days, the students will look up the *actual* high and low temperatures.

1. Go to a website such as http://www.weather.com/ to find the predicted high and low temperatures for a 10-day period for a city and state of your choice.

2. Record the predicted high and low temperatures for each of the 10 days. Record these values in the second column of each table.

Answers will vary.

Day	Predicted High	Actual High	Difference (error)
1			
2			
3			
4			
5			
6			
7			
8			
9			
10			

Day	Predicted Low	Actual Low	Difference (error)
1			
2			
3			
4			
5			
6			
7			
8			
9			
10			

3. For the next 10 days, record the actual high and low temperatures for your chosen city for that day. Record these values in the third column of each table.

4. For each day, compute the difference between the predicted and actual temperature and record the results in the fourth column of each table. We will call this difference the *error*.

$$\text{Error} = (\text{predicted temperature}) - (\text{actual temperature})$$

5. If the error is *negative*, does this mean that the weather service overestimated or underestimated the temperature?

6. If the error is *positive*, does this mean that the weather service overestimated or underestimated the temperature?

7. Find the mean (average) error for the high temperature predictions.

8. Find the mean (average) error for the low temperature predictions.

9. Which set of predictions was most accurate, the high temperature predictions or the low temperature predictions?

Chapter 10   Summary

Section 10.1   Real Numbers and the Real Number Line

Key Concepts

The numbers . . . $-3, -2, -1, 0, 1, 2, 3, \ldots$ and so on are called **integers**. The negative integers lie to the left of zero on the number line.

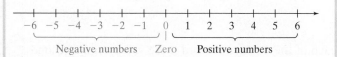

A number that can be written as a ratio of two integers is called a **rational number** (division by zero is excluded).

Irrational numbers are numbers that cannot be written as a ratio of integers.

The set of all rational numbers and irrational numbers together forms the **real numbers**.

The **absolute value** of a number a is denoted $|a|$. The value of $|a|$ is the distance between a and 0 on the number line.

Two numbers that are the same distance from zero on the number line, but on opposite sides of zero are called **opposites**.

The double-negative property states that the opposite of a negative number is a positive number. That is, $-(-a) = a$, for $a > 0$.

Examples

Example 1

The temperature 5° below zero can be represented by a negative number: $-5°$.

Example 2

The following numbers are rational because they can be written as a ratio of two integers.

a. $\dfrac{1}{2}$ is a ratio of 1 and 2

b. -0.75 is a ratio of -3 and 4

c. -4 is a ratio of -4 and 1

Example 3

a. $|5| = 5$ b. $|-13| = 13$ c. $|0| = 0$

Example 4

The opposite of 12 is $-(12) = -12$.

Example 5

The opposite of -23 is $-(-23) = 23$.

Section 10.2   Addition of Real Numbers

Key Concepts

To add integers by using a number line, locate the first addend on the number line. Then to add a positive number, move to the right on the number line. To add a negative number, move to the left on the number line.

Integers can be added by using the following rules:

Adding Numbers with the Same Sign

To add two numbers with the same sign, add their absolute values and apply the common sign.

Adding Numbers with Different Signs

To add two numbers with different signs, subtract the smaller absolute value from the larger absolute value. Then apply the sign of the number having the larger absolute value.

Examples

Example 1

Add $-2 + (-4)$ by using the number line.

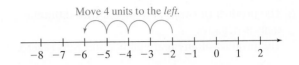

Move 4 units to the *left*.

$-2 + (-4) = -6$

Example 2

a. $5 + 2 = 7$

b. $-5 + (-2) = -7$

Example 3

a. $6 + (-5) = 1$

b. $(-6) + 5 = -1$

Section 10.3   Subtraction of Real Numbers

Key Concepts

Subtraction of Real Numbers

If a and b are real numbers, then

$$a - b = a + (-b)$$

To perform subtraction, follow these steps:

1. Leave the first number (the minuend) unchanged.
2. Change the subtraction sign to an addition sign.
3. Add the opposite of the second number (the subtrahend).

Examples

Example 1

a. $3 - 9 = 3 + (-9) = -6$

b. $-3 - 9 = -3 + (-9) = -12$

c. $3 - (-9) = 3 + (9) = 12$

d. $-3 - (-9) = -3 + (9) = 6$

Section 10.4 Multiplication and Division of Real Numbers

Key Concepts

Multiplication of Real Numbers

1. The product of two real numbers with the same sign is positive.
2. The product of two real numbers with different signs is negative.
3. The product of any real number and zero is zero.

The product of an *even* number of negative factors is *positive*.

The product of an *odd* number of negative factors is *negative*.

When evaluating an exponential expression, attention must be paid when parentheses are used. That is, $(-2)^4 = (-2)(-2)(-2)(-2) = 16$, while $-2^4 = -1 \cdot (2)(2)(2)(2) = -16$.

Two numbers are **reciprocals** if their product is 1.

Division of Real Numbers

1. The quotient of two real numbers with the same sign is positive.
2. The quotient of two real numbers with different signs is negative.

Division by zero is undefined.
Zero divided by a nonzero number is 0.

Examples

Example 1

a. $-8(-3) = 24$

b. $8(-3) = -24$

c. $-8(0) = 0$

Example 2

a. $(-5)(-4)(-1)(-3) = 60$

b. $(-2)(-1)(-6)(-3)(-2) = -72$

Example 3

a. $\left(-\dfrac{2}{3}\right)^2 = \left(-\dfrac{2}{3}\right)\left(-\dfrac{2}{3}\right) = \dfrac{4}{9}$

b. $-\left(\dfrac{2}{3}\right)^2 = -1 \cdot \left(\dfrac{2}{3}\right)\left(\dfrac{2}{3}\right) = -\dfrac{4}{9}$

Example 4

The reciprocal of -4 is $-\dfrac{1}{4}$ because

$-4 \cdot -\dfrac{1}{4} = 1$

Example 5

a. $-36 \div (-9) = 4$ b. $\dfrac{42}{-6} = -7$

Example 6

a. $\dfrac{-15}{0}$ is undefined.   b. $\dfrac{0}{-3} = 0$

Section 10.5   Order of Operations

Key Concepts

Order of Operations

1. Perform all operations inside parentheses and other grouping symbols first.
2. Simplify any expressions containing exponents or square roots.
3. Perform multiplication or division in the order that they appear from left to right.
4. Perform addition or subtraction in the order that they appear from left to right.

Examples

Example 1

$$-15 - 2(8-11)^2 = -15 - 2(-3)^2$$
$$= -15 - 2 \cdot 9$$
$$= -15 - 18$$
$$= -33$$

Example 2

$$\frac{1}{5} \div \left(-\frac{7}{20}\right) \cdot \left(-\frac{9}{2}\right) = \frac{1}{5} \cdot \left(-\frac{20}{7}\right) \cdot \left(-\frac{9}{2}\right)$$
$$= \frac{1}{\underset{1}{5}} \cdot \left(-\frac{\overset{4}{20}}{7}\right) \cdot \left(-\frac{9}{2}\right)$$
$$= -\frac{4}{7} \cdot \left(-\frac{9}{\underset{1}{2}}\right)$$
$$= \frac{18}{7}$$

Chapter 10   Review Exercises

Section 10.1

For Exercises 1–4, write an integer that represents each numerical value.

1. The population of Detroit, Michigan, decreased by 10,227 from 2005 to 2010. (*Source*: U.S. Census Bureau, www.census.gov)
   −10,227

2. The country's deficit is $5 billion.
   −$5 billion

3. The temperature rose 15° in one day.
   15°

4. Cecelia's bank account earned $10 in interest.
   $10

For Exercises 5–8, locate the numbers on the number line.

5. −2   6. $-5\frac{1}{3}$ 7. 0   8. 3.8

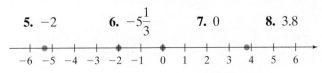

For Exercises 9–12, determine the opposite and the absolute value for each number.

9. −4
   4, 4

10. $-\frac{1}{2}$
    $\frac{1}{2}, \frac{1}{2}$

11. 3.5
    −3.5, 3.5

12. 6
    −6, 6

13. Evaluate.

   a. −(−9)
      9

   b. −|−9|
      −9

14. Evaluate.

   a. −(1.5)
      −1.5

   b. −|1.5|
      −1.5

For Exercises 15–20, place the correct symbol, > or <, between the two numbers.

15. $-\frac{5}{6}$ ☐ −1   >

16. −0.5 ☐ 0.5   <

17. |3| ☐ −|3| >

18. 8 ☐ |−9| <

19. −2.8 ☐ −2   <

20. $\left|-\frac{3}{2}\right|$ ☐ $-\frac{1}{2}$ >

Writing   ←→ Translating Expression   Geometry   Scientific Calculator   Video   NS E

Section 10.2

For Exercises 21–24, add the integers by using the number line.

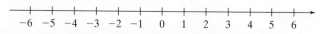

21. $6 + (-2)$ 4   **22.** $-3 + 6$ 3

23. $-3 + (-2)$ –5   **24.** $-3 + 0$ –3

25. State the rule for adding two numbers with the same sign.   To add two numbers with the same sign, add their absolute values and apply the common sign.

26. State the rule for adding two numbers with different signs.   To add two numbers with different signs, subtract the smaller absolute value from the larger absolute value. Then apply the sign of the number having the larger absolute value.

For Exercises 27–34, add the real numbers.

27. $35 + (-22)$ **28.** $-105 + 90$
13   –15

29. $-29 + (-41)$ **30.** $3.22 + (-4.1)$
–70   –0.88

31. $-6.5 + (-4.16)$ **32.** $\left(-1\dfrac{3}{4}\right) + \left(2\dfrac{1}{2}\right)$ $\frac{3}{4}$
–10.66

33. $\left(-\dfrac{1}{5}\right) + \left(-\dfrac{7}{10}\right)$ $-\frac{9}{10}$ **34.** $2 + \left(-\dfrac{7}{3}\right)$ $-\frac{1}{3}$

For Exercises 35–40, translate each phrase to a mathematical expression. Then simplify the expression.

35. The sum of 23 and -35 $23 + (-35); -12$

36. 57 plus -10 $57 + (-10); 47$

37. The total of -5, -13, and 20 $-5 + (-13) + 20; 2$

38. -42 increased by 12 $-42 + 12; -30$

39. 3 more than -12 $-12 + 3; -9$

40. -89 plus -22 $-89 + (-22); -111$

For Exercises 41 and 42, add.

41. $-3 + (-10) + 12 + 14 + (-10)$ 3

42. $9 + (-15) + 2 + (-7) + (-4)$ –15

Section 10.3

43. State the steps for subtracting two numbers.
1. Leave the first number (the minuend) unchanged.
2. Change the subtraction sign to an addition sign.
3. Add the opposite of the second number (the subtrahend).

For Exercises 44–51, subtract the real numbers.

44. $4 - (-23)$ 27   **45.** $19 - 44$ –25

46. $-2 - (-24)$ 22   **47.** $-289 - 130$ –419

48. $-2.9 - 4.5$ –7.4   **49.** $3.8 - 4.5$ –0.7

50. $\left(-\dfrac{5}{3}\right) - \left(-\dfrac{5}{12}\right)$ $-\frac{5}{4}$ **51.** $0 - \left(-\dfrac{20}{21}\right)$ $\frac{20}{21}$

For Exercises 52–55, translate the mathematical statement to an English phrase. Answers will vary.

52. $4 - 6$
For example: The difference of 4 and 6

53. $23 - (-6)$
For example: 23 minus negative 6

54. $-2 - 14$
For example: 14 subtracted from -2

55. $-25 - (-7)$
For example: Subtract -7 from -25.

56. The temperature in Fargo, North Dakota, rose from $-6°F$ to $-1°F$. By how many degrees did the temperature rise?   The temperature rose 5°F.

57. Sam's balance in his checking account was $-\$40$, so he deposited \$132. What is his present balance?   Sam's balance is now \$92.

58. Find the average of the golf scores: $-3, 4, 0, 9, -2, -1, 0, 5, -3$ (These scores are the number of holes above or below par.)   The average is 1 above par.

Section 10.4

For Exercises 59–70, multiply or divide as indicated.

59. $6(-3)$ –18   **60.** $\dfrac{-12}{4}$ –3

61. $\dfrac{-900}{-60}$ 15   **62.** $(-7)(-8)$ 56

63. $-2.8 \div 0.04$ –70   **64.** $(-62.6)(2.5)$ –156.5

65. $\left(-\dfrac{2}{3}\right)\left(-\dfrac{21}{8}\right)$ $\frac{7}{4}$ **66.** $\left(-2\dfrac{1}{8}\right) \div \left(1\dfrac{1}{4}\right)$ $-\frac{17}{10}$ or $-1\frac{7}{10}$

67. $\left(-\dfrac{1}{5}\right) \div 0$ Undefined   **68.** $\dfrac{0}{-5}$ 0

69. $(-1)(-8)(2)(1)(-2)$ –32   **70.** $\dfrac{-9}{-5}$ $\frac{9}{5}$ or $1\frac{4}{5}$

For Exercises 71–76, simplify.

71. $(-6)^2$ 36   **72.** -6^2 –36

73. $\left(-\dfrac{3}{4}\right)^3$ $-\frac{27}{64}$ **74.** $-\left(\dfrac{3}{4}\right)^3$ $-\frac{27}{64}$

75. $(-1)^{10}$ 1   **76.** $(-1)^{21}$ –1

77. What is the sign of the product of three negative factors?   Negative

78. What is the sign of the product of four negative factors?   Positive

For Exercises 79–82, translate the English phrase into a mathematical expression. Then simplify.

79. The quotient of −45 and −15 $-45 \div (-15)$; 3

80. The product of −4 and 19 $-4 \cdot 19$; −76

81. 30 times −5 $30(-5)$; −150

82. −136 divided by −8 $-136 \div (-8)$; 17

Section 10.5

For Exercises 83–91, simplify by using the order of operations.

83. $28 \div (-7) \cdot 3 - (-1)$ −11

84. $(-4)^3 \div 8 - (-6)$ −2

85. $|10 - (-3)^2| \cdot (-11) + 4$ −7

86. $[-9 - (-7)]^3 \cdot 3 \div (-6)$ 4

87. $18 - (-5)^2 + 14 \div 2$ 0

88. $\left(\dfrac{1}{15}\right) \div \left(-\dfrac{7}{10}\right) \cdot \left(\dfrac{3}{2}\right) + \left(-\dfrac{6}{7}\right)$ −1

89. $\left(-\dfrac{3}{8}\right)^2 - \left(-\dfrac{1}{2}\right)^3$ $\dfrac{17}{64}$

90. $6 - [5 - (2 - 8)]$ −5

91. $\dfrac{3 - |2 + (-7)|}{3^2 - 5^2}$ $\dfrac{1}{8}$

92. Find the average temperature for one week: $2°, 4°, -6°, -1°, 0°, -4°, -2°$ −1°

Chapter 10   Test

1. Write an integer that represents the numerical value.

   a. Dwayne lost $220 during his last trip to Las Vegas.   −$220

   b. Garth Brooks has 26 more platinum albums than Elvis Presley.   26

For Exercises 2–4, refer to these numbers: $-3, -\frac{3}{5}, 0, \sqrt{7}, 4, -1, \frac{4}{7}, -\pi$

2. List all the numbers that are integers. $-3, 0, 4, -1$

3. List all the rational numbers. $-3, -\frac{3}{5}, 0, 4, -1, \frac{4}{7}$

4. List all the irrational numbers. $\sqrt{7}, -\pi$

For Exercises 5–10, place the correct symbol, $>$ or $<$, between the two numbers.

5. $-5 \boxed{} -2$ <

6. $|-5| \boxed{} |-2|$ >

7. $0 \boxed{} -2.4$ >

8. $\dfrac{4}{5} \boxed{} -\dfrac{2}{3}$ >

9. $-|-9| \boxed{} 9$ <

10. $-|33.1| \boxed{} |-33.1|$ <

For Exercises 11–18, add or subtract as indicated.

11. $9 + (-14)$ −5

12. $-23 + (-5)$ −28

13. $-4 - (-13)$ 9

14. $-30 - 11$ −41

15. $-1.5 + 2.1$ 0.6

16. $0.5 - 2.8$ −2.3

17. $-\dfrac{2}{3} - \dfrac{4}{7}$ $-\dfrac{26}{21}$ or $-1\dfrac{5}{21}$

18. $\dfrac{5}{4} + \left(-\dfrac{7}{8}\right)$ $\dfrac{3}{8}$

For Exercises 19–26, multiply or divide as indicated.

19. $6(-12)$ −72

20. $(-11)(-8)$ 88

21. $\dfrac{-24}{-12}$ 2

22. $\dfrac{54}{-3}$ −18

23. $\dfrac{-44}{0}$ Undefined

24. $(-91)(0)$ 0

25. $\dfrac{3}{10} \div \left(-\dfrac{4}{5}\right)$ $-\dfrac{3}{8}$ **26.** $\dfrac{-13}{-6}$ $\dfrac{13}{6}$

27. **a.** What is the sign of the product of an even number of negative factors? Positive

b. What is the sign of the product of an odd number of negative factors? Negative

28. Simplify the exponential expressions.

a. $(-8)^2$ **b.** -8^2 **c.** $(-4)^3$ **d.** -4^3
64 -64 -64 -64

For Exercises 29–34, translate to a mathematical expression. Then simplify the expression.

29. The product of -3 and -7 $-3(-7)$; 21

30. 8 more than -13 $-13 + 8$; -5

31. Subtract -4 from 18. $18 - (-4)$; 22

32. The quotient of 6 and $-\dfrac{2}{3}$ $6 \div \left(-\dfrac{2}{3}\right)$; -9

33. -8.1 increased by 5 $-8.1 + 5$; -3.1

34. The total of -3, 15, -6, and -1
$-3 + 15 + (-6) + (-1)$; 5

For Exercises 35–42, simplify.

35. $-14 + 22 - (-5) + (-10)$ 3

36. $(-3)(-1)(-4)(-1)(-5)$ -60

37. $-20 \div (-2)^2 + (-14)$ -19

38. $12 \cdot (-6) + [20 - (-12)] - 15$ -55

39. $-\dfrac{2}{15} + \left(-\dfrac{20}{21} \cdot \dfrac{7}{5}\right)$ $-\dfrac{22}{15}$ **40.** $\left(-\dfrac{1}{3}\right)^2 \div \left(\dfrac{5}{6} - \dfrac{1}{9}\right)$ $\dfrac{2}{13}$

41. $16 - 2[5 - (1 - 4)]$ 0 **42.** $\dfrac{15 - 2|3 - 9|}{8 - 2^2}$ $\dfrac{3}{4}$

43. Find the average temperature:
$4°, -3°, -1°, 5°, -2°, 0°, 4°$ 1°

Chapters 1–10 Cumulative Review Exercises

For Exercises 1–4, add, subtract, or multiply as indicated.

1. 3490
$\underline{+123}$
3613

2. 2901
$\underline{-332}$
2569

3. 23
34
98
$\underline{+22}$
177

4. 790
$\underline{\times 24}$
$18,960$

5. Write the prime factorization of 720.
$2 \cdot 2 \cdot 2 \cdot 2 \cdot 3 \cdot 3 \cdot 5$ or $2^4 \cdot 3^2 \cdot 5$

6. Write a fraction that represents the shaded region of the figure.

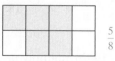

$\dfrac{5}{8}$

7. On a quiz, Harold missed 3 out of 14 questions. Write a fraction representing the fraction of the quiz questions that he answered *correctly*.
Harold got $\dfrac{11}{14}$ of the quiz correct.

8. Amy has a box that contains 20 oz of snack crackers. How many individual packages will she get if she puts $2\frac{1}{2}$ oz in each package?
Amy will have 8 packages.

9. Find the LCM of 16, 40, and 10. 80

10. Simplify. $\dfrac{3}{16} + \dfrac{33}{40} - \dfrac{7}{10}$ $\dfrac{5}{16}$

11. Add. $3\dfrac{3}{5} + 2\dfrac{13}{15}$ $6\dfrac{7}{15}$

12. Subtract. $16\dfrac{1}{2} - 12\dfrac{13}{14}$ $3\dfrac{4}{7}$

13. Round the numbers to the indicated place.

a. 34.2298 Thousandths **b.** 9.0314 Tenths
34.230 9.0

14. Convert cents to dollars. 209¢ $2.09

15. Multiply. 204.55(2.4) 490.92

16. Divide. $402.5 \div 3.5$ 115

17. Write a ratio of the shortest side to the longest side of the rectangle.

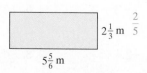

$2\frac{1}{3}$ m $\dfrac{2}{5}$

$5\frac{5}{6}$ m

18. A DC-10 aircraft used 9964 gal of fuel in 4 hr. Find the unit rate in gallons per hour.
The aircraft used 2491 gal/hr.

19. On a map 1 in. represents 6 mi. What is the distance between two cities that measure $3\frac{1}{2}$ in. on the map?   21 mi

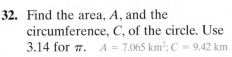 **20.** Given that the two triangles are similar, find sides x and y. $x = 2.7$ cm; $y = 7$ cm

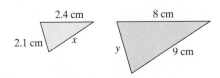

For Exercises 21–23, solve the percent equations.

21. What is 32% of 600?   192

22. What percent of 300 is 336?   112%

23. 15 is 6% of what?   250

24. A pair of shoes was discounted 20%. If the original price was $86, what is the sale price?
The sale price is $68.80.

For Exercises 25–28, convert the units of measure.

25. 2 ft 4 in. = _____ in.   28 in.

26. 20 qt = _____ gal   5 gal

27. 60 mL = _____ L   0.06 L

28. 30 oz = _____ lb $1\frac{7}{8}$ lb

 29. A car travels 6 mi due north and then turns and travels 8 mi due east. What is the distance of the car from the point of origin?
The distance is 10 mi.

 For Exercises 30 and 31, find the area.

30. Parallelogram   16.5 yd²

31. Square $5\frac{1}{16}$ m²

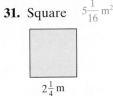

3.3 yd

5 yd

$2\frac{1}{4}$ m

32. Find the area, A, and the circumference, C, of the circle. Use 3.14 for π. $A = 7.065$ km²; $C = 9.42$ km

1.5 km

For Exercises 33–35, use the following data.

The number of miles walked in one day by 10 selected people is given.

4   4   4   3   6   4   6   5   3   4

33. Complete the frequency distribution for the data.

Number of Miles	Tally	Frequency (Number of Walkers)
3	‖	2
4	‖‖‖	5
5			1
6	‖	2

34. Construct a horizontal bar graph from the frequency distribution in Exercise 33. Label the vertical axis with the number of miles and the horizontal axis with the frequency.

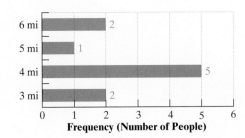

35. What is the mean number of miles walked per day?   4.3 mi

36. Refer to the circle graph. If the monthly budget for a small business is $1200, how much will be spent on postage? $216

Budget for an Office

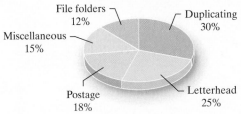

File folders 12%
Miscellaneous 15%
Duplicating 30%
Postage 18%
Letterhead 25%

For Exercises 37–40, simplify.

37. $43 - (-12)$ 55

38. $-12 + (-5) - 3 - (-8)$ -12

39. $(-4)^2 - 6^2$ -20

40. $\frac{8}{9} \cdot \left(\frac{1}{3} - \frac{5}{6}\right)^3 \div \left(-\frac{2}{3}\right)$ $\frac{1}{6}$

 Writing    Translating Expression   Geometry   Scientific Calculator   Video   NS E

Solving Equations

CHAPTER OUTLINE

Chapter 11

In this chapter, we learn how to simplify algebraic expressions by clearing parentheses and combining like terms. Then we move on to solving linear equations and using equations to solve applications problems.

Review Your Skills

To prepare for solving equations, practice performing the order of operations with positive and negative numbers. To complete the puzzle, first answer the questions and fill in the appropriate box. Then fill the grid so that every row, every column, and every 2×3 box contains the digits 1 through 6.

A. $(7 - 9)(-1 - 2)$

B. $3(1 - 4) + 12$

C. $-36 \div 4 \div (-9)$

D. $-14 \div 7(-2)$

E. $-4 - 2(5 - 8)$

F. $-2[5 - (-5)] \div (4 - 9)$

			A		
1	4	2	6	5	3
5	6	B 3	4	2	1
6	C 1	5	3	4	2
3	2	D 4	1	6	5
4	5	1	2	3	6
E 2	3	6	5	1	F 4

Section 11.1   Properties of Real Numbers

1. Algebraic Expressions

We begin our study of algebra with a few key terms. Recall that a **variable** is a letter or symbol that can represent any number. **Constants** are values that never change. Here are some examples of variables and constants.

Variables	Constants
x, y, z, A, l, w	$3, -1, \frac{2}{3}, \pi$

An algebraic **expression** is a collection of variables and numbers under operations such as addition, subtraction, multiplication, and division. Here are some examples of expressions.

$$2x, \quad 4 + y, \quad 3t - 7, \quad \frac{y}{8}$$

Algebraic expressions are often used in applications.

Skill Practice

1. Smoked turkey costs $6.99 per pound. Write an expression that represents the cost of p pounds.
2. The width of a basketball court is 44 ft shorter than its length, l. Write an expression that represents the width.
3. Six tons of gravel is to be carried away by n trucks. Write an expression for the amount carried by each truck. Assume that each truck carries an equal amount of gravel.

Classroom Examples: p. 643,
Exercises 4 and 8

Example 1   Using Algebraic Expressions in Applications

a. At a discount CD store, each CD costs $7.99. Suppose n is the number of CDs that a customer buys. Write an expression that represents the cost for n CDs.

b. The length of a rectangle is 5 in. longer than the width, w. Write an expression that represents the length of the rectangle.

c. A rope that is L ft long is to be cut into five pieces of equal length. Write an expression that represents the length of each piece.

Solution:

a. The cost of 1 CD is $7.99.

The cost of 2 CDs is $7.99(2) = $15.98.

The cost of 3 CDs is $7.99(3) = $23.97.

The cost of n CDs is $7.99($n$) or simply 7.99n$.

b. The length of a rectangle is 5 in. more than the width. The phrase *more than* implies addition. Thus, the length is represented by

$$\text{Length} = w + 5$$

c. For this scenario, drawing a figure may be helpful (Figure 11-1). The original length L must be cut (divided) into five pieces of equal length. Thus, the length (in feet) of each piece is given by

$$\text{Length of each piece} = \frac{L}{5}$$

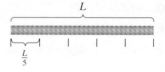

Figure 11-1

---

2. Evaluating Expressions

The value of an expression depends on the values of the variables within the expression. When we substitute numerical values for the variables within an expression, we call this *evaluating the expression*.

Answers

1. 6.99p$ 2. $l - 44$ 3. $\dfrac{6}{n}$

| Example 2 | **Evaluating Expressions** |

Evaluate each expression for the given value of the variable(s).

a. $3x^2$ for $x = -5$

b. $4a - 8b$ for $a = -6$, $b = \frac{1}{2}$

Solution:

To evaluate the expressions, we substitute the given number for the variable. We recommend using parentheses so that you can "see" where to insert the given numerical values.

a. $3x^2 = 3(\ )^2$ Replace the variable with parentheses.

     $= 3(-5)^2$ Substitute $x = -5$.

     $= 3(25)$ Simplify exponents.

     $= 75$

b. $4a - 8b = 4(\ ) - 8(\ )$ Replace the variables with parentheses.

     $= 4(-6) - 8\left(\frac{1}{2}\right)$ Substitute $a = -6$ and $b = \frac{1}{2}$.

     $= -24 - 4$ Multiply before subtracting.

     $= -28$

Skill Practice

Evaluate each expression for the given value of the variable(s).

4. $-4y^2$ for $y = -3$

5. $-6t + 4v$ for $t = -1$ and $v = -\frac{1}{2}$

Classroom Example: p. 643, Exercise 20

Instructor Note: The common error is to write $3 \cdot -5^2$ when substituting -5 for x. If students use parentheses, they are less likely to make that error.

3. Properties of Real Numbers

Several important properties of whole numbers were introduced in Sections 1.2 and 1.5 involving addition and multiplication. These properties also hold for real numbers and are summarized in Tables 11-1 through 11-3.

Table 11-1   **Commutative Properties of Real Numbers**

Property	In Symbols/Examples	Comments/Notes
Commutative property of addition	$a + b = b + a$ Ex.: $-4 + 7 = 7 + (-4)$ $x + 3 = 3 + x$	The order in which two real numbers are added does not affect the sum.
Commutative property of multiplication	$a \cdot b = b \cdot a$ Ex.: $-5 \cdot 9 = 9 \cdot (-5)$ $8y = y \cdot 8$	The order in which two real numbers are multiplied does not affect the product.

Table 11-2   **Associative Properties of Real Numbers**

Property	In Symbols/Examples	Comments/Notes
Associative property of addition	$(a + b) + c = a + (b + c)$ Ex.: $(5 + 8) + 1 = 5 + (8 + 1)$ $(t + n) + 3 = t + (n + 3)$	The manner in which three real numbers are grouped under addition does not affect the sum.
Associative property of multiplication	$(a \cdot b) \cdot c = a \cdot (b \cdot c)$ Ex.: $(-2 \cdot 3) \cdot 6 = -2 \cdot (3 \cdot 6)$ $3 \cdot (m \cdot n) = (3 \cdot m) \cdot n$	The manner in which three real numbers are grouped under multiplication does not affect the product.

Answers

4. -36 **5.** 4

Table 11-3 **Distributive Property of Multiplication over Addition**

Property	In Symbols/Examples	Comments/Notes
Distributive property of multiplication over addition*	$a \cdot (b + c) = a \cdot b + a \cdot c$ Ex.: $-3(2 + x) = -3(2) + (-3)(x)$ $\qquad = -6 + (-3)x$ $\qquad = -6 - 3x$	Each term inside the parentheses is multiplied by the factor outside the parentheses.

*Note that the distributive property of multiplication over addition is sometimes referred to as just the *distributive property*.

Example 3 demonstrates the use of the commutative properties.

Example 3 **Applying the Commutative Properties of Real Numbers**

Apply the commutative property of addition or multiplication to rewrite the expression.

a. $6 + p$　　**b.** $y(7)$　　**c.** $-5 + n$　　**d.** xy

Solution:

a. $6 + p = p + 6$　　　　Commutative property of addition

b. $y(7) = 7y$　　　　　　Commutative property of multiplication

c. $-5 + n = n + (-5)$　　Commutative property of addition

$\qquad = n - 5$

d. $xy = yx$　　　　　　　Commutative property of multiplication

Recall from Section 1.3 that subtraction is not a commutative operation. However, if we rewrite the difference of two numbers $a - b$ as $a + (-b)$, then we can apply the commutative property of addition. For example,

$$x - 9 = x + (-9) \qquad \text{Rewrite as addition of the opposite.}$$

$$= -9 + x \qquad \text{Apply the commutative property of addition.}$$

Example 4 demonstrates the associative properties of addition and multiplication.

Example 4 **Applying the Associative Properties of Real Numbers**

Use the associative property of addition or multiplication to rewrite each expression. Then simplify the expression.

a. $5(7w)$　　**b.** $1.2 + (4.5 + y)$　　**c.** $-\dfrac{2}{5}\left(-\dfrac{5}{2}z\right)$

Solution:

a. $5(7w) = (5 \cdot 7)w$　　　　　　　　Apply the associative property of multiplication.

$\qquad = 35w$　　　　　　　　　　　Simplify.

b. $1.2 + (4.5 + y) = (1.2 + 4.5) + y$　　Apply the associative property of addition.

$\qquad = 5.7 + y$　　　　　　　　Simplify.

c. $-\dfrac{2}{5}\left(-\dfrac{5}{2}z\right) = \left(-\dfrac{2}{5}\cdot-\dfrac{5}{2}\right)z$ Apply the associative property of multiplication.

$\qquad\qquad = 1 \cdot z$ Simplify. Note that the factors within the parentheses are reciprocals. Therefore, their product is 1.

$\qquad\qquad = z$

Note that in most cases, a detailed application of the associative properties will not be given. Instead the process will be written in one step, such as

$$5(7w) = 35w \qquad 1.2 + (4.5 + y) = 5.7 + y \qquad -\dfrac{2}{5}\left(-\dfrac{5}{2}z\right) = z$$

Example 5 demonstrates the use of the distributive property.

Example 5 **Applying the Distributive Property**

Apply the distributive property.

a. $3(x + 4)$ **b.** $2(3y - 5z + 1)$ **c.** $\dfrac{2}{3}\left(6p + \dfrac{1}{4}\right)$

Solution:

a. $3(x + 4) = 3(x) + 3(4)$ Apply the distributive property.

$\qquad\qquad = 3x + 12$ Simplify.

b. $2(3y - 5z + 1) = 2(3y + (-5z) + 1)$ First write the subtraction as addition of the opposite.

$\qquad\qquad = 2(3y + (-5z) + 1)$ Apply the distributive property.

$\qquad\qquad = 2(3y) + 2(-5z) + 2(1)$

$\qquad\qquad = 6y + (-10z) + 2$ Simplify.

$\qquad\qquad = 6y - 10z + 2$

> **TIP:** In Example 5(b), we rewrote the expression by writing the subtraction as addition of the opposite. Often this step is not shown, and fewer steps are shown overall. For example:
>
> $$2(3y - 5z + 1) = 2(3y) + 2(-5z) + 2(1)$$
> $$= 6y - 10z + 2$$

c. $\dfrac{2}{3}\left(6p + \dfrac{1}{4}\right) = \dfrac{2}{3}(6p) + \dfrac{2}{3}\left(\dfrac{1}{4}\right)$ Apply the distributive property.

$\qquad\qquad = \dfrac{2}{3}\left(\dfrac{6p}{1}\right) + \dfrac{2}{3}\left(\dfrac{1}{4}\right)$ Write the whole number as an improper fraction.

$\qquad\qquad = \dfrac{2}{3}\left(\dfrac{\overset{2}{6p}}{1}\right) + \dfrac{\overset{1}{2}}{3}\left(\dfrac{1}{\underset{2}{4}}\right)$ Multiply fractions.

$\qquad\qquad = 4p + \dfrac{1}{6}$ Simplify.

Example 6 **Applying the Distributive Property**

Apply the distributive property.

a. $-8(2 - 5y)$ **b.** $-(-4a + b + 3c)$

Solution:

a. $-8(2 - 5y)$

$= -8[2 + (-5y)]$ Write the subtraction as addition of the opposite.

$= -8[2 + (-5y)]$ Apply the distributive property.

$= -8(2) + (-8)(-5y)$

$= -16 + 40y$ Simplify.

b. $-(-4a + b + 3c)$

$= -1 \cdot (-4a + b + 3c)$ The negative sign preceding the parentheses indicates that we take the opposite of the expression within parentheses. This is equivalent to multiplying the expression within parentheses by -1.

$= -1(-4a) + (-1)(b) + (-1)(3c)$ Apply the distributive property.

$= 4a - b - 3c$ Simplify.

> **TIP:** Notice that a negative factor outside the parentheses changes the signs of all terms to which it is multiplied.
>
> $$-1 \cdot (-4a + b + 3c)$$
> $$= +4a - b - 3c$$

Section 11.1 Practice Exercises

Study Skills Exercise

For additional exercises, see Classroom Activities 11.1A–11.1B in the *Student's Resource Manual* at www.mhhe.com/moh.

When beginning a study of algebra, some students do not understand the concept of a variable. A variable is a letter that represents an unknown value. When trying to find what number added to 5 equals -12, we write $5 + x = -12$. Rewrite the following, using variables:

What number times 4 equals 6? _____

What number divided by 5 equals 3? _____

Vocabulary and Key Concepts

1. a. A ___variable___ is a letter or symbol that can represent any number.

b. The ___commutative___ property of addition states that $a + b = b + a$. For example, the expression $6 + x$ can be written as ___$x + 6$___.

c. The commutative property of multiplication states that $a \cdot b = $ ___$b \cdot a$___. For example, the expression $x(6)$ can be written as ___$6x$___.

d. The __associative__ property of addition states that $(a + b) + c = a + (b + c)$. For example, the expression $-5 + (7 + x)$ can be written as __$(-5 + 7) + x$__.

e. The associative property of multiplication states that $(a \cdot b) \cdot c =$ __$a \cdot (b \cdot c)$__. For example, the expression $-5(7x)$ can be written as __$(-5 \cdot 7)x$__.

f. The distributive property of multiplication over addition states that $a \cdot (b + c) =$ __$a \cdot b + a \cdot c$__. Using this property, the expression $4(x + 5)$ simplifies as __$4x + 20$__.

Concept 1: Algebraic Expressions

2. Shawna makes $54 more per week than her brother Cory. Write an expression for Shawna's weekly income if Cory's weekly income is represented by w. $w + 54$

3. Maria needs to buy 8 wine glasses. Write an expression for the cost of 8 glasses at p dollars each. **(See Example 1.)** $8p$

4. Carolyn sells homemade candles. Write an expression for her total revenue if she sells 5 candles for r dollars each. $5r$

5. Jonathan is 4 in. taller than his brother. Write an expression for Jonathan's height if his brother is t in. tall. $t + 4$

6. It takes Perry $\frac{1}{2}$ hr longer than David to mow the lawn. If it takes David l hr to mow the lawn, write an expression for the amount of time it takes Perry to mow the lawn. $l + \frac{1}{2}$

7. A sedan travels 6 mph slower than a sports car. Write an expression for the speed of the sedan if the sports car travels v mph. $v - 6$

8. Marcus's daughter is 30 years younger than he is. Write an expression for his daughter's age if Marcus is A years old. $A - 30$

9. A piece of ribbon is cut into n pieces of equal length. If the length of the ribbon is 4 yd, write an expression for the length of each piece. $\frac{4}{n}$

10. A party is planned for 20 people. If there is p oz of punch available, write an expression for the amount of punch for each person, assuming that each person drinks an equal amount. $\frac{p}{20}$

11. The price of gas has doubled over the last 3 years. If gas cost g dollars per gallon 3 years ago, write an expression for the current price per gallon. $2g$

12. Suppose that the amount of rain that fell on Monday was twice the amount that fell on Sunday. Write an expression for the amount of rain on Monday, if Sunday's amount was t in. $2t$

Concept 2: Evaluating Expressions

For Exercises 13–20, evaluate the expression for the given values. **(See Example 2.)**

13. $-6x$ for

  a. $x = 2$ -12

  b. $x = -5$ 30

14. $-2y^2$ for

  a. $y = 3$ -18

  b. $y = -3$ -18

15. $3p + 5q$ for

  a. $p = 2, q = -\frac{1}{5}$ 5

  b. $p = -5, q = 0$ -15

16. $9c - 2d$ for

  a. $c = -1, d = \frac{1}{2}$ -10

  b. $c = 3, d = -2$ 31

17. $-a^2$ for

  a. $a = -7$ -49

  b. $a = 7$ -49

18. $-b^3$ for

  a. $b = -3$ 27

  b. $b = 3$ -27

19. $-4(r - s)^2$ for

  a. $r = 8, s = 6$ -16

  b. $r = 3, s = -1$ -64

20. $-5(u + v)^2$ for

  a. $u = 10, v = -7$ -45

  b. $u = 0, v = -2$ -20

For Exercises 21–24, evaluate the expression when $x = -2$, $y = \dfrac{2}{3}$, $z = 4$, and $w = -\dfrac{1}{2}$.

21. $y(x - 4)$ -4

22. $w(-x - 4)$ 1

23. $z^2 - x + 6$ 24

24. $x^3 - w - \dfrac{3}{2}$ -9

For Exercises 25–28, evaluate the expression when $a = 12$, $b = -3$, and $c = -2$.

25. $bc \div a$ $\dfrac{1}{2}$

26. $5b - c$ -13

27. $b^2 - c^2$ 5

28. $b^2 + c^2$ 13

For Exercises 29–32, find the area A or perimeter P.

29. $P = 2l + 2w$ for $l = 6$ in. and $w = 2.3$ in. (perimeter of a rectangle) $P = 16.6$ in.

30. $A = lw$ for $l = \dfrac{3}{2}$ ft and $w = 4$ ft (area of a rectangle) $A = 6$ ft^2

31. $A = \pi r^2$ for $\pi = \dfrac{22}{7}$ m and $r = \dfrac{7}{2}$ m (area of a circle) $A = \dfrac{77}{2}$ m^2

32. $A = \dfrac{1}{2}bh$ for $b = \dfrac{4}{5}$ yd and $h = \dfrac{10}{11}$ yd (area of a triangle) $A = \dfrac{4}{11}$ yd^2

Concept 3: Properties of Real Numbers

For Exercises 33–44, apply the commutative property of addition or multiplication to rewrite each expression. **(See Example 3.)**

33. $5 + w$ $w + 5$

34. $t + 2$ $2 + t$

35. $-\dfrac{1}{3} + b$ $b + \left(-\dfrac{1}{3}\right)$ or $b - \dfrac{1}{3}$

36. $-\dfrac{1}{2} + c$ $c + \left(-\dfrac{1}{2}\right)$ or $c - \dfrac{1}{2}$

37. $r(2)$ $2r$

38. $a(-4)$ $-4a$

39. $t(-s)$ $-st$

40. $d(-c)$ $-cd$

41. xy yx

42. ab ba

43. $7 - p$ $-p + 7$

44. $8 - q$ $-q + 8$

For Exercises 45–56, apply the associative property of addition or multiplication to rewrite each expression. Then simplify the expression. **(See Example 4.)**

45. $-2(6b)$
$(-2 \cdot 6)b$; $-12b$

46. $-3(2c)$
$(-3 \cdot 2)c$; $-6c$

47. $3 + (8 + t)$
$(3 + 8) + t$; $11 + t$

48. $7 + (5 + p)$
$(7 + 5) + p$; $12 + p$

49. $-4.2 + (2.5 + r)$
$(-4.2 + 2.5) + r$; $-1.7 + r$

50. $1.1 + (-0.8 + w)$
$(1.1 + (-0.8)) + w$; $0.3 + w$

51. $3(6x)$
$(3 \cdot 6)x$; $18x$

52. $9(5k)$
$(9 \cdot 5)k$; $45k$

53. $-\dfrac{4}{7}\left(-\dfrac{7}{4}d\right)$
$\left(-\dfrac{4}{7} \cdot -\dfrac{7}{4}\right)d$; d

54. $\dfrac{5}{6}\left(\dfrac{6}{5}m\right)$
$\left(\dfrac{5}{6} \cdot \dfrac{6}{5}\right)m$; m

55. $-9 + (-12 + h)$
$(-9 + (-12)) + h$; $-21 + h$

56. $-11 + (-4 + s)$
$(-11 + (-4)) + s$; $-15 + s$

For Exercises 57–76, apply the distributive property. **(See Examples 5 and 6.)**

57. $4(x + 8)$
$4x + 32$

58. $5(3 + w)$
$15 + 5w$

59. $-2(p + 4)$
$-2p - 8$

60. $-6(k + 2)$
$-6k - 12$

61. $-10(t - 3)$
$-10t + 30$

62. $-7(p - 4)$
$-7p + 28$

63. $-5(-2 + x)$
$10 - 5x$

64. $-8(-3 + y)$
$24 - 8y$

65. $4(a + 4b - c)$
$4a + 16b - 4c$

66. $2(3q - r + s)$
$6q - 2r + 2s$

67. $4\left(\dfrac{2}{3} + g\right)$
$\dfrac{8}{3} + 4g$

68. $8\left(\dfrac{5}{6} + m\right)$
$\dfrac{20}{3} + 8m$

69. $-(3 - n)$
$-3 + n$

70. $-(13 - t)$
$-13 + t$

71. $-(-a - 8)$
$a + 8$

72. $-(-d - 10)$
$d + 10$

73. $-(3x + 9 - 5y)$
$-3x - 9 + 5y$

74. $-(a - 8b + 4c)$
$-a + 8b - 4c$

75. $-(-5q - 2s - 3t)$
$5q + 2s + 3t$

76. $-(-10p - 12q + 3)$
$10p + 12q - 3$

Mixed Exercises

For Exercises 77–96, apply the appropriate property to simplify the expression.

77. $6(2x)$
$12x$

78. $-3(12k)$
$-36k$

79. $6(2 + x)$
$12 + 6x$

80. $-3(12 + k)$
$-36 - 3k$

81. $-6(-1 - k)$
$6 + 6k$

82. $-4(-8 + h)$
$32 - 4h$

83. $-6 + (-1 - k)$
$-7 - k$

84. $-4 + (-8 + h)$
$-12 + h$

85. $-8 + (4 - p)$
$-4 - p$

86. $3 + (25 - m)$
$28 - m$

87. $-8(4 - p)$
$-32 + 8p$

88. $3(25 - m)$
$75 - 3m$

89. $8\left(\dfrac{1}{2}a\right)$
$4a$

90. $-20\left(\dfrac{1}{5}b\right)$
$-4b$

91. $8\left(\dfrac{1}{2} + a\right)$
$4 + 8a$

92. $-20\left(\dfrac{1}{5} + b\right)$
$-4 - 20b$

93. $\dfrac{5}{9}(9 + y)$

$5 + \dfrac{5}{9}y$

94. $-\dfrac{3}{4}(8 - b)$

$-6 + \dfrac{3}{4}b$

95. $\dfrac{5}{9}(9y)$

$5y$

96. $-\dfrac{3}{4}(8b)$

$-6b$

1. Definition of *Like* Terms

An algebraic expression is the sum of one or more terms. A **term** is a number or a product or quotient of numbers and variables. For example, the expression

$$-8x^3 + xy - 40 \qquad \text{can be written as} \qquad -8x^3 + xy + (-40)$$

This expression consists of the terms $-8x^3$, xy, and -40. The terms $-8x^3$ and xy are called **variable terms**, and the term -40 is called a **constant term**.

It is important to distinguish between a term and the factors within a term. For example, the quantity xy is one term, and the values x and y are factors within the term. The constant factor in a term is called the **coefficient** of the term.

Term	Coefficient of the Term
$-8x^3$	-8
xy or $1xy$	1
-40	-40
$\dfrac{x}{4}$ or $\dfrac{1}{4}x$	$\dfrac{1}{4}$

Concepts

1. Definition of *Like* Terms
2. Combining *Like* Terms
3. Clearing Parentheses and Combining *Like* Terms

Avoiding Mistakes

Variables without a coefficient explicitly written have a coefficient of 1. Thus, the term xy is equal to $1xy$. The 1 is understood.

Terms are said to be *like* terms if they each have the same variables and the corresponding variables are raised to the same powers. For example,

Like Terms	Unlike Terms	
$-4x$ and $6x$	$-4x$ and $6y$	(different variables)
$18ab$ and $4ba$	$18ab$ and $4a$	(different variables)
$7m^2n^5$ and $3m^2n^5$	$7m^2n^5$ and $3mn^5$	(different powers of m)
$5p$ and $-3p$	$5p$ and 3	(different variables)
8 and 10	8 and $10x$	(different variables)

Example 1 Identifying Terms, Coefficients, and *Like* Terms

a. List the terms of the expression: $1.4x^3 - 6x^2 + x + 5$

b. Identify the coefficient of each term: $1.4x^3 - 6x^2 + x + 5$

c. Which two terms are *like* terms? $-6x, 5, -3y,$ and $4x$

Solution:

a. The expression $1.4x^3 - 6x^2 + x + 5$ can be written as
$1.4x^3 + (-6x^2) + x + 5$

Therefore, the terms are $1.4x^3, -6x^2, x,$ and 5.

b. The coefficients are $1.4, -6, 1,$ and 5.

c. The terms $-6x$ and $4x$ are *like* terms.

2. Combining *Like* Terms

Two terms may be combined if they are *like* terms. To add or subtract *like* terms, we use the distributive property, as shown in Example 2.

Example 2 Using the Distributive Property to Add and Subtract *Like* Terms

Add or subtract as indicated.

a. $8y + 6y$ **b.** $-15w + 4w - w$

Solution:

a. $8y + 6y = (8 + 6)y$ Apply the distributive property.

$\qquad\quad = 14y$ Simplify.

b. $-15w + 4w - w = -15w + 4w - 1w$ First note that $w = 1w$.

$\qquad\qquad\qquad = (-15 + 4 - 1)w$ Apply the distributive property.

$\qquad\qquad\qquad = (-12)w$ Simplify within parentheses.

$\qquad\qquad\qquad = -12w$

Although the distributive property is used to add and subtract *like* terms, it is tedious to write each step. Observe that adding or subtracting *like* terms is a matter of adding or subtracting the coefficients and leaving the variable factors unchanged. This can be shown in one step.

$$8y + 6y = 14y \qquad \text{and} \qquad -15w + 4w - 1w = -12w$$

This shortcut will be used throughout the text.

| Example 3 | Adding and Subtracting *Like* Terms |

Simplify by combining *like* terms.

$$-3x + 8y + 4x - 19 - 10y$$

Solution:

$$-3x + 8y + 4x - 19 - 10y$$

$= -3x + 4x + 8y - 10y - 19$	Use the commutative and associative properties of addition to group the *like* terms together.
$= 1x - 2y - 19$	Combine *like* terms.
$= x - 2y - 19$	Note that $1x = x$. Also note that the remaining terms cannot be combined because they are not *like* terms. The variable factors are different.

Skill Practice

Simplify.

6. $4a - 10b - a + 16b + 9$

Classroom Example: p. 649, Exercise 44

| Example 4 | Adding and Subtracting *Like* Terms |

Simplify by combining *like* terms.

a. $\dfrac{2}{5}m + \dfrac{1}{8}n - \dfrac{1}{5}m + \dfrac{3}{8}n$ **b.** $0.2a - 1.4 + 1.4a - 6b - 2.1$

Solution:

a. $\dfrac{2}{5}m + \dfrac{1}{8}n - \dfrac{1}{5}m + \dfrac{3}{8}n$

$= \dfrac{2}{5}m - \dfrac{1}{5}m + \dfrac{1}{8}n + \dfrac{3}{8}n$	Group *like* terms together.
$= \dfrac{1}{5}m + \dfrac{4}{8}n$	Combine *like* terms.
$= \dfrac{1}{5}m + \dfrac{1}{2}n$	Simplify fractions.

b. $0.2a - 1.4 + 1.4a - 6b - 2.1$

| $= 0.2a + 1.4a - 6b - 1.4 - 2.1$ | Group *like* terms together. |
| $= 1.6a - 6b - 3.5$ | Combine *like* terms. |

Skill Practice

Simplify.

7. $\dfrac{5}{9} - \dfrac{2}{3}w + \dfrac{5}{3}w - \dfrac{4}{9}$

8. $6.3x - 4.1y - 2.4$ $+ 2.1y + 1.1$

Classroom Example: p. 650, Exercise 58

3. Clearing Parentheses and Combining *Like* Terms

Notice that when the distributive property is applied, the original parentheses are dropped. This is often called *clearing parentheses*.

Answers

6. $3a + 6b + 9$ **7.** $w + \dfrac{1}{9}$

8. $6.3x - 2y - 1.3$

Skill Practice

Simplify.

9. $8 - 6(w + 4)$

Classroom Example: p. 650, Exercise 70

Example 5 Clearing Parentheses and Combining *Like* Terms

Simplify by clearing parentheses and combining *like* terms. $6 - 3(2y + 9)$

Solution:

$6 - 3(2y + 9)$ The order of operations indicates that we must perform multiplication before subtraction.

It is also important to understand that a factor of -3 (not 3) will be multiplied by all terms within the parentheses. To see why, rewrite the subtraction in terms of addition of the opposite.

$6 - 3(2y + 9) = 6 + (-3)(2y + 9)$ Rewrite subtraction as addition of the opposite.

$= 6 + (-3)(2y) + (-3)(9)$ Apply the distributive property.

$= 6 + (-6y) + (-27)$ Simplify.

$= -6y + 6 + (-27)$ Group *like* terms together.

$= -6y - 21$ Combine *like* terms.

Skill Practice

Simplify.

10. $-5(10 - m) - 2(m + 1)$

Classroom Example: p. 650, Exercise 74

Example 6 Clearing Parentheses and Combining *Like* Terms

Simplify by clearing parentheses and combining *like* terms.
$-8(x - 4) - 5(x + 7)$

Solution:

$-8(x - 4) - 5(x + 7)$

$= -8[x + (-4)] + (-5)(x + 7)$ Rewrite subtraction as addition of the opposite.

$= -8[x + (-4)] + (-5)(x + 7)$ Apply the distributive property.

$= -8(x) + (-8)(-4) + (-5)(x) + (-5)(7)$

$= -8x + 32 - 5x - 35$ Simplify.

$= -8x - 5x + 32 - 35$ Group *like* terms together.

$= -13x - 3$ Combine *like* terms.

Answers

9. $-6w - 16$ **10.** $3m - 52$

Section 11.2 Practice Exercises

Study Skills Exercise

For additional exercises, see Classroom Activities 11.2A–11.2B in the *Student's Resource Manual* at www.mhhe.com/moh.

Two important concepts in this section are *terms* and *factors*. Consider the expression $2x + 5y$. The quantities $2x$ and $5y$ are terms of the expression. Now consider the single term expression $2xy$. In this expression, $2, x,$ and y are factors. Write in your own words the difference between a term and a factor.

Writing   Translating Expression   Geometry   Scientific Calculator   Video   NSE

Vocabulary and Key Concepts

1. **a.** A ___term___ is a number or the product or quotient of numbers and variables.

   b. The term $-5xy$ is called a (constant/variable) term, whereas the term -5 is called a (constant/variable) term.   variable; constant

   c. The numerical factor within a term is called the ___coefficient___ of the term.

   d. Two terms are ___like___ terms if they each have the same variables and the corresponding variables are raised to the same powers.

Review Exercises

2. Evaluate the expression for $a = 4$ and $b = -5$.

   a. $6(a + 3b)$ −66

   b. $6a + 18b$ −66

For Exercises 3–8, simplify the expression, using the associative or distributive properties to clear parentheses.

3. $6(p + 3)$ $6p + 18$

4. $(-7p + 2) + 10$ $-7p + 12$

5. $4(-6q)$ $-24q$

6. $-3(t - 2)$ $-3t + 6$

7. $13 + (-4 - h)$ $9 - h$

8. $-(x - 20y - 14z)$ $-x + 20y + 14z$

Concept 1: Definition of *Like* Terms

For Exercises 9–16, identify each term as a variable term or a constant term. **(See Example 1.)**

9. $2a$ variable term

10. -4 constant term

11. 8 constant term

12. $-8k$ variable term

13. $-4pq$ variable term

14. $8st$ variable term

15. $10h^2$ variable term

16. 3 constant term

For Exercises 17–24, identify the coefficients for each term. **(See Example 1.)**

17. $6p - 4q$ $6, -4$

18. $-5a^3 - 2a$ $-5, -2$

19. $-14h + 12$ $-14, 12$

20. $8x + 9$ $8, 9$

21. $x - y$ $1, -1$

22. $p - q$ $1, -1$

23. $5t - 8s - 3$ $5, -8, -3$

24. $6g - 16h - 2$ $6, -16, -2$

For Exercises 25–36, determine if the two terms are *like* terms or unlike terms. **(See Example 1.)**

25. $3a, -2a$ *Like* terms

26. $8b, 12b$ *Like* terms

27. $4x, 4y$ Unlike terms

28. $-9k, -9h$ Unlike terms

29. $7xy, -3yx$ *Like* terms

30. $-5ab, ba$ *Like* terms

31. $6a, 13a^2$ Unlike terms

32. $20k^3, 3k$ Unlike terms

33. $14, 14y$ Unlike terms

34. $25x, 25$ Unlike terms

35. $17, -32$ *Like* terms

36. $8, -22$ *Like* terms

Concept 2: Combining *Like* Terms

For Exercises 37–62, combine the *like* terms. **(See Examples 2–4.)**

37. $6rs + 8rs$ $14rs$

38. $4x + 21x$ $25x$

39. $-4h + 12h$ $8h$

40. $9p - 13p$ $-4p$

41. $4x^2 + 9 - x^2$ $3x^2 + 9$

42. $13t^2 - t^2 + 4$ $12t^2 + 4$

43. $10x - 12y - 4x - 3y$ $6x - 15y$

44. $14a - 5b + 3a - b$ $17a - 6b$

45. $-6k - 9k + 12k$ $-3k$

46. $-11p + 23p - p$ $11p$

47. $-8uv + 6u + 12uv$ $4uv + 6u$

48. $9pq - 9p + 13pq$ $22pq - 9p$

49. $6 - 14m - 15 - 2m$ $-16m - 9$

50. $1 - 8n + 5 - 3n$ $-11n + 6$

51. $18 - 3a + 5b - 6a + 2$ $-9a + 5b + 20$

52. $13 + w - 5z - 4 + 7w$ $8w - 5z + 9$

53. $-5p^2 + 6p - p^2 + 7 - 8p$ $-6p^2 - 2p + 7$

54. $-3q^2 - 10q + q^2 - 15 + 5q$ $-2q^2 - 5q - 15$

55. $\dfrac{1}{2}y + \dfrac{3}{2}y - \dfrac{5}{6}$ $2y - \dfrac{5}{6}$

56. $-\dfrac{4}{5}p + \dfrac{2}{5}p + \dfrac{4}{7}$ $-\dfrac{2}{5}p + \dfrac{4}{7}$

57. $\dfrac{3}{4}a + 3 - \dfrac{1}{8}a + 6$ $\dfrac{5}{8}a + 9$

58. $\frac{1}{3}b - 4 + \frac{2}{9}b - 4$ $\frac{5}{9}b - 8$ **59.** $2.3x^2 + 4.1x - 5.3x^2 - 6x$ **60.** $1.2y - 0.4y^2 - 0.3y - 1.5y^2$
                                                            $-3x^2 - 1.9x$ $-1.9y^2 + 0.9y$

61. $4.4 - 0.9a + 3.2$ $-0.9a + 7.6$ **62.** $9.7 - 8.8b - 3.2$ $-8.8b + 6.5$

Concept 3: Clearing Parentheses and Combining *Like* Terms

For Exercises 63–86, clear parentheses and combine *like* terms. **(See Examples 5 and 6.)**

63. $5(t - 6) + 2$ $5t - 28$ **64.** $7(a - 4) + 8$ $7a - 20$ **65.** $-3(2x + 1) - 13$ $-6x - 16$

66. $-2(4b + 3) - 10$ $-8b - 16$ **67.** $4 + 6(y - 3)$ $6y - 14$ **68.** $11 + 2(p - 8)$ $2p - 5$

69. $21 - 7(3 - q)$ $7q$ **70.** $10 - 5(2 - 5m)$ $25m$ **71.** $-3 - (2n + 1)$ $-2n - 4$

72. $-13 - (6s + 5)$ $-6s - 18$ **73.** $-2(a + 3b) - (4a - 5b)$ **74.** $-(2m - 7n) - 3(6m - n)$
                                                            $-6a - b$ $-20m + 10n$

75. $10(x + 5) - 3(2x + 9)$ **76.** $6(y - 9) - 5(2y - 5)$ **77.** $-(12z + 1) + 2(7z - 5)$
   $4x + 23$ $-4y - 29$ $2z - 11$

78. $-(8w + 5) + 3(w - 15)$ **79.** $3(w + 3) - (4w + y) - 3y$ **80.** $2(s + 6) - (8s - t) + 6t$
   $-5w - 50$ $-w - 4y + 9$ $-6s + 7t + 12$

81. $20a - 4(b + 3a) - 5b$ **82.** $16p - 3(2p - q) + 7q$ **83.** $6 - (3m - n) - 2(m + 8) + 5n$
   $8a - 9b$ $10p + 10q$ $-5m + 6n - 10$

84. $12 - (5u + v) - 4(u - 6) + 2v$ **85.** $15 + 2(w - 4) - (2w - 5z) + 7z$ **86.** $7 + 3(2a - 5) - (6a - 8b) - 2b$
   $-9u + v + 36$ $12z + 7$ $6b - 8$

Expanding Your Skills

For Exercises 87–94, clear parentheses and combine *like* terms in expressions involving fractions and decimals.

87. $6\left(\frac{1}{2}x - \frac{2}{3}\right) - 4\left(\frac{5}{2}x + \frac{3}{4}\right)$ $-7x - 7$ **88.** $-12\left(\frac{5}{6}p + \frac{1}{4}\right) + 9\left(\frac{2}{9}p - \frac{1}{3}\right)$ $-8p - 6$

89. $\frac{2}{3}(9y + 6) - \frac{3}{2}(18y - 16)$ $-21y + 28$ **90.** $-\frac{1}{4}(4w - 8) + \frac{1}{2}(4w + 10)$ $w + 7$

91. $10(0.2q - 3) - 100(0.04q - 0.5)$ $-2q + 20$ **92.** $100(0.14b + 0.2) - 10(1.3b - 4)$ $b + 60$

93. $100(1.04a - 2.1b) - 10(21.1a + 0.3b)$ **94.** $10(-7.2x - y) + 1000(0.023x + 0.004y)$ $-49x - 6y$
   $-107a - 213b$

Section 11.3 Addition and Subtraction Properties of Equality

Concepts

1. Definition of an Equation
2. Addition and Subtraction Properties of Equality

1. Definition of an Equation

An **equation** is a statement that indicates that two quantities are equal. The following are equations.

$$x = 7 \qquad z + 3 = 8 \qquad -6p = 18$$

All equations have an equal sign. Furthermore, notice that the equal sign separates the equation into two parts, the left-hand side and the right-hand side. A **solution** to an equation is a value of the variable that makes the equation a true statement. Substituting a solution to an equation for the variable makes the right-hand side equal to the left-hand side.

Equation	Solution	Check
$x = 7$	7	$x = 7$

$$7 = 7 \checkmark$$

Substitute 7 for x.
Right-hand side equals left-hand side.

$z + 3 = 8$	5	$z + 3 = 8$

$$5 + 3 = 8$$

$$8 = 8 \checkmark$$

Substitute 5 for z.
Right-hand side equals left-hand side.

$-6p = 18$	-3	$-6p = 18$

$$-6(-3) = 18$$

$$18 = 18 \checkmark$$

Substitute -3 for p.
Right-hand side equals left-hand side.

Concept Connections

Which of the following are equations?
1. $-2w = 6$
2. $4 = t - 3$
3. $x + 8$

Example 1 **Determining Whether a Number Is a Solution to an Equation**

Determine whether the given number is a solution to the equation.

a. $2x - 9 = 3$; 6 **b.** $8 = 8p - 4$; $-\frac{1}{2}$

Solution:

a. $2x - 9 = 3$

$2(6) - 9 \overset{?}{=} 3$ Substitute 6 for x.

$12 - 9 \overset{?}{=} 3$ Simplify.

$3 = 3 \checkmark$ The right-hand side equals the left-hand side.
Thus, 6 is a solution to the equation $2x - 9 = 3$.

b. $8 = 8p - 4$

$8 \overset{?}{=} 8\left(-\frac{1}{2}\right) - 4$ Substitute $-\frac{1}{2}$ for p.

$8 \overset{?}{=} -4 - 4$ Simplify.

$8 \neq -8$ The right-hand side does not equal the left-hand side.
Thus, $-\frac{1}{2}$ is *not* a solution to the equation $8 = 8p - 4$.

Skill Practice

Determine whether the given number is a solution to the equation.
4. $2 + 3x = 23$; 7
5. $-4x + 1 = 9$; 2

Classroom Example: p. 655, Exercise 10

In the study of algebra, you will encounter a variety of equations. In this chapter, we focus on a specific type of equation called a linear equation in one variable.

Linear Equation in One Variable

A **linear equation in one variable** is an equation that can be written in the form $ax + b = c$. In this equation a, b, and c are real numbers, $a \neq 0$, and x is the variable.

Note: A linear equation in one variable contains only one variable and the exponent on the variable is 1.

Answers

1. Equation 2. Equation
3. Not an equation
4. Yes, 7 is a solution.
5. No, 2 is not a solution.

2. Addition and Subtraction Properties of Equality

Given the equation $x = 3$, we can easily determine that the solution is 3. The solution to the equation $2x + 14 = 20$ is also 3. These two equations are called **equivalent equations** because they have the same solution. However, while the solution to $x = 3$ is obvious, the solution to $2x + 14 = 20$ is not. Our goal in this chapter is to learn how to *solve* equations.

To solve an equation, we use algebraic principles to write an equation such as $2x + 14 = 20$ in an equivalent but simpler form, such as $x = 3$. The addition and subtraction properties of equality are the first tools we will learn to solve an equation.

> **The Addition and Subtraction Properties of Equality**
>
> Let a, b, and c represent algebraic expressions.
>
> 1. The **addition property of equality**:   If $a = b$
>   then $a + c = b + c$
>
> 2. The **subtraction property of equality**:   If $a = b$
>   then $a - c = b - c$

The addition and subtraction properties of equality indicate that adding or subtracting the same quantity to each side of an equation results in an equivalent equation. This is true because if two quantities are increased (or decreased) by the same amount, then the resulting quantities will also be equal (Figure 11-2).

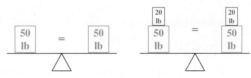

Figure 11-2

Example 2   Applying the Addition Property of Equality

Solve the equations and check the solution.

   a. $x - 6 = 18$ **b.** $3.8 = -4.1 + x$

Solution:

To solve an equation, the goal is to isolate the variable on one side of the equation. That is, we want to create an equivalent equation of the form $x = $ number. To accomplish this, we can use the fact that the sum of a number and its opposite is zero.

a. $x - 6 = 18$

   $x - 6 + 6 = 18 + 6$ To isolate x, add 6 to both sides, because $-6 + 6 = 0$.

   $x + 0 = 24$ Simplify.

   $x = 24$ The variable is isolated (by itself) on the left-hand side of the equation. The solution is 24.

Check: $x - 6 = 18$ Original equation

$(24) - 6 \stackrel{?}{=} 18$ Substitute 24 for x.

$18 = 18$ ✓   The right-hand side = the left-hand side.

b. $3.8 = -4.1 + x$

$3.8 + 4.1 = -4.1 + 4.1 + x$ To isolate x, add 4.1 to both sides, because $-4.1 + 4.1 = 0$.

$7.9 = 0 + x$ Simplify.

$7.9 = x$ The solution is 7.9.

Check: $3.8 = -4.1 + x$ Original equation

$3.8 \stackrel{?}{=} -4.1 + (7.9)$ Substitute 7.9 for x.

$3.8 = 3.8$ ✓

In Example 3, we apply the subtraction property of equality. This indicates that we can subtract the same quantity from both sides of the equation to obtain an equivalent equation.

Example 3   **Applying the Subtraction Property of Equality**

Solve the equations and check.

a. $z + 11 = 14$ **b.** $-8 = 2 + q$

Solution:

a. $z + 11 = 14$

$z + 11 - 11 = 14 - 11$ Subtract 11 from both sides, because $11 - 11 = 0$.

$z + 0 = 3$ Simplify.

$z = 3$ The solution is 3.

Check: $z + 11 = 14$ Original equation

$(3) + 11 \stackrel{?}{=} 14$ Substitute 3 for z.

$14 = 14$ ✓

b. $-8 = 2 + q$

$-8 - 2 = 2 - 2 + q$ Subtract 2 from both sides, because $2 - 2 = 0$.

$-10 = 0 + q$ Simplify.

$-10 = q$ The solution is -10.

Check: $-8 = 2 + q$ Original equation

$-8 \stackrel{?}{=} 2 + (-10)$ Substitute -10 for q.

$-8 = -8$ ✓

Answers
9. 13   **10.** -17

Skill Practice

Solve the equation.

11. $\frac{4}{5} + c = -\frac{1}{4}$

12. $-9.13 = y - 2.27$

Classroom Examples: p. 656,
Exercises 62 and 64

Example 4 **Applying the Addition and Subtraction Properties of Equality**

Solve the equations.

a. $\frac{3}{10} + m = -\frac{2}{3}$ **b.** $8.54 = p + 1.96$

Solution:

a.

$$\frac{3}{10} + m = -\frac{2}{3}$$ To isolate m, we subtract $\frac{3}{10}$ because $\frac{3}{10} - \frac{3}{10} = 0$.

$$\frac{3}{10} - \frac{3}{10} + m = -\frac{2}{3} - \frac{3}{10}$$ Subtract $\frac{3}{10}$ from both sides.

$$0 + m = -\frac{2 \cdot 10}{3 \cdot 10} - \frac{3 \cdot 3}{10 \cdot 3}$$ To subtract the fractions, first obtain a common denominator. The LCD is 30.

$$m = -\frac{20}{30} - \frac{9}{30}$$

$$m = -\frac{29}{30}$$ The solution is $-\frac{29}{30}$ and checks in the original equation.

b.

$$8.54 = p + 1.96$$ To isolate p, we subtract 1.96 because $1.96 - 1.96 = 0$.

$$8.54 - 1.96 = p + 1.96 - 1.96$$ Subtract 1.96 from each side.

$$6.58 = p + 0$$

$$6.58 = p$$ The solution is 6.58 and checks in the original equation.

Answers

11. $-\frac{21}{20}$ **12.** -6.86

Section 11.3 Practice Exercises

Study Skills Exercise

For additional exercises, see Classroom Activities 11.3A–11.3B in the *Student's Resource Manual* at www.mhhe.com/moh.

Up to this point we have been simplifying expressions. We will now begin solving equations. Consider the two lists:

Expressions	Equations
$3x + 2y$	$5x + 2 = 6$
$6(8 + x) + 2$	$2(x - 5) = 14$
$7y$	$7 = y$

Explain the difference between an expression and an equation.

Vocabulary and Key Concepts

1. a. A _____linear_____ equation in one variable is an equation that can be written in the form $ax + b = c$.

b. A _____solution_____ to an equation is a value of the variable that makes the equation a true statement.

c. The equations $3x = 12$ and $x = 4$ are called ____equivalent____ equations because they have the same solution.

d. The ____addition____ property of equality tells us that adding the same quantity to both sides of an equation, results in an equivalent equation.

e. The ____subtraction____ property of equality tells us that subtracting the same quantity from both sides of an equation, results in an equivalent equation.

Writing   Translating Expression   Geometry   Scientific Calculator   Video   NS E

Review Exercises

2. Explain why the terms in the given sum cannot be combined: $3x + 7x^2$

The exponents on the variable x are different. Therefore, the terms are not *like* terms and cannot be combined.

For Exercises 3–8, simplify the expression.

3. $-10a + 3b - 3a + 13b$
$-13a + 16b$

4. $4 - 23y + 11 - 16y$
$-39y + 15$

5. $-(-8h + 2k - 13)$
$8h - 2k + 13$

6. $3(-4m + 3) - 12$
$-12m - 3$

7. $5z - 8(z - 3) - 20$
$-3z + 4$

8. $-(7p - 12) - 10(1 - p) + 6$
$3p + 8$

Concept 1: Definition of an Equation

For Exercises 9–20, determine whether the given number is a solution to the equation. **(See Example 1.)**

9. $5x + 3 = -2$; -1
-1 is a solution.

10. $3y - 2 = 4$; 2
2 is a solution.

11. $10 = p - 16$; 26
26 is a solution.

12. $-14 = q - 1$; -13
-13 is a solution.

13. $-z + 8 = 20$; 12
12 is not a solution.

14. $-7 - w = -10$; -3
-3 is not a solution.

15. $6m - 3 = -6$; $-\dfrac{1}{2}$
$-\dfrac{1}{2}$ is a solution.

16. $-12n + 2 = -1$; $\dfrac{1}{4}$
$\dfrac{1}{4}$ is a solution.

17. $13 = 13 + 6t$; 0
0 is a solution.

18. $-\dfrac{1}{5} = r - \dfrac{1}{5}$; 0
0 is a solution.

19. $25 = -5q - 5$; 4
4 is not a solution.

20. $39 = -7p - 4$; 5
5 is not a solution.

Concept 2: Addition and Subtraction Properties of Equality

For Exercises 21–26, fill in the blank with the appropriate number.

21. $13 + (-13) =$ _____ 0

22. $6 +$ _____ $= 0$ -6

23. _____ $+ (-7) = 0$ 7

24. $1 + (-1) =$ _____ 0

25. $3.2 +$ _____ $= 0$ -3.2

26. _____ $+ (-0.3) = 0$ 0.3

For Exercises 27–38, solve the equation using the addition property of equality. **(See Example 2.)**

27. $g - 23 = 14$ 37

28. $h - 12 = 30$ 42

29. $-4 + k = 12$ 16

30. $-16 + m = 4$ 20

31. $-18 = n - 3$ -15

32. $-9 = t - 6$ -3

33. $-\dfrac{5}{6} + p = \dfrac{1}{3}$ $\dfrac{7}{6}$

34. $-\dfrac{3}{4} + q = \dfrac{3}{2}$ $\dfrac{9}{4}$

35. $k - 4.3 = -1.2$ 3.1

36. $a - 0.04 = -2.04$ -2

37. $13 = -21 + w$ 34

38. $2 = -17 + w$ 19

For Exercises 39–44, fill in the blank with the appropriate number.

39. $52 -$ _____ $= 0$ 52

40. $2 - 2 =$ _____ 0

41. $18 - 18 =$ _____ 0

42. _____ $- 15 = 0$ 15

43. _____ $- 100 = 0$ 100

44. $21 -$ _____ $= 0$ 21

For Exercises 45–56, solve the equation using the subtraction property of equality. **(See Example 3.)**

45. $x + 34 = 6$ -28

46. $y + 12 = 4$ -8

47. $17 + b = 20$ 3

48. $5 + c = 14$ 9

49. $-32 = t + 14$ -46

50. $-23 = k + 11$ -34

51. $8.2 = 21.8 + m$ -13.6

52. $16.01 = 20.88 + n$ -4.87

53. $a + \dfrac{3}{5} = -\dfrac{7}{10}$ $-\dfrac{13}{10}$

54. $b + \dfrac{1}{4} = -\dfrac{3}{8}$ $-\dfrac{5}{8}$

55. $21 = 14 + w$ 7

56. $9 = 8 + u$ 1

Mixed Exercises

For Exercises 57–77, solve the equation by using the appropriate property. **(See Example 4.)**

57. $1 + p = 0$ -1

58. $r - 12 = 13$ 25

59. $-34 + t = -40$ -6

60. $7 + q = 4$ -3

61. $\frac{2}{3} = y - \frac{5}{12}$ $\frac{13}{12}$

62. $\frac{7}{11} = z + \frac{3}{11}$ $\frac{4}{11}$

63. $-2.5 = -1.1 + m$ -1.4

64. $-4.1 = -3.5 + n$ -0.6

65. $w - 23 = -11$ 12

66. $p - 10 = -9$ 1

67. $x + 21 = 16$ -5

68. $y + 18 = -4$ -22

69. $-2 = a - 15$ 13

70. $-1 = b - 49$ 48

71. $4.01 + p = 3.22$ -0.79

72. $2.8 + q = 6.1$ 3.3

73. $t + \frac{3}{8} = 2$ $\frac{13}{8}$

74. $r - \frac{4}{7} = -1$ $-\frac{3}{7}$

75. $27 = z - 22$ 49

76. $109 = x + 49$ 60

77. $-70 = -55 + w$ -15

Expanding Your Skills

For Exercises 78–83, first simplify each side of the equation. Then solve the equation.

78. $5h - 4h + 4 = 3$
  -1

79. $10x - 9x - 11 = 15$
  26

80. $9 + (-2) = 4 + t$
  3

81. $-13 + 15 = p + 5$
  -3

82. $3(r - 2) - 2r = 6 + (-2)$
  10

83. $4(k + 2) - 3k = -6 + 9$
  -5

Section 11.4	Multiplication and Division Properties of Equality

Concepts

1. Multiplication and Division Properties of Equality
2. Using the Properties of Equality

1. Multiplication and Division Properties of Equality

Adding or subtracting the same quantity to both sides of an equation results in an equivalent equation. In a similar way, multiplying or dividing both sides of an equation by the same nonzero quantity also results in an equivalent equation. This is stated formally as the multiplication and division properties of equality.

> **The Multiplication and Division Properties of Equality**
>
> Let a, b, and c represent algebraic expressions, $c \neq 0$.
>
> **1.** The **multiplication property of equality:** If $a = b$
>   then $c \cdot a = c \cdot b$
>
> **2.** The **division property of equality:** If $a = b$
>   then $\dfrac{a}{c} = \dfrac{b}{c}$

To understand the multiplication property of equality, suppose we start with a true equation such as $10 = 10$. If both sides of the equation are multiplied by a constant such as 3, the result is also a true statement (Figure 11-3).

$$10 = 10$$
$$3 \cdot 10 = 3 \cdot 10$$
$$30 = 30$$

Figure 11-3

To solve an equation in the variable x, the goal is to write the equation in the form $x =$ number. In particular, notice that we desire the coefficient of x to be 1. That is, we want to write the equation as $1 \cdot x =$ number. To solve an equation such as $3x = 12$, we can multiply both sides of the equation by the reciprocal of the x-term coefficient. In this case, multiply both sides by the reciprocal of 3, which is $\frac{1}{3}$.

$$3x = 12$$

$$\frac{1}{3} \cdot (3x) = \frac{1}{3} \cdot (12) \qquad \text{Multiply by the reciprocal of 3, which is } \frac{1}{3}.$$

$$1 \cdot x = 4 \qquad \text{The coefficient of the } x\text{-term is now 1.}$$

$$x = 4 \qquad \text{Simplify. The solution is 4.}$$

> **TIP:** Recall that the product of a number and its reciprocal is 1. For example:
>
> $$\frac{1}{5} \cdot (5) = 1$$
>
> $$\frac{3}{2} \cdot \frac{2}{3} = 1$$
>
> $$-\frac{7}{2} \cdot \left(-\frac{2}{7}\right) = 1$$

The division property of equality can also be used to solve the equation $3x = 12$ by dividing both sides by the coefficient of the x-term. In this case, divide both sides by 3 to make the coefficient of x equal to 1.

$$3x = 12$$

$$\frac{3x}{3} = \frac{12}{3} \qquad \text{Divide by the coefficient of } x \text{ which is 3.}$$

$$1 \cdot x = 4 \qquad \text{The coefficient on the } x\text{-term is now 1.}$$

$$x = 4 \qquad \text{Simplify. The solution is 4.}$$

> **TIP:** Recall that the quotient of a nonzero real number and itself is 1. For example:
>
> $$\frac{5}{5} = 1 \quad \text{and} \quad \frac{-3.5}{-3.5} = 1$$

Example 1 Applying the Multiplication and Division Properties of Equality

Solve the equations by using the multiplication or division property of equality.

a. $10x = 50$ **b.** $28 = -4p$

Solution:

a. $10x = 50$

$$\frac{10x}{10} = \frac{50}{10} \qquad \text{To obtain a coefficient of 1 for the } x\text{-term, divide both sides by 10.}$$

$$1x = 5 \qquad \text{Simplify.}$$

$$x = 5 \qquad \text{The solution is 5.}$$

$$\underline{\text{Check:}} \qquad 10x = 50 \qquad \text{Original equation}$$

$$10(5) \overset{?}{=} 50 \qquad \text{Substitute 5 for } x.$$

$$50 = 50 \ ✔$$

> **Skill Practice**
>
> Solve.
> **1.** $4x = 32$
> **2.** $18 = -2w$
>
> Classroom Examples: p. 661, Exercise 20

Answers
1. 8 **2.** −9

> **TIP:** In Example 1(a) we could also have multiplied both sides by $\frac{1}{10}$ to obtain a coefficient of 1 for the *x*-term.
>
> $$\frac{1}{10}(10x) = \frac{1}{10}(50)$$
> $$1x = 5$$

b. $28 = -4p$

$\dfrac{28}{-4} = \dfrac{-4p}{-4}$ To obtain a coefficient of 1 for the *x*-term, divide both sides by -4. This is also equivalent to multiplying by $-\frac{1}{4}$.

$-7 = 1p$ Simplify.

$-7 = p$ The solution is -7 and checks in the original equation.

Skill Practice

Solve.

3. $1.9 = -m$

4. $4.1z = 28.29$

Classroom Examples: p. 661, Exercises 24 and 30

| **Example 2** | **Applying the Multiplication and Division Properties of Equality** |

Solve the equation by using the multiplication or division property of equality.

a. $-y = 3.4$ **b.** $23.18 = 6.1w$

Solution:

a. $-y = 3.4$ Note that $-y$ is the same as $-1 \cdot y$.

    $-1y = 3.4$

    $\dfrac{-1 \cdot y}{-1} = \dfrac{3.4}{-1}$ To obtain a coefficient of 1 for the *y*-term, divide both sides by -1.

    $1y = -3.4$ Simplify.

    $y = -3.4$ The solution is -3.4 and checks in the original equation.

> **TIP:** In Example 2(a), we could have also multiplied both sides by -1 to obtain a coefficient of 1 for *y*.
>
> $$(-1)(-y) = (-1)3.4$$
> $$y = -3.4$$

b. $23.18 = 6.1w$

    $\dfrac{23.18}{6.1} = \dfrac{6.1w}{6.1}$ To obtain a coefficient of 1 on the *w*-term, divide both sides by 6.1.

    $3.8 = 1w$ Simplify.

    $3.8 = w$ The solution is 3.8 and checks in the original equation.

Answers

3. -1.9 **4.** 6.9

Example 3   Applying the Multiplication and Division
                Properties of Equality

Solve the equation by using the multiplication or division property of equality.

a. $-\dfrac{2}{3}p = -\dfrac{4}{7}$ **b.** $5 = \dfrac{d}{8}$

Solution:

a. $-\dfrac{2}{3}p = -\dfrac{4}{7}$

$-\dfrac{3}{2}\left(-\dfrac{2}{3}p\right) = -\dfrac{3}{2}\left(-\dfrac{4}{7}\right)$ To obtain a coefficient of 1 for the p-term,
multiply by the reciprocal of $-\frac{2}{3}$ which is $-\frac{3}{2}$.

$1p = \dfrac{12}{14}$ Simplify.

$p = \dfrac{6}{7}$ The solution is $\frac{6}{7}$.

> **TIP:** For Example 3(a) you may have first thought of dividing by $-\frac{2}{3}$. Recall from Section 2.5 that dividing by a fraction is the same as multiplying by its reciprocal.

b. $5 = \dfrac{d}{8}$

$5 = \dfrac{1}{8}d$ The expression $\frac{d}{8}$ is equivalent to $\frac{1}{8}d$.

$8(5) = 8\left(\dfrac{1}{8}d\right)$ To obtain a coefficient of 1 on the d-term, multiply
both sides by the reciprocal of $\frac{1}{8}$, which is 8.

$40 = 1d$ Simplify.

$40 = d$ The solution is 40.

2. Using the Properties of Equality

It is important to distinguish between cases where the addition or subtraction property of equality should be used to isolate a variable versus where the multiplication or division property of equality should be used. Compare the equations:

$$4 + x = 12 \quad \text{and} \quad 4x = 12$$

To solve each equation, we want to isolate the variable. In the equation $4 + x = 12$, we subtract 4 from both sides so that the constant being added to x is zero. For the equation $4x = 12$, we divide by 4 (or multiply by the reciprocal $\frac{1}{4}$) so that the coefficient on x is 1.

$$4 + x = 12 \qquad \text{and} \qquad 4x = 12$$

$$4 - 4 + x = 12 - 4 \qquad\qquad \dfrac{4x}{4} = \dfrac{12}{4}$$

$$0 + x = 8 \qquad\qquad 1x = 3$$

$$x = 8 \qquad\qquad x = 3$$

In general, if the operation between a term and the variable is addition or subtraction, we apply the subtraction or addition property of equality. If the variable

Classroom Examples: p. 661,
Exercises 32 and 38

> **TIP:** When applying the multiplication or division properties, we will generally use the following conventions:
> - If the coefficient of the variable term is expressed as a fraction, multiply both sides by its reciprocal.
> - Otherwise, divide both sides by the coefficient itself.

Answers

5. $-\dfrac{3}{5}$ **6.** 36

is multiplied or divided by a constant factor then apply the division or multiplication property of equality.

In Example 4, we practice distinguishing which property of equality to use.

Example 4 Solving Linear Equations

Solve the equations.

a. $\dfrac{m}{12} = -3$ **b.** $3.2 = x + 19.5$ **c.** $6 = -4t$ **d.** $y - \dfrac{5}{9} = \dfrac{2}{3}$

Solution:

a. $\dfrac{m}{12} = -3$ The operation between m and 12 is division. To obtain a coefficient of 1 for the m-term, multiply both sides by 12.

$12\left(\dfrac{m}{12}\right) = 12(-3)$ Multiply both sides by 12.

$m = -36$ Simplify both sides. The solution -36 checks in the original equation.

b. $3.2 = x + 19.5$ The operation between x and 19.5 is addition. To isolate the x-term, we can subtract 19.5 from both sides because $19.5 - 19.5 = 0$.

$3.2 - 19.5 = x + 19.5 - 19.5$ Subtract 19.5 from both sides.

$-16.3 = x$ Simplify. The solution -16.3 checks in the original equation.

c. $6 = -4t$ The operation between t and -4 is multiplication. To obtain a coefficient of 1 on the t-term, we can divide both sides by -4.

$\dfrac{6}{-4} = \dfrac{-4t}{-4}$ Divide both sides by -4.

$-\dfrac{6}{4} = t$

$-\dfrac{3}{2} = t$ Simplify. The solution $-\frac{3}{2}$ checks in the original equation.

d. $y - \dfrac{5}{9} = \dfrac{2}{3}$ The operation between y and $\frac{5}{9}$ is subtraction. We can add $\frac{5}{9}$ to both sides to isolate y.

$y - \dfrac{5}{9} + \dfrac{5}{9} = \dfrac{2}{3} + \dfrac{5}{9}$ Add $\frac{5}{9}$ to both sides.

$y = \dfrac{2 \cdot 3}{3 \cdot 3} + \dfrac{5}{9}$ Obtain a common denominator. The LCD is 9.

$y = \dfrac{6}{9} + \dfrac{5}{9}$ Add the fractions.

$y = \dfrac{11}{9}$ The solution $\frac{11}{9}$ checks in the original equation.

Skill Practice

Solve.

7. $\dfrac{t}{5} = -8$

8. $-4.6 + x = 12.9$

9. $5 = -2p$

10. $z + \dfrac{1}{3} = \dfrac{5}{6}$

Classroom Examples: p. 662,
Exercises 58, 70, and 72

Instructor Note: Remind students that $\frac{m}{12} = \frac{1}{12} \cdot m$.

Answers

7. -40 **8.** 17.5

9. $-\dfrac{5}{2}$ **10.** $\dfrac{1}{2}$

Section 11.4   Practice Exercises

For additional exercises, see Classroom Activities 11.4A–11.4B in the *Student's Resource Manual* at www.mhhe.com/moh.

Study Skills Exercise

> One way to know that you really understand a concept is to try to explain it to someone else. In your own words, explain when you would apply the multiplication property of equality or the division property of equality.

Vocabulary and Key Concepts

1. a. The _____multiplication_____ property of equality tells us that multiplying both sides of an equation by the same nonzero quantity, results in an equivalent equation.

b. The _____division_____ property of equality tells us that dividing both sides of an equation by the same nonzero quantity, results in an equivalent equation.

Review Exercises

2. Determine whether 5 is a solution to the equation.

a. $2x + 3 = 13$ Yes   **b.** $2x = 10$ Yes

For Exercises 3–10, solve the equation.

3. $p - 12 = 33$
45

4. $-8 = 10 + k$
−18

5. $16 = h - 5$
21

6. $-4 + w = 22$
26

7. $p - 6 = -19$
−13

8. $\dfrac{1}{6} = -\dfrac{11}{6} + m$
2

9. $n + \dfrac{1}{2} = -\dfrac{2}{3}$ $-\dfrac{7}{6}$

10. $2.4 + z = -12$
−14.4

Concept 1: Multiplication and Division Properties of Equality

For Exercises 11–18, fill in the blank with the appropriate number.

11. $3 \cdot \underline{\hspace{1cm}} = 1$
$\dfrac{1}{3}$

12. $-6 \cdot \underline{\hspace{1cm}} = 1$
$-\dfrac{1}{6}$

13. $-\dfrac{4}{7} \cdot \underline{\hspace{1cm}} = 1$ $-\dfrac{7}{4}$

14. $\dfrac{3}{10} \cdot \underline{\hspace{1cm}} = 1$ $\dfrac{10}{3}$

15. $-7 \div \underline{\hspace{1cm}} = 1$
−7

16. $2 \div \underline{\hspace{1cm}} = 1$
2

17. $5.1 \div \underline{\hspace{1cm}} = 1$
5.1

18. $-6.8 \div \underline{\hspace{1cm}} = 1$
−6.8

For Exercises 19–50, solve the equation by using the multiplication or division property of equality. **(See Examples 1–3.)**

19. $14b = -42$
−3

20. $-6p = 12$
−2

 21. $-8k = 56$
−7

22. $5y = -25$
−5

23. $-t = -13$
13

24. $-h = -17$
17

25. $\dfrac{2}{3}m = 14$
21

26. $\dfrac{5}{9}n = 40$
72

 27. $\dfrac{b}{7} = -3$
−21

28. $\dfrac{a}{4} = -12$
−48

29. $-2.8 = -0.7t$
4

30. $-3.3 = -3r$
1.1

31. $-\dfrac{u}{2} = -15$
30

32. $-\dfrac{v}{10} = -4$
40

33. $6 = -18w$
$-\dfrac{1}{3}$

34. $4 = -32g$
$-\dfrac{1}{8}$

35. $1.3x = 5.33$
4.1

36. $8.1y = 17.82$
2.2

37. $\dfrac{5}{4}k = -\dfrac{1}{2}$ $-\dfrac{2}{5}$

38. $-\dfrac{11}{12}h = -\dfrac{1}{6}$ $\dfrac{2}{11}$

39. $0 = \dfrac{3}{8}m$ 0

40. $0 = \dfrac{1}{10}n$ 0

41. $-\dfrac{9}{4}x = -\dfrac{3}{5}$ $\dfrac{4}{15}$

42. $-\dfrac{15}{14}y = \dfrac{1}{2}$ $-\dfrac{7}{15}$

43. $100 = 5k$ 20

44. $95 = 19h$ 5

45. $31 = -p$ -31

46. $-6 = -z$ 6

47. $3p = \dfrac{5}{2}$ $\dfrac{5}{6}$

48. $2q = \dfrac{7}{5}$ $\dfrac{7}{10}$

 49. $-4a = 0$ 0

50. $-7b = 0$ 0

Concept 2: Using the Properties of Equality

For Exercises 51–82, solve the equation. **(See Example 4.)**

51. $4 + x = -12$
-16

52. $6 + z = -18$
-24

53. $4y = -12$
-3

54. $6p = -18$
-3

55. $q - 4 = -12$
-8

56. $p - 6 = -18$
-12

57. $\dfrac{h}{4} = -12$
-48

58. $\dfrac{w}{6} = -18$
-108

59. $\dfrac{2}{3} + t = 1$
$\dfrac{1}{3}$

60. $\dfrac{3}{4} + q = 1$
$\dfrac{1}{4}$

61. $-9a = -12$
$\dfrac{4}{3}$

62. $-8b = -44$
$\dfrac{11}{2}$

63. $7 = r - 23$
30

64. $11 = s - 4$
15

65. $-\dfrac{y}{3} = 5$
-15

66. $-\dfrac{h}{5} = 1$
-5

67. $2p = \dfrac{5}{6}$
$\dfrac{5}{12}$

68. $4q = \dfrac{3}{5}$
$\dfrac{3}{20}$

 69. $-\dfrac{3}{7}x = \dfrac{9}{10}$ $-\dfrac{21}{10}$

70. $-\dfrac{2}{11}y = \dfrac{4}{15}$ $-\dfrac{22}{15}$

71. $t - 12.9 = 15$
27.9

72. $c - 4.11 = 1.2$
5.31

73. $5 + u = 3.2$
-1.8

74. $3 + v = 1.7$
-1.3

75. $50 = a + 72$
-22

76. $23 = w + 41$
-18

77. $-1 = b - 16$
15

78. $-5 = y - 8$
3

79. $-12 = 30x$
$-\dfrac{2}{5}$

80. $-10 = 12h$
$-\dfrac{5}{6}$

81. $-6 = -\dfrac{1}{2}q$
12

82. $4 = -\dfrac{1}{6}k$
-24

Expanding Your Skills

For Exercises 83–88, first simplify each side of the equation. Then solve the equation.

83. $5x - 2x = -15$
-5

84. $13y - 10y = -18$
-6

85. $3p + 4p = 25 - 4$
3

86. $2q + 3q = 54 - 9$
9

87. $-2(a + 3) - 6a + 6 = 8$
-1

88. $-(b - 11) - 3b - 11 = -16$
4

Section 11.5 Solving Equations with Multiple Steps

Concepts

1. Solving Equations with Multiple Steps
2. Solving Linear Equations Involving Parentheses

1. Solving Equations with Multiple Steps

In Sections 11.3 and 11.4 we studied a one-step process to solve linear equations. We used the addition, subtraction, multiplication, and division properties of equality. In this section we combine these properties to solve equations that require multiple steps. This is shown in Example 1.

✏ Writing   ←→ Translating Expression   ◁ Geometry   🖩 Scientific Calculator   ▶ Video   NS&E

Classroom Example: p. 668, Exercise 10

Example 1 Solving a Linear Equation

Solve. $2x - 3 = 15$

Solution:

Remember that our goal is to isolate x. Therefore, in this equation, we first isolate the *term* containing x. This can be done by adding 3 to both sides.

$2x - 3 + 3 = 15 + 3$	Add 3 to both sides.
$2x = 18$	The term containing x is now isolated (by itself). The resulting equation now requires only one step to solve.
$\dfrac{2x}{2} = \dfrac{18}{2}$	Divide both sides by 2 to make the coefficient on x equal to 1.
$x = 9$	Simplify. The solution is 9.

$$\text{Check:} \quad 2x - 3 = 15 \qquad \text{Original equation}$$
$$2(9) - 3 \overset{?}{=} 15 \qquad \text{Substitute 9 for } x.$$
$$18 - 3 = 15 \checkmark$$

Skill Practice

Solve.

1. $3x + 7 = 25$

Instructor Note: When solving an equation, the order of operations is reversed.

As Example 1 shows, we will generally apply the addition (or subtraction) property of equality to isolate the variable term first. Then we will apply the multiplication (or division) property of equality to obtain a coefficient of 1 on the variable term.

Example 2 Solving a Linear Equation

Solve. $14 = \dfrac{y}{4} + 8$

Solution:

$14 = \dfrac{y}{4} + 8$	
$14 - 8 = \dfrac{y}{4} + 8 - 8$	Subtract 8 from both sides. This will isolate the term containing the variable y.
$6 = \dfrac{y}{4}$	Simplify.
$6(4) = \dfrac{y}{4}(4)$	Multiply both sides by 4.
$24 = y$	Simplify. The solution is 24.

$$\text{Check: } 14 = \dfrac{y}{4} + 8 \qquad \text{Original equation}$$
$$14 \overset{?}{=} \dfrac{(24)}{4} + 8 \qquad \text{Substitute 24 for } y.$$
$$14 = 6 + 8 \checkmark$$

Skill Practice

Solve.

2. $-13 = -9 + \dfrac{y}{2}$

Classroom Example: p. 668, Exercise 24

Answers

1. 6   2. -8

Example 3 Solving a Linear Equation

Solve. $2z - 9.2 = 2.6$

Solution:

$$2z - 9.2 = 2.6$$

$2z - 9.2 + 9.2 = 2.6 + 9.2$ Add 9.2 to both sides. This will isolate the term containing the variable z.

$2z = 11.8$ Simplify.

$\dfrac{2z}{2} = \dfrac{11.8}{2}$ Divide both sides by 2.

$z = 5.9$ The solution is 5.9 and checks in the original equation.

In Example 4, the variable x appears on both sides of the equation. In this case, apply the addition or subtraction properties of equality to collect the variable terms on one side of the equation and the constant terms on the other side.

Example 4 Solving a Linear Equation with Variables on Both Sides

Solve. $4x + 5 = -2x - 13$

Solution:

To isolate x, we must first "move" all x-terms to one side of the equation. For example, suppose we add $2x$ to both sides. This would "remove" the x-term from the right-hand side because $-2x + 2x = 0$. The term $2x$ is then combined with $4x$ on the left-hand side.

$$4x + 5 = -2x - 13$$

$4x + 2x + 5 = -2x + 2x - 13$ Add $2x$ to both sides.

$6x + 5 = -13$ Simplify. Next, we want to isolate the term containing x.

$6x + 5 - 5 = -13 - 5$ Subtract 5 from both sides to isolate the x-term.

$6x = -18$ Simplify.

$\dfrac{6x}{6} = \dfrac{-18}{6}$ Divide both sides by 6 to obtain a coefficient of 1 for the x-term.

$x = -3$ The solution is -3 and checks in the original equation.

Answers

3. 4.5   **4.** 10

TIP: Note that the variable may be isolated on either side of the equation. In Example 4, for instance, we could have isolated the *x*-terms on the right-hand side of the equation.

$$4x + 5 = -2x - 13$$

$$4x - 4x + 5 = -2x - 4x - 13$$ Subtract $4x$ from both sides. This "removes" the x-term from the left-hand side.

$$5 = -6x - 13$$

$$5 + 13 = -6x - 13 + 13$$ Add 13 to both sides to isolate the x-term.

$$18 = -6x$$ Simplify.

$$\frac{18}{-6} = \frac{-6x}{-6}$$ Divide both sides by -6.

$$-3 = x$$ This is the same solution as in Example 4.

Often we can simplify both sides of an equation before applying the properties of equality. This is demonstrated in Example 5.

Example 5 Solving a Linear Equation by Simplifying First

Solve. $6 - 8y + 3 = y + 3y + 6$

Solution:

Notice that *like* terms can be combined on both sides of the equation first, to make the equation simpler.

$$6 - 8y + 3 = y + 3y + 6$$

$$9 - 8y = 4y + 6$$ Combine *like* terms. On the left-hand side $6 + 3 = 9$. On the right-hand side, $y + 3y = 4y$.

$$9 - 8y - 4y = 4y - 4y + 6$$ We can collect all variable terms on the left-hand side by subtracting $4y$ from both sides.

$$9 - 12y = 6$$ Simplify.

$$9 - 9 - 12y = 6 - 9$$ To isolate the y-term on the left, subtract 9 from both sides.

$$-12y = -3$$ Simplify.

$$\frac{-12y}{-12} = \frac{-3}{-12}$$ Divide both sides by -12 to make the coefficient on the y-term 1.

$$y = \frac{1}{4}$$ The solution is $\frac{1}{4}$ and checks in the original equation.

Skill Practice

Solve.
5. $14 - 3w + 2$
   $= 4w + 21 - 2w$

Classroom Example: p. 668, Exercise 44

2. Solving Linear Equations Involving Parentheses

In Examples 1–5, we used multiple steps to solve equations. We also learned how to collect the variable terms on one side of the equation so that the variable could be isolated. The following guidelines summarize the steps to solve a linear equation.

Answer

5. -1

Solving a Linear Equation in One Variable

Step 1 Simplify both sides of the equation.

  - Clear parentheses if necessary.
  - Combine *like* terms if necessary.

Step 2 Use the addition or subtraction property of equality to collect the variable terms on one side of the equation.

Step 3 Use the addition or subtraction property of equality to collect the constant terms on the *other* side of the equation.

Step 4 Use the multiplication or division property of equality to make the coefficient of the variable term equal to 1.

Step 5 Check the answer in the original equation.

Skill Practice

Solve.

6. $6(z + 4) - 9$
$= -12 - 3z$

Classroom Example: p. 668, Exercise 48

Example 6   **Solving a Linear Equation**

Solve. $2(y - 6) + 32 = 8 - 4y$

Solution:

$2(y - 6) + 32 = 8 - 4y$

$2y - 12 + 32 = 8 - 4y$ **Step 1:**   Simplify both sides of the equation. Clear parentheses.

$2y + 20 = 8 - 4y$ Combine *like* terms on the left-hand side. Note that $-12 + 32 = 20$.

$2y + 4y + 20 = 8 - 4y + 4y$ **Step 2:**   Add $4y$ to both sides to collect the variable terms on the left.

$6y + 20 = 8$ Simplify.

$6y + 20 - 20 = 8 - 20$ **Step 3:**   Subtract 20 from both sides to collect the constants on the right.

$6y = -12$ Simplify.

$\dfrac{6y}{6} = \dfrac{-12}{6}$ **Step 4:**   Divide both sides by 6 to obtain a coefficient of 1 on the y-term.

$y = -2$ Simplify. The solution is -2.

Check: $2(y - 6) + 32 = 8 - 4y$ **Step 5:**   Check the solution in the original equation.

$2(-2 - 6) + 32 \overset{?}{=} 8 - 4(-2)$ Substitute -2 for y.

$2(-8) + 32 \overset{?}{=} 8 - (-8)$

$-16 + 32 \overset{?}{=} 16$

$16 = 16$ ✓   The solution checks.

Answer

6. -3

Example 7 Solving a Linear Equation

Solve. $2x + 3x + 2 = -4(3 - x)$

Solution:

$2x + 3x + 2 = -4(3 - x)$

$5x + 2 = -12 + 4x$ **Step 1:** Simplify both sides of the equation. On the left, combine *like* terms. On the right, clear parentheses.

$5x - 4x + 2 = -12 + 4x - 4x$ **Step 2:** Subtract $4x$ from both sides to collect the variable terms on the left.

$x + 2 = -12$ Simplify.

$x + 2 - 2 = -12 - 2$ **Step 3:** Subtract 2 from both sides to collect the constants on the right.

$x = -14$ **Step 4:** The coefficient on the x-term is already 1. The solution is -14.

Check: $2x + 3x + 2 = -4(3 - x)$ **Step 5:** Check in the original equation.

$2(-14) + 3(-14) + 2 \overset{?}{=} -4[3 - (-14)]$ Substitute -14 for x.

$-28 - 42 + 2 \overset{?}{=} -4(17)$

$-70 + 2 \overset{?}{=} -68$

$-68 = -68 \checkmark$ The solution checks.

Skill Practice

Solve.
7. $2 - y - 4$
$= 6 - 2(y - 8)$

Classroom Example: p. 669,
Exercise 56

Answer
7. 24

Section 11.5   Practice Exercises

For additional exercises, see Classroom Activities 11.5A–11.5B in the *Student's Resource Manual* at www.mhhe.com/moh.

Study Skills Exercise

When you are solving multistep equations, it is recommended that you write an explanation for each step along the way. In the following example, an equation is solved with each step shown. Your job is to write an explanation for each step.

                                        Explanation

$-7x + 2 = 4(x - 5)$

$-7x + 2 = 4x - 20$

$-7x - 4x + 2 = 4x - 4x - 20$

$-11x + 2 = -20$

$-11x + 2 - 2 = -20 - 2$

$-11x = -22$

$\dfrac{-11x}{-11} = \dfrac{-22}{-11}$

$x = 2$ The solution is 2.

 Writing   Translating Expression   Geometry   Scientific Calculator   Video   NS&E

Vocabulary and Key Concepts

1. Consider the equation $3x - 6 = 18$. According to the recommended procedure to solve a linear equation, should we add 6 to both sides first, or should we divide both sides by 3 first?   Add 6 to both sides first.

Review Exercises

For Exercises 2–8, solve the equation.

2. $4c = -\dfrac{1}{3}$ $-\dfrac{1}{12}$

3. $\dfrac{1}{3}b = -4$ -12

4. $-\dfrac{1}{5} + t = \dfrac{6}{5}$ $\dfrac{7}{5}$

5. $-\dfrac{3}{8} = w + \dfrac{1}{4}$ $-\dfrac{5}{8}$

6. $-p = -\dfrac{7}{10}$ $\dfrac{7}{10}$

7. $-8h = 0$ 0

8. $5 + q = 0$ -5

Concept 1: Solving Equations with Multiple Steps

For Exercises 9–28, solve the equation. **(See Examples 1–3.)**

9. $3m + 2 = 14$ 4

10. $-2n + 5 = -15$ 10

11. $-8c - 12 = 36$ -6

12. $5t - 1 = -11$ -2

13. $1 = -4z + 21$ 5

14. $-4 = -3p + 14$ 6

15. $9 = 12x - 7$ $\dfrac{4}{3}$

16. $-7 = 5y - 8$ $\dfrac{1}{5}$

17. $3.4 - 2d = 8.2$ -2.4

18. $2.9 - 4g = 23.3$ -5.1

19. $-0.57 = 15h + 16.23$ -1.12

20. $1.9 = 8k + 4.06$ -0.27

21. $\dfrac{b}{3} - 12 = -9$ 9

22. $\dfrac{c}{5} + 2 = 4$ 10

23. $-9 = \dfrac{w}{2} - 3$ -12

24. $-16 = \dfrac{t}{4} - 14$ -8

25. $3x + \dfrac{1}{2} = \dfrac{5}{4}$ $\dfrac{1}{4}$

26. $9z - \dfrac{3}{8} = \dfrac{9}{16}$ $\dfrac{5}{48}$

27. $10 - y = 37$ -27

28. $25 - c = -3$ 28

For Exercises 29–46, solve the equation. **(See Examples 4 and 5.)**

29. $8 + 4b = 2 + 2b$ -3

30. $2w + 10 = 5w - 5$ 5

31. $7 - 5t = 3t - 2$ $\dfrac{9}{8}$

32. $4 - 2p = 8 + 5p$ $-\dfrac{4}{7}$

33. $4 - 3d = 5d - 4$ 1

34. $-3k + 14 = -4 + 3k$ 3

35. $12p = 3p + 21$ $\dfrac{7}{3}$

36. $2x + 10 = 4x$ 5

37. $-z - 2 = -2z$ 2

38. $9y = -y + 25$ $\dfrac{5}{2}$

39. $1 + \dfrac{1}{4}p = 2 + \dfrac{3}{4}p$ -2

40. $\dfrac{4}{3} + \dfrac{2}{3}q = -\dfrac{5}{3} - \dfrac{1}{3}q - 4$ -7

41. $4 + 2a - 7 = 3a + a + 3$ -3

42. $4b + 2b - 7 = 2 + 4b + 5$ 7

43. $-8w + 8 + 3w = 2 - 6w + 2$ -4

44. $-12 + 5m + 10 = -2m - 10 - m$ -1

45. $6y + 2y - 2 = 14 + 3y - 12$ $\dfrac{4}{5}$

46. $-7t - 20 + 7 = -7 - 3t$ $-\dfrac{3}{2}$

Concept 2: Solving Linear Equations Involving Parentheses

For Exercises 47–60, solve the equation. **(See Examples 6 and 7.)**

47. $3n - 4(n - 1) = 16$ -12

48. $4p - 3(p + 2) = 18$ 24

49. $9q - 5(q - 3) = 5q$ 15

50. $6h - 2(h + 6) = 10h$ -2

51. $2(1 - m) = 5 - 3m$ 3

52. $3(2 - g) = 12 - g$ -3

53. $-4(k - 2) + 14 = 3k - 20$ 6   **54.** $-3(x + 4) - 9 = -2x + 12$ -33 **55.** $3z - 9 = 3(5z - 1)$ $-\dfrac{1}{2}$

56. $4y - 9 = 8(y - 2)$ $\dfrac{7}{4}$ **57.** $6w + 2(w - 1) = 14 - (3w + 1)$ **58.** $-3t - 3(t - 4) = 2 - (2t - 1)$
   $\dfrac{15}{11}$ $\dfrac{9}{4}$

59. $6(u - 1) + 5u + 1 = 5(u + 6) - u$ **60.** $2(2v + 3) + 8v = 6(v - 1) + 3v$
   5   -4

Problem Recognition Exercises

Equations versus Expressions

For Exercises 1–6, identify the problem as an expression or as an equation.

1. $-5 + 4x - 6x = 7$ equation   **2.** $-8(4 - 7x) + 4$ expression   **3.** $4 - 6(2x - 3) + 1$ expression

4. $10 - x = 2x + 19$ equation   **5.** $6 - 3(x + 4) = 6$ equation   **6.** $9 - 6(x + 1)$ expression

For Exercises 7–30, identify as an expression or an equation. Then simplify the expression or solve the equation.

7. $5t = 20$ equation; 4   **8.** $6x - 2 = 36$ equation; $\dfrac{19}{3}$ **9.** $4(x - 5) + 12$ expression; $4x - 8$

10. $16 - 2k + 2 + k$ **11.** $5 + t = 20$ equation; 15   **12.** $0 = 7y - 3y + 8$ equation; -2
   expression; $-k + 18$

13. $5(t - 3) = 20$ equation; 7   **14.** $11 = \dfrac{s}{4} + 3$ equation; 32   **15.** $\dfrac{2}{9}(9x - 5) + \dfrac{1}{9}$ expression; $2x - 1$

16. $5 - 3(2t + 7)$ expression; $-6t - 16$ **17.** $5x - 3 = 20$ equation; $\dfrac{23}{5}$ **18.** $6x = 36$ equation; 6

19. $-14 = \dfrac{r}{6} - 12$ equation; -12 **20.** $16 - 2k + 2 = 0$ equation; 9   **21.** $4 - 2(4x + 5)$ expression; $-8x - 6$

22. $\dfrac{1}{7}(y + 21) + \dfrac{6}{7}y$ expression; $y + 3$ **23.** $2.3u + 0.2 = -1.2u + 7.2$ **24.** $6 + x = 36$ equation; 30
   equation; 2

25. $5 + 3p - 2 = 0$ equation; -1 **26.** $\dfrac{1}{5}b + \dfrac{7}{10} = \dfrac{3}{5}$ equation; $-\dfrac{1}{2}$ **27.** $0 = 2x + 5x + 1$ equation; $-\dfrac{1}{7}$

28. $7.5w - 2.7 = 1.4w + 15.6$ **29.** $5 + 7p - 2p + 9$ **30.** $3(y + 2) - 21$ expression; $3y - 15$
   equation; 3   expression; $5p + 14$

Section 11.6   Applications and Problem Solving

Concepts

1. Problem-Solving Flowchart
2. Translating Verbal Statements into Equations
3. Applications of Linear Equations

1. Problem-Solving Flowchart

Linear equations can be used to solve many real-world applications. In Section 1.8, we introduced guidelines for problem solving. In this section, we expand on these guidelines to enable us to write equations to solve applications. Consider the problem-solving flowchart.

Problem-Solving Flowchart for Word Problems

Instructor Note: Stress the importance of labeling the variable, including units of measure.

Step 1 — Read the problem completely.
- Familiarize yourself with the problem. Estimate the answer if possible.

Step 2 — Assign labels to unknown quantities.
- Identify the unknown quantity or quantities. Let a variable represent one of the unknowns. Draw a picture and write down relevant formulas, when needed.

Step 3 — Write an equation in words.
- Verbalize what quantities must be equal.

Step 4 — Write a mathematical equation.
- Replace the verbal equation with a mathematical equation using x or some other variable.

Step 5 — Solve the equation.
- Solve for the variable by using the steps for solving linear equations.

Step 6 — Interpret the results and write the final answer in words.
- Once you have obtained a numerical value for the variable, recall what it represents in the context of the problem. Can this value be used to determine other unknowns in the problem? Write an answer to the word problem *in words*.

Avoiding Mistakes

It is always a good idea to check your answer in the context of the problem to determine if your answer is reasonable.

2. Translating Verbal Statements into Equations

We begin solving word problems with practice translating between an English sentence and an algebraic equation. First, spend a minute to recall some of the key words that represent addition, subtraction, multiplication, and division. See Table 11-4.

Table 11-4

Addition: $a + b$	Subtraction: $a - b$
the sum of a and b	*the difference of a and b*
a plus b	*a minus b*
b added to a	*b subtracted from a*
b more than a	*a decreased by b*
a increased by b	*b less than a*
the total of a and b	

Multiplication: $a \cdot b$	Division: $a \div b$
the product of a and b	*the quotient of a and b*
a times b	*a divided by b*
a multiplied by b	*b divided into a*
	the ratio of a and b
	a over b
	a per b

Example 1 Translating a Sentence into a Mathematical Equation

A number decreased by 7 is 12. Find the number.

Solution:

Step 1: Read the problem completely.

Let x represent the number.   **Step 2:** Label the variable.

A number decreased by 7 is 12.   **Step 3:** Write the equation in words.
$x \quad - \quad 7 = 12$

Step 4: Translate to a mathematical equation.

$x - 7 = 12$ **Step 5:** Solve the equation.

$x - 7 + 7 = 12 + 7$ Add 7 to both sides.

$x = 19$

The number is 19.   **Step 6:** Interpret the answer in words.

Example 2 Translating a Sentence into a Mathematical Equation

Two times the sum of a number and 8 results in 38.

Solution:

Step 1: Read the problem completely.

Classroom Example: p. 676, Exercise 22

Let x represent the number.

Two times the sum of a number and 8 results in 38.

two times   results in
$$2 \cdot \underbrace{(x + 8)}_{\substack{\text{the sum of a} \\ \text{number and 8}}} = \underset{38}{38}$$

$$2(x + 8) = 38$$

$$2x + 16 = 38$$

$$2x + 16 - 16 = 38 - 16$$

$$2x = 22$$

$$\frac{2x}{2} = \frac{22}{2}$$

$$x = 11$$

The number is 11.

Step 2: Label the variable.

Step 3: Write the equation in words.

Step 4: Translate to a mathematical equation.

Step 5: Solve the equation.

Clear parentheses.

Subtract 16 from both sides.

Simplify.

Divide both sides by 2.

Step 6: Interpret the answer in words.

3. Applications of Linear Equations

In Examples 3–6, we solve application problems by using linear equations.

Example 3   Applying a Linear Equation to Carpentry

A carpenter must cut an 8-ft board into two pieces to build a brace for a picnic table. If one piece is to be 4 times as long as the other piece, how long should each piece be?

Solution:

We can let x represent the length of either piece. However, if we choose x to be the length of the shorter piece, then the longer piece has to be $4x$ (4 times as long).

    Let x = length of the shorter piece.
Then $4x$ = length of the longer piece.

Step 1: Read the problem completely.

Classroom Example: p. 676, Exercise 26

Step 2: Label the variables. Draw a picture.

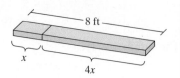

$$\begin{pmatrix} \text{Length of} \\ \text{one piece} \end{pmatrix} + \begin{pmatrix} \text{length of} \\ \text{other piece} \end{pmatrix} = \begin{pmatrix} \text{total} \\ \text{length} \end{pmatrix}$$
$$x \quad\quad + \quad\quad 4x \quad\quad = \quad\quad 8$$

Step 3: Write an equation in words.

Step 4: Write a mathematical equation.

Answers

2. The number is 7.
3. One piece should be 23 ft, and the other should be 69 ft.

$$x + 4x = 8$$

Step 5: Solve the equation.

$$5x = 8$$

Combine *like* terms.

$$\frac{5x}{5} = \frac{8}{5}$$

Divide both sides by 5.

$$x = \frac{8}{5} \text{ or } 1.6$$

Recall that x represents the length of the shorter piece. Therefore, the shorter piece is 1.6 ft. The longer piece is given by $4x$ or $4(1.6 \text{ ft}) = 6.4 \text{ ft}$.

The pieces are 1.6 ft and 6.4 ft.

Step 6: Interpret the results in words.

> ### Avoiding Mistakes
>
> It is good practice to verify that an answer is reasonable. In Example 3, the two boards should total 8 ft. We have 1.6 ft + 6.4 ft = 8 ft, as desired.

Example 4 **Applying a Linear Equation**

One model of a Panasonic high-definition plasma TV sells for $1500 more than a certain model made by Planar. The combined cost for these two models is $8500. Find the cost for each model. (*Source: Consumer Reports*)

Solution:

Step 1: Read the problem completely.

Let x represent the cost of the Planar TV.

Step 2: Label the variables.

Then $x + 1500$ represents the cost of the Panasonic.

$$\begin{pmatrix} \text{Cost of} \\ \text{Planar} \end{pmatrix} + \begin{pmatrix} \text{cost of} \\ \text{Panasonic} \end{pmatrix} = \begin{pmatrix} \text{total} \\ \text{cost} \end{pmatrix}$$

Step 3: Write an equation in words.

$$x \quad + \quad x + 1500 \quad = 8500$$

Step 4: Write a mathematical equation.

$$x + x + 1500 = 8500$$

Step 5: Solve the equation.

$$2x + 1500 = 8500$$

Combine *like* terms.

$$2x + 1500 - 1500 = 8500 - 1500$$

Subtract 1500 from both sides.

$$2x = 7000$$

$$\frac{2x}{2} = \frac{7000}{2}$$

Divide both sides by 2.

$$x = 3500$$

The Planar model TV costs $3500.
The Panasonic model is represented by $x + 1500 = \$3500 + \$1500 = \$5000$.

> ### Skill Practice
>
> 4. A kit of cordless 18-volt tools made by Craftsman costs $310 less than a similar kit made by DeWalt. The combined cost for both models is $690. Find the cost for each model. (*Source: Consumer Reports*)

Classroom Example: p. 676, Exercise 28

Answer
4. The Craftsman model costs $190, and the DeWalt model costs $500.

TIP: In Example 4, we could have let x represent the cost of *either* the Planar model TV or the Panasonic model.

Suppose we had let x represent the cost of the Panasonic model. Then $x - 1500$ is the cost of the Planar model (the Planar model is *less* expensive).

$$\begin{pmatrix} \text{Cost of} \\ \text{Planar} \end{pmatrix} + \begin{pmatrix} \text{cost of} \\ \text{Panasonic} \end{pmatrix} = \begin{pmatrix} \text{total} \\ \text{cost} \end{pmatrix}$$
$$x - 1500 + \qquad x \qquad = 8500$$
$$2x - 1500 = 8500$$
$$2x - 1500 + 1500 = 8500 + 1500$$
$$2x = 10{,}000$$
$$x = 5000$$

Therefore, the Panasonic model costs \$5000 as expected.
The Planar model costs $x - 1500$ or \$5000 − \$1500 = \$3500.

Skill Practice

5. The perimeter of a tennis court is 228 ft. If the length is 42 ft longer than the width, find the dimensions of the court.

Classroom Example: p. 677, Exercise 30

Example 5 **Applying a Linear Equation to Geometry**

The perimeter of the soccer field at Giants Stadium is 338 m. If the length is 37 m longer than the width, find the dimensions of the field.

Solution:

Step 1: Read the problem completely.

Let w represent the width.
Then $w + 37$ represents the length.

Step 2: Label the variables and write down relevant formulas. Draw a figure to help you.

Recall that the formula for the perimeter of a rectangle is given by $P = 2l + 2w$.

$$P = 2l + 2w$$

Step 3: The perimeter formula for a rectangle can be used as the equation.

$$338 = 2(w + 37) + 2(w)$$

Step 4: Write a mathematical equation, using the labeled variables.

$$338 = 2w + 74 + 2w$$

Step 5: Solve the equation. Clear parentheses.

$$338 = 4w + 74$$

Combine *like* terms.

$$338 - 74 = 4w + 74 - 74$$

Subtract 74 from both sides.

$$264 = 4w$$

$$\frac{264}{4} = \frac{4w}{4}$$

Divide by 4 on both sides.

$$66 = w$$

Step 6: Interpret the answer in words.

The width is 66 m.

Answer

5. The length is 78 ft, and the width is 36 ft.

The length is given by $w + 37 = 66\text{ m} + 37\text{ m} = 103\text{ m}$.

Example 6 **Using a Linear Equation in a Consumer Application**

Joanne has a cellular phone plan in which she pays $39.95 per month for 450 min of air time. Additional minutes beyond 450 are charged at a rate of $0.40 per minute. If Joanne's bill comes to $87.95, how many minutes did she use beyond 450 min?

Solution:

	Step 1:	Read the problem.
Let x represent the number of minutes beyond 450.	**Step 2:**	Label the variable.

Then $0.40x$ represents the cost for x additional minutes.

$$\left(\begin{matrix}\text{Monthly}\\ \text{fee}\end{matrix}\right) + \left(\begin{matrix}\text{cost of}\\ \text{additional minutes}\end{matrix}\right) = \left(\begin{matrix}\text{total}\\ \text{cost}\end{matrix}\right)$$

		Step 3:	Write an equation in words.
39.95	$+$ $0.40x$ $=$ 87.95	**Step 4:**	Write a mathematical equation.

$$39.95 + 0.40x = 87.95 \qquad \textbf{Step 5:} \quad \text{Solve the equation.}$$

$$39.95 - 39.95 + 0.40x = 87.95 - 39.95 \qquad \text{Subtract 39.95.}$$

$$0.40x = 48.00$$

$$\frac{0.40x}{0.40} = \frac{48.00}{0.40} \qquad \text{Divide by 0.40.}$$

$$x = 120$$

Joanne talked for 120 min beyond her allotted 450 min.

Skill Practice

6. D.J. signs up for a new credit card that earns travel miles with a certain airline. She initially earns 15,000 travel miles by signing up for the new card. Then for each dollar spent she earns 2.5 travel miles. If at the end of one year she has 38,500 travel miles, how many dollars did she charge on the credit card?

Classroom Example: p. 677, Exercise 32

Answer
6. D.J. charged $9400.

Section 11.6 Practice Exercises

Study Skills Exercise

For additional exercises, see Classroom Activities 11.6A–11.6B in the *Student's Resource Manual* at www.mhhe.com/moh.

In solving an application it is very important first to read and understand what is being asked in the problem. One way to do this is to read the problem several times. Another is to read it out loud so you can hear yourself. Still another is to try to rewrite the problem in your own words. Which of these methods do you think will help you in understanding an application?

Review Exercises

1. Use substitution to determine if 3 is a solution to the equation $4x + 1 = 11$. No

For Exercises 2–8, solve the equation.

2. $3t - 15 = -22$ $-\dfrac{7}{3}$

3. $\dfrac{b}{5} - 5 = -14$ -45

4. $2x + 22 = 6x - 2$ 6

 Writing ⬅➡ Translating Expression △ Geometry 🖩 Scientific Calculator ▶ Video NS E

5. $4(r + 4) - 12 = 18 - r$ $\dfrac{14}{5}$ **6.** $-(y - 9) + 5(y + 3) = -2(3y + 7)$ $-\dfrac{19}{5}$

7. $4.4p - 2.6 = 1.2p - 5$ -0.75 **8.** $\dfrac{2}{3}w - \dfrac{1}{6} = \dfrac{1}{3}w + \dfrac{5}{6}$ 3

Concept 2: Translating Verbal Statements into Equations

For Exercises 9–24, **a.** Write an equation that represents the problem and **b.** Solve the problem. (See Examples 1 and 2.)

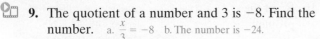

9. The quotient of a number and 3 is -8. Find the number. a. $\dfrac{x}{3} = -8$ b. The number is -24.

10. The difference of -2 and a number is -14. Find the number. a. $-2 - x = -14$ b. The number is 12.

11. A number subtracted from -30 results in 42. Find the number. a. $-30 - x = 42$ b. The number is -72.

12. A number increased by 13 results in -100. Find the number. a. $x + 13 = -100$ b. The number is -113.

13. The total of 30 and a number is 13. Find the number. a. $30 + x = 13$ b. The number is -17.

14. Sixty is -5 times a number. Find the number. a. $60 = -5x$ b. The number is -12.

15. Five less than the quotient of a number and 4 is equal to -12. Find the number. a. $\dfrac{x}{4} - 5 = -12$ b. The number is -28.

16. Eight decreased by the product of a number and 3 is equal to 5. Find the number. a. $8 - 3x = 5$ b. The number is 1.

17. One-half increased by a number is 4. Find the number. a. $\dfrac{1}{2} + x = 4$ b. The number is $\dfrac{7}{2}$ or $3\dfrac{1}{2}$.

18. Five-thirds decreased by a number is 1. Find the number. a. $\dfrac{5}{3} - x = 1$ b. The number is $\dfrac{2}{3}$.

19. The product of -12 and a number is the same as the sum of the number and 26. Find the number. a. $-12x = x + 26$ b. The number is -2.

20. The difference of a number and 16 is the same as the product of the number and -3. Find the number. a. $x - 16 = -3x$ b. The number is 4.

21. Ten times the total of a number and 5.1 is 56. Find the number. a. $10(x + 5.1) = 56$ b. The number is 0.5.

22. Three times the difference of a number and 5 is 15. Find the number. a. $3(x - 5) = 15$ b. The number is 10.

23. The product of 3 and a number is the same as 10 less than twice the number. a. $3x = 2x - 10$ b. The number is -10.

24. Six less than a number is the same as 3 more than twice the number. a. $x - 6 = 2x + 3$ b. The number is -9.

Concept 3: Applications of Linear Equations (Mixed Exercises)

For Exercises 25–44, solve the problem by using the problem-solving flowchart found on page 670.

25. Zachary has a piece of electrical wire that is 12 m long. He wants to cut it into two pieces, one that is 3 times as long as the other. How long should he cut each piece? **(See Example 3.)** The pieces should be 3 m and 9 m.

26. A 9-ft-long piece of weather stripping is cut into two pieces. If one piece is twice the length of the other piece, how long is each piece? The pieces are 3 ft and 6 ft.

27. In the 1990s the musical group Metallica had 6 fewer hits than the group Boyz II Men. If together the groups had 26 hits, how many hits did each group have? **(See Example 4.)** Metallica had 10 hits while Boyz II Men had 16 hits.

28. A two-piece set of luggage costs $150. If sold individually, the large bag costs $40 more than the small bag. What are the individual costs for each bag? The large bag costs $95, and the small bag costs $55.

29. The perimeter of a rectangular soccer field is 460 yd. The length is 30 yd longer than the width. What are the dimensions of the field? **(See Example 5.)** The soccer field is 100 yd by 130 yd.

30. The width of an Olympic size swimming pool is one-half its length. If the perimeter is 150 m, what are the dimensions of the pool? The length is 50 m, and the width is 25 m.

31. A cellular phone company charges $49.95 each month, which includes 500 min. For minutes used beyond the first 500, the charge is $0.25 per minute. Jim's bill came to $62.45. How many minutes over 500 min did he use? **(See Example 6.)** Jim used 50 min over the 500 min.

32. U-Rent-It car company charges $19 per day, which includes 200 mi of driving. If a person travels more than 200 mi, a fee of $0.30 per mile is charged for each mile over 200 mi. If Mr. Cain's bill for one day came to $62.50, how many miles did he travel over 200 mi? Mr. Cain traveled 145 mi over 200 mi.

33. Felicia has a piece of ribbon that is 4 ft long. She wants to cut the ribbon to make two pieces so that one piece is twice as long as the other. How long should each piece be? The pieces are $1\frac{1}{3}$ ft and $2\frac{2}{3}$ ft long.

34. A pipe, 6 m long, needs to be cut into two sections. If one section must be 3 times longer than the other, how long must each section be? The sections should be $1\frac{1}{2}$ m and $4\frac{1}{2}$ m long.

35. In a football game, the Tampa Bay Buccaneers won the game with 6 points more than twice the number of points earned by their opponent, the Oakland Raiders. If there was a total of 69 points in the game, how many points did each team score? Tampa Bay had 48 points and Oakland had 21 points.

36. In the first Superbowl in 1967, the Green Bay Packers won the game with 5 points more than 3 times the number of points earned by their opponents, the Kansas City Chiefs. If there was a total of 45 points in the game, how many points did each team score? Green Bay had 35 points and Kansas City had 10 points.

37. An apartment complex charges a refundable security deposit when renting an apartment. The deposit is $350 less than the monthly rent. If Charlene paid a total of $950 for her first month's rent and security deposit, how much is her monthly rent? How much is the security deposit? Charlene's rent is $650 a month with a security deposit of $300.

38. José buys a shirt and a pair of pants for $71.90. If the shirt costs $20 less than the pants, what is the cost of the shirt? What is the cost of the pants? The shirt costs $25.95 and the pants cost $45.95.

39. Stefan is paid a salary of $480 a week at his job. When he works overtime, he receives $18 an hour. If his weekly paycheck came to $588, how many hours of overtime did he put in that week? Stefan worked 6 hr of overtime.

40. Victoria is paid a salary of $480 a week at her job. She worked 8 hr of overtime during the holidays, and her weekly paycheck came to $672. What is her overtime pay per hour? Victoria makes $24 an hour for overtime.

41. Raul signed up for his classes for the spring semester. He signed up for 4 credit-hours more in the spring than he did in the fall. If he has a total of 28 hr in the two semesters, how many hours did he take in the fall? How many hours did he take in the spring? Raul took 12 hr in the fall and 16 hr in the spring.

42. A computer with a monitor costs $899. If the computer costs $241 more than the monitor, what is the price of the computer? What is the price of the monitor? The computer costs $570 and the monitor costs $329.

43. Ann-Marie got two offers for jobs as a salesperson. One job pays a weekly salary of $300, and the other pays on commission. If commission pays 24% of sales, how much merchandise must Ann-Marie sell each week to match the weekly salary? Ann-Marie must sell $1250 of merchandise each week.

44. Mercedes had a calling card worth $20. On this card, long-distance calls cost $0.05 more per minute than local calls. If Mercedes made 35 min of local calls and 52 min of long-distance calls, what was she paying per minute for each type of call? Local calls are $0.20 per minute, and long-distance calls are $0.25 per minute.

Group Activity

Deciphering a Coded Message

Materials: Pencil and paper

Estimated Time: 20 minutes

Group Size: Pairs

Cryptography is the study of coding and decoding messages. One type of coding process assigns a number to each letter of the alphabet and to the space character. For example:

A	B	C	D	E	F	G	H	I	J	K	L	M	N
1	2	3	4	5	6	7	8	9	10	11	12	13	14

O	P	Q	R	S	T	U	V	W	X	Y	Z	space
15	16	17	18	19	20	21	22	23	24	25	26	27

According to the number assigned to each letter, the message "Do the Math" would be coded as follows:

D O _ T H E _ M A T H
4 / 15 / 27 / 20 / 8 / 5 / 27 / 13 / 1 / 20 / 8

Now suppose each letter is encoded by applying a formula such as $x + 3 = y$, where x is the original number of the letter and y is the code number of the letter. For example, the letter A would be coded by $1 + 3 = 4$, B would be coded $2 + 3 = 5$, and so on.

Using this encoding, we have

Message: D O _ T H E _ M A T H

Original: 4 / 15 / 27 / 20 / 8 / 5 / 27 / 13 / 1 / 20 / 8

Coded form: 7 / 18 / 30 / 23 / 11 / 8 / 30 / 16 / 4 / 23 / 11

To decode this message, the receiver would need to reverse the operation by solving for x, that is, use the formula $x = y - 3$.

1. Each pair of students will encode the following message by adding 3 to each number:

   Life is too short for long division.

   > 15 / 12 / 9 / 8 / 30 / 12 / 22 / 30 / 23 / 18 / 18 / 30 / 22 / 11 / 18 / 21 / 23 / 30 /
   > 9 / 18 / 21 / 30 / 15 / 18 / 17 / 10 / 30 / 7 / 12 / 25 / 12 / 22 / 12 / 18 / 17

2. Each pair of students will decode the following message by subtracting 3 from each number.

   17 / 4 / 23 / 24 / 21 / 4 / 15 / 30 / 17 / 24 / 16 / 5 / 8 / 21 / 22 / 30 / 4 / 21 / 8 / 30 /
   10 / 18 / 18 / 7 / 30 / 9 / 18 / 21 / 30 / 28 / 18 / 24 / 21 / 30 / 11 / 8 / 4 / 15 / 23 / 11

   > NATURAL_NUMBERS_ARE_GOOD_FOR_YOUR_HEALTH

Instructor Note: This activity can be extended by changing the formula.

Chapter 11   Summary

Section 11.1   Properties of Real Numbers

Key Concepts

An algebraic **expression** is a collection of **variables** and **constants** under operations such as addition, subtraction, multiplication, and division.

To evaluate an expression, first replace the variable with parentheses. Then insert the values and simplify using the order of operations.

The Properties of Real Numbers

Commutative property of addition:

$a + b = b + a$

Commutative property of multiplication:

$a \cdot b = b \cdot a$

Associative property of addition:

$(a + b) + c = a + (b + c)$

Associative property of multiplication:

$(a \cdot b) \cdot c = a \cdot (b \cdot c)$

Distributive property of multiplication over addition:

$a(b + c) = a \cdot b + a \cdot c$

Examples

Example 1

$3x + 8$ is an algebraic expression, x is a variable, and 8 is a constant.

Example 2

Evaluate $4x - 5y$ for $x = -\dfrac{1}{2}$ and $y = 3$.

$$4x - 5y = 4(\ \ ) - 5(\ \ )$$
$$= 4\left(-\frac{1}{2}\right) - 5(3)$$
$$= -2 - 15$$
$$= -17$$

Example 3

$5 + w = w + 5$ is an example of the commutative property of addition.

Example 4

$3(4y) = (3 \cdot 4)y$ is an example of the associative property of multiplication.

Example 5

$-2(7 + t) = -2(7) + (-2)t$ is an example of the distributive property of multiplication over addition.

Section 11.2   Simplifying Expressions

Key Concepts

An algebraic expression is the sum of one or more terms. A **term** is a number or a product or quotient of numbers and variables. If a term contains a variable, it is called a **variable term**. A term with no variable is called a **constant term**. The **coefficient** of a term is the numerical factor of the term.

Terms that have exactly the same variable factors are called *like* **terms**.

Like terms can be combined by applying the distributive property.

To simplify an expression, first clear parentheses by using the distributive property. Group *like* terms together. Then combine *like* terms.

Examples

Example 1

In the expression $12x + 3$,

$12x$ is a variable term.

The term 3 is a constant term.

12 is the coefficient of the term $12x$.

Example 2

$5h$ and $-2h$ are *like* terms because the variable part h is the same.

$6t$ and $6t^2$ are not *like* terms because the variable parts t and t^2 are not exactly the same.

Example 3

$$3x + 15x - 7x = (3 + 15 - 7)x$$
$$= 11x$$

Example 4

Simplify: $3(k - 4) - (6k + 10) + 14$

$3(k - 4) - (6k + 10) + 14$

$= 3k - 12 - 6k - 10 + 14$ Clear parentheses.

$= 3k - 6k - 12 - 10 + 14$ Group *like* terms together.

$= -3k - 8$ Combine *like* terms.

Section 11.3   Addition and Subtraction Properties of Equality

Key Concepts

An **equation** is a statement that indicates that two quantities are equal.

A **solution** to an equation is a value of the variable that makes the equation a true statement.

Definition of a Linear Equation in One Variable

Let a, b, and c be real numbers such that $a \neq 0$. A **linear equation in one variable** is an equation that can be written in the form $ax + b = c$.

Two equations that have the same solution are called **equivalent equations**.

The Addition and Subtraction Properties of Equality

Let a, b, and c represent algebraic expressions.

1. The **addition property of equality**:
   If $a = b$
   then $a + c = b + c$

2. The **subtraction property of equality**:
   If $a = b$
   then $a - c = b - c$

Examples

Example 1

$3x + 4 = 6$ is an equation while $3x + 4$ is an expression.

Example 2

The number -4 is a solution to the equation $5x + 7 = -13$ because when we substitute -4 for x, we get a true statement.

$$5(-4) + 7 \stackrel{?}{=} -13$$
$$-20 + 7 \stackrel{?}{=} -13$$
$$-13 = -13 \checkmark$$

Example 3

The equation $5x + 7 = -13$ is equivalent to the equation $x = -4$ because they both have the same solution of -4.

Example 4

To solve the equation $t - 12 = -3$, use the addition property of equality.

$$t - 12 = -3$$
$$t - 12 + 12 = -3 + 12$$
$$t = 9 \qquad \text{The solution is 9.}$$

Example 5

To solve the equation $-1 = p + 2$, use the subtraction property of equality.

$$-1 = p + 2$$
$$-1 - 2 = p + 2 - 2$$
$$-3 = p \qquad \text{The solution is } -3.$$

Section 11.4 — Multiplication and Division Properties of Equality

Key Concepts

The Multiplication and Division Properties of Equality

Let a, b, and c represent algebraic expressions, $c \neq 0$.

1. The **multiplication property of equality**:

   If $\qquad a = b$

   then $\quad c \cdot a = c \cdot b$

2. The **division property of equality**:

   If $\qquad a = b$

   then $\quad \dfrac{a}{c} = \dfrac{b}{c}$

When applying the multiplication or division properties of equality to obtain a coefficient of 1 for the variable term, we will generally use the following conventions.

- If the coefficient of the variable term is expressed as a fraction, multiply both sides by its reciprocal.
- Otherwise, divide both sides by the coefficient itself.

We can determine which property to use to solve an equation as follows: Note the operation on the variable. Then use the property of equality of the inverse operation.

Examples

Example 1

Solve. $\quad 3a = -18$

$$\frac{3a}{3} = \frac{-18}{3} \qquad \text{Divide both sides by 3.}$$

$$a = -6 \qquad \text{The solution is } -6.$$

Example 2

Solve. $\quad -\dfrac{3}{5}b = 9$

$$\left(-\frac{5}{3}\right)\left(-\frac{3}{5}\right)b = \left(-\frac{5}{3}\right) \cdot 9 \qquad \begin{array}{l}\text{Multiply both sides by} \\ \text{the reciprocal of } -\frac{3}{5}.\end{array}$$

$$b = -15 \qquad \text{The solution is } -15.$$

Example 3

Solve. $\quad \dfrac{w}{2} = -11$

$$2\left(\frac{w}{2}\right) = 2(-11) \qquad \text{Multiply both sides by 2.}$$

$$w = -22 \qquad \text{The solution is } -22.$$

Example 4

Solve. $\quad 4x = 20 \quad$ and $\quad 4 + x = 20$

$$4x = 20 \qquad\qquad 4 + x = 20$$

$$\frac{4x}{4} = \frac{20}{4} \qquad 4 - 4 + x = 20 - 4$$

$$x = 5 \qquad\qquad x = 16$$

The solution is 5. $\qquad$ The solution is 16.

Section 11.5 Solving Equations with Multiple Steps

Key Concepts

Steps to Solve a Linear Equation in One Variable

1. Simplify both sides of the equation.
   - Clear parentheses if necessary.
   - Combine *like* terms if necessary.
2. Use the addition or subtraction property of equality to collect the variable terms on one side of the equation.
3. Use the addition or subtraction property of equality to collect the constant terms on the *other* side of the equation.
4. Use the multiplication or division property of equality to make the coefficient of the variable term equal to 1.
5. Check the answer in the original equation.

Examples

Example 1

Solve.

$$2(2x - 5) - 3 = 3(x - 1) - x$$

$$4x - 10 - 3 = 3x - 3 - x \qquad \text{Clear parentheses.}$$

$$4x - 13 = 2x - 3 \qquad \text{Combine } like \text{ terms.}$$

$$4x - 2x - 13 = 2x - 2x - 3 \qquad \text{Subtract } 2x.$$

$$2x - 13 = -3 \qquad \text{Simplify.}$$

$$2x - 13 + 13 = -3 + 13 \qquad \text{Add 13.}$$

$$2x = 10 \qquad \text{Simplify.}$$

$$\frac{2x}{2} = \frac{10}{2} \qquad \text{Divide by 2.}$$

$$x = 5 \qquad \text{Simplify.}$$

The solution is 5.

Section 11.6 Applications and Problem Solving

Key Concepts

Problem-Solving Flowchart for Word Problems

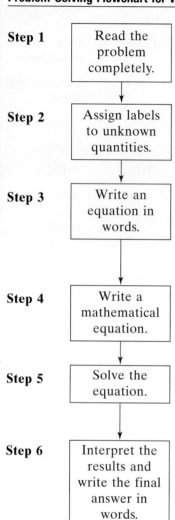

Step 1 — Read the problem completely.

Step 2 — Assign labels to unknown quantities.

Step 3 — Write an equation in words.

Step 4 — Write a mathematical equation.

Step 5 — Solve the equation.

Step 6 — Interpret the results and write the final answer in words.

Examples

Example 1

Subtract 5 times a number from 14. The result is −6. Find the number.

Let n represent the unknown number.

$$\underbrace{14}_{\text{from } 14} \quad \underbrace{-\,5n}_{\text{subtract } 5n} \quad \underbrace{=\,-6}_{\text{result is } -6}$$

$$14 - 5n = -6$$

$$14 - 14 - 5n = -6 - 14$$

$$-5n = -20$$

$$\frac{-5n}{-5} = \frac{-20}{-5}$$

$$n = 4$$

The number is 4.

Example 2

An electrician needs to cut a 10-ft wire into two pieces so that one piece is 4 times as long as the other. How long should each piece be?

Let x represent one piece of the wire.
The other piece will be $4x$ ft long.

The two pieces added will be 10 ft.

$$x + 4x = 10$$

$$5x = 10$$

$$\frac{5x}{5} = \frac{10}{5}$$

$$x = 2$$

One piece of wire is 2 ft.
The other piece is 4(2 ft) = 8 ft.

Chapter 11   Review Exercises

Section 11.1

1. a. Michael is 8 years older than his sister. Write an expression for Michael's age if his sister is a years old. $a + 8$

   b. What is Michael's age if his sister is 35 years old?   43 years old

2. Evaluate the expression $-\left(\dfrac{4}{x}\right)^2$ for

   a. $x = 3$ $-\dfrac{16}{9}$ **b.** $x = 4$ -1

   c. $x = 2$ -4 **d.** $x = -2$ -4

3. Evaluate the expression $-2(x + y)^2$ for $x = 6$ and $y = -9$. -18

 4. Find the volume, V, for a sphere with $r = \dfrac{3}{2}$ meters, using the formula $V = \dfrac{4}{3}\pi r^3$. Use $\pi \approx \dfrac{22}{7}$. $\dfrac{99}{7}$ m³

5. Apply the commutative property of addition or multiplication to rewrite the expression.

   a. $t - 5$ **b.** $h \cdot 3$
      $-5 + t$ $3h$

6. Apply the associative property of addition or multiplication to rewrite the expression. Then simplify the expression.

   a. $-4(2 \cdot p)$ **b.** $(m + 10) - 12$
      $(-4 \cdot 2)p; \ -8p$ $m + (10 - 12); \ m - 2$

7. Apply the distributive property. $3(2b + 5)$
   $6b + 15$

8. Apply the distributive property.
   $-(-4k + 8m - 12)$
   $4k - 8m + 12$

Section 11.2

For Exercises 9 and 10, list the coefficients of each term.

9. $3a^2 - 5a + 12$ $3, -5, 12$

10. $-6xy - y + 2x + 1$ $-6, -1, 2, 1$

For Exercises 11–14, determine if the two terms are *like* terms or unlike terms.

11. $5t^2, 5t$ Unlike terms   **12.** $4h, -2h$ *Like* terms

13. $21, -5$ *Like* terms   **14.** $-8, -8k$ Unlike terms

For Exercises 15–18, combine *like* terms. Clear parentheses if necessary.

15. $6y + 8x - 2y - 2x + 10$ $6x + 4y + 10$

16. $12a - 5 + 9b - 5a + 14$ $7a + 9b + 9$

17. $4(u - 3v) - 5u + v$ $-u - 11v$

18. $-5(p + 4) + 6(p + 1) - 2$ $p - 16$

Section 11.3

For Exercises 19 and 20, determine if -3 is a solution to the equation.

19. $5x + 10 = -5$ -3 is a solution.

20. $-3(x - 1) = -9 + x$ -3 is not a solution.

21. Explain how to decide whether to use the addition property of equality or the subtraction property of equality to solve an equation. If a constant is being added to the variable term, use the subtraction property. If a constant is being subtracted from the variable term, use the addition property.

For Exercises 22–30, solve the equation, using either the addition or the subtraction property of equality.

22. $r + 23 = -12$ -35 **23.** $k - 3 = -15$ -12

24. $10 = p - 4$ 14 **25.** $21 = q - 3$ 24

26. $4.1 + m = 5.2$ 1.1 **27.** $-3.1 + n = 1.9$ 5

28. $a - \dfrac{2}{3} = \dfrac{5}{6}$ $\dfrac{3}{2}$ **29.** $b + \dfrac{1}{5} = -\dfrac{9}{10}$ $-\dfrac{11}{10}$

30. $\dfrac{3}{4} + h = 2$ $\dfrac{5}{4}$

Section 11.4

For Exercises 31–42, solve the equation by using either the multiplication or the division property of equality.

31. $4d = -28$ -7 **32.** $-3c = -12$ 4

33. $\dfrac{t}{2} = -13$ -26 **34.** $\dfrac{p}{5} = 7$ 35

35. $-\dfrac{4}{5}y = -16$ 20 **36.** $\dfrac{2}{3}x = -14$ -21

37. $1.4 = -0.7m$ -2

38. $-3.6 = 0.9n$ -4

39. $\frac{1}{3}w = \frac{3}{7}$ $\frac{9}{7}$

40. $-\frac{1}{4}s = \frac{2}{3}$ $-\frac{8}{3}$

41. $-42 = -7p$ 6

42. $51 = -3b$ -17

Section 11.5

For Exercises 43–54, solve the equation.

43. $9x + 7 = -2$ -1

44. $8y - 3 = 13$ 2

45. $45 = 6m - 3$ 8

46. $-25 = 2n - 1$ -12

47. $4 = \frac{3}{5}m - 2$ 10

48. $\frac{5}{8}p - 1 = 14$ 24

49. $5x + 12 = 4x - 16$ -28

50. $-4t - 2 = -3t + 5$ -7

51. $6(w - 2) + 15 = 3(w + 3) - 2$ $\frac{4}{3}$

52. $-4(h - 5) + h = 7(h + 1) - 5$ $\frac{9}{5}$

53. $-(5a + 3) - 3(a - 2) = 24 - a$ -3

54. $-(4b - 7) = 2(b + 3) - 4b + 13$ -6

Section 11.6

For Exercises 55–58, **a.** write an equation that represents the problem and **b.** solve the problem.

55. The product of a number and -6 results in the sum of the number and 2. Find the number.
a. $-6x = x + 2$ **b.** $-\frac{2}{7}$

56. The total of -9, 3, and twice a number is -2. Find the number.   **a.** $-9 + 3 + 2x = -2$ **b.** 2

57. One-third decreased by a number is 2. Find the number.   **a.** $\frac{1}{3} - x = 2$ **b.** $-\frac{5}{3}$

58. Two less than the quotient of a number and 8 is $\frac{1}{4}$. Find the number.
a. $\frac{x}{8} - 2 = \frac{1}{4}$ **b.** 18

59. Actor Johnny Depp has had 16 fewer film and TV appearances than Tom Hanks. Together they had 116 appearances. How many appearances did each star have?
Johnny Depp had 50 appearances and Tom Hanks had 66 appearances.

60. Marty has a cell phone plan that costs $44.98 per month and includes 400 text messages. If Marty goes over 400 text messages, it costs $0.10 per message. If Marty's bill one month was $47.48, how many text messages did Marty make over the allotted 400?
Marty made 25 text messages over the allotted 400.

61. The top of a rectangular table has length 32 in. more than the width. If the perimeter of the table is 224 in., find the width and length of the table.
The width is 40 in. and the length is 72 in.

Chapter 11   Test

1. A high school student sells magazine subscriptions at $19.95 each. Write an expression that represents the total amount made for m magazines. $19.95m$

2. Evaluate the expression for $x = 4$ and $y = -1$.
$-x^2 + y^2$ -15

3. Find the area A in the formula
$A = 2lw + 2lh + 2wh$ for $l = 3$ ft, $w = 2.5$ ft, and $h = 1.75$ ft (surface area of a rectangular solid). $A = 34.25 \text{ ft}^2$

For Exercises 4–8, name the property demonstrated. Choose from the commutative property of addition, commutative property of multiplication, associative property of addition, associative property of multiplication, and distributive property of multiplication over addition.

4. $-5(9x) = (-5 \cdot 9)x$
Associative property of multiplication

5. $-5x + 9 = 9 - 5x$
Commutative property of addition

6. $-3 + (u + v) = (-3 + u) + v$
Associative property of addition

7. $-4(b + 2) = -4b - 8$
Distributive property of multiplication over addition

8. $g(-6) = -6g$
Commutative property of multiplication

For Exercises 9–14, simplify the expressions.

9. $4(a + 9) - 12$ $4a + 24$

10. $-3(6b) + 5b + 8$ $-13b + 8$

11. $14y + 2(y - 9) + 21$ $16y + 3$

 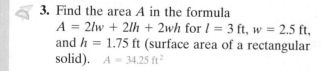

12. $2 + (5 - w) + 3(-2w)$ $7 - 7w$

13. $-(3x - 5) + 2x - 1$ $-x + 4$

14. $6y + 2 - 5(y - 4)$ $y + 22$

15. Explain the difference between an expression and an equation.
An expression is a collection of terms. An equation has an equal sign that indicates that two expressions are equal.

16. Identify the following as either an expression or an equation.

a. $4x + 5$
Expression
b. $4x + 5 = 2$
Equation
c. $-3(p + 5)$
Expression
d. $2(q - 3) = 6$
Equation
e. $2(q - 3) + 6$
Expression
f. $6(4y + 3) + (y - 2)$
Expression

For Exercises 17–28, solve the equation.

17. $-6x = 12$ -2

18. $-6 + x = 12$ 18

19. $\dfrac{x}{-6} = 12$ -72

20. $12x = -6$ $-\dfrac{1}{2}$

21. $12 = -3p + 9$ -1

22. $1.8m = 0.36$ 0.2

23. $p + \dfrac{1}{16} = -\dfrac{3}{4}$ $-\dfrac{13}{16}$

24. $-\dfrac{5}{12} = -\dfrac{5}{6}n$ $\dfrac{1}{2}$

25. $5h - 2 = -h + 16$ 3

26. $\dfrac{x}{7} = -12$ -84

27. $-2(q - 5) = 6q + 10$ 0

28. $-(4k - 2) - k = 2(k - 6)$ 2

29. The product of -2 and a number is the same as the total of 15 and the number. Find the number.
The number is -5.

30. The budget for making the movie *The Twilight Saga: New Moon* was one-fifth the budget for making *Harry Potter: The Half-Blood Prince*. Together the budgets totaled $300 million. What was the budget for *The Twilight Saga: New Moon*?
The budget for *New Moon* was $50 million.

31. A 300-cm copper pipe is cut into two sections. One section is 4 times longer than the other. What are the lengths of the two sections in centimeters?
The lengths are 60 cm and 240 cm.

32. Sela's electric bill was $90.03. She pays a monthly fee of $9.95 and is charged $0.104 per kWh (kilowatt-hour). How many kilowatt-hours did she use that month?
Sela used 770 kWh.

Chapters 1–11 Cumulative Review Exercises

1. Identify the place value of the underlined digit.

a. 34,911
Hundreds
b. 209,001
Ten-thousands
c. 5,901,888
Hundred-thousands

For Exercises 2–4, round the number to the indicated place value.

2. 45,921; thousands 46,000

3. 1,285,000; ten-thousands 1,290,000

4. 25,449; hundreds 25,400

5. Divide 39,190 by 46. Identify the dividend, divisor, whole part of the quotient, and remainder.
Dividend is 39,190; divisor is 46; quotient is 851; remainder is 44.

6. Identify each number as prime or composite.

a. 59
Prime
b. 91
Composite
c. 39
Composite

7. Write the reciprocal of the number $\dfrac{8}{23}$. $\dfrac{23}{8}$

8. Multiply. $\dfrac{2}{13} \cdot \dfrac{39}{5}$ $\dfrac{6}{5}$

9. Divide. $\dfrac{21}{10} \div \dfrac{75}{8}$ $\dfrac{28}{125}$

10. Simplify. $\dfrac{1300}{10,000}$ $\dfrac{13}{100}$

11. Add and subtract. $\dfrac{9}{25} - \dfrac{1}{10} + \dfrac{4}{15}$ $\dfrac{79}{150}$

12. Find the area of the triangle.
7 ft^2

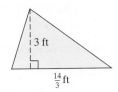

13. Convert to an improper fraction. $3\dfrac{7}{8}$ $\dfrac{31}{8}$

14. Convert to a mixed fraction. $\dfrac{28}{5}$ $5\dfrac{3}{5}$

15. Simplify. Write the answer as a mixed number.
$2\dfrac{1}{4} + 5\dfrac{5}{6} \cdot 1\dfrac{1}{10}$ $8\dfrac{2}{3}$

For Exercises 16–20, simplify.

16. 3.1(4.5) 13.95

17. 0.08 ÷ (0.16) 0.5

18. 23.991 + 3.2 + 4.03
31.221

19. 78.002 − 34.25
43.752

20. 4.2 − (2.0 − 1.2)² 3.56

21. Sarah cleans three apartments in a weekend. The apartments have 5, 6, and 4 rooms, respectively. If she earns $300 for the weekend, how much does she make per room? Sarah makes $20 per room.

For Exercises 22–24, solve the proportion.

22. $\dfrac{2}{9} = \dfrac{1.8}{p}$
8.1

23. $\dfrac{n}{1\frac{1}{2}} = \dfrac{8}{15}$
$4\frac{4}{5}$

24. $\dfrac{24}{15} = \dfrac{m}{35}$
56

25. One type of cat treats, Kitty Treats, contains 2 oz and sells for $1.84. Cat Goodies sells for $2.25 but contains 2.5 oz. Find the unit prices to determine the better buy.
Kitty Treats costs $0.92 per ounce. Cat Goodies costs $0.90 per ounce. Cat Goodies is the better buy.

For Exercises 26–30, complete the table.

	Decimal	Fraction	Percent
26.	0.15	$\frac{3}{20}$	15%
27.	0.125	$\frac{1}{8}$	12.5%
28.	1.1	$\frac{11}{10}$	110%
29.	$0.\overline{2}$	$\frac{2}{9}$	$22.\overline{2}$%
30.	0.002	$\frac{1}{500}$	0.2

31. A 1-gal jug of fruit punch will be poured into 6-fl-oz glasses. How many complete glasses can be filled? 21 glasses can be filled.

32. A Formula One Grand Prix race was won with a time of 1:41:45. Convert this to minutes.
101.75 min

For Exercises 33–35, convert the metric measurements.

33. 680 cc = _____ L 0.68 L

34. 3.2 km = _____ m 3200 m

35. 45 g = _____ cg 4500 cg

36. What is the area for a round table with a 40-in. diameter? Use 3.14 for π. 1256 in.²

37. A page in a book is $8\frac{1}{2}$ in. by 10 in. However, there is a $\frac{1}{2}$-in. margin around the edges of the page. How much area can be used for printed material?
The area is 67.5 in.²

38. Find the volume of a can of paint if the can is a cylinder with radius 6.5 in. and height 7.5 in. Use 3.14 for π, and round the answer to the nearest whole unit. 995 in.²

39. Find the circumference of the circle. Use 3.14 for π. 18.84 yd

40. Find the length of side x. 24 mm

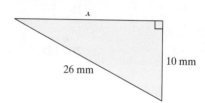

For Exercises 41–43, refer to the list of data depicting the number of turkey subs sold each day for two weeks at Subs-R-Us.

15	12	8	5	6	12	10
20	7	5	8	9	11	12

41. Find the mean. 10

42. Find the median. 9.5

43. Find the mode. 12

44. The table shows the gas mileage (mpg) for several different types of vehicles.

Vehicle	Miles per gallon (mpg)
Hybrid	50
Sedan	29
SUV	9
Truck	21
Sports car	11
Station wagon	20

Construct a bar graph with horizontal bars. The length of each bar corresponds to the miles per gallon for each type of vehicle.

Miles per Gallon Comparison

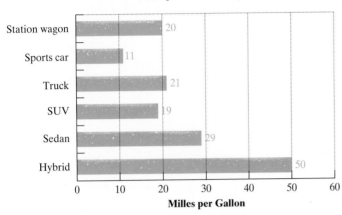

For Exercises 46–49, perform the indicated operations.

46. $-129 - (-132)$ 3   **47.** $16 \div (-4) \cdot 3$ −12

48. $4(5 - 11) + (-1)$ −25   **49.** $5 - 23 + 12 - 3$ −9

For Exercises 50 and 51, simplify the expression.

50. $-2(x + 14) + 15 - 3x$ −5x − 13

51. $3y - (5y + 6) - 12$ −2y − 18

For Exercises 52–55, solve the equation.

52. $4p + 5 = -11$ −4

53. $9(t - 1) - 7t + 2 = t - 15$ −8

54. $2x = -4(x - 3)$ 2

55. $16 = 7 - (3x - 1)$ $-\dfrac{8}{3}$

45. The circle graph represents activities of high school students after school.

After-School Activites

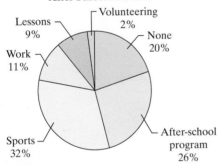

a. If a school has about 520 students, how many are enrolled in sports?   Approximately 166 students

b. If a school has about 650 students, how many do not participate in any after-school activities?   130 students

Additional Topics Appendix

Energy and Power

1. Units of Energy

In this section, we discuss common units of energy and power. **Energy** is defined as the amount of work that a physical system is capable of performing. Energy is stored in fossil fuels (coal, wood, oil, and natural gas), the food we eat, and even our own bodies.

Three common units of energy that we discuss here are the

Foot-pound (ft·lb)

British thermal unit (Btu)

Kilocalorie (or simply Calorie, Cal)

One foot-pound (1 ft·lb) is equal to the amount of energy required to lift 1 lb a distance of 1 ft (see Figure A-1).

Figure A-1

Notice that the foot-pound (ft·lb) is a product of feet and pounds. Thus, to compute the amount of energy necessary to lift a weight a certain distance, we have

$$\text{Energy} = \begin{pmatrix} \text{distance} \\ \text{in feet} \end{pmatrix} \cdot \begin{pmatrix} \text{weight} \\ \text{in pounds} \end{pmatrix}$$

Example 1 Computing Energy in Foot-pounds

How many foot-pounds of energy is required to lift a steel beam weighing 475 lb to the top of a 64-ft roof?

Solution:

$$\text{Energy} = \begin{pmatrix} \text{distance} \\ \text{in feet} \end{pmatrix} \cdot \begin{pmatrix} \text{weight} \\ \text{in pounds} \end{pmatrix}$$

$$= (64 \text{ ft}) \cdot (475 \text{ lb})$$

$$= 30,400 \text{ ft·lb}$$

Concepts

1. Units of Energy
2. Units of Power

Skill Practice

1. How many foot-pounds of energy is required to lift a 250-lb slab of concrete to the top of a 40-ft bridge?

Classroom Example: p. A-4, Exercise 4

Answer

1. 10,000 ft·lb

A-1

The British thermal unit (Btu) is often used to measure energy consumption for heating and air conditioning systems. For example, a portable heater might have a rating of 60,000 Btu. To convert between foot-pounds and British thermal units, we have

$$1 \text{ Btu} \approx 778 \text{ ft·lb}$$

Skill Practice

Convert.

2. 500 Btu ≈ _____ ft·lb

3. 2723 ft·lb ≈ _____ Btu

Classroom Examples: p. A-4,
Exercises 10 and 12

Example 2   **Converting Units of Energy**

a. 1250 Btu ≈ _____ ft·lb

b. 5057 ft·lb ≈ _____ Btu

Solution:

a. $1250 \text{ Btu} \approx \dfrac{1250 \text{ Btu}}{1} \cdot \dfrac{778 \text{ ft·lb}}{1 \text{ Btu}}$ Recall that 1 Btu ≈ 778 ft·lb.

$$= 972{,}500 \text{ ft·lb}$$

b. $5057 \text{ ft·lb} \approx \dfrac{5057 \text{ ft·lb}}{1} \cdot \dfrac{1 \text{ Btu}}{778 \text{ ft·lb}}$

$$= 6.5 \text{ Btu}$$

Heat is also a type of energy. In the metric system, heat is measured in calories or kilocalories. One kilocalorie (1 kcal) is equal to 1000 cal. However, kilocalories are often called **Calories** (spelled with a capital C and abbreviated as Cal). It is these larger units of Calories that are used frequently in dietary science.

Skill Practice

4. Antonio burns 600 Cal/hr bicycling at a fast pace. How many Calories will he burn if he rides for 1 hr 20 min?

Classroom Example: p. A-4,
Exercise 26

Example 3   **Computing Calories Used in Exercise**

Walking at a moderate pace burns 480 Cal/hr. If Molly walks a half-marathon in 3 hr 15 min, how many calories does she burn?

Solution:

We will assume that the rate of 480 Cal/hr burned remains constant. Therefore, we can set up a proportion to solve this problem.

First note that 3 hr 15 min is equal to 3.25 hr. Thus,

$$\begin{array}{l} \text{number of Calories} \rightarrow \\ \text{number of hours} \rightarrow \end{array} \dfrac{480}{1} = \dfrac{x}{3.25} \begin{array}{l} \leftarrow \text{number of Calories} \\ \leftarrow \text{number of hours} \end{array}$$

$$(480)(3.25) = (1) \cdot x \qquad \text{Cross-multiply to form the cross products.}$$

$$1560 = x \qquad \text{Simplify.}$$

Molly will burn 1560 Cal walking a half-marathon.

2. Units of Power

Power is the rate at which energy is released. One way to express power is in foot-pounds per second $\left(\dfrac{\text{ft·lb}}{\text{sec}} \right)$.

Answers

2. 389,000 ft·lb

3. 3.5 Btu   **4.** 800 Cal

Example 4 **Finding Power**

a. Find the power if a 50-lb box is lifted 10 ft in 4 sec.

b. Find the power if a 50-lb box is lifted 10 ft in 2 sec.

Solution:

a. Because power is expressed in units of $\frac{\text{ft·lb}}{\text{sec}}$ we have

$$\text{Power} = \frac{(10 \text{ ft})(50 \text{ lb})}{4 \text{ sec}} = \frac{500 \text{ ft·lb}}{4 \text{ sec}} = 125 \frac{\text{ft·lb}}{\text{sec}}$$

b. To find the power used to lift the same box the same distance in one-half the time, we have

$$\text{Power} = \frac{(10 \text{ ft})(50 \text{ lb})}{2 \text{ sec}} = \frac{500 \text{ ft·lb}}{2 \text{ sec}} = 250 \frac{\text{ft·lb}}{\text{sec}}$$

Notice that more power is required to lift the box more quickly.

Skill Practice

5. Find the power if an 80-lb box is lifted 20 ft in 10 sec.

6. Find the power if an 80-lb box is lifted 20 ft in 5 sec.

Classroom Example: p. A-5, Exercise 32

A unit of power that is used in the U.S. Customary System is the horsepower.

$$1 \text{ horsepower (hp)} = 550 \frac{\text{ft·lb}}{\text{sec}}$$

A measure of 1 hp loosely indicates that a horse can move 550 lb a distance of 1 ft in 1 sec.

Example 5 **Convert Units of Power**

A Mercury engine is rated 200 hp. Convert this to $\frac{\text{ft·lb}}{\text{sec}}$.

Solution:

$$200 \text{ hp} = \frac{200 \text{ hp}}{1} \cdot \frac{550 \frac{\text{ft·lb}}{\text{sec}}}{1 \text{ hp}} \qquad \text{Recall that 1 hp} = 550 \frac{\text{ft·lb}}{\text{sec}}.$$

$$= 110,000 \frac{\text{ft·lb}}{\text{sec}}$$

Skill Practice

7. A lawn mower is rated 6.5 hp. Convert this to $\frac{\text{ft·lb}}{\text{sec}}$.

Classroom Example: p. A-5, Exercise 42

Answers

5. $160 \frac{\text{ft·lb}}{\text{sec}}$ 6. $320 \frac{\text{ft·lb}}{\text{sec}}$

7. $3575 \frac{\text{ft·lb}}{\text{sec}}$

Section A.1 Practice Exercises

Vocabulary and Key Concepts

For additional exercises, see Classroom Activities A.1A–A.1B in the *Student's Resource Manual* at www.mhhe.com/moh.

1. _____Energy_____ is defined as the amount of work that a physical system performs.

Concept 1: Units of Energy

For Exercises 2–7, find the energy in foot-pounds.

2. Find the energy required to lift a 3800-lb car 6 ft.
   22,800 ft·lb

3. Find the energy required to lift a 3000-lb car 5 ft.
   (See Example 1.)   15,000 ft·lb

 Writing   Translating Expression   Geometry    Scientific Calculator   Video   NS&E

4. Find the energy required to lift 200 lb a distance of 2 yd.   1200 ft·lb

5. Find the energy required to lift 50 lb a distance of 1.5 yd.   225 ft·lb

6. How much energy would be required to lift 2.5 tons a distance of 3 ft?   15,000 ft·lb

7. How much energy would be required to lift 1.5 tons a distance of 4 ft?   12,000 ft·lb

For Exercises 8–11, find the foot-pound equivalent for the heater or air conditioner. **(See Example 2.)**

8. "Toasty" Patio Heater (40,000 Btu)
31,120,000 ft·lb

9. Portable battery powered Heatpro space heater (3000 Btu)   2,334,000 ft·lb

10. Air conditioner with remote (14,000 Btu)
10,892,000 ft·lb

11. Window air conditioner (8000 Btu)
6,224,000 ft·lb

For Exercises 12 and 13, convert the units of energy. Round to the nearest whole unit, if necessary. **(See Example 2.)**

12. 4000 ft·lb ≈ _____ Btu   5 Btu

13. 53,000 ft·lb ≈ _____ Btu   68 Btu

14. The amount of energy from 1 gal of gasoline is approximately 96,472,000 ft·lb. Find the amount of energy in Btu.   124,000 Btu

15. The amount of energy from 1 barrel (bbl) (42 gal) of crude oil is approximately 4,512,400,000 ft·lb. Find the amount of energy in Btu.   5,800,000 Btu

16. The amount of energy from 1 gal of propane is 90,000 Btu. Find the amount of energy in foot-pounds.   70,020,000 ft·lb

17. The amount of energy from 1 ft^3 of natural gas is 1026 Btu. Find the amount of energy in foot-pounds.   798,228 ft·lb

For Exercises 18–23, convert to hours. Write the answer in both fraction form and decimal form, rounded to the nearest hundredth.

18. 40 min = _____ hr
$\frac{2}{3}$ hr or 0.67 hr

19. 45 min = _____ hr
$\frac{3}{4}$ hr or 0.75 hr

20. 1 hr 15 min = _____ hr
$\frac{5}{4}$ hr or 1.25 hr

21. 2 hr 30 min = _____ hr
$\frac{5}{2}$ hr or 2.5 hr

22. 2 hr 24 min = _____ hr
$\frac{12}{5}$ hr or 2.4 hr

23. 1 hr 6 min = _____ hr
$\frac{11}{10}$ hr or 1.1 hr

For Exercises 24–29, determine the number of Calories burned. Use the information given in the table. Round to the nearest whole unit, if necessary. **(See Example 3.)**

24. Bicycling at 14–16 mph for 1 hr 20 min.   933 Cal

25. Running a 10-min mile pace for 45 min.   443 Cal

26. Walking for 2 hr 45 min.   770 Cal

27. Playing lacrosse for 2 hr 30 min.   1250 Cal

28. Mowing the lawn for 48 min.   304 Cal

29. Playing basketball for 1 hr 40 min.   933 Cal

Activity	Cal/hr
Basketball game	560
Bicycling, 14–16 mph	700
Mowing the lawn	380
Lacrosse	500
Running, 10-min mile pace	590
Walking, moderate pace	280

Concept 2: Units of Power

30. Find the power in foot-pounds per second of lifting 25 lb a distance of 5 ft in 5 sec. $25 \frac{\text{ft·lb}}{\text{sec}}$

31. Find the power in foot-pounds per second of lifting 40 lb a distance of 3 ft in 2 sec. **(See Example 4.)** $60 \frac{\text{ft·lb}}{\text{sec}}$

32. A machine can raise 200 lb a distance of 1 yd in 6 sec. Find the power in foot-pounds per second. $100 \frac{\text{ft·lb}}{\text{sec}}$

33. An engine can raise 300 lb a distance of 1.5 yd in 10 sec. Find the power in foot-pounds per second. $135 \frac{\text{ft·lb}}{\text{sec}}$

34. Find the power in foot-pounds per second of raising 1 ton a distance of 3 ft in 15 sec. $400 \frac{\text{ft·lb}}{\text{sec}}$

35. Find the power in foot-pounds per second of raising 1 ton a distance of 3 ft in 30 sec. $200 \frac{\text{ft·lb}}{\text{sec}}$

For Exercises 36–39, convert foot-pounds per second to horsepower.

36. $550 \frac{\text{ft·lb}}{\text{sec}} = $ _____ hp 1 hp

37. $1100 \frac{\text{ft·lb}}{\text{sec}} = $ _____ hp 2 hp

38. $4950 \frac{\text{ft·lb}}{\text{sec}} = $ _____ hp 9 hp

39. $6050 \frac{\text{ft·lb}}{\text{sec}} = $ _____ hp 11 hp

40. A new LS2 6.0-L small-block V-8 is the standard engine in the 2005 Corvette C6. It delivers peak output levels of 400 hp. Convert this to foot-pounds per second. $220,000 \frac{\text{ft·lb}}{\text{sec}}$

41. The 2003 Porsche 911 is rated at 315 hp. Convert this to foot-pounds per second. **(See Example 5.)** $173,250 \frac{\text{ft·lb}}{\text{sec}}$

42. A six-speed car will produce up to 550 hp. Convert this to foot-pounds per second. $302,500 \frac{\text{ft·lb}}{\text{sec}}$

43. A Dodge Ram pickup has a 215-hp engine. Convert this to foot-pounds per second. $118,250 \frac{\text{ft·lb}}{\text{sec}}$

Expanding Your Skills

Electric energy is often measured by the watt-hour (Wh). For example, a 60-W lightbulb will emit 60 Wh of energy in 1 hr. For Exercises 44 and 45 use the fact that 1 kilowatt-hour (kWh) = 1000 Wh.

44. A television is rated at 250 W. Suppose the television is on for 3 hr/day.

  a. How many watt-hours are used over a 30-day period? 22,500 Wh

  b. How many kilowatt-hours are used over a 30-day period? 22.5 kWh

  c. If the power company charges $0.11 per kilowatt-hour, what is the cost to run the television for a 30-day period? $2.48

45. A hot water heater is rated at 3500 W. Suppose the water heater comes on for 6 hr/day.

  a. How many watt-hours are used over a 30-day period? 630,000 Wh

  b. How many kilowatt-hours are used over a 30-day period? 630 kWh

  c. If the power company charges $0.082 per kilowatt-hour, what is the cost to run the hot water heater for a 30-day period? $51.66

 Writing   Translating Expression   Geometry   Scientific Calculator   Video   NS E

Section A.2   Scientific Notation

Concept Connections

Fill in the blank.

1. $1000 = 10^{\square}$
2. $\dfrac{1}{1000} = 10^{\square}$
3. $0.01 = 10^{\square}$
4. $100 = 10^{\square}$

1. Scientific Notation

In Section 1.7, we studied powers of 10. Recall:

$10^6 = 10 \cdot 10 \cdot 10 \cdot 10 \cdot 10 \cdot 10 = 1,000,000$

$10^5 = 10 \cdot 10 \cdot 10 \cdot 10 \cdot 10 = 100,000$

$10^4 = 10 \cdot 10 \cdot 10 \cdot 10 = 10,000$

$10^3 = 10 \cdot 10 \cdot 10 = 1000$

$10^2 = 10 \cdot 10 = 100$

$10^1 = 10$

$10^0 = 1$

Notice that as the exponents are decreased by 1, the numbers on the right are one-tenth of the number from the row above.

If we continue this pattern, then 10^0 would equal 1.

$10^{-1} = 0.1$

$10^{-2} = 0.01$

$10^{-3} = 0.001$

$10^{-4} = 0.0001$

$10^{-5} = 0.00001$

$10^{-6} = 0.000001$

If this pattern is continued for negative exponents, we get the fractions $\frac{1}{10}, \frac{1}{100}, \frac{1}{1000}$, and so on. In decimal form, we have $0.1, 0.01, 0.001, \ldots$.

In many applications in mathematics, business, and science, it is necessary to work with very large or very small numbers. For example,

- America's teens spend $153,000,000,000 yearly in online shopping.
- A piece of paper is 0.002 in. thick.

The value $153,000,000,000 can be expressed as $1.53 \times 100,000,000,000$

$$= 1.53 \times 10^{11}$$

The value 0.002 can be expressed as 2×0.001

$$= 2 \times 10^{-3}$$

The numbers 1.53×10^{11} and 2×10^{-3} are written in scientific notation. Using **scientific notation**, we express a number as the product of two factors. One factor is a number greater than or equal to 1 but less than 10. The other factor is a power of 10.

> ### Scientific Notation
> A positive number written in scientific notation is written as $a \times 10^n$, where a is a number greater than or equal to 1, but less than 10, and n is an integer.

2. Converting to Scientific Notation

To write a number in scientific notation, follow these guidelines.

Move the decimal point so that its new location is to the right of the first nonzero digit. Count the number of places that the decimal point is moved. Then

Answers

1. 3   **2.** −3   **3.** −2   **4.** 2

1. If the original number is *greater than or equal to 10:*

   The exponent for the power of 10 is *positive* and is equal to the number of places that the decimal point was moved.

$$7{,}500{,}000 = 7.5 \times 10^{6} \qquad 30{,}000 = 3 \times 10^{4}$$

Move the
decimal point 6
places to the *left.*

Move the
decimal point 4
places to the *left.*

2. If the original number is *between 0 and 1:*

   The exponent for the power of 10 is *negative.* Its absolute value is equal to the number of places the decimal point was moved.

$$0.000089 = 8.9 \times 10^{-5} \qquad 0.00000004 = 4 \times 10^{-8}$$

Move the decimal
point 5 places to
the *right.*

Move the decimal
point 8 places to
the *right.*

3. If the original number is *between 1 and 10*

   The exponent on 10 is 0.

$$2.5 = 2.5 \times 10^{0} \qquad \textit{Note: } 10^{0} = 1.$$

   For this case, scientific notation is not needed.

Example 1 **Writing Numbers in Scientific Notation**

Write the number in scientific notation.

a. 93,000,000 mi   (the distance between the Earth and the Sun)

b. 0.000 000 000 753 kg   (the mass of a dust particle)

c. 300,000,000 m/sec   (the speed of light)

d. 0.00017 m   (length of the smallest insect in the world)

Solution:

a. 93,000,000 mi   The number is greater than 10. Move the decimal
                  point left 7 places.

   $= 9.3 \times 10^{7}$ mi   For a number greater than 10, the exponent is
                  positive.

b. 0.000 000 000 753 kg   The number is between 0 and 1. Move the decimal
                  point to the right 10 places.

   $= 7.53 \times 10^{-10}$ kg   For a number between 0 and 1, the exponent is
                  negative.

c. 300,000,000 m/sec   The number is greater than 10. Move the decimal
                  point to the left 8 places.

   $= 3 \times 10^{8}$ m/sec   For a number greater than 10, the exponent is
                  positive.

d. 0.00017 m   The number is between 0 and 1. Move the decimal
                  point to the right 4 places.

   $= 1.7 \times 10^{-4}$ m   For a number between 0 and 1, the exponent is
                  negative.

3. Converting Scientific Notation to Standard Form

To convert from scientific notation to standard form, follow these guidelines.

1. If the exponent on 10 is *positive,* move the decimal point to the right the same number of places as the exponent. Add zeros as necessary.

2. If the exponent on 10 is *negative,* move the decimal point to the left the same number of places as the exponent. Add zeros as necessary.

Skill Practice

Convert to standard form.

9. 2.79×10^{-8}
10. 8.603×10^{5}
11. 1×10^{-6}
12. 6×10^{1}

Classroom Examples: p. A-9, Exercises 48 and 52

Answers

9. 0.0000000279
10. 860,300
11. 0.000001
12. 60

Example 2 Converting Scientific Notation to Standard Form

Convert to decimal notation.

a. 3.52×10^{-5} b. 4.6×10^{4} c. 9×10^{-12} d. 1×10^{15}

Solution:

a. $3.52 \times 10^{-5} = 0.0000352$ The exponent is negative. Move the decimal point to the left 5 places.

b. $4.6 \times 10^{4} = 46{,}000$ The exponent is positive. Move the decimal point to the right 4 places.

c. $9 \times 10^{-12} = 0.000\,000\,000\,009$ The exponent is negative. Move the decimal point to the left 12 places.

d. $1 \times 10^{15} = 1{,}000{,}000{,}000{,}000{,}000$ The exponent is positive. Move the decimal point to the right 15 places.

Section A.2 Practice Exercises

Vocabulary and Key Concepts

For additional exercises, see Classroom Activity A.2A in the *Student's Resource Manual* at www.mhhe.com/moh.

1. A positive number is written in ___scientific___ ___notation___ if it is written as a product of two factors. The first factor is a number greater than or equal to 1 but less than 10, and the second factor is a power of 10.

Concept 1: Scientific Notation

For Exercises 2–9, write each power of 10 in exponential notation.

2. 10,000 10^{4} 3. 100,000 10^{5} 4. 1000 10^{3} 5. 1,000,000 10^{6}

6. 0.001 10^{-3} 7. 0.01 10^{-2} 8. 0.0001 10^{-4} 9. 0.1 10^{-1}

For Exercises 10–17, identify which of the expressions are in correct scientific notation. (Answer yes or no.)

10. 43×10^{3} No   11. 82×10^{-4} No   12. 6.1×10^{-1} Yes   13. 2.34×10^{4} Yes

14. 2×10^{10} Yes   15. 8×10^{-5} Yes   16. 0.02×10^{4} No   17. 0.052×10^{-3} No

 Writing   Translating Expression   Geometry   Scientific Calculator   Video   NSE

Concept 2: Converting to Scientific Notation

For Exercises 18–21, convert the number to scientific notation. **(See Example 1.)**

18. The Andromeda Galaxy contains more than 230,000,000,000 stars. 2.3×10^{11} stars

19. The distance from Earth to Alpha Centauri is approximately 25,000,000,000,000 mi. 2.5×10^{13} mi

20. The diameter of an atom is 0.000 000 2 mm. 2×10^{-7} mm

21. The length of a flea is 0.0625 in. 6.25×10^{-2} in.

For Exercises 22–37, convert to scientific notation.

22. 20,000,000 2×10^{7}

23. 5000 5×10^{3}

24. 8,100,000 8.1×10^{6}

25. 62,000 6.2×10^{4}

26. 0.003 3×10^{-3}

27. 0.0009 9×10^{-4}

28. 0.025 2.5×10^{-2}

29. 0.58 5.8×10^{-1}

30. 142,000 1.42×10^{5}

31. 25,500,000 2.55×10^{7}

32. 0.0000491 4.91×10^{-5}

33. 0.000116 1.16×10^{-4}

34. 0.082 8.2×10^{-2}

35. 0.15 1.5×10^{-1}

36. 4920 4.92×10^{3}

37. 13,400 1.34×10^{4}

Concept 3: Converting Scientific Notation to Standard Form

For Exercises 38–53, convert to standard form. **(See Example 2.)**

38. 6×10^{3}
6000

39. 3×10^{4}
30,000

40. 8×10^{-2}
0.08

41. 2×10^{-5}
0.00002

42. 4.4×10^{-1}
0.44

43. 2.1×10^{-3}
0.0021

44. 3.7×10^{4}
37,000

45. 5.5×10^{3}
5500

46. 3.26×10^{2}
326

47. 6.13×10^{7}
61,300,000

48. 1.29×10^{-2}
0.0129

49. 4.04×10^{-4}
0.000404

50. 2.003×10^{-6}
0.000002003

51. 5.02×10^{-5}
0.0000502

52. 9.001×10^{8}
900,100,000

53. 7.07×10^{6}
7,070,000

Expanding Your Skills

For Exercises 54–61, write the mass of the planet in scientific notation if it is not already in scientific notation.

	Planet	Mass (kg)	Scientific Notation
54.	Mercury	0.33×10^{24}	3.3×10^{23}
55.	Venus	4.87×10^{24}	Already in scientific notation
56.	Earth	5.98×10^{24}	Already in scientific notation
57.	Mars	0.64×10^{24}	6.4×10^{23}
58.	Jupiter	1899×10^{24}	1.898×10^{27}
59.	Saturn	586.5×10^{24}	5.865×10^{26}
60.	Uranus	86.8×10^{24}	8.68×10^{25}
61.	Neptune	102.4×10^{24}	1.024×10^{26}

Section A.3   Rectangular Coordinate System

Concepts

1. Rectangular Coordinate System
2. Plotting Points in a Rectangular Coordinate System
3. Graphing Ordered Pairs in an Application

1. Rectangular Coordinate System

In Section 9.1 we created line graphs that illustrated how one variable relates to another. For example, Table A-1 represents the daily revenue for a popular movie for its opening two weeks. In table form, the information is difficult to picture and interpret. However, Figure A-2 shows a graph of these data. From the graph we can speculate that days 1–3 and 8–10 were weekends because revenue was up. We can also see that as time went on, revenue dropped.

Table A-1

Day number	Revenue ($ Millions)
1	59.8
2	51.4
3	39.9
4	10.3
5	8
6	6.8
7	5.9
8	17.1
9	24.9
10	15.8
11	3.6
12	3.5
13	3.1
14	3

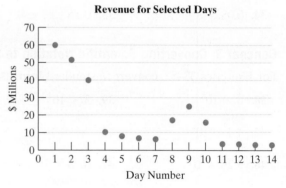

Figure A-2

Figure A-2 represents the variables of time and revenue. In general, to picture two variables, we use a graph with two number lines drawn at right angles to each other (Figure A-3). This forms a **rectangular coordinate system**. The horizontal line is called the **x-axis**, and the vertical line is called the **y-axis**. The point where the lines intersect is called the **origin**. On the x-axis, the numbers to the right of the origin are positive, and the numbers to the left are negative. On the y-axis, the numbers above the origin are positive, and the numbers below the origin are negative. The x- and y-axes divide the graphing area into four regions called **quadrants**.

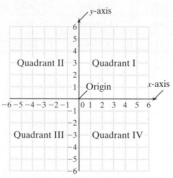

Figure A-3

2. Plotting Points in a Rectangular Coordinate System

Points graphed in a rectangular coordinate system are defined by two numbers as an **ordered pair** (x, y). The first number (called the **x-coordinate** or first coordinate) is the horizontal position from the origin. The second number (called the **y-coordinate** or second coordinate) is the vertical position from the origin. Example 1 shows how points are plotted in a rectangular coordinate system.

Example 1   **Plotting Points in a Rectangular Coordinate System**

Plot the points.

a. $(3, 4)$ **b.** $(-4, 2)$ **c.** $(2, -4)$ **d.** $(-5, -3)$

Solution:

a. The ordered pair $(3, 4)$ indicates that $x = 3$ and $y = 4$.

$$\underset{x}{\downarrow} \quad \underset{y}{\downarrow}$$

Because the x-coordinate is positive, start at the origin and move 3 units in the positive x direction (to the right). Then, because the y-coordinate is positive, move 4 units in the positive y direction (upward). Draw a dot at the final location. See Figure A-4.

The point $(3, 4)$ is located in Quadrant I.

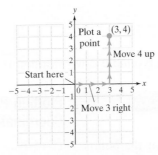

Figure A-4

b. The ordered pair $(-4, 2)$ indicates that $x = -4$ and $y = 2$.

$$\underset{x}{\downarrow} \quad \underset{y}{\downarrow}$$

Because the x-coordinate is *negative*, we start at the origin and move 4 units in the *negative x* direction (left). Then, because the y-coordinate is positive, move 2 units in the positive y direction (upward). Draw a dot at the final location. See Figure A-5.

The point $(-4, 2)$ is located in Quadrant II.

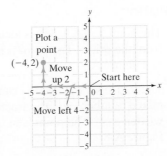

Figure A-5

c. The ordered pair $(2, -4)$ indicates that $x = 2$ and $y = -4$.

$$\underset{x}{\downarrow} \quad \underset{y}{\downarrow}$$

Because the x-coordinate is positive, start at the origin and move 2 units in the positive x direction (to the right). Then, because the y-coordinate is *negative*, move 4 units in the *negative y* direction (downward). Draw a dot at the final location. See Figure A-6.

The point $(2, -4)$ is located in Quadrant IV.

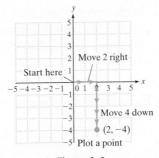

Figure A-6

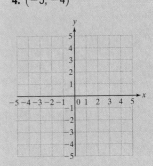

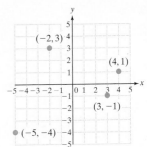

Classroom Examples: pp. A-14–A-15,
Exercises 22 and 28

Concept Connections

5. Answer true or false: The ordered pair $(6, -1)$ is the same as $(-1, 6)$.

TIP: Notice that changing the order of the x- and y-coordinates changes the location of the point. In Example 1(b), the point $(-4, 2)$ is in Quadrant II. In Example 1(c), the point $(2, -4)$ is in Quadrant IV (Figure A-6). This is why points are represented by *ordered* pairs. The order is important.

---

d. The ordered pair $(-5, -3)$ indicates that $x = -5$ and $y = -3$.

$$\underset{x}{\downarrow} \quad \underset{y}{\downarrow}$$

Because the x-coordinate is *negative*, start at the origin and move 5 units in the *negative* x direction (to the left). Then, because the y-coordinate is *negative*, move 3 units in the *negative y* direction (downward). Draw a dot at the final location. See Figure A-7.

The point $(-5, -3)$ is located in Quadrant III.

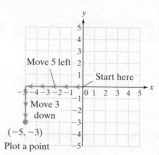

Figure A-7

In Example 2, we plot points whose coordinates are fractions, decimals, or zero.

Skill Practice

Plot the points.

6. $(2.1, -3.5)$
7. $\left(-\frac{9}{4}, \frac{9}{2}\right)$
8. $(5, 0)$
9. $(0, 2.5)$

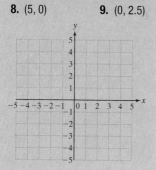

Instructor Note: Point out to students that *every* point on the x-axis can be written as $(x, 0)$ and *every* point on the y-axis can be written as $(0, y)$.

Example 2 **Plotting Points in a Rectangular Coordinate System**

Plot the points.

a. $(-4.6, -3.8)$ **b.** $\left(-\dfrac{5}{2}, \dfrac{13}{3}\right)$ **c.** $(3, 0)$ **d.** $(0, -4)$

Solution:

a. The point $(-4.6, -3.8)$ is located 4.6 units to the left and 3.8 units down from the origin. The point is in Quadrant III. See Figure A-8.

b. The improper fraction $-\frac{5}{2}$ can be written as the mixed number $-2\frac{1}{2}$. The fraction $\frac{13}{3}$ can be written as $4\frac{1}{3}$. The point is in Quadrant II. See Figure A-8.

c. In the ordered pair $(3, 0)$, the y-coordinate is zero. Therefore, we move neither upward nor downward from the origin. This indicates that the point is on the x-axis. See Figure A-8.

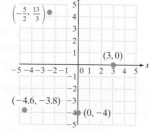

Figure A-8

d. In the ordered pair $(0, -4)$, the x-coordinate is zero. Therefore, we move to neither the left nor the right of the origin. This indicates that the point is on the y-axis. See Figure A-8.

Answers

5. False

6–9.

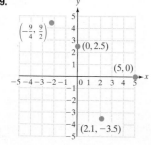

In Example 3, we practice reading ordered pairs from a graph.

| Example 3 | Reading Ordered Pairs from a Graph |

Estimate the coordinates of points A, B, C, D, E, and F from Figure A-9.

Solution:

    Point A is at $(-5, -1)$.

    Point B is at $(-4, 4)$.

    Point C is at $(2\frac{1}{2}, 0)$.

    Point D is at $(5, 2)$.

    Point E is at $(3\frac{1}{2}, -1\frac{1}{2})$.

    Point F is at $(0, 0)$.

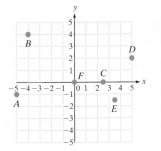

Figure A-9

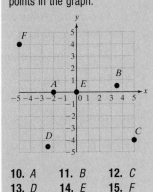
3. Graphing Ordered Pairs in an Application

Graphing ordered pairs in an application is essentially the same process as creating a graph.

| Example 4 | Graphing Ordered Pairs in an Application |

The data given in Table A-2 represent the gas mileage for a subcompact car for various speeds. Let x represent the speed of the car, and let y represent the corresponding gas mileage.

a. Write the table values as ordered pairs.

b. Graph the ordered pairs.

Table A-2

Speed x (mph)	Gas Mileage y (mpg)
25	27
30	29
35	33
40	34
45	35
50	33
55	30
60	27
65	24
70	21

Solution:

a. The ordered pairs are given by $(25, 27)$, $(30, 29)$, $(35, 33)$, $(40, 34)$, $(45, 35)$, $(50, 33)$, $(55, 30)$, $(60, 27)$, $(65, 24)$, and $(70, 21)$.

b.

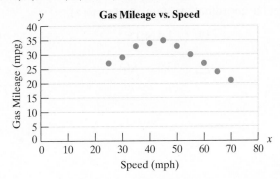

From the graph we see that the most efficient speed is approximately 45 mph. That is where the gas mileage is at its peak.

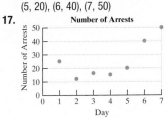

Section A.3   Practice Exercises

Vocabulary and Key Concepts

For additional exercises, see Classroom Activities A.3A–A.3B in the *Student's Resource Manual* at www.mhhe.com/moh.

1. a. In a rectangular coordinate system, two number lines are drawn at right angles to each other. The horizontal line is called the ___x___-axis, and the vertical line is called the ___y-axis___.

b. A point in a rectangular coordinate system is defined by a(n) ___ordered___ pair, (x, y).

c. In a rectangular coordinate system, the point where the x- and y-axes intersect is called the ___origin___ and is represented by the ordered pair ___$(0, 0)$___.

d. The x- and y-axes divide the coordinate plane into four regions called ___quadrants___.

e. A point with a positive x-coordinate and a ___negative___ y-coordinate is in Quadrant IV.

f. In Quadrant ___III___, both the x- and y-coordinates are negative.

Concept 1: Rectangular Coordinate System

Use the words in Exercises 2–8 to label the rectangular coordinate system.

2. x-axis

3. y-axis

4. Origin

5. Quadrant I

6. Quadrant II

7. Quadrant III

8. Quadrant IV

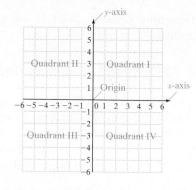

Concept 2: Plotting Points in a Rectangular Coordinate System

For Exercises 9–14, plot the points on the rectangular coordinate system. **(See Example 1.)**

9. $(-1, 4)$

10. $(4, -1)$

11. $(2, 2)$

12. $(3, -3)$

13. $(-5, -2)$

14. $(-3, 2)$

15. Explain how to plot the point $(-1.8, 3.1)$.   First move to the left 1.8 units from the origin. Then go up 3.1 units. Place a dot at the final location. The point is in Quadrant II.

16. Explain how to plot the point $\left(\dfrac{15}{2}, \dfrac{15}{7}\right)$.   Write the fractions as mixed numbers $\left(7\dfrac{1}{2}, 2\dfrac{1}{7}\right)$.

Move $7\dfrac{1}{2}$ units to the right of the origin and $2\dfrac{1}{7}$ units up. Place a dot at the final location. The point is in Quadrant I.

For Exercises 17–22, plot the points on the rectangular coordinate system. **(See Example 2.)**

17. $(-0.6, 1.1)$

18. $(2.3, 4.9)$

19. $(-1.4, 4.1)$

20. $(5.1, -3.8)$

21. $(0.9, -1.1)$

22. $(-3.3, -4.6)$

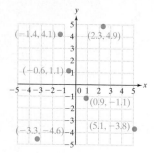

For Exercises 23–28, plot the points on the rectangular coordinate system. **(See Example 2.)**

23. $\left(\dfrac{1}{2}, \dfrac{5}{2}\right)$

24. $\left(-\dfrac{10}{3}, \dfrac{8}{3}\right)$

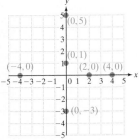 **25.** $\left(-\dfrac{13}{6}, -1\right)$

26. $\left(\dfrac{24}{5}, -\dfrac{8}{5}\right)$

27. $\left(\dfrac{13}{4}, \dfrac{29}{8}\right)$

28. $\left(\dfrac{15}{7}, -\dfrac{17}{6}\right)$

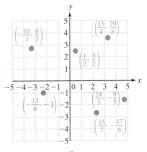

For Exercises 29–34, plot the points on the rectangular coordinate system. **(See Example 2.)**

29. $(0, -3)$

30. $(2, 0)$

31. $(4, 0)$

32. $(0, 5)$

33. $(0, 1)$

34. $(-4, 0)$

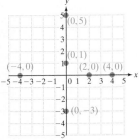

For Exercises 35–46, for each point, identify in which quadrant or on which axis it lies.

35. $(32, -44)$ Quadrant IV

36. $(-12, 25)$ Quadrant II

37. $(-10, -5)$ Quadrant III

38. $(100, 82)$ Quadrant I

39. $(54.9, 0)$ *x*-axis

40. $(0, -23.33)$ *y*-axis

41. $\left(0, \dfrac{55}{17}\right)$ *y*-axis

42. $\left(-\dfrac{3}{4}, 0\right)$ *x*-axis

43. $(-27, 3)$ Quadrant II

44. $(5, -75)$ Quadrant IV

45. $(35, 66)$ Quadrant I

46. $\left(-\dfrac{2}{17}, -\dfrac{1}{50}\right)$ Quadrant III

For Exercises 47–55, estimate the coordinate of points *A, B, C, D, E, F, G, H* and *I*. **(See Example 3.)**

47. *A* $(0, 3)$

48. *B* $(-1, 0)$

49. *C* $(2, 3)$

50. *D* $(-4, 2)$

51. *E* $(-5, -2)$

52. *F* $(1, -1)$

53. *G* $(4, -2)$

54. *H* $(5, \frac{1}{2})$

55. *I* $(-2, -5)$

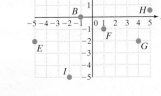

Concept 3: Graphing Ordered Pairs in an Application

For Exercises 56–61, write the table values as ordered pairs. Then graph the ordered pairs.

56. The table gives the average monthly temperature in Anchorage, Alaska, for 1 yr. Let January represent $x = 1$, February represent $x = 2$, etc.

Month x	Temperature y (°C)
1	−7
2	−3
3	1
4	6
5	12
6	17
7	18
8	18
9	14
10	6
11	−1
12	−7

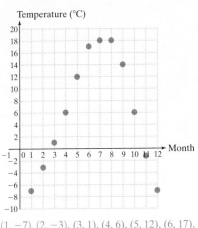

$(1, -7), (2, -3), (3, 1), (4, 6), (5, 12), (6, 17),$
$(7, 18), (8, 18), (9, 14), (10, 6),$
$(11, -1), (12, -7)$

57. The table gives the average monthly temperature in Moscow for 1 yr. Let January represent $x = 1$, February represent $x = 2$, etc. **(See Example 4.)**

Month x	Temperature y (°C)
1	−6
2	−5
3	1
4	11
5	18
6	22
7	24
8	22
9	15
10	8
11	1
12	−4

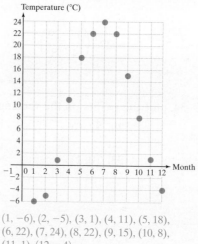

$(1, -6), (2, -5), (3, 1), (4, 11), (5, 18),$
$(6, 22), (7, 24), (8, 22), (9, 15), (10, 8),$
$(11, 1), (12, -4)$

58. This table gives the value of a $20,000 car as it depreciates over 6 years.

Year after Purchase, x	Value of the Car, y ($)
1	15,000
2	13,200
3	11,616
4	9,900
5	8,491
6	7,218

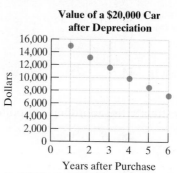

Value of a $20,000 Car after Depreciation

$(1, 15000), (2, 13200), (3, 11616), (4, 9900), (5, 8491), (6, 7218)$

59. The table gives the value each year after purchase of a used car bought for $15,000.

Year after Purchase x	Value of Car y ($)
1	13,200
2	11,352
3	9,649
4	8,201
5	6,971
6	5,925

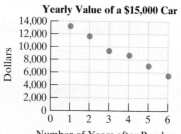

Yearly Value of a $15,000 Car

$(1, 13200), (2, 11352), (3, 9649), (4, 8201), (5, 6971), (6, 5925)$

 Writing   Translating Expression   Geometry   Scientific Calculator   Video   NS E

60. The table gives the gas mileage for vehicles of different weights.

Weight (lb)	Miles per Gallon
3504	18
2833	22
4060	17
2050	38
2205	36
2625	28

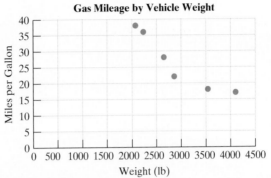

(3504, 18), (2833, 22), (4060, 17), (2050, 38), (2205, 36), (2625, 28)

61. This table gives the gas mileage for a VW Beetle at different speeds.

Speed x (mph)	Gas Mileage y (mpg)
30	33
40	35
50	40
60	37
70	30

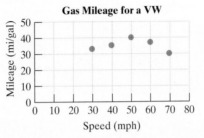

(30, 33), (40, 35), (50, 40), (60, 37), (70, 30)

Student Answer Appendix

Chapter 1

Chapter Opener Puzzle

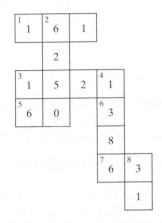

Section 1.1 Practice Exercises, pp. 6–8

1. a. periods   **b.** hundreds   **c.** thousands
3. 7: ones; 5: tens; 4: hundreds; 3: thousands;
1: ten-thousands; 2: hundred-thousands; 8: millions
5. Tens   **7.** Ones   **9.** Hundreds   **11.** Thousands
13. Hundred-thousands   **15.** Billions
17. Ten-thousands   **19.** Millions   **21.** Ten-millions
23. Billions   **25.** 5 tens + 8 ones; $5 \times 10 + 8 \times 1$
27. 5 hundreds + 3 tens + 9 ones;
$5 \times 100 + 3 \times 10 + 9 \times 1$
29. 5 hundreds + 3 ones; $5 \times 100 + 3 \times 1$
31. 1 ten-thousand + 2 hundreds + 4 tens + 1 one;
$1 \times 10,000 + 2 \times 100 + 4 \times 10 + 1 \times 1$
33. 524   **35.** 150   **37.** 1,906   **39.** 85,007
41. Ones, thousands, millions, billions
43. Two hundred forty-one
45. Six hundred three
47. Thirty-one thousand, five hundred thirty
49. One hundred thousand, two hundred thirty-four
51. Nine thousand, five hundred thirty-five
53. Twenty thousand, three hundred twenty
55. One thousand, three hundred seventy-seven
57. 6,005   **59.** 672,000   **61.** 1,484,250
63.

```
    d              a       c                   b
 ───●──┼──┼──┼──┼──┼──●──┼──●──┼──┼──┼──┼──●──→
    0  1  2  3  4  5  6  7  8  9 10 11 12 13
```

65. 10   **67.** 4   **69.** 8 is greater than 2, or 2 is less than 8
71. 3 is less than 7, or 7 is greater than 3   **73.** <
75. >   **77.** <   **79.** >   **81.** <   **83.** <
85. False   **87.** 99   **89.** There is no greatest whole number.   **91.** 7   **93.** 964

Section 1.2 Practice Exercises, pp. 16–20

1. a. addends   **b.** sum   **c.** commutative   **d.** 4; 4
e. associative   **f.** polygon   **g.** perimeter
3. 3 hundreds + 5 tens + 1 one; $3 \times 100 + 5 \times 10 + 1 \times 1$

5. 1 hundred + 7 ones; $1 \times 100 + 7 \times 1$ **7.** 4012
9.

+	0	1	2	3	4	5	6	7	8	9
0	0	1	2	3	4	5	6	7	8	9
1	1	2	3	4	5	6	7	8	9	10
2	2	3	4	5	6	7	8	9	10	11
3	3	4	5	6	7	8	9	10	11	12
4	4	5	6	7	8	9	10	11	12	13
5	5	6	7	8	9	10	11	12	13	14
6	6	7	8	9	10	11	12	13	14	15
7	7	8	9	10	11	12	13	14	15	16
8	8	9	10	11	12	13	14	15	16	17
9	9	10	11	12	13	14	15	16	17	18

11. Addends: 2, 8; sum: 10
13. Addends: 11, 10; sum: 21
15. Addends: 5, 8, 2; sum: 15
17. 74   **19.** 58   **21.** 48   **23.** 19   **25.** 588
27. 798   **29.** 237   **31.** 198   **33.** 84   **35.** 115
37. 937   **39.** 850   **41.** 41   **43.** 29   **45.** 1003
47. 836   **49.** 24,004   **51.** 132,658   **53.** 21 + 30
55. 13 + 8   **57.** 23 + (9 + 10)   **59.** (41 + 3) + 22
61. The sum of any number and 0 is that number.
a. 423   **b.** 25   **c.** 67   **63.** 100 + 42; 142
65. 23 + 81; 104   **67.** 76 + 2; 78   **69.** 1320 + 448; 1768
71. For example: The sum of 54 and 24
73. For example: 88 added to 12
75. For example: The total of 4, 23, and 77
77. For example: 10 increased by 8   **79.** 276 people
81. 58,493,000 viewers   **83.** $45,500   **85.** $245
87. 13,538 participants   **89.** 821,024 nonteachers
91. 104 cm   **93.** 110 m   **95.** 42 yd   **97.** 288 ft

Section 1.2 Calculator Connections, p. 21

99. 9,536,940   **100.** 908,788   **101.** 8,163,940
102. 192,780   **103.** 21,491,394   **104.** 5,257,179
105. $148,500,000   **106.** 128,107,616 votes

Section 1.3 Practice Exercises, pp. 28–31

1. minuend; subtrahend; difference
3. 1151   **5.** 899   **7.** 0 < 10
9. Minuend: 12; subtrahend: 8; difference: 4
11. Minuend: 21; subtrahend: 12; difference: 9
13. Minuend: 9; subtrahend: 6; difference: 3
15. 18 + 9 = 27   **17.** 27 + 75 = 102   **19.** 5
21. 3   **23.** 6   **25.** 45   **27.** 61   **29.** 1126
31. 321   **33.** 10,004   **35.** 1103   **37.** 17   **39.** 49
41. 104   **43.** 521   **45.** 23   **47.** 4764   **49.** 1403
51. 2217   **53.** 378   **55.** 713   **57.** 30,941
59. 5,662,119   **61.** 78 − 23; 55   **63.** 78 − 6; 72
65. 422 − 100; 322   **67.** 1090 − 72; 1018
69. 50 − 13; 37   **71.** 103 − 35; 68

73. For example: 93 minus 27

75. For example: Subtract 85 from 165.

77. The expression $7 - 4$ means 7 minus 4, yielding a difference of 3. The expression $4 - 7$ means 4 minus 7 which results in a difference of -3. (This is a mathematical skill we have not yet learned.)   **79.** $33   **81.** 55 more hits

83. 8 plants   **85.** 2479 times   **87.** 13 m   **89.** 10 yd

91. 30 thousand marriages   **93.** 119 thousand marriages

Section 1.3 Calculator Connections, p. 31

95. 4,447,302   **96.** 897,058,513   **97.** 2,906,455

98. 49,408 mi^2 **99.** 17,139 mi^2 **100.** The difference in land area between Colorado and Rhode Island is 102,673 mi^2.

101. 13,093 mi^2

Section 1.4 Practice Exercises, pp. 36–38

1. rounding   **3.** 26   **5.** 5007   **7.** Ten-thousands

9. If the digit in the tens place is 0, 1, 2, 3, or 4, then change the tens and ones digits to 0. If the digit in the tens place is 5, 6, 7, 8, or 9, increase the digit in the hundreds place by 1 and change the tens and ones digits to 0.

11. 340   **13.** 730   **15.** 9400   **17.** 8500   **19.** 35,000

21. 3000   **23.** 10,000   **25.** 109,000   **27.** 490,000

29. $77,000,000   **31.** 239,000 mi   **33.** 160   **35.** 180

37. 500   **39.** 2100   **41.** $151,000,000

43. $11,000,000 more   **45.** $10,000,000

47. a. year 4; $3,500,000   **b.** year 6; $2,000,000

49. Massachusetts; 79,000 students   **51.** 71,000 students

53. Answers may vary.   **55.** 10,000 mm   **57.** 440 in.

Section 1.5 Practice Exercises, pp. 47–50

1. a. factors; product   **b.** commutative   **c.** associative

d. 0; 0   **e.** 7; 7   **f.** distributive   **g.** area   **h.** $l \times w$

3. 1,010,000   **5.** 5400   **7.** 6×5; 30

9. 3×9; 27   **11.** Factors: 13, 42; product: 546

13. Factors: 3, 5, 2; product: 30

15. For example: 5×12; $5 \cdot 12$; 5(12)

17. d   **19.** e   **21.** c   **23.** 8×14

25. $(6 \times 2) \times 10$ **27.** $(5 \times 7) + (5 \times 4)$

29. 144   **31.** 52   **33.** 655   **35.** 1376

37. 11,280   **39.** 23,184   **41.** 378,126   **43.** 448

45. 1632   **47.** 864   **49.** 2431   **51.** 6631

53. 19,177   **55.** 186,702   **57.** 21,241,448

59. 4,047,804   **61.** 24,000   **63.** 2,100,000

65. 72,000,000   **67.** 36,000,000   **69.** 60,000,000

71. 2,400,000,000   **73.** $1000   **75.** $2,720,000

77. 4000 minutes   **79.** $1665   **81.** 287,500 sheets

83. 372 mi   **85.** 276 ft^2 **87.** 5329 cm^2

89. 105,300 mi^2 **91. a.** 2400 in.^{2} **b.** 42 windows

c. 100,800 in.^{2} **93.** 128 ft^2

Section 1.6 Practice Exercises, pp. 58–61

1. a. dividend; divisor, quotient   **b.** 1   **c.** 5   **d.** 0

e. undefined   **f.** remainder

3. 4944   **5.** 1253   **7.** 664,210   **9.** 902

11. Dividend: 72; divisor: 8; quotient: 9

13. Dividend: 64; divisor: 8; quotient: 8

15. Dividend: 45; divisor: 9; quotient: 5

17. You cannot divide a number by zero (the quotient is undefined). If you divide zero by a number (other than zero), the quotient is always zero.

19. 15   **21.** 0   **23.** Undefined   **25.** 1

27. Undefined   **29.** 0   **31.** $2 \times 3 = 6$, $2 \times 6 \neq 3$

33. Multiply the quotient and the divisor to get the dividend.

35. 13   **37.** 41   **39.** 486   **41.** 409   **43.** 203

45. 822   **47.** Correct   **49.** Incorrect; 253 R2

51. Correct   **53.** Incorrect; 25 R3   **55.** 7 R5

57. 10 R2   **59.** 27 R1   **61.** 197 R2   **63.** 42 R4

65. 1557 R1   **67.** 751 R6   **69.** 835 R2   **71.** 479 R9

73. 43 R19   **75.** 308   **77.** 1259 R26   **79.** 229 R96

81. 302   **83.** $497 \div 71$; 7   **85.** $877 \div 14$; 62 R9

87. $42 \div 6$; 7   **89.** 14 classrooms

91. 5 cases; 8 cans left over   **93.** 52 mph   **95.** 22 lb

97. $1200 \div 20 = 60$; approximately 60 words per minute

99. Yes, they can all attend if they sit in the second balcony.

101. a. 12 loads   **b.** 2 oz

Section 1.6 Calculator Connections, p. 61

103. 7,665,000,000 bbl   **104.** 13,000 min

105. $888 billion   **106.** Each crate weighs 255 lb.

Chapter 1 Problem Recognition Exercises, p. 62

1. a. 65   **b.** 676   **c.** 39   **d.** 4   **2. a.** 6   **b.** 85   **c.** 1734

d. 119   **3. a.** 293,712   **b.** 5122   **c.** 87 R18   **d.** 5006

4. a. 1112   **b.** 10 R86   **c.** 1340   **d.** 139,764

5. a. 197   **b.** 156   **6. a.** 6225   **b.** 6004

7. 4180   **8.** 41,800   **9.** 418,000   **10.** 4,180,000

11. 35,000   **12.** 3500   **13.** 350   **14.** 35

15. 246,000   **16.** 2,820,000   **17.** 20,000   **18.** 540,000

Section 1.7 Practice Exercises, pp. 67–70

1. a. base; 4   **b.** powers   **c.** square root; 81

d. order; operations   **e.** variable; constants   **f.** mean

3. True   **5.** False   **7.** True   **9.** 9^4 **11.** 2^7

13. 3^6 **15.** $4^4 \cdot 2^3$ **17.** $8 \cdot 8 \cdot 8 \cdot 8$

19. $4 \cdot 4 \cdot 4 \cdot 4 \cdot 4 \cdot 4 \cdot 4 \cdot 4$ **21.** 8   **23.** 9   **25.** 27

27. 125   **29.** 32   **31.** 81   **33.** 1   **35.** 1

37. The number 1 raised to any power equals 1.   **39.** 1000

41. 100,000   **43.** 2   **45.** 6   **47.** 10   **49.** 0

51. No, addition and subtraction should be performed in the order in which they appear from left to right.   **53.** 26

55. 1   **57.** 49   **59.** 3   **61.** 2   **63.** 53   **65.** 8

67. 45   **69.** 24   **71.** 4   **73.** 40   **75.** 90   **77.** 25

79. 4   **81.** 81   **83.** 18   **85.** 0   **87.** 5   **89.** 6

91. 3   **93.** 201   **95.** 109   **97.** 18

99. The average mileage rating was 34 mpg.

101. 121 mm per month   **103.** 24   **105.** 70

Section 1.7 Calculator Connections, p. 70

107. 24,336   **108.** 174,724   **109.** 248,832

110. 1,500,625   **111.** 79,507   **112.** 357,911

113. 8028   **114.** 293,834   **115.** 66,049   **116.** 1728

117. 35   **118.** 43

Section 1.8 Practice Exercises, pp. 76–80

1. $4 \div 0$ **3.** $71 + 14$; 85   **5.** $2 \cdot 14$; 28

7. $102 - 32$; 70   **9.** $10 \cdot 13$; 130

11. $24 \div 6$; 4   **13.** $5 + 13 + 25$; 43

15. For example: sum, added to, increased by, more than, plus, total of   **17.** For example: difference, minus, decreased by, less, subtract   **19.** Denali is 6074 ft higher than White Mountain Peak.   **21.** 18,960,000 barrels per day

23. The whole screen has 12,096 pixels.

25. There will be 120 classes of Prealgebra.

27. There will be 9 gal used.

29. Jeannette will pay $42,236 for 1 year.

31. The Prius can go 1100 mi.

33. The maximum capacity is 3150 seats.

35. Jackson's monthly payment was $390.

37. Each trip will take 2 hr.   **39.** Perimeter

41. The cost will be $86.   **43.** The cost is $1020.

45. There will be $36 left in Gina's account.

47. The total bill was $154,032.

49. a. Latayne will receive $48.   **b.** She can buy 6 CDs.

51. Michael Jordan scored 33,454 points with the Bulls.

53. a. One bottle will last for 30 days.   **b.** The owner should order a refill no later than September 28.

55. a. The distance is 360 mi.   **b.** 14 in. represents 840 mi.

57. 104 boxes will be filled completely with 2 books left over.

59. a. Marc needs five $20 bills.   **b.** He will receive $16 in change.   **61.** He earned $520.

Chapter 1 Review Exercises, pp. 88–92

1. Ten-thousands   **2.** Hundred-thousands   **3.** 92,046

4. 503,160   **5.** 3 millions + 4 hundred-thousands + 8 hundreds + 2 tens; $3 \times 1,000,000 + 4 \times 100,000 + 8 \times 100 + 2 \times 10$ **6.** 3 ten-thousands + 5 hundreds + 5 tens + 4 ones; $3 \times 10,000 + 5 \times 100 + 5 \times 10 + 4 \times 1$

7. Two hundred forty-five

8. Thirty thousand, eight hundred sixty-one   **9.** 3602

10. 800,039

11.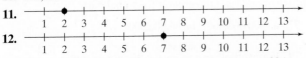

12.

13. True   **14.** False   **15.** Addends: 105, 119; sum: 224

16. Addends: 53, 21; sum: 74   **17.** 71   **18.** 54

19. 17,410   **20.** 70,642   **21. a.** Commutative property
b. Associative property   **c.** Commutative property

22. 403 + 79; 482   **23.** 44 + 92; 136   **24.** 36 + 7; 43

25. 23 + 6; 29   **26. a.** 96 cars   **b.** 66 Fords

27. 45,797 thousand seniors   **28.** 177 m

29. Minuend: 14; subtrahend: 8; difference: 6

30. Minuend: 102; subtrahend: 78; difference: 24

31. 26   **32.** 20   **33.** 121   **34.** 1090   **35.** 31,019

36. 34,188   **37.** 38−31; 7   **38.** 111 − 15; 96

39. 251−42; 209   **40.** 90 − 52; 38   **41.** 71,892,438 tons

42. $7,200,000   **43.** 2336 thousand visitors

44. 5,000,000   **45.** 9,330,000   **46.** 800,000

47. 1500   **48.** 13,000,000 people   **49.** 163,000 m^3

50. Factors: 32, 12; product: 384

51. Factors: 33, 40; product: 1320   **52. a.** Yes   **b.** Yes
c. No   **53.** c   **54.** e   **55.** d   **56.** a

57. b   **58.** 6106   **59.** 52,224   **60.** 3,000,000

61. $429   **62.** 7714 lb

63. 7; divisor: 6, dividend: 42, quotient: 7

64. 13; divisor: 4, dividend: 52, quotient: 13   **65.** 3

66. 1   **67.** Undefined   **68.** 0

69. Multiply the quotient and the divisor to get the dividend.

70. Multiply the whole number part of the quotient and the divisor, and then add the remainder to get the dividend.

71. 58   **72.** 41 R7   **73.** 52 R3   **74.** $\frac{72}{4}$; 18

75. $9\overline{)108}$; 12   **76.** 26 photos with 1 left over

77. a. 4 T-shirts   **b.** 5 hats   **78.** 8^5 **79.** 2$^4 \cdot$ 5^3

80. 125   **81.** 256   **82.** 1   **83.** 1,000,000   **84.** 8

85. 12   **86.** 7   **87.** 75   **88.** 90   **89.** 15   **90.** 28

91. 55   **92.** 8   **93.** $89   **94.** 8 houses per month

95. a. The Cincinnati Zoo has 13,000 more animals than the San Diego Zoo.   **b.** The San Diego Zoo has 50 more species than the Cincinnati Zoo.   **96. a.** 21 mi   **b.** 840 mi

97. He will receive $19,600,000 per year.

98. a. She should purchase 48 plants.   **b.** The plants will cost $144.   **c.** The fence will cost $80.   **d.** Aletha's total cost will be $224.

Chapter 1 Test, pp. 92–93

1. a. Hundreds   **b.** Thousands   **c.** Millions
d. Ten-thousands   **2. a.** 4,065,000   **b.** Twenty-one million, three hundred twenty-five thousand   **c.** Twelve million, two hundred eighty-seven thousand   **d.** 729,000   **e.** Eleven million, four hundred ten thousand

3. a. 14 > 6   **b.** 72 < 81   **4.** 129   **5.** 328

6. 113   **7.** 227   **8.** 2842   **9.** 447   **10.** 21 R9

11. 546   **12.** 8103   **13.** 20   **14.** 1,500,000,000

15. 336   **16.** 0   **17.** Undefined

18. a. The associative property of multiplication; the expression shows a change in grouping.
b. The commutative property of multiplication; the expression shows a change in the order of the factors.

19. a. 4900   **b.** 12,000   **c.** 8,000,000

20. There were approximately 1,430,000 people.

21. 4   **22.** 24   **23.** 48   **24.** 33

25. Jennifer has a higher average of 29. Brittany has an average of 28.

26. a. 442 thousand users   **b.** The largest increase was between year 3 and year 4. The increase was 15,430 thousand.

27. The North Side Fire Department is the busiest with an average of five calls per week.

28. 156 mm   **29.** Perimeter: 350 ft; area: 6016 ft^2

30. 4,560,000 m^2

Chapter 2

Chapter Opener Puzzle

			A 1	2	4
3	5	6	1	2	4
B 1	2	3	C 4	D 6	E 5
6	4	2	5	3	1
2	1	F 4	6	5	3
G 5	3	1	H 2	4	I 6
4	6	5	3	J 1	2

Section 2.1 Practice Exercises, pp. 102–106

1. a. fractions **b.** numerator; denominator **c.** proper
d. improper **e.** mixed
3. Numerator: 2; denominator: 3
5. Numerator: 12; denominator: 11
7. $6 \div 1; 6$ **9.** $2 \div 2; 1$ **11.** $0 \div 3; 0$
13. $2 \div 0$; undefined   **15.** $\frac{3}{4}$ **17.** $\frac{5}{9}$ **19.** $\frac{1}{6}$ **21.** $\frac{3}{8}$
23. $\frac{3}{4}$ **25.** $\frac{1}{8}$ **27.** $\frac{41}{103}$ **29.** $\frac{10}{21}$ **31.** Proper
33. Improper   **35.** Improper   **37.** Proper
39. $\frac{5}{2}$ **41.** $\frac{12}{4}$ **43.** $\frac{9}{8}$ **45.** $\frac{7}{4}; 1\frac{3}{4}$ **47.** $\frac{13}{8}; 1\frac{5}{8}$
49. $\frac{7}{4}$ **51.** $\frac{38}{9}$ **53.** $\frac{24}{7}$ **55.** $\frac{29}{4}$ **57.** $\frac{137}{12}$
59. $\frac{171}{8}$ **61.** 19   **63.** 7   **65.** $4\frac{5}{8}$ **67.** $7\frac{4}{5}$
69. $2\frac{7}{10}$ **71.** $5\frac{7}{9}$ **73.** $12\frac{1}{11}$ **75.** $3\frac{5}{6}$ **77.** $44\frac{1}{7}$
79. $1056\frac{1}{5}$ **81.** $810\frac{3}{11}$ **83.** $12\frac{7}{15}$

85.
$\frac{3}{4}$
⊢—⊢—●—⊢—⊢
0   1

87.
$\frac{1}{3}$
⊢—●—⊢—⊢—⊢
0   1

89.
$\frac{2}{3}$
⊢—⊢—●—⊢—⊢
0   1

91.
$1\frac{1}{6}$
⊢—⊢—⊢—●—⊢—⊢—⊢—⊢—⊢—⊢—⊢—⊢
0   1   2

93.
$1\frac{2}{3}$
⊢—⊢—⊢—⊢—⊢—⊢—●—⊢—⊢
0   1   2

95. False   **97.** True

Section 2.2 Practice Exercises, pp. 111–113

1. a. factor **b.** prime **c.** composite **d.** prime
3. $\frac{8}{12}; \frac{4}{12}$ **5.** $\frac{5}{4}; \frac{3}{4}$ **7.** $\frac{7}{12}$; proper   **9.** $4\frac{3}{5}$
11. For example: $2 \cdot 4$ and $1 \cdot 8$
13. For example: $4 \cdot 6$ and $2 \cdot 2 \cdot 2 \cdot 3$
15.

Product	36	42	30	15	81
Factor	12	7	30	15	27
Factor	3	6	1	1	3
Sum	15	13	31	16	30

17. A whole number is divisible by 2 if it is an even number.
19. A whole number is divisible by 3 if the sum of its digits
is divisible by 3.   **21. a.** No **b.** Yes **c.** Yes **d.** No
23. a. No **b.** No **c.** No **d.** No
25. a. Yes **b.** Yes **c.** No **d.** No
27. a. Yes **b.** No **c.** Yes **d.** Yes
29. Yes   **31.** Prime   **33.** Composite   **35.** Composite
37. Prime   **39.** Neither   **41.** Composite
43. Prime   **45.** Composite
47. There are two whole numbers that are neither prime
nor composite, 0 and 1.   **49.** False

51. 2, 3, 5, 7, 11, 13, 17, 19, 23, 29, 31, 37, 41, 43, 47
53. No, 9 is not a prime number.   **55.** Yes
57. $2 \cdot 5 \cdot 7$ **59.** $2 \cdot 2 \cdot 5 \cdot 13$ or $2^2 \cdot 5 \cdot 13$
61. $3 \cdot 7 \cdot 7$ or $3 \cdot 7^2$ **63.** $2 \cdot 3 \cdot 23$
65. $2 \cdot 2 \cdot 2 \cdot 7 \cdot 11$ or $2^3 \cdot 7 \cdot 11$ **67.** Prime
69. 1, 2, 3, 4, 6, 12   **71.** 1, 2, 4, 8, 16, 32
73. 1, 3, 9, 27, 81   **75.** 1, 2, 3, 4, 6, 8, 12, 16, 24, 48
77. No   **79.** Yes   **81.** Yes   **83.** No   **85.** Yes
87. No   **89.** Yes   **91.** No

Section 2.3 Practice Exercises, pp. 118–121

1. lowest   **3.** $5 \cdot 29$ **5.** $2 \cdot 2 \cdot 23$ or $2^2 \cdot 23$
7. $5 \cdot 17$ **9.** $3 \cdot 5 \cdot 13$
11.
13.
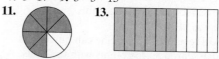

15. False   **17.** $\neq$ **19.** $=$ **21.** $=$
23. $\neq$ **25.** $\frac{1}{2}$ **27.** $\frac{1}{3}$ **29.** $\frac{9}{5}$ **31.** $\frac{5}{4}$ **33.** $\frac{4}{5}$
35. 1   **37.** 2   **39.** 1   **41.** $\frac{3}{4}$ **43.** 3   **45.** $\frac{7}{10}$
47. $\frac{13}{15}$ **49.** $\frac{77}{39}$ **51.** $\frac{2}{5}$ **53.** $\frac{2}{7}$ **55.** 0
57. Undefined   **59.** $\frac{3}{5}$ **61.** $\frac{3}{4}$ **63.** $\frac{5}{3}$ **65.** $\frac{21}{11}$
67. $\frac{17}{100}$ **69.** Heads: $\frac{5}{12}$; tails: $\frac{7}{12}$ **71. a.** $\frac{3}{13}$ **b.** $\frac{10}{13}$
73. a. Jonathan: $\frac{5}{7}$; Jared: $\frac{6}{7}$ **b.** Jared sold the greater
fractional part.   **75. a.** Raymond: $\frac{10}{11}$; Travis: $\frac{9}{11}$
b. Raymond read the greater fractional part.
77. a. 300,000,000 **b.** 36,000,000 **c.** $\frac{3}{25}$
79. For example: $\frac{6}{8}, \frac{9}{12}, \frac{12}{16}$ **81.** For example: $\frac{6}{9}, \frac{4}{6}, \frac{2}{3}$

Section 2.3 Calculator Connections, p. 121

83. $\frac{8}{9}$ **84.** $\frac{13}{14}$ **85.** $\frac{41}{51}$ **86.** $\frac{21}{10}$ **87.** $\frac{29}{30}$
88. $\frac{13}{7}$ **89.** $\frac{3}{2}$ **90.** $\frac{31}{19}$

Section 2.4 Practice Exercises, pp. 128–132

1. a. one-tenth **b.** $\frac{1}{2}bh$
3. Numerator: 10; denominator: 14; $\frac{5}{7}$
5. Numerator: 25; denominator: 15; $\frac{5}{3}$ **7.**
9. 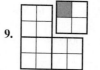   **11.** $\frac{1}{8}$ **13.** 6   **15.** $\frac{3}{16}$ **17.** $\frac{14}{81}$
19. $\frac{24}{35}$ **21.** $\frac{8}{11}$ **23.** $\frac{24}{5}$ **25.** $\frac{65}{36}$ **27.** $\frac{2}{15}$ **29.** $\frac{5}{8}$

31. $\frac{35}{4}$ **33.** $\frac{8}{3}$ **35.** $\frac{4}{5}$ **37.** 8   **39.** 12   **41.** $\frac{30}{7}$

43. 10   **45.** $\frac{5}{3}$ **47.** $\frac{3}{8}$ **49.** 24   **51.** $\frac{1}{1000}$

53. $\frac{1}{1,000,000}$ **55.** $\frac{1}{81}$ **57.** $\frac{27}{8}$ **59.** 27   **61.** $\frac{1}{225}$

63. 2   **65.** $\frac{2}{9}$ **67.**   **69.**

71. 44 cm^2 **73.** 32 m^2 **75.** 4 yd^2 **77.** $\frac{1}{4}$ cm^2

79. $\frac{195}{256}$ in.^{2} **81.** 48 yd^2 **83.** 9 cm^2

85. The amount left is 10 gal.

87. Trey ate $\frac{1}{8}$ of the pizza for breakfast.

89. Corrine will prepare $4\frac{1}{8}$ lb.   **91.** 6,550,000

93. First place: $800; second place: $300; third place: $100

95. a. $\frac{1}{36}$ **b.** $\frac{1}{6}$ **97.** $\frac{1}{5}$ **99.** $\frac{8}{9}$ **101.** $\frac{1}{32}$

103. They are the same.

Section 2.5 Practice Exercises, pp. 138–141

1. reciprocals   **3.** $\frac{18}{5}$ **5.** 2   **7.** $\frac{5}{3}$ **9.** 1   **11.** 1

13. $\frac{8}{7}$ **15.** $\frac{9}{10}$ **17.** $\frac{1}{4}$ **19.** No reciprocal exists.

21. $\frac{1}{3}$ **23.** multiplying   **25.** $\frac{8}{25}$ **27.** $\frac{35}{26}$ **29.** $\frac{35}{9}$

31. 5   **33.** 1   **35.** $\frac{21}{2}$ **37.** $\frac{3}{5}$ **39.** $\frac{1}{4}$ **41.** 20

43. 16   **45.** $\frac{90}{13}$ **47.** 20   **49.** $\frac{7}{2}$ **51.** $\frac{5}{36}$

53. 8   **55.** $\frac{2}{5}$ **57.** $\frac{40}{3}$ **59.** 2   **61.** $\frac{55}{56}$ **63.** $\frac{3}{2}$

65. $\frac{2}{3} \cdot 6$ multiplies $\frac{2}{3}$ by $\frac{6}{1}$, and $\frac{2}{3} \div 6$ multiplies $\frac{2}{3}$ by $\frac{1}{6}$. So $\frac{2}{3} \cdot 6 = 4$ and $\frac{2}{3} \div 6 = \frac{1}{9}$.   **67.** $\frac{3}{7}$ **69.** $\frac{7}{6}$ **71.** $\frac{7}{32}$

73. $\frac{9}{400}$ **75.** 49   **77.** $\frac{7}{16}$ **79.** 18

81. Li wrapped 54 packages.   **83.** 24 cups of juice

85. The stack will be 12 in. high.

87. a. 27 commercials in 1 hr   **b.** 648 commercials in 1 day

89. a. Ricardo's mother will pay $16,000.

b. Ricardo will have to pay $8000.   **c.** He will have to finance $216,000.   **91. a.** She plans to sell $\frac{3}{4}$ acre.

b. She will keep $\frac{3}{2}$ or $1\frac{1}{2}$ acres.

93. She can prepare 14 samples.

95. 12 ft, because $30 \div \frac{5}{2} = 12$.   **97.** Less   **99.** More

Chapter 2 Problem Recognition Exercises, p. 142

1. a. $\frac{16}{5}$ **b.** $\frac{16}{5}$ **c.** $\frac{20}{9}$ **d.** $\frac{9}{20}$

2. a. $\frac{40}{7}$ **b.** $\frac{40}{7}$ **c.** $\frac{35}{18}$ **d.** $\frac{18}{35}$

3. a. $\frac{27}{2}$ **b.** $\frac{27}{2}$ **c.** $\frac{32}{3}$ **d.** $\frac{3}{32}$

4. a. 9   **b.** 9   **c.** 25   **d.** $\frac{1}{25}$

5. a. $\frac{25}{36}$ **b.** 1   **c.** 1   **d.** $\frac{25}{36}$

6. a. 0   **b.** 0   **c.** Undefined   **d.** 0

7. a. $\frac{8}{189}$ **b.** $\frac{7}{96}$ **c.** $\frac{2}{21}$ **d.** $\frac{21}{128}$

8. a. $\frac{7}{27}$ **b.** $\frac{7}{12}$ **c.** $\frac{3}{7}$ **d.** $\frac{27}{28}$

9. a. $\frac{27}{20}$ **b.** $\frac{108}{5}$ **c.** $\frac{3}{80}$ **d.** $\frac{3}{5}$

10. a. $\frac{2}{5}$ **b.** $\frac{1}{250}$ **c.** 160   **d.** $\frac{8}{5}$

11. a. $\frac{2}{3}$ **b.** $\frac{2}{3}$ **c.** $\frac{2}{3}$ **d.** $\frac{3}{2}$

12. a. $\frac{3}{5}$ **b.** $\frac{5}{3}$ **c.** 60   **d.** 60

13. a. 32   **b.** 2   **c.** 2   **d.** 32

14. a. $\frac{1}{14}$ **b.** $\frac{2}{7}$ **c.** $\frac{1}{14}$ **d.** $\frac{2}{7}$

15. a. $\frac{8}{3}$ **b.** 96   **c.** $\frac{1}{9}$ **d.** 144

16. a. $\frac{1}{6}$ **b.** $\frac{3}{8}$ **c.** $\frac{2}{9}$ **d.** $\frac{9}{8}$

Section 2.6 Practice Exercises, pp. 147–149

1. improper   **3.** $\frac{26}{9}$ **5.** $\frac{12}{11}$ **7.** $\frac{2}{9}$ **9.** $\frac{17}{5}$

11. $\frac{11}{7}$ **13.** $12\frac{5}{6}$ **15.** $9\frac{3}{4}$ **17.** $7\frac{2}{5}$ **19.** $1\frac{2}{3}$

21. 38   **23.** $27\frac{2}{3}$ **25.** $72\frac{1}{2}$ **27.** 0   **29.** $7\frac{1}{2}$

31. $2\frac{4}{25}$ **33.** $\frac{34}{55}$ **35.** $4\frac{5}{12}$ **37.** $2\frac{6}{17}$ **39.** 2

41. 0   **43.** 17   **45.** $4\frac{2}{3}$ **47.** $1\frac{3}{4}$

49. Tabitha earned $38.   **51.** $642\frac{1}{2}$ lb

53. a. 7 weeks old   **b.** $8\frac{1}{2}$ weeks old

55. a. Lucy earned $72 more than Ricky.   **b.** Together they earned $922.

57. 2   **59.** $5\frac{1}{3}$ **61.** $1\frac{4}{5}$ **63.** 0

65. $1\frac{1}{6}$ **67.** 0   **69.** $1\frac{1}{2}$ **71.** Undefined   **73.** $2\frac{5}{8}$

75. $2\frac{3}{8}$ **77.** The total cost is $168.

Section 2.6 Calculator Connections, p. 149

79. $318\frac{1}{4}$ **80.** $3\frac{1}{15}$ **81.** $17\frac{18}{19}$ **82.** $466\frac{1}{5}$

83. $2\frac{99}{146}$ **84.** $2\frac{404}{753}$ **85.** $480\frac{1}{8}$ **86.** $280\frac{5}{27}$

Chapter 2 Review Exercises, pp. 156–159

1. $\frac{1}{2}$ **2.** $\frac{4}{7}$ **3. a.** $\frac{5}{3}$ **b.** Improper

4. a. $\frac{1}{6}$ **b.** Proper **5.** $\frac{7}{15}$ **6.** $\frac{23}{8}$ or $2\frac{7}{8}$ **7.** $\frac{7}{6}$ or $1\frac{1}{6}$

8. $\frac{43}{7}$ **9.** $\frac{57}{5}$ **10.** 17 **11.** $5\frac{2}{9}$ **12.** $1\frac{2}{21}$

13.–15.

16. $134\frac{3}{7}$ **17.** $60\frac{11}{13}$ **18.** $21, 51, 1200$

19. $55, 140, 260, 1200$ **20.** $58, 124, 140, 260, 1200$
21. Prime **22.** Composite **23.** Neither
24. Neither **25.** $2 \cdot 2 \cdot 2 \cdot 2 \cdot 2 \cdot 2$ or 2^6
26. $2 \cdot 3 \cdot 5 \cdot 11$ **27.** $2 \cdot 2 \cdot 3 \cdot 3 \cdot 5 \cdot 5$ or $2^2 \cdot 3^2 \cdot 5^2$
28. $1, 2, 3, 4, 6, 8, 12, 16, 24, 48$
29. $1, 2, 4, 5, 8, 10, 16, 20, 40, 80$ **30.** $\neq$ **31.** $=$

32. $\frac{1}{4}$ **33.** $\frac{2}{7}$ **34.** $\frac{3}{2}$ **35.** $\frac{7}{3}$ **36.** 1 **37.** 2

38. $\frac{4}{5}$ **39.** $\frac{7}{10}$ **40.** $\frac{14}{15}, \frac{1}{15}$ **41. a.** $\frac{3}{5}$ **b.** $\frac{2}{5}$

42. $\frac{6}{35}$ **43.** $\frac{32}{9}$ **44.** 63 **45.** 15 **46.** $\frac{1}{5}$ **47.** $\frac{12}{7}$

48. $\frac{1}{10,000}$ **49.** $\frac{1}{625}$ **50.** $\frac{1}{1000}$ **51.** $\frac{1}{17}$

52. $A = \frac{1}{2}bh$ **53.** $A = lw$ **54.** 51 ft²

55. $\frac{10}{3}$ or $3\frac{1}{3}$ m² **56.** 40 yd²

57. Maximus requires $\frac{7}{2}$ or $3\frac{1}{2}$ yd of lumber.
58. There are 900 African American students.
59. There are 300 Asian American students.
60. There are 300 Hispanic female students.
61. There are 750 Caucasian male students.

62. 1 **63.** 1 **64.** $\frac{2}{7}$ **65.** $\frac{1}{7}$

66. Reciprocal does not exist. **67.** 6 **68.** $\frac{1}{5}$

69. multiplying **70.** $\frac{16}{9}$ **71.** $\frac{7}{5}$ **72.** $\frac{1}{21}$ **73.** $\frac{1}{6}$

74. $\frac{8}{3}$ **75.** 14 **76.** $\frac{1}{64}$ **77.** $\frac{4}{5}$ **78.** $\frac{18}{5}$

79. $\frac{1}{52}$ **80.** $\frac{4}{5} \times 20$; 16 **81.** $18 \div \frac{2}{3}$; 27

82. 36 bags of candy **83.** Amelia earned $576.

84. The area is $\frac{640}{3}$ or $213\frac{1}{3}$ ft².

85. Yes. $9 \div \frac{3}{8} = 24$ so he will have 24 pieces, which is more than enough for his class.

86. $23\frac{7}{15}$ **87.** $23\frac{2}{3}$ **88.** 8 **89.** $22\frac{1}{2}$ **90.** 0

91. $1\frac{1}{2}$ **92.** $\frac{10}{11}$ **93.** $4\frac{1}{2}$ **94.** $2\frac{3}{11}$ **95.** $\frac{3}{5}$

96. 0 **97.** It will take $3\frac{1}{8}$ gal.

98. There will be 10 pieces.

Chapter 2 Test, pp. 159–160

1. a. $\frac{5}{8}$ **b.** Proper **2. a.** $\frac{7}{3}$ **b.** Improper

3. $\frac{11}{2}; 5\frac{1}{2}$

4. $\frac{7}{7}$ is an improper fraction because the numerator is greater than or equal to the denominator.

5. a. $3\frac{2}{3}$ **b.** $\frac{34}{9}$ **6.**

7.

8.

9.

10. a. Composite **b.** Neither **c.** Prime
d. Neither **e.** Prime **f.** Composite
11. a. $1, 3, 5, 9, 15, 45$ **b.** $3 \cdot 3 \cdot 5$ or $3^2 \cdot 5$
12. a. Add the digits of the number. If the sum is divisible by 3, then the original number is divisible by 3.
b. Yes. **13. a.** No **b.** Yes **c.** Yes **d.** No **14.** $=$

15. $\neq$ **16.** $\frac{10}{7}$ or $1\frac{3}{7}$ **17.** $\frac{6}{7}$

18. a. Christine: $\frac{3}{5}$; Brad: $\frac{4}{5}$

b. Brad has the greater fractional part completed.

19. $\frac{19}{69}$ **20.** $\frac{25}{2}$ or $12\frac{1}{2}$ **21.** $\frac{4}{9}$ **22.** $\frac{1}{2}$ **23.** $\frac{4}{15}$

24. $\frac{3}{4}$ **25.** $\frac{4}{35}$ **26.** $9\frac{3}{5}$ **27.** $\frac{13}{12}$

28. $\frac{44}{3}$ or $14\frac{2}{3}$ cm² **29.** $20 \div \frac{1}{4}$

30. 48 quarter-pounders

31. 5 dogs are female pure breeds.

32. They can build on a maximum of $\frac{2}{5}$ acre.

Chapters 1–2 Cumulative Review Exercises, pp. 160–161

1.

	Height (ft)	
Mountain	Standard Form	Words
Mt. Foraker (Alaska)	17,400	Seventeen thousand, four hundred
Mt. Kilimanjaro (Tanzania)	19,340	Nineteen thousand, three hundred forty
El Libertador (Argentina)	22,047	Twenty-two thousand, forty-seven
Mont Blanc (France-Italy)	15,771	Fifteen thousand, seven hundred seventy-one

2. 1430 **3.** 139 **4.** 214,344 **5.** 24 **6.** 1863
7. 18 R2 **8.** 120,000,000,000 **9.** 184 **10.** 6
11. 22 **12.** 16 **13.** 4 **14.** d **15.** c **16.** b
17. e **18.** a **19. a.** $\frac{4}{7}$ **b.** $\frac{7}{3}$ or $2\frac{1}{3}$
20. a. Proper **b.** Improper **c.** Improper
21. a. 1, 2, 3, 5, 6, 10, 15, 30 **b.** $2 \cdot 3 \cdot 5$
22. a. $\frac{12}{7}$ **b.** $\frac{2}{5}$ **23.** $\frac{119}{171}$ **24.** $\frac{5}{6}$
25. Yes. $\frac{8}{13} \cdot \frac{5}{16} = \frac{5}{26}$ and $\frac{5}{16} \cdot \frac{8}{13} = \frac{5}{26}$
26. Yes. $\left(\frac{1}{2} \cdot \frac{2}{9}\right) \cdot \frac{5}{3} = \frac{1}{9} \cdot \frac{5}{3} = \frac{5}{27}$ and
$\frac{1}{2} \cdot \left(\frac{2}{9} \cdot \frac{5}{3}\right) = \frac{1}{2} \cdot \frac{10}{27} = \frac{5}{27}$
27. $\frac{6}{25}$ **28.** $\frac{11}{9}$ or $1\frac{2}{9}$ m² **29.** 50 ft²
30. $\frac{3}{40}$ of the students are males from out of state.

Chapter 3
Chapter Opener Puzzles

A crossword puzzle with the following entries:
2 (across) NUMERATOR, with 1 (down) D...
5 (across) RECIPROCAL
6 (across) IMPROPER
Down clues include: DENOMINATOR, PRIME, PROPER

Section 3.1 Practice Exercises, pp. 167–170

1. like **3.** 8 ft **5.** 20 m **7.** 7 fourths
9. 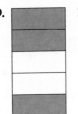 **11.** $\frac{13}{11}$ **13.** $\frac{9}{5}$ **15.** 1
17. $\frac{2}{3}$ **19.** $\frac{13}{10}$ **21.** $\frac{5}{2}$
23. Bethany has $\frac{5}{2}$ or $2\frac{1}{2}$ cups of bleach and water mixture.
25. 11 baskets **27.** 6 fifths
29.
31. $\frac{3}{8}$ **33.** $\frac{3}{2}$ **35.** 2 **37.** $\frac{2}{3}$ **39.** $\frac{2}{5}$ **41.** $\frac{1}{4}$
43. $\frac{1}{4}$ g is left. **45.** $\frac{3}{2}$ **47.** $\frac{12}{5}$ **49.** 1 **51.** $\frac{4}{5}$
53. $\frac{5}{2}$ **55.** $\frac{7}{4}$ **57.** $\frac{81}{100}$ **59.** $\frac{5}{3}$ **61.** $\frac{9}{5}$ **63.** $\frac{16}{7}$
65. $\frac{13}{3}$ **67.** $\frac{1}{3}$ **69.** $\frac{12}{7}$ or $1\frac{5}{7}$ m **71.** $\frac{7}{2}$ or $3\frac{1}{2}$ in.
73. There was $\frac{1}{2}$ gal left over. **75.** He used $\frac{3}{8}$ L.
77. a. Thilan walked $5\frac{1}{2}$ mi total. **b.** He walked an average of $\frac{11}{12}$ mi per day.
79. Perimeter: 2 ft; area: $\frac{15}{64}$ ft²
81. Perimeter: $\frac{70}{3}$ or $23\frac{1}{3}$ yd; area: $\frac{286}{9}$ or $31\frac{7}{9}$ yd²
83. $\frac{3}{5} + \frac{2}{5}$; 1 **85.** $\frac{11}{15} - \frac{8}{15}$; $\frac{1}{5}$

Section 3.2 Practice Exercises, pp. 176–179

1. a. multiple **b.** least common multiple **c.** least common denominator
3. $\frac{1}{2}$ **5.** $\frac{5}{3}$ **7.** 6
9. a. 48, 72, 240 **b.** 4, 8, 12
11. a. 72, 360, 108 **b.** 6, 12, 9
13. 50 **15.** 48 **17.** 120 **19.** 72 **21.** 60
23. 75 **25.** 120 **27.** 210 **29.** 540 **31.** 60
33. 240 **35.** 180 **37.** 180
39. The shortest length of floor space is 60 in. (5 ft).
41. It will take 120 hr (5 days) for the satellites to be lined up again.

43. $\frac{14}{21}$ **45.** $\frac{10}{16}$ **47.** $\frac{12}{16}$ **49.** $\frac{12}{15}$ **51.** $\frac{49}{42}$

53. $\frac{121}{99}$ **55.** $\frac{15}{39}$ **57.** $\frac{11,000}{4000}$ **59.** $\frac{15}{70}$ **61.** $>$

63. $<$ **65.** $=$ **67.** $<$ **69.** $\frac{7}{8}$ **71.** $\frac{2}{3}, \frac{3}{4}, \frac{7}{8}$

73. $\frac{1}{4}, \frac{5}{16}, \frac{3}{8}$ **75.** $\frac{13}{12}, \frac{17}{15}, \frac{4}{3}$

77. The longest cut is above the left eye. The shortest cut is on the right hand.

79. The greatest amount is $\frac{2}{3}$ lb of turkey. The least amount is $\frac{3}{5}$ lb of ham.

81. a and b

Section 3.3 Practice Exercises, pp. 186–188

1. a. is   **b.** is not   **3.** $\frac{12}{14}$ **5.** $\frac{14}{21}$ **7.** $\frac{25}{5}$ **9.** $\frac{8}{4}$

11. $\frac{80}{100}$ **13.** $\frac{5}{40}$ **15.** $\frac{19}{24}$ **17.** $\frac{1}{6}$ **19.** $\frac{1}{4}$ **21.** $\frac{83}{42}$

23. $\frac{27}{40}$ **25.** $\frac{1}{3}$ **27.** $\frac{25}{36}$ **29.** $\frac{5}{8}$ **31.** $\frac{25}{8}$ **33.** $\frac{8}{3}$

35. $\frac{17}{3}$ **37.** $\frac{2}{7}$ **39.** $\frac{89}{100}$ **41.** $\frac{1}{100}$ **43.** $\frac{391}{1000}$

45. $\frac{9}{8}$ **47.** $\frac{23}{60}$ **49.** $\frac{9}{16}$ **51.** $\frac{1}{36}$ **53.** $\frac{7}{12}$ **55.** $\frac{7}{3}$

57. $\frac{4}{3}$ **59.** $\frac{38}{35}$ **61.** $\frac{13}{125}$ **63.** $\frac{23}{24}$

65. Inez added $1\frac{1}{8}$ cups.

67. The storm delivered $\frac{5}{32}$ in. of rain.

69. The trough now holds the original amount of 5 gal.

71. a. $\frac{13}{36}$ **b.** $\frac{23}{36}$ **73.** $2\frac{3}{5}$ m

75. $a = \frac{3}{8}$ ft, $b = \frac{3}{8}$ ft; perimeter: 3 ft   **77.** b

Section 3.4 Practice Exercises, pp. 194–198

1. $\frac{19}{15}$ **3.** $\frac{13}{6}$ **5.** $\frac{24}{5}$ **7.** $\frac{1}{2}$ **9.** $7\frac{4}{11}$ **11.** $15\frac{3}{7}$

13. $15\frac{9}{16}$ **15.** $10\frac{13}{15}$ **17.** 5 **19.** 2 **21.** $3\frac{1}{5}$

23. $8\frac{2}{3}$ **25.** $14\frac{1}{2}$ **27.** $23\frac{1}{8}$ **29.** $19\frac{17}{48}$ **31.** $9\frac{7}{8}$

33. $42\frac{2}{7}$ **35.** $11\frac{3}{5}$ **37.** $2\frac{2}{15}$ **39.** $12\frac{1}{6}$ **41.** $2\frac{5}{14}$

43. $\frac{3}{3}$ **45.** $\frac{12}{12}$ **47.** $11\frac{1}{2}$ **49.** $1\frac{3}{4}$ **51.** $7\frac{13}{14}$

53. $3\frac{1}{6}$ **55.** $2\frac{7}{9}$ **57.** $2\frac{3}{17}$ **59.** $6\frac{5}{14}$ **61.** $7\frac{7}{24}$

63. $6\frac{2}{15}$ **65.** $\frac{11}{16}$ **67.** $9\frac{7}{36}$ **69.** $\frac{29}{32}$ **71.** $10\frac{20}{21}$

73. $\frac{32}{35}$ **75.** $7\frac{13}{72}$ **77.** $7\frac{3}{4}$ in.   **79.** The index finger is longer.   **81.** The total is $16\frac{11}{12}$ hr.   **83.** $3\frac{5}{12}$ ft

85. $\frac{1}{32}$ in.   **87.** The printing area width is 6 in.

89. There is $2\frac{5}{6}$ hr remaining.   **91.** The blinds will hang $\frac{1}{3}$ ft below the window.   **93. a.** $3\frac{3}{8}$ L   **b.** $\frac{5}{8}$ L

95. 4 **97.** $4\frac{1}{4}$

Section 3.4 Calculator Connections, p. 199

98. $\frac{211}{168}$ or $1\frac{43}{168}$ **99.** $\frac{11}{30}$ **100.** $\frac{37}{132}$ **101.** $\frac{137}{391}$

102. $\frac{2509}{54}$ or $46\frac{25}{54}$ **103.** $\frac{2171}{84}$ or $25\frac{71}{84}$ **104.** $\frac{402}{77}$ or $5\frac{17}{77}$

105. $\frac{213}{68}$ or $3\frac{9}{68}$

Chapter 3 Problem Recognition Exercises, pp. 199–200

1. a. $\frac{9}{5}$ or $1\frac{4}{5}$ **b.** $\frac{14}{25}$ **c.** $\frac{7}{2}$ or $3\frac{1}{2}$ **d.** 1

2. a. $\frac{10}{9}$ or $1\frac{1}{9}$ **b.** $\frac{8}{5}$ or $1\frac{3}{5}$ **c.** $\frac{13}{6}$ or $2\frac{1}{6}$ **d.** $\frac{1}{2}$

3. a. $\frac{17}{4}$ or $4\frac{1}{4}$ **b.** $\frac{5}{4}$ or $1\frac{1}{4}$ **c.** $\frac{11}{6}$ or $1\frac{5}{6}$ **d.** $\frac{33}{8}$ or $4\frac{1}{8}$

4. a. $\frac{221}{18}$ or $12\frac{5}{18}$ **b.** $\frac{26}{17}$ or $1\frac{9}{17}$ **c.** $\frac{3}{2}$ or $1\frac{1}{2}$ **d.** $\frac{43}{6}$ or $7\frac{1}{6}$

5. a. $\frac{29}{8}$ or $3\frac{5}{8}$ **b.** $\frac{3}{2}$ or $1\frac{1}{2}$ **c.** $\frac{32}{3}$ or $10\frac{2}{3}$ **d.** $\frac{35}{8}$ or $4\frac{3}{8}$

6. a. $\frac{11}{6}$ or $1\frac{5}{6}$ **b.** $\frac{5}{3}$ or $1\frac{2}{3}$ **c.** $\frac{17}{3}$ or $5\frac{2}{3}$ **d.** $\frac{22}{3}$ or $7\frac{1}{3}$

7. a. $\frac{53}{15}$ or $3\frac{8}{15}$ **b.** $\frac{73}{13}$ or $4\frac{13}{15}$ **c.** $\frac{14}{5}$ or $2\frac{4}{5}$ **d.** $\frac{63}{10}$ or $6\frac{3}{10}$

8. a. $\frac{25}{18}$ or $1\frac{7}{18}$ **b.** $\frac{50}{9}$ or $5\frac{5}{9}$ **c.** $\frac{7}{9}$ **d.** $\frac{43}{9}$ or $4\frac{7}{9}$

9. a. 1 **b.** $\frac{106}{45}$ or $2\frac{16}{45}$ **c.** $\frac{81}{25}$ or $3\frac{6}{25}$ **d.** $\frac{56}{45}$ or $1\frac{11}{45}$

10. a. 1 **b.** $\frac{58}{21}$ or $2\frac{16}{21}$ **c.** $\frac{40}{21}$ or $1\frac{19}{21}$ **d.** $\frac{49}{9}$ or $5\frac{4}{9}$

Section 3.5 Practice Exercises, pp. 205–208

1. $2\frac{6}{7}$ **3.** $12\frac{2}{9}$ **5.** $2\frac{2}{3}$ **7.** $3\frac{13}{36}$ **9.** $\frac{67}{13}$ **11.** $\frac{39}{10}$

13. $5\frac{4}{5}$ **15.** $1\frac{11}{19}$ **17.** $2\frac{1}{4}$ **19.** $3\frac{2}{3}$ **21.** $7\frac{1}{2}$ **23.** $4\frac{2}{7}$

25. 13 **27.** $\frac{13}{25}$ **29.** $1\frac{10}{27}$ **31.** 1 **33.** $1\frac{3}{7}$

35. a. The difference is $\frac{3}{10}$ sec.   **b.** The average is $3\frac{3}{5}$ sec.

37. a. The total weight loss is 51 lb.   **b.** The average is $8\frac{1}{2}$ lb.

c. The difference is $6\frac{1}{2}$ lb.   **39.** The stock dropped $\$3\frac{7}{8}$.

41. George will receive $26,750.

43. Each piece is $3\frac{13}{16}$ ft.   **45.** $2\frac{1}{4}$ lb of cheese was eaten.

47. 20 loaves can be made.   **49.** The new rate is $7\frac{1}{4}$ points.

51. Stephanie will need $11\frac{1}{4}$ yd for the dresses.

53. Wilma has $1\frac{1}{12}$ lb left.   **55.** Joan saves $152\frac{1}{2}$ gal.

57. She needs $15\frac{1}{3}$ ft more.   **59.** The perimeter is 100 in.

61. Matt needs $76\frac{1}{3}$ ft of gutter.

63. The area of the whole roof is $1022\frac{7}{16}$ ft².

65. a. The area is $247\frac{1}{2}$ m².   **b.** They will need 65 m.

67. $152\frac{3}{4}$ m²

Chapter 3 Review Exercises, pp. 213–215

1. 8 books   **2.** 18 cm   **3.** 12 mi   **4.** 11 CDs
5. Fractions with the same denominators are considered like fractions.
6. For example: like fractions: $\frac{4}{7}, \frac{2}{7}$; unlike fractions: $\frac{1}{9}, \frac{3}{16}$.

7. $\frac{3}{2}$ **8.** $\frac{2}{3}$ **9.** $\frac{1}{2}$ **10.** 1   **11.** $\frac{9}{7}$ **12.** 3

13. $\frac{3}{4}$ **14.** $\frac{20}{9}$ **15.** $\frac{11}{13}$ **16.** $\frac{1}{7}$ **17.** 12 in. or 1 ft

18. a. $\frac{28}{5}$ in.   **b.** $\frac{12}{5}$ in.
19. a. 7, 14, 21, 28   **b.** 13, 26, 39, 52   **c.** 22, 44, 66, 88
20. 6 and 8 have many common multiples including 24, 48, and 72. Of all the common multiples, 24 is the least.
21. a. 1, 2, 4, 5, 10, 20, 25, 50, 100   **b.** 1, 5, 13, 65
c. 1, 2, 5, 7, 10, 14, 35, 70   **22. a.** $2 \cdot 2 \cdot 5 \cdot 5$ **b.** $5 \cdot 13$
c. $2 \cdot 5 \cdot 7$ **23.** 150   **24.** 1584   **25.** 420
26. 96   **27.** They will meet on the 12th day.

28. $\frac{15}{48}$ **29.** $\frac{63}{35}$ **30.** $\frac{35}{60}$ **31.** $\frac{170}{150}$ **32.** <

33. >   **34.** =   **35.** $\frac{8}{15}, \frac{72}{105}, \frac{7}{10}, \frac{27}{35}$ **36.** $\frac{17}{24}$

37. $\frac{29}{100}$ **38.** $\frac{1}{25}$ **39.** $\frac{1}{2}$ **40.** $\frac{47}{11}$ **41.** $\frac{43}{20}$

42. $\frac{4}{15}$ **43.** $\frac{1}{34}$ **44.** $\frac{37}{1000}$ **45.** $\frac{17}{40}$ **46.** $\frac{12}{7}$

47. $\frac{1}{15}$ **48.** $\frac{3}{5}$ **49. a.** $\frac{35}{4}$ or $8\frac{3}{4}$ m   **b.** $\frac{315}{128}$ or $2\frac{59}{128}$ m²

50. a. $\frac{23}{3}$ or $7\frac{2}{3}$ yd   **b.** $\frac{7}{2}$ or $3\frac{1}{2}$ yd²   **51.** $11\frac{11}{63}$

52. $14\frac{7}{16}$ **53.** $2\frac{5}{8}$ **54.** $1\frac{11}{12}$ **55.** $3\frac{1}{24}$ **56.** $2\frac{8}{15}$

57. $12\frac{5}{14}$ **58.** $6\frac{3}{16}$ **59.** $3\frac{2}{5}$ **60.** $3\frac{3}{14}$ **61.** $63\frac{15}{16}$

62. $50\frac{1}{2}$ **63.** 8; $8\frac{5}{18}$ **64.** 11; $10\frac{14}{15}$ **65.** 50; $50\frac{9}{40}$

66. 23; $22\frac{71}{75}$ **67.** Corry drove a total of $8\frac{1}{6}$ hr.

68. Denise will have $\frac{7}{8}$ acre left.   **69.** $12\frac{2}{5}$ **70.** $1\frac{1}{4}$

71. $\frac{4}{27}$ **72.** 18   **73.** 12   **74.** $9\frac{1}{3}$
75. The appraised value is $144,000.
76. There are $1\frac{1}{4}$ lb of nuts in each bag.

Chapter 3 Test, p. 216

1. $\frac{7}{5}$ **2.** $\frac{1}{2}$ **3.** When subtracting like fractions, keep the same denominator and subtract the numerators. When multiplying fractions, multiply the denominators as well as the numerators.
4. a. 24, 48, 72, 96   **b.** 1, 2, 3, 4, 6, 8, 12, 24
c. $2 \cdot 2 \cdot 2 \cdot 3$ or $2^3 \cdot 3$

5. 240   **6.** $\frac{35}{63}$ **7.** $\frac{33}{63}$ **8.** $\frac{36}{63}$ **9.** $\frac{11}{21}, \frac{5}{9}, \frac{4}{7}$

10. $\frac{9}{16}$ **11.** $\frac{1}{3}$ **12.** $\frac{1}{3}$ **13.** $\frac{2}{3}$ **14.** $17\frac{3}{8}$

15. $2\frac{1}{11}$ **16.** $60\frac{5}{12}$ **17.** $1\frac{1}{2}$ **18.** $\frac{25}{6}$ or $4\frac{1}{6}$

19. 7   **20.** $\frac{12}{295}$ **21.** $\frac{10}{3}$ or $3\frac{1}{3}$ **22.** 1 lb is needed.
23. The Ford Expedition can tow 8950 lb.
24. Area: $25\frac{2}{25}$ m²; perimeter: $20\frac{1}{5}$ m
25. Justin has $10,500 for cabinets.
26. The difference is $4\frac{2}{3}$ ft.

Chapters 1–3 Cumulative Review Exercises, p. 217

1. Twenty-three million, four hundred thousand, eight hundred six
2. 96   **3.** 48   **4.** 1728   **5.** 3   **6.** 1,500,000,000
7. $4^2 \cdot 5^4 \cdot 8^2$ **8.** 36   **9.** 17, 19, 23, 29, 31
10. $2 \cdot 5 \cdot 7$ **11.** Numerator: 21; denominator: 17
12. $\frac{4}{16}$ or $\frac{1}{4}$ **13.** $\frac{17}{22}$ had pepperoni and $\frac{5}{22}$ did not have pepperoni.   **14. a.** Improper   **b.** Proper   **c.** Improper
15. b   **16. a.** Composite   **b.** Composite   **c.** Prime

17. $2 \cdot 2 \cdot 2 \cdot 3 \cdot 3 \cdot 5$ or $2^3 \cdot 3^2 \cdot 5$ **18.** $\frac{1}{5}$ **19.** $\frac{3}{8}$

20. $\frac{4}{7}$ **21.** $\frac{3}{4}$ **22.** $\frac{33}{16}$ **23.** $\frac{2}{5}$ **24.** $\frac{305}{22}$ or $13\frac{19}{22}$

25. $\frac{26}{17}$ or $1\frac{9}{17}$ **26.** $\frac{10}{3}$ or $3\frac{1}{3}$
27. The distance around is approximately 88 cm.

28. $4\frac{1}{3}$ yd   **29.** $7\frac{7}{8}$ m²   **30. a.** $2\frac{3}{10}$ **b.** $6\frac{4}{5}$

Chapter 4

Chapter Opener Puzzle

Mathematicians shop at the $\frac{d}{1}\frac{e}{2}\frac{c}{3}\frac{i}{4}\frac{m}{5}\frac{a}{6}\frac{l}{7}\frac{l}{8}$

Section 4.1 Practice Exercises, pp. 226–229

1. a. decimal

   b. tenths; hundredths; thousandths

3. 100   **5.** 10,000   **7.** $\frac{1}{100}$ **9.** $\frac{1}{10,000}$

11. Tenths   **13.** Hundredths   **15.** Tens

17. Ten-thousandths   **19.** Thousandths   **21.** Ones

23. Nine-tenths   **25.** Twenty-three hundredths

27. Thirty-three thousandths

29. Four hundred seven ten-thousandths

31. Three and twenty-four hundredths

33. Five and nine-tenths   **35.** Fifty-two and three-tenths

37. Six and two hundred nineteen thousandths

39. 8472.014   **41.** 700.07   **43.** 2,469,000.506

45. $3\frac{7}{10}$ **47.** $2\frac{4}{5}$ **49.** $\frac{1}{4}$ **51.** $\frac{11}{20}$ **53.** $20\frac{203}{250}$

55. $15\frac{1}{2000}$ **57.** $\frac{42}{5}$ **59.** $\frac{157}{50}$ **61.** $\frac{47}{2}$ **63.** $\frac{1191}{100}$

65. 34.2, 34.25, 34.29, 34.3   **67.** 0.042, 0.043, $\frac{4}{10}$, 0.42, 0.43

69. <   **71.** >   **73.** >   **75.** <   **77.** a, b

79. 0.3444, 0.3493, 0.3558, 0.3585, 0.3664

81. These numbers are equivalent, but they represent different levels of accuracy.

83. 7.1   **85.** 49.9   **87.** 33.42   **89.** 9.096

91. 21.0   **93.** 7.000   **95.** 0.0079   **97.** 0.0036 mph

	Number	Hundreds	Tens	Tenths	Hun-dredths	Thou-sandths
99.	971.0948	1000	970	971.1	971.09	971.095
101.	21.9754	0	20	22.0	21.98	21.975

103. 0.972

Section 4.2 Practice Exercises, pp. 235–239

1. b, c   **3.** b, c   **5.** 23.5   **7.** 8.603   **9.** 2.8300

11. 63.2   **13.** 8.951   **15.** 15.991   **17.** 79.8005

19. 31.0148   **21.** 62.6032   **23.** 100.414   **25.** 128.44

27. 82.063   **29.** 14.24   **31.** 3.68   **33.** 12.32

35. 5.677   **37.** 1.877   **39.** 57.368   **41.** 21.6646

43. 14.765   **45.** 159.558   **47.** 15.347   **49.** 6.581

51. 19.912   **53.** 10.3327   **55.** 5.9156   **57.** 9.001

59. a. 321.724 days   **b.** 156.73 days

61. a. The water is rising 1.7 in./hr.   **b.** At 1:00 P.M. the level will be 11 in.   **c.** At 3:00 P.M. the level will be 14.4 in.

63.

Check No.	Description	Payment	Deposit	Balance
				$ 245.62
2409	Electric bill	$ 52.48		193.14
2410	Groceries	72.44		120.70
2411	Department store	108.34		12.36
	Paycheck		$1084.90	1097.26
2412	Restaurant	23.87		1073.39
	Transfer from savings		200	1273.39

65. 1.35 million cells per microliter

67. The pile containing the two nickels and two pennies is higher.

69. $x = 8.9$ in.; $y = 15.4$ in.; the perimeter is 98.8 in.

71. $x = 2.075$ ft; $y = 2.59$ ft; the perimeter is 22.17 ft.

73. 27.2 mi   **75.** 7 mm

Section 4.2 Calculator Connections, p. 239

77. IBM decreased by $1.99 per share.

78. FedEx increased by $6.56 per share.

79. Between March and April, FedEx increased the most, by $6.36 per share.

80. Between February and March, IBM increased the most, by $3.04 per share.

81. Between January and February, FedEx decreased the most, by $2.78 per share.

82. Between January and February, IBM decreased the most, by $6.92 per share.

Section 4.3 Practice Exercises, pp. 245–247

1. front   **3.** 1000   **5.** 0.01   **7.** 0.4   **9.** 3.6   **11.** 8

13. 0.18   **15.** 17.904   **17.** 0.028   **19.** 100   **21.** 30

23. 0.07   **25.** 0.2   **27.** 37.35   **29.** 4.176

31. 4.736   **33.** 2.891   **35.** 114.88   **37.** 2.248

39. 0.00144   **41.** $(0.3)^2 = 0.09$, which is not equal to 0.9.

43. 0.0036   **45.** 6.25   **47.** 0.16   **49.** 1.69

51. 0.001   **53.** 0.0016   **55.** The decimal point will move to the right two places.

57. a. 51   **b.** 510   **c.** 5100   **d.** 51,000

59. The decimal point will move to the left one place.

61. 3490   **63.** 96,590   **65.** 0.933   **67.** 0.05403

69. 20.01   **71.** 0.00005   **73.** 324¢   **75.** 37¢

77. $3.47   **79.** $20.41   **81. a.** $1   **b.** $1.50

83. 2,600,000   **85.** 400,000   **87.** $20,549,000,000

89. a. 201.6 lb of gasoline   **b.** 640 lb of CO_2

91. The bill was $423.61.   **93.** $48.81 can be saved.

95. 0.00115 km^2 **97.** The area is 333 yd^2.

99. a. 0.09   **b.** 0.3   **101.** 0.1   **103.** 0.6

Section 4.4 Practice Exercises, pp. 255–258

1. a. repeating   **b.** terminating

3. 5280   **5.** 3.776   **7.** 2.02   **9.** 0.9   **11.** 0.18

13. 0.53   **15.** 21.1   **17.** 1.96   **19.** 0.035

21. 16.84   **23.** 0.12   **25.** 0.16   **27.** $5.\overline{3}$ **29.** $3.1\overline{6}$

31. $2.\overline{15}$ **33.** 503   **35.** 9.92   **37.** 56   **39.** 2.975
41. $208.\overline{3}$ **43.** 48.5   **45.** 1100   **47.** 42,060
49. The decimal point will move to the left two places.
51. 0.03923   **53.** 9.802   **55.** 0.00027
57. 0.00102   **59. a.** 2.4 **b.** 2.44 **c.** 2.444
61. a. 1.9 **b.** 1.89 **c.** 1.889
63. a. 3.6 **b.** 3.63 **c.** 3.626   **65.** 0.26   **67.** 14.8
69. 20.667   **71.** 35.67   **73.** 111.3
75. Unreasonable; $960   **77.** Unreasonable; $140,000
79. The monthly payment is $42.50.
81. a. 13 bulbs would be needed (rounded up to the nearest whole unit).   **b.** $9.75   **c.** The energy efficient fluorescent bulb would be more cost effective.
83. Babe Ruth's batting average was 0.342.
85. 2.2 mph   **87.** 47.265   **89.** b, d

Section 4.4 Calculator Connections, p. 258

91. 1149686.166   **92.** 3411.4045   **93.** 1914.0625
94. 69,568.83693   **95.** 95.6627907   **96.** 293.5070423
97. Answers will vary.   **98.** Answers will vary.
99. a. 0.37   **b.** Yes the claim is accurate. The decimal, 0.37 is greater than $0.\overline{3}$, which is equal to $\frac{1}{3}$.
100. 272 people per square mile
101. a. 1,600,000 mi per day   **b.** $66,666.\overline{6}$ mph
102. When we say that 1 year is 365 days, we are ignoring the 0.256 day each year. In 4 years, that amount is $4 \times 0.256 = 1.024$, which is another whole day. This is why we add one more day to the calendar every 4 years.

Chapter 4 Problem Recognition Exercises, p. 259

1. a. 223.04 **b.** 12,304 **c.** 23.04 **d.** 1.2304 **e.** 123.05 **f.** 1.2304 **g.** 12,304 **h.** 123.03
2. a. 6078.3 **b.** 5,078,300 **c.** 4078.3 **d.** 5.0783 **e.** 5078.301 **f.** 5.0783 **g.** 5,078,300 **h.** 5078.299
3. a. 7.191 **b.** 7.191   **4. a.** 730.4634 **b.** 730.4634
5. a. 52.64 **b.** 52.64   **6. a.** 59.384 **b.** 59.384
7. a. 86.4 **b.** 5.4   **8. a.** 185 **b.** 46.25
9. a. 80 **b.** 448   **10. a.** 54 **b.** 496.8
11. 1   **12.** 1   **13.** 4000   **14.** 6,400,000
15. 200,000   **16.** 2700   **17.** 1,350,000,000
18. 1,700,000   **19.** 4.4001   **20.** 76.7001

Section 4.5 Practice Exercises, pp. 265–268

1. 0.9   **3.** 0.141   **5.** $\frac{3}{5}$ **7.** $\frac{7}{20}$
9. 4.25   **11.** $\frac{4}{10}$; 0.4
13. $\frac{98}{100}$; 0.98   **15.** 0.28   **17.** 0.632   **19.** 0.875
21. 3.2   **23.** 5.25   **25.** 1.2   **27.** 0.75
29. 3.3125   **31.** 7.45   **33.** $0.8\overline{8}$ **35.** $3.\overline{8}$
37. $0.4\overline{6}$ **39.** $0.52\overline{7}$ **41.** $0.\overline{54}$ **43.** $0.\overline{126}$
45. $1.1\overline{36}$ **47.** 0.143   **49.** 0.08   **51.** 0.9 **53.** 0.71
55. 1.2   **57. a.** $0.\overline{1}$ **b.** $0.\overline{2}$ **c.** $0.\overline{4}$ **d.** $0.\overline{5}$
If we memorize that $\frac{1}{9} = 0.\overline{1}$, then $\frac{2}{9} = 2 \cdot \frac{1}{9} = 2 \cdot 0.\overline{1} = 0.\overline{2}$, and so on.

59.

	Decimal Form	Fraction Form
a.	0.45	$\frac{9}{20}$
b.	1.625	$\frac{13}{8}$ or $1\frac{5}{8}$
c.	$0.\overline{7}$	$\frac{7}{9}$
d.	$0.\overline{45}$	$\frac{5}{11}$

61.

	Decimal Form	Fraction Form
a.	$0.\overline{3}$	$\frac{1}{3}$
b.	2.125	$\frac{17}{8}$ or $2\frac{1}{8}$
c.	$0.8\overline{63}$	$\frac{19}{22}$
d.	1.68	$\frac{42}{25}$

63.

Stock	Closing Price ($) (Decimal)	Closing Price ($) (Fraction)
McGraw-Hill	69.25	$69\frac{1}{4}$
Walgreens	44.95	$44\frac{19}{20}$
Home Depot	38.50	$38\frac{1}{2}$
General Electric	37.44	$37\frac{11}{25}$

65. =   **67.** <   **69.** >   **71.** <   **73.** =   **75.** <
77. $\frac{1}{10}, 0.\overline{1}, \frac{1}{5}$

79. $1.75, 1.\overline{7}, 1.8$

81. $\frac{9}{9} = 1$ **83.** 7

Section 4.6 Practice Exercises, pp. 274–277

1. a. $\frac{17}{20}$ **b.** 4.6   **3.** 313.72   **5.** $\frac{107}{27}$ **7.** $\frac{5}{4}$ **9.** 6.96
11. 6.25   **13.** 10   **15.** 8.77   **17.** 25.75   **19.** 2
21. 12.98   **23.** 4   **25.** 12.1   **27.** 67.35   **29.** 25.05
31. 23.4   **33.** 1.28   **35.** 10.83   **37.** 2.84
39. $0.93\overline{5}$ **41.** $4.4\overline{3}$

43. a. 471 mi   **b.** 62.8 mph
45. Jorge will be charged $98.75.
47. She has 24.3 g left for dinner.
49. Caren should get $4.77 in change.
51. Duncan's average is 78.75.
53. The average snowfall per month is 14.54 in.
55. Answers will vary.
57. a. 29.8   **b.** Overweight   **59.** 3.475   **61.** 0.52

Section 4.6 Calculator Connections, p. 278

63. a. 237 shares   **b.** $13.90 will be left over.
64. a. Approximately 921,800 homes could be powered.
b. Approximately 342,678 additional homes could be powered.
65. a. Marty will have to finance $120,000.   **b.** There are 360 months in 30 yr.   **c.** He will pay $287,409.60   **d.** He will pay $167,409.60 in interest.
66. a. Gwen needs to finance $94,000.   **b.** There are 180 months in 15 yr.   **c.** Gwen will pay $152,820.00.   **d.** She will pay $58,820.00 in interest.
67. Each person will get approximately $13,410.10.
68. The average price is $110.28.

Chapter 4 Review Exercises, pp. 285–288

1. The 3 is in the tens place, 2 is in the ones place, 1 is in the tenths place, and 6 is in the hundredths place.
2. The 2 is in the ones place, 0 is in the tenths place, 7 is in the hundredths place, and 9 is in the thousandths place.
3. Five and seven-tenths   **4.** Ten and twenty-one hundredths   **5.** Fifty-one and eight thousandths
6. One hundred nine and one-hundredth   **7.** 33,015.047
8. 100.01   **9.** $4\frac{4}{5}$ **10.** $\frac{1}{40}$ **11.** $\frac{13}{10}$ **12.** $\frac{27}{4}$
13. >   **14.** <   **15.** 4.3875, 4.3953, 4.4839, 4.5000, 4.5142
16. 89.92   **17.** 34.890
18. a. The amount in the box is less than the advertised amount.   **b.** The amount rounds to 12.5 oz.
19. a, b   **20.** b, c   **21.** 49.743   **22.** 273.22   **23.** 5.45
24. 1.902   **25.** 197.96   **26.** 38.993   **27.** 7.809   **28.** 82.265
29. $x = 4.5$ in., $y = 5.07$ in.; the perimeter is 201 in.
30. a. Between days 1 and 2, the increase was $0.194.
b. Between days 3 and 4, the decrease was $0.209.
31. 3.74 in.   **32.** 8.19   **33.** 74.113   **34.** 264.44
35. 346.5   **36.** 85,490   **37.** 100.34   **38.** 0.9201
39. 1.0422   **40.** 28,100,000   **41.** 432,000
42. a. Eight batteries cost $15.96 on sale.   **b.** A customer can save $2.03.   **43.** The call will cost $5.75.
44. Area = 940 ft²; perimeter = 127 ft
45. a. 7280 people   **b.** 18,580 people   **46.** 17.1
47. 42.8   **48.** $4.1\overline{3}$ **49.** $8.7\overline{6}$ **50.** 27   **51.** 0.03
52. 4.9393   **53.** 9.0234   **54.** 553,800   **55.** 260
56.

	$8.\overline{6}$	**$52.\overline{52}$**	**$0.\overline{409}$**
Tenths	8.7	52.5	0.4
Hundredths	8.67	52.53	0.41
Thousandths	8.667	52.525	0.409
Ten-thousandths	8.6667	52.5253	0.4094

57. 11.62   **58.** 11.97   **59. a.** $0.50 per roll   **b.** $0.57 per roll   **c.** The 12-pack is better.

60. $\frac{6}{10}$; 0.6   **61.** $\frac{35}{100}$; 0.35   **62.** $\frac{54}{1000}$; 0.054   **63.** 2.4
64. 3.52   **65.** 0.192   **66.** 0.4375   **67.** $0.58\overline{3}$ **68.** $1.52\overline{7}$
69. $4.3\overline{18}$ **70.** $0.\overline{153846}$ **71.** 0.29   **72.** 0.87   **73.** 3.67
74. 2.83   **75.** $\frac{2}{9}$ **76.** $1\frac{2}{3}$ **77.** $3\frac{1}{3}$ **78.** $5\frac{7}{9}$
79.

Stock	Closing Price ($) (Decimal)	Closing Price ($) (Fraction)
Ford	13.02	$13\frac{1}{50}$
Microsoft	30.50	$30\frac{1}{2}$
Citibank	4.37	$4\frac{37}{100}$

80. >   **81.** =   **82.** <   **83.** 0.28   **84.** 0.713
85. 5   **86.** 125.6   **87.** 78.5   **88.** 25.12
89. $89.90 will be saved by buying the combo package.
90. Marvin must drive 34 mi more.

Chapter 4 Test, pp. 289–290

1. a. Tens place   **b.** Hundredths place
2. Five hundred nine and twenty-four thousandths
3. $1\frac{13}{50}$; $\frac{63}{50}$ **4.** 0.4419, 0.4484, 0.4489, 0.4495
5. b is correct.   **6.** 52.832   **7.** 21.29   **8.** 126.45
9. 5.08   **10.** 1.22   **11.** 12.2243   **12.** $120.\overline{6}$
13. 439.81   **14.** 4.592   **15.** 57,923   **16.** 8012
17. 0.002931   **18. a.** 61.4°F   **b.** 1.4°F
19. a. 50,500,000 votes   **b.** 51,000,000 votes   **c.** The difference is approximately 500,000 in favor of Al Gore.
20. a. 67.5 in.²   **b.** 75.5 in.²   **c.** 157.3 in.²
21. He made $3094.75.   **22.** She will pay approximately $37.50 per month.   **23.** He will use 10 gal of gas.
24.

Year	Time in Seconds (Decimal)	Time in Seconds (Fraction)
1998	38.24	$38\frac{6}{25}$
2002	38.23	$38\frac{23}{100}$
2006	37.30	$37\frac{3}{10}$
2010	38.21	$38\frac{21}{100}$

25. $3.2, 3\frac{1}{2}, 3.\overline{5}$

26. 9.57   **27.** 47.25   **28. a.** 38.8 mi   **b.** 5.5 mi/day

Chapters 1–4 Cumulative Review Exercises, pp. 290–291

1. 14   **2.** 4039   **3.** 4840   **4.** 3872   **5.** 2,415,000
6. Dividend: 4530; divisor: 225; whole-number part of the quotient: 20; remainder: 30   **7.** To check a division problem, multiply the whole-number part of the quotient and the divisor. Then add the remainder to get the dividend. That is, $20 \times 225 + 30 = 4530$.   **8.** The difference between sales for Wal-Mart and Sears is $181,956 million.

9. $\dfrac{6}{55}$ **10.** $\dfrac{4}{7}$ **11.** $\dfrac{49}{100}$ **12.** 2   **13.** $\dfrac{2}{3}$ **14.** 0

15. There is $9000 left.   **16.** $\dfrac{2}{5}$ **17.** $\dfrac{97}{100}$ **18.** $\dfrac{38}{11}$

19. $\dfrac{33}{7}$ **20.** $\dfrac{3}{2}$ **21.** Area: $\dfrac{15}{64}$ ft^2; perimeter: 2 ft

22. The average is $1\dfrac{3}{16}$ km.   **23.** 174.13

24. 668.79   **25.** 75.275   **26.** 16   **27.** 339.12
28. 46.48   **29. a.** 3.75248   **b.** 3.75248   **c.** Commutative property of multiplication
30.

Bone	Length (in.) (Decimal)	Length (in.) (Mixed Number)
Femur	19.875	$19\frac{7}{8}$
Fibula	15.9375	$15\frac{15}{16}$
Humerus	14.375	$14\frac{3}{8}$
Innominate bone (hip)	7.5	$7\frac{1}{2}$

Chapter 5

Chapter Opener Puzzles

4	1	2	5	3
2	3	4	1	5
1	ᵃ2	5	3	ᶜ4
3	5	1	4	2
ᵈ5	4	ᵇ3	2	1

a. 20.25   **b.** 202.5   **c.** 0.2025   **d.** 2.025

Section 5.1 Practice Exercises, pp. 298–300

1. ratio   **3.** $5:6$ and $\dfrac{5}{6}$ **5.** 11 to 4 and $\dfrac{11}{4}$

7. $1:2$ and 1 to 2   **9. a.** $\dfrac{3}{2}$ **b.** $\dfrac{2}{3}$ **c.** $\dfrac{3}{5}$

11. a. $\dfrac{21}{52}$ **b.** $\dfrac{21}{31}$ **13.** $\dfrac{2}{3}$ **15.** $\dfrac{1}{5}$ **17.** $\dfrac{4}{1}$ **19.** $\dfrac{11}{5}$

21. $\dfrac{6}{5}$ **23.** $\dfrac{1}{2}$ **25.** $\dfrac{3}{2}$ **27.** $\dfrac{6}{7}$ **29.** $\dfrac{8}{9}$ **31.** $\dfrac{7}{1}$

33. $\dfrac{1}{8}$ **35.** $\dfrac{5}{4}$ **37.** $\dfrac{4}{11}$ **39. a.** $\dfrac{6}{16} = \dfrac{3}{8}$ **b.** $\dfrac{\frac{1}{2}}{1\frac{1}{3}} = \dfrac{3}{8}$

41. $\dfrac{1}{11}$ **43.** $\dfrac{10}{1}$ **45.** $\dfrac{15}{32}$ **47.** $\dfrac{20}{61}$ **49.** $\dfrac{2}{3}$

51. $\dfrac{1}{4}$ **53.** 13 units   **55. a.** 1.5   **b.** $1.\overline{6}$ **c.** 1.6
d. 1.625; yes   **57.** Answers will vary.

Section 5.2 Practice Exercises, pp. 304–307

1. a. rate   **b.** unit   **3.** $3:5$ and $\dfrac{3}{5}$ **5.** $\dfrac{4}{3}$ **7.** $\dfrac{9}{17}$ **9.** $\dfrac{\$32}{5 \text{ ft}^2}$

11. $\dfrac{117 \text{ mi}}{2 \text{ hr}}$ **13.** $\dfrac{\$29}{4 \text{ hr}}$ **15.** $\dfrac{1 \text{ page}}{2 \text{ sec}}$ **17.** $\dfrac{65 \text{ calories}}{4 \text{ crackers}}$

19. $\dfrac{\$15}{2 \text{ trays}}$ **21.** a, c, d   **23.** 113 mi/day
25. 96 km/hr   **27.** $55 per payment   **29.** $0.69/lb
31. $256,000 per person   **33.** 14.3 m/sec
35. $0.219 per oz   **37.** $0.995 per liter
39. $52.50 per tire   **41.** $5.417 per bodysuit
43. a. $0.334/oz   **b.** $0.334/oz   **c.** Both sizes cost the same amount per ounce.
45. The larger can is $0.123 per ounce. The smaller can is $0.164 per ounce. The larger can is the better buy.
47. Coca-Cola: 3.25 g/fl oz; Mello Yello: 3.92 g/fl oz; Ginger Ale: 3 g/fl oz; Mello Yello has the greatest amount per fluid oz.
49. 130,000 platelets per microliter; Since the patient's platelet count is above 20,000 per microliter, the patient does *not* have a life-threatening condition.
51. 43,000 prisoners per year   **53. a.** $0.76 per month
b. $1.90 per month   **c.** IBM

Section 5.2 Calculator Connections, pp. 307–308

55. a. 9.9 wins/year   **b.** 8.6 wins/year   **c.** Shula
56. a. 2.1 wins/loss   **b.** 1.5 wins/loss   **c.** Shula
57. a. $0.38 per ounce   **b.** $0.18 per ounce
c. $0.19 per ounce; The best buy is Dial.
58. The unit prices are $0.181 per ounce, $0.280 per ounce, and $0.255 per ounce. The best buy is the 48-oz jar.
59. a. $0.401/oz   **b.** $0.167/oz
c. $0.322/oz; The best buy is the 12-oz can.
60. a. $0.062/oz   **b.** $0.023/oz; The case of twelve 12-fl-oz cans for $3.33 is the better buy.

Section 5.3 Practice Exercises, pp. 313–315

1. a. equation   **b.** proportion   **3.** $\dfrac{1}{15}$ **5.** $\dfrac{3 \text{ apples}}{1 \text{ pie}}$

7. $\dfrac{22 \text{ mi}}{3 \text{ gal}}$ **9.** $\dfrac{4}{16} = \dfrac{5}{20}$ **11.** $\dfrac{25}{15} = \dfrac{10}{6}$ **13.** $\dfrac{2}{3} = \dfrac{4}{6}$

15. $\dfrac{30}{25} = \dfrac{12}{10}$ **17.** $\dfrac{\$6.25}{1 \text{ hr}} = \dfrac{\$187.50}{30 \text{ hr}}$ **19.** $\dfrac{1 \text{ in.}}{7 \text{ mi}} = \dfrac{5 \text{ in.}}{35 \text{ mi}}$

21. No   **23.** Yes   **25.** Yes   **27.** Yes   **29.** Yes
31. Yes   **33.** No   **35.** Divide by 2   **37.** Divide by 5
39. Divide by 8   **41.** Divide by 0.6   **43.** Yes
45. No   **47.** $x = 4$ **49.** $x = 3$ **51.** $p = 75$
53. $n = 12$ **55.** $t = 12$ **57.** $y = 36$ **59.** $x = 3$
61. $m = \dfrac{15}{2}$ or $7\frac{1}{2}$ or 7.5   **63.** $k = 30$ **65.** $h = 2.5$

67. $x = 4$ **69.** $z = \dfrac{1}{80}$

Chapter 5 Problem Recognition Exercises, p. 315

1. a. Proportion; $\frac{15}{2}$ **b.** Product of fractions; $\frac{15}{32}$

2. a. Product of fractions; $\frac{3}{25}$ **b.** Proportion; 4

3. a. Product of fractions; $\frac{3}{49}$ **b.** Proportion; 4

4. a. Proportion; 2   **b.** Product of fractions; $\frac{6}{25}$

5. a. Proportion; 9   **b.** Product of fractions; 32

6. a. Product of fractions; 8   **b.** Proportion; $\frac{98}{5}$

7. a. 14   **b.** $\frac{5}{2}$ **c.** $\frac{3}{5}$ **d.** $\frac{18}{245}$

8. a. $\frac{3}{25}$ **b.** $\frac{3}{5}$ **c.** $\frac{16}{3}$ **d.** $\frac{112}{15}$

9. a. 4   **b.** $\frac{98}{5}$ **c.** $\frac{48}{35}$ **d.** $\frac{49}{25}$

10. a. 18   **b.** $\frac{29}{3}$ **c.** $\frac{11}{18}$ **d.** 22

Section 5.4 Practice Exercises, pp. 321–325

1. a. similar   **b.** proportional   **3.** $=$ **5.** $\neq$

7. $n = \frac{20}{3}$ or $6\frac{2}{3}$ or $6.\overline{6}$ **9.** $k = 6$ **11.** $y = 4.9$

13. Pam can drive 610 mi on 10 gal of gas.
15. 78 kg of crushed rock will be required.
17. The actual distance is about 80 mi.
19. There are 3800 male students.
21. Heads would come up about 315 times.
23. There would be approximately 3 earned runs for a 9-inning game.   **25.** Pierre can buy 684€.
27. 45 visits would be a result of falls.
29. $\frac{2}{3}$ cup of water   **31. a.** 195 e-mails   **b.** 585 min or 9.75 hr
33. 0.98 megabyte or 980 kilobytes   **35.** 29.08 in.
37. 252 mL of acid
39. There are approximately 357 bass in the lake.
41. There are approximately 4000 bison in the park.
43. $x = 24$ cm, $y = 36$ cm   **45.** $x = 1$ yd, $y = 10.5$ yd
47. $x = 15$ cm, $y = 4$ in.   **49.** The flagpole is 12 ft high.
51. The platform is 2.4 m high.   **53.** $x = 17.5$ in.
55. $x = 6$ ft, $y = 8$ ft   **57.** $x = 21$ ft; $y = 21$ ft; $z = 53.2$ ft

Section 5.4 Calculator Connections, p. 325

59. There were approximately 166,005 crimes committed.
60. The Washington Monument is approximately 555 ft tall.
61. Approximately 15,400 women would be expected to have breast cancer.
62. Approximately 295,000 men would be expected to have prostate disease.

Chapter 5 Review Exercises, pp. 330–332

1. 5 to 4 and $\frac{5}{4}$ **2.** 3 : 1 and $\frac{3}{1}$ **3.** 8 : 7 and 8 to 7

4. a. $\frac{2}{3}$ **b.** $\frac{3}{2}$ **c.** $\frac{3}{5}$ **5. a.** $\frac{4}{5}$ **b.** $\frac{5}{4}$ **c.** $\frac{5}{9}$

6. a. $\frac{12}{52}$ **b.** $\frac{12}{40}$ **7.** $\frac{4}{1}$ **8.** $\frac{7}{5}$ **9.** $\frac{2}{5}$ **10.** $\frac{1}{4}$

11. $\frac{9}{2}$ **12.** $\frac{4}{13}$ **13.** $\frac{4}{3}$ **14.** $\frac{170}{13}$

15. a. This year's enrollment is 1520 students.   **b.** $\frac{4}{19}$

16. $\frac{19}{12}$ **17.** $\frac{1}{5}$ **18.** $\frac{24}{49}$ **19.** $\frac{4 \text{ hot dogs}}{9 \text{ min}}$

20. $\frac{2 \text{ mi}}{17 \text{ min}}$ **21.** $\frac{650 \text{ tons}}{9 \text{ ft}}$ **22.** $\frac{473 \text{ crimes}}{5000 \text{ people}}$

23. All unit rates have a denominator of 1, and reduced rates may not.   **24.** 33 mph   **25.** 4° per hour
26. 90 times/sec   **27.** 11 min/lawn   **28.** $0.599 per ounce
29. $6.667 per towel   **30. a.** $0.262/oz   **b.** $0.280/oz; The 32-oz bottle is the better buy.   **31. a.** $0.175/oz
b. $0.159/oz; The 44-oz jar is the better buy.
32. $0.499 per ounce   **33.** The difference is about 25¢ per roll or $0.25 per roll.   **34.** 0.6275 in./hr
35. a. There was an increase of 120,000 hybrid vehicles.
b. There will be 10,000 additional hybrid vehicles per month.
36. a. There was an increase of 63 lb.   **b.** Americans increased the amount of vegetables in their diet by 3.5 lb per year.

37. $\frac{16}{14} = \frac{12}{10\frac{1}{2}}$ **38.** $\frac{8}{20} = \frac{6}{15}$ **39.** $\frac{5}{3} = \frac{10}{6}$

40. $\frac{4}{3} = \frac{20}{15}$ **41.** $\frac{\$11}{1 \text{ hr}} = \frac{\$88}{8 \text{ hr}}$ **42.** $\frac{2 \text{ in.}}{5 \text{ mi}} = \frac{6 \text{ in.}}{15 \text{ mi}}$

43. No   **44.** Yes   **45.** Yes   **46.** No   **47.** Yes
48. No   **49.** No   **50** Yes   **51.** $x = 4$ **52.** $y = 27$
53. $b = 3$ **54.** $p = 2$ **55.** $h = 13.6$ **56.** $k = 0.9$
57. The human equivalent is 84 years.
58. Lavu can buy 42,750 yen.
59. Alabama had approximately 4,600,000 people.
60. The tax would be $6.96.   **61.** $x = 10$ in., $y = 62.1$ in.
62. The building is 8 m high.   **63.** $x = 1.6$ yd, $y = 1.8$ yd
64. $x = 10.8$ cm, $y = 30$ cm

Chapter 5 Test, pp. 332–333

1. 25 to 521, 25 : 521, $\frac{25}{521}$ **2. a.** $\frac{17}{23}$ **b.** $\frac{17}{6}$ **3.** $\frac{2}{15}$

4. $\frac{11}{6}$ **5.** $\frac{5}{8}$ **6. a.** $\frac{21}{125}$ **b.** $\frac{9}{125}$ **c.** The poverty ratio was greater in New Mexico.   **7. a.** $\frac{\frac{1}{2}}{1\frac{1}{2}} = \frac{1}{3}$ **b.** $\frac{30}{90} = \frac{1}{3}$

8. $\frac{85 \text{ mi}}{2 \text{ hr}}$ **9.** $\frac{10 \text{ lb}}{3 \text{ weeks}}$ **10.** $\frac{1 \text{ g}}{2 \text{ cookies}}$

11. 21.45 g/cm³   **12.** 2.29 oz/lb   **13.** $0.22 per ounce
14. $0.50 per ring   **15.** Generic: $0.044 per caplet; Advil: $0.179 capsule. The generic pain reliever is the better buy.
16. They form equal ratios or rates.

17. $\frac{42}{15} = \frac{28}{10}$ **18.** $\frac{20 \text{ pages}}{12 \text{ min}} = \frac{30 \text{ pages}}{18 \text{ min}}$

19. $\dfrac{\$15}{1\ hr} = \dfrac{\$75}{5\ hr}$ **20.** No   **21.** $p = 35$ **22.** $x = 12.5$
23. $n = 5$ **24.** $y = 6$ **25.** It will take 7.5 min.
26. Cherise spends 30 hr each week on homework outside of class.   **27.** There are approximately 27 goldfish in her pond.
28. $x = 1\frac{1}{2}$ mi, $y = 8$ mi   **29.** 16 cm

Chapters 1–5 Cumulative Review Exercises, pp. 334–335

1. Five hundred three thousand, forty-two
2. Approximately 1400   **3.** 22,600,000   **4.** 22 R 3
5. 22.1875   **6.** 6   **7.**

8. $\dfrac{7}{5}$ **9.** $\dfrac{39}{14}$ **10.** $\dfrac{9}{25}$

11. Bruce has $4\frac{1}{2}$ in. of sandwich left.   **12.** 2   **13.** $\dfrac{35}{9}$

14. $\dfrac{9}{13}$ **15.** Emil needs $13\frac{1}{12}$ ft of wallpaper border.

16. It sold $61\frac{11}{16}$ acres, and $20\frac{9}{16}$ acres were left.

17. There are 59 ninths.   **18.** One thousand four and seven hundred one thousandths   **19.** 28.057

20. $\dfrac{109}{25}$ **21.** 4392.3   **22.** 2.379   **23.** 130.9 cm

24. $\dfrac{61}{44}$ or 61 : 44   **25.** $\dfrac{13}{1}$ **26.** $\dfrac{7}{50}$; Approximately 7 out of 50 deaths are due to cancer.   **27.** 125 people/mi^2

28. a. Yes   **b.** No   **29.** $x = 4.5$
30. Jim can drive 100 mi on 4 gal.

Chapter 6
Chapter Opener Puzzle

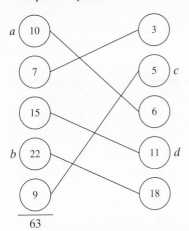

Section 6.1 Practice Exercises, pp. 343–345

1. percent   **3.** 48%   **5.** 50%   **7.** 25%   **9.** 2%

11. 70%   **13.** Replace the symbol % by $\times \dfrac{1}{100}$ (or $\div$ 100).

Then reduce the fraction to lowest terms.

15. $\dfrac{3}{100}$ **17.** $\dfrac{21}{25}$ **19.** $\dfrac{1}{4}$ **21.** $\dfrac{17}{500}$ **23.** $\dfrac{23}{20}$ or $1\frac{3}{20}$

25. $\dfrac{7}{4}$ or $1\frac{3}{4}$ **27.** $\dfrac{1}{200}$ **29.** $\dfrac{1}{400}$ **31.** $\dfrac{2}{3}$ **33.** $\dfrac{49}{200}$

35. Replace the % symbol by $\times$ 0.01 (or $\div$ 100).
37. 0.72   **39.** 0.66   **41.** 0.129   **43.** 0.4105
45. 2.01   **47.** 0.0075   **49.** 0.1625   **51.** 0.622
53. 25%   **55.** 100%   **57.** 150%   **59.** d   **61.** b

63. a   **65.** d   **67.** b   **69.** c   **71.** 0.076; $\dfrac{19}{250}$

73. 0.043; $\dfrac{43}{1000}$ **75.** 0.02; $\dfrac{1}{50}$ **77.** 0.35; $\dfrac{7}{20}$

79. $40\% = 0.4$ or $\dfrac{2}{5}$; $42\% = 0.42$ or $\dfrac{21}{50}$; $59\% = 0.59$

or $\dfrac{59}{100}$; $73\% = 0.73$ or $\dfrac{73}{100}$

Section 6.2 Practice Exercises, pp. 350–353

1. a. False   **b.** True   **c.** True   **3.** $\dfrac{13}{10}$ or $1\frac{3}{10}$

5. $\dfrac{1}{200}$ **7.** $0.06\overline{3}$ **9.** 0.003   **11.** 162%   **13.** 26%

15. 125%   **17.** 77%   **19.** 27%   **21.** 19%
23. 175%   **25.** 12.4%   **27.** 0.6%   **29.** 101.4%
31. 71%   **33.** 95%   **35.** 87.5% or $87\frac{1}{2}$%
37. 81.25% or $81\frac{1}{4}$%   **39.** $83.\overline{3}$% or $83\frac{1}{3}$%
41. $44.\overline{4}$% or $44\frac{4}{9}$%   **43.** 25%   **45.** 10%
47. $66.\overline{6}$% or $66\frac{2}{3}$%   **49.** 175%   **51.** 135%
53. $122.\overline{2}$% or $122\frac{2}{9}$%   **55.** $166.\overline{6}$% or $166\frac{2}{3}$%
57. 42.9%   **59.** 7.7%   **61.** 45.5%   **63.** 86.7%
65. The fraction $\frac{1}{2} = 0.5$ and $\frac{1}{2}\% = 0.5\% = 0.005$.
67. $25\% = 0.25$ and $0.25\% = 0.0025$
69. a, c   **71.** a, c

73.

	Fraction	Decimal	Percent
a.	$\frac{1}{4}$	0.25	25%
b.	$\frac{23}{25}$	0.92	92%
c.	$\frac{3}{20}$	0.15	15%
d.	$\frac{8}{5}$ or $1\frac{3}{5}$	1.6	160%
e.	$\frac{1}{100}$	0.01	1%
f.	$\frac{1}{200}$	0.005	0.5%

75.

	Fraction	Decimal	Percent
a.	$\frac{7}{50}$	0.14	14%
b.	$\frac{87}{100}$	0.87	87%
c.	1	1	100%
d.	$\frac{1}{3}$	$0.\overline{3}$	$33.\overline{3}$% or $33\frac{1}{3}$%
e.	$\frac{1}{500}$	0.002	0.2%
f.	$\frac{19}{20}$	0.95	95%

77. $1.4 > 100\%$ **79.** $0.052 < 50\%$

Section 6.3 Practice Exercises, pp. 359–363

1. a. percent **b.** cross   **3.** 130%
5. 37.5% or $37\frac{1}{2}$%   **7.** 1%   **9.** $\frac{1}{50}$ **11.** 0.82
13. 1   **15.** Yes   **17.** No   **19.** Yes
21. 45

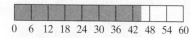

0   6 12 18 24 30 36 42 48 54 60
23. 32

0   4   8   12 16 20 24 28 32 36 40
25. Amount: 12; base: 20; $p = 60$ **27.** Amount: 99; base: 200; $p = 49.5$ **29.** Amount: 50; base: 40; $p = 125$
31. $\frac{10}{100} = \frac{12}{120}$ **33.** $\frac{80}{100} = \frac{72}{90}$ **35.** $\frac{104}{100} = \frac{21{,}684}{20{,}850}$
37. 108 employees   **39.** 0.2   **41.** 560
43. Pedro pays $20,160 in taxes.
45. Approximately 219 of the 304 teens were not wearing seat belts.   **47.** 36   **49.** 230 lb   **51.** 1350
53. Albert makes $1600 per month.   **55.** Amiee has a total of 35 e-mails.   **57.** 35%   **59.** 120%   **61.** 87.5%
63. She answered 72.5% correctly.   **65.** 20%
67. 26.7%   **69.** 70 mm of rain fell in August.
71. Approximately 1900 freshmen were admitted.
73. Terry made approximately 36.5% of his three-point shots.   **75. a.** 106 five-person households own dogs.
b. 23 three-person households own dogs.
77. 73 were Chevys.   **79.** There were 180 total vehicles.
81. $331.20   **83.** $11.60   **85.** $6.30

Section 6.4 Practice Exercises, pp. 368–371

1. Divide both sides of the equation by 26 to get $x = 2.5$.
3. $x = 9$ **5.** $x = 300$ **7.** $x = 825$ **9.** $x = 420$
11. $x = (0.35)(700); x = 245$
13. $(0.0055)(900) = x; x = 4.95$
15. $x = (1.33)(600); x = 798$
17. 50% equals one-half of the number. So multiply the number by $\frac{1}{2}$.
19. $2 \times 14 = 28$ **21.** $\frac{1}{2} \times 40 = 20$
23. There is 3.84 oz of sodium hypochlorite.
25. Marino completed approximately 5015 passes.
27. $18 = 0.4x; x = 45$ **29.** $0.92x = 41.4; x = 45$
31. $3.09 = 1.03x; x = 3$
33. There were 1175 subjects tested.
35. At that time, the population was about 280 million.
37. 13%   **39.** 108%   **41.** 0.5%   **43.** 17%
45. $x \cdot 480 = 120; x = 25\%$
47. $666 = x \cdot 740; x = 90\%$
49. $x \cdot 300 = 400; x = 133.3\%$
51. 70% of the hot dogs were sold.
53. a. There are 80 total employees. **b.** 12.5% missed 3 days of work. **c.** 75% missed 1 to 5 days of work.
55. There were 35 million total hospital stays that year.

57. Approximately 12.6% of Florida's panthers live in Everglades National Park.   **59.** 416 parents would be expected to have started saving for their children's education.
61. The total cost is $2400.   **63.** 15.6 min of commercials would be expected.   **65.** 6,350,000 people ages 25–34 made over $10/hr.   **67.** There is a total of 16,000,000 workers in the 16–24 age group.
69. a. 200 beats per minute.   **b.** Between 120 and 170 beats per minute.

Chapter 6 Problem Recognition Exercises, p. 372

1. 8.2   **2.** 4.1   **3.** 16.4   **4.** 41   **5.** 164   **6.** 12.3
7. Greater than   **8.** Less than   **9.** Greater than
10. Greater than   **11.** 3000   **12.** 24%   **13.** 4.8
14. 15%   **15.** 70   **16.** 36   **17.** 6.3   **18.** 250
19. 300%   **20.** 0.7   **21.** 75,000   **22.** 37.5%
23. 25   **24.** 135   **25.** 100   **26.** 6000   **27.** 0.8%
28. 125%   **29.** 2.6   **30.** 20   **31.** 75%   **32.** 6.05
33. $133\frac{1}{3}$%   **34.** 400%

Section 6.5 Practice Exercises, pp. 379–382

1. a. $\left(\begin{array}{c}\text{Sales}\\\text{tax}\end{array}\right) = \left(\begin{array}{c}\text{sales tax}\\\text{rate}\end{array}\right) \cdot \left(\begin{array}{c}\text{cost of}\\\text{merchandise}\end{array}\right)$

b. $(\text{Commission}) = \left(\begin{array}{c}\text{commission}\\\text{rate}\end{array}\right) \cdot \left(\begin{array}{c}\text{total}\\\text{sales}\end{array}\right)$

c. $(\text{Discount}) = \left(\begin{array}{c}\text{discount}\\\text{rate}\end{array}\right) \cdot \left(\begin{array}{c}\text{original}\\\text{price}\end{array}\right)$

d. $(\text{Markup}) = \left(\begin{array}{c}\text{markup}\\\text{rate}\end{array}\right) \cdot \left(\begin{array}{c}\text{original}\\\text{price}\end{array}\right)$

3. 12   **5.** 28   **7.** 26,000   **9.** 24%   **11.** 8.8

	Cost of Merchandise	Sales Tax Rate	Amount of Tax	Total Cost
13.	$ 20.00	5%	$1.00	$ 21.00
15.	$ 12.50	4%	$0.50	$ 13.00
17.	$110.00	2.5%	$2.75	$112.75
19.	$ 55.00	6%	$3.30	$ 58.30

21. The total bill is $71.66.   **23.** The tax rate is 7%.
25. The price is $44.50.

	Total Sales	Commission Rate	Amount of Commission
27.	$ 20,000.00	5%	$ 1000.00
29.	$125,000.00	8%	$10,000.00
31.	$ 5400.00	10%	$ 540.00

33. Zach made $3360 in commission.   **35.** Rodney's commission rate is 15%.   **37.** Her sales were $1,400,000.
39. Kabir's commission totaled $5810.00.

	Original Price	Discount Rate	Amount of Discount	Sale Price
41.	$175.00	15%	$ 26.25	$148.75
43.	$900.00	30%	$270.00	$630.00
45.	$110.00	30%	$ 33.00	$ 77.00
47.	$ 58.40	40%	$ 23.36	$ 35.04

49. a. The discount is $55.   **b.** The discounted yearly membership will cost $495.   **51.** The discount rate is 20%.
53. The set of dishes is not free. After the first discount, the price was 50% or one-half of $112, which is $56. Then the second discount is 50% or one-half of $56, which is $28.
55. The discount is $47.00, and the discount rate is 20%.

	Original Price	Markup Rate	Amount of Markup	Retail Price
57.	$ 92.00	5%	$ 4.60	$ 96.60
59.	$110.00	8%	$ 8.80	$118.80
61.	$325.00	30%	$ 97.50	$422.50
63.	$ 45.00	20%	$ 9.00	$ 54.00

65. a. The markup is $27.00.   **b.** The retail price is $177.00.
c. The total price is $189.39.   **67.** The markup rate is 25%.
69. The markup rate is 54%.

Section 6.6 Practice Exercises, pp. 385–387

1. a. $\left(\dfrac{\text{Percent}}{\text{increase}}\right) = \left(\dfrac{\text{amount of increase}}{\text{original value}}\right) \times 100\%$

b. $\left(\dfrac{\text{Percent}}{\text{decrease}}\right) = \left(\dfrac{\text{amount of decrease}}{\text{original value}}\right) \times 100\%$

3. a. The total price will be $68.25.   **b.** The total price with the 20% discount would be $54.60.   **c.** Damien will save $13.65.   **5.** Pablo's commission is $31.50.
7. Multiply the decimal by 100% by moving the decimal point 2 places to the right and attaching the % sign.
9. 5%   **11.** 12%   **13. a.** Increase   **b.** 11
15. a. Decrease   **b.** 10   **17 a.** Decrease   **b.** 9
19. a. Increase   **b.** 12   **21.** c   **23.** 75%
25. 100%   **27.** 10%   **29.** 3%   **31.** a   **33.** 5%
35. 68%   **37.** 25.3%   **39.** 15%

Section 6.6 Calculator Connections, p. 388

	Country	Population in 2005 (Millions)	Population in 2010 (Millions)	Change (Millions)	Percent Increase or Decrease
41.	Mexico	110.8	112.3	1.5	1.4% increase
42.	France	61.4	62.6	1.2	2.0% increase
43.	Bulgaria	8.11	7.78	0.33	4.1% decrease
44.	Trinidad	1.075	1.047	0.028	2.6% decrease

45. a. An increase of 6.35 million people   **b.** Between 2008 and 2009, there was a 71.2% increase in unemployment.
c. A decrease of 0.43 million people   **d.** Between 2009 and 2010, there was a 2.8% decrease in unemployment.
46. 8.7%

Section 6.7 Practice Exercises, pp. 394–396

1. a. Simple; principal   **b.** $I = Prt$ **c.** Compound

d. $A = P \cdot \left(1 + \dfrac{r}{n}\right)^{n \cdot t}$

	U.S. National Parks	Visitors in 2000 (Thousands)	Visitors in 2004 (Thousands)	Change	Percent Increase or Decrease
3.	Bryce Canyon, UT	1099	987	112	10% decrease
5.	Denali, AK	364	404	40	11% increase

7. $900, $6900   **9.** $1212, $6262   **11.** $2160, $14,160
13. $1890.00, $12,390.00   **15. a.** $350   **b.** $2850
17. a. $48   **b.** $448   **19.** $12,360   **21.** $5625
23. There are 6 total compounding periods.
25. There are 24 total compounding periods.
27. a. $560   **b.**

Year	Interest Earned	Total Amount in Account
1	$20.00	$520.00
2	20.80	540.80
3	21.63	**562.43**

29. a. $8960   **b.** $8998.91

Year	Interest Earned	Total Amount in Account
1	$320.00	$8320.00
2	332.80	8652.80
3	346.11	**8998.91**

c. $38.91
31. A = total amount in the account;
P = principal; r = annual interest rate;
n = number of compounding periods per year;
t = time in years

Section 6.7 Calculator Connections, p. 397

33. $6230.91   **34.** $14,725.49   **35.** $6622.88
36. $4373.77   **37.** $10,934.43   **38.** $9941.60
39. $16,019.47   **40.** $10,555.99

Chapter 6 Review Exercises, pp. 404–407

1. 75%   **2.** 33%   **3.** 125%   **4.** 50%   **5.** b, c
6. c, d   **7.** f   **8.** d   **9.** a   **10.** b
11. c   **12.** e   **13.** e   **14.** c   **15.** f   **16.** a
17. d   **18.** b   **19.** $\dfrac{21}{50}$; 0.42   **20.** $\dfrac{1}{5}$; 0.20
21. 0.0615   **22.** 0.0529   **23.** $\dfrac{183}{2000}$ **24.** $\dfrac{101}{2000}$
25. 17%   **26.** 44%   **27.** 80%   **28.** 175%
29. 12%   **30.** 110%   **31.** 0.5%   **32.** 40%
33. 87.5%   **34.** 76%   **35.** 60%   **36.** 10%

	Fraction	Decimal	Percent
37.	$\dfrac{9}{20}$	0.45	45%
38.	1	1	100%
39.	$\dfrac{3}{50}$	0.06	6%
40.	$\dfrac{6}{5}$	1.2	120%
41.	$\dfrac{9}{1000}$	0.009	0.9%
42.	$\dfrac{3}{4}$	0.75	75%

43. Amount: 67.50; base: 150; $p = 45$
44. Amount: 360; base: 3000; $p = 12$
45. Amount: 30.24; base: 144; $p = 21$
46. Amount: 31.8; base: 30; $p = 106$
47. $\dfrac{6}{8} = \dfrac{75}{100}$ **48.** $\dfrac{27}{180} = \dfrac{15}{100}$ **49.** $\dfrac{840}{420} = \dfrac{200}{100}$
50. $\dfrac{6}{2000} = \dfrac{0.3}{100}$ **51.** 6   **52.** 3.68   **53.** 12.5%
54. 0.32%   **55.** 39   **56.** 20
57. Approximately 11 people would be no-shows.
58. 850 people were surveyed.
59. Victoria spends 40% on rent.
60. There are 65 cars.   **61.** $0.18 \cdot 900 = x$; $x = 162$
62. $x = 0.29 \cdot 404$; $x = 117.16$
63. $18.90 = x \cdot 63$; $x = 30\%$
64. $x \cdot 250 = 86$; $x = 34.4\%$
65. $30 = 0.25 \cdot x$; $x = 120$
66. $26 = 1.30 \cdot x$; $x = 20$
67. The original price is $68.00.
68. Veronica read 55% of the novel.
69. Elaine can consume 720 fat calories.
70. a. 39,000,000   **b.** 80,800,000
71. The sales tax is $76.74.
72. The sales tax rate is 7%.
73. a. The tax is $7.22.   **b.** The tax rate is 8%.
74. The total amount for 4 nights will be $1053.00.
75. The commission rate was approximately 10.6%.
76. Andre earned $163 in commission.
77. Sela will earn $131 that day.
78. The commission rate is 3.5%.
79. The discount is $8.69. The sale price is $20.26.
80. The discount is $174.70. The final price is $1522.30.
81. The markup rate is 30%.
82. The baskets will sell for $59 each.
83. a. Increase   **b.** 25%   **84. a.** Decrease   **b.** 80%
85. 118.3%   **86.** 43.75%   **87.** 123%   **88.** 5233%
89. $1224, $11,424   **90.** $1400; $8400
91. Jean-Luc will have to pay $2687.50.
92. Kyle's brother will owe him $840.
93.

Year	Interest	Total
1	$240.00	$6240.00
2	249.60	6489.60
3	259.58	**6749.18**

94.

Compound Periods	Interest	Total
Period 1 (end of first 6 months)	$150.00	$10,150.00
Period 2 (end of year 1)	152.25	10,302.25
Period 3 (end of 18 months)	154.53	10,456.78
Period 4 (end of year 2)	156.85	10,613.63

95. $995.91   **96.** $2624.17   **97.** $16,976.32
98. $9813.88

Chapter 6 Test, pp. 408–409

1. 22%   **2.**

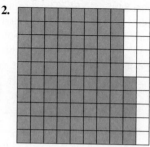

3. a. 0.054; $\dfrac{27}{500}$ **b.** 0.0015; $\dfrac{3}{2000}$ **c.** 1.70; $\dfrac{17}{10}$
4. a. $\dfrac{1}{100}$ **b.** $\dfrac{1}{4}$ **c.** $\dfrac{1}{3}$ **d.** $\dfrac{1}{2}$ **e.** $\dfrac{2}{3}$ **f.** $\dfrac{3}{4}$ **g.** 1   **h.** $\dfrac{3}{2}$
5. 0.028; $\dfrac{7}{250}$ **6.** 0.099; $\dfrac{99}{1000}$
7. Multiply the fraction by 100%.   **8.** 60%
9. 0.4%   **10.** 175%   **11.** 71.4%
12. Multiply the decimal by 100%.   **13.** 32%
14. 5.2%   **15.** 130%   **16.** 0.6%   **17.** 36
18. 19.2   **19.** 350   **20.** 200   **21.** 90%   **22.** 50%
23. a. 730 mg   **b.** 98.6%   **24.** 390 m^3 **25.** 420 m^3
26. a. The amount of sales tax is $2.10.   **b.** The sales tax rate is 7%.
27. Charles will earn $610.
28. The discount rate of this product is 60%.
29. 29.6%   **30. a.** $1200   **b.** $6200   **31.** $31,268.76

Chapters 1–6 Cumulative Review Exercises, pp. 409–411

1. Millions place
2.

| Country | Area (mi²) | |
	Standard Form	Words
a. United States	3,539,245	Three million, five hundred thirty-nine thousand, two hundred forty-five
b. Saudi Arabia	830,000	Eight hundred thirty thousand
c. Falkland Islands	4,700	Four thousand, seven hundred
d. Colombia	401,044	Four hundred one thousand, forty-four

3. 3,488,200   **4.** 87   **5.** 3185   **6.** 11
7. a. Improper   **b.** Improper   **c.** Proper   **d.** Proper
8. $\dfrac{4}{3}$ or $1\dfrac{1}{3}$ **9.** 24   **10.** $\dfrac{3}{2}$ or $1\dfrac{1}{2}$ **11.** $\dfrac{2}{3}$
12. $\dfrac{15}{32}$ yd^2 **13.** 9 km   **14.** $\dfrac{473}{1000}$ **15.** $15\dfrac{1}{10}$
16. $\dfrac{459}{2}$ or $229\dfrac{1}{2}$ in.^{2}
17. a. 18, 36, 54, 72   **b.** 1, 2, 3, 6, 9, 18   **c.** $2 \cdot 3^2$
18. a. $\dfrac{5}{2}$ **b.** $\dfrac{5}{6}$ **19.** 0.375   **20.** $1.\overline{3}$ **21.** $0.\overline{7}$
22. 0.75   **23.** 65.3%   **24.** 42.1%   **25.** 0.085

26. 8500   **27.** 8.5   **28.** 850,000   **29.** $p = 20$
30. $p = 3.75$ **31.** $p = 6\frac{1}{2}$ **32.** $p = 27$
33. It will take $2\frac{1}{2}$ hr.
34. The unit price is $0.25 per ounce.
35. It will take about 7.2 min.
36. The DC-10 flew 514 mph.   **37.** 13%
38. a. 5 million   **b.** Approximately 0.045 million people per year or 45,000 people per year
39. Kevin will have $15,080.
40. There is $91,473.02 paid in interest.

Chapter 7

Chapter Opener Puzzle

Section 7.1 Practice Exercises, pp. 419–423

1. a. measure   **b.** conversion   **3.** 1 mi   **5.** 3 ft
7. $\frac{1}{3}$ yd   **9.** b   **11.** a   **13.** c   **15.** 6 ft   **17.** 72 in.
19. 10,560 ft   **21.** 8 yd   **23.** $\frac{3}{4}$ ft   **25.** $\frac{1}{3}$ mi   **27.** b
29. a   **31.** 3 yd   **33.** 42 in.   **35.** $2\frac{1}{4}$ mi   **37.** 18 ft
39. $4\frac{2}{3}$ yd   **41.** 563,200 yd   **43.** $4\frac{3}{4}$ yd   **45.** 72 in.
47. 50,688 in.   **49.** 0.2 mi   **51.** 101,376 in.
53. a. 128 in.   **b.** $10\frac{2}{3}$ ft   **55. a.** 3.5 ft   **b.** 42 in.
57. 10'8"   **59.** 12 ft   **61.** 8'6"   **63.** 1 ft 9 in.
65. 8 ft 10 in.   **67.** 28 ft   **69.** 3'2"   **71.** 6 ft 1 in.
73. $5\frac{1}{2}$ ft   **75.** 18 pieces of border are needed.
77. 50 ft 8 in. is needed.   **79.** The plumber used 7'2" of pipe.   **81.** 7 ft is left over.   **83.** Each piece is 2 ft 3 in. long.   **85.** The total length is 46'.   **87.** 4 rolls
89. 6 yd² **91.** 3 ft² **93.** 720 in.² **95.** 27 ft²

Section 7.2 Practice Exercises, pp. 427–429

1. c

	Object	in.	ft	yd	mi
3.	Length of a hallway	144 in.	12 ft	4 yd	
5.	Height of a tree	216 in.	18 ft	6 yd	
7.	Perimeter of a backyard	1,800 in.	150 ft	50 yd	

9. 2 pt   **11.** 16 oz   **13.** 365 days   **15.** 4 qt   **17.** d
19. b   **21.** 730 days   **23.** $1\frac{1}{2}$ hr   **25.** 3 min   **27.** 3 days
29. 1 hr   **31.** 1512 hr   **33.** 80.5 min   **35.** 175.25 min
37. Gil ran for 5 hr 35 min.   **39.** The total time was 1 hr 34 min.   **41.** 2 lb   **43.** 4000 lb   **45.** 64 oz
47. $1\frac{1}{2}$ tons or 1.5 tons   **49.** 10 lb 8 oz   **51.** 8 lb 2 oz
53. 6 lb 8 oz   **55.** The total weight is 312 lb 8 oz.
57. The truck will have to make 2 trips.   **59.** 2 c
61. 24 qt   **63.** 16 c   **65.** $\frac{1}{2}$ gal   **67.** 16 fl oz   **69.** 6 tsp
71. Yes, 3 c is 24 fl oz, so the 48-fl-oz jar will suffice.
73. The unit price for the 24-fl-oz jar is about $0.112 per ounce, and the unit price for the 1-qt jar is about $0.103 per ounce; therefore the 1-qt jar is the better buy.

Section 7.2 Calculator Connections, p. 429

	Object	fl oz	c	pt	qt	gal
75.	Bottle of canola oil	32 fl oz	4 c	2 pt	1 qt	0.25 gal
76.	Can of soda	12 fl oz	1.5 c	0.75 pt	0.375 qt	0.09375 gal
77.	Laundry detergent	128 fl oz	16 c	8 pt	4 qt	1 gal
78.	Container of gasoline	640 fl oz	80 c	40 pt	20 qt	5 gal
79.	Bottle of Gatorade	16 fl oz	2 c	1 pt	0.5 qt	0.125 gal
80.	Container of orange juice	24 fl oz	3 c	1.5 pt	0.75 qt	0.1875 gal
81.	Bottle of spring water	8 fl oz	1 c	0.5 pt	0.25 qt	0.0625 gal
82.	Milkshake	16 fl oz	2 c	1 pt	0.5 qt	0.125 gal
83.	Jug of maple syrup	64 fl oz	8 c	4 pt	2 qt	0.5 gal

Section 7.3 Practice Exercises, pp. 434–438

1. a. metric   **b.** prefix   **c.** meter; m   **3.** 1.25 mi
5. 3 lb   **7.** 1440 min   **9.** 56 oz   **11.** b, f, g
13. 3.2 cm or 32 mm   **15.** 2.1 cm or 21 mm
17. a. 5 cm   **b.** 2 cm   **c.** 14 cm   **d.** 10 cm²
19. a   **21.** d   **23.** d   **25.** $\frac{1 \text{ km}}{1000 \text{ m}}$ **27.** $\frac{1 \text{ m}}{100 \text{ cm}}$
29. $\frac{1 \text{ m}}{10 \text{ dm}}$ **31.** 2.43 km   **33.** 10.3 m   **35.** 50,000 mm
37. 4000 m   **39.** 43.1 mm   **41.** 0.3328 km

43. 0.345 m   **45.** 250 m   **47.** 400.3 dm   **49.** 0.007 cm
51. 2091 cm   **53.** 2.538 km   **55.** 0.27 km
57. No, she needs 1.04 m of framing.   **59.** It will take
13 tiles.   **61.** There can be 24 parking spaces.
63. 300 cm^2 **65.** 41,000 cm^2

Section 7.4 Practice Exercises, pp. 443–447

1. a. gram; g   **b.** liter; L   **c.** cubic   **d.** microgram

	Object	mm	cm	m	km
3.	Distance between Orlando and Miami	670,000,000	67,000,000	670,000	670
5.	Length of a screw	25	2.5	0.025	0.000025
7.	Thickness of a dime	1.35	0.135	0.00135	0.00000135

9. 0.539 kg   **11.** 2500 g   **13.** 33.4 mg   **15.** 9 kg
17. 4.5 g

	Object	mg	cg	g	kg
19.	Bag of cat food	1,580,000	158,000	1580	1.58
21.	Can of tuna	170,000	17,000	170	0.17
23.	Box of raisins	425,000	42,500	425	0.425
25.	Dose of acetaminophen	325	32.5	0.325	0.000325

27. <   **29.** =   **31.** <   **33.** Cubic centimeter
35. 3.2 L   **37.** 700 cL   **39.** 0.42 dL   **41.** 64 mL
43. 40 cc

	Object	mL	cL	L	kL
45.	1 Tablespoon	15	1.5	0.015	0.000015
47.	Bottle of vinegar	355	35.5	0.355	0.000355
49.	Bottle of soda pop	2,000	200	2	0.002
51.	Capacity of a cooler	37,700	3,770	37.7	0.0377

53. c   **55.** b   **57.** c, d   **59.** 11.2014 dm   **61.** 0.6 g
63. 0.019 kL   **65.** Stacy gets 9.45 g per week.
67. The price is $1.65 per liter.   **69.** A 6-pack contains
4.26 L.   **71.** 520 mg of sodium per 1-qt bottle
73. 5.25 g of the drug would be given in 1 wk.   **75.** 2 mL
77. 500 people   **79.** 9.6 mg   **81. a.** 400 mg
b. 8000 mg or 8 g   **83.** 10 mcg   **85.** 200 mcg
87. 1 mg   **89.** 3 mL   **91.** 3.6 g   **93.** 3.3 metric tons
95. 10,900 kg

Chapter 7 Problem Recognition Exercises, p. 447

1. 9 qt   **2.** 2.2 m   **3.** 12 oz   **4.** 300 mL   **5.** 4 yd

6. 6030 g   **7.** $\frac{3}{4}$ ft   **8.** 2640 ft   **9.** 3 tons

10. 4 qt   **11.** $\frac{1}{2}$ T   **12.** 0.021 km   **13.** 36 cc

14. 4 lb   **15.** 4.322 kg   **16.** 5000 mm   **17.** 2.5 c
18. 8.5 min   **19.** 0.5 gal   **20.** 3.25 c   **21.** 5460 g

22. 902 cL   **23.** 16,016 yd   **24.** 3 lb   **25.** 3240 lb
26. 4600 m   **27.** 2.5 days   **28.** 8 mL

Section 7.5 Practice Exercises, pp. 453–456

1. a. Fahrenheit; 32; 212   **b.** Celsius; 0; 100
3. d, f   **5.** b, e   **7.** c, f   **9.** b, g   **11.** b
13. a   **15.** 5.1 cm   **17.** 8.8 yd   **19.** 122 m
21. 1.1 m   **23.** 15.2 cm   **25.** 2.7 kg   **27.** 0.4 oz
29. 1.2 lb   **31.** 1980 kg   **33.** 5.7 L   **35.** 4 fl oz
37. 32 fl oz   **39.** The box of sugar costs $0.100 per ounce,
and the packets cost $0.118 per ounce. The 2-lb box is the
better buy.   **41.** 18 mi is about 28.98 km. Therefore the
30-km race is longer than 18 mi.
43. 97 lb is approximately 43.65 kg.   **45.** The price is
approximately $7.22 per gallon.   **47.** A hockey puck is
1 in. thick.   **49.** Tony weighs about 222 lb.
51. 45 cc is 1.5 fl oz.   **53.** 40.8 ft   **55.** 77°F
57. 20°C   **59.** 86°F   **61.** 7232°F   **63.** It is a hot day.
The temperature is 95°F.
65. In Italy, the Celsius scale is used. Converting 25°C to
Fahrenheit gives 77°F which would be a warm day.
67. $F = \frac{9}{5}C + 32 = \frac{9}{5} \cdot 100 + 32 = 9(20) + 32$
$= 180 + 32 = 212$ **69.** The Navigator weighs
approximately 2.565 metric tons.   **71.** The average weight
of the blue whale is approximately 240,000 lb.

Chapter 7 Review Exercises, pp. 462–463

1. 4 ft   **2.** 39 in.   **3.** 3520 yd   **4.** $1\frac{1}{4}$ mi

5. $1\frac{1}{3}$ mi   **6.** 2640 ft   **7.** 72 in.   **8.** 0.1 mi

9. 9 ft 3 in.   **10.** 6'4"   **11.** 2'10"   **12.** 3 ft 9 in.
13. 21'   **14.** 9 ft 4 in.   **15.** 2 ft 1 in.   **16.** 3 yd 1 ft

17. $7\frac{1}{2}$ ft   **18.** There is 102 ft or 34 yd of wire left.

19. 3 days   **20.** 360 sec   **21.** 80 oz   **22.** 168 hr

23. $1\frac{1}{2}$ c   **24.** 500 lb   **25.** $1\frac{3}{4}$ tons   **26.** 2.5 hr

27. 0.5 hr   **28.** 16 pt   **29.** $\frac{3}{4}$ lb   **30.** 4 gal

31. 144.5 min   **32.** The total weight was 11 lb 13 oz.

33. 375 lb will go to each location.   **34.** There are $2\frac{1}{4}$ gal
of soda.   **35.** b   **36.** a   **37.** c   **38.** d
39. 520 mm   **40.** 9.1 cm   **41.** 2338 m   **42.** 0.093 km
43. 3.4 m   **44.** 0.21 dam   **45.** 0.04 m   **46.** 300 cm
47. 1200 mm   **48.** 402.3 km   **49.** The difference is
3688 m.   **50.** 22.6 cm   **51.** 610 cg   **52.** 0.42 kg
53. 3.212 g   **54.** 70 g   **55.** 50 mg   **56.** 100 cg
57. 0.3 L   **58.** 240 L   **59.** 8.3 L   **60.** 124 cc
61. 22.5 cL   **62.** 490 L   **63.** Perimeter: 6.5 m; area:
2.5 m^2 **64.** 5 glasses can be filled.   **65.** The difference
is 64.8 kg.   **66.** No, the board is 25 cm too short.
67. a. 3.2 mg   **b.** 44.8 mg   **68.** 450 mcg
69. There is 1.2 cc or 1.2 mL of fluid left.   **70.** Clayton
took 7.5 g.   **71.** 15.75 cm   **72.** 2.5 fl oz   **73.** 5 oz
74. 5.26 qt   **75.** 1.04 m   **76.** 45 kg   **77.** 74.53 mi
78. 5.7 L   **79.** 45 cc   **80.** 11,250 kg
81. The difference in height is 38.2 cm.
82. There are approximately 6.72 servings.

83. The total amount of cough syrup is approximately 0.42 L.
84. The marathon is approximately 26.2 mi.

85. $C = \dfrac{5}{9}(F - 32)$ **86.** 82.2°C to 85°C

87. $F = \dfrac{9}{5}C + 32$ **88.** 46.4°F

Chapter 7 Test, pp. 464–465

1. c, d, g, j **2.** f, h, i **3.** a, b, e **4.** $8\dfrac{1}{3}$ yd

5. 5.5 tons **6.** 10 mi **7.** 10 fl oz of liquid
8. 20 min **9.** 9′ **10.** 4′2″ **11.** He lost 7 oz.
12. 19 ft 7 in. **13.** 75.25 min **14.** 2.4 cm or 24 mm
15. c **16.** 1.158 km **17.** 15 mL **18. a.** Cubic
centimeters **b.** 235 cc **c.** 1000 cc **19.** 41,100 cg
20. 7 servings **21.** 2.1 qt **22.** 109 yd **23.** 2.8 mi
24. 2929 m **25.** 50.8 cm tall and 96.52-cm wingspan
26. 11 lb **27.** 190.6°C **28.** 35.6°F **29.** 28 mg
30. 1750 mcg per week

Chapter 7 Cumulative Review Exercises, pp. 465–466

1. a. 2000 **b.** 42,100 **2.** 56 cm **3.** 180 cm²
4. 4 **5. a.** Ford Motor Company spends the most. That
amount is $7400 million or $7,400,000,000. **b.** The difference
between IBM and Motorola is $302 million or $302,000,000.
c. The total amount spent is $26,917 million or $26,917,000,000.

6. $\dfrac{6}{39}$ **7.** The number 32,542 is not divisible by 3

because the sum of the digits (16) is not divisible by 3.
8. 2 · 2 · 3 · 3 · 3 **9.** 540 in.²
10. $\frac{1}{4}$ of the recipe would call for $\frac{3}{4}$ c of oatmeal. This is less
than 1 c so Keesha does have enough. **11.** 10

12. $\dfrac{7}{5}$ **13.** $9\dfrac{1}{2}$ **14.** $18\dfrac{8}{9}$ **15.** $2\dfrac{6}{17}$ **16.** $3\dfrac{5}{6}$

	Fraction	Decimal
17.	$\dfrac{1}{3}$	$0.\overline{3}$
18.	$\dfrac{9}{20}$	0.45
19.	$\dfrac{5}{4}$	1.25
20.	$\dfrac{7}{2}$	3.5
21.	$\dfrac{3}{8}$	0.375
22.	$\dfrac{1}{25}$	0.04

23. a. $\dfrac{6}{5}$ **b.** $\dfrac{6}{11}$ **24.** 40 bottles **25.** 90 cars

26. 6.7 beds per nurse **27.** No **28.** 80%

29. $x = \dfrac{16}{3}$ yd **30.** 2290 trees **31.** 27 people

32. 6% **33.** $15,000 in sales **34.** $1020 in interest

35. 5.8 kg **36.** 12.9 lb **37.** 182.9 cm **38.** 6 ft

39. 7 pt **40.** 3.3 L

Chapter 8

Chapter Opener Puzzles

1. e **2.** d **3.** c **4.** a **5.** f **6.** b

Section 8.1 Practice Exercises, pp. 473–477

1. a. point **b.** line **c.** segment **d.** $P; Q$ **e.** angle;
vertex **f.** right; 180 **g.** protractor **h.** acute; obtuse
i. complementary; supplementary **j.** parallel
k. perpendicular
3. A line extends forever in both directions. A line
segment is a portion of a line between and including two
endpoints. **5.** Ray **7.** Point **9.** Line
11. For example:

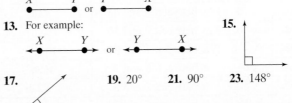

13. For example:

15.

17. **19.** 20° **21.** 90° **23.** 148°

25. Right **27.** Obtuse **29.** Acute **31.** Straight
33. 10° **35.** 63° **37.** 60.5° **39.** 1° **41.** 100°
43. 53° **45.** 142.6° **47.** 1°
49. No, because the sum of two angles that are both greater
than 90° will be more than 180°.
51. Yes. For two angles to add to 90°, the angles themselves
must both be less than 90°. **53.** A 90° angle
55. **57.**

59. $m(\angle a) = 41°; m(\angle b) = 139°; m(\angle c) = 139°$
61. $m(\angle a) = 26°; m(\angle b) = 112°;$
$m(\angle c) = 26°; m(\angle d) = 42°$
63. The two lines are perpendicular. **65.** Vertical angles
67. a, c or b, h or e, g or f, d **69.** a, e or f, b
71. $m(\angle a) = 55°; m(\angle b) = 125°;$
$m(\angle c) = 55°; m(\angle d) = 55°;$
$m(\angle e) = 125°; m(\angle f) = 55°;$
$m(\angle g) = 125°$
73. $m(\angle a) = 120°; m(\angle b) = 60°; m(\angle c) = 120°;$
$m(\angle d) = 120°; m(\angle e) = 60°; m(\angle f) = 120°;$
$m(\angle g) = 60°$
75. True **77.** True **79.** False **81.** True
83. True **85.** 70° **87.** 90° **89. a.** 48° **b.** 48°
c. 132° **91.** 180° **93.** 120°

Section 8.2 Practice Exercises, pp. 482–485

1. a. 180 **b.** acute; right; obtuse **c.** equilateral
d. isosceles **e.** scalene **f.** hypotenuse; legs
g. Pythagorean; c^2
3. Yes **5.** No **7.** No **9.** $m(\angle a) = 54°$
11. $m(\angle b) = 78°$ **13.** $m(\angle a) = 60°, m(\angle b) = 80°$
15. $m(\angle a) = 40°, m(\angle b) = 72°$ **17.** c, f **19.** b, d
21. b, c, e **23.** 7 **25.** 49 **27.** 16 **29.** 4
31. 6 **33.** 36 **35.** 81 **37.** 9 **39.** $c = 5$ m
41. $b = 12$ yd **43.** Leg = 10 ft

45. Hypotenuse = 40 in.   **47.** The brace is 20 in. long.
49. The height is 9 km.
51. The car is 25 mi from the starting point.
53. 24 m   **55.** 30 km
57. c = 5 in.; perimeter = 28 in.   **59.** Perimeter = 72 ft

Section 8.2 Calculator Connections, pp. 486–487

	Square Root	Estimate	Calculator Approximation (Round to 3 Decimal Places)
	$\sqrt{50}$	is between 7 and 8	7.071
61.	$\sqrt{10}$	is between 3 and 4	3.162
62.	$\sqrt{90}$	is between 9 and 10	9.487
63.	$\sqrt{116}$	is between 10 and 11	10.770
64.	$\sqrt{65}$	is between 8 and 9	8.062
65.	$\sqrt{5}$	is between 2 and 3	2.236
66.	$\sqrt{48}$	is between 6 and 7	6.928

67. 20.682   **68.** 56.434   **69.** 1116.244   **70.** 7100.423
71. 0.7   **72.** 0.5   **73.** 0.748   **74.** 0.906
75. b = 21 ft   **76.** a = 16 cm
77. Hypotenuse = 11.180 mi
78. Hypotenuse = 8.246 m   **79.** Leg = 18.439 in.
80. Leg = 9.950 ft   **81.** The diagonal length is 1.41 ft.
82. The length of the diagonal is 134.16 ft.
83. The length of the diagonal is 35.36 ft.

Section 8.3 Practice Exercises, pp. 494–498

1. a. perimeter **b.** area
3. a. acute triangle **b.** scalene triangle
5. a. right triangle **b.** isosceles triangle
7. a. obtuse triangle **b.** isosceles triangle
9. a, b, c, d, e, h   **11.** a, b, e   **13.** a, b, e, h
15. 80 cm   **17.** 260 mm   **19.** 10.7 m
21. 10 ft 6 in.   **23.** 5 ft or 60 in.
25. x = 550 mm; y = 3 dm; perimeter = 26 dm or 2600 mm
27. 280 ft of rain gutters is needed.   **29.** 576 yd^2
31. 54 m^2 **33.** 656 in.^{2} **35.** 18.4 ft^2 **37.** 12.375 ft^2
39. 148.5 yd^2 **41.** 280 mm^2 **43.** 60 in.^{2}
45. The area to be carpeted is 382.5 ft^2. The area to be tiled is 13.5 ft^2.   **47.** The area of the sign is 16.5 yd^2.
49. a. The area is 483 ft^2. **b.** They will need 2 paint kits.
51. The area is increased by 9 times.
53. False   **55.** True

Section 8.4 Practice Exercises, pp. 503–506

1. a. radius **b.** diameter **c.** circumference
d. diameter **e.** 3.14; $\frac{22}{7}$ **f.** Either formula can be used.
3. 1260 cm^2 **5.** 630 cm^2
7. Yes. Since a rectangle is a special type of parallelogram (one that contains four right angles), the area formula for a parallelogram applies to a rectangle.
9. 12 in.   **11.** 3 m   **13.** 4 in.   **15.** 8.3 m   **17.** c
19. π is the circumference divided by the diameter. That is, $\pi = \frac{C}{d}$.   **21. a.** 4π m **b.** 12.56 m

23. a. 20π cm **b.** 62.8 cm   **25. a.** 4.2π cm **b.** 13.188 cm
27. a. 5π km **b.** 15.7 km   **29.** 18.84 cm   **31.** 14.13 in.
33. 6.908 cm   **35. a.** 49π m^2 **b.** 154 m^2
37. a. 441π in.^{2} **b.** 1386 in.^{2}
39. a. 156.25π mm^2 **b.** 491 mm^2
41. a. 38.44π ft^2 **b.** 121 ft^2 **43.** 2.72 ft^2
45. 55.04 in.^{2} **47.** 18.28 in.^{2} **49.** 113.04 mm^2
51. 222.39 in.^{2} **53.** 16.642 mi   **55.** 2826 ft^2
57. a. ≈804 mi^2 **b.** ≈79 mi^2
59. a. 81.64 in. **b.** 147 times   **61. a.** 21 in. **b.** 43.8 ft

Section 8.4 Calculator Connections, p. 507

62. Area ≈ 517.1341 cm^2; circumference ≈ 80.6133 cm
63. Area ≈ 81.7128 ft^2; circumference ≈ 32.0442 ft
64. Area ≈ 70.8822 in.^{2}; circumference ≈ 29.8451 in.
65. Area ≈ 8371.1644 mm^2; circumference ≈ 324.3380 mm
66.

Diameter	Cost	Area	Cost per in.^{2}
8 in.	$ 6.50	50.27 in.^{2}	$ 0.129
12 in.	12.40	113.10 in.^{2}	0.110

The 12-in. is the better buy.

Chapter 8 Problem Recognition Exercises, p. 508

1. Area = 25 ft^2; perimeter = 20 ft
2. Area = 144 m^2; perimeter = 48 m
3. Area = 12 m^2 or 120,000 cm^2; perimeter 14 m or 1400 cm   **4.** Area = 1ft^2 or 144 in.^{2}; perimeter = 5 ft or 60 in.   **5.** Area = $\frac{1}{3}$ yd^2 or 3 ft^2; perimeter = 3 yd or 9 ft
6. Area = 0.473 km^2 or 473,000 m^2; perimeter = 3.24 km or 3240 m   **7.** Area = 6 yd^2; perimeter = 12 yd
8. Area = 30 cm^2; perimeter = 30 cm
9. Area = 44 m^2; perimeter = 32 m
10. Area = 88 in.^{2}; perimeter = 40 in.
11. Area ≈ 28.26 yd^2; circumference ≈ 18.84 yd
12. Area ≈ 1256 cm^2; circumference ≈ 125.6 cm
13. Area ≈ 154 cm^2; circumference ≈ 44 cm
14. Area ≈ 616 ft^2; circumference ≈ 88 ft
15. Area ≈ 38.28 ft^2; perimeter ≈ 26.28 ft

Section 8.5 Practice Exercises, pp. 513–517

1. a. s^3 **b.** lwh **c.** $\pi r^2 h$ **d.** cone; sphere
3. C ≈ 25.12 in; A ≈ 50.24 in.^{2} **5.** 187.52 cm^2
7. b, d   **9.** Area = 1 ft^2; volume = 1 ft^3
11. Area = 1 km^2; volume = 1 km^3 **13.** 2.744 cm^3
15. 48 ft^3 **17.** 12.56 mm^3 **19.** 3052.08 yd^3
21. 235.5 cm^3 **23.** 452.16 ft^3 **25.** 289 in.^{3}
27. 314 ft^3 **29.** 10 ft^3 **31. a.** 2575 ft^3 **b.** 19,260 gal
33. 502 mm^3 **35.** 32 ft^3 **37.** 56 in.^{3}
39. $\frac{11}{36}$ ft^3 or 0.306 ft^3 or 528 in.^{3} **41.** 109.3 in.^{3}
43. 450 ft^3 **45.** 84.78 in.^{3} **47.** 50,240 cm^3

Chapter 8 Review Exercises, pp. 523–526

1. d   **2.** a   **3.** c   **4.** b
5. The measure of an acute angle is between 0° and 90°.
6. The measure of an obtuse angle is between 90° and 180°.

7. The measure of a straight angle is 180°.

8. The measure of a right angle is 90°.

9. a. 57° **b.** 147° **10. a.** 70° **b.** 160° **11.** 60°

12. 90° **13.** 175° **14.** 180° **15.** b **16.** b, c

17. a, c **18.** 62° **19.** 118° **20.** 118° **21.** 62°

22. 62° **23.** 118° **24.** 118° **25.** $m(\angle x) = 40°$

26. $m(\angle x) = 80°$; $m(\angle y) = 32°$

27. An obtuse triangle has one obtuse angle.

28. An equilateral triangle has three sides of equal length and three angles of equal measure.

29. A right triangle has a right (90°) angle.

30. An acute triangle has three acute angles.

31. An isosceles triangle has two sides of equal length and two angles of equal measure.

32. A scalene triangle has no sides or angles of equal measure.

33. 5 **34.** 7 **35.** 10 **36.** 8

37. The sum of the squares of the legs of a right triangle equals the square of the hypotenuse. **38.** $b = 7$ cm

39. $c = 20$ ft **40.** 13 m of string is extended.

41. They both have sides of equal length, but a square also has four right angles.

42. A parallelogram must have both pairs of opposite sides parallel.

43. A square is a rectangle with four sides of equal length.

44. A rectangle is a parallelogram with four right angles.

45. 90 cm **46.** 17.3 m **47.** 56 mi **48.** 400 yd

49. 42 ft **50.** 15.5 ft **51.** 20 in.² **52.** 51 ft²

53. 7056 ft² **54.** The area is 36 m². The perimeter is 36 m.

55. 90 mm **56.** 6.4 ft **57.** 22.5 mm **58.** 1.6 ft

59. $C = 50.24$ m; $A = 200.96$ m²

60. $C = 13.2$ yd; $A = 13.86$ yd²

61. $C = 440$ in.; $A = 15,400$ in.²

62. $C = 125.6$ ft; $A = 1256$ ft²

63. 134.88 in.² **64. a.** 28.26 cm² **b.** 7.065 cm² **c.** No

65. 5.57 yd² **66.** 25,000 cm³ **67.** 226.08 ft³

68. 14,130 in.³ **69.** 37.68 km³ **70.** 995 in.³

71. 113 in.³ **72.** 335 in.³ **73.** 28,500 in.³ **74.** $2\frac{1}{3}$ ft³

Chapter 8 Test, pp. 527–528

1. d **2.** c **3.** 74° **4.** 33° **5.** 103°

6. $\frac{5}{2}$ ft or $2\frac{1}{2}$ ft **7.** $\frac{55}{7}$ ft or $7\frac{6}{7}$ ft **8.** 70,650 ft²

9. 48 ft² **10. a.** 2 **b.** 16 **11.** Obtuse

12. Acute **13.** Right **14.** Straight

15. $m(\angle x) = 125°$, $m(\angle y) = 55°$ **16.** They are each 45°.

17. 49° **18.** 180° **19.** $m(\angle A) = 80°$ **20.** 12 ft

21. 100 m **22.** d **23.** c **24.** f **25.** b **26.** a

27. e **28.** 96 in. **29.** 3 rolls are needed.

30. The area is 72 in.² **31.** The area of the rectangular pizza is 96 in.² The area of the round pizza is approximately 113.04 in.² The round pizza is larger by about 17 in.²

32. The volume is about 151 ft³. **33.** The volume is 1260 in.³ **34.** The volume is 2002 cm³.

Chapters 1–8 Cumulative Review Exercises, pp. 529–530

1. 3835 **2.** 0 **3.** Undefined **4.** 666,000

5. 2,511,000 **6.** $\frac{3}{5}, \frac{2}{3}, \frac{5}{6}$ **7.** There is $10\frac{1}{2}$ fl oz left.

8. 18 **9.** $\frac{1}{18}$ **10.** $\frac{6}{5}$ **11.** 60 **12.** $\frac{67}{60}$ **13.** $\frac{67}{60}$

14. $16\frac{1}{2}$ **15.** $\frac{46}{9}$ **16.** Four glasses cost $47.96.

17. Geraldo will save the cost of one shirt which is $13.49.

	Fraction	Decimal
18.	$\frac{3}{8}$	0.375
19.	$\frac{2}{9}$	$0.\overline{2}$
20.	$\frac{1}{50}$	0.02

21. $\frac{2}{3}$ **22.** $n = 37.35$ **23.** 17 pizzas **24.** 34 mpg

25. $3436 per hour **26.** 52.8 **27.** 72 **28.** 130%

29. 20% markup **30.** 16% discount **31.** 10 ft

32. Yes, $4\frac{1}{2}$ ft is 54 in. **33.** There is a total of $1\frac{1}{4}$ c or 10 fl oz of liquid. **34.** 100 kph ≈ 62 mph **35.** 41°F

36. 13.3 m **37.** 11 ft **38.** 1256 cm²

39. 3 yd² or 27 ft² **40.** 452 in.³

Chapter 9

Chapter Opener Puzzle

a. 6-month **b.** 0.46% **c.** 1.02% **d.** 0.60% **e.** 1-year

Section 9.1 Practice Exercises, pp. 538–543

1. a. Statistics **b.** table; cells **c.** pictograph

3. Mt. Kosciusko; Australia **5.** 2514 ft **7.** 3.6 yr

9. 2.8 yr **11.** Men

13.

	Dog	Cat	Neither
Boy	4	1	3
Girl	3	4	5

15. a. The 18- to 29-year age group has the greatest percentage of Internet users.

b.

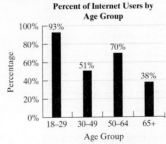

Percent of Internet Users by Age Group

17.

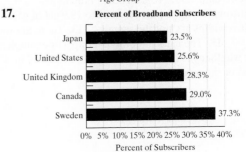

Percent of Broadband Subscribers

19. a. One icon represents 100 servings sold.   **b.** About 450 servings   **c.** Sunday   **21. a.** Barnes & Noble/B. Dalton has approximately $4.5 billion in book sales. **b.** There is approximately $11.5 billion in book sales. **23.** 48.4%   **25.** The trend for women over 65 in the labor force shows a slight increase.   **27.** For example: 18% **29.** The most cars were sold in 2008. 22,400 cars were sold. **31.** 4800 cars **33.** The greatest increase was between 2007 and 2008. **35. a.**

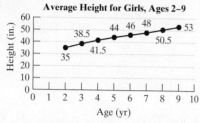

Average Height for Girls, Ages 2–9

b. Approximately 56 in. **37.** There are 14 servings per container, which means that there is 8 g × 14 = 112 g of fat in one container. **39.** The daily value of fat is approximately 61.5 g.

Section 9.2 Practice Exercises pp. 545–549

1. a. frequency   **b.** histogram **3.** There are 72 data.   **5.** 9–12 **7.**

Class Intervals (Age in Years)	Tally	Frequency (Number of Professors)
56–58	II	2
59–61	I	1
62–64	I	1
65–67	₩ II	7
68–70	₩	5
71–73	IIII	4

a. The class of 65–67 has the most values. **b.** There are 20 values represented in the table. **c.** Of the professors, 25% retire when they are 68 to 70 years old. **9.**

Class Intervals (Amount in Gal)	Tally	Frequency (Number of Customers)
8.0–9.9	IIII	4
10.0–11.9	I	1
12.0–13.9	₩	5
14.0–15.9	IIII	4
16.0–17.9		0
18.0–19.9	II	2

a. The 12.0–13.9 class has the highest frequency. **b.** There are 16 data values represented in the table. **c.** Of the customers, 12.5% purchased 18 to 19.9 gal of gas. **11.** The class widths are not the same.   **13.** There are too few classes.   **15.** The class intervals overlap. For example, it is unclear whether the data value 12 should be placed in the first class or the second class.

17.

Class Interval (Height, in.)	Frequency (Number of Students)
62–63	2
64–65	3
66–67	4
68–69	4
70–71	4
72–73	3

19.

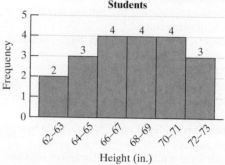

Heights of Valencia College Students

21.

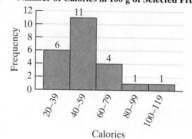

Number of Calories in 100 g of Selected Fruits

Section 9.3 Practice Exercises, pp. 553–555

1. circle; sectors **3.** 64,000   **5.** 640   **7.** 25%   **9.** 2.5 times **11.** There were 15.6 million viewers represented. **13.** *The Young and the Restless* has 2 times as many viewers.   **15.** Of the viewers, approximately 17% watch *General Hospital.*   **17.** There are 960 Latina CDs. **19.** There are 640 CDs that are classical or jazz. **21.** 13.5 million Wii systems were sold. **23.** 5.5 million Play Station 3 systems were sold. **25.**   **27.**   **29.**

31.   **33.**

35.

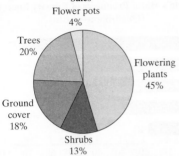

Sunshine Nursery Distribution of Sales

Flower pots 4%
Trees 20%
Flowering plants 45%
Ground cover 18%
Shrubs 13%

37. a.

	Expenses	Percent	Number of Degrees
Tuition	$9000	75%	270°
Books	600	5%	18°
Housing	2400	20%	72°

b.

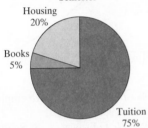

College Expenses for a Semester

Housing 20%
Books 5%
Tuition 75%

Section 9.4 Practice Exercises, pp. 561–565

1. a. mean **b.** median **c.** mean **d.** mode **e.** weighted
3. 6   **5.** 4   **7.** 15.2   **9.** 8.76 in.   **11.** 5.8 hr
13. a. 397 Cal **b.** 386 Cal **c.** There is only an 11-Cal difference in the means.   **15. a.** 86.5% **b.** 81%
c. The low score of 59% decreased Zach's average by 5.5%.
17. 17   **19.** 110.5   **21.** 52.5   **23.** 3.93 deaths per 1000
25. 58 years old   **27.** 51.7 million passengers   **29.** 4
31. No mode   **33.** 21, 24   **35.** $300   **37.** 5.2%
39. These data are bimodal: $2.49 and $2.51.
41. Mean: 85.5%; median: 94.5%; The median gave Jonathan a better overall score.   **43.** Mean: $250; median: $256; mode: There is no mode.   **45.** Mean: $942,500; median: $848,500; mode: $850,000

47.

Age (yr)	Number of Students	Product
16	7	112
17	9	153
18	6	108
19	3	57
Total:	25	430

The mean age is 17.2 years.

49. The mean number of students per class is approximately 28.   **51.** 3.59   **53.** 2.73

Section 9.5 Practice Exercises, pp. 569–572

1. a. experiment **b.** sample **c.** probability
d. complement **e.** 1
3. Mean: 17.2; median: 16; no mode   **5.** Mean: 8.875; median: 8.5; mode: 8   **7.** Mean: 16.5; median: 18.5; mode: 20
9. {1, 2, 3, 4, 5, 6, 7, 8, 9, 10}   **11.** {2, 3, 4, 5, 6, 7, 8, 9, 10, 11, 12}
13. 3 ways   **15.** c, d, g, h   **17.** $\frac{2}{6} = \frac{1}{3}$
19. $\frac{3}{6} = \frac{1}{2}$ **21.** $\frac{5}{8}$ **23.** $\frac{1}{8}$ **25.** 1
27. An impossible event is one in which the probability is 0.
29. $\frac{12}{52} = \frac{3}{13}$ **31.** $\frac{12}{16} = \frac{3}{4}$
33. a. $\frac{18}{120} = \frac{3}{20}$ **b.** $\frac{27}{120} = \frac{9}{40}$ **c.** 30%
35. a. $\frac{21}{60} = \frac{7}{20}$ **b.** 50% **37. a.** $\frac{7}{29}$ **b.** $\frac{11}{29}$ **c.** 62%
39. $1 - \frac{2}{11} = \frac{9}{11}$ **41.** $100\% - 1.2\% = 98.8\%$

Chapter 9 Review Exercises, pp. 577–580

1. Godiva   **2.** Breyers   **3.** Blue Bell has 2 times more sodium than Edy's Grand.   **4.** There is a 10-g difference.   **5.** 374 acres   **6.** The difference is 260 acres.
7. The difference is 4 acres.   **8.** The greatest increase was between 1950 and 1960.   **9.** 1 icon represents 50 tornadoes.   **10.** 300   **11.** June   **12.** 75   **13.** 2010
14. 4900   **15.** Increasing   **16.** ≈7000
17.

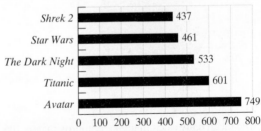

High Grossing Movies in the United States

Shrek 2   437
Star Wars   461
The Dark Night   533
Titanic   601
Avatar   749

Dollars (in millions)

18.

Class Intervals (Age)	Frequency
18–21	4
22–25	5
26–29	4
30–33	3
34–37	1
38–41	1
42–45	2

19.

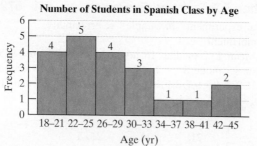

Number of Students in Spanish Class by Age

20. There are 24 types of subs.

21. $\frac{2}{3}$ of the subs contain beef.

22. $\frac{1}{3}$ of the subs do not contain beef.

23. a.

Education Level	Number of People	Percent	Number of Degrees
Grade school	10	5%	18°
High school	50	25%	90°
Some college	60	30%	108°
Four-year degree	40	20%	72°
Postgraduate	40	20%	72°

b.

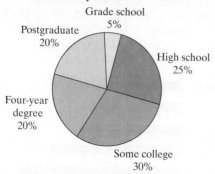

Percent by Education Level

24. Mean: 17.5; median: 18; mode: 20

25. The mean daily calcium intake is approximately 1060 mg.

26. The median is 20,562 seats.   **27.** 4

28. The mean age is 10.95 years.

Age (yr)	Number of Children	Product
10	8	80
11	6	66
12	5	60
13	1	13

29. {blue, green, brown, black, gray, white}

30. $\frac{1}{6}$ **31.** a, c, d, e, g   **32. a.** $\frac{1}{2}$ **b.** $\frac{1}{2}$ **c.** 0

Chapter 9 Test, pp. 581–583

1.

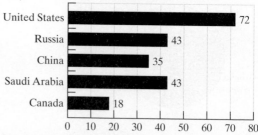

World's Major Producers of Primary Energy (Quadrillions of Btu)

2. The year 1820 had the greatest percent of workers employed in farm occupations. This was 72%.

3.

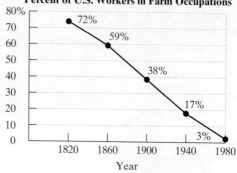

Percent of U.S. Workers in Farm Occupations

4. Approximately 10% of U.S. workers were employed in farm occupations in the year 1960.

5. $1000   **6.** $4500   **7.** February   **8.** Seattle

9. 1.73 in.   **10.** May

11.

Number of Minutes Used Monthly	Tally	Frequency
51–100	ⱧⱧ I	6
101–150	II	2
151–200	III	3
201–250	II	2
251–300	IIII	4
301–350	III	3

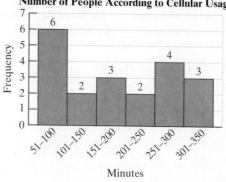

Number of People According to Cellular Usage

12. 66 people would have carpet. **13.** 40 people would have tile. **14.** 270 people would have something other than linoleum. **15.** 19,173 ft **16.** 19,340 ft **17.** There is no mode. **18.** Mean: $14.60; median: $15; mode $16

19. a. {1, 2, 3, 4, 5, 6, 7, 8} **b.** $\frac{1}{8}$ **c.** $\frac{1}{2}$ **d.** $\frac{1}{4}$

20. a. $\frac{2}{7}$ **b.** $\frac{5}{7}$ **21.** 3.09 **22.** c

Chapters 1–9 Cumulative Review Exercises, pp. 583–585

1. a. Millions **b.** Ten-thousands **c.** Hundreds
2. 12,645 **3.** $700 \times 1200 = 840,000$ **4.** Divisor: 23; dividend: 651; quotient: 28; remainder: 7
5. $\frac{3}{8}$ **6.** $\frac{2}{3}$ **7.** $\frac{5}{2}$ **8.** $\frac{3}{2}$ **9.** $\frac{1}{3}$ **10.** $\frac{37}{100}$
11. 2 **12.** $\frac{1}{6}$

13.

Stock	Yesterday's Closing Price ($)	Increase/ Decrease	Today's Closing Price ($)
RylGold	13.28	0.27	13.55
NetSolve	9.51	−0.17	9.34
Metals USA	14.35	0.10	14.45
PAM Transpt	18.09	0.09	18.18
Steel Tch	21.63	−0.37	21.26

14. 6841.2 **15.** 6.8412 **16.** 68,412 **17. a.** 0.75 million km^2 or equivalently, 750,000 km^2 **b.** 20.3%
18. Quick Cut Lawn Company's rate is 0.55 hr per customer. Speedy Lawn Company's rate is 0.5 hr per customer. Speedy Lawn Company is faster.
19. 125 min or 2 hr 5 min **20.** $x = 5$ m, $y = 22.4$ m
21. 122 people **22.** 17.02 million **23.** 65%
24. $1404 **25.** 29 in. **26.** 18 qt **27.** 9 yd 1 ft
28. 9.64 km or 9640 m **29.** 4 lb 3 oz **30.** Obtuse
31. Right **32.** Acute **33.** Area: 8 ft^2 **34.** 66 m^3
35.

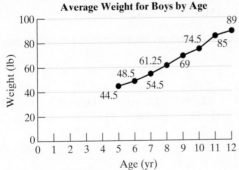

Average Weight for Boys by Age

36. Mean: 105; median: 123 **37.** 3
38. {yellow, blue, red, green}
39. $\frac{1}{4}$ **40.** $\frac{3}{4}$

Chapter 10

Chapter Opener Puzzle

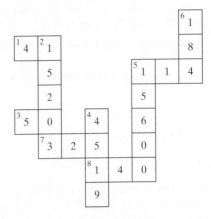

Section 10.1 Practice Exercises, pp. 594–596

1. a. positive; negative **b.** integers **c.** rational; irrational **d.** absolute **e.** opposites **3.** −86 m
5. $3800 **7.** −$500 **9.** −14 lb **11.** 140,000
13.
15.
17.
19.
21.
23.
25. Rational **27.** Rational **29.** Rational
31. Irrational **33.** Irrational **35.** Rational
37. > **39.** > **41.** < **43.** < **45.** >
47. > **49.** < **51.** < **53.** 2 **55.** 4.5
57. $\frac{5}{2}$ **59.** 0 **61.** 3.2 **63.** 21
65. a. −8 **b.** |−12| **67. a.** 7.8 **b.** |7.8| **69.** $\left|-\frac{4}{5}\right|$
71. Neither, they are equal. **73.** −5 **75.** 12
77. $\frac{1}{6}$ **79.** $-\frac{2}{11}$ **81.** −8.1 **83.** 1.14
85. −6 **87.** −(−2) **89.** |7| **91.** |−3|
93. −|14| **95.** −|−30| **97.** −2 **99.** −5.3
101. 15 **103.** 4.7 **105.** $-\frac{12}{17}$ **107.** $\frac{3}{8}$

Section 10.2 Practice Exercises, pp. 602–604

1. a. 0 **b.** negative; positive **c.** Subtract the smaller absolute value from the larger absolute value. The sum takes on the sign of the addend with the greater absolute value.
3. > **5.** = **7.** < **9.** −2 **11.** 2 **13.** −8
15. 6 **17.** −7 **19.** −4
21. To add two numbers with the same sign, add their absolute values and apply the common sign.

23. 15   **25.** −73   **27.** −124   **29.** 89   **31.** 52
33. −22   **35.** −24   **37.** 45   **39.** 0   **41.** 0
43. 9   **45.** −26   **47.** −41   **49.** −150   **51.** −17
53. −41   **55.** 2   **57.** −30   **59.** 10   **61.** −8
63. 0   **65.** −12   **67.** 26   **69.** −38.3   **71.** −2
73. $-3\frac{1}{2}$ **75.** −10.2   **77.** 1.4   **79.** $\frac{3}{16}$ **81.** $-\frac{7}{12}$
83. $-1\frac{1}{2}$ **85.** Sum, added to, increased by, more than, plus, total
87. $89 + (-11); 78$ **89.** $-2 + (-4) + 14 + 20; 28$
91. $-12 + (-4.5); -16.5$ **93.** $-\frac{1}{3} + 2; \frac{5}{3}$
95. $-\frac{1}{5} + 1; \frac{4}{5}$ **97.** −6°F   **99.** −$16.11
101. $320.32   **103.** For example: $-6 + (-8)$
105. For example: $5 + (-5)$

Section 10.2 Calculator Connections, p. 604
106. −120   **107.** −566   **108.** −68.221   **109.** −332.5
110. 711   **111.** 339

Section 10.3 Practice Exercises, pp. 609–611

1. a. $(-b)$ **b.** $-5 + 4$ **3.** −47   **5.** $\frac{1}{36}$ **7.** $-\frac{41}{36}$
9. −4   **11.** $2 + (-9); -7$ **13.** $4 + 3; 7$
15. $-3 + (-15); -18$ **17.** $-11 + 13; 2$
19. 52   **21.** −33   **23.** −12   **25.** 8   **27.** 0
29. 161   **31.** −34   **33.** −22   **35.** −26   **37.** −1
39. 32   **41.** −15
43. Minus, difference, decreased, less than, subtract from
45. $14 - 23; -9$ **47.** $5 - 12; -7$ **49.** $105 - 110; -5$
51. $320 - (-20); 340$ **53.** $-35 - 24; -59$
55. $-34 - 21; -55$ **57.** −8.3   **59.** −4.2   **61.** 5.5
63. 8.3   **65.** $\frac{5}{6}$ **67.** $\frac{2}{5}$ **69.** $-1\frac{1}{2}$ **71.** $-\frac{7}{4}$
73. 0   **75.** −1   **77.** 16   **79.** 1   **81.** 52   **83.** 5.2
85. 6423°F   **87.** The balance was $18,085.51.
89. The difference is 0.18 point.
91. His new balance is −$375.   **93.** 64,827 ft
95. The range is $3° - (-8°) = 11°$.
97. For example, $4 - 10$ **99.** $-11, -15, -19$
101. $-1, -\frac{4}{3}, -\frac{5}{3}$ **103.** Positive   **105.** Positive
107. Negative   **109.** Negative

Section 10.3 Calculator Connections, p. 612
111. −413   **112.** −433   **113.** 66.77   **114.** 20.06
115. 112.8   **116.** 129.7

Chapter 10 Problem Recognition Exercises, p. 612

1. −12   **2.** −2   **3.** −12   **4.** −2   **5.** 55   **6.** −35
7. −35   **8.** 55   **9.** 21.4   **10.** −83.8   **11.** −83.8
12. 21.4   **13.** 2   **14.** −41   **15.** −41   **16.** 2
17. 4   **18.** 4   **19.** 20   **20.** 20   **21.** $-\frac{9}{8}$ or $-1\frac{1}{8}$

22. $-\frac{11}{8}$ or $-1\frac{3}{8}$ **23.** $-\frac{9}{8}$ or $-1\frac{1}{8}$ **24.** $\frac{9}{8}$ or $1\frac{1}{8}$
25. $-\frac{17}{18}$ **26.** $-\frac{11}{18}$ **27.** $-\frac{11}{18}$ **28.** $\frac{11}{18}$
29. $-\frac{13}{4}$ or $-3\frac{1}{4}$ **30.** $\frac{7}{2}$ or $3\frac{1}{2}$ **31.** $-\frac{9}{2}$ or $-4\frac{1}{2}$
32. 4   **33.** $\frac{9}{4}$ or $2\frac{1}{4}$ **34.** $-\frac{9}{5}$ or $-1\frac{4}{5}$
35. −1.999   **36.** −1.987   **37.** 0   **38.** 0
39. −112   **40.** 28

Section 10.4 Practice Exercises, pp. 619–622
1. a. positive; negative   **b.** positive; negative
c. reciprocals   **3.** 19   **5.** −44   **7.** 17   **9.** −15
11. −48   **13.** 45   **15.** −72   **17.** 3.84   **19.** −2.4
21. −7.7   **23.** 0   **25.** $\frac{4}{7}$ **27.** $-\frac{1}{7}$ **29.** $-\frac{5}{2}$ or $-2\frac{1}{2}$
31. $\frac{13}{3}$ or $4\frac{1}{3}$ **33.** 0   **35.** 4.9   **37.** $-3(-1); 3$
39. $-5 \cdot 3; -15$ **41.** $1.3(-3); -3.9$ **43.** $3(-12);$
-36 customers   **45.** $4(-8); -32$ yd   **47.** 400
49. −88   **51.** 0   **53.** 1   **55.** −1   **57.** −100
59. 100   **61.** −27   **63.** −27   **65.** −0.008
67. $-\frac{8}{27}$ **69.** 36   **71.** −36   **73.** −3   **75.** −7
77. $\frac{5}{3}$ **79.** $\frac{2}{3}$ **81.** Undefined   **83.** 0   **85.** 4
87. 1.3   **89.** $-\frac{10}{7}$ **91.** Undefined   **93.** $\frac{5}{8}$
95. $-\frac{3}{4}$ **97.** $-100 \div 20; -5$ **99.** $-32 \div (-64); \frac{1}{2}$
101. $-52 \div 13; -4$ **103.** 2   **105.** −48   **107.** 3
109. 3   **111.** 20   **113.** −15   **115.** $-\frac{2}{5}$ **117.** $\frac{7}{6}$
119. $(-2)^{50}$ **121.** $(5)^{41}$ **123.** Negative
125. Positive

Section 10.4 Calculator Connections, p. 622
127. −359,723   **128.** 594,125   **129.** 5290.18
130. 54   **131.** −629   **132.** −33

Chapter 10 Problem Recognition Exercises, p. 622

1. 20   **2.** −75   **3.** 10   **4.** −3   **5.** 72
6. −34   **7.** 18   **8.** −38   **9.** −80   **10.** 80
11. −80   **12.** 80   **13.** −16   **14.** −72   **15.** −5
16. 25   **17.** 24   **18.** 24   **19.** −24   **20.** −24
21. $-\frac{27}{50}$ **22.** $-\frac{2}{3}$ **23.** $-\frac{23}{45}$ **24.** $\frac{77}{45}$ **25.** $-\frac{13}{9}$
26. $\frac{6}{7}$ **27.** $-\frac{13}{4}$ or $-3\frac{1}{4}$ **28.** $-\frac{29}{36}$ **29.** 55.1
30. −9.44   **31.** −93.91   **32.** −0.02   **33.** $\frac{12}{11}$
34. $-\frac{1}{6}$ **35.** 0   **36.** Undefined   **37.** −21,000
38. 40,000   **39.** 0   **40.** 1

Section 10.5 Practice Exercises, pp. 625–626

1. 44   **3.** $-\dfrac{3}{4}$ **5.** 3.08   **7.** -120 **9.** 1   **11.** -44

13. 29   **15.** -150 **17.** -3.3 **19.** 64   **21.** 16

23. -20 **25.** -16 **27.** 11   **29.** 2   **31.** -5

33. -8 **35.** 8   **37.** -96 **39.** -2 **41.** $-\dfrac{3}{2}$

43. $\dfrac{1}{3}$ **45.** $5\dfrac{1}{2}$ **47.** 14   **49.** -5 **51.** 38

53. -2 **55.** $\dfrac{1}{5}$ **57.** -1 **59.** $\dfrac{3}{7}$ **61.** $-\dfrac{1}{3}$

63. $-2°$ **65.** -3 **67.** $-129.28°F$ **69.** 3   **71.** -1

Chapter 10 Review Exercises, pp. 631–633

1. $-10,227$ **2.** $-\$5$ billion   **3.** $15°$ **4.** \$10

5. – 8.

9. 4, 4   **10.** $\dfrac{1}{2}, \dfrac{1}{2}$ **11.** $-3.5, 3.5$ **12.** $-6, 6$

13. a. 9   **b.** -9 **14. a.** -1.5 **b.** -1.5 **15.** $>$

16. $<$ **17.** $>$ **18.** $<$ **19.** $<$ **20.** $>$

21. 4   **22.** 3   **23.** -5 **24.** -3

25. To add two numbers with the same sign, add their absolute values and apply the common sign.

26. To add two numbers with different signs, subtract the smaller absolute value from the larger absolute value. Then apply the sign of the number having the larger absolute value.

27. 13   **28.** -15 **29.** -70 **30.** -0.88 **31.** -10.66

32. $\dfrac{3}{4}$ **33.** $-\dfrac{9}{10}$ **34.** $-\dfrac{1}{3}$ **35.** $23 + (-35); -12$

36. $57 + (-10); 47$ **37.** $-5 + (-13) + 20; 2$

38. $-42 + 12; -30$ **39.** $-12 + 3; -9$

40. $-89 + (-22); -111$ **41.** 3   **42.** -15

43. 1. Leave the first number (the minuend) unchanged.
2. Change the subtraction sign to an addition sign.
3. Add the opposite of the second number (the subtrahend).

44. 27   **45.** -25 **46.** 22   **47.** -419 **48.** -7.4

49. -0.7 **50.** $-\dfrac{5}{4}$ **51.** $\dfrac{20}{21}$

52. For example: The difference of 4 and 6
53. For example: 23 minus negative 6
54. For example: 14 subtracted from -2
55. For example: Subtract -7 from -25.
56. The temperature rose 5°F.
57. Sam's balance is now \$92.
58. The average is 1 above par.

59. -18 **60.** -3 **61.** 15   **62.** 56   **63.** -70

64. -156.5 **65.** $\dfrac{7}{4}$ **66.** $-\dfrac{17}{10}$ or $-1\dfrac{7}{10}$

67. Undefined   **68.** 0   **69.** -32 **70.** $\dfrac{9}{5}$ or $1\dfrac{4}{5}$

71. 36   **72.** -36 **73.** $-\dfrac{27}{64}$ **74.** $-\dfrac{27}{64}$ **75.** 1

76. -1 **77.** Negative   **78.** Positive
79. $-45 \div (-15); 3$ **80.** $-4 \cdot 19; -76$

81. $30(-5); -150$ **82.** $-136 \div (-8); 17$ **83.** -11

84. -2 **85.** -7 **86.** 4   **87.** 0   **88.** -1

89. $\dfrac{17}{64}$ **90.** -5 **91.** $\dfrac{1}{8}$ **92.** $-1°$

Chapter 10 Test, pp. 633–634

1. a. $-\$220$ **b.** 26   **2.** $-3, 0, 4, -1$

3. $-3, -\dfrac{3}{5}, 0, 4, -1, \dfrac{4}{7}$ **4.** $\sqrt{7}, -\pi$ **5.** $<$

6. $>$ **7.** $>$ **8.** $>$ **9.** $<$ **10.** $<$

11. -5 **12.** -28 **13.** 9   **14.** -41 **15.** 0.6

16. -2.3 **17.** $-\dfrac{26}{21}$ or $-1\dfrac{5}{21}$ **18.** $\dfrac{3}{8}$

19. -72 **20.** 88   **21.** 2   **22.** -18 **23.** Undefined

24. 0   **25.** $-\dfrac{3}{8}$ **26.** $\dfrac{13}{6}$ **27. a.** Positive   **b.** Negative

28. a. 64   **b.** -64 **c.** -64 **d.** -64 **29.** $-3(-7); 21$

30. $-13 + 8; -5$ **31.** $18 - (-4); 22$

32. $6 \div \left(-\dfrac{2}{3}\right); -9$ **33.** $-8.1 + 5; -3.1$

34. $-3 + 15 + (-6) + (-1); 5$ **35.** 3   **36.** -60

37. -19 **38.** -55 **39.** $-\dfrac{22}{15}$ **40.** $\dfrac{2}{13}$

41. 0   **42.** $\dfrac{3}{4}$ **43.** $1°$

Chapters 1–10 Cumulative Review Exercises, pp. 634–635

1. 3613   **2.** 2569   **3.** 177   **4.** 18,960

5. $2 \cdot 2 \cdot 2 \cdot 2 \cdot 3 \cdot 3 \cdot 5$ or $2^4 \cdot 3^2 \cdot 5$ **6.** $\dfrac{5}{8}$

7. Harold got $\dfrac{11}{14}$ of the quiz correct.

8. Amy will have 8 packages.   **9.** 80   **10.** $\dfrac{5}{16}$

11. $6\dfrac{7}{15}$ **12.** $3\dfrac{4}{7}$ **13. a.** 34.230   **b.** 9.0

14. \$2.09   **15.** 490.92   **16.** 115   **17.** $\dfrac{2}{5}$

18. The aircraft used 2491 gal/hr.   **19.** 21 mi
20. $x = 2.7$ cm; $y = 7$ cm   **21.** 192   **22.** 112%

23. 250   **24.** The sale price is \$68.80.   **25.** 28 in.

26. 5 gal   **27.** 0.06 L   **28.** $1\dfrac{7}{8}$ lb

29. The distance is 10 mi.   **30.** 16.5 yd^2 **31.** $5\dfrac{1}{16}$ m^2

32. $A = 7.065$ km^2; $C = 9.42$ km

33.

Number of Miles	Tally	Frequency (Number of Walkers)
3	‖	2
4	‖‖|	5
5			1
6	‖	2

34.

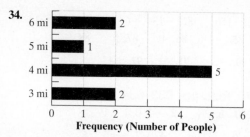

Frequency (Number of People)

35. 4.3 mi   **36.** $216   **37.** 55   **38.** −12

39. −20   **40.** $\dfrac{1}{6}$

Chapter 11

Chapter Opener Puzzle

1	4	2	A 6	5	3
5	6	B 3	4	2	1
6	C 1	5	3	4	2
3	2	D 4	1	6	5
4	5	1	2	3	6
E 2	3	6	5	1	F 4

Section 11.1 Practice Exercises, pp. 642–645

1. a. variable   **b.** commutative; $x + 6$ **c.** $b \cdot a$; $6x$
d. associative; $(-5 + 7) + x$ **e.** $a \cdot (b \cdot c)$; $(-5 \cdot 7)x$
f. $a \cdot b + a \cdot c$; $4x + 20$ **3.** $8p$ **5.** $t + 4$ **7.** $v - 6$
9. $\dfrac{4}{n}$ **11.** $2g$ **13. a.** −12   **b.** 30   **15. a.** 5   **b.** −15
17. a. −49   **b.** −49   **19. a.** −16   **b.** −64   **21.** −4
23. 24   **25.** $\dfrac{1}{2}$ **27.** 5   **29.** $P = 16.6$ in.   **31.** $A = \dfrac{77}{2}$ m^2
33. $w + 5$ **35.** $b + (-\frac{1}{3})$ or $b - \frac{1}{3}$ **37.** $2r$ **39.** $-st$
41. yx **43.** $-p + 7$ **45.** $(-2 \cdot 6)b$; $-12b$
47. $(3 + 8) + t$; $11 + t$ **49.** $(-4.2 + 2.5) + r$; $-1.7 + r$
51. $(3 \cdot 6)x$; $18x$ **53.** $\left(-\dfrac{4}{7} \cdot -\dfrac{7}{4}\right)d$; d
55. $(-9 + (-12)) + h$; $-21 + h$ **57.** $4x + 32$
59. $-2p - 8$ **61.** $-10t + 30$ **63.** $10 - 5x$
65. $4a + 16b - 4c$ **67.** $\dfrac{8}{3} + 4g$ **69.** $-3 + n$
71. $a + 8$ **73.** $-3x - 9 + 5y$ **75.** $5q + 2s + 3t$
77. $12x$ **79.** $12 + 6x$ **81.** $6 + 6k$ **83.** $-7 - k$
85. $-4 - p$ **87.** $-32 + 8p$ **89.** $4a$ **91.** $4 + 8a$
93. $5 + \dfrac{5}{9}y$ **95.** $5y$

Section 11.2 Practice Exercises, pp. 648–650

1. a. term   **b.** variable; constant   **c.** coefficient   **d.** like
3. $6p + 18$ **5.** $-24q$ **7.** $9 - h$ **9.** variable term
11. constant term   **13.** variable term

15. variable term   **17.** 6, −4   **19.** −14, 12
21. 1, −1   **23.** 5, −8, −3   **25.** Like terms
27. Unlike terms   **29.** Like terms
31. Unlike terms   **33.** Unlike terms   **35.** Like terms
37. $14rs$ **39.** $8h$ **41.** $3x^2 + 9$ **43.** $6x - 15y$
45. $-3k$ **47.** $4uv + 6u$ **49.** $-16m - 9$
51. $-9a + 5b + 20$ **53.** $-6p^2 - 2p + 7$
55. $2y - \dfrac{5}{6}$ **57.** $\dfrac{5}{8}a + 9$ **59.** $-3x^2 - 1.9x$
61. $-0.9a + 7.6$ **63.** $5t - 28$ **65.** $-6x - 16$
67. $6y - 14$ **69.** $7q$ **71.** $-2n - 4$ **73.** $-6a - b$
75. $4x + 23$ **77.** $2z - 11$ **79.** $-w - 4y + 9$
81. $8a - 9b$ **83.** $-5m + 6n - 10$ **85.** $12z + 7$
87. $-7x - 7$ **89.** $-21y + 28$ **91.** $-2q + 20$
93. $-107a - 213b$

Section 11.3 Practice Exercises, pp. 654–656

1. a. linear   **b.** solution   **c.** equivalent   **d.** addition
e. subtraction   **3.** $-13a + 16b$ **5.** $8h - 2k + 13$
7. $-3z + 4$ **9.** −1 is a solution.   **11.** 26 is a solution.
13. 12 is not a solution.   **15.** $-\dfrac{1}{2}$ is a solution.
17. 0 is a solution.   **19.** 4 is not a solution.   **21.** 0
23. 7   **25.** −3.2   **27.** 37   **29.** 16   **31.** −15
33. $\dfrac{7}{6}$ **35.** 3.1   **37.** 34   **39.** 52   **41.** 0   **43.** 100
45. −28   **47.** 3   **49.** −46   **51.** −13.6   **53.** $-\dfrac{13}{10}$
55. 7   **57.** −1   **59.** −6   **61.** $\dfrac{13}{12}$ **63.** −1.4
65. 12   **67.** −5   **69.** 13   **71.** −0.79   **73.** $\dfrac{13}{8}$
75. 49   **77.** −15   **79.** 26   **81.** −3   **83.** −5

Section 11.4 Practice Exercises, pp. 661–662

1. a. multiplication   **b.** division
3. 45   **5.** 21   **7.** −13   **9.** $-\dfrac{7}{6}$ **11.** $\dfrac{1}{3}$
13. $-\dfrac{7}{4}$ **15.** −7   **17.** 5.1   **19.** −3   **21.** −7
23. 13   **25.** 21   **27.** −21   **29.** 4   **31.** 30
33. $-\dfrac{1}{3}$ **35.** 4.1   **37.** $-\dfrac{2}{5}$ **39.** 0   **41.** $\dfrac{4}{15}$
43. 20   **45.** −31   **47.** $\dfrac{5}{6}$ **49.** 0   **51.** −16
53. −3   **55.** −8   **57.** −48   **59.** $\dfrac{1}{3}$ **61.** $\dfrac{4}{3}$
63. 30   **65.** −15   **67.** $\dfrac{5}{12}$ **69.** $-\dfrac{21}{10}$ **71.** 27.9
73. −1.8   **75.** −22   **77.** 15   **79.** $-\dfrac{2}{5}$ **81.** 12
83. −5   **85.** 3   **87.** −1

Section 11.5 Practice Exercises, pp. 667–669

1. Add 6 to both sides first.
3. −12   **5.** $-\dfrac{5}{8}$ **7.** 0   **9.** 4   **11.** −6
13. 5   **15.** $\dfrac{4}{3}$ **17.** −2.4   **19.** −1.12   **21.** 9

23. -12 **25.** $\frac{1}{4}$ **27.** -27 **29.** -3 **31.** $\frac{9}{8}$

33. 1 **35.** $\frac{7}{3}$ **37.** 2 **39.** -2 **41.** -3

43. -4 **45.** $\frac{4}{5}$ **47.** -12 **49.** 15 **51.** 3

53. 6 **55.** $-\frac{1}{2}$ **57.** $\frac{15}{11}$ **59.** 5

Chapter 11 Problem Recognition Exercises, p. 669

1. Equation   **2.** Expression   **3.** Expression
4. Equation   **5.** Equation   **6.** Expression
7. equation; 4 **8.** equation; $\frac{19}{3}$ **9.** expression; $4x - 8$
10. expression; $-k + 18$ **11.** equation; 15
12. equation; -2 **13.** equation; 7 **14.** equation; 32
15. expression; $2x - 1$ **16.** expression; $-6t - 16$
17. equation; $\frac{23}{5}$ **18.** equation; 6 **19.** equation; -12
20. equation; 9 **21.** expression; $-8x - 6$
22. expression; $y + 3$ **23.** equation; 2
24. equation; 30 **25.** equation; -1
26. equation; $-\frac{1}{2}$ **27.** equation; $-\frac{1}{7}$ **28.** equation; 3
29. expression; $5p + 14$ **30.** expression; $3y - 15$

Section 11.6 Practice Exercises, pp. 675–677

1. No   **3.** -45 **5.** $\frac{14}{5}$ **7.** -0.75
9. a. $\frac{x}{3} = -8$ **b.** The number is -24.
11. a. $-30 - x = 42$ **b.** The number is -72.
13. a. $30 + x = 13$ **b.** The number is -17.
15. a. $\frac{x}{4} - 5 = -12$ **b.** The number is -28.
17. a. $\frac{1}{2} + x = 4$ **b.** The number is $\frac{7}{2}$ or $3\frac{1}{2}$.
19. a. $-12x = x + 26$ **b.** The number is -2.
21. a. $10(x + 5.1) = 56$ **b.** The number is 0.5.
23. a. $3x = 2x - 10$ **b.** The number is -10.
25. The pieces should be 3 m and 9 m.
27. Metallica had 10 hits while Boyz II Men had 16 hits.
29. The soccer field is 100 yd by 130 yd.
31. Jim used 50 min over the 500 min.
33. The pieces are $1\frac{1}{3}$ ft and $2\frac{2}{3}$ ft long.
35. Tampa Bay had 48 points and Oakland had 21 points.
37. Charlene's rent is $650 a month with a security deposit of $300.   **39.** Stefan worked 6 hr of overtime.
41. Raul took 12 hr in the fall and 16 hr in the spring.
43. Ann-Marie must sell $1250 of merchandise each week.

Chapter 11 Review Exercises, pp. 685–686

1. a. $a + 8$ **b.** 43 years old   **2. a.** $-\frac{16}{9}$ **b.** -1
c. -4 **d.** -4 **3.** -18 **4.** $\frac{99}{7}$ m^3 **5. a.** $-5 + t$

b. $3h$ **6. a.** $(-4 \cdot 2)p; -8p$ **b.** $m + (10 - 12); m - 2$
7. $6b + 15$ **8.** $4k - 8m + 12$ **9.** $3, -5, 12$
10. $-6, -1, 2, 1$ **11.** Unlike terms   **12.** *Like* terms
13. *Like* terms   **14.** Unlike terms
15. $6x + 4y + 10$ **16.** $7a + 9b + 9$ **17.** $-u - 11v$
18. $p - 16$ **19.** -3 is a solution.   **20.** -3 is not a solution.   **21.** If a constant is being added to the variable term, use the subtraction property. If a constant is being subtracted from the variable term, use the addition property.
22. -35 **23.** -12 **24.** 14 **25.** 24 **26.** 1.1 **27.** 5
28. $\frac{3}{2}$ **29.** $-\frac{11}{10}$ **30.** $\frac{5}{4}$ **31.** -7 **32.** 4
33. -26 **34.** 35 **35.** 20 **36.** -21 **37.** -2
38. -4 **39.** $\frac{9}{7}$ **40.** $-\frac{8}{3}$ **41.** 6 **42.** -17
43. -1 **44.** 2 **45.** 8 **46.** -12 **47.** 10 **48.** 24
49. -28 **50.** -7 **51.** $\frac{4}{3}$ **52.** $\frac{9}{5}$ **53.** -3
54. -6 **55. a.** $-6x = x + 2$ **b.** $-\frac{2}{7}$
56. a. $-9 + 3 + 2x = -2$ **b.** 2 **57. a.** $\frac{1}{3} - x = 2$
b. $-\frac{5}{3}$ **58. a.** $\frac{x}{8} - 2 = \frac{1}{4}$ **b.** 18
59. Johnny Depp had 50 appearances and Tom Hanks had 66 appearances.   **60.** Marty made 25 text messages over the allotted 400.   **61.** The width is 40 in. and the length is 72 in.

Chapter 11 Test, pp. 686–687

1. $19.95m$ **2.** -15 **3.** $A = 34.25$ ft^2
4. Associative property of multiplication
5. Commutative property of addition
6. Associative property of addition
7. Distributive property of multiplication over addition
8. Commutative property of multiplication
9. $4a + 24$ **10.** $-13b + 8$ **11.** $16y + 3$
12. $7 - 7w$ **13.** $-x + 4$ **14.** $y + 22$ **15.** An expression is a collection of terms. An equation has an equal sign that indicates that two expressions are equal.
16. a. Expression   **b.** Equation   **c.** Expression
d. Equation   **e.** Expression   **f.** Expression
17. -2 **18.** 18 **19.** -72 **20.** $-\frac{1}{2}$ **21.** -1
22. 0.2 **23.** $-\frac{13}{16}$ **24.** $\frac{1}{2}$ **25.** 3 **26.** -84
27. 0 **28.** 2 **29.** The number is -5.
30. The budget for *New Moon* was $50 million.
31. The lengths are 60 cm and 240 cm.
32. Sela used 770 kWh.

Chapters 1–11 Cumulative Review Exercises, pp. 687–689

1. a. Hundreds   **b.** Ten-thousands   **c.** Hundred-thousands   **2.** 46,000   **3.** 1,290,000   **4.** 25,400
5. Dividend is 39,190; divisor is 46; quotient is 851; remainder is 44.
6. a. Prime   **b.** Composite   **c.** Composite
7. $\frac{23}{8}$ **8.** $\frac{6}{5}$ **9.** $\frac{28}{125}$ **10.** $\frac{13}{100}$ **11.** $\frac{79}{150}$

12. 7 ft^2 **13.** $\dfrac{31}{8}$ **14.** $5\dfrac{3}{5}$ **15.** $8\dfrac{2}{3}$ **16.** 13.95

17. 0.5   **18.** 31.221   **19.** 43.752   **20.** 3.56

21. Sarah makes \$20 per room.   **22.** 8.1

23. $\dfrac{4}{5}$ **24.** 56

25. Kitty Treats costs \$0.92 per ounce. Cat Goodies costs \$0.90 per ounce. Cat Goodies is the better buy.

	Decimal	Fraction	Percent
26.	0.15	$\dfrac{3}{20}$	15%
27.	0.125	$\dfrac{1}{8}$	12.5%
28.	1.1	$\dfrac{11}{10}$	110%
29.	$0.\overline{2}$	$\dfrac{2}{9}$	$22.\overline{2}\%$
30.	0.002	$\dfrac{1}{500}$	0.2

31. 21 glasses can be filled.   **32.** 101.75 min   **33.** 0.68 L

34. 3200 m   **35.** 4500 cg   **36.** 1256 in.^{2}

37. The area is 67.5 in.^{2} **38.** 995 in.^{2} **39.** 18.84 yd

40. 24 mm   **41.** 10   **42.** 9.5   **43.** 12

44.

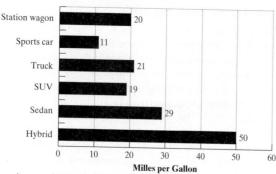

Miles per Gallon Comparison

Station wagon 20, Sports car 11, Truck 21, SUV 19, Sedan 29, Hybrid 50

Milles per Gallon

45. a. Approximately 166 students   **b.** 130 students

46. 3   **47.** −12   **48.** −25   **49.** −9

50. $-5x - 13$ **51.** $-2y - 18$ **52.** −4   **53.** −8

54. 2   **55.** $-\dfrac{8}{3}$

Additional Topics Appendix

Section A.1 Practice Exercises pp. A-3–A-5

1. Energy   **3.** 15,000 ft·lb   **5.** 225 ft·lb

7. 12,000 ft·lb   **9.** 2,334,000 ft·lb   **11.** 6,224,000 ft·lb

13. 68 Btu   **15.** 5,800,000 Btu   **17.** 798,228 ft·lb

19. $\dfrac{3}{4}$ hr or 0.75 hr   **21.** $\dfrac{5}{2}$ hr or 2.5 hr

23. $\dfrac{11}{10}$ hr or 1.1 hr   **25.** 443 Cal   **27.** 1250 Cal

29. 933 Cal   **31.** $60 \dfrac{\text{ft·lb}}{\text{sec}}$ **33.** $135 \dfrac{\text{ft·lb}}{\text{sec}}$

35. $200 \dfrac{\text{ft·lb}}{\text{sec}}$ **37.** 2 hp   **39.** 11 hp

41. $173,250 \dfrac{\text{ft·lb}}{\text{sec}}$ **43.** $118,250 \dfrac{\text{ft·lb}}{\text{sec}}$

45. a. 630,000 Wh   **b.** 630 kWh   **c.** \$51.66

Section A.2 Practice Exercises pp. A-8–A-9

1. scientific notation   **3.** 10^5 **5.** 10^6 **7.** 10^{-2}

9. 10^{-1} **11.** No   **13.** Yes   **15.** Yes   **17.** No

19. 2.5×10^{13} mi   **21.** 6.25×10^{-2} in.   **23.** 5×10^3

25. 6.2×10^4 **27.** 9×10^{-4} **29.** 5.8×10^{-1}

31. 2.55×10^7 **33.** 1.16×10^{-4} **35.** 1.5×10^{-1}

37. 1.34×10^4 **39.** 30,000   **41.** 0.00002   **43.** 0.0021

45. 5500   **47.** 61,300,000   **49.** 0.000404

51. 0.0000502   **53.** 7,070,000

	Planet	Mass	Scientific Notation
55.	Venus	4.87×10^{24}	Already in scientific notation
57.	Mars	0.64×10^{24}	6.4×10^{23}
59.	Saturn	586.5×10^{24}	5.865×10^{26}
61.	Neptune	102.4×10^{24}	1.024×10^{26}

Section A.3 Practice Exercises pp. A-14–A-17

1. a. x; y-axis   **b.** ordered   **c.** origin; $(0, 0)$ **d.** quadrants

e. negative   **f.** III

3., 5., 7.

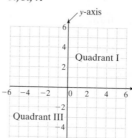

9., 11., 13.

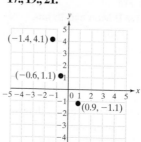

15. First move to the left 1.8 units from the origin. Then go up 3.1 units. Place a dot at the final location. The point is in Quadrant II.

17., 19., 21.

23., 25., 27.

29., 31., 33.

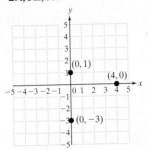

35. Quadrant IV   **37.** Quadrant III   **39.** *x*-axis
41. *y*-axis   **43.** Quadrant II   **45.** Quadrant I
47. $(0, 3)$ **49.** $(2, 3)$ **51.** $(-5, -2)$ **53.** $(4, -2)$
55. $(-2, -5)$

57.

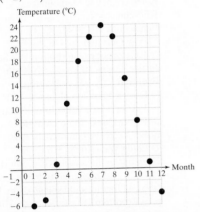

$(1, -6), (2, -5), (3, 1), (4, 11), (5, 18),$
$(6, 22), (7, 24), (8, 22), (9, 15), (10, 8),$
$(11, 1), (12, -4)$

59.

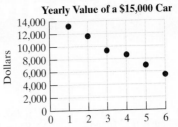

$(1, 13200), (2, 11352), (3, 9649), (4, 8201), (5, 6971),$
$(6, 5925)$

61.

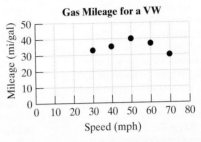

$(30, 33), (40, 35), (50, 40), (60, 37), (70, 30)$

Photo Credits

Application Index

Consumer Applications

Cooking

Distance/Speed/Time

Statistics and Demographics

Subject Index

Geometry Formulas

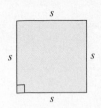

Perimeter of a Square
$P = s + s + s + s$
$P = 4s$

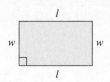

Perimeter of a Rectangle
$P = l + l + w + w$
$P = 2l + 2w$

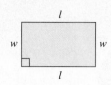

Area of a Rectangle
$A = \text{length} \times \text{width}$
$A = lw$

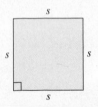

Area of a Square
$A = \text{length} \times \text{width}$
$A = s \cdot s$
$A = s^2$

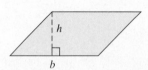

Area of a Parallelogram
$A = \text{base} \times \text{height}$
$A = bh$

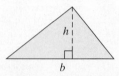

Area of a Triangle
$A = \frac{1}{2} \times \text{base} \times \text{height}$
$A = \frac{1}{2}bh$

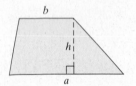

Area of a Trapezoid
$A = \frac{1}{2} \times (\text{sum of the parallel sides}) \times \text{height}$
$A = \frac{1}{2} \cdot (a + b) \cdot h$

Circumference of a Circle
The circumference, C, of a circle is
given by: $C = \pi d$ or $C = 2\pi r$

Area of a Circle
The area, A, of a circle
is given by: $A = \pi r^2$

Rectangular Solid
$V = lwh$

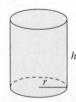

Right Circular Cylinder
$V = \pi r^2 h$

Sphere
$V = \frac{4}{3}\pi r^3$